The Feynman LECTURES ON PHYSICS

MAINLY ELECTROMAGNETISM AND MATTER

RICHARD P. FEYNMAN

Richard Chace Tolman Professor of Theoretical Physics
California Institute of Technology

ROBERT B. LEIGHTON

Professor of Physics
California Institute of Technology

MATTHEW SANDS

Professor
Stanford University

ADDISON-WESLEY PUBLISHING COMPANY
Reading, Massachusetts
Menlo Park, California · London · Amsterdam · Don Mills, Ontario · Sydney

---

Printed in the United States of America

Library of Congress Catalog Card No. 63-20717

Fifth printing, July 1975

ISBN 0-201-02011-4-H
ISBN 0-201-02117-X-P
ABCDEFGHIJ-MU-7987

Feynman's Preface

These are the lectures in physics that I gave last year and the year before to the freshman and sophomore classes at Caltech. The lectures are, of course, not verbatim—they have been edited, sometimes extensively and sometimes less so. The lectures form only part of the complete course. The whole group of 180 students gathered in a big lecture room twice a week to hear these lectures and then they broke up into small groups of 15 to 20 students in recitation sections under the guidance of a teaching assistant. In addition, there was a laboratory session once a week.

The special problem we tried to get at with these lectures was to maintain the interest of the very enthusiastic and rather smart students coming out of the high schools and into Caltech. They have heard a lot about how interesting and exciting physics is—the theory of relativity, quantum mechanics, and other modern ideas. By the end of two years of our previous course, many would be very discouraged because there were really very few grand, new, modern ideas presented to them. They were made to study inclined planes, electrostatics, and so forth, and after two years it was quite stultifying. The problem was whether or not we could make a course which would save the more advanced and excited student by maintaining his enthusiasm.

The lectures here are not in any way meant to be a survey course, but are very serious. I thought to address them to the most intelligent in the class and to make sure, if possible, that even the most intelligent student was unable to completely encompass everything that was in the lectures—by putting in suggestions of applications of the ideas and concepts in various directions outside the main line of attack. For this reason, though, I tried very hard to make all the statements as accurate as possible, to point out in every case where the equations and ideas fitted into the body of physics, and how—when they learned more—things would be modified. I also felt that for such students it is important to indicate what it is that they should—if they are sufficiently clever—be able to understand by deduction from what has been said before, and what is being put in as something new. When new ideas came in, I would try either to deduce them if they were deducible, or to explain that it *was* a new idea which hadn't any basis in terms of things they had already learned and which was not supposed to be provable—but was just added in.

At the start of these lectures, I assumed that the students knew something when they came out of high school—such things as geometrical optics, simple chemistry ideas, and so on. I also didn't see that there was any reason to make the lectures in a definite order, in the sense that I would not be allowed to mention something until I was ready to discuss it in detail. There was a great deal of mention of things to come, without complete discussions. These more complete discussions would come later when the preparation became more advanced. Examples are the discussions of inductance, and of energy levels, which are at first brought in in a very qualitative way and are later developed more completely.

At the same time that I was aiming at the more active student, I also wanted to take care of the fellow for whom the extra fireworks and side applications are merely disquieting and who cannot be expected to learn most of the material in the lecture at all. For such students I wanted there to be at least a central core or backbone of material which he *could* get. Even if he didn't understand everything in a lecture, I hoped he wouldn't get nervous. I didn't expect him to understand everything, but only the central and most direct features. It takes, of course, a certain intelligence on his part to see which are the central theorems and central ideas, and which are the more advanced side issues and applications which he may understand only in later years.

In giving these lectures there was one serious difficulty: in the way the course was given, there wasn't any feedback from the students to the lecturer to indicate how well the lectures were going over. This is indeed a very serious difficulty, and I don't know how good the lectures really are. The whole thing was essentially an experiment. And if I did it again I wouldn't do it the same way—I hope I *don't* have to do it again! I think, though, that things worked out—so far as the physics is concerned—quite satisfactorily in the first year.

In the second year I was not so satisfied. In the first part of the course, dealing with electricity and magnetism, I couldn't think of any really unique or different way of doing it—of any way that would be particularly more exciting than the usual way of presenting it. So I don't think I did very much in the lectures on electricity and magnetism. At the end of the second year I had originally intended to go on, after the electricity and magnetism, by giving some more lectures on the properties of materials, but mainly to take up things like fundamental modes, solutions of the diffusion equation, vibrating systems, orthogonal functions, . . . developing the first stages of what are usually called "the mathematical methods of physics." In retrospect, I think that if I were doing it again I would go back to that original idea. But since it was not planned that I would be giving these lectures again, it was suggested that it might be a good idea to try to give an introduction to the quantum mechanics—what you will find in Volume III.

It is perfectly clear that students who will major in physics can wait until their third year for quantum mechanics. On the other hand, the argument was made that many of the students in our course study physics as a background for their primary interest in other fields. And the usual way of dealing with quantum mechanics makes that subject almost unavailable for the great majority of students because they have to take so long to learn it. Yet, in its real applications—especially in its more complex applications, such as in electrical engineering and chemistry—the full machinery of the differential equation approach is not actually used. So I tried to describe the principles of quantum mechanics in a way which wouldn't require that one first know the mathematics of partial differential equations. Even for a physicist I think that is an interesting thing to try to do—to present quantum mechanics in this reverse fashion—for several reasons which may be apparent in the lectures themselves. However, I think that the experiment in the quantum mechanics part was not completely successful—in large part because I really did not have enough time at the end (I should, for instance, have had three or four more lectures in order to deal more completely with such matters as energy bands and the spatial dependence of amplitudes). Also, I had never presented the subject this way before, so the lack of feedback was particularly serious. I now believe the quantum mechanics should be given at a later time. Maybe I'll have a chance to do it again someday. Then I'll do it right.

The reason there are no lectures on how to solve problems is because there were recitation sections. Although I did put in three lectures in the first year on how to solve problems, they are not included here. Also there was a lecture on inertial guidance which certainly belongs after the lecture on rotating systems, but which was, unfortunately, omitted. The fifth and sixth lectures are actually due to Matthew Sands, as I was out of town.

The question, of course, is how well this experiment has succeeded. My own point of view—which, however, does not seem to be shared by most of the people who worked with the students—is pessimistic. I don't think I did very well by the students. When I look at the way the majority of the students handled the problems on the examinations, I think that the system is a failure. Of course, my friends point out to me that there were one or two dozen students who—very surprisingly—understood almost everything in all of the lectures, and who were quite active in working with the material and worrying about the many points in an excited and interested way. These people have now, I believe, a first-rate background in physics—and they are, after all, the ones I was trying to get at. But then, "The power of instruction is seldom of much efficacy except in those happy dispositions where it is almost superfluous." (Gibbon)

Still, I didn't want to leave any student completely behind, as perhaps I did. I think one way we could help the students more would be by putting more hard work into developing a set of problems which would elucidate some of the ideas in the lectures. Problems give a good opportunity to fill out the material of the lectures and make more realistic, more complete, and more settled in the mind the ideas that have been exposed.

I think, however, that there isn't any solution to this problem of education other than to realize that the best teaching can be done only when there is a direct individual relationship between a student and a good teacher—a situation in which the student discusses the ideas, thinks about the things, and talks about the things. It's impossible to learn very much by simply sitting in a lecture, or even by simply doing problems that are assigned. But in our modern times we have so many students to teach that we have to try to find some substitute for the ideal. Perhaps my lectures can make some contribution. Perhaps in some small place where there are individual teachers and students, they may get some inspiration or some ideas from the lectures. Perhaps they will have fun thinking them through—or going on to develop some of the ideas further.

RICHARD P. FEYNMAN

June, 1963

Foreword

For some forty years Richard P. Feynman focussed his curiosity on the mysterious workings of the physical world, and bent his intellect to searching out the order in its chaos. Now, he has given two years of his ability and his energy to his Lectures on Physics for beginning students. For them he has distilled the essence of his knowledge, and has created in terms they can hope to grasp a picture of the physicist's universe. To his lectures he has brought the brilliance and clarity of his thought, the originality and vitality of his approach, and the contagious enthusiasm of his delivery. It was a joy to behold.

The first year's lectures formed the basis for the first volume of this set of books. We have tried in this the second volume to make some kind of a record of a part of the second year's lectures—which were given to the sophomore class during the 1962–1963 academic year. The rest of the second year's lectures will make up Volume III.

Of the second year of lectures, the first two-thirds were devoted to a fairly complete treatment of the physics of electricity and magnetism. Its presentation was intended to serve a dual purpose. We hoped, first, to give the students a complete view of one of the great chapters of physics—from the early gropings of Franklin, through the great synthesis of Maxwell, on to the Lorentz electron theory of material properties, and ending with the still unsolved dilemmas of the electromagnetic self-energy. And we hoped, second, by introducing at the outset the calculus of vector fields, to give a solid introduction to the mathematics of field theories. To emphasize the general utility of the mathematical methods, related subjects from other parts of physics were sometimes analyzed together with their electric counterparts. We continually tried to drive home the generality of the mathematics. ("The same equations have the same solutions.") And we emphasized this point by the kinds of exercises and examinations we gave with the course.

Following the electromagnetism there are two chapters each on elasticity and fluid flow. In the first chapter of each pair, the elementary and practical aspects are treated. The second chapter on each subject attempts to give an overview of the whole complex range of phenomena which the subject can lead to. These four chapters can well be omitted without serious loss, since they are not at all a necessary preparation for Volume III.

The last quarter, approximately, of the second year was dedicated to an introduction to quantum mechanics. This material has been put into the third volume.

In this record of the Feynman Lectures we wished to do more than provide a transcription of what was said. We hoped to make the written version as clear an exposition as possible of the ideas on which the original lectures were based. For some of the lectures this could be done by making only minor adjustments of the wording in the original transcript. For others of the lectures a major reworking and rearrangement of the material was required. Sometimes we felt we should add some new material to improve the clarity or balance of the presentation. Throughout the process we benefitted from the continual help and advice of Professor Feynman.

The translation of over 1,000,000 spoken words into a coherent text on a tight schedule is a formidable task, particularly when it is accompanied by the other onerous burdens which come with the introduction of a new course—preparing for recitation sections, and meeting students, designing exercises and examinations, and grading them, and so on. Many hands—and heads—were involved. In some instances we have, I believe, been able to render a faithful image—or a tenderly retouched portrait—of the original Feynman. In other instances we have fallen far short of this ideal. Our successes are owed to all those who helped. The failures, we regret.

As explained in detail in the Foreword to Volume I, these lectures were but one aspect of a program initiated and supervised by the Physics Course Revision Committee (R. B. Leighton, Chairman, H. V. Neher, and M. Sands) at the California Institute of Technology, and supported financially by the Ford Foundation. In addition, the following people helped with one aspect or another of the preparation of textual material for this second volume: T. K. Caughey, M. L. Clayton, J. B. Curcio, J. B. Hartle, T. W. H. Harvey, M. H. Israel, W. J. Karzas, R. W. Kavanagh, R. B. Leighton, J. Mathews, M. S. Plesset, F. L. Warren, W. Whaling, C. H. Wilts, and B. Zimmerman. Others contributed indirectly through their work on the course: J. Blue, G. F. Chapline, M. J. Clauser, R. Dolen, H. H. Hill, and A. M. Title. Professor Gerry Neugebauer contributed in all aspects of our task with a diligence and devotion far beyond the dictates of duty.

The story of physics you find here would, however, not have been, except for the extraordinary ability and industry of Richard P. Feynman.

MATTHEW SANDS

March, 1964

Contents

1

Electromagnetism

Review: Chapter 12, Vol. I, *Characteristics of Force*

1-1 Electrical forces

Consider a force like gravitation which varies predominantly inversely as the square of the distance, but which is about a *billion-billion-billion-billion* times stronger. And with another difference. There are two kinds of "matter," which we can call positive and negative. Like kinds repel and unlike kinds attract—unlike gravity where there is only attraction. What would happen?

A bunch of positives would repel with an enormous force and spread out in all directions. A bunch of negatives would do the same. But an evenly mixed bunch of positives and negatives would do something completely different. The opposite pieces would be pulled together by the enormous attractions. The net result would be that the terrific forces would balance themselves out almost perfectly, by forming tight, fine mixtures of the positive and the negative, and between two separate bunches of such mixtures there would be practically no attraction or repulsion at all.

There is such a force: the electrical force. And all matter is a mixture of positive protons and negative electrons which are attracting and repelling with this great force. So perfect is the balance, however, that when you stand near someone else you don't feel any force at all. If there were even a little bit of unbalance you would know it. If you were standing at arm's length from someone and each of you had *one percent* more electrons than protons, the repelling force would be incredible. How great? Enough to lift the Empire State Building? No! To lift Mount Everest? No! The repulsion would be enough to lift a "weight" equal to that of the entire earth!

With such enormous forces so perfectly balanced in this intimate mixture, it is not hard to understand that matter, trying to keep its positive and negative charges in the finest balance, can have a great stiffness and strength. The Empire State Building, for example, swings only eight feet in the wind because the electrical forces hold every electron and proton more or less in its proper place. On the other hand, if we look at matter on a scale small enough that we see only a few atoms, any small piece will not, usually, have an equal number of positive and negative charges, and so there will be strong residual electrical forces. Even when there are equal numbers of both charges in two neighboring small pieces, there may still be large net electrical forces because the forces between individual charges vary inversely as the square of the distance. A net force can arise if a negative charge of one piece is closer to the positive than to the negative charges of the other piece. The attractive forces can then be larger than the repulsive ones and there can be a net attraction between two small pieces with no excess charges. The force that holds the atoms together, and the chemical forces that hold molecules together, are really electrical forces acting in regions where the balance of charge is not perfect, or where the distances are very small.

You know, of course, that atoms are made with positive protons in the nucleus and with electrons outside. You may ask: "If this electrical force is so terrific, why don't the protons and electrons just get on top of each other? If they want to be in an intimate mixture, why isn't it still more intimate?" The answer has to do with the quantum effects. If we try to confine our electrons in a region that is very close to the protons, then according to the uncertainty principle they must have some mean square momentum which is larger the more we try to confine them. It is this motion, required by the laws of quantum mechanics, that keeps the electrical attraction from bringing the charges any closer together.

There is another question: "What holds the nucleus together"? In a nucleus there are several protons, all of which are positive. Why don't they push themselves apart? It turns out that in nuclei there are, in addition to electrical forces, nonelectrical forces, called nuclear forces, which are greater than the electrical forces and which are able to hold the protons together in spite of the electrical repulsion. The nuclear forces, however, have a short range—their force falls off much more rapidly than $1/r^2$. And this has an important consequence. If a nucleus has too many protons in it, it gets too big, and it will not stay together. An example is uranium, with 92 protons. The nuclear forces act mainly between each proton (or neutron) and its nearest neighbor, while the electrical forces act over larger distances, giving a repulsion between each proton and all of the others in the nucleus. The more protons in a nucleus, the stronger is the electrical repulsion, until, as in the case of uranium, the balance is so delicate that the nucleus is almost ready to fly apart from the repulsive electrical force. If such a nucleus is just "tapped" lightly (as can be done by sending in a slow neutron), it breaks into two pieces, each with positive charge, and these pieces fly apart by electrical repulsion. The energy which is liberated is the energy of the atomic bomb. This energy is usually called "nuclear" energy, but it is really "electrical" energy released when electrical forces have overcome the attractive nuclear forces.

Lower case Greek letters and commonly used capitals

α		alpha
β		beta
γ	Γ	gamma
δ	Δ	delta
ϵ		epsilon
ζ		zeta
η		eta
θ	Θ	theta
ι		iota
κ		kappa
λ	Λ	lambda
μ		mu
ν		nu
ξ	Ξ	xi (ksi)
o		omicron
π	Π	pi
ρ		rho
σ	Σ	sigma
τ		tau
υ	Υ	upsilon
ϕ	Φ	phi
χ		chi (khi)
ψ	Ψ	psi
ω	Ω	omega

We may ask, finally, what holds a negatively charged electron together (since it has no nuclear forces). If an electron is all made of one kind of substance, each part should repel the other parts. Why, then, doesn't it fly apart? But does the electron have "parts"? Perhaps we should say that the electron is just a point and that electrical forces only act between *different* point charges, so that the electron does not act upon itself. Perhaps. All we can say is that the question of what holds the electron together has produced many difficulties in the attempts to form a complete theory of electromagnetism. The question has never been answered. We will entertain overselves by discussing this subject some more in later chapters.

As we have seen, we should expect that it is a combination of electrical forces and quantum-mechanical effects that will determine the detailed structure of materials in bulk, and, therefore, their properties. Some materials are hard, some are soft. Some are electrical "conductors"—because their electrons are free to move about; others are "insulators"—because their electrons are held tightly to individual atoms. We shall consider later how some of these properties come about, but that is a very complicated subject, so we will begin by looking at the electrical forces only in simple situations. We begin by treating only the laws of electricity—including magnetism, which is really a part of the same subject.

We have said that the electrical force, like a gravitational force, decreases inversely as the square of the distance between charges. This relationship is called Coulomb's law. But it is not precisely true when charges are moving—the electrical forces depend also on the motions of the charges in a complicated way. One part of the force between moving charges we call the *magnetic* force. It is really one aspect of an electrical effect. That is why we call the subject "electromagnetism."

There is an important general principle that makes it possible to treat electromagnetic forces in a relatively simple way. We find, from experiment, that the force that acts on a particular charge—no matter how many other charges there are or how they are moving—depends only on the position of that particular charge, on the velocity of the charge, and on the amount of charge. We can write the force $\boldsymbol{F}$ on a charge q moving with a velocity $\boldsymbol{v}$ as

$$\boldsymbol{F} = q(\boldsymbol{E} + \boldsymbol{v} \times \boldsymbol{B}). \tag{1.1}$$

We call $\boldsymbol{E}$ the *electric field* and $\boldsymbol{B}$ the *magnetic field* at the location of the charge. The important thing is that the electrical forces from all the other charges in the universe can be summarized by giving just these two vectors. Their values will depend on *where* the charge is, and may change with *time*. Furthermore, if we replace that charge with another charge, the force on the new charge will be just in proportion to the amount of charge so long as all the rest of the charges in the world do not change their positions or motions. (In real situations, of course, each charge produces forces on all other charges in the neighborhood and may cause these other charges to move, and so in some cases the fields *can* change if we replace our particular charge by another.)

We know from Vol. I how to find the motion of a particle if we know the force on it. Equation (1.1) can be combined with the equation of motion to give

$$\frac{d}{dt}\left[\frac{m\boldsymbol{v}}{(1 - v^2/c^2)^{1/2}}\right] = \boldsymbol{F} = q(\boldsymbol{E} + \boldsymbol{v} \times \boldsymbol{B}). \tag{1.2}$$

So if $\boldsymbol{E}$ and $\boldsymbol{B}$ are given, we can find the motions. Now we need to know how the $\boldsymbol{E}$'s and $\boldsymbol{B}$'s are produced.

One of the most important simplifying principles about the way the fields are produced is this: Suppose a number of charges moving in some manner would produce a field $\boldsymbol{E}_1$, and another set of charges would produce $\boldsymbol{E}_2$. If both sets of charges are in place at the same time (keeping the same locations and motions they had when considered separately), then the field produced is just the sum

$$\boldsymbol{E} = \boldsymbol{E}_1 + \boldsymbol{E}_2. \tag{1.3}$$

This fact is called *the principle of superposition* of fields. It holds also for magnetic fields.

This principle means that if we know the law for the electric and magnetic fields produced by a *single* charge moving in an arbitrary way, then all the laws of electrodynamics are complete. If we want to know the force on charge A we need only calculate the $\boldsymbol{E}$ and $\boldsymbol{B}$ produced by each of the charges B, C, D, etc., and then add the $\boldsymbol{E}$'s and $\boldsymbol{B}$'s from all the charges to find the fields, and from them the forces acting on charge A. If it had only turned out that the field produced by a single charge was simple, this would be the neatest way to describe the laws of electrodynamics. We have already given a description of this law (Chapter 28, Vol. I) and it is, unfortunately, rather complicated.

It turns out that the form in which the laws of electrodynamics are simplest are not what you might expect. It is *not* simplest to give a formula for the force that one charge produces on another. It is true that when charges are standing still the Coulomb force law is simple, but when charges are moving about the relations are complicated by delays in time and by the effects of acceleration, among others. As a result, we do not wish to present electrodynamics only through the force laws between charges; we find it more convenient to consider another point of view—a point of view in which the laws of electrodynamics appear to be the most easily manageable.

1–2 Electric and magnetic fields

First, we must extend, somewhat, our ideas of the electric and magnetic vectors, $\boldsymbol{E}$ and $\boldsymbol{B}$. We have defined them in terms of the forces that are felt by a charge. We wish now to speak of electric and magnetic fields *at a point* even when there is no charge present. We are saying, in effect, that since there are forces "acting on" the charge, there is still "something" there when the charge is removed. If a charge located at the point (x, y, z) at the time t feels the force $\boldsymbol{F}$ given by Eq. (1.1) we associate the vectors $\boldsymbol{E}$ and $\boldsymbol{B}$ with *the point* in space (x, y, z). We may think of $\boldsymbol{E}(x, y, z, t)$ and $\boldsymbol{B}(x, y, z, t)$ as giving the forces that *would be* experienced at the time t by a charge located at (x, y, z), *with the condition* that placing the charge there *did not disturb* the positions or motions of all the other charges responsible for the fields.

Following this idea, we associate with *every* point (x, y, z) in space two vectors $\boldsymbol{E}$ and $\boldsymbol{B}$, which may be changing with time. The electric and magnetic fields are, then, viewed as *vector functions* of x, y, z, and t. Since a vector is specified by its components, each of the fields $\boldsymbol{E}(x, y, z, t)$ and $\boldsymbol{B}(x, y, z, t)$ represent three mathematical functions of x, y, z, and t.

It is precisely because $\boldsymbol{E}$ (or $\boldsymbol{B}$) can be specified at every point in space that it is called a "field." A "field" is any physical quantity which takes on different values at different points in space. Temperature, for example, is a field—in this case a scalar field, which we write as $T(x, y, z)$. The temperature could also vary in time, and we would say the temperature field is time-dependent, and write $T(x, y, z, t)$. Another example is the "velocity field" of a flowing liquid. We write $\boldsymbol{v}(x, y, z, t)$ for the velocity of the liquid at each point in space at the time t. It is a vector field.

Returning to the electromagnetic fields—although they are produced by charges according to complicated formulas, they have the following important characteristic: the relationships between the values of the fields at *one point* and the values at a *nearby point* are very simple. With only a few such relationships in the form of differential equations we can describe the fields completely. It is in terms of such equations that the laws of electrodynamics are most simply written.

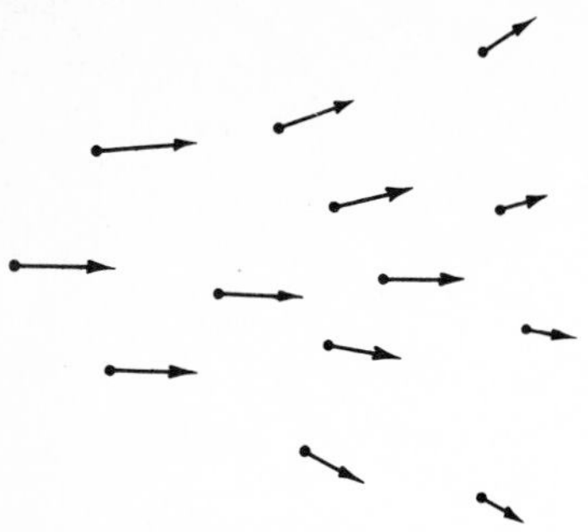

Fig. 1–1. A vector field may be represented by drawing a set of arrows whose magnitudes and directions indicate the values of the vector field at the points from which the arrows are drawn.

There have been various inventions to help the mind visualize the behavior of fields. The most correct is also the most abstract: we simply consider the fields as mathematical functions of position and time. We can also attempt to get a mental picture of the field by drawing vectors at many points in space, each of which gives the field strength and direction at that point. Such a representation is shown in Fig. 1–1. We can go further, however, and draw lines which are everywhere tangent to the vectors—which, so to speak, follow the arrows and keep track of the direction of the field. When we do this we lose track of the *lengths* of the vectors, but we can keep track of the strength of the field by drawing the lines far apart when the field is weak and close together when it is strong. We adopt the convention that the *number of lines per unit area* at right angles to the lines is proportional to the *field strength*. This is, of course, only an approximation, and it will require, in general, that new lines sometimes start up in order to keep the number up to the strength of the field. The field of Fig. 1–1 is represented by field lines in Fig. 1–2.

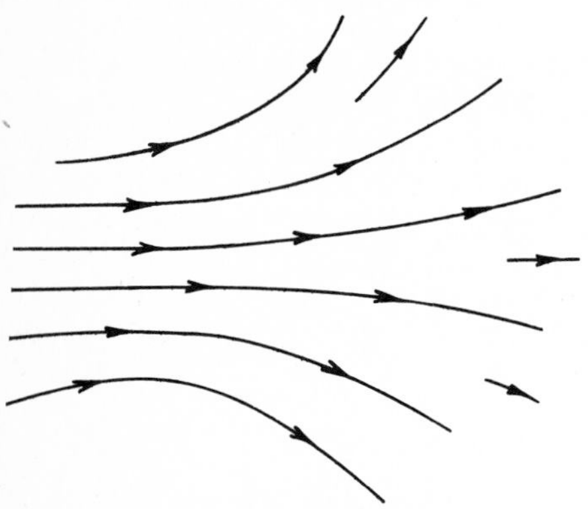

Fig. 1–2. A vector field can be represented by drawing lines which are tangent to the direction of the field vector at each point, and by drawing the density of lines proportional to the magnitude of the field vector.

1–3 Characteristics of vector fields

There are two mathematically important properties of a vector field which we will use in our description of the laws of electricity from the field point of view. Suppose we imagine a closed surface of some kind and ask whether we are losing "something" from the inside; that is, does the field have a quality of "outflow"? For instance, for a velocity field we might ask whether the velocity is always outward on the surface or, more generally, whether more fluid flows out (per unit time) than comes in. We call the net amount of fluid going out through the surface per unit time the "flux of velocity" through the surface. The flow through an element of a surface is just equal to the component of the velocity perpendicular to the surface times the area of the surface. For an arbitrary closed surface, the *net outward flow*—or *flux*—is the average outward normal component of the velocity, times the area of the surface:

$$\text{Flux} = (\text{average normal component})\cdot(\text{surface area}). \tag{1.4}$$

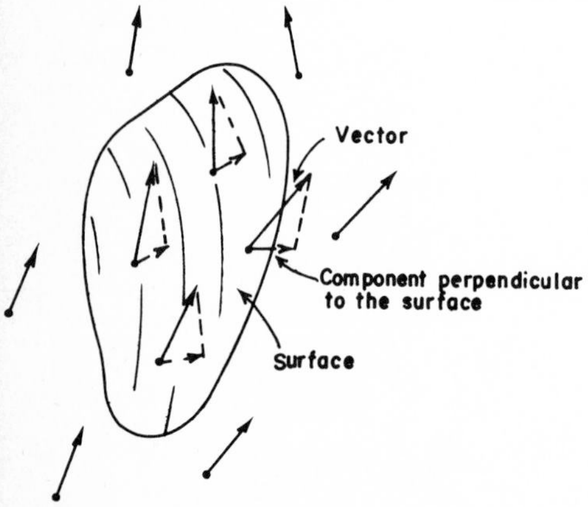

Fig. 1–3. The flux of a vector field through a surface is defined as the average value of the normal component of the vector times the area of the surface.

In the case of an electric field, we can mathematically define something analogous to an outflow, and we again call it the flux, but of course it is not the flow of any substance, because the electric field is not the velocity of anything. It turns out, however, that the mathematical quantity which is the average normal component of the field still has a useful significance. We speak, then, of the *electric flux*—also defined by Eq. (1.4). Finally, it is also useful to speak of the flux not only through a completely closed surface, but through any bounded surface. As before, the flux through such a surface is defined as the average normal component of a vector times the area of the surface. These ideas are illustrated in Fig. 1–3.

There is a second property of a vector field that has to do with a line, rather than a surface. Suppose again that we think of a velocity field that describes the flow of a liquid. We might ask this interesting question: Is the liquid circulating?

By that we mean: Is there a net rotational motion around some loop? Suppose that we instantaneously freeze the liquid everywhere except inside of a tube which is of uniform bore, and which goes in a loop that closes back on itself as in Fig. 1–4. Outside of the tube the liquid stops moving, but inside the tube it may keep on moving because of the momentum in the trapped liquid—that is, if there is more momentum heading one way around the tube than the other. We define a quantity called the *circulation* as the resulting speed of the liquid in the tube times its circumference. We can again extend our ideas and define the "circulation" for any vector field (even when there isn't anything moving). For any vector field the *circulation around any imagined closed curve* is defined as the average tangential component of the vector (in a consistent sense) multiplied by the circumference of the loop (Fig. 1–5).

$$\text{Circulation} = (\text{average tangential component})\cdot(\text{distance around}). \tag{1.5}$$

You will see that this definition does indeed give a number which is proportional to the circulation velocity in the quickly frozen tube described above.

With just these two ideas—flux and circulation—we can describe all the laws of electricity and magnetism at once. You may not understand the significance of the laws right away, but they will give you some idea of the way the physics of electromagnetism will be ultimately described.

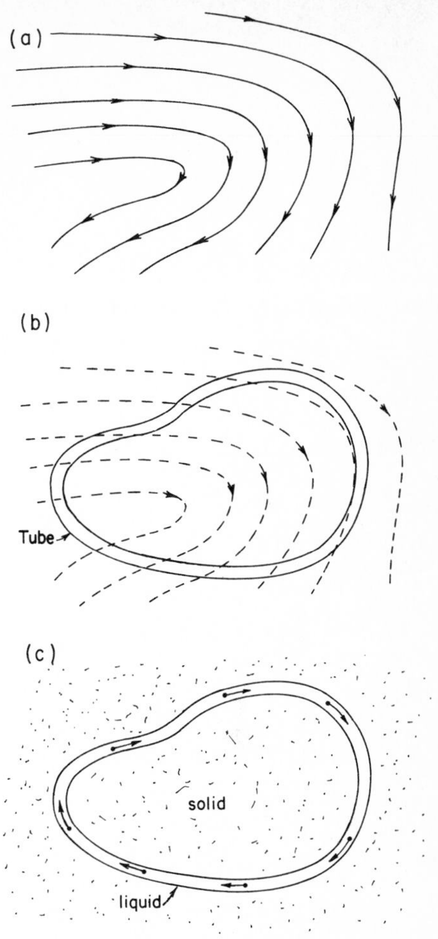

Fig. 1–4. (a) The velocity field in a liquid. Imagine a tube of uniform cross section that follows an arbitrary closed curve as in (b). If the liquid were suddenly frozen everywhere except inside the tube, the liquid in the tube would circulate as shown in (c).

1–4 The laws of electromagnetism

The first law of electromagnetism describes the flux of the electric field:

$$\text{The flux of } \boldsymbol{E} \text{ through any closed surface} = \frac{\text{the net charge inside}}{\epsilon_0}, \tag{1.6}$$

where ϵ_0 is a convenient constant. (The constant ϵ_0 is usually read as "epsilon-zero" or "epsilon-naught".) If there are no charges inside the surface, even though there are charges nearby outside the surface, the *average* normal component of $\boldsymbol{E}$ is zero, so there is no net flux through the surface. To show the power of this type of statement, we can show that Eq. (1.6) is the same as Coulomb's law, provided only that we also add the idea that the field from a single charge is spherically symmetric. For a point charge, we draw a sphere around the charge. Then the average normal component is just the value of the magnitude of $\boldsymbol{E}$ at any point, since the field must be directed radially and have the same strength for all points on the sphere. Our rule now says that the field at the surface of the sphere, times the area of the sphere—that is, the outgoing flux—is proportional to the charge inside. If we were to make the radius of the sphere bigger, the area would increase as the square of the radius. The average normal component of the electric field times that area must still be equal to the same charge inside, and so the field must decrease as the square of the distance—we get an "inverse square" field.

If we have an arbitrary stationary curve in space and measure the circulation of the electric field around the curve, we will find that it is not, in general, zero (although it is for the Coulomb field). Rather, for electricity there is a second law that states: for any surface S (not closed) whose edge is the curve C,

$$\text{Circulation of } \boldsymbol{E} \text{ around } C = -\frac{d}{dt}(\text{flux of } \boldsymbol{B} \text{ through } S). \tag{1.7}$$

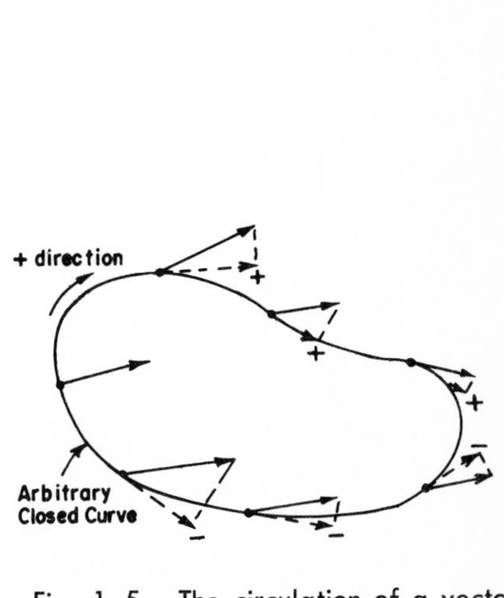

Fig. 1–5. The circulation of a vector field is the average tangential component of the vector (in a consistent sense) times the circumference of the loop.

We can complete the laws of the electromagnetic field by writing two corresponding equations for the magnetic field $\boldsymbol{B}$.

$$\text{Flux of } \boldsymbol{B} \text{ through any closed surface} = 0. \tag{1.8}$$

For a surface S bounded by the curve C,

$$c^2(\text{circulation of } \boldsymbol{B} \text{ around } C) = \frac{d}{dt}(\text{flux of } \boldsymbol{E} \text{ through } S) + \frac{\text{flux of electric current through } S}{\epsilon_0}. \tag{1.9}$$

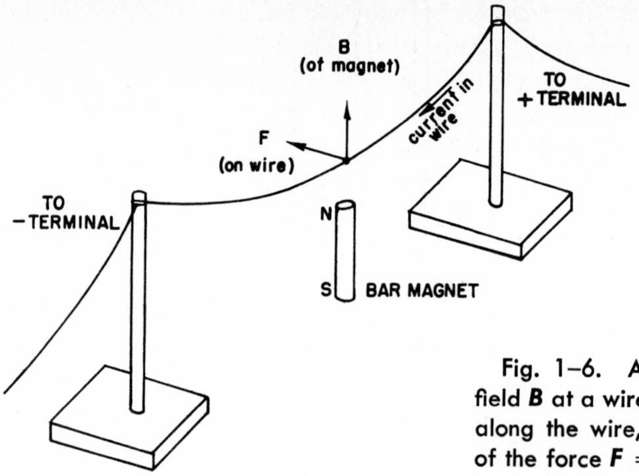

Fig. 1–6. A bar magnet gives a field $\boldsymbol{B}$ at a wire. When there is a current along the wire, the wire moves because of the force $\boldsymbol{F} = q\boldsymbol{v} \times \boldsymbol{B}$.

The constant c^2 that appears in Eq. (1.9) is the square of the velocity of light. It appears because magnetism is in reality a relativistic effect of electricity. The constant ϵ_0 has been stuck in to make the units of electric current come out in a convenient way.

Equations (1.6) through (1.9), together with Eq. (1.1), are all the laws of electrodynamics*. As you remember, the laws of Newton were very simple to write down, but they had a lot of complicated consequences and it took us a long time to learn about them all. These laws are not nearly as simple to write down, which means that the consequences are going to be more elaborate and it will take us quite a lot of time to figure them all out.

We can illustrate some of the laws of electrodynamics by a series of small experiments which show qualitatively the interrelationships of electric and magnetic fields. You have experienced the first term of Eq. (1.1) when combing your hair, so we won't show that one. The second part of Eq. (1.1) can be demonstrated by passing a current through a wire which hangs above a bar magnet, as shown in Fig. 1–6. The wire will move when a current is turned on because of the force $\boldsymbol{F} = q\boldsymbol{v} \times \boldsymbol{B}$. When a current exists, the charges inside the wire are moving, so they have a velocity $\boldsymbol{v}$, and the magnetic field from the magnet exerts a force on them, which results in pushing the wire sideways.

When the wire is pushed to the left, we would expect that the magnet must feel a push to the right. (Otherwise we could put the whole thing on a wagon and have a propulsion system that didn't conserve momentum!) Although the force is too small to make movement of the bar magnet visible, a more sensitively supported magnet, like a compass needle, will show the movement.

How does the wire push on the magnet? The current in the wire produces a magnetic field of its own that exerts forces on the magnet. According to the last

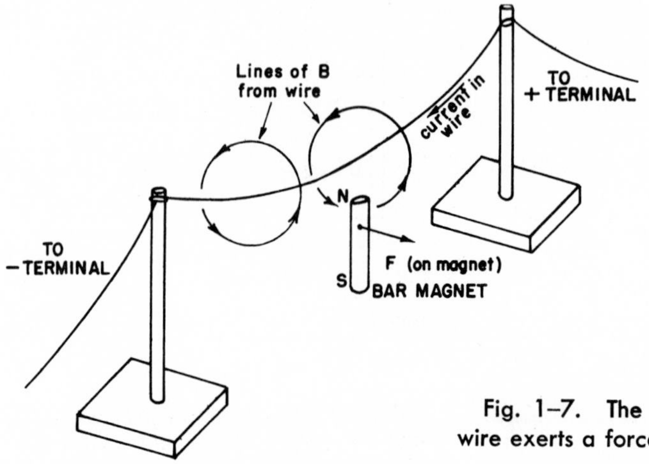

Fig. 1–7. The magnetic field of the wire exerts a force on the magnet.

* We need only to add a remark about some conventions for the *sign* of the circulation.

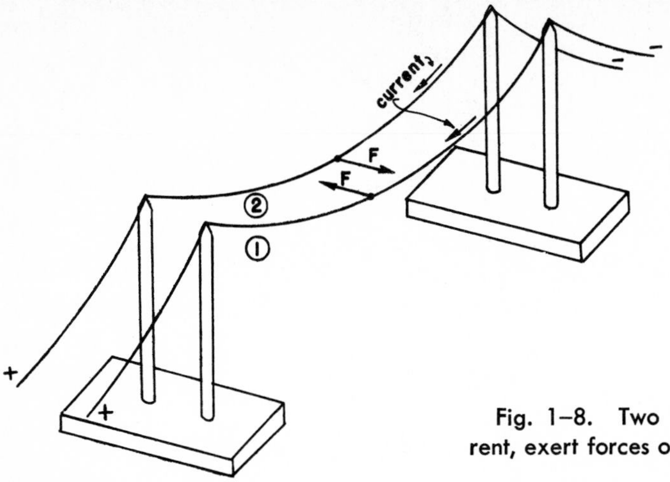

Fig. 1–8. Two wires, carrying current, exert forces on each other.

term in Eq. (1.9), a current must have a *circulation* of $\boldsymbol{B}$—in this case, the lines of $\boldsymbol{B}$ are loops around the wire, as shown in Fig. 1–7. This $\boldsymbol{B}$-field is responsible for the force on the magnet.

Equation (1.9) tells us that for a fixed current through the wire the circulation of $\boldsymbol{B}$ is the same for *any* curve that surrounds the wire. For curves—say circles—that are farther away from the wire, the circumference is larger, so the tangential component of $\boldsymbol{B}$ must decrease. You can see that we would, in fact, expect $\boldsymbol{B}$ to decrease linearly with the distance from a long straight wire.

Now, we have said that a current through a wire produces a magnetic field, and that when there is a magnetic field present there is a force on a wire carrying a current. Then we should also expect that if we make a magnetic field with a current in one wire, it should exert a force on another wire which also carries a current. This can be shown by using two hanging wires as shown in Fig. 1–8. When the currents are in the same direction, the two wires attract, but when the currents are opposite, they repel.

In short, electrical currents, as well as magnets, make magnetic fields. But wait, what is a magnet, anyway? If magnetic fields are produced by moving charges, is it not possible that the magnetic field from a piece of iron is really the result of currents? It appears to be so. We can replace the bar magnet of our experiment with a coil of wire, as shown in Fig. 1–9. When a current is passed through the coil—as well as through the straight wire above it—we observe a motion of the wire exactly as before, when we had a magnet instead of a coil. In other words, the current in the coil imitates a magnet. It appears, then, that a piece of iron acts as though it contains a perpetual circulating current. We can, in fact, understand magnets in terms of permanent currents in the atoms of the iron. The force on the magnet in Fig. 1–7 is due to the second term in Eq. (1.1).

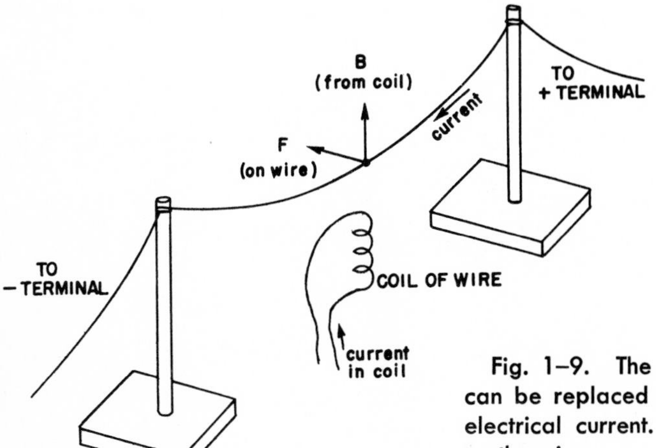

Fig. 1–9. The bar magnet of Fig. 1–6 can be replaced by a coil carrying an electrical current. A similar force acts on the wire.

Where do the currents come from? One possibility would be from the motion of the electrons in atomic orbits. Actually, that is not the case for iron, although it is for some materials. In addition to moving around in an atom, an electron also spins about on its own axis—something like the spin of the earth—and it is the current from this spin that gives the magnetic field in iron. (We say "something like the spin of the earth" because the question is so deep in quantum mechanics that the classical ideas do not really describe things too well.) In most substances, some electrons spin one way and some spin the other, so the magnetism cancels out, but in iron—for a mysterious reason which we will discuss later—many of the electrons are spinning with their axes lined up, and that is the source of the magnetism.

Since the fields of magnets are from currents, we do not have to add any extra term to Eqs. (1.8) or (1.9) to take care of magnets. We just take *all* currents, including the circulating currents of the spinning electrons, and then the law is right. You should also notice that Eq. (1.8) says that there are no magnetic "charges" analogous to the electrical charges appearing on the right side of Eq. (1.6). None has been found.

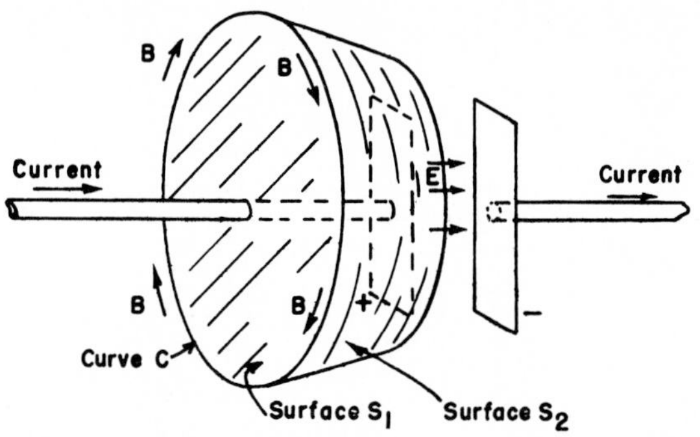

Fig. 1–10. The circulation of $\boldsymbol{B}$ around the curve C is given either by the current passing through the surface S_1, or by the rate of change of the flux of $\boldsymbol{E}$ through the surface S_2.

The first term on the right-hand side of Eq. (1.9) was discovered theoretically by Maxwell and is of great importance. It says that changing *electric* fields produce magnetic effects. In fact, without this term the equation would not make sense, because without it there could be no currents in circuits that are not complete loops. But such currents do exist, as we can see in the following example. Imagine a capacitor made of two flat plates. It is being charged by a current that flows toward one plate and away from the other, as shown in Fig. 1–10. We draw a curve C around one of the wires and fill it in with a surface which crosses the wire, as shown by the surface S_1 in the figure. According to Eq. (1.9), the circulation of $\boldsymbol{B}$ around C is given by the current in the wire (times c^2). But what if we fill in the curve with a *different* surface S_2, which is shaped like a bowl and passes between the plates of the capacitor, staying always away from the wire? There is certainly no current through this surface. But, surely, just changing the location of an imaginary surface is not going to change a real magnetic field! The circulation of $\boldsymbol{B}$ must be what it was before. The first term on the right-hand side of Eq. (1.9) does, indeed, combine with the second term to give the same result for the two surfaces S_1 and S_2. For S_2 the circulation of $\boldsymbol{B}$ is given in terms of the rate of change of the flux of $\boldsymbol{E}$ between the plates of the capacitor. And it works out that the changing $\boldsymbol{E}$ is related to the current in just the way required for Eq. (1.9) to be correct. Maxwell saw that it was needed, and he was the first to write the complete equation.

With the setup shown in Fig. 1–6 we can demonstrate another of the laws of electromagnetism. We disconnect the ends of the hanging wire from the battery and connect them to a galvanometer which tells us when there is a current through the wire. When we *push* the wire sideways through the magnetic field of the magnet, we observe a current. Such an effect is again just another consequence of Eq. (1.1)—the electrons in the wire feel the force $\boldsymbol{F} = q\boldsymbol{v} \times \boldsymbol{B}$. The electrons have a sidewise velocity because they move with the wire. This $\boldsymbol{v}$ with a vertical $\boldsymbol{B}$ from the magnet results in a force on the electrons directed *along* the wire, which starts the electrons moving toward the galvanometer.

Suppose, however, that we leave the wire alone and move the magnet. We guess from relativity that it should make no difference, and indeed, we observe a similar current in the galvanometer. How does the magnetic field produce forces on charges at rest? According to Eq. (1.1) there must be an electric field. A moving magnet must make an electric field. How that happens is said quantitatively by Eq. (1.7). This equation describes many phenomena of great practical interest, such as those that occur in electric generators and transformers.

The most remarkable consequence of our equations is that the combination of Eq. (1.7) and Eq. (1.9) contains the explanation of the radiation of electromagnetic effects over large distances. The reason is roughly something like this: suppose that somewhere we have a magnetic field which is increasing because, say, a current is turned on suddenly in a wire. Then by Eq. (1.7) there must be a circulation of an electric field. As the electric field builds up to produce its circulation, then according to Eq. (1.9) a magnetic circulation will be generated. But the building up of *this* magnetic field will produce a new circulation of the electric field, and so on. In this way fields work their way through space without the need of charges or currents except at their source. That is the way we *see* each other! It is all in the equations of the electromagnetic fields.

1–5 What are the fields?

We now make a few remarks on our way of looking at this subject. You may be saying: "All this business of fluxes and circulations is pretty abstract. There are electric fields at every point in space; then there are these 'laws.' But what is *actually* happening? Why can't you explain it, for instance, by whatever it *is* that goes between the charges." Well, it depends on your prejudices. Many physicists used to say that direct action with nothing in between was inconceivable. (How could they find an idea inconceivable when it had already been conceived?) They would say: "Look, the only forces we know are the direct action of one piece of matter on another. It is impossible that there can be a force with nothing to transmit it." But what really happens when we study the "direct action" of one piece of matter right against another? We discover that it is not one piece right against the other; they are slightly separated, and there are electrical forces acting on a tiny scale. Thus we find that we are going to explain so-called direct-contact action in terms of the picture for electrical forces. It is certainly not sensible to try to insist that an electrical force has to look like the old, familiar, muscular push or pull, when it will turn out that the muscular pushes and pulls are going to be interpreted as electrical forces! The only sensible question is what is the *most convenient* way to look at electrical effects. Some people prefer to represent them as the interaction at a distance of charges, and to use a complicated law. Others love the field lines. They draw field lines all the time, and feel that writing $\boldsymbol{E}$'s and $\boldsymbol{B}$'s is too abstract. The field lines, however, are only a crude way of describing a field, and it is very difficult to give the correct, quantitative laws directly in terms of field lines. Also, the ideas of the field lines do not contain the deepest principle of electrodynamics, which is the superposition principle. Even though we know how the field lines look for one set of charges and what the field lines look like for another set of charges, we don't get any idea about what the field line patterns will look like when both sets are present together. From the mathematical standpoint, on the other hand, superposition is easy—we simply add the two vectors. The field lines have some advantage in giving a vivid picture, but they also have some disadvantages. The direct interaction way of thinking has great advantages when thinking of electrical charges at rest, but has great disadvantages when dealing with charges in rapid motion.

The best way is to use the abstract field idea. That it is abstract is unfortunate, but necessary. The attempts to try to represent the electric field as the motion of some kind of gear wheels, or in terms of lines, or of stresses in some kind of material have used up more effort of physicists than it would have taken simply to get the right answers about electrodynamics. It is interesting that the correct equations for the behavior of light in crystals were worked out by McCullough in 1843. But people said to him: "Yes, but there is no real material whose mechanical properties could possibly satisfy those equations, and since light is an oscillation that must vibrate in *something*, we cannot believe this abstract equation business." If people had been more open-minded, they might have believed in the right equations for the behavior of light a lot earlier than they did.

In the case of the magnetic field we can make the following point: Suppose that you finally succeeded in making up a picture of the magnetic field in terms of some kind of lines or of gear wheels running through space. Then you try to explain what happens to two charges moving in space, both at the same speed and parallel to each other. Because they are moving, they will behave like two currents and will have a magnetic field associated with them (like the currents in the wires of Fig. 1–8). An observer who was riding along with the two charges, however, would see both charges as stationary, and would say that there is *no* magnetic field. The "gear wheels" or "lines" disappear when you ride along with the object! All we have done is to invent a *new* problem. How can the gear wheels disappear?! The people who draw field lines are in a similar difficulty. Not only is it not possible to say whether the field lines move or do not move with charges—they may disappear completely in certain coordinate frames

What we are saying, then, is that magnetism is really a relativistic effect. In the case of the two charges we just considered, travelling parallel to each other, we would expect to have to make relativistic corrections to their motion, with terms of order v^2/c^2. These corrections must correspond to the magnetic force. But what about the force between the two wires in our experiment (Fig. 1–8). There the magnetic force is the *whole* force. It didn't look like a "relativistic correction." Also, if we estimate the velocities of the electrons in the wire (you can do this yourself), we find that their average speed along the wire is about 0.01 centimeter per second. So v^2/c^2 is about 10^{-25}. Surely a negligible "correction." But no! Although the magnetic force is, in this case, 10^{-25} of the "normal" electrical force between the moving electrons, remember that the "normal" electrical forces have disappeared because of the almost perfect balancing out—because the wires have the same number of protons as electrons. The balance is much more precise than one part in 10^{25}, and the small relativistic term which we call the magnetic force is the only term left. It becomes the dominant term.

It is the near-perfect cancellation of electrical effects which allowed relativity effects (that is, magnetism) to be studied and the correct equations—to order v^2/c^2—to be discovered, even though physicists didn't *know* that's what was happening. And that is why, when relativity was discovered, the electromagnetic laws didn't need to be changed. They—unlike mechanics—were already correct to a precision of v^2/c^2.

1–6 Electromagnetism in science and technology

Let us end this chapter by pointing out that among the many phenomena studied by the Greeks there were two very strange ones: that if you rubbed a piece of amber you could lift up little pieces of papyrus, and that there was a strange rock from the island of Magnesia which attracted iron. It is amazing to think that these were the only phenomena known to the Greeks in which the effects of electricity or magnetism were apparent. The reason that these were the only phenomena that appeared is due primarily to the fantastic precision of the balancing of charges that we mentioned earlier. Study by scientists who came after the Greeks uncovered one new phenomena after another that were really some aspect of these amber and/or lodestone effects. Now we realize that the phenomena of chemical interaction and, ultimately, of life itself are to be understood in terms of electromagnetism.

At the same time that an understanding of the subject of electromagnetism was being developed, technical possibilities that defied the imagination of the people that came before were appearing: it became possible to signal by telegraph over long distances, and to talk to another person miles away without any connections between, and to run huge power systems—a great water wheel, connected by filaments over hundreds of miles to another engine that turns in response to the master wheel—many thousands of branching filaments—ten thousand engines in ten thousand places running the machines of industries and homes—all turning because of the knowledge of the laws of electromagnetism.

Today we are applying even more subtle effects. The electrical forces, enormous as they are, can also be very tiny, and we can control them and use them in very many ways. So delicate are our instruments that we can tell what a man is doing by the way he affects the electrons in a thin metal rod hundreds of miles away. All we need to do is to use the rod as an antenna for a television receiver!

From a long view of the history of mankind—seen from, say, ten thousand years from now—there can be little doubt that the most significant event of the 19th century will be judged as Maxwell's discovery of the laws of electrodynamics. The American Civil War will pale into provincial insignificance in comparison with this important scientific event of the same decade.

2

Differential Calculus of Vector Fields

2–1 Understanding physics

Review: Chapter 11, Vol. I, *Vectors*

The physicist needs a facility in looking at problems from several points of view. The exact analysis of real physical problems is usually quite complicated, and any particular physical situation may be too complicated to analyze directly by solving the differential equation. But one can still get a very good idea of the behavior of a system if one has some feel for the character of the solution in different circumstances. Ideas such as the field lines, capacitance, resistance, and inductance are, for such purposes, very useful. So we will spend much of our time analyzing them. In this way we will get a feel as to what should happen in different electromagnetic situations. On the other hand, none of the heuristic models, such as field lines, is really adequate and accurate for all situations. There is only one precise way of presenting the laws, and that is by means of differential equations. They have the advantage of being fundamental and, so far as we know, precise. If you have learned the differential equations you can always go back to them. There is nothing to unlearn.

It will take you some time to understand what should happen in different circumstances. You will have to solve the equations. Each time you solve the equations, you will learn something about the character of the solutions. To keep these solutions in mind, it will be useful also to study their meaning in terms of field lines and of other concepts. This is the way you will really "understand" the equations. That is the difference between mathematics and physics. Mathematicians, or people who have very mathematical minds, are often led astray when "studying" physics because they lose sight of the physics. They say: "Look, these differential equations—the Maxwell equations—are all there is to electrodynamics; it is admitted by the physicists that there is nothing which is not contained in the equations. The equations are complicated, but after all they are only mathematical equations and if I understand them mathematically inside out, I will understand the physics inside out." Only it doesn't work that way. Mathematicians who study physics with that point of view—and there have been many of them—usually make little contribution to physics and, in fact, little to mathematics. They fail because the actual physical situations in the real world are so complicated that it is necessary to have a much broader understanding of the equations.

What it means really to understand an equation—that is, in more than a strictly mathematical sense—was described by Dirac. He said: "I understand what an equation means if I have a way of figuring out the characteristics of its solution without actually solving it." So if we have a way of knowing what should happen in given circumstances without actually solving the equations, then we "understand" the equations, as applied to these circumstances. A physical understanding is a completely unmathematical, imprecise, and inexact thing, but absolutely necessary for a physicist.

Ordinarily, a course like this is given by developing gradually the physical ideas—by starting with simple situations and going on to more and more complicated situations. This requires that you continuously forget things you previously learned—things that are true in certain situations, but which are not true in general. For example, the "law" that the electrical force depends on the square of the distance is not *always* true. We prefer the opposite approach. We prefer to take first the *complete* laws, and then to step back and apply them to simple situations, developing the physical ideas as we go along. And that is what we are going to do.

Our approach is completely opposite to the historical approach in which one develops the subject in terms of the experiments by which the information was obtained. But the subject of physics has been developed over the past 200 years by some very ingenious people, and as we have only a limited time to acquire our knowledge, we cannot possibly cover everything they did. Unfortunately one of the things that we shall have a tendency to lose in these lectures is the historical, experimental development. It is hoped that in the laboratory some of this lack can be corrected. You can also fill in what we must leave out by reading the Encyclopedia Brittanica, which has excellent historical articles on electricity and on other parts of physics. You will also find historical information in many textbooks on electricity and magnetism.

2–2 Scalar and vector fields—T and h

We begin now with the abstract, mathematical view of the theory of electricity and magnetism. The ultimate idea is to explain the meaning of the laws given in Chapter 1. But to do this we must first explain a new and peculiar notation that we want to use. So let us forget electromagnetism for the moment and discuss the mathematics of vector fields. It is of very great importance, not only for electromagnetism, but for all kinds of physical circumstances. Just as ordinary differential and integral calculus is so important to all branches of physics, so also is the differential calculus of vectors. We turn to that subject.

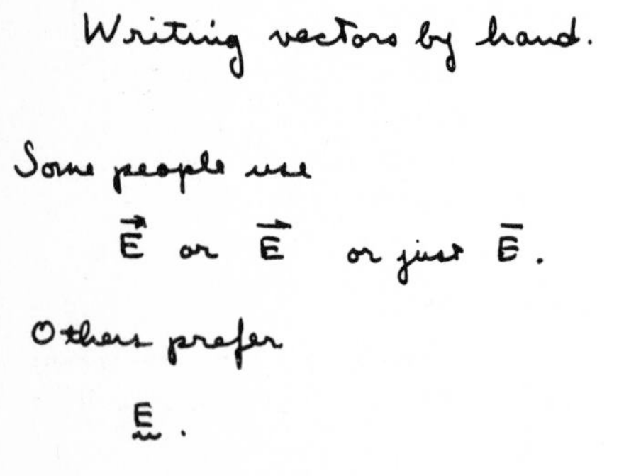

Listed below are a few facts from the algebra of vectors. It is assumed that you already know them.

$$\boldsymbol{A} \cdot \boldsymbol{B} = \text{scalar} = A_x B_x + A_y B_y + A_z B_z \tag{2.1}$$

$$\boldsymbol{A} \times \boldsymbol{B} = \text{vector} \tag{2.2}$$
$$(\boldsymbol{A} \times \boldsymbol{B})_z = A_x B_y - A_y B_x$$
$$(\boldsymbol{A} \times \boldsymbol{B})_x = A_y B_z - A_z B_y$$
$$(\boldsymbol{A} \times \boldsymbol{B})_y = A_z B_x - A_x B_z$$

$$\boldsymbol{A} \times \boldsymbol{A} = 0 \tag{2.3}$$

$$\boldsymbol{A} \cdot (\boldsymbol{A} \times \boldsymbol{B}) = 0 \tag{2.4}$$

$$\boldsymbol{A} \cdot (\boldsymbol{B} \times \boldsymbol{C}) = (\boldsymbol{A} \times \boldsymbol{B}) \cdot \boldsymbol{C} \tag{2.5}$$

$$\boldsymbol{A} \times (\boldsymbol{B} \times \boldsymbol{C}) = \boldsymbol{B}(\boldsymbol{A} \cdot \boldsymbol{C}) - \boldsymbol{C}(\boldsymbol{A} \cdot \boldsymbol{B}) \tag{2.6}$$

Also we will want to use the two following equalities from the calculus:

$$\Delta f(x, y, z) = \frac{\partial f}{\partial x} \Delta x + \frac{\partial f}{\partial y} \Delta y + \frac{\partial f}{\partial z} \Delta z, \tag{2.7}$$

$$\frac{\partial^2 f}{\partial x\, \partial y} = \frac{\partial^2 f}{\partial y\, \partial x}. \tag{2.8}$$

The first equation (2.7) is, of course, true only in the limit that Δx, Δy, and Δz go toward zero.

The simplest possible physical field is a scalar field. By a field, you remember, we mean a quantity which depends upon position in space. By a *scalar field* we merely mean a field which is characterized at each point by a single number—a scalar. Of course the number may change in time, but we need not worry about that for the moment. We will talk about what the field looks like at a given instant. As an example of a scalar field, consider a solid block of material which has been heated at some places and cooled at others, so that the temperature of the body varies from point to point in a complicated way. Then the temperature will be a function of x, y, and z, the position in space measured in a rectangular coordinate system. Temperature is a scalar field.

Writing vectors by hand.

Some people use

$$\vec{E} \quad or \quad \overrightarrow{E} \quad or\ just\ \bar{E}.$$

Other prefer

$$\underset{\sim}{E}.$$

We like the following way:

A B C D E F G

H I J K L M N

O P Q R S T U

V W X Y Z

Small letters are harder:

a b c d e f g

h i j k l m n

o p q r s t u

v w x y z

You can invent your own.

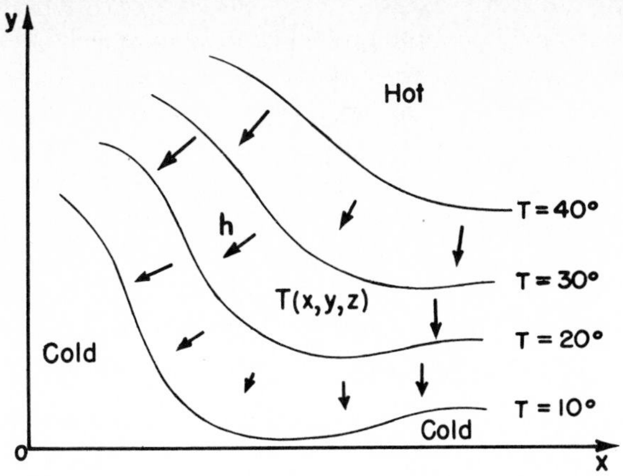

Fig. 2–1. Temperature T is an example of a scalar field. With each point (x, y, z) in space there is associated a number $T(x, y, z)$. All points on the surface marked $T = 20°$ (shown as a curve at $z = 0$) are at the same temperature. The arrows are samples of the heat flow vector $\boldsymbol{h}$.

One way of thinking about scalar fields is to imagine "contours" which are imaginary surfaces drawn through all points for which the field has the same value, just as contour lines on a map connect points with the same height. For a temperature field the contours are called "isothermal surfaces" or isotherms. Figure 2–1 illustrates a temperature field and shows the dependence of T on x and y when $z = 0$. Several isotherms are drawn.

There are also vector fields. The idea is very simple. A vector is given for each point in space. The vector varies from point to point. As an example, consider a rotating body. The velocity of the material of the body at any point is a vector which is a function of position (Fig. 2–2). As a second example, consider the flow of heat in a block of material. If the temperature in the block is high at one place and low at another, there will be a flow of heat from the hotter places to the colder. The heat will be flowing in different directions in different parts of the block. The heat flow is a directional quantity which we call $\boldsymbol{h}$. Its magnitude is a measure of how much heat is flowing. Examples of the heat flow vector are also shown in Fig. 2–1.

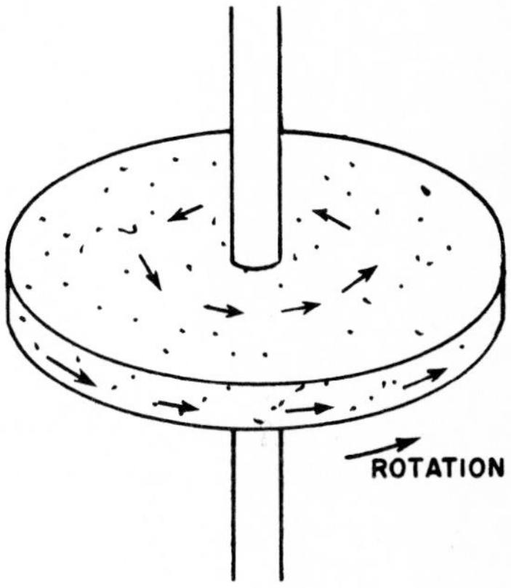

Fig. 2–2. The velocity of the atoms in a rotating object is an example of a vector field.

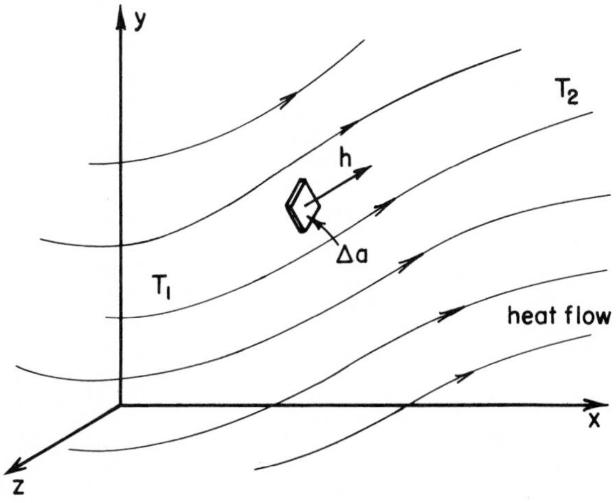

Fig. 2–3. Heat flow is a vector field. The vector $\boldsymbol{h}$ points along the direction of the flow. Its magnitude is the energy transported per unit time across a surface element oriented perpendicular to the flow, divided by the area of the surface element.

Let's make a more precise definition of $\boldsymbol{h}$: The magnitude of the vector heat flow at a point is the amount of thermal energy that passes, per unit time and per unit area, through an infinitesimal surface element at right angles to the direction of flow. The vector points in the direction of flow (see Fig. 2–3). In symbols: If ΔJ is the thermal energy that passes per unit time through the surface element Δa, then

$$\boldsymbol{h} = \frac{\Delta J}{\Delta a}\,\boldsymbol{e}_f, \tag{2.9}$$

where $\boldsymbol{e}_f$ is a *unit vector* in the direction of flow.

The vector $\boldsymbol{h}$ can be defined in another way—in terms of its components. We ask how much heat flows through a small surface at *any* angle with respect to the flow. In Fig. 2–4 we show a small surface Δa_2 inclined with respect to Δa_1, which is perpendicular to the flow. The *unit vector* $\boldsymbol{n}$ is normal to the surface Δa_2. The

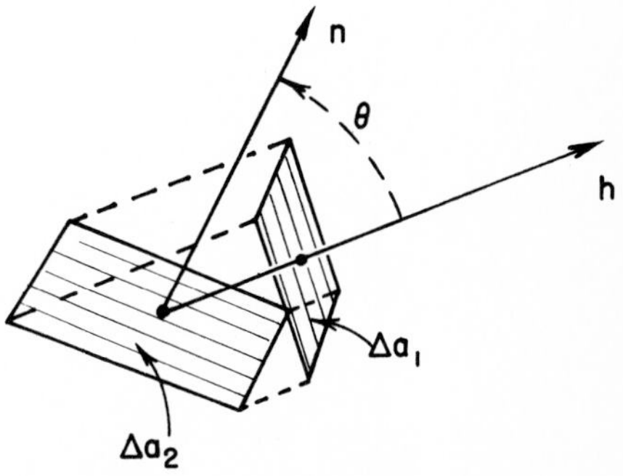

Fig. 2–4. The heat flow through Δa_2 is the same as through Δa_1.

angle θ between $\boldsymbol{n}$ and $\boldsymbol{h}$ is the same as the angle between the surfaces (since $\boldsymbol{h}$ is normal to Δa_1). Now what is the heat flow *per unit area* through Δa_2? The flow through Δa_2 is the same as through Δa_1; only the areas are different. In fact, $\Delta a_1 = \Delta a_2 \cos\theta$. The heat flow through Δa_2 is

$$\frac{\Delta J}{\Delta a_2} = \frac{\Delta J}{\Delta a_1}\cos\theta = \boldsymbol{h}\cdot\boldsymbol{n}. \tag{2.10}$$

We interpret this equation: the heat flow (per unit time and per unit area) through *any* surface element whose unit normal is $\boldsymbol{n}$, is given by $\boldsymbol{h}\cdot\boldsymbol{n}$. Equally, we could say: the component of the heat flow perpendicular to the surface element Δa_2 is $\boldsymbol{h}\cdot\boldsymbol{n}$. We can, if we wish, consider that these statements *define* $\boldsymbol{h}$. We will be applying the same ideas to other vector fields.

2–3 Derivatives of fields—the gradient

When fields vary in time, we can describe the variation by giving their derivatives with respect to t. We want to describe the variations with position in a similar way, because we are interested in the relationship between, say, the temperature in one place and the temperature at a nearby place. How shall we take the derivative of the temperature with respect to position? Do we differentiate the temperature with respect to x? Or with respect to y, or z?

Useful physical laws do not depend upon the orientation of the coordinate system. They should, therefore, be written in a form in which either both sides are scalars or both sides are vectors. What is the derivative of a scalar field, say $\partial T/\partial x$? Is it a scalar, or a vector, or what? It is neither a scalar nor a vector, as you can easily appreciate, because if we took a different x-axis, $\partial T/\partial x$ would certainly be different. But notice: We have three possible derivatives: $\partial T/\partial x$, $\partial T/\partial y$, and $\partial T/\partial z$. Since there are three kinds of derivatives and we know that it takes three numbers to form a vector, perhaps these three derivatives are the components of a vector:

$$\left(\frac{\partial T}{\partial x}, \frac{\partial T}{\partial y}, \frac{\partial T}{\partial z}\right) \stackrel{?}{=} \text{a vector}. \tag{2.11}$$

Of course it is not generally true that *any* three numbers form a vector. It is true only if, when we rotate the coordinate system, the components of the vector transform among themselves in the correct way. So it is necessary to analyze how these derivatives are changed by a rotation of the coordinate system. We shall show that (2.11) is indeed a vector. The derivatives do transform in the correct way when the coordinate system is rotated.

We can see this in several ways. One way is to ask a question whose answer is independent of the coordinate system, and try to express the answer in an "invariant" form. For instance, if $S = \boldsymbol{A}\cdot\boldsymbol{B}$, and if $\boldsymbol{A}$ and $\boldsymbol{B}$ are vectors, we know—because we proved it in Chapter 11 of Vol. I—that S is a scalar. We *know* that S is a scalar without investigating whether it changes with changes in coordinate systems. It *can't*, because it's a dot product of two vectors. Similarly, if we *know* that $\boldsymbol{A}$ is a vector, and we have three numbers B_1, B_2, and B_3, and we find out that

$$A_xB_1 + A_yB_2 + A_zB_3 = S, \tag{2.12}$$

where S is the same for any coordinate system, then it *must* be that the three numbers B_1, B_2, B_3 are the components B_x, B_y, B_z of some vector $\boldsymbol{B}$.

Now let's think of the temperature field. Suppose we take two points P_1 and P_2, separated by the small interval $\Delta\boldsymbol{R}$. The temperature at P_1 is T_1 and at P_2 is T_2, and the difference $\Delta T = T_2 - T_1$. The temperatures at these real, physical points certainly do not depend on what axis we choose for measuring the coordinates. In particular, ΔT is a number independent of the coordinate system. It is a scalar.

If we choose some convenient set of axes, we could write $T_1 = T(x, y, z)$ and $T_2 = T(x + \Delta x, y + \Delta y, z + \Delta z)$, where Δx, Δy, and Δz are the components of the vector $\Delta \boldsymbol{R}$ (Fig. 2–5). Remembering Eq. (2.7), we can write

$$\Delta T = \frac{\partial T}{\partial x}\,\Delta x + \frac{\partial T}{\partial y}\,\Delta y + \frac{\partial T}{\partial z}\,\Delta z. \tag{2.13}$$

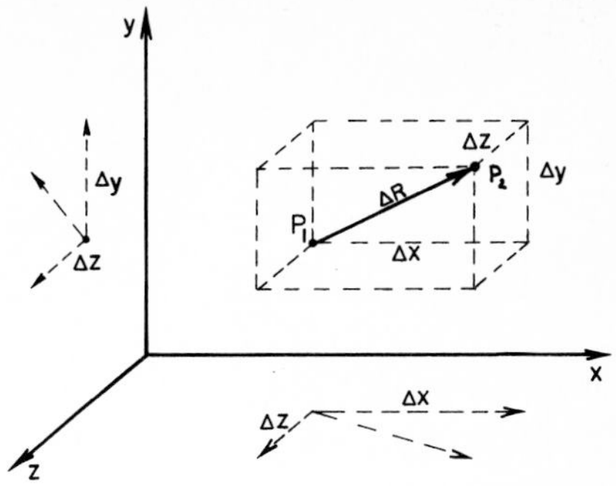

Fig. 2–5. The vector $\Delta \boldsymbol{R}$, whose components are Δx, Δy, and Δz.

The left side of Eq. (2.13) is a scalar. The right side is the sum of three products with Δx, Δy, and Δz, which are the components of a vector. It follows that the three numbers

$$\frac{\partial T}{\partial x},\ \frac{\partial T}{\partial y},\ \frac{\partial T}{\partial z}$$

are also the x-, y-, and z-components of a vector. We write this new vector with the symbol ∇T. The symbol ∇ (called "del") is an upside-down Δ, and is supposed to remind us of differentiation. People read ∇T in various ways: "del-T," or "gradient of T," or "grad T;"

$$\text{grad}\ T = \nabla T = \left(\frac{\partial T}{\partial x}, \frac{\partial T}{\partial y}, \frac{\partial T}{\partial z}\right).^{*} \tag{2.14}$$

Using this notation, we can rewrite Eq. (2.13) in the more compact form

$$\Delta T = \nabla T \cdot \Delta \boldsymbol{R}. \tag{2.15}$$

In words, this equation says that the difference in temperature between two nearby points is the dot product of the gradient of T and the vector displacement between the points. The form of Eq. (2.15) also illustrates clearly our proof above that ∇T is indeed a vector.

Perhaps you are still not convinced? Let's prove it in a different way. (Although if you look carefully, you may be able to see that it's really the same proof in a longer-winded form!) We shall show that the components of ∇T transform in just the same way that components of $\boldsymbol{R}$ do. If they do, ∇T is a vector according to our original definition of a vector in Chapter 11 of Vol. I. We take a new coordinate system x', y', z', and in this new system we calculate $\partial T/\partial x'$, $\partial T/\partial y'$, and $\partial T/\partial z'$. To make things a little simpler, we let $z = z'$, so that we can forget about the z-coordinate. (You can check out the more general case for yourself.)

We take an $x'y'$-system rotated an angle θ with respect to the xy-system, as in Fig. 2–6(a). For a point (x, y) the coordinates in the prime system are

$$x' = x\cos\theta + y\sin\theta, \tag{2.16}$$

$$y' = -x\sin\theta + y\cos\theta. \tag{2.17}$$

Or, solving for x and y,

$$x = x'\cos\theta - y'\sin\theta, \tag{2.18}$$

$$y = x'\sin\theta + y'\cos\theta. \tag{2.19}$$

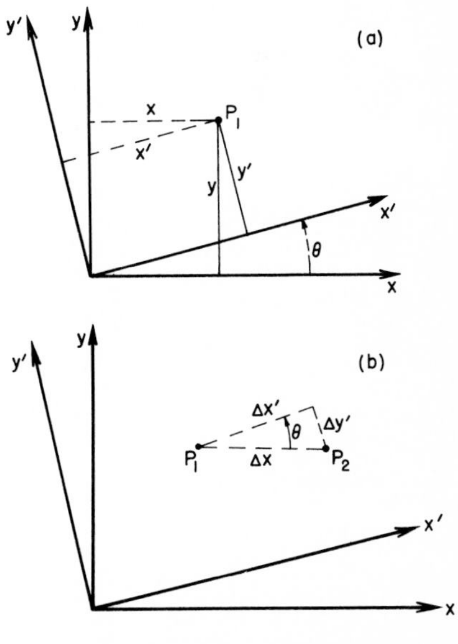

Fig. 2–6. (a) Transformation to a rotated coordinate system. (b) Special case of an interval $\Delta \boldsymbol{R}$ parallel to the x-axis.

If any pair of numbers transforms with these equations in the same way that x and y do, they are the components of a vector.

Now let's look at the difference in temperature between the two nearby points P_1 and P_2, chosen as in Fig. 2–6(b). If we calculate with the x- and y-coordinates, we would write

$$\Delta T = \frac{\partial T}{\partial x}\,\Delta x \tag{2.20}$$

—since Δy is zero.

---

* In our notation, the expression (a, b, c) represents a vector with components a, b, and c. If you like to use the unit vectors $\boldsymbol{i}$, $\boldsymbol{j}$, and $\boldsymbol{k}$, you may write

$$\nabla T = \boldsymbol{i}\,\frac{\partial T}{\partial x} + \boldsymbol{j}\,\frac{\partial T}{\partial y} + \boldsymbol{k}\,\frac{\partial T}{\partial z}.$$

What would a computation in the prime system give? We would have written

$$\Delta T = \frac{\partial T}{\partial x'}\,\Delta x' + \frac{\partial T}{\partial y'}\,\Delta y'. \tag{2.21}$$

Looking at Fig. 2–6(b), we see that

$$\Delta x' = \Delta x \cos\theta \tag{2.22}$$

and

$$\Delta y' = -\Delta x \sin\theta, \tag{2.23}$$

since Δy is negative when Δx is positive. Substituting these in Eq. (2.21), we find that

$$\Delta T = \frac{\partial T}{\partial x'}\,\Delta x\cos\theta - \frac{\partial T}{\partial y'}\,\Delta x\sin\theta \tag{2.24}$$

$$= \left(\frac{\partial T}{\partial x'}\cos\theta - \frac{\partial T}{\partial y'}\sin\theta\right)\Delta x. \tag{2.25}$$

Comparing Eq. (2.25) with (2.20), we see that

$$\frac{\partial T}{\partial x} = \frac{\partial T}{\partial x'}\cos\theta - \frac{\partial T}{\partial y'}\sin\theta. \tag{2.26}$$

This equation says that $\partial T/\partial x$ is obtained from $\partial T/\partial x'$ and $\partial T/\partial y'$, just as x is obtained from x' and y' in Eq. (2.18). So $\partial T/\partial x$ is the x-component of a vector. The same kind of arguments would show that $\partial T/\partial y$ and $\partial T/\partial z$ are y- and z-components. So $\boldsymbol{\nabla} T$ is definitely a vector. It is a vector field derived from the scalar field T.

2–4 The operator $\boldsymbol{\nabla}$

Now we can do something that is extremely amusing and ingenious—and characteristic of the things that make mathematics beautiful. The argument that grad T, or $\boldsymbol{\nabla} T$, is a vector did not depend upon *what* scalar field we were differentiating. All the arguments would go the same if T were replaced by *any scalar field.* Since the transformation equations are the same no matter what we differentiate, we could just as well omit the T and replace Eq. (2.26) by the operator equation

$$\frac{\partial}{\partial x} = \frac{\partial}{\partial x'}\cos\theta - \frac{\partial}{\partial y'}\sin\theta. \tag{2.27}$$

We leave the operators, as Jeans said, "hungry for something to differentiate."

Since the differential operators themselves transform as the components of a vector should, we can call them components of a *vector operator*. We can write

$$\boldsymbol{\nabla} = \left(\frac{\partial}{\partial x}, \frac{\partial}{\partial y}, \frac{\partial}{\partial z}\right), \tag{2.28}$$

which means, of course,

$$\nabla_x = \frac{\partial}{\partial x}, \qquad \nabla_y = \frac{\partial}{\partial y}, \qquad \nabla_z = \frac{\partial}{\partial z}. \tag{2.29}$$

We have abstracted the gradient away from the T—that is the wonderful idea.

You must always remember, of course, that $\boldsymbol{\nabla}$ is an operator. Alone, it means nothing. If $\boldsymbol{\nabla}$ by itself means nothing, what does it mean if we multiply *it* by a scalar—say T—to get the product $T\boldsymbol{\nabla}$? (One can always multiply a vector by a scalar.) It still does not mean anything. Its x-component is

$$T\frac{\partial}{\partial x}, \tag{2.30}$$

which is not a number, but is still some kind of operator. However, according to the algebra of vectors we would still call $T\boldsymbol{\nabla}$ a vector.

Now let's multiply ∇ by a scalar on the other side, so that we have the product (∇T). In ordinary algebra

$$TA = AT, \tag{2.31}$$

but we have to remember that operator algebra is a little different from ordinary vector algebra. With operators we must always keep the sequence right, so that the operations make proper sense. You will have no difficulty if you just remember that the operator ∇ obeys the same convention as the derivative notation. What is to be differentiated must be placed on the right of the ∇. The order is important.

Keeping in mind this problem of order, we understand that $T\nabla$ is an operator, but the product ∇T is no longer a hungry operator; the operator is completely satisfied. It is indeed a physical vector having a meaning. It represents the spatial rate of change of T. The x-component of ∇T is how fast T changes in the x-direction. What is the direction of the vector ∇T? We know that the rate of change of T in any direction is the component of ∇T in that direction (see Eq. 2.15). It follows that the direction of ∇T is that in which it has the largest possible component—in other words, the direction in which T changes the fastest. The gradient of T has the direction of the steepest uphill slope (in T).

2–5 Operations with ∇

Can we do any other algebra with the vector operator ∇? Let us try combining it with a vector. We can combine two vectors by making a dot product. We could make the products

$$(\text{a vector}) \cdot \nabla, \qquad \text{or} \qquad \nabla \cdot (\text{a vector}).$$

The first one doesn't mean anything yet, because it is still an operator. What it might ultimately mean would depend on what it is made to operate on. The second product is some scalar field. ($\boldsymbol{A} \cdot \boldsymbol{B}$ is always a scalar.)

Let's try the dot product of ∇ with a vector field we know, say $\boldsymbol{h}$. We write out the components:

$$\nabla \cdot \boldsymbol{h} = \nabla_x h_x + \nabla_y h_y + \nabla_z h_z \tag{2.32}$$

or

$$\nabla \cdot \boldsymbol{h} = \frac{\partial h_x}{\partial x} + \frac{\partial h_y}{\partial y} + \frac{\partial h_z}{\partial z} \cdot \tag{2.33}$$

The sum is invariant under a coordinate transformation. If we were to choose a different system (indicated by primes), we would have*

$$\nabla' \cdot \boldsymbol{h} = \frac{\partial h_{x'}}{\partial x'} + \frac{\partial h_{y'}}{\partial y'} + \frac{\partial h_{z'}}{\partial z'}, \tag{2.34}$$

which is the *same* number as would be gotten from Eq. (2.33), even though it looks different. That is,

$$\nabla' \cdot \boldsymbol{h} = \nabla \cdot \boldsymbol{h} \tag{2.35}$$

for every point in space. So $\nabla \cdot \boldsymbol{h}$ is a scalar field, which must represent some physical quantity. You should realize that the combination of derivatives in $\nabla \cdot \boldsymbol{h}$ is rather special. There are all sorts of other combinations like $\partial h_y/\partial x$, which are neither scalars nor components of vectors.

The scalar quantity $\nabla \cdot$ (a vector) is extremely useful in physics. It has been given the name the *divergence*. For example,

$$\nabla \cdot \boldsymbol{h} = \text{div}\, \boldsymbol{h} = \text{"divergence of } \boldsymbol{h}\text{."} \tag{2.36}$$

As we did for ∇T, we can ascribe a physical significance to $\nabla \cdot \boldsymbol{h}$. We shall, however, postpone that until later.

---

* We think of $\boldsymbol{h}$ as a *physical* quantity that depends on position in space, and not strictly as a mathematical function of three variables. When $\boldsymbol{h}$ is "differentiated" with respect to x, y, and z, or with respect to x', y', and z', the mathematical expression for $\boldsymbol{h}$ must first be expressed as a function of the appropriate variables.

First, we wish to see what else we can cook up with the vector operator ∇. What about a cross product? We must expect that

$$\nabla \times \boldsymbol{h} = \text{a vector}. \tag{2.37}$$

It is a vector whose components we can write by the usual rule for cross products (see Eq. 2.2):

$$(\nabla \times \boldsymbol{h})_z = \nabla_x h_y - \nabla_y h_x = \frac{\partial h_y}{\partial x} - \frac{\partial h_x}{\partial y}. \tag{2.38}$$

Similarly,

$$(\nabla \times \boldsymbol{h})_x = \nabla_y h_z - \nabla_z h_y = \frac{\partial h_z}{\partial y} - \frac{\partial h_y}{\partial z} \tag{2.39}$$

and

$$(\nabla \times \boldsymbol{h})_y = \nabla_z h_x - \nabla_x h_z = \frac{\partial h_x}{\partial z} - \frac{\partial h_z}{\partial x}. \tag{2.40}$$

The combination $\nabla \times \boldsymbol{h}$ is called "the *curl* of $\boldsymbol{h}$." The reason for the name and the physical meaning of the combination will be discussed later.

Summarizing, we have three kinds of combinations with ∇:

$$\begin{aligned} \nabla T &= \text{grad } T = \text{a vector}, \\ \nabla \cdot \boldsymbol{h} &= \text{div } \boldsymbol{h} = \text{a scalar}, \\ \nabla \times \boldsymbol{h} &= \text{curl } \boldsymbol{h} = \text{a vector}. \end{aligned}$$

Using these combinations, we can write about the spatial variations of fields in a convenient way—in a way that is general, in that it doesn't depend on any particular set of axes.

As an example of the use of our vector differential operator ∇, we write a set of vector equations which contain the same laws of electromagnetism that we gave in words in Chapter 1. They are called Maxwell's equations.

Maxwell's Equations

$$\begin{aligned} &(1) \quad \nabla \cdot \boldsymbol{E} = \frac{\rho}{\epsilon_0} \\ &(2) \quad \nabla \times \boldsymbol{E} = -\frac{\partial \boldsymbol{B}}{\partial t} \\ &(3) \quad \nabla \cdot \boldsymbol{B} = 0 \\ &(4) \quad c^2 \nabla \times \boldsymbol{B} = \frac{\partial \boldsymbol{E}}{\partial t} + \frac{\boldsymbol{j}}{\epsilon_0} \end{aligned} \tag{2.41}$$

where ρ (rho), the "electric charge density," is the amount of charge per unit volume, and $\boldsymbol{j}$, the "electric current density," is the rate at which charge flows through a unit area per second. These four equations contain the complete classical theory of the electromagnetic field. You see what an elegantly simple form we can get with our new notation!

2–6 The differential equation of heat flow

Let us give another example of a law of physics written in vector notation. The law is not a precise one, but for many metals and a number of other substances that conduct heat it is quite accurate. You know that if you take a slab of material and heat one face to temperature T_2 and cool the other to a different temperature T_1, the heat will flow through the material from T_2 to T_1 [Fig. 2–7(a)]. The heat flow is proportional to the area A of the faces, and to the temperature difference. It is also inversely proportional to d, the distance between the plates. (For a given temperature difference, the thinner the slab the greater the heat flow.) Letting J be the thermal energy that passes per unit time through the slab, we write

$$J = \kappa(T_2 - T_1)\frac{A}{d}. \tag{2.42}$$

The constant of proportionality κ (kappa) is called the *thermal conductivity*.

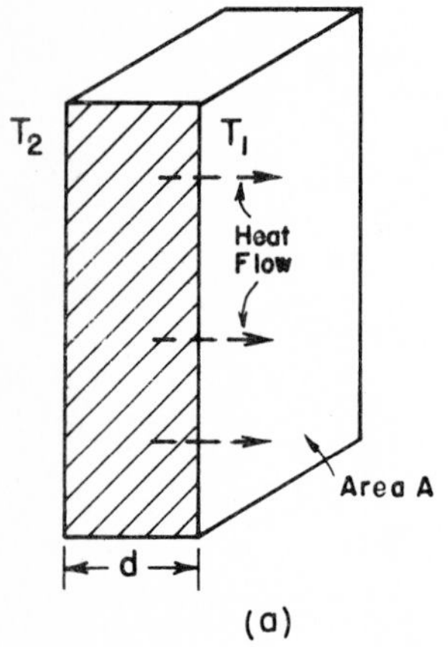

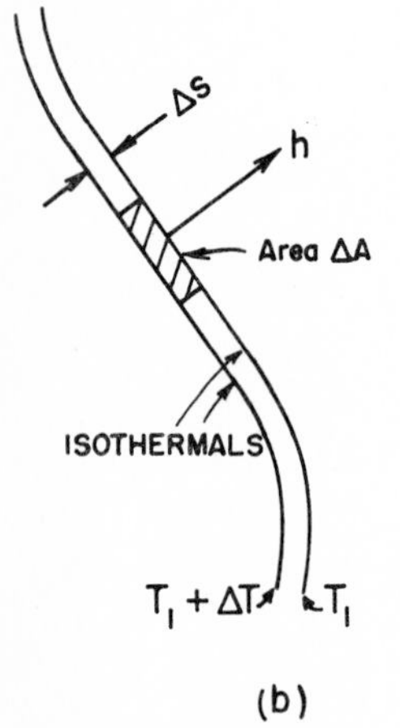

Fig. 2–7. (a) Heat flow through a slab. (b) An infinitesimal slab parallel to an isothermal surface in a large block.

What will happen in a more complicated case? Say in an odd-shaped block of material in which the temperature varies in peculiar ways? Suppose we look at a tiny piece of the block and imagine a slab like that of Fig. 2–7(a) on a miniature scale. We orient the faces parallel to the isothermal surfaces, as in Fig. 2–7(b), so that Eq. (2.42) is correct for the small slab.

If the area of the small slab is ΔA, the heat flow per unit time is

$$\Delta J = \kappa\,\Delta T\,\frac{\Delta A}{\Delta s}, \tag{2.43}$$

where Δs is the thickness of the slab. Now $\Delta J/\Delta A$ we have defined earlier as the magnitude of $\boldsymbol{h}$, whose direction is the heat flow. The heat flow will be from $T_1 + \Delta T$ toward T_1, and so it will be perpendicular to the isotherms, as drawn in Fig. 2–7(b). Also, $\Delta T/\Delta s$ is just the rate of change of T with position. And since the position change is perpendicular to the isotherms, our $\Delta T/\Delta s$ is the maximum rate of change. It is, therefore, just the magnitude of ∇T. Now since the direction of ∇T is opposite to that of $\boldsymbol{h}$, we can write (2.43) as a vector equation:

$$\boldsymbol{h} = -\kappa\,\nabla T. \tag{2.44}$$

(The minus sign is necessary because heat flows "downhill" in temperature.) Equation (2.44) is the differential equation of heat conduction in bulk materials. You see that it is a proper vector equation. Each side is a vector if κ is just a number. It is the generalization to arbitrary cases of the special relation (2.42) for rectangular slabs. Later we should learn to write all sorts of elementary physics relations like (2.42) in the more sophisticated vector notation. This notation is useful not only because it makes the equations *look* simpler. It also shows most clearly the *physical content* of the equations without reference to any arbitrarily chosen coordinate system.

2–7 Second derivatives of vector fields

So far we have had only first derivatives. Why not second derivatives? We could have several combinations:

$$\begin{aligned}
&\text{(a)}\quad \nabla\cdot(\nabla T)\\
&\text{(b)}\quad \nabla\times(\nabla T)\\
&\text{(c)}\quad \nabla(\nabla\cdot\boldsymbol{h})\\
&\text{(d)}\quad \nabla\cdot(\nabla\times\boldsymbol{h})\\
&\text{(e)}\quad \nabla\times(\nabla\times\boldsymbol{h})
\end{aligned} \tag{2.45}$$

You can check that these are all the possible combinations.

Let's look first at the second one, (b). It has the same form as

$$\boldsymbol{A}\times(\boldsymbol{A}T) = (\boldsymbol{A}\times\boldsymbol{A})T = 0,$$

since $\boldsymbol{A}\times\boldsymbol{A}$ is always zero. So we should have

$$\text{curl (grad } T) = \nabla\times(\nabla T) = 0. \tag{2.46}$$

We can see how this equation comes about if we go through once with the components:

$$\begin{aligned}
[\nabla\times(\nabla T)]_z &= \nabla_x(\nabla T)_y - \nabla_y(\nabla T)_x\\
&= \frac{\partial}{\partial x}\left(\frac{\partial T}{\partial y}\right) - \frac{\partial}{\partial y}\left(\frac{\partial T}{\partial x}\right),
\end{aligned} \tag{2.47}$$

which is zero (by Eq. 2.8). It goes the same for the other components. So $\nabla\times(\nabla T) = 0$, for any temperature distribution—in fact, for *any* scalar function.

Now let us take another example. Let us see whether we can find another zero. The dot product of a vector with a cross product which contains that vector is zero:

$$\boldsymbol{A}\cdot(\boldsymbol{A}\times\boldsymbol{B}) = 0, \tag{2.48}$$

because $\boldsymbol{A}\times\boldsymbol{B}$ is perpendicular to $\boldsymbol{A}$, and so has no components in the direction $\boldsymbol{A}$. The same combination appears in (d) of (2.45), so we have

$$\boldsymbol{\nabla}\cdot(\boldsymbol{\nabla}\times\boldsymbol{h}) = \text{div}\,(\text{curl}\,\boldsymbol{h}) = 0. \tag{2.49}$$

Again, it is easy to show that it is zero by carrying through the operations with components.

Now we are going to state two mathematical theorems that we will not prove. They are very interesting and useful theorems for physicists to know.

In a physical problem we frequently find that the curl of some quantity—say of the vector field $\boldsymbol{A}$—is zero. Now we have seen (Eq. 2.46) that the curl of a gradient is zero, which is easy to remember because of the way the vectors work. It could certainly be, then, that $\boldsymbol{A}$ is the gradient of some quantity, because then its curl would necessarily be zero. The interesting theorem is that if the curl $\boldsymbol{A}$ is zero, then $\boldsymbol{A}$ is *always* the gradient of *something*—there is some scalar field ψ (psi) such that $\boldsymbol{A}$ is equal to grad ψ. In other words, we have the

THEOREM:

$$\begin{array}{ll} \text{If} & \boldsymbol{\nabla}\times\boldsymbol{A} = 0 \\ \text{there is a} & \psi \\ \text{such that} & \boldsymbol{A} = \boldsymbol{\nabla}\psi. \end{array} \tag{2.50}$$

There is a similar theorem if the divergence of $\boldsymbol{A}$ is zero. We have seen in Eq. (2.49) that the divergence of a curl of something is always zero. If you come across a vector field $\boldsymbol{D}$ for which div $\boldsymbol{D}$ is zero, then you can conclude that $\boldsymbol{D}$ is the curl of some vector field $\boldsymbol{C}$.

THEOREM:

$$\begin{array}{ll} \text{If} & \boldsymbol{\nabla}\cdot\boldsymbol{D} = 0 \\ \text{there is a} & \boldsymbol{C} \\ \text{such that} & \boldsymbol{D} = \boldsymbol{\nabla}\times\boldsymbol{C}. \end{array} \tag{2.51}$$

In looking at the possible combinations of two $\boldsymbol{\nabla}$ operators, we have found that two of them always give zero. Now we look at the ones that are *not* zero. Take the combination $\boldsymbol{\nabla}\cdot(\boldsymbol{\nabla}T)$, which was first on our list. It is not, in general, zero. We write out the components:

$$\boldsymbol{\nabla}T = \nabla_xT + \nabla_yT + \nabla_zT.$$

Then

$$\begin{aligned} \boldsymbol{\nabla}\cdot(\boldsymbol{\nabla}T) &= \nabla_x(\nabla_xT) + \nabla_y(\nabla_yT) + \nabla_z(\nabla_zT) \\ &= \frac{\partial^2T}{\partial x^2} + \frac{\partial^2T}{\partial y^2} + \frac{\partial^2T}{\partial z^2}, \end{aligned} \tag{2.52}$$

which would, in general, come out to be some number. It is a scalar field.

You see that we do not need to keep the parentheses, but can write, without any chance of confusion,

$$\boldsymbol{\nabla}\cdot(\boldsymbol{\nabla}T) = \boldsymbol{\nabla}\cdot\boldsymbol{\nabla}T = (\boldsymbol{\nabla}\cdot\boldsymbol{\nabla})T = \nabla^2T. \tag{2.53}$$

We look at ∇^2 as a new operator. It is a scalar operator. Because it appears often in physics, it has been given a special name—the *Laplacian*.

$$\text{Laplacian} = \nabla^2 = \frac{\partial^2}{\partial x^2} + \frac{\partial^2}{\partial y^2} + \frac{\partial^2}{\partial z^2}. \tag{2.54}$$

Since the Laplacian is a scalar operator, we may operate with it on a vector—by which we mean the same operation on each component in rectangular coordinates:

$$\nabla^2 \boldsymbol{h} = (\nabla^2 h_x, \nabla^2 h_y, \nabla^2 h_z).$$

Let's look at one more possibility: $\nabla \times (\nabla \times \boldsymbol{h})$, which was (e) in the list (2.45). Now the curl of the curl can be written differently if we use the vector equality (2.6):

$$\boldsymbol{A} \times (\boldsymbol{B} \times \boldsymbol{C}) = \boldsymbol{B}(\boldsymbol{A} \cdot \boldsymbol{C}) - \boldsymbol{C}(\boldsymbol{A} \cdot \boldsymbol{B}). \tag{2.55}$$

In order to use this formula, we should replace $\boldsymbol{A}$ and $\boldsymbol{B}$ by the operator ∇ and put $\boldsymbol{C} = \boldsymbol{h}$. If we do that, we get

$$\nabla \times (\nabla \times \boldsymbol{h}) = \nabla(\nabla \cdot \boldsymbol{h}) - \boldsymbol{h}(\nabla \cdot \nabla) \ldots ???$$

Wait a minute! Something is wrong. The first two terms are vectors all right (the operators are satisfied), but the last term doesn't come out to anything. It's still an operator. The trouble is that we haven't been careful enough about keeping the order of our terms straight. If you look again at Eq. (2.55), however, you see that we could equally well have written it as

$$\boldsymbol{A} \times (\boldsymbol{B} \times \boldsymbol{C}) = \boldsymbol{B}(\boldsymbol{A} \cdot \boldsymbol{C}) - (\boldsymbol{A} \cdot \boldsymbol{B})\boldsymbol{C}. \tag{2.56}$$

The order of terms looks better. Now let's make our substitution in (2.56). We get

$$\nabla \times (\nabla \times \boldsymbol{h}) = \nabla(\nabla \cdot \boldsymbol{h}) - (\nabla \cdot \nabla)\boldsymbol{h}. \tag{2.57}$$

This form looks all right. It is, in fact, correct, as you can verify by computing the components. The last term is the Laplacian, so we can equally well write

$$\nabla \times (\nabla \times \boldsymbol{h}) = \nabla(\nabla \cdot \boldsymbol{h}) - \nabla^2 \boldsymbol{h}. \tag{2.58}$$

We have had something to say about all of the combinations in our list of double ∇'s, except for (c), $\nabla(\nabla \cdot \boldsymbol{h})$. It is a possible vector field, but there is nothing special to say about it. It's just some vector field which may occasionally come up.

It will be convenient to have a table of our conclusions:

$$\begin{aligned}
&\text{(a)} \quad \nabla \cdot (\nabla T) = \nabla^2 T = \text{a scalar field}\\
&\text{(b)} \quad \nabla \times (\nabla T) = 0\\
&\text{(c)} \quad \nabla(\nabla \cdot \boldsymbol{h}) = \text{a vector field}\\
&\text{(d)} \quad \nabla \cdot (\nabla \times \boldsymbol{h}) = 0\\
&\text{(e)} \quad \nabla \times (\nabla \times \boldsymbol{h}) = \nabla(\nabla \cdot \boldsymbol{h}) - \nabla^2 \boldsymbol{h}\\
&\text{(f)} \quad (\nabla \cdot \nabla)\boldsymbol{h} = \nabla^2 \boldsymbol{h} = \text{a vector field}
\end{aligned} \tag{2.59}$$

You may notice that we haven't tried to invent a new vector operator $(\nabla \times \nabla)$. Do you see why?

2–8 Pitfalls

We have been applying our knowledge of ordinary vector algebra to the algebra of the operator ∇. We have to be careful, though, because it is possible to go astray. There are two pitfalls which we will mention, although they will not come up in this course. What would you say about the following expression, that involves the two scalar functions ψ and ϕ (phi):

$$(\nabla\psi) \times (\nabla\phi)?$$

You might want to say: it must be zero because it's just like

$$(Aa) \times (Ab),$$

which is zero because the cross product of two *equal* vectors $A \times A$ is always zero. But in our example the two operators ∇ are not equal! The first one operates on one function, ψ; the other operates on a different function, ϕ. So although we represent them by the same symbol ∇, they must be considered as different operators. Clearly, the direction of $\nabla\psi$ depends on the function ψ, so it is not likely to be parallel to $\nabla\phi$.

$$(\nabla\psi) \times (\nabla\phi) \neq 0 \quad \text{(generally)}.$$

Fortunately, we won't have to use such expressions. (What we have said doesn't change the fact that $\nabla \times \nabla\psi = 0$ for any scalar field, because here both ∇'s operate on the same function.)

Pitfall number two (which, again, we need not get into in our course) is the following: The rules that we have outlined here are simple and nice when we use rectangular coordinates. For example, if we have $\nabla^2 \boldsymbol{h}$ and we want the x-component, it is

$$(\nabla^2 \boldsymbol{h})_x = \left(\frac{\partial^2}{\partial x^2} + \frac{\partial^2}{\partial y^2} + \frac{\partial^2}{\partial z^2}\right) h_x = \nabla^2 h_x. \tag{2.60}$$

The same expression would *not* work if we were to ask for the *radial* component of $\nabla^2 \boldsymbol{h}$. The radial component of $\nabla^2 \boldsymbol{h}$ is not equal to $\nabla^2 h_r$. The reason is that when we are dealing with the algebra of vectors, the directions of the vectors are all quite definite. But when we are dealing with vector fields, their directions are different at different places. If we try to describe a vector field in, say, polar coordinates, what we call the "radial" direction varies from point to point. So we can get into a lot of trouble when we start to differentiate the components. For example, even for a *constant* vector field, the radial component changes from point to point.

It is usually safest and simplest just to stick to rectangular coordinates and avoid trouble, but there is one exception worth mentioning: Since the Laplacian ∇^2, is a scalar, we can write it in any coordinate system we want to (for example, in polar coordinates). But since it is a differential operator, we should use it only on vectors whose components are in a fixed direction—that means rectangular coordinates. So we shall express all of our vector fields in terms of their x-, y-, and z-components when we write our vector differential equations out in components.

3

Vector Integral Calculus

3–1 Vector integrals; the line integral of $\nabla\Psi$

We found in Chapter 2 that there were various ways of taking derivatives of fields. Some gave vector fields; some gave scalar fields. Although we developed many different formulas, everything in Chapter 2 could be summarized in one rule: the operators $\partial/\partial x$, $\partial/\partial y$, and $\partial/\partial z$ are the three components of a vector operator ∇. We would now like to get some understanding of the significance of the derivatives of fields. We will then have a better feeling for what a vector field equation means.

We have already discussed the meaning of the gradient operation (∇ on a scalar). Now we turn to the meanings of the divergence and curl operations. The interpretation of these quantities is best done in terms of certain vector integrals and equations relating such integrals. These equations cannot, unfortunately, be obtained from vector algebra by some easy substitution, so you will just have to learn them as something new. Of these integral formulas, one is practically trivial, but the other two are not. We will derive them and explain their implications. The equations we shall study are really mathematical theorems. They will be useful not only for interpreting the meaning and the content of the divergence and the curl, but also in working out general physical theories. These mathematical theorems are, for the theory of fields, what the theorem of the conservation of energy is to the mechanics of particles. General theorems like these are important for a deeper understanding of physics. You will find, though, that they are not very useful for solving problems—except in the simplest cases. It is delightful, however, that in the beginning of our subject there will be many simple problems which can be solved with the three integral formulas we are going to treat. We will see, however, as the problems get harder, that we can no longer use these simple methods.

We take up first an integral formula involving the gradient. The relation contains a very simple idea: Since the gradient represents the rate of change of a field quantity, if we integrate that rate of change, we should get the total change. Suppose we have the scalar field $\psi(x, y, z)$. At any two points (1) and (2), the function ψ will have the values $\psi(1)$ and $\psi(2)$, respectively. [We use a convenient notation, in which (2) represents the point (x_2, y_2, z_2) and $\psi(2)$ means the same thing as $\psi(x_2, y_2, z_2)$.] If Γ(gamma) is any curve joining (1) and (2), as in Fig. 3–1, the following relation is true:

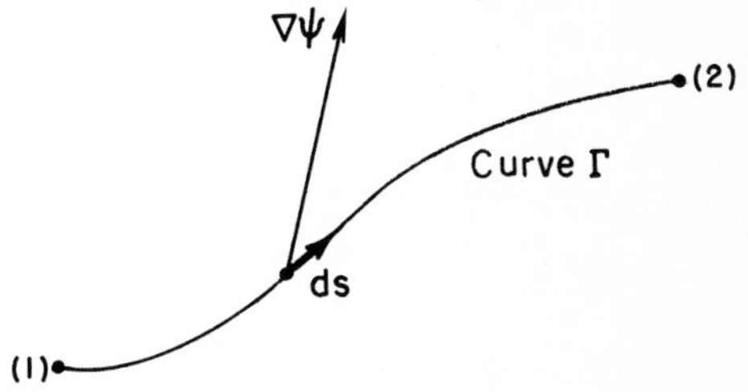

Fig. 3–1. The terms used in Eq. (3.1). The vector $\nabla\psi$ is evaluated at the line element ds.

THEOREM 1.

$$\psi(2) - \psi(1) = \int_{\substack{(1)\\ \text{along } \Gamma}}^{(2)} (\nabla\psi)\cdot d\mathbf{s}. \tag{3.1}$$

The integral is a *line integral*, from (1) to (2) along the curve Γ, of the dot product of $\nabla\psi$—a vector—with $d\mathbf{s}$—another vector which is an infinitesimal line element of the curve Γ (directed away from (1) and toward (2)).

First, we should review what we mean by a line integral. Consider a scalar function $f(x, y, z)$, and the curve Γ joining two points (1) and (2). We mark off the curve at a number of points and join these points by straight-line segments, as shown in Fig. 3–2. Each segment has the length Δs_i, where i is an index that runs 1, 2, 3, ... By the line integral

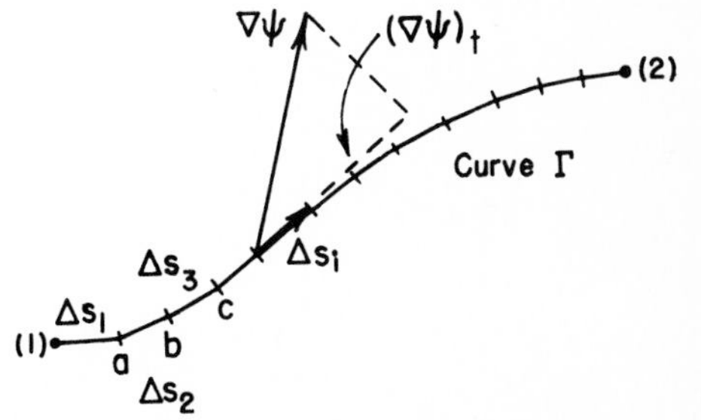

Fig. 3–2. The line integral is the limit of a sum.

$$\int_{\substack{(1)\\ \text{along } \Gamma}}^{(2)} f\, ds$$

we mean the limit of the sum

$$\sum f_i \, \Delta s_i,$$

where f_i is the value of the function at the ith segment. The limiting value is what the sum approaches as we add more and more segments (in a sensible way, so that the largest $\Delta s_i \to 0$).

The integral in our theorem, Eq. (3.1), means the same thing, although it looks a little different. Instead of f, we have another scalar—the component of $\nabla\psi$ in the direction of $\Delta\mathbf{s}$. If we write $(\nabla\psi)_t$ for this tangential component, it is clear that

$$(\nabla\psi)_t \, \Delta s = (\nabla\psi) \cdot \Delta\mathbf{s}. \tag{3.2}$$

The integral in Eq. (3.1) means the sum of such terms.

Now let's see why Eq. (3.1) is true. In Chapter 1, we showed that the component of $\nabla\psi$ along a small displacement $\Delta\mathbf{R}$ was the rate of change of ψ in the direction of $\Delta\mathbf{R}$. Consider the line segment $\Delta\mathbf{s}$ from (1) to point a in Fig. 3–2. According to our definition,

$$\Delta\psi_1 = \psi(a) - \psi(1) = (\nabla\psi)_1 \cdot \Delta\mathbf{s}_1. \tag{3.3}$$

Also, we have

$$\psi(b) - \psi(a) = (\nabla\psi)_2 \cdot \Delta\mathbf{s}_2, \tag{3.4}$$

where, of course, $(\nabla\psi)_1$ means the gradient evaluated at the segment $\Delta\mathbf{s}_1$, and $(\nabla\psi)_2$, the gradient evaluated at $\Delta\mathbf{s}_2$. If we add Eqs. (3.3) and (3.4), we get

$$\psi(b) - \psi(1) = (\nabla\psi)_1 \cdot \Delta\mathbf{s}_1 + (\nabla\psi)_2 \cdot \Delta\mathbf{s}_2. \tag{3.5}$$

You can see that if we keep adding such terms, we get the result

$$\psi(2) - \psi(1) = \sum (\nabla\psi)_i \cdot \Delta\mathbf{s}_i. \tag{3.6}$$

The left-hand side doesn't depend on how we choose our intervals—if (1) and (2) are kept always the same—so we can take the limit of the right-hand side. We have therefore proved Eq. (3.1).

You can see from our proof that just as the equality doesn't depend on how the points $a, b, c, \ldots,$ are chosen, similarly it doesn't depend on what we choose for the curve Γ to join (1) and (2). Our theorem is correct for *any* curve from (1) to (2).

One remark on notation: You will see that there is no confusion if we write, for convenience,

$$(\nabla\psi) \cdot d\mathbf{s} = \nabla\psi \cdot d\mathbf{s}. \tag{3.7}$$

With this notation, our theorem is

THEOREM 1.

$$\psi(2) - \psi(1) = \underset{\substack{\text{any curve from} \\ \text{(1) to (2)}}}{\int_{(1)}^{(2)}} \nabla\psi \cdot d\mathbf{s}. \tag{3.8}$$

3–2 The flux of a vector field

Before we consider our next integral theorem—a theorem about the divergence—we would like to study a certain idea which has an easily understood physical significance in the case of heat flow. We have defined the vector $\mathbf{h}$, which represents the heat that flows through a unit area in a unit time. Suppose that inside a block of material we have some closed surface S which encloses the volume V (Fig. 3–3). We would like to find out how much heat is flowing out of this *volume*. We can, of course, find it by calculating the total heat flow out of the *surface S*.

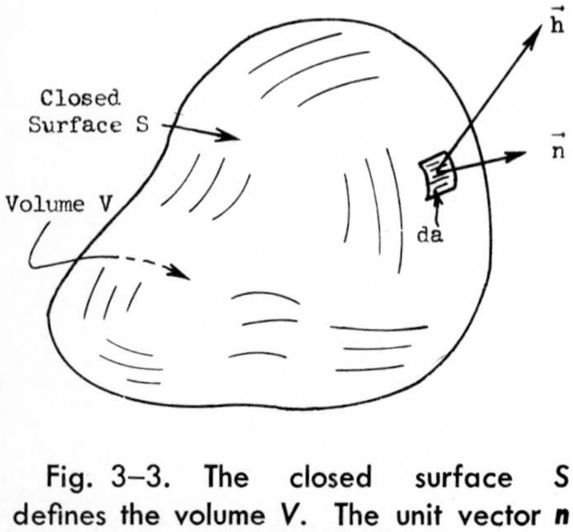

Fig. 3–3. The closed surface S defines the volume V. The unit vector **n** is the outward normal to the surface element *da*, and **h** is the heat-flow vector at the surface element.

We write da for the area of an element of the surface. The symbol stands for a two-dimensional differential. If, for instance, the area happened to be in the xy-plane we would have

$$da = dx\,dy.$$

Later we shall have integrals over volume and for these it is convenient to consider a differential volume that is a little cube. So when we write dV we mean

$$dV = dx\,dy\,dz.$$

Some people like to write d^2a instead of da to remind themselves that it is kind of a second-order quantity. They would also write d^3V instead of dV. We will use the simpler notation, and assume that you can remember that an area has two dimensions and a volume has three.

The heat flow out through the surface element da is the area times the component of $\boldsymbol{h}$ perpendicular to da. We have already defined $\boldsymbol{n}$ as a unit vector pointing outward at right angles to the surface (Fig. 3–3). The component of $\boldsymbol{h}$ that we want is

$$h_n = \boldsymbol{h} \cdot \boldsymbol{n}. \tag{3.9}$$

The heat flow out through da is then

$$\boldsymbol{h} \cdot \boldsymbol{n}\, da. \tag{3.10}$$

To get the total heat flow through any surface we sum the contributions from all the elements of the surface. In other words, we integrate (3.10) over the whole surface:

$$\text{Total heat flow outward through } S = \int_S \boldsymbol{h} \cdot \boldsymbol{n}\, da. \tag{3.11}$$

We are also going to call this surface integral "the flux of $\boldsymbol{h}$ through the surface." Originally the word flux meant flow, so that the surface integral just means the flow of $\boldsymbol{h}$ through the surface. We may think: $\boldsymbol{h}$ is the "current density" of heat flow and the surface integral of it is the total heat current directed out of the surface; that is, the thermal energy per unit time (joules per second).

We would like to generalize this idea to the case where the vector does not represent the flow of anything; for instance, it might be the electric field. We can certainly still integrate the normal component of the electric field over an area if we wish. Although it is not the flow of anything, we still call it the "flux." We say

$$\text{Flux of } \boldsymbol{E} \text{ through the surface } S = \int_S \boldsymbol{E} \cdot \boldsymbol{n}\, da. \tag{3.12}$$

We generalize the word "flux" to mean the "surface integral of the normal component" of a vector. We will also use the same definition even when the surface considered is not a closed one, as it is here.

Returning to the special case of heat flow, let us take a situation in which *heat is conserved.* For example, imagine some material in which after an initial heating no further heat energy is generated or absorbed. Then, if there is a net heat flow out of a closed surface, the heat content of the volume inside must decrease. So, in circumstances in which heat would be conserved, we say that

$$\int_S \boldsymbol{h} \cdot \boldsymbol{n}\, da = -\frac{dQ}{dt}, \tag{3.13}$$

where Q is the heat inside the surface. The heat flux out of S is equal to minus the rate of change with respect to time of the total heat Q inside of S. This interpretation is possible because we are speaking of heat flow and also because we supposed that the heat was conserved. We could not, of course, speak of the total heat inside the volume if heat were being generated there.

Now we shall point out an interesting fact about the flux of any vector. You may think of the heat flow vector if you wish, but what we say will be true for any vector field $\boldsymbol{C}$. Imagine that we have a closed surface S that encloses the volume V. We now separate the volume into two parts by some kind of a "cut," as in Fig. 3–4. Now we have two closed surfaces and volumes. The volume V_1 is enclosed in the surface S_1, which is made up of part of the original surface S_a and of the surface of the cut, S_{ab}. The volume V_2 is enclosed by S_2, which is made up of the rest of the original surface S_b and closed off by the cut S_{ab}. Now consider the

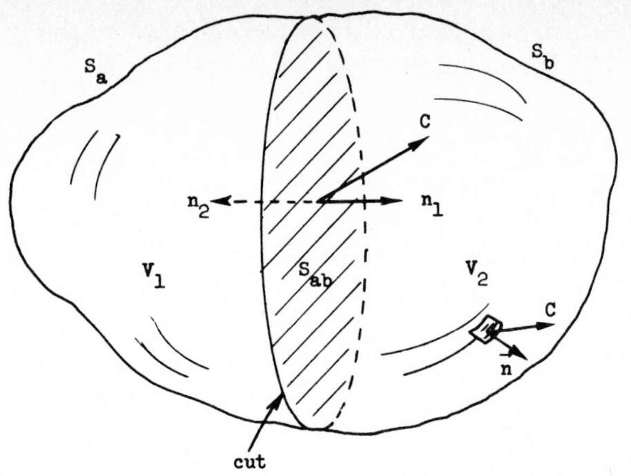

Fig. 3–4. A volume V contained inside the surface S is divided into two pieces by a "cut" at the surface S_{ab}. We now have the volume V_1 enclosed in the surface $S_1 = S_a + S_{ab}$ and the volume V_2 enclosed in the surface $S_2 = S_b + S_{ab}$.

following question: Suppose we calculate the flux out through surface S_1 and add to it the flux through surface S_2. Does the sum equal the flux through the whole surface that we started with? The answer is yes. The flux through the part of the surfaces S_{ab} common to both S_1 and S_2 just exactly cancels out. For the flux of the vector $\boldsymbol{C}$ out of V_1, we can write

$$\text{Flux through } S_1 = \int_{S_a} \boldsymbol{C} \cdot \boldsymbol{n}\, da + \int_{S_{ab}} \boldsymbol{C} \cdot \boldsymbol{n}_1\, da, \tag{3.14}$$

and for the flux out of V_2,

$$\text{Flux through } S_2 = \int_{S_b} \boldsymbol{C} \cdot \boldsymbol{n}\, da + \int_{S_{ab}} \boldsymbol{C} \cdot \boldsymbol{n}_2\, da. \tag{3.15}$$

Note that in the second integral we have written $\boldsymbol{n}_1$ for the outward normal for S_{ab} when it belongs to S_1, and $\boldsymbol{n}_2$ when it belongs to S_2, as shown in Fig. 3–4. Clearly, $\boldsymbol{n}_1 = -\boldsymbol{n}_2$, so that

$$\int_{S_{ab}} \boldsymbol{C} \cdot \boldsymbol{n}_1\, da = -\int_{S_{ab}} \boldsymbol{C} \cdot \boldsymbol{n}_2\, da. \tag{3.16}$$

If we now add Eqs. (3.14) and (3.15), we see that the sum of the fluxes through S_1 and S_2 is just the sum of two integrals which, taken together, give the flux through the original surface $S = S_a + S_b$.

We see that the flux through the complete outer surface S can be considered as the sum of the fluxes from the two pieces into which the volume was broken. We can similarly subdivide again—say by cutting V_1 into two pieces. You see that the same arguments apply. So for *any* way of dividing the original volume, it must be generally true that the flux through the outer surface, which is the original integral, is equal to a sum of the fluxes out of all the little interior pieces.

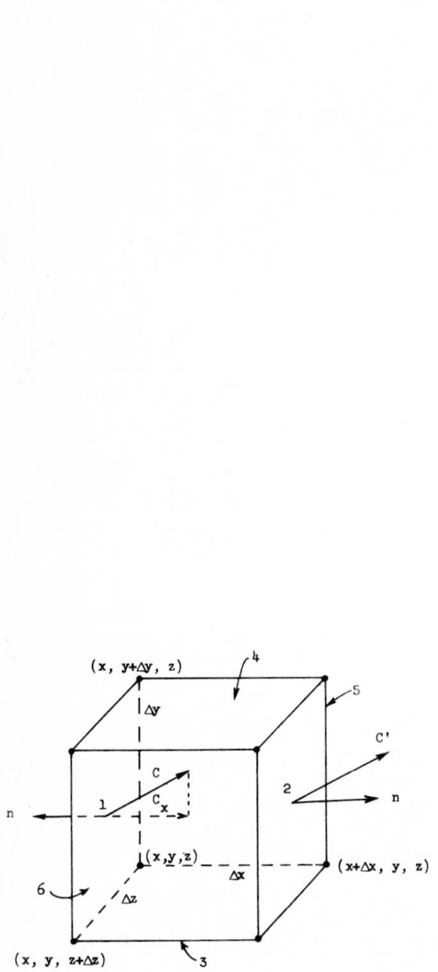

Fig. 3–5. Computation of the flux of C out of a small cube.

3–3 The flux from a cube; Gauss' theorem

We now take the special case of a small cube* and find an interesting formula for the flux out of it. Consider a cube whose edges are lined up with the axes as in Fig. 3–5. Let us suppose that the coordinates of the corner nearest the origin are x, y, z. Let Δx be the length of the cube in the x-direction, Δy be the length in the y-direction, and Δz be the length in the z-direction. We wish to find the flux of a vector field $\boldsymbol{C}$ through the surface of the cube. We shall do this by making a sum of the fluxes through each of the six faces. First, consider the face marked 1 in the figure. The flux *outward* on this face is the negative of the x-component of $\boldsymbol{C}$, integrated over the area of the face. This flux is

$$-\int C_x\, dy\, dz.$$

Since we are considering a *small* cube, we can approximate this integral by the

* The following development applies equally well to any rectangular parallelepiped.

value of C_x at the center of the face—which we call the point (1)—multiplied by the area of the face, $\Delta y\,\Delta z$:

$$\text{Flux out of 1} = -C_x(1)\,\Delta y\,\Delta z.$$

Similarly, for the flux out of face 2, we write

$$\text{Flux out of 2} = C_x(2)\,\Delta y\,\Delta z.$$

Now $C_x(1)$ and $C_x(2)$ are, in general, slightly different. If Δx is small enough, we can write

$$C_x(2) = C_x(1) + \frac{\partial C_x}{\partial x}\,\Delta x.$$

There are, of course, more terms, but they will involve $(\Delta_x)^2$ and higher powers, and so will be negligible if we consider only the limit of small Δx. So the flux through face 2 is

$$\text{Flux out of 2} = \left[C_x(1) + \frac{\partial C_x}{\partial x}\,\Delta x\right]\Delta y\,\Delta z.$$

Summing the fluxes for faces 1 and 2, we get

$$\text{Flux out of 1 and 2} = \frac{\partial C_x}{\partial x}\,\Delta x\,\Delta y\,\Delta z.$$

The derivative should really be evaluated at the center of face 1; that is, at $[x, y + (\Delta y/2), z + (\Delta z/2)]$. But in the limit of an infinitesimal cube, we make a negligible error if we evaluate it at the corner (x, y, z).

Applying the same reasoning to each of the other pairs of faces, we have

$$\text{Flux out of 3 and 4} = \frac{\partial C_y}{\partial y}\,\Delta x\,\Delta y\,\Delta z$$

and

$$\text{Flux out of 5 and 6} = \frac{\partial C_z}{\partial z}\,\Delta x\,\Delta y\,\Delta z.$$

The total flux through all the faces is the sum of these terms. We find that

$$\int_{\text{cube}} \boldsymbol{C}\cdot\boldsymbol{n}\,da = \left(\frac{\partial C_x}{\partial x} + \frac{\partial C_y}{\partial y} + \frac{\partial C_z}{\partial z}\right)\Delta x\,\Delta y\,\Delta z,$$

and the sum of the derivatives is just $\boldsymbol{\nabla}\cdot\boldsymbol{C}$. Also, $\Delta x\,\Delta y\,\Delta z = \Delta V$, the volume of the cube. So we can say that *for an infinitesimal cube*

$$\int_{\text{surface}} \boldsymbol{C}\cdot\boldsymbol{n}\,da = (\boldsymbol{\nabla}\cdot\boldsymbol{C})\,\Delta V. \tag{3.17}$$

We have shown that the outward flux from the surface of an infinitesimal cube is equal to the divergence of the vector multiplied by the volume of the cube. We now see the "meaning" of the divergence of a vector. The divergence of a vector at the point P is the flux—the outgoing "flow" of $\boldsymbol{C}$—*per unit volume*, in the neighborhood of P.

We have connected the divergence of $\boldsymbol{C}$ to the flux of $\boldsymbol{C}$ out of each infinitesimal volume. For any finite volume we can use the fact we proved above—that the total flux from a volume is the sum of the fluxes out of each part. We can, that is, integrate the divergence over the entire volume. This gives us the theorem that the integral of the normal component of any vector over any closed surface can also be written as the integral of the divergence of the vector over the volume enclosed by the surface. This theorem is named after Gauss.

GAUSS' THEOREM.

$$\int_S \boldsymbol{C}\cdot\boldsymbol{n}\,da = \int_V \boldsymbol{\nabla}\cdot\boldsymbol{C}\,dV, \tag{3.18}$$

where S is any closed surface and V is the volume inside it.

3–4 Heat conduction; the diffusion equation

Let's consider an example of the use of this theorem, just to get familiar with it. Suppose we take again the case of heat flow in, say, a metal. Suppose we have a simple situation in which all the heat has been previously put in and the body is just cooling off. There are no sources of heat, so that heat is conserved. Then how much heat is there inside some chosen volume at any time? It must be *decreasing* by just the amount that flows out of the surface of the volume. If our volume is a little cube, we would write, following Eq. (3.17),

$$\text{Heat out} = \int_{\text{cube}} \boldsymbol{h} \cdot \boldsymbol{n}\, da = \boldsymbol{\nabla} \cdot \boldsymbol{h}\, \Delta V. \tag{3.19}$$

But this must equal the rate of loss of the heat inside the cube. If q is the heat per unit volume, the heat in the cube is $q\,\Delta V$, and the rate of *loss* is

$$-\frac{d}{dt}(q\,\Delta V) = -\frac{dq}{dt}\,\Delta V. \tag{3.20}$$

Comparing (3.19) and (3.20), we see that

$$-\frac{dq}{dt} = \boldsymbol{\nabla} \cdot \boldsymbol{h}. \tag{3.21}$$

Take careful note of the form of this equation; the form appears often in physics. It expresses a conservation law—here the conservation of heat. We have expressed the same physical fact in another way in Eq. (3.13). Here we have the *differential* form of a conservation equation, while Eq. (3.13) is the *integral* form.

We have obtained Eq. (3.21) by applying Eq. (3.13) to an infinitesimal cube. We can also go the other way. For a big volume V bounded by S, Gauss' law says that

$$\int_S \boldsymbol{h} \cdot \boldsymbol{n}\, da = \int \boldsymbol{\nabla} \cdot \boldsymbol{h}\, dV. \tag{3.22}$$

Using (3.21), the integral on the right-hand side is found to be just $-dQ/dt$, and again we have Eq. (3.13).

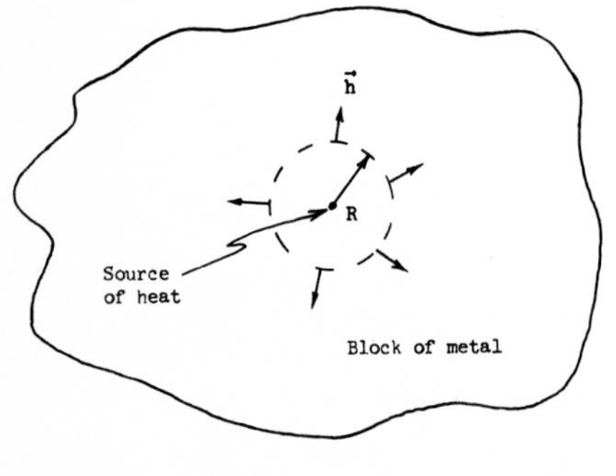

Fig. 3–6. In the region near a point source of heat, the heat flow is radially outward.

Now let's consider a different case. Imagine that we have a block of material and that inside it there is a very tiny hole in which some chemical reaction is taking place and generating heat. Or we could imagine that there are some wires running into a tiny resistor that is being heated by an electric current. We shall suppose that the heat is generated practically at a point, and let W represent the energy liberated per second at that point. We shall suppose that in the rest of the volume heat is conserved, and that the heat generation has been going on for a long time—so that now the temperature is no longer changing anywhere. The problem is: What does the heat vector $\boldsymbol{h}$ look like at various places in the metal? How much heat flow is there at each point?

We know that if we integrate the normal component of $\boldsymbol{h}$ over a closed surface that encloses the source, we will always get W. All the heat that is being generated at the point source must flow out through the surface, since we have supposed that the flow is steady. We have the difficult problem of finding a vector field which, when integrated over any surface, always gives W. We can, however, find the field rather easily by taking a somewhat special surface. We take a sphere of radius R, centered at the source, and assume that the heat flow is radial (Fig. 3–6). Our intuition tells us that $\boldsymbol{h}$ should be radial if the block of material is large and we don't get too close to the edges, and it should also have the same magnitude at all points on the sphere. You see that we are adding a certain amount of guesswork—usually called "physical intuition"—to our mathematics in order to find the answer.

When $\boldsymbol{h}$ is radial and spherically symmetric, the integral of the normal component of $\boldsymbol{h}$ over the area is very simple, because the normal component is just the magnitude of $\boldsymbol{h}$ and is constant. The area over which we integrate is $4\pi R^2$. We have then that

$$\int_S \boldsymbol{h} \cdot \boldsymbol{n}\, da = h \cdot 4\pi R^2 \tag{3.23}$$

(where h is the magnitude of $\boldsymbol{h}$). This integral should equal W, the rate at which heat is produced at the source. We get

$$h = \frac{W}{4\pi R^2},$$

or

$$\boldsymbol{h} = \frac{W}{4\pi R^2}\, \boldsymbol{e}_r, \tag{3.24}$$

where, as usual, $\boldsymbol{e}_r$ represents a unit vector in the radial direction. Our result says that $\boldsymbol{h}$ is proportional to W and varies inversely as the square of the distance from the source.

The result we have just obtained applies to the heat flow in the vicinity of a point source of heat. Let's now try to find the equations that hold in the most general kind of heat flow, keeping only the condition that heat is conserved. We will be dealing only with what happens at places outside of any sources or absorbers of heat.

The differential equation for the conduction of heat was derived in Chapter 2. According to Eq. (2.44),

$$\boldsymbol{h} = -\kappa\, \boldsymbol{\nabla} T. \tag{3.25}$$

(Remember that this relationship is an approximate one, but fairly good for some materials like metals.) It is applicable, of course, only in regions of the material where there is no generation or absorption of heat. We derived above another relation, Eq. (3.21), that holds when heat is conserved. If we combine that equation with (3.25), we get

$$-\frac{dq}{dt} = \boldsymbol{\nabla} \cdot \boldsymbol{h} = -\boldsymbol{\nabla} \cdot (\kappa\, \boldsymbol{\nabla} T),$$

or

$$\frac{dq}{dt} = \kappa\, \boldsymbol{\nabla} \cdot \boldsymbol{\nabla} T = \kappa\, \nabla^2 T, \tag{3.26}$$

if κ is a constant. You remember that q is the amount of heat in a unit volume and $\boldsymbol{\nabla} \cdot \boldsymbol{\nabla} = \nabla^2$ is the Laplacian operator

$$\nabla^2 = \frac{\partial^2}{\partial x^2} + \frac{\partial^2}{\partial y^2} + \frac{\partial^2}{\partial z^2}.$$

If we now make one more assumption we can obtain a very interesting equation. We assume that the temperature of the material is proportional to the heat content per unit volume—that is, that the material has a definite specific heat. When this assumption is valid (as it often is), we can write

$$\Delta q = c_v\, \Delta T$$

or

$$\frac{dq}{dt} = c_v \frac{dT}{dt}. \tag{3.27}$$

The rate of change of heat is proportional to the rate of change of temperature. The constant or proportionality c_v is, here, the specific heat per unit *volume* of the material. Using Eq. (3.27) with (3.26), we get

$$\frac{dT}{dt} = \frac{\kappa}{c_v} \nabla^2 T. \tag{3.28}$$

We find that the *time* rate of change of T—at every point—is proportional to the Laplacian of T, which is the second derivative of its spatial dependence. We have a differential equation—in x, y, z, and t—for the temperature T.

The differential equation (3.28) is called the *heat diffusion equation.* It is often written as

$$\frac{dT}{dt} = D\,\nabla^2 T, \tag{3.29}$$

where D is called the *diffusion* constant, and is here equal to κ/c_v.

The diffusion equation appears in many physical problems—in the diffusion of gases, in the diffusion of neutrons, and in others. We have already discussed the physics of some of these phenomena in Chapter 43 of Vol. I. Now you have the complete equation that describes diffusion in the most general possible situation. At some later time we will take up ways of solving the diffusion equation to find how the temperature varies in particular cases. We turn back now to consider other theorems about vector fields.

3–5 The circulation of a vector field

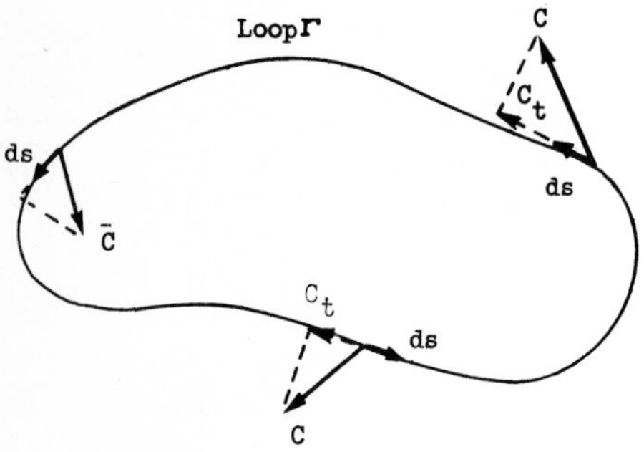

Fig. 3–7. The circulation of $\boldsymbol{C}$ around the curve Γ is the line integral of C_t, the tangential component of $\boldsymbol{C}$.

We wish now to look at the curl in somewhat the same way we looked at the divergence. We obtained Gauss' theorem by considering the integral over a surface, although it was not obvious at the beginning that we were going to be dealing with the divergence. How did we know that we were supposed to integrate over a surface in order to get the divergence? It was not at all clear that this would be the result. And so with an apparent equal lack of justification, we shall calculate something else about a vector and show that it is related to the curl. This time we calculate what is called the circulation of a vector field. If $\boldsymbol{C}$ is any vector field, we take its component along a curved line and take the integral of this component all the way around a complete loop. The integral is called the *circulation* of the vector field around the loop. We have already considered a line integral of $\nabla\psi$ earlier in this chapter. Now we do the same kind of thing for *any* vector field $\boldsymbol{C}$.

Let Γ be any closed loop in space—imaginary, of course. An example is given in Fig. 3–7. The line integral of the tangential component of $\boldsymbol{C}$ around the loop is written as

$$\oint_\Gamma C_t\,ds = \oint_\Gamma \boldsymbol{C}\cdot d\boldsymbol{s}. \tag{3.30}$$

You should note that the integral is taken all the way around, not from one point to another as we did before. The little circle on the integral sign is to remind us that the integral is to be taken all the way around. This integral is called the circulation of the vector field around the curve Γ. The name came originally from considering the circulation of a liquid. But the name—like flux—has been extended to apply to any field even when there is no material "circulating."

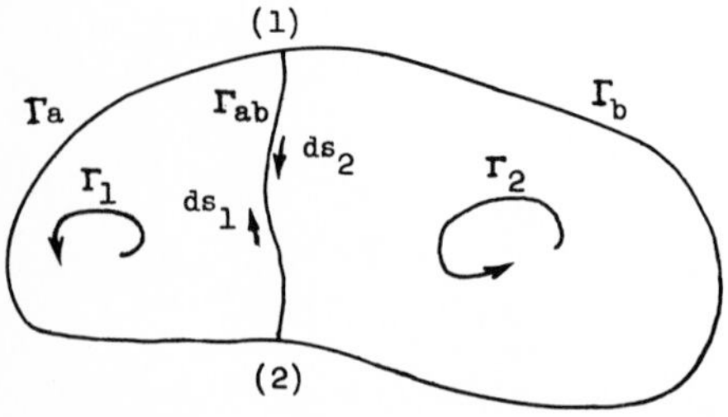

Fig. 3–8. The circulation around the whole loop is the sum of the circulations around the two loops: $\Gamma_1 = \Gamma_a + \Gamma_{ab}$ and $\Gamma_2 = \Gamma_b + \Gamma_{ab}$.

Playing the same kind of game we did with the flux, we can show that the circulation around a loop is the sum of the circulations around two partial loops. Suppose we break up our curve of Fig. 3–7 into two loops, by joining two points (1) and (2) on the original curve by some line that cuts across as shown in Fig. 3–8. There are now two loops, Γ_1 and Γ_2. Γ_1 is made up of Γ_a, which is that part of the original curve to the left of (1) and (2), plus Γ_{ab}, the "short cut." Γ_2 is made up of the rest of the original curve plus the short cut.

The circulation around Γ_1 is the sum of an integral along Γ_a and along Γ_{ab}. Similarly, the circulation around Γ_2 is the sum of two parts, one along Γ_b and the other along Γ_{ab}. The integral along Γ_{ab} will have, for the curve Γ_2, the opposite sign from what it has for Γ_1, because the direction of travel is opposite—we must take both our line integrals with the same "sense" of rotation.

Following the same kind of argument we used before, you can see that the sum of the two circulations will give just the line integral around the original curve Γ. The parts due to Γ_{ab} cancel. The circulation around the one part plus the circulation around the second part equals the circulation about the outer line. We can continue the process of cutting the original loop into any number of smaller loops. When we add the circulations of the smaller loops, there is always a cancellation of the parts on their adjacent portions, so that the sum is equivalent to the circulation around the original single loop.

Now let us suppose that the original loop is the boundary of some surface. There are, of course, an infinite number of surfaces which all have the original loops as the boundary. Our results will not, however, depend on which surface we choose. First, we break our original loop into a number of small loops that all lie on the surface we have chosen, as in Fig. 3–9. No matter what the shape of the surface, if we choose our small loops small enough, we can assume that each of the small loops will enclose an area which is essentially flat. Also, we can choose our small loops so that each is very nearly a square. Now we can calculate the circulation around the big loop Γ by finding the circulations around all of the little squares and then taking their sum.

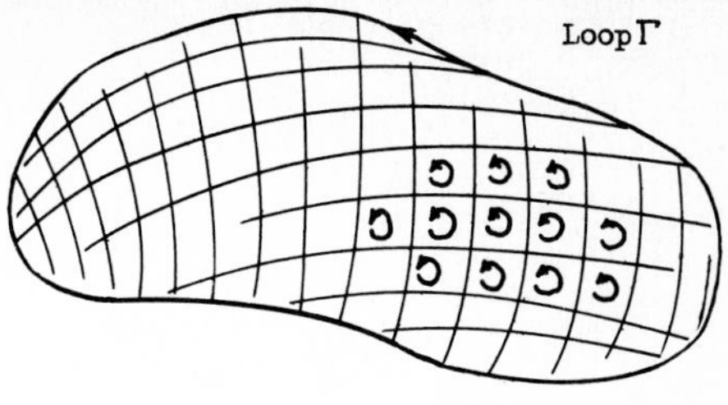

Fig. 3–9. Some surface bounded by the loop Γ is chosen. The surface is divided into a number of small areas, each approximately a square. The circulation around Γ is the sum of the circulations around the little loops.

3–6 The circulation around a square; Stokes' theorem

How shall we find the circulation for each little square? One question is, how is the square oriented in space? We could easily make the calculation if it had a special orientation. For example, if it were in one of the coordinate planes. Since we have not assumed anything as yet about the orientation of the coordinate axes, we can just as well choose the axes so that the one little square we are concentrating on at the moment lies in the xy-plane, as in Fig. 3–10. If our result is expressed in vector notation, we can say that it will be the same no matter what the particular orientation of the plane.

We want now to find the circulation of the field $\boldsymbol{C}$ around our little square. It will be easy to do the line integral if we make the square small enough that the vector $\boldsymbol{C}$ doesn't change much along any one side of the square. (The assumption is better the smaller the square, so we are really talking about infinitesimal squares.) Starting at the point (x, y)—the lower left corner of the figure—we go around in the direction indicated by the arrows. Along the first side—marked (1)—the tangential component is $C_x(1)$ and the distance is Δx. The first part of the integral is $C_x(1)\,\Delta x$. Along the second leg, we get $C_y(2)\,\Delta y$. Along the third, we get $-C_x(3)\,\Delta x$, and along the fourth, $-C_y(4)\,\Delta y$. The minus signs are required because we want the tangential component in the direction of travel. The whole line integral is then

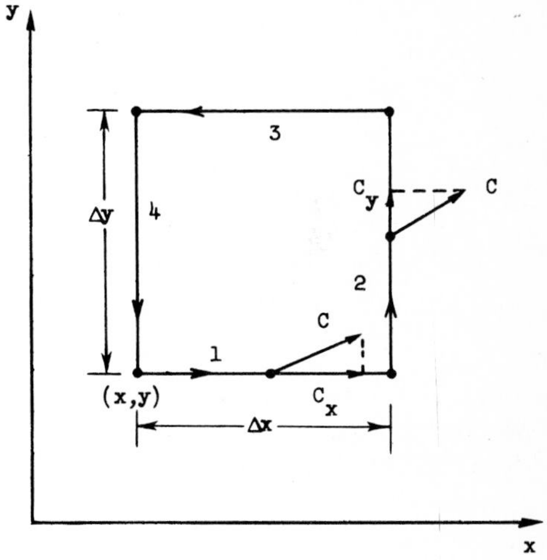

Fig. 3–10. Computing the circulation of **C** around a small square.

$$\oint \boldsymbol{C}\cdot d\boldsymbol{s} = C_x(1)\,\Delta x + C_y(2)\,\Delta y - C_x(3)\,\Delta x - C_y(4)\,\Delta y. \tag{3.31}$$

Now let's look at the first and third pieces. Together they are

$$[C_x(1) - C_x(3)]\,\Delta x. \tag{3.32}$$

You might think that to our approximation the difference is zero. That is true to the first approximation. We can be more accurate, however, and take into account the rate of change of C_x. If we do, we may write

$$C_x(3) = C_x(1) + \frac{\partial C_x}{\partial y}\,\Delta y. \tag{3.33}$$

If we included the next approximation, it would involve terms in $(\Delta y)^2$, but since we will ultimately think of the limit as $\Delta y \to 0$, such terms can be neglected. Putting (3.33) together with (3.32), we find that

$$[C_x(1) - C_x(3)]\,\Delta y = -\frac{\partial C_x}{\partial y}\,\Delta x\,\Delta y. \tag{3.34}$$

The derivative can, to our approximation, be evaluated at (x, y).

Similarly, for the other two terms in the circulation, we may write

$$C_y(2)\,\Delta y - C_y(4)\,\Delta y = \frac{\partial C_y}{\partial x}\,\Delta x\,\Delta y. \tag{3.35}$$

The circulation around our square is then

$$\left(\frac{\partial C_y}{\partial x} - \frac{\partial C_x}{\partial y}\right)\Delta x\,\Delta y, \tag{3.36}$$

which is interesting, because the two terms in the parentheses are just the z-component of the curl. Also, we note that $\Delta x\,\Delta y$ is the area of our square. So we can write our circulation (3.36) as

$$(\nabla \times \boldsymbol{C})_z\,\Delta a.$$

But the z-component really means the component *normal* to the surface element. We can, therefore, write the circulation around a differential square in an invariant vector form:

$$\oint \boldsymbol{C}\cdot d\boldsymbol{s} = (\nabla \times \boldsymbol{C})_n\,\Delta a = (\nabla \times \boldsymbol{C})\cdot \boldsymbol{n}\,\Delta a. \tag{3.37}$$

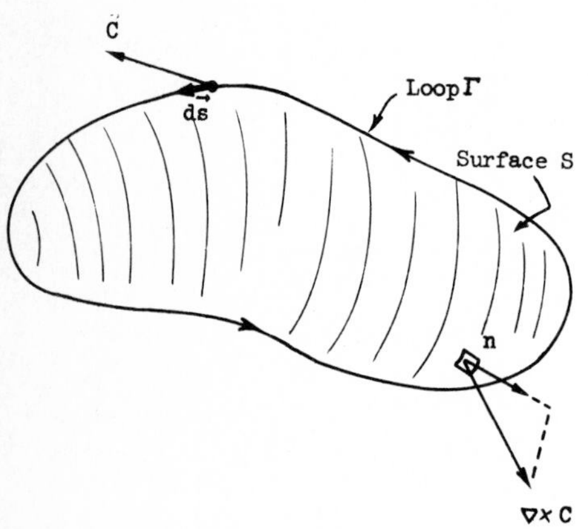

Fig. 3–11. The circulation of **C** around Γ is the surface integral of the normal component of $\nabla \times \boldsymbol{C}$.

Our result is: the circulation of any vector $\boldsymbol{C}$ around an infinitesimal square is the component of the curl of $\boldsymbol{C}$ normal to the surface, times the area of the square.

The circulation around any loop Γ can now be easily related to the curl of the vector field. We fill in the loop with any convenient surface S, as in Fig. 3–11, and add the circulations around a set of infinitesimal squares in this surface. The sum can be written as an integral. Our result is a very useful theorem called Stokes' theorem (after Mr. Stokes).

STOKES' THEOREM.

$$\oint_{\Gamma} \boldsymbol{C}\cdot d\boldsymbol{s} = \int_S (\nabla \times \boldsymbol{C})_n\,da, \tag{3.38}$$

where S is any surface bounded by Γ.

We must now speak about a convention of signs. In Fig. 3–10 the z-axis would point *toward* you in a "usual"—that is, "right-handed"—system of axes. When we took our line integral with a "positive" sense of rotation, we found that the circulation was equal to the z-component of $\nabla \times \boldsymbol{C}$. If we had gone around the other way, we would have gotten the opposite sign. Now how shall we know, in general, what direction to choose for the positive direction of the "normal" component of $\nabla \times \boldsymbol{C}$? The "positive" normal must always be related to the sense of rotation, as in Fig. 3–10. It is indicated for the general case in Fig. 3–11.

One way of remembering the relationship is by the "right-hand rule." If you make the fingers of your *right* hand go around the curve Γ, with the fingertips pointed in the direction of the positive sense of $d\boldsymbol{s}$, then your thumb points in the direction of the *positive* normal to the surface S.

3–7 Curl-free and divergence-free fields

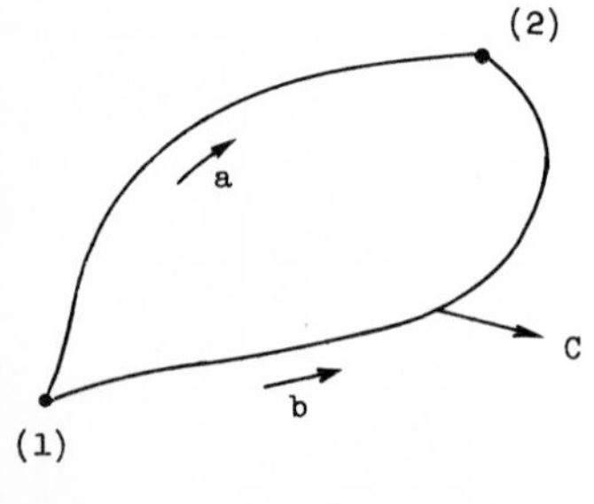

Fig. 3–12. If $\nabla \times \boldsymbol{C}$ is zero, the circulation around the closed curve Γ is zero. The line integral of $\boldsymbol{C}\cdot d\boldsymbol{s}$ from (1) to (2) along a must be the same as the line integral along b.

We would like, now, to consider some consequences of our new theorems. Take first the case of a vector whose curl is *everywhere* zero. Then Stokes' theorem says that the circulation around any loop is zero. Now if we choose two points (1) and (2) on a closed curve (Fig. 3–12), it follows that the line integral of the tangential component from (1) to (2) is independent of which of the two possible paths is taken. We can conclude that the integral from (1) to (2) can depend only on the location of these points—that is to say, it is some function of position only. The same logic was used in Chapter 14 of Vol. I, where we proved that if the integral around a closed loop of some quantity is always zero, then that integral can be represented as the difference of a function of the position of the two ends. This fact allowed us to invent the idea of a potential. We proved, furthermore, that the vector field was the gradient of this potential function (see Eq. 14.13 of Vol. I).

It follows that any vector field whose curl is zero is equal to the gradient of some scalar function. That is, if $\nabla \times \boldsymbol{C} = 0$, everywhere, there is some ψ (psi) for which $\boldsymbol{C} = \nabla\psi$—a useful idea. We can, if we wish, describe this special kind of vector field by means of a scalar field.

Let's show something else. Suppose we have *any* scalar field ϕ (phi). If we take its gradient, $\nabla\phi$, the integral of this vector around any closed loop must be zero. Its line integral from point (1) to point (2) is $[\phi(2) - \phi(1)]$. If (1) and (2) are the same points, our Theorem 1, Eq. (3.8), tells us that the line integral is zero:

$$\oint_{\text{loop}} \nabla\phi \cdot ds = 0.$$

Using Stokes' theorem, we can conclude that

$$\int \nabla \times (\nabla\phi)\, da = 0$$

over *any* surface. But if the integral is zero over *any* surface, the integrand must be zero. So

$$\nabla \times (\nabla\phi) = 0, \quad \text{always.}$$

We proved the same result in Section 2–7 by vector algebra.

Let's look now at a special case in which we fill in a *small* loop Γ with a *large* surface S, as indicated in Fig. 3–13. We would like, in fact, to see what happens when the loop shrinks down to a point, so that the surface boundary disappears—the surface becomes closed. Now if the vector $\boldsymbol{C}$ is everywhere finite, the line integral around Γ must go to zero as we shrink the loop—the integral is roughly proportional to the circumference of Γ, which goes to zero. According to Stokes' theorem, the surface integral of $(\nabla \times \boldsymbol{C})_n$ must also vanish. Somehow, as we close the surface we add in contributions that cancel out what was there before. So we have a new theorem:

$$\int_{\substack{\text{any closed}\\ \text{surface}}} (\nabla \times \boldsymbol{C})_n\, da = 0. \tag{3.39}$$

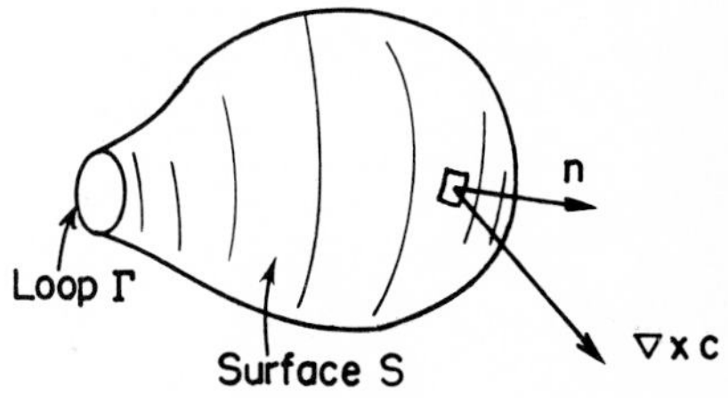

Fig. 3–13. Going to the limit of a closed surface, we find that the surface integral of $(\nabla \times \boldsymbol{C})_n$ must vanish.

Now this is interesting, because we already have a theorem about the surface integral of a vector field. Such a surface integral is equal to the volume integral of the divergence of the vector, according to Gauss' theorem (Eq. 3.18). Gauss' theorem, applied to $\nabla \times \boldsymbol{C}$, says

$$\int_{\substack{\text{closed}\\ \text{surface}}} (\nabla \times \boldsymbol{C})_n\, da = \int_{\substack{\text{volume}\\ \text{inside}}} \nabla \cdot (\nabla \times \boldsymbol{C})\, dV. \tag{3.40}$$

So we conclude that the second integral must also be zero:

$$\int_{\substack{\text{any}\\ \text{volume}}} \nabla \cdot (\nabla \times \boldsymbol{C})\, dV = 0, \tag{3.41}$$

and this is true for any vector field $\boldsymbol{C}$ whatever. Since Eq. (3.41) is true for *any volume*, it must be true that at *every point* in space the integrand is zero. We have

$$\nabla \cdot (\nabla \times \boldsymbol{C}) = 0, \quad \text{always.}$$

But this is the same result we got from vector algebra in Section 2–7. Now we begin to see how everything fits together.

3–8 Summary

Let us summarize what we have found about the vector calculus. These are really the salient points of Chapters 2 and 3:

1. The operators $\partial/\partial x$, $\partial/\partial y$, and $\partial/\partial z$ can be considered as the three components of a vector operator ∇, and the formulas which result from vector algebra by treating this operator as a vector are correct:

$$\nabla = \left(\frac{\partial}{\partial x}, \frac{\partial}{\partial y}, \frac{\partial}{\partial z}\right).$$

2. The difference of the values of a scalar field at two points is equal to the line integral of the tangential component of the gradient of that scalar along any curve at all between the first and second points:

$$\psi(2) - \psi(1) = \int_{\substack{(1)\\ \text{any curve}}}^{(2)} \nabla\psi \cdot d\mathbf{s}. \tag{3.42}$$

3. The surface integral of the normal component of an arbitrary vector over a closed surface is equal to the integral of the divergence of the vector over the volume interior to the surface:

$$\int_{\substack{\text{closed}\\ \text{surface}}} \mathbf{C}\cdot\mathbf{n}\,da = \int_{\substack{\text{volume}\\ \text{inside}}} \nabla\cdot\mathbf{C}\,dV. \tag{3.43}$$

4. The line integral of the tangential component of an arbitrary vector around a closed loop is equal to the surface integral of the normal component of the curl of that vector over any surface which is bounded by the loop.

$$\int_{\text{boundary}} \mathbf{C}\cdot d\mathbf{s} = \int_{\text{surface}} (\nabla\times\mathbf{C})\cdot\mathbf{n}\,da. \tag{3.44}$$

Electrostatics

Review: Chapters 13 and 14, Vol. I, *Work and Potential Energy*

4–1 Statics

We begin now our detailed study of the theory of electromagnetism. All of electromagnetism is contained in the Maxwell equations.

Maxwell's equations:

$$\nabla \cdot \boldsymbol{E} = \frac{\rho}{\epsilon_0}, \tag{4.1}$$

$$\nabla \times \boldsymbol{E} = -\frac{\partial \boldsymbol{B}}{\partial t}, \tag{4.2}$$

$$c^2 \nabla \times \boldsymbol{B} = \frac{\partial \boldsymbol{E}}{\partial t} + \frac{\boldsymbol{j}}{\epsilon_0}, \tag{4.3}$$

$$\nabla \cdot \boldsymbol{B} = 0. \tag{4.4}$$

The situations that are described by these equations can be very complicated. We will consider first relatively simple situations, and learn how to handle them before we take up more complicated ones. The easiest circumstance to treat is one in which nothing depends on the time—called the *static* case. All charges are permanently fixed in space, or if they do move, they move as a steady flow in a circuit (so ρ and $\boldsymbol{j}$ are constant in time). In these circumstances, all of the terms in the Maxwell equations which are time derivatives of the field are zero. In this case, the Maxwell equations become:

Electrostatics:

$$\nabla \cdot \boldsymbol{E} = \frac{\rho}{\epsilon_0}, \tag{4.5}$$

$$\nabla \times \boldsymbol{E} = 0. \tag{4.6}$$

Magnetostatics:

$$\nabla \times \boldsymbol{B} = \frac{\boldsymbol{j}}{\epsilon_0 c^2}, \tag{4.7}$$

$$\nabla \cdot \boldsymbol{B} = 0. \tag{4.8}$$

$$\epsilon_0 c^2 = \frac{10^7}{4\pi}$$

$$\frac{1}{4\pi\epsilon_0} \approx 9 \times 10^9$$

$$[\epsilon_0] = \text{coulomb}^2/\text{newton}\cdot\text{meter}^2$$

You will notice an interesting thing about this set of four equations. It can be separated into two pairs. The electric field $\boldsymbol{E}$ appears only in the first two, and the magnetic field $\boldsymbol{B}$ appears only in the second two. The two fields are not interconnected. This means that *electricity and magnetism are distinct phenomena so long as charges and currents are static.* The interdependence of $\boldsymbol{E}$ and $\boldsymbol{B}$ does not appear until there are changes in charges or currents, as when a condensor is charged, or a magnet moved. Only when there are sufficiently rapid changes, so that the time derivatives in Maxwell's equations become significant, will $\boldsymbol{E}$ and $\boldsymbol{B}$ depend on each other.

Now if you look at the equations of statics you will see that the study of the two subjects we call electrostatics and magnetostatics is ideal from the point of view of learning about the mathematical properties of vector fields. Electrostatics is a neat example of a vector field with *zero curl* and a *given divergence.* Magnetostatics is a neat example of a field with *zero divergence* and a *given curl.* The more conventional—and you may be thinking, more satisfactory—way of presenting the theory of electromagnetism is to start first with electrostatics and thus to learn about the divergence. Magnetostatics and the curl are taken up later. Finally, electricity and magnetism are put together. We have chosen to start with the complete theory of vector calculus. Now we shall apply it to the special case of electrostatics, the field of $\boldsymbol{E}$ given by the first pair of equations.

We will begin with the simplest situations—ones in which the positions of all charges are specified. If we had only to study electrostatics at this level (as we shall do in the next two chapters), life would be very simple—in fact, almost trivial. Everything can be obtained from Coulomb's law and some integration, as you will see. In many real electrostatic problems, however, we do not *know*, initially, where the charges are. We know only that they have distributed themselves in ways that depend on the properties of matter. The positions that the charges take up depend on the $\boldsymbol{E}$ field, which in turn depends on the positions of the charges. Then things can get quite complicated. If, for instance, a charged body is brought near a conductor or insulator, the electrons and protons in the conductor or insulator will move around. The charge density ρ in Eq. (4.5) may have one part that we know about, from the charge that we brought up; but there will be other parts from charges that have moved around in the conductor. And all of the charges must be taken into account. One can get into some rather subtle and interesting problems. So although this chapter is to be on electrostatics, it will not cover the more beautiful and subtle parts of the subject. It will treat only the situation where we can assume that the positions of all the charges are known. Naturally, you should be able to do that case before you try to handle the other ones.

4–2 Coulomb's law; superposition

It would be logical to use Eqs. (4.5) and (4.6) as our starting points. It will be easier, however, if we start somewhere else and come back to these equations. The results will be equivalent. We will start with a law that we have talked about before, called Coulomb's law, which says that between two charges at rest there is a force directly proportional to the product of the charges and inversely proportional to the square of the distance between. The force is along the straight line from one charge to the other.

Coulomb's law:

$$\boldsymbol{F}_1 = \frac{1}{4\pi\epsilon_0}\frac{q_1q_2}{r_{12}^2}\boldsymbol{e}_{12} = -\boldsymbol{F}_2. \tag{4.9}$$

$\boldsymbol{F}_1$ is the force *on* charge q_1, $\boldsymbol{e}_{12}$ is the unit vector in the direction *to* q_1 *from* q_2, and r_{12} is the distance between q_1 and q_2. The force $\boldsymbol{F}_2$ on q_2 is equal and opposite to $\boldsymbol{F}_1$.

The constant of proportionality, for historical reasons, is written as $1/4\pi\epsilon_0$. In the system of units which we use—the mks system—it is defined as exactly 10^{-7} times the speed of light squared. Now since the speed of light is approximately 3×10^8 meters per second, the constant is approximately 9×10^9, and the unit turns out to be newton·meter^2 per coulomb^2 or volt·meter per coulomb.

$$\begin{aligned}\frac{1}{4\pi\epsilon_0} &= 10^{-7}c^2 \quad \text{(by definition)}\\ &= 9.0 \times 10^9 \text{ (by experiment)}.\end{aligned} \tag{4.10}$$

$$\text{Unit: newton·meter}^2/\text{coulomb}^2,$$

$$\text{or volt·meter/coulomb}.$$

When there are more than two charges present—the only really interesting times—we must supplement Coulomb's law with one other fact of nature: the force on any charge is the vector sum of the Coulomb forces from each of the other charges. This fact is called "the principle of superposition." That's all there is to electrostatics. If we combine the Coulomb law and the principle of superposition, there is nothing else. Equations (4.5) and (4.6)—the electrostatic equations—say no more and no less.

When applying Coulomb's law, it is convenient to introduce the idea of an electric field. We say that the field $\boldsymbol{E}(1)$ is the force *per unit charge* on q_1 (due to all other charges). Dividing Eq. (4.9) by q_1, we have, for one other charge besides q_1,

$$\boldsymbol{E}(1) = \frac{1}{4\pi\epsilon_0} \frac{q_2}{r_{12}^2} \boldsymbol{e}_{12}. \tag{4.11}$$

Also, we consider that $\boldsymbol{E}(1)$ describes something about the point (1) even if q_1 were not there—assuming that all other charges keep their same positions. We say: $\boldsymbol{E}(1)$ is the electric field *at* the point (1).

The electric field $\boldsymbol{E}$ is a vector, so by Eq. (4.11) we really mean three equations—one for each component. Writing out explicitly the x-component, Eq. (4.11) means

$$E_x(x_1, y_1, z_1) = \frac{q_2}{4\pi\epsilon_0} \frac{x_1 - x_2}{[(x_1 - x_2)^2 + (y_1 - y_2)^2 + (z_1 - z_2)^2]^{3/2}}, \tag{4.12}$$

and similarly for the other components.

If there are many charges present, the field $\boldsymbol{E}$ at any point (1) is a sum of the contributions from each of the other charges. Each term of the sum will look like (4.11) or (4.12). Letting q_j be the magnitude of the jth charge, and $\boldsymbol{r}_{1j}$ the displacement from q_j to the point (1), we write

$$\boldsymbol{E}(1) = \sum_j \frac{1}{4\pi\epsilon_0} \frac{q_j}{r_{1j}^2} \boldsymbol{e}_{1j}. \tag{4.13}$$

Which means, of course,

$$E_x(x_1, y_1, z_1) = \sum_j \frac{1}{4\pi\epsilon_0} \frac{q_j(x_1 - x_j)}{[(x_1 - x_j)^2 + (y_1 - y_j)^2 + (z_1 - z_j)^2]^{3/2}}, \tag{4.14}$$

and so on.

Often it is convenient to ignore the fact that charges come in packages like electrons and protons, and think of them as being spread out in a continuous smear—or in a "distribution," as it is called. This is O.K. so long as we are not interested in what is happening on too small a scale. We describe a charge distribution by the "charge density," $\rho(x, y, z)$. If the amount of charge in a small volume ΔV_2 located at the point (2) is Δq_2, then ρ is defined by

$$\Delta q_2 = \rho(2)\,\Delta V_2. \tag{4.15}$$

To use Coulomb's law with such a description, we replace the sums of Eqs. (4.13) or (4.14) by integrals over all volumes containing charges. Then we have

$$\boldsymbol{E}(1) = \frac{1}{4\pi\epsilon_0} \int_{\substack{\text{all}\\\text{space}}} \frac{\rho(2)\boldsymbol{e}_{12}\,dV_2}{r_{12}^2}. \tag{4.16}$$

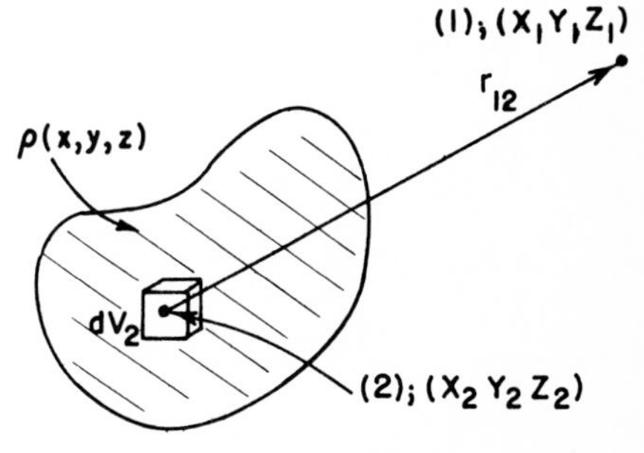

Fig. 4–1. The electric field $\boldsymbol{E}$ at point (1), from a charge distribution, is obtained from an integral over the distribution. Point (1) could also be inside the distribution.

Some people prefer to write

$$\boldsymbol{e}_{12} = \frac{\boldsymbol{r}_{12}}{r_{12}},$$

where $\boldsymbol{r}_{12}$ is the vector displacement *to* (1) *from* (2), as shown in Fig. 4–1. The integral for $\boldsymbol{E}$ is then written as

$$\boldsymbol{E}(1) = \frac{1}{4\pi\epsilon_0} \int_{\substack{\text{all}\\\text{space}}} \frac{\rho(2)\boldsymbol{r}_{12}\,dV_2}{r_{12}^3}. \tag{4.17}$$

When we want to calculate something with these integrals, we usually have to write them out in explicit detail. For the x-component of either Eq. (4.16) or (4.17), we would have

$$E_x(x_1, y_1, z_1) = \int_{\substack{\text{all}\\\text{space}}} \frac{(x_1 - x_2)\rho(x_2, y_2, z_2)\,dx_2\,dy_2\,dz_2}{4\pi\epsilon_0[(x_1 - x_2)^2 + (y_1 - y_2)^2 + (z_1 - z_2)^2]^{3/2}}. \tag{4.18}$$

We are not going to use this formula much. We write it here only to emphasize the fact that we have completely solved all the electrostatic problems in which we know the locations of all of the charges. Given the charges, what are the fields? *Answer:* Do this integral. So there is nothing to the subject; it is just a case of doing complicated integrals over three dimensions—strictly a job for a computing machine!

With our integrals we can find the fields produced by a sheet of charge, from a line of charge, from a spherical shell of charge, or from any specified distribution. It is important to realize, as we go on to draw field lines, to talk about potentials, or to calculate divergences, that we already have the answer here. It is merely a matter of it being sometimes easier to do an integral by some clever guesswork than by actually carrying it out. The guesswork requires learning all kinds of strange things. In practice, it might be easier to forget trying to be clever and always to do the integral directly instead of being so smart. We are, however, going to try to be smart about it. We shall go on to discuss some other features of the electric field.

4–3 Electric potential

First we take up the idea of electric potential, which is related to the work done in carrying a charge from one point to another. There is some distribution of charge, which produces an electric field. We ask about how much work it would take to carry a small charge from one place to another. The work done *against* the electrical forces in carrying a charge along some path is the *negative* of the component of the electrical force in the direction of the motion, integrated along the path. If we carry a charge from point a to point b,

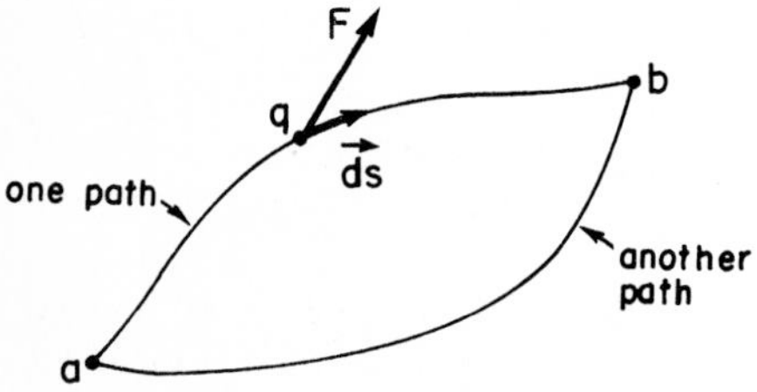

Fig. 4–2. The work done in carrying a charge from a to b is the negative of the integral of $\boldsymbol{F} \cdot d\boldsymbol{s}$ along the path taken.

$$W = -\int_a^b \boldsymbol{F} \cdot d\boldsymbol{s},$$

where $\boldsymbol{F}$ is the electrical force *on* the charge at each point, and $d\boldsymbol{s}$ is the differential vector displacement along the path. (See Fig. 4–2.)

It is more interesting for our purposes to consider the work that would be done in carrying *one unit* of charge. Then the force on the charge is numerically the same as the electric field. Calling the work done against electrical forces in this case W(unit), we write

$$W(\text{unit}) = -\int_a^b \boldsymbol{E} \cdot d\boldsymbol{s}. \tag{4.19}$$

Now, in general, what we get with this kind of an integral depends on the path we take. But if the integral of (4.19) depended on the path from a to b, we could get work out of the field by carrying the charge to b along one path and then back to a on the other. We would go to b along the path for which W is smaller and *back* along the other, getting *out* more work than we put *in*.

There is nothing impossible, in principle, about getting energy out of a field. We shall, in fact, encounter fields where it is possible. It could be that as you move a charge you produce forces on the other part of the "machinery." If the "machinery" moved against the force it would lose energy, thereby keeping the total energy in the world constant. For *electrostatics*, however, there is no such "machinery." We know what the forces back on the sources of the field are. They are the Coulomb forces on the charges responsible for the field. If the other charges are fixed in position—as we assume in *electrostatics* only—these back forces can do no work on them. There is no way to get energy from them—provided, of course, that the principle of energy conservation works for electrostatic situations. We believe that it will work, but let's just show that it must follow from Coulomb's law of force.

We consider first what happens in the field due to a single charge q. Let point a be at the distance r_1 from q, and point b at r_2. Now we carry a different charge, which we will call the "test" charge, and whose magnitude we choose to be one unit, from a to b. Let's start with the easiest possible path to calculate. We carry our test charge first along the arc of a circle, then along a radius, as shown in part (a) of Fig. 4–3. Now on that particular path it is child's play to find the work done (otherwise we wouldn't have picked it). First, there is no work done at all on the path from a to a'. The field is radial (from Coulomb's law), so it is at right angles to the direction of motion. Next, on the path from a' to b, the field is in the direction of motion and varies as $1/r^2$. Thus the work done on the test charge in carrying it from a to b would be

$$-\int_a^b \boldsymbol{E}\cdot d\boldsymbol{s} = -\frac{q}{4\pi\epsilon_0}\int_{a'}^b \frac{dr}{r^2} = -\frac{q}{4\pi\epsilon_0}\left(\frac{1}{r_a}-\frac{1}{r_b}\right). \tag{4.20}$$

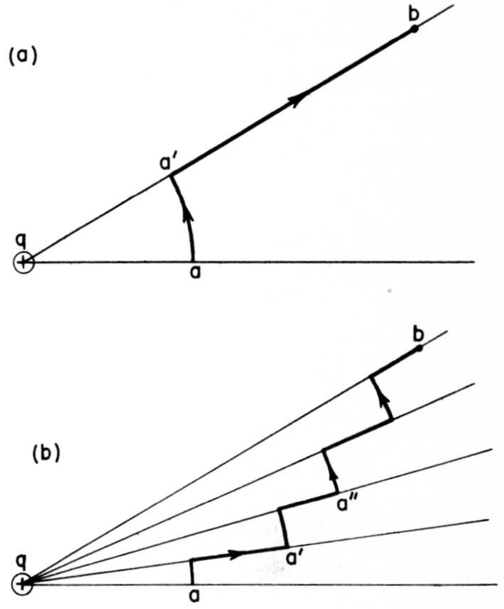

Fig. 4–3. In carrying a test charge from a to b the same work is done along either path.

Now let's take another easy path. For instance, the one shown in part (b) of Fig. 4–3. It goes for awhile along an arc of a circle, then radially for awhile, then along an arc again, then radially, and so on. Every time we go along the circular parts, we do no work. Every time we go along the radial parts, we must just integrate $1/r^2$. Along the first radial stretch, we integrate from r_a to $r_{a'}$, then along the next radial stretch from $r_{a'}$ to $r_{a''}$, and so on. The sum of all these integrals is the same as a single integral directly from r_a to r_b. We get the same answer for this path that we did for the first path we tried. It is clear that we would get the same answer for *any* path which is made up of an arbitrary number of the same kinds of pieces.

What about smooth paths? Would we get the same answer? We discussed this point previously in Chapter 13 of Vol. I. Applying the same arguments used there, we can conclude that work done in carrying a unit charge from a to b is independent of the path.

$$\left.\begin{matrix} W(\text{unit}) \\ a \to b \end{matrix}\right\} = -\int_{\substack{a \\ \text{any} \\ \text{path}}}^b \boldsymbol{E}\cdot d\boldsymbol{s}.$$

Since the work done depends only on the endpoints, it can be represented as the difference between two numbers. We can see this in the following way. Let's choose a reference point P_0 and agree to evaluate our integral by using a path that always goes *by way of* point P_0. Let $\phi(a)$ stand for the work done against the field in going *from* P_0 to point a, and let $\phi(b)$ be the work done in going *from* P_0 to point b (Fig. 4–4). The work in going *to* P_0 from a (on the way to b) is the negative of $\phi(a)$, so we have that

$$-\int_a^b \boldsymbol{E}\cdot d\boldsymbol{s} = \phi(b) - \phi(a). \tag{4.21}$$

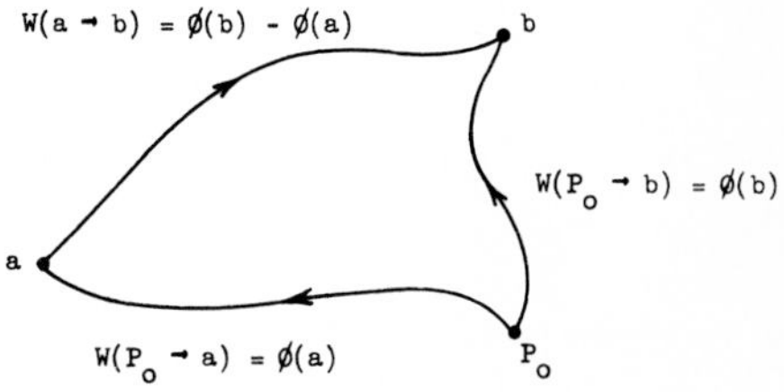

Fig. 4–4. The work done in going along any path from a to b is the negative of the work from some point P_0 to a plus the work from P_0 to b.

Since only the difference in the function ϕ at two points is ever involved, we do not really have to specify the location of P_0. Once we have chosen some reference point, however, a number ϕ is determined for *any* point in space; ϕ is then a *scalar field*. It is a function of x, y, z. We call this scalar function the *electrostatic potential* at any point.

Electrostatic potential:

$$\phi(P) = -\int_{P_0}^P \boldsymbol{E}\cdot d\boldsymbol{s}. \tag{4.22}$$

For convenience, we will often take the reference point at infinity. Then, for a single charge at the origin, the potential ϕ is given for any point (x, y, z)—using Eq. (4.20):

$$\phi(x, y, z) = \frac{q}{4\pi\epsilon_0}\frac{1}{r}. \tag{4.23}$$

The electric field from several charges can be written as the sum of the electric field from the first, from the second, from the third, etc. When we integrate the sum to find the potential we get a sum of integrals. Each of the integrals is the potential from one of the charges. We conclude that the potential ϕ from a lot of charges is the sum of the potentials from all the individual charges. There is a superposition principle also for potentials. Using the same kind of arguments by which we found the electric field from a group of charges and for a distribution of charges, we can get the complete formulas for the potential ϕ at a point we call (1):

$$\phi(1) = \sum_j \frac{1}{4\pi\epsilon_0} \frac{q_j}{r_{1j}}, \tag{4.24}$$

$$\phi(1) = \frac{1}{4\pi\epsilon_0} \int \frac{\rho(2)\, dV_2}{r_{12}}. \tag{4.25}$$

Remember that the potential ϕ has a physical significance: it is the potential energy which a unit charge would have if brought to the specified point in space from some reference point.

4–4 $E = -\nabla\phi$

Who cares about ϕ? Forces on charges are given by $\boldsymbol{E}$, the electric field. The point is that $\boldsymbol{E}$ can be obtained easily from ϕ—it is as easy, in fact, as taking a derivative. Consider two points, one at x and one at $(x + dx)$, but both at the same y and z, and ask how much work is done in carrying a unit charge from one point to the other. The path is along the horizontal line from x to $x + dx$. The work done is the difference in the potential at the two points:

$$\Delta W = \phi(x + \Delta x, y, z) - \phi(x, y, z) = \frac{\partial\phi}{\partial x}\Delta x.$$

But the work done against the field for the same path is

$$\Delta W = -\int \boldsymbol{E} \cdot d\boldsymbol{s} = -E_x\, \Delta x.$$

We see that

$$E_x = -\frac{\partial\phi}{\partial x}. \tag{4.26}$$

Similarly, $E_y = -\partial\phi/\partial y$, $E_z = -\partial\phi/\partial z$, or, summarizing with the notation of vector analysis,

$$\boldsymbol{E} = -\nabla\phi. \tag{4.27}$$

This equation is the differential form of Eq. (4.22). Any problem with specified charges can be solved by computing the potential from (4.24) or (4.25) and using (4.27) to get the field. Equation (4.27) also agrees with what we found from vector calculus: that for any scalar field ϕ

$$\int_a^b \nabla\phi \cdot d\boldsymbol{s} = \phi(b) - \phi(a). \tag{4.28}$$

According to Eq. (4.25) the scalar potential ϕ is given by a three-dimensional integral similar to the one we had for $\boldsymbol{E}$. Is there any advantage to computing ϕ rather than $\boldsymbol{E}$? Yes. There is only one integral for ϕ, while there are three integrals for $\boldsymbol{E}$—because it is a vector. Furthermore, $1/r$ is usually a little easier to integrate than x/r^3. It turns out in many practical cases that it is easier to calculate ϕ and then take the gradient to find the electric field, than it is to evaluate the three integrals for $\boldsymbol{E}$. It is merely a practical matter.

There is also a deeper physical significance to the potential ϕ. We have shown that $\boldsymbol{E}$ of Coulomb's law is obtained from $\boldsymbol{E} = -\text{grad}\,\phi$, when ϕ is given by (4.22). But if $\boldsymbol{E}$ is equal to the gradient of a scalar field, then we know from the vector calculus that the curl of $\boldsymbol{E}$ must vanish:

$$\nabla \times \boldsymbol{E} = 0. \tag{4.29}$$

But that is just our second fundamental equation of electrostatics, Eq. (4.6). We have shown that Coulomb's law gives an $\boldsymbol{E}$ field that staisfies that condition. So far, everything is all right.

We had really proved that $\nabla \times \boldsymbol{E}$ was zero before we defined the potential. We had shown that the work done around a closed path is zero. That is, that

$$\oint \boldsymbol{E} \cdot d\boldsymbol{s} = 0$$

for *any* path. We saw in Chapter 3 that for any such field $\nabla \times \boldsymbol{E}$ must be zero everywhere. The electric field in electrostatics is an example of a curl-free field.

You can practice your vector calculus by proving that $\nabla \times \boldsymbol{E}$ is zero in a different way—by computing the components of $\nabla \times \boldsymbol{E}$ for the field of a point charge, as given by Eq. (4.11). If you get zero, the superposition principle says you would get zero for the field of any charge distribution.

We should point out an important fact. For any *radial* force the work done is independent of the path, and there exists a potential. If you think about it, the entire argument we made above to show that the work integral was independent of the path depended only on the fact that the force from a single charge was radial and spherically symmetric. It did not depend on the fact that the dependence on distance was as $1/r^2$—there could have been any r dependence. The existence of a potential, and the fact that the curl of $\boldsymbol{E}$ is zero, comes really only from the *symmetry* and *direction* of the electrostatic forces. Because of this, Eq. (4–28)—or (4.29)—can contain only part of the laws of electricity.

4–5 The flux of *E*

We will now derive a field equation that depends specifically and directly on the fact that the force law is inverse square. That the field varies inversely as the square of the distance seems, for some people, to be "only natural," because "that's the way things spread out." Take a light source with light streaming out: the amount of light that passes through a surface cut out by a cone with its apex at the source is the same no matter at what radius the surface is placed. It must be so if there is to be conservation of light energy. The amount of light per unit area—the intensity—must vary inversely as the area cut by the cone, i.e., inversely as the square of the distance from the source. Certainly the electric field should vary inversely as the square of the distance for the same reason! But there is no such thing as the "same reason" here. Nobody can say that the electric field measures the flow of something like light which must be conserved. *If* we had a "model" of the electric field in which the electric field vector represented the direction and speed—say the current—of some kind of little "bullets" which were flying out, *and* if our model required that these bullets were conserved, that none could ever disappear once it was shot out of a charge, then we might say that we can "see" that the inverse square law is necessary. On the other hand, there would necessarily be some mathematical way to express this physical idea. If the electric field *were* like conserved bullets going out, then it would vary inversely as the square of the distance and we would be able to describe that behavior by an equation—which is purely mathematical. Now there is no harm in thinking this way, so long as we do not say that the electric field *is made* out of bullets, but realize that we are using a model to help us find the right mathematics.

Suppose, indeed, that we imagine for a moment that the electric field did represent the flow of something that was conserved—everywhere, that is, except at charges. (It has to start somewhere!) We imagine that whatever it is flows out of a charge into the space around. If $\boldsymbol{E}$ were the vector of such a flow (as $\boldsymbol{h}$ is for heat flow), it would have a $1/r^2$ dependence near a point source. Now we wish to use this model to find out how to state the inverse square law in a deeper or more abstract way, rather than simply saying "inverse square." (You may wonder why we should want to avoid the direct statement of such a simple law, and want instead to imply the same thing sneakily in a different way. Patience! It will turn out to be useful.)

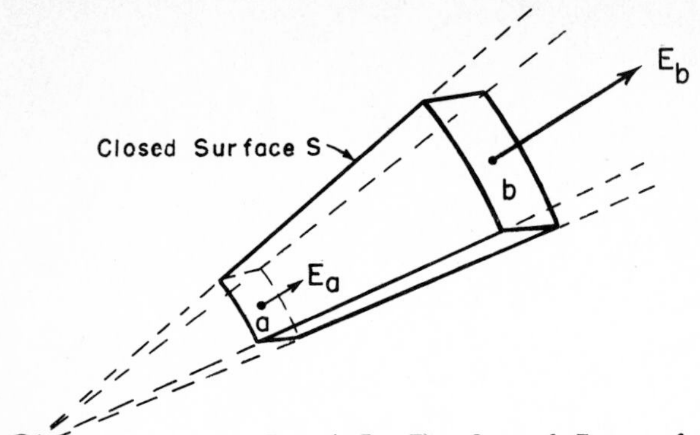

Fig. 4–5. The flux of ***E*** out of the surface S is zero.

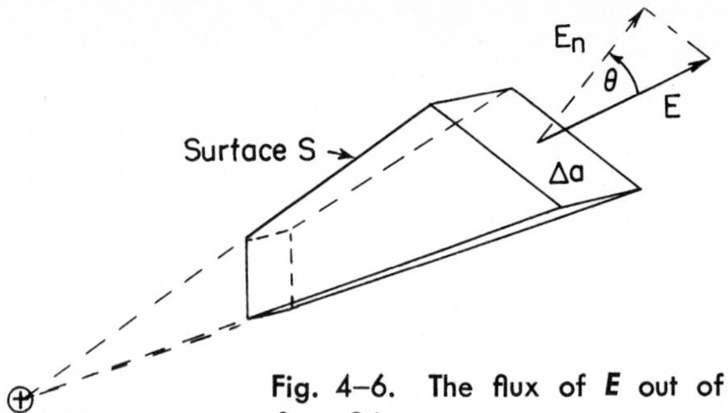

Fig. 4–6. The flux of ***E*** out of the surface S is zero.

We ask: What is the "flow" of $\boldsymbol{E}$ out of an arbitrary closed surface in the neighborhood of a point charge? First let's take an easy surface—the one shown in Fig. 4–5. If the $\boldsymbol{E}$ field is like a flow, the net flow out of this box should be zero. That is what we get if by the "flow" from this surface we mean the surface integral of the normal component of $\boldsymbol{E}$—that is, the flux of $\boldsymbol{E}$. On the radial faces, the normal component is zero. On the spherical faces, the normal component E_n is just the magnitude of $\boldsymbol{E}$—minus for the smaller face and plus for the larger face. The magnitude of $\boldsymbol{E}$ decreases as $1/r^2$, but the surface area is proportional to r^2, so the product is independent of r. The flux of $\boldsymbol{E}$ into face a is just cancelled by the flux out of face b. The total flow out of S is zero, which is to say that for this surface

$$\int_S E_n\,da = 0. \tag{4.30}$$

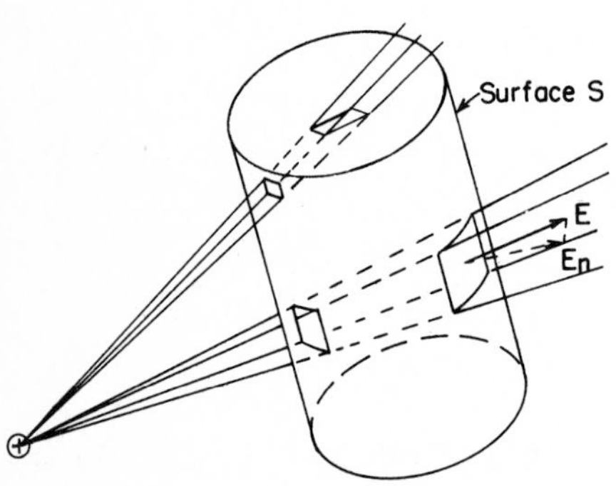

Fig. 4–7. Any volume can be thought of as completely made up of infinitesimal truncated cones. The flux of ***E*** from one end of each conical segment is equal and opposite to the flux from the other end. The total flux from the surface S is therefore zero.

Next we show that the two end surfaces may be tilted with respect to the radial line without changing the integral (4.30). Although it is true in general, for our purposes it is only necessary to show that this is true when the end surfaces are small, so that they subtend a small angle from the source—in fact, an infinitesimal angle. In Fig. 4–6 we show a surface S whose "sides" are radial, but whose "ends" are tilted. The end surfaces are not small in the figure, but you are to imagine the situation for very small end surfaces. Then the field $\boldsymbol{E}$ will be sufficiently uniform over the surface that we can use just its value at the center. When we tilt the surface by an angle θ, the area is increased by the factor $1/\cos\theta$. But E_n, the component of $\boldsymbol{E}$ normal to the surface, is decreased by the factor $\cos\theta$. The product $E_n\,\Delta a$ is unchanged. The flux out of the whole surface S is still zero.

Now it is easy to see that the flux out of a volume enclosed by *any* surface S must be zero. Any volume can be thought of as made up of pieces, like that in Fig. 4–6. The surface will be subdivided completely into pairs of end surfaces, and since the fluxes in and out of these end surfaces cancel by pairs, the total flux out of the surface will be zero. The idea is illustrated in Fig. 4–7. We have the completely general result that the total flux of $\boldsymbol{E}$ out of *any* surface S in the field of a point charge is zero.

But notice! Our proof works only if the surface *S does not surround* the charge. What would happen if the point charge were *inside* the surface? We could still divide our surface into pairs of areas that are matched by radial lines through the charge, as shown in Fig. 4–8. The fluxes through the two surfaces are still equal—by the same arguments as before—only now they have the *same* sign. The flux out of a surface that *surrounds* a charge is *not* zero. Then what is it? We can find out by a little trick. Suppose we "remove" the charge from the "inside" by surrounding the charge by a little surface S' totally inside the original surface S, as shown in Fig. 4–9. Now the volume enclosed *between* the two surfaces S and S' has no charge in it. The total flux out of this volume (including that through S') is zero, by the arguments we have given above. The arguments tell us, in fact, that the flux *into* the volume through S' is the same as the flux outward through S.

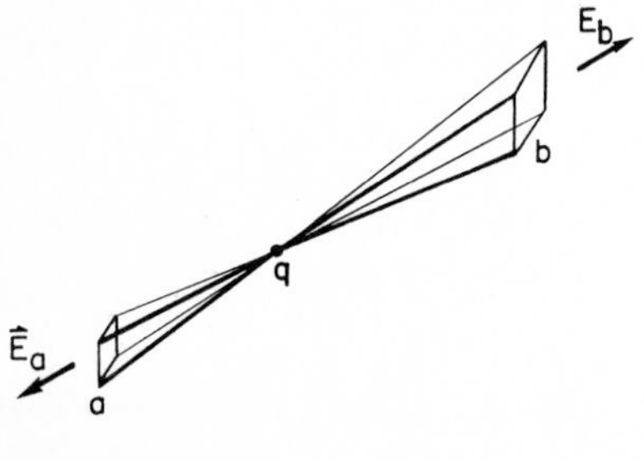

Fig. 4–8. If a charge is inside a surface, the flux out is not zero.

We can choose any shape we wish for S', so let's make it a sphere centered on the charge, as in Fig. 4–10. Then we can easily calculate the flux through it. If the radius of the little sphere is r, the value of $\boldsymbol{E}$ everywhere on its surface is

$$\frac{1}{4\pi\epsilon_0}\frac{q}{r^2},$$

and is directed always normal to the surface. We find the total flux through S' if we multiply this normal component of $\boldsymbol{E}$ by the surface area:

$$\text{Flux through the suface } S' = \left(\frac{1}{4\pi\epsilon_0}\frac{q}{r^2}\right)(4\pi r^2) = \frac{q}{\epsilon_0}, \tag{4.31}$$

a number independent of the radius of the sphere! We know then that the flux outward through S is also q/ϵ_0—a value independent of the shape of S so long as the charge q is inside.

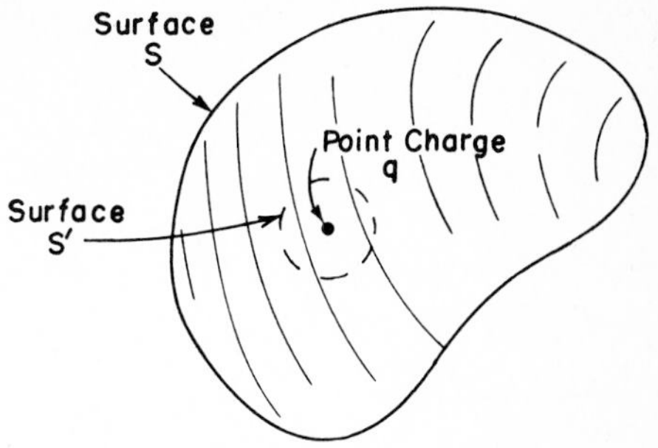

Fig. 4–9. The flux through S is the same as the flux through S'.

We can write our conclusions as follows:

$$\int_{\text{any surface } S} E_n\,da = \begin{cases} 0; & q \text{ outside } S \\ \dfrac{q}{\epsilon_0}; & q \text{ inside } S \end{cases} \tag{4.32}$$

Let's return to our "bullet" analogy and see if it makes sense. Our theorem says that the net flow of bullets through a surface is zero if the surface does not enclose the gun that shoots the bullets. If the gun is enclosed in a surface, whatever size and shape it is, the number of bullets passing through is the same—it is given by the rate at which bullets are generated at the gun. It all seems quite reasonable for conserved bullets. But does the model tell us anything more than we get simply by writing Eq. (4.32)? No one has succeeded in making these "bullets" do anything else but produce this one law. After that, they produce nothing but errors. That is why today we prefer to represent the electromagnetic field purely abstractly.

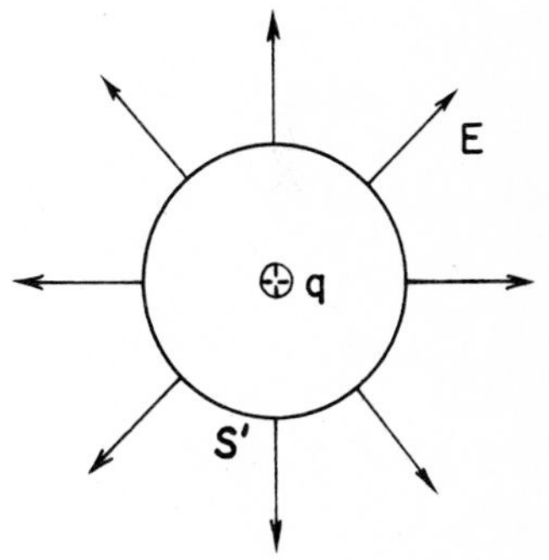

Fig. 4–10. The flux through a spherical surface containing a point charge q is q/ϵ_0.

4–6 Gauss' law; the divergence of $\boldsymbol{E}$

Our nice result, Eq. (4.32), was proved for a single point charge. Now suppose that there are two charges, a charge q_1 at one point and a charge q_2 at another. The problem looks more difficult. The electric field whose normal component we integrate for the flux is the field due to both charges. That is, if $\boldsymbol{E}_1$ represents the electric field that would have been produced by q_1 alone, and $\boldsymbol{E}_2$ represents the electric field produced by q_2 alone, the total electric field is $\boldsymbol{E} = \boldsymbol{E}_1 + \boldsymbol{E}_2$. The flux through any closed surface S is

$$\int_S (E_{1n} + E_{2n})\,da = \int_S E_{1n}\,da + \int_S E_{2n}\,da. \tag{4.33}$$

The flux with both charges present is the flux due to a single charge plus the flux due to the other charge. If both charges are outside S, the flux through S is zero. If q_1 is inside S but q_2 is outside, then the first integral gives q_1/ϵ_0 and the second integral gives zero. If the surface encloses both charges, each will give its contribution and we have that the flux is $(q_1 + q_2)/\epsilon_0$. The general rule is clearly that the total flux out of a closed surface is equal to the total charge *inside*, divided by ϵ_0.

Our result is an important general law of the electrostatic field, called Gauss' law.

Gauss' law:

$$\int_{\substack{\text{any closed}\\ \text{surface } S}} E_n\,da = \frac{\text{sum of charges inside}}{\epsilon_0}, \tag{4.34}$$

or

$$\int_{\substack{\text{any closed}\\ \text{surface } S}} \boldsymbol{E}\cdot\boldsymbol{n}\,da = \frac{Q_{\text{int}}}{\epsilon_0}, \tag{4.35}$$

where

$$Q_{\text{int}} = \sum_{\text{inside } S} q_i. \tag{4.36}$$

If we describe the location of charges in terms of a charge density ρ, we can consider that each infinitesimal volume dV contains a "point" charge $\rho\, dV$. The sum over all charges is then the integral

$$Q_{\text{int}} = \int_{\substack{\text{volume}\\ \text{inside } S}} \rho\, dV. \tag{4.37}$$

From our derivation you see that Gauss' law follows from the fact that the exponent in Coulomb's law is exactly two. A $1/r^3$ field, or any $1/r^n$ field with $n \neq 2$, would not give Gauss' law. So Gauss' law is just an expression, in a different form, of the Coulomb law of forces between two charges. In fact, working back from Gauss' law, you can derive Coulomb's law. The two are quite equivalent so long as we keep in mind the rule that the forces between charges is radial.

We would now like to write Gauss' law in terms of derivatives. To do this, we apply Gauss' law to an infinitesimal cubical surface. We showed in Chapter 3 that the flux of $\boldsymbol{E}$ out of such a cube is $\boldsymbol{\nabla} \cdot \boldsymbol{E}$ times the volume dV of the cube. The charge inside of dV, by the definition of ρ, is equal to $\rho\, dV$, so Gauss' law gives

$$\boldsymbol{\nabla} \cdot \boldsymbol{E}\, dV = \frac{\rho\, dV}{\epsilon_0},$$

or

$$\boldsymbol{\nabla} \cdot \boldsymbol{E} = \frac{\rho}{\epsilon_0}. \tag{4.38}$$

The differential form of Gauss' law is the first of our fundamental field equations of electrostatics, Eq. (4.5). We have now shown that the two equations of electrostatics, Eqs. (4.5) and (4.6), are equivalent to Coulomb's law of force. We will now consider one example of the use of Gauss' law. (We will come later to many more examples.)

4–7 Field of a sphere of charge

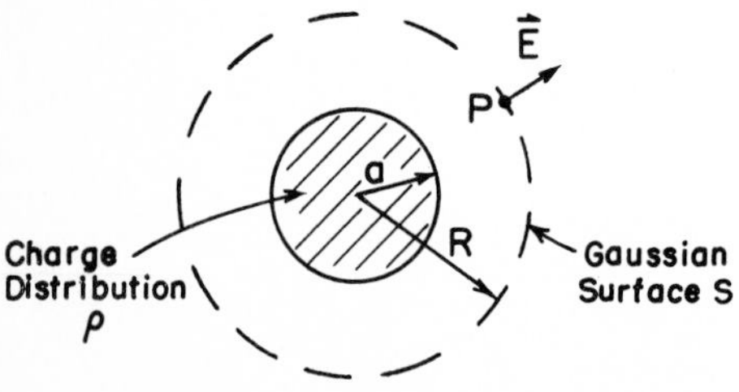

Fig. 4–11. Using Gauss' law to find the field of a uniform sphere of charge.

One of the difficult problems we had when we studied the theory of gravitational attractions was to prove that the force produced by a solid sphere of matter was the same at the surface of the sphere as it would be if all the matter were concentrated at the center. For many years Newton didn't make public his theory of gravitation, because he couldn't be sure this theorem was true. We proved the theorem in Chapter 13 of Vol. I by doing the integral for the potential and then finding the gravitational force by using the gradient. Now we can prove the theorem in a most simple fashion. Only this time we will prove the corresponding theorem for a uniform sphere of electrical charge. (Since the laws of electrostatics are the same as those of gravitation, the same proof could be done for the gravitational field.)

We ask: What is the electric field $\boldsymbol{E}$ at a point P anywhere outside the surface of a sphere filled with a uniform distribution of charge? Since there is no "special" direction, we can assume that $\boldsymbol{E}$ is everywhere directed away from the center of the sphere. We consider an imaginary surface that is spherical and concentric with the sphere of charge, and that passes through the point P (Fig. 4–11). For this surface, the flux outward is

$$\int E_n\, da = E \cdot 4\pi R^2.$$

Gauss' law tells us that this flux is equal to the total charge Q of the sphere (over ϵ_0):

$$E \cdot 4\pi R^2 = \frac{Q}{\epsilon_0},$$

or

$$E = \frac{1}{4\pi\epsilon_0} \frac{Q}{r^2}, \tag{4.39}$$

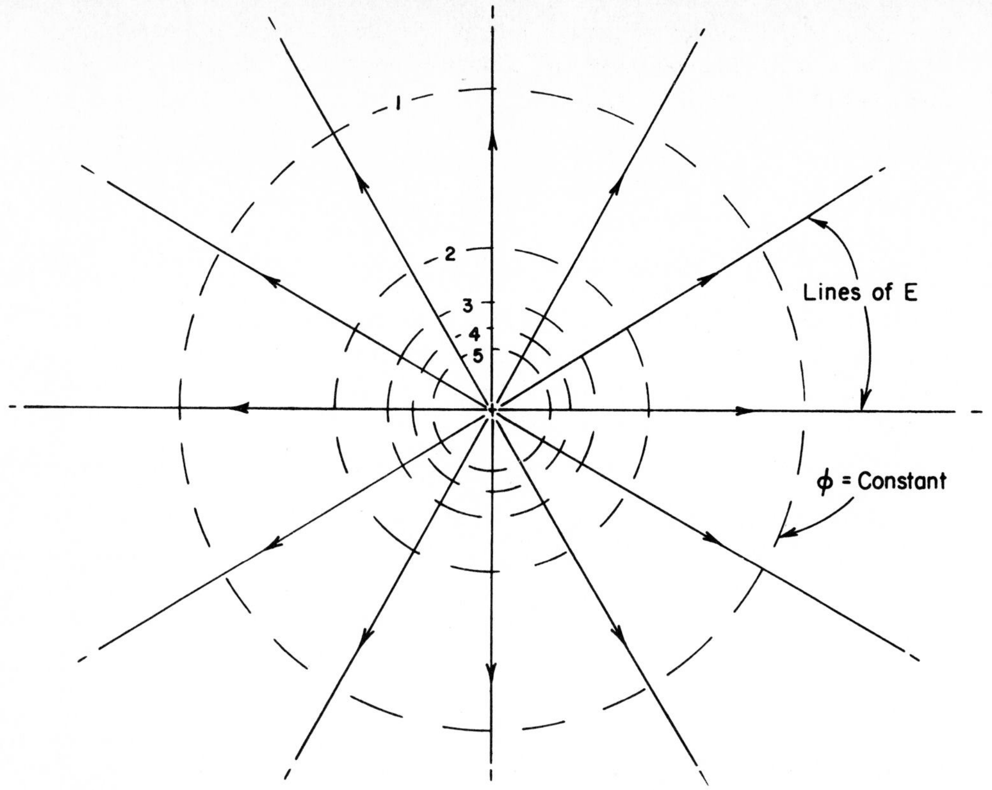

Fig. 4–12. Field lines and equipotential surfaces for a positive point charge.

which is the same formula we would have for a point charge Q. We have proved Newton's problem more easily than by doing the integral. It is, of course, a false kind of easiness—it has taken you some time to be able to understand Gauss' law, so you may think that no time has really been saved. But after you have used the theorem more and more, it begins to pay. It is a question of efficiency.

4–8 Field lines; equipotential surfaces

We would like now to give a geometrical description of the electrostatic field. The two laws of electrostatics, one that the flux is proportional to the charge inside and the other that the electric field is the gradient of a potential, can also be represented geometrically. We illustrate this with two examples.

First, we take the field of a point charge. We draw lines in the direction of the field—lines which are always tangent to the field, as in Fig. 4–12. These are called *field lines*. The lines show everywhere the direction of the electric vector. But we also wish to represent the magnitude of the vector. We can make the rule that the strength of the electric field will be represented by the "density" of the lines. By the density of the lines we mean the number of lines per unit area through a surface perpendicular to the lines. With these two rules we can have a picture of the electric field. For a point charge, the density of the lines must decrease as $1/r^2$. But the area of a spherical surface perpendicular to the lines at any radius r *increases* as r^2, so if we always keep the same *number* of lines for *all* distances from the charge, the *density* will remain in proportion to the magnitude of the field. We can guarantee that there are the same number of lines at every distance if we insist that the lines be *continuous*—that once a line is started from the charge, it never stops. In terms of the field lines, Gauss' law says that lines should start only at plus charges and stop at minus charges. The number which *leave* a charge q must be equal to q/ϵ_0.

Now, we can find a similar geometrical picture for the potential ϕ. The easiest way to represent the potential is to draw surfaces on which ϕ is a constant. We call them *equipotential* surfaces—surfaces of equal potential. Now what is the geometri-

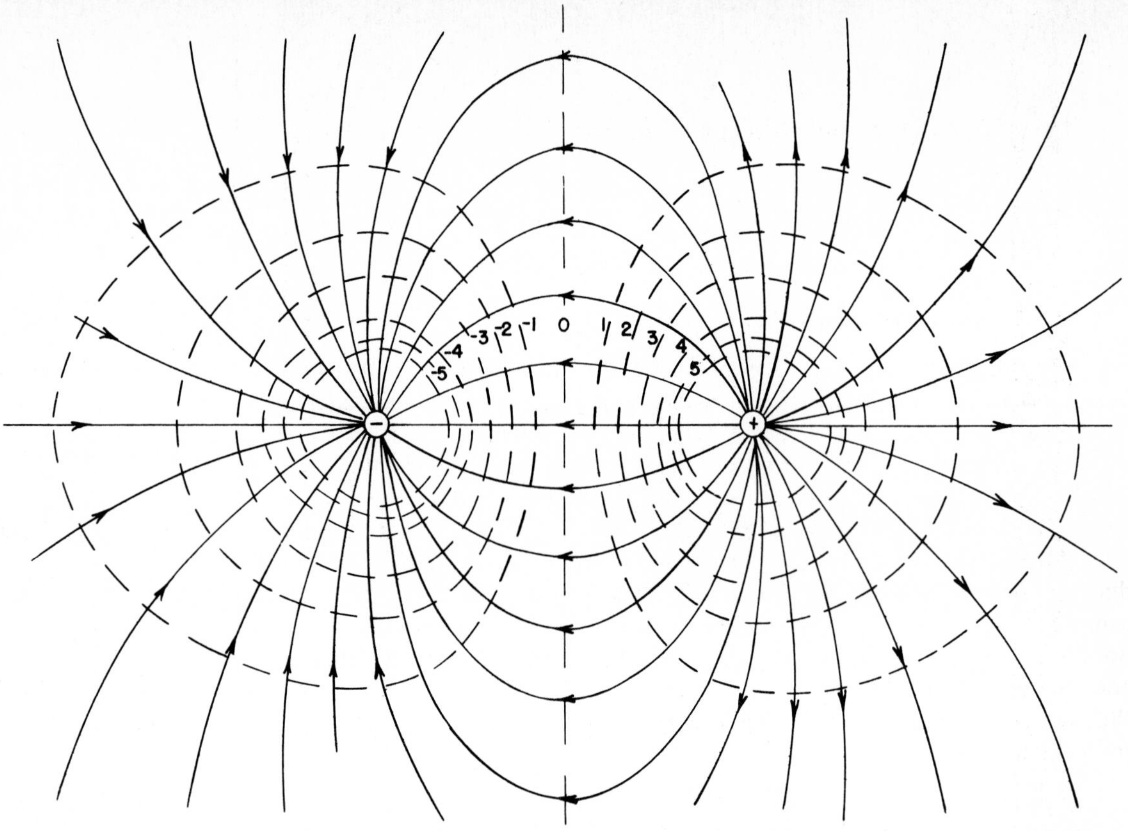

Fig. 4–13. Field lines and equipotentials for two equal and opposite point charges.

cal relationship of the equipotential surfaces to the field lines? The electric field is the gradient of the potential. The gradient is in the direction of the most rapid change of the potential, and is therefore perpendicular to an equipotential surface. If $\boldsymbol{E}$ were *not* perpendicular to the surface, it would have a component *in* the surface. The potential would be changing in the surface, but then it wouldn't be an equipotential. The equipotential surfaces must then be everywhere at right angles to the electric field lines.

For a point charge all by itself, the equipotential surfaces are spheres centered at the charge. We have shown in Fig. 4–12 the intersection of these spheres with a plane through the charge.

As a second example, we consider the field near two equal charges, a positive one and a negative one. To get the field is easy. The field is the superposition of the fields from each of the two charges. So, we can take two pictures like Fig. 4–12 and superimpose them—impossible! Then we would have field lines crossing each other, and that's not possible, because $\boldsymbol{E}$ can't have *two* directions at the same point. The disadvantage of the field-line picture is now evident. By geometrical arguments it is impossible to analyze in a very simple way where the new lines go. From the two independent pictures, we can't get the combined picture. The principle of superposition, a simple and deep principle about electric fields, does not have, in the field-line picture, an easy representation.

The field-line picture has its uses, however, so we might still like to draw the picture for a pair of equal (and opposite) charges. If we calculate the fields from Eq. (4.13) and the potentials from (4.23), we can draw the field lines and equipotentials. Figure 4–13 shows the result. But we first had to solve the problem mathematically!

A Note about Units

Quantity	*Unit*
F	newton
Q	coulomb
L	meter
W	joule
$\rho \sim Q/L^3$	coulomb/meter^3
$1/\epsilon_0 \sim FL^2/Q^2$	newton·meter^2/coulomb^2
$E \sim F/Q$	newton/coulomb
$\phi \sim W/Q$	joule/coulomb = volt
$E \sim \phi/L$	volt/meter
$1/\epsilon_0 \sim EL^2/Q$	volt·meter/coulomb

5

Application of Gauss' Law

5–1 Electrostatics is Gauss' law plus . . .

There are two laws of electrostatics: that the flux of the electric field from a volume is proportional to the charge inside—Gauss' law, and that the circulation of the electric field is zero—$\boldsymbol{E}$ is a gradient. From these two laws, all the predictions of electrostatics follow. But to say these things mathematically is one thing; to use them easily, and with a certain amount of ingenuity, is another. In this chapter we will work through a number of calculations which can be made with Gauss' law directly. We will prove theorems and describe some effects, particularly in conductors, that can be understood very easily from Gauss' law. Gauss' law by itself cannot give the solution of any problem because the other law must be obeyed too. So when we use Gauss' law for the solution of particular problems, we will have to add something to it. We will have to presuppose, for instance, some idea of how the field looks—based, for example, on arguments of symmetry. Or we may have to introduce specifically the idea that the field is the gradient of a potential.

5–2 Equilibrium in an electrostatic field

Consider first the following question: When can a point charge be in stable mechanical equilibrium in the electric field of other charges? As an example, imagine three negative charges at the corners of an equilateral triangle in a horizontal plane. Would a positive charge placed at the center of the triangle remain there? (It will be simpler if we ignore gravity for the moment, although including it would not change the results.) The force on the positive charge is zero, but is the equilibrium stable? Would the charge return to the equilibrium position if displaced slightly? The answer is no.

There are *no* points of stable equilibrium in *any* electrostatic field—except right on top of another charge. Using Gauss' law, it is easy to see why. First, for a charge to be in equilibrium at any particular point P_0, the field must be zero. Second, if the equilibrium is to be a stable one, we require that if we move the charge away from P_0 in *any* direction, there should be a restoring force directed opposite to the displacement. The electric field at *all* nearby points must be pointing inward—toward the point P_0. But that is in violation of Gauss' law if there is no charge at P_0, as we can easily see.

Consider a tiny imaginary surface that encloses P_0, as in Fig. 5–1. If the electric field everywhere in the vicinity is pointed toward P_0, the surface integral of the normal component is certainly not zero. For the case shown in the figure, the flux through the surface must be a negative number. But Gauss' law says that the flux of electric field through any surface is proportional to the total charge inside. If there is no charge at P_0, the field we have imagined violates Gauss' law. It is impossible to balance a positive charge in empty space—at a point where there is not some negative charge. A positive charge *can* be in equilibrium if it is in the middle of a distributed negative charge. Of course, the negative charge distribution would have to be held in place by other than electrical forces!

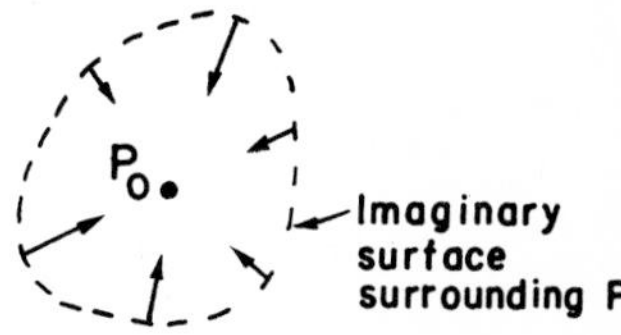

Fig. 5–1. If P_0 were a position of stable equilibrium for a positive charge, the electric field everywhere in the neighborhood would point toward P_0.

Our result has been obtained for a point charge. Does the same conclusion hold for a complicated arrangement of charges held together in fixed relative positions—with rods, for example? We consider the question for two equal charges fixed on a rod. Is it possible that this combination can be in equilibrium in some electrostatic field? The answer is again no. The *total* force on the rod cannot be restoring for displacements in every direction.

Call $\boldsymbol{F}$ the total force on the rod in any position—$\boldsymbol{F}$ is then a vector field. Following the argument used above, we conclude that at a position of stable equilibrium, the divergence of $\boldsymbol{F}$ must be a negative number. But the total force on the rod is the first charge times the field at its position, plus the second charge times the field at its position:

$$\boldsymbol{F} = q_1\boldsymbol{E}_1 + q_2\boldsymbol{E}_2. \tag{5.1}$$

The divergence of $\boldsymbol{F}$ is given by

$$\nabla \cdot \boldsymbol{F} = q_1 (\nabla \cdot \boldsymbol{E}_1) + q_2 (\nabla \cdot \boldsymbol{E}_2).$$

If each of the two charges q_1 and q_2 is in free space, both $\nabla \cdot \boldsymbol{E}_1$ and $\nabla \cdot \boldsymbol{E}_2$ are zero, and $\nabla \cdot \boldsymbol{F}$ is zero—not negative, as would be required for equilibrium. You can see that an extension of the argument shows that no rigid combination of any number of charges can have a position of stable equilibrium in an electrostatic field in free space.

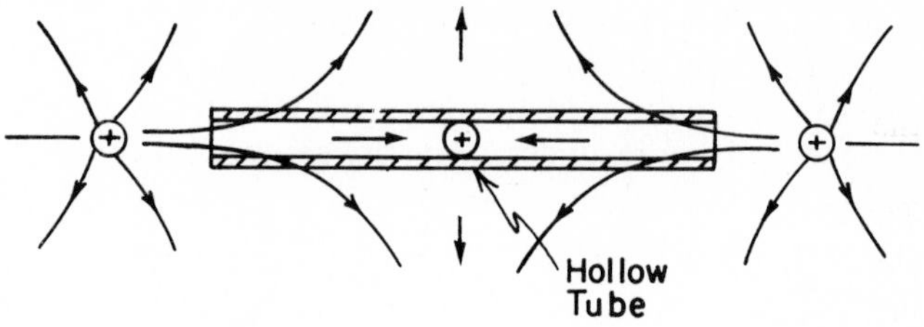

Fig. 5–2. A charge can be in equilibrium if there are mechanical constraints.

Now we have not shown that equilibrium is forbidden if there are pivots or other mechanical constraints. As an example, consider a hollow tube in which a charge can move back and forth freely, but not sideways. Now it is very easy to devise an electric field that points inward at both ends of the tube if it is allowed that the field may point laterally outward near the center of the tube. We simply place positive charges at each end of the tube, as in Fig. 5–2. There can now be an equilibrium point even though the divergence of $\boldsymbol{E}$ is zero. The charge, of course, would not be in stable equilibrium for sideways motion were it not for "non-electrical" forces from the tube walls.

5–3 Equilibrium with conductors

There is no stable spot in the field of a system of fixed charges. What about a system of charged conductors? Can a system of charged conductors produce a field that will have a stable equilibrium point for a point charge? (We mean at a point other than on a conductor, of course.) You know that conductors have the property that charges can move freely around in them. Perhaps when the point charge is displaced slightly, the other charges on the conductors will move in a way that will give a restoring force to the point charge? The answer is still no—although the proof we have just given doesn't show it. The proof for this case is more difficult, and we will only indicate how it goes.

First, we note that when charges redistribute themselves on the conductors, they can only do so if their motion decreases their total potential energy. (Some energy is lost to heat as they move in the conductor.) Now we have already shown that if the charges producing a field are *stationary*, there is, near any zero point P_0 in the field, some direction for which moving a point charge away from P_0 will *decrease* the energy of the system (since the force is *away* from P_0). Any readjustment of the charges on the conductors can only lower the potential energy still more, so (by the principle of virtual work) their motion will only *increase* the force in that particular direction away from P_0, and not reverse it.

Our conclusions do not mean that it is not possible to balance a charge by electrical forces. It is possible if one is willing to control the locations or the sizes of the supporting charges with suitable devices. You know that a rod standing on its point in a gravitational field is unstable, but this does not prove that it cannot be balanced on the end of a finger. Similarly, a charge can be held in one spot by electric fields if they are *variable*. But not with a passive—that is, a *static*—system.

5–4 Stability of atoms

If charges cannot be held stably in position, it is surely not proper to imagine matter to be made up of static *point* charges (electrons and protons) governed only by the laws of electrostatics. Such a static configuration is impossible; it would collapse!

It was once suggested that the positive charge of an atom could be distributed uniformly in a sphere, and the negative charges, the electrons, could be at rest inside the positive charge, as shown in Fig. 5–3. This was the first atomic model, proposed by Thompson. But Rutherford concluded from the experiment of Geiger and Marsden that the positive charges were very much concentrated, in what he called the nucleus. Thompson's static model had to be abandoned. Rutherford and Bohr then suggested that the equilibrium might be dynamic, with the electrons revolving in orbits, as shown in Fig. 5–4. The electrons would be kept from falling in toward the nucleus by their orbital motion. We already know at least one difficulty with this picture. With such motion, the electrons would be accelerating (because of the circular motion) and would, therefore, be radiating energy. They would lose the kinetic energy required to stay in orbit, and would spiral in toward the nucleus. Again unstable!

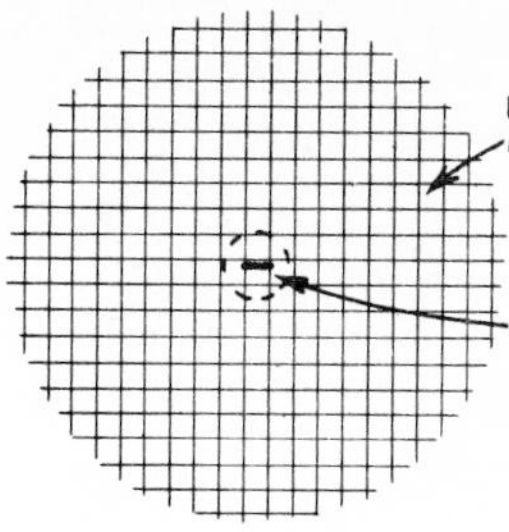

Fig. 5–3. The Thompson model of an atom.

The stability of the atoms is now explained in terms of quantum mechanics. The electrostatic forces pull the electron as close to the nucleus as possible, but the electron is compelled to stay spread out in space over a distance given by the uncertainty principle. If it were confined in too small a space, it would have a great uncertainty in momentum. But that means that it would have a high expected energy—which it would use to escape from the electrical attraction. The net result is an electrical equilibrium not too different from the idea of Thompson—only it is the *negative* charge that is spread out (because the mass of the electron is so much smaller than the mass of the proton).

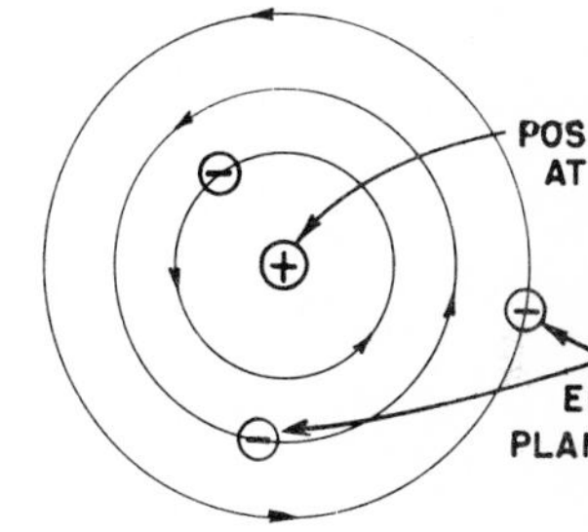

Fig. 5–4. The Rutherford-Bohr model of an atom.

5–5 The field of a line charge

Gauss' law can be used to solve a number of electrostatic field problems involving a special symmetry—usually spherical, cylindrical, or planar symmetry. In the remainder of this chapter we will apply Gauss' law to a few such problems. The ease with which these problems can be solved may give the misleading impression that the method is very powerful, and that one should be able to go on to many other problems. It is unfortunately not so. One soon exhausts the list of problems that can be solved easily with Gauss' law. In later chapters we will develop more powerful methods for investigating electrostatic fields.

As our first example, we consider a system with cylindrical symmetry. Suppose that we have a very long, uniformly charged rod. By this we mean that electric charges are distributed uniformly along an indefinitely long straight line, with the charge λ per unit length. We wish to know the electric field. The problem can, of course, be solved by integrating the contribution to the field from every part of the line. We are going to do it without integrating, by using Gauss' law and some guesswork. First, we surmise that the electric field will be directed radially outward from the line. Any axial component from charges on one side would be accompanied by an equal axial component from charges on the other side. The result could only be a radial field. It also seems reasonable that the field should have the same magnitude at all points equidistant from the line. This is obvious. (It may not be easy to prove, but it is true if space is symmetric—as we believe it is.)

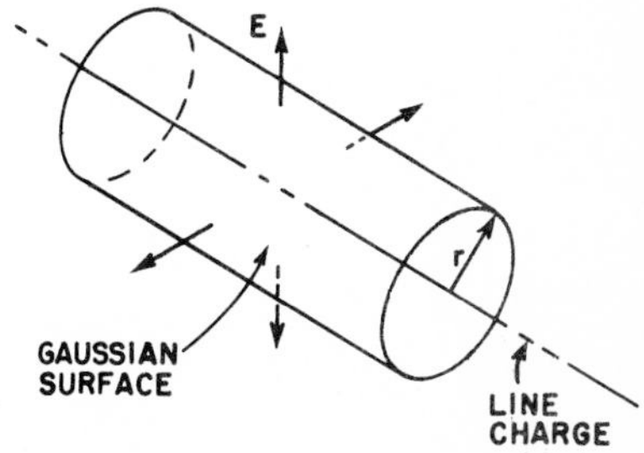

Fig. 5–5. A cylindrical gaussian surface coaxial with a line charge.

We can use Gauss' law in the following way. We consider an *imaginary* surface in the shape of a cylinder coaxial with the line, as shown in Fig. 5–5. According to Gauss' law, the total flux of $\boldsymbol{E}$ from this surface is equal to the charge inside divided by ϵ_0. Since the field is assumed to be normal to the surface, the normal component is the magnitude of the field. Let's call it E. Also, let the radius of the cylinder be r, and its length be taken as one unit, for convenience. The flux through the cylindrical surface is equal to E times the area of the surface, which is $2\pi r$. The flux through the two end faces is zero because the electric field is tangential to them. The total charge inside our surface is just λ, because the length of the line inside is one unit. Gauss' law then gives

$$E \cdot 2\pi r = \lambda/\epsilon_0,$$

$$E = \frac{\lambda}{2\pi\epsilon_0 r}. \tag{5.2}$$

The electric field of a line charge depends inversely on the *first* power of the distance from the line.

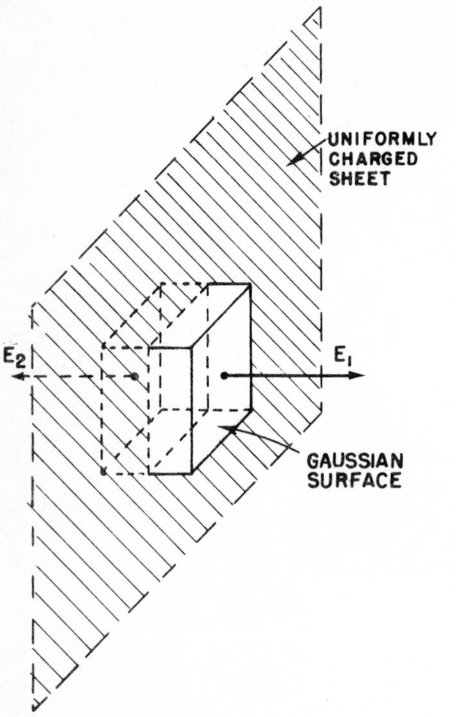

Fig. 5–6. The electric field near a uniformly charged sheet can be found by applying Gauss' law to an imaginary box.

5–6 A sheet of charge; two sheets

As another example, we will calculate the field from a uniform plane sheet of charge. Suppose that the sheet is infinite in extent and that the charge per unit area is σ. We are going to take another guess. Considerations of symmetry lead us to believe that the field direction is everywhere normal to the plane, and *if we have no field from any other charges in the world*, the fields must be the same (in magnitude) on each side. This time we choose for our Gaussian surface a rectangular box that cuts through the sheet, as shown in Fig. 5–6. The two faces parallel to the sheet will have equal areas, say A. The field is normal to these two faces, and parallel to the other four. The total flux is E times the area of the first face, plus E times the area of the opposite face—with no contribution from the other four faces. The total charge enclosed in the box is σA. Equating the flux to the charge inside, we have

$$EA + EA = \frac{\sigma A}{\epsilon_0},$$

from which

$$E = \frac{\sigma}{2\epsilon_0}, \tag{5.3}$$

a simple but important result.

You may remember that the same result was obtained in an earlier chapter by an integration over the entire surface. Gauss' law gives us the answer, in this instance, much more quickly (although it is not as generally applicable as the earlier method).

We emphasize that this result applies *only* to the field due to the charges on the sheet. If there are other charges in the neighborhood, the total field near the sheet would be the sum of (5.3) and the field of the other charges. Gauss' law would then tell us only that

$$E_1 + E_2 = \frac{\sigma}{\epsilon_0}, \tag{5.4}$$

where E_1 and E_2 are the fields directed outward on each side of the sheet.

The problem of two parallel sheets with equal and opposite charge densities, $+\sigma$ and $-\sigma$, is equally simple if we assume again that the outside world is quite symmetric. Either by superposing two solutions for a single sheet or by constructing a gaussian box that includes both sheets, it is easily seen that the field is zero *outside* of the two sheets (Fig. 5–7a). By considering a box that includes only one surface or the other, as in (b) or (c) of the figure, it can be seen that the field between the sheets must be twice what it is for a single sheet. The result is

$$E \text{ (between the sheets)} = \sigma/\epsilon_0, \tag{5.5}$$

$$E \text{ (outside)} = 0. \tag{5.6}$$

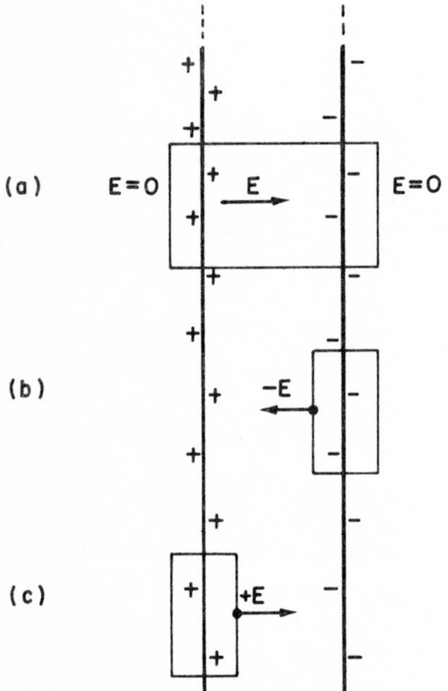

Fig. 5–7. The field between two charged sheets is σ/ϵ_0.

5–7 A sphere of charge; a spherical shell

We have already (in Chapter 4) used Gauss' law to find the field outside a uniformly charged spherical region. The same method can also give us the field at points *inside* the sphere. For example, the computation can be used to obtain a good approximation to the field inside an atomic nucleus. In spite of the fact that the protons in a nucleus repel each other, they are, because of the strong nuclear forces, spread nearly uniformly throughout the body of the nucleus.

Suppose that we have a sphere of radius R filled uniformly with charge. Let ρ be the charge per unit volume. Again using arguments of symmetry, we assume the field to be radial and equal in magnitude at all points at the same distance from the center. To find the field at the distance r from the center, we take a spherical gaussian surface of radius r ($r < R$), as shown in Fig. 5–8. The flux out of this surface is

$$4\pi r^2 E.$$

The charge inside our gaussian surface is the volume inside times ρ, or

$$\tfrac{4}{3}\pi r^3 \rho.$$

Using Gauss' law, it follows that the magnitude of the field is given by

$$E = \frac{\rho r}{3\epsilon_0} \qquad (r < R). \tag{5.7}$$

You can see that this formula gives the proper result for $r = R$. The electric field is *proportional* to the radius and is directed radially outward.

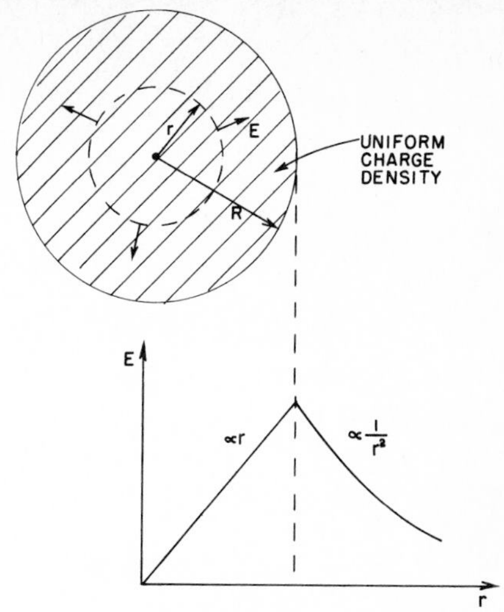

Fig. 5–8. Gauss' law can be used to find the field inside a uniformly charged sphere.

The arguments we have just given for a uniformly charged sphere can be applied also to a thin spherical shell of charge. Assuming that the field is everywhere radial and is spherically symmetric, one gets immediately from Gauss' law that the field outside the shell is like that of a point charge, while the field everywhere inside the shell is zero. (A gaussian surface inside the shell will contain no charge.)

5–8 Is the field of a point charge exactly $1/r^2$?

If we look in a little more detail at *how* the field inside the shell gets to be zero, we can see more clearly why it is that Gauss' law is true only because the coulomb force depends exactly on the square of the distance. Consider any point P inside a uniform spherical shell of charge. Imagine a small cone whose apex is at P and which extends to the surface of the sphere, where it cuts out a small surface area Δa_1, as in Fig. 5–9. An exactly symmetric cone diverging from the opposite side of P would cut out the surface area Δa_2. If the distances from P to these two elements of area are r_1 and r_2, the areas are in the ratio

$$\frac{\Delta a_2}{\Delta a_1} = \frac{r_2^2}{r_1^2}.$$

(You can show this by geometry for any point P inside the sphere.)

If the surface of the sphere is uniformly charged, the charge Δq on each of the elements of area is proportional to the area, so

$$\frac{\Delta q_2}{\Delta q_1} = \frac{\Delta a_2}{\Delta a_1}.$$

Coulomb's law then says that the magnitudes of the fields produced at P by these two surface elements are in the ratio

$$\frac{E_2}{E_1} = \frac{q_2/r_2^2}{q_1/r_1^2} = 1.$$

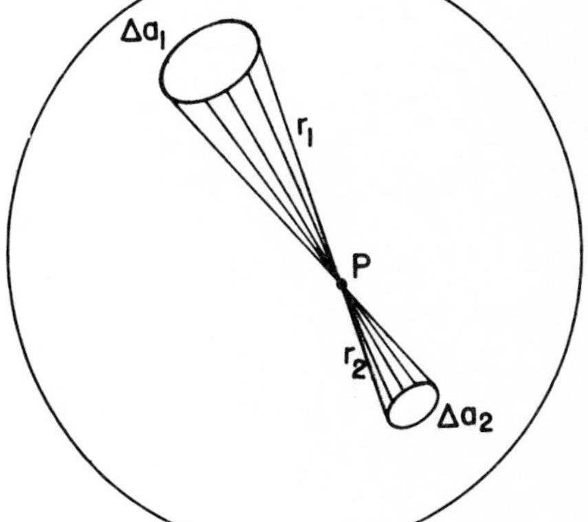

Fig. 5–9. The field is zero at any point P inside a spherical shell of charge.

The fields cancel exactly. Since all parts of the surface can be paired off in the same way, the total field at P is zero. But you can see that it would not be so if the exponent of r in Coulomb's law were not exactly two.

The validity of Gauss' law depends upon the inverse square law of Coulomb. If the force law were not exactly the inverse square, it would not be true that the field inside a uniformly charged sphere would be exactly zero. For instance, if the force varied more rapidly, like, say, the inverse cube of r, that portion of the surface which is nearer to an interior point would produce a field which is larger than that which is farther away, resulting in a radial inward field for a positive surface charge. These conclusions suggest an elegant way of finding out whether the inverse square law is precisely correct. We need only determine whether or not the field inside of a uniformly charged spherical shell is precisely zero.

It is lucky that such a method exists. It is usually difficult to measure a physical quantity to high precision—a one percent result may not be too difficult, but how would one go about measuring, say, Coulomb's law to an accuracy of one part in a billion? It is almost certainly not possible with the best available techniques to measure the *force* between two charged objects with such an accuracy. But by determining only that the electric fields inside a charged sphere are *smaller* than some value we can make a highly accurate measurement of the correctness of Gauss' law, and hence of the inverse square dependence of Coulomb's law. What one does, in effect, is *compare* the force law to an ideal inverse square. Such comparisons of things that are equal, or nearly so, are usually the bases of the most precise physical measurements.

How shall we observe the field inside a charged sphere? One way is to try to charge an object by touching it to the inside of a spherical conductor. You know that if we touch a small metal ball to a charged object and then touch it to an electrometer the meter will become charged and the pointer will move from zero (Fig. 5–10a). The ball picks up charge because there are electric fields outside the charged sphere that cause charges to run onto (or off) the little ball. If you do the same experiment by touching the little ball to the *inside* of the charged sphere, you find that no charge is carried to the electrometer. With such an experiment you can easily show that the field inside is, at most, a few percent of the field outside, and that Gauss' law is at least approximately correct.

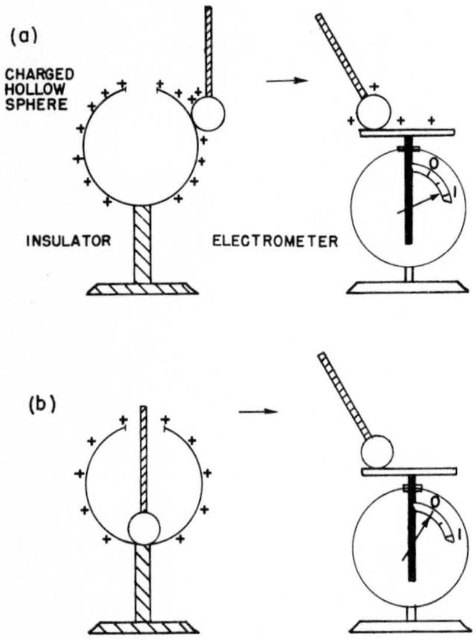

Fig. 5–10. The electric field is zero inside a closed conducting shell.

It appears that Benjamin Franklin was the first to notice that the field inside a conducting shell is zero. The result seemed strange to him. When he reported his observation to Priestley, the latter suggested that it might be connected with an inverse square law, since it was known that a spherical shell of matter produced no gravitational field inside. But Coulomb didn't measure the inverse square dependence until 18 years later, and Gauss' law came even later still.

Gauss' law has been checked carefully by putting an electrometer inside a large sphere and observing whether any deflections occur when the sphere is charged to a high voltage. A null result is always obtained. Knowing the geometry of the apparatus and the sensitivity of the meter, it is possible to compute the minimum field that would be observed. From this number it is possible to place an upper limit on the deviation of the exponent from two. If we write that the electrostatic force depends on $r^{-2+\epsilon}$, we can place an upper bound on ϵ. By this method Maxwell determined that ϵ was less than $1/10{,}000$. The experiment was repeated and improved upon in 1936 by Plimpton and Laughton. They found that Coulomb's exponent differs from two by less than one part in a billion.

Now that brings up an interesting question: How accurate do we know this Coulomb law to be in various circumstances? The experiments we just described measure the dependence of the field on distance for distances of some tens of centimeters. But what about the distances inside an atom—in the hydrogen atom, for instance, where we believe the electron is attracted to the nucleus by the same inverse square law? It is true that quantum mechanics must be used for the mechanical part of the behavior of the electron, but the force is the usual electrostatic one. In the formulation of the problem, the potential energy of an electron must be known as a function of distance from the nucleus, and Coulomb's law gives a potential which varies inversely with the first power of the distance. How accurately is the exponent known for such small distances? As a result of very careful measurements in 1947 by Lamb and Retherford on the relative positions of the energy levels of hydrogen, we know that the exponent is correct again to one part in a billion on the atomic scale—that is, at distances of the order of one angstrom (10^{-8} centimeter).

The accuracy of the Lamb-Retherford measurement was possible again because of a physical "accident." Two of the states of a hydrogen atom are expected to have almost indentical energies *only* if the potential varies exactly as $1/r$. A measurement was made of the very slight *difference* in energies by finding the frequency ω of the photons that are emitted or absorbed in the transition from one state to the other, using for the energy difference $\Delta E = \hbar\omega$. Computations showed that ΔE would have been noticeably different from what was observed if the exponent in the force law $1/r^2$ differed from 2 by as much as one part in a billion.

Is the same exponent correct at still shorter distances? From measurements in nuclear physics it is found that there are electrostatic forces at typical nuclear distances—at about 10^{-13} centimeter—and that they still vary approximately as the inverse square. We shall look at some of the evidence in a later chapter. Coulomb's law is, we know, still valid, at least to some extent, at distances of the order of 10^{-13} centimeter.

How about 10^{-14} centimeter? This range can be investigated by bombarding protons with very energetic electrons and observing how they are scattered. Results to date seem to indicate that the law fails at these distances. The electrical force seems to be about 10 times too weak at distances less than 10^{-14} centimeter. Now there are two possible explanations. One is that the Coulomb law does not work at such small distances; the other is that our objects, the electrons and protons, are not point charges. Perhaps either the electron or proton, or both, is some kind of a smear. Most physicists prefer to think that the charge of the proton is smeared. We know that protons interact strongly with mesons. This implies that a proton will, from time to time, exist as a neutron with a π^+ meson around it. Such a configuration would act—on the average—like a little sphere of positive charge. We know that the field from a sphere of charge does not vary as $1/r^2$ all the way into the center. It is quite likely that the proton charge is smeared, but the theory of pions is still quite incomplete, so it may also be that Coulomb's law fails at very small distances. The question is still open.

One more point: The inverse square law is valid at distances like one meter and also at 10^{-10}m; but is the coefficient $1/4\pi\epsilon_0$ the same? The answer is yes; at least to an accuracy of 15 parts in a million.

We go back now to an important matter that we slighted when we spoke of the experimental verification of Gauss' law. You may have wondered how the experiment of Maxwell or of Plimpton and Laughton could give such an accuracy unless the spherical conductor they used was a perfect sphere. An accuracy of one part in a billion is really something to achieve, and you might well ask whether they could make a sphere which was that precise. There are certain to be slight irregularities in any real sphere and if there are irregularities, will they not produce fields inside? We wish to show now that it is not necessary to have a perfect sphere. It is possible, in fact, to show that there is no field inside a closed conducting shell of *any* shape. In other words, the experiments depended on $1/r^2$, but had nothing to do with the surface being a sphere (except that with a sphere it is easier to calculate what the fields *would* be if Coulomb had been wrong), so we take up that subject now. To show this, it is necessary to know some of the properties of electrical conductors.

5–9 The fields of a conductor

An electrical conductor is a solid that contains many "free" electrons. The electrons can move around freely *in* the material, but cannot leave the surface. In a metal there are so many free electrons that any electric field will set large numbers of them into motion. Either the current of electrons so set up must be continually kept moving by external sources of energy, or the motion of the electrons will cease as they discharge the sources producing the initial field. In "electrostatic" situations, we do not consider continuous sources of current (they will be considered later when we study magnetostatics), so the electrons move only until they have arranged themselves to produce zero electric field everywhere inside the conductor. (This usually happens in a small fraction of a second.) If there were any field left, this field would urge still more electrons to move; the only electrostatic solution is that the field is everywhere zero inside.

Now consider the *interior* of a charged conducting object. (By "interior" we mean in the *metal* itself.) Since the metal is a conductor, the interior field must be zero, and so the gradient of the potential ϕ is zero. That means that ϕ does not vary from point to point. Every conductor is an equipotential *region*, and its surface is an equipotential surface. Since in a conducting material the electric field is everywhere zero, the divergence of $\boldsymbol{E}$ is zero, and by Gauss' law the charge density in the *interior* of the conductor must be zero.

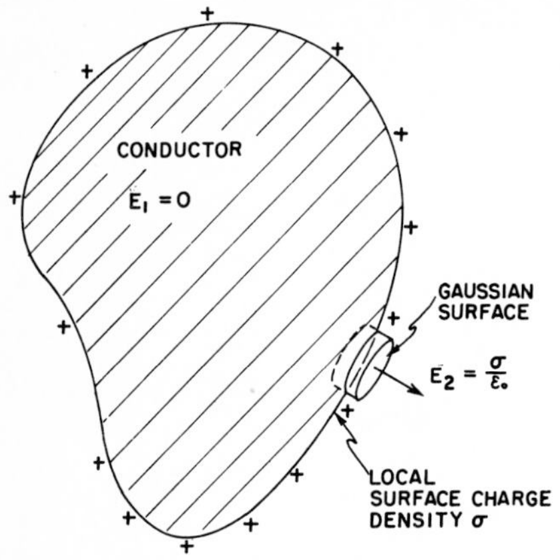

Fig. 5–11. The electric field just outside the surface of a conductor is proportional to the local surface density of charge.

If there can be no charges in a conductor, how can it ever be charged? What do we mean when we say a conductor is "charged"? Where are the charges? The answer is that they reside at the surface of the conductor, where there are strong forces to keep them from leaving—they are not completely "free." When we study solid-state physics, we shall find that the excess charge of any conductor is on the average within one or two atomic layers of the surface. For our present purposes, it is accurate enough to say that if any charge is put on, or *in*, a conductor it all accumulates on the surface; there is no charge in the interior of a conductor.

We note also that the electric field *just outside* the surface of a conductor must be normal to the surface. There can be no tangential component. If there were a tangential component, the electrons would move *along* the surface; there are no forces preventing that. Saying it another way: we know that the electric field lines must always go at right angles to an equipotential surface.

We can also, using Gauss' law, relate the field strength just outside a conductor to the local density of the charge at the surface. For a gaussian surface, we take a small cylindrical box half inside and half outside the surface, like the one shown in Fig. 5–11. There is a contribution to the total flux of $\boldsymbol{E}$ only from the side of the box outside the conductor. The field just outside the surface of a conductor is then

Outside a conductor:

$$E = \frac{\sigma}{\epsilon_0}, \tag{5.8}$$

where σ is the *local* surface charge density.

Why does a sheet of charge on a conductor produce a different field than *just* a sheet of charge? In other words, why is (5.8) twice as large as (5.3)? The reason, of course, is that we have *not* said for the conductor that there are no "other" charges around. There must, in fact, be some to make $\boldsymbol{E} = 0$ in the conductor. The charges in the immediate neighborhood of a point P on the surface do, in fact, give a field $E_{\text{local}} = \sigma_{\text{local}}/2\epsilon_0$ both inside and outside the surface. But all the rest of the charges on the conductor "conspire" to produce an additional field at the point P equal in magnitude to E_{local}. The total field inside goes to zero and the field outside to $2E_{\text{local}} = \sigma/\epsilon_0$.

5–10 The field in a cavity of a conductor

We return now to the problem of the hollow container—a conductor with a cavity. There is no field in the *metal*, but what about in the *cavity*? We shall show that if the cavity is *empty* then there are no fields in it, *no matter what the shape* of the conductor or the cavity—say for the one in Fig. 5–12. Consider a gaussian surface, like S in Fig. 5–12, that encloses the cavity but stays everywhere in the conducting material. Everywhere on S the field is zero, so there is no flux through S and the *total* charge inside S is zero. For a spherical shell, one could then argue from symmetry that there could be *no* charge inside. But, in general, we can only say that there are equal amounts of positive and negative charge on the inner surface of the conductor. There *could* be a positive surface charge on one part and a negative one somewhere else, as indicated in Fig. 5–12. Such a thing cannot be ruled out by Gauss' law.

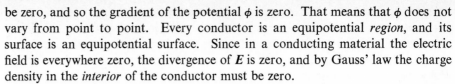

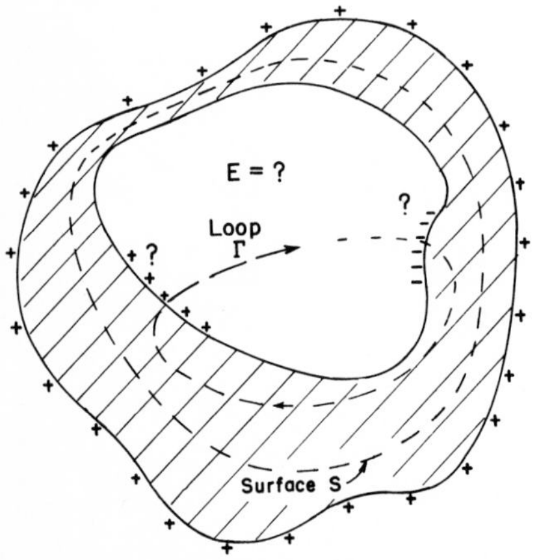

Fig. 5–12. What is the field in an empty cavity of a conductor, for any shape?

What really happens, of course, is that any equal and opposite charges on the inner surface would slide around to meet each other, cancelling out completely. We can show that they must cancel completely by using the law that the circulation of $\boldsymbol{E}$ is always zero (electrostatics). Suppose there were charges on some parts of the inner surface. We know that there would have to be an equal number of opposite charges somewhere else. Now any lines of $\boldsymbol{E}$ would have to start on the positive charges and end on the negative charges (since we are considering only the case that there are no free charges in the cavity). Now imagine a loop Γ that crosses the cavity along a line of force from some positive charge to some negative charge, and returns to its starting point via the conductor (as in Fig. 5–12). The integral along such a line of force from the positive to the negative charges would not be zero. The integral through the metal is zero, since $\boldsymbol{E} = 0$. So we would have

$$\oint \boldsymbol{E} \cdot d\boldsymbol{s} \neq 0???$$

But the line integral of $\boldsymbol{E}$ around any closed loop in an electrostatic field is always zero. So there can be no fields inside the empty cavity, nor any charges on the inside surface.

You should notice carefully one important qualification we have made. We have always said "inside an *empty*" cavity. If some charges are *placed* at some fixed locations in the cavity—as on an insulator or on a small conductor insulated from the main one—then there *can* be fields in the cavity. But then that is not an "empty" cavity.

We have shown that if a cavity is completely enclosed by a conductor, no static distribution of charges *outside* can ever produce any fields inside. This explains the principle of "shielding" electrical equipment by placing it in a metal can. The same arguments can be used to show that no static distribution of charges *inside* a closed conductor can produce any fields *outside*. Shielding works both ways! In electrostatics—but not in varying fields—the fields on the two sides of a closed conducting shell are completely independent.

Now you see why it was possible to check Coulomb's law to such a great precision. The shape of the hollow shell used doesn't matter. It doesn't need to be spherical; it could be square! If Gauss' law is exact, the field inside is always zero. Now you also understand why it is safe to sit inside the high-voltage terminal of a million-volt van de Graaff generator, without worrying about getting a shock—because of Gauss' law.

6

The Electric Field in Various Circumstances

Review: Chapter 23, Vol. I, *Resonance*

6–1 Equations of the electrostatic potential

This chapter will describe the behavior of the electric field in a number of different circumstances. It will provide some experience with the way the electric field behaves, and will describe some of the mathematical methods which are used to find this field.

We begin by pointing out that the whole mathematical problem is the solution of two equations, the Maxwell equations for electrostatics:

$$\nabla \cdot \boldsymbol{E} = \frac{\rho}{\epsilon_0}, \tag{6.1}$$

$$\nabla \times \boldsymbol{E} = 0. \tag{6.2}$$

In fact, the two can be combined into a single equation. From the second equation, we know at once that we can describe the field as the gradient of a scalar (see Section 3–7):

$$\boldsymbol{E} = -\nabla\phi. \tag{6.3}$$

We may, if we wish, completely describe any particular electric field in terms of its potential ϕ. We obtain the differential equation that ϕ must obey by substituting Eq. (6.3) into (6.1), to get

$$\nabla \cdot \nabla\phi = -\frac{\rho}{\epsilon_0}. \tag{6.4}$$

The divergence of the gradient of ϕ is the same as ∇^2 operating on ϕ:

$$\nabla \cdot \nabla\phi = \nabla^2\phi = \frac{\partial^2\phi}{\partial x^2} + \frac{\partial^2\phi}{\partial y^2} + \frac{\partial^2\phi}{\partial z^2}, \tag{6.5}$$

so we write Eq. (6.4) as

$$\nabla^2\phi = -\frac{\rho}{\epsilon_0}. \tag{6.6}$$

The operator ∇^2 is called the Laplacian, and Eq. (6.6) is called the Poisson equation. The entire subject of electrostatics, from a mathematical point of view, is merely a study of the solutions of the single equation (6.6). Once ϕ is obtained by solving Eq. (6.6) we can find $\boldsymbol{E}$ immediately from Eq. (6.3).

We take up first the special class of problems in which ρ is given as a function of x, y, z. In that case the problem is almost trivial, for we already know the solution of Eq. (6.6) for the general case. We have shown that if ρ is known at every point, the potential at point (1) is

$$\phi(1) = \int \frac{\rho(2)\,dV_2}{4\pi\epsilon_0 r_{12}}, \tag{6.7}$$

where $\rho(2)$ is the charge density, dV_2 is the volume element at point (2), and r_{12} is the distance between points (1) and (2). The solution of the *differential* equation (6.6) is reduced to an *integration* over space. The solution (6.7) should be especially noted, because there are many situations in physics that lead to equations like

$$\nabla^2\,(\text{something}) = (\text{something else}),$$

and Eq. (6.7) is a prototype of the solution for any of these problems.

The solution of electrostatic field problems is thus completely straightforward when the positions of all the charges are known. Let's see how it works in a few examples.

6–2 The electric dipole

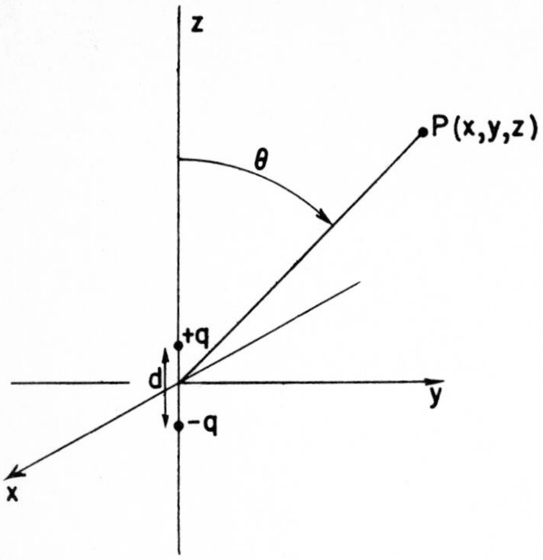

Fig. 6–1. A dipole: two charges $+q$ and $-q$ the distance d apart.

First, take two point charges, $+q$ and $-q$, separated by the distance d. Let the z-axis go through the charges, and pick the origin halfway between, as shown in Fig. 6–1. Then, using (4.24), the potential from the two charges is given by

$$\phi(x, y, z) = \frac{1}{4\pi\epsilon_0}\left[\frac{q}{\sqrt{[z - (d/2)]^2 + x^2 + y^2}} + \frac{-q}{\sqrt{[z + (d/2)]^2 + x^2 + y^2}}\right]. \tag{6.8}$$

We are not going to write out the formula for the electric field, but we can always calculate it once we have the potential. So we have solved the problem of two charges.

There is an important special case in which the two charges are very close together—which is to say that we are interested in the fields only at distances from the charges large in comparison with their separation. We call such a close pair of charges a *dipole*. Dipoles are very common.

A "dipole" antenna can often be approximated by two charges separated by a small distance—if we don't ask about the field too close to the antenna. (We are usually interested in antennas with *moving* charges; then the equations of statics do not really apply, but for some purposes they are an adequate approximation.)

More important perhaps, are atomic dipoles. If there is an electric field in any material, the electrons and protons feel opposite forces and are displaced relative to each other. In a conductor, you remember, some of the electrons move to the surfaces, so that the field inside becomes zero. In an insulator the electrons cannot move very far; they are pulled back by the attraction of the nucleus. They do, however, shift a little bit. So although an atom, or molecule, remains neutral in an external electric field, there is a very tiny separation of its positive and negative charges and it becomes a microscopic dipole. If we are interested in the fields of these atomic dipoles in the neighborhood of ordinary-sized objects, we are normally dealing with distances large compared with the separations of the pairs of charges.

In some molecules the charges are somewhat separated even in the absence of external fields, because of the form of the molecule. In a water molecule, for example, there is a net negative charge on the oxygen atom and a net positive charge on each of the two hydrogen atoms, which are not placed symmetrically but as in Fig. 6–2. Although the charge of the whole molecule is zero, there is a charge distribution with a little more negative charge on one side and a little more positive charge on the other. This arrangement is certainly not as simple as two point charges, but when seen from far away the system acts like a dipole. As we shall see a little later, the field at large distances is not sensitive to the fine details.

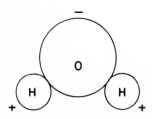

Fig. 6–2. The water molecule H_2O. The hydrogen atoms have slightly less than their share of the electron cloud; the oxygen, slightly more.

Let's look, then, at the field of two opposite charges with a small separation d. If d becomes zero, the two charges are on top of each other, the two potentials cancel, and there is no field. But if they are not exactly on top of each other, we can get a good approximation to the potential by expanding the terms of (6.8) in a power series in the small quantity d (using the binomial expansion). Keeping terms only to first order in d, we can write

$$\left(z - \frac{d}{2}\right)^2 \approx z^2 - zd.$$

It is convenient to write

$$x^2 + y^2 + z^2 = r^2.$$

Then

$$\left(z - \frac{d}{2}\right)^2 + x^2 + y^2 \approx r^2 - zd = r^2\left(1 - \frac{zd}{r^2}\right),$$

and

$$\frac{1}{\sqrt{[z - (d/2)]^2 + x^2 + y^2}} \approx \frac{1}{\sqrt{r^2[1 - (zd/r^2)]}} \approx \frac{1}{r}\left(1 - \frac{zd}{r^2}\right)^{-1/2}.$$

Using the binomial expansion again for $[1 - (zd/r^2)]^{-1/2}$—and throwing away terms with higher powers than the square of d—we get

$$\frac{1}{r}\left(1 + \frac{1}{2}\frac{zd}{r^2}\right).$$

Similarly,

$$\frac{1}{\sqrt{[z + (d/2)]^2 + x^2 + y^2}} \approx \frac{1}{r}\left(1 - \frac{1}{2}\frac{zd}{r^2}\right).$$

The difference of these two terms gives for the potential

$$\phi(x, y, z) = \frac{1}{4\pi\epsilon_0}\frac{z}{r^3}\,qd. \tag{6.9}$$

The potential, and hence the field, which is its derivative, is proportional to qd, the product of the charge and the separation. This product is defined as the *dipole moment* of the two charges, for which we will use the symbol p (do *not* confuse with momentum!):

$$p = qd. \tag{6.10}$$

Equation (6.9) can also be written as

$$\phi(x, y, z) = \frac{1}{4\pi\epsilon_0}\frac{p\cos\theta}{r^2}, \tag{6.11}$$

since $z/r = \cos\theta$, where θ is the angle between the axis of the dipole and the radius vector to the point (x, y, z)—see Fig. 6–1. The *potential* of a dipole decreases as $1/r^2$ for a given direction from the axis (whereas for a point charge it goes as $1/r$). The electric field $\boldsymbol{E}$ of the dipole will then decrease as $1/r^3$.

We can put our formula into a vector form if we define $\boldsymbol{p}$ as a vector whose magnitude is p and whose direction is along the axis of the dipole, pointing from q_- toward q_+. Then

$$p\cos\theta = \boldsymbol{p}\cdot\boldsymbol{e}_r, \tag{6.12}$$

where $\boldsymbol{e}_r$ is the unit radial vector (Fig. 6–3). We can also represent the point (x, y, z) by $\boldsymbol{r}$. Then

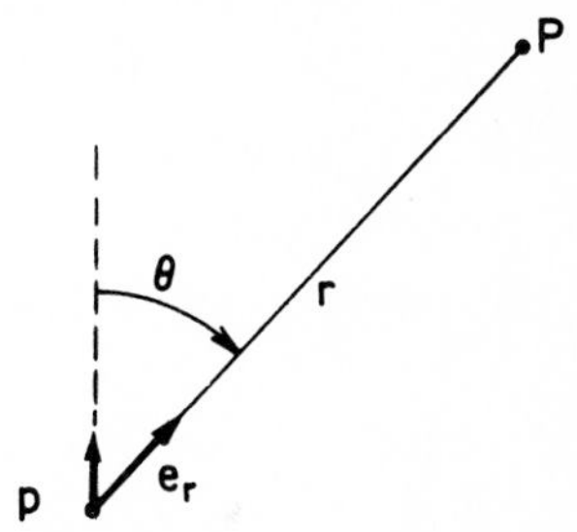

Fig. 6–3. Vector notation for a dipole.

Dipole potential:

$$\phi(\boldsymbol{r}) = \frac{1}{4\pi\epsilon_0}\frac{\boldsymbol{p}\cdot\boldsymbol{e}_r}{r^2} = \frac{1}{4\pi\epsilon_0}\frac{\boldsymbol{p}\cdot\boldsymbol{r}}{r^3}. \tag{6.13}$$

This formula is valid for a dipole with any orientation and position if $\boldsymbol{r}$ represents the vector from the dipole to the point of interest.

If we want the electric field of the dipole we can get it by taking the gradient of ϕ. For example, the z-component of the field is $-\partial\phi/\partial z$. For a dipole oriented along the z-axis we can use (6.9):

$$-\frac{\partial\phi}{\partial z} = -\frac{p}{4\pi\epsilon_0}\frac{\partial}{\partial z}\left(\frac{z}{r^3}\right) = -\frac{p}{4\pi\epsilon_0}\left(\frac{1}{r^3} - \frac{3z^2}{r^5}\right),$$

or

$$E_z = \frac{p}{4\pi\epsilon_0}\frac{3\cos^2\theta - 1}{r^3}. \tag{6.14}$$

The x- and y-components are

$$E_x = \frac{p}{4\pi\epsilon_0}\frac{3zx}{r^5}, \qquad E_y = \frac{p}{4\pi\epsilon_0}\frac{3zy}{r^5}.$$

These two can be combined to give one component directed *perpendicular* to the z-axis, which we will call the transverse component $E_\perp$:

$$E_\perp = \sqrt{E_x^2 + E_y^2} = \frac{p}{4\pi\epsilon_0}\frac{3z}{r^5}\sqrt{x^2 + y^2}$$

or

$$E_\perp = \frac{p}{4\pi\epsilon_0}\frac{3\cos\theta\sin\theta}{r^3}. \tag{6.15}$$

The transverse component $E_\perp$ is in the x-y plane and points directly away from the *axis* of the dipole. The total field, of course, is

$$E = \sqrt{E_z^2 + E_\perp^2}.$$

The dipole field varies inversely as the cube of the distance from the dipole. On the axis, at $\theta = 0$, it is twice as strong as at $\theta = 90°$. At both of these special angles the electric field has only a z-component, but of opposite sign at the two places (Fig. 6–4).

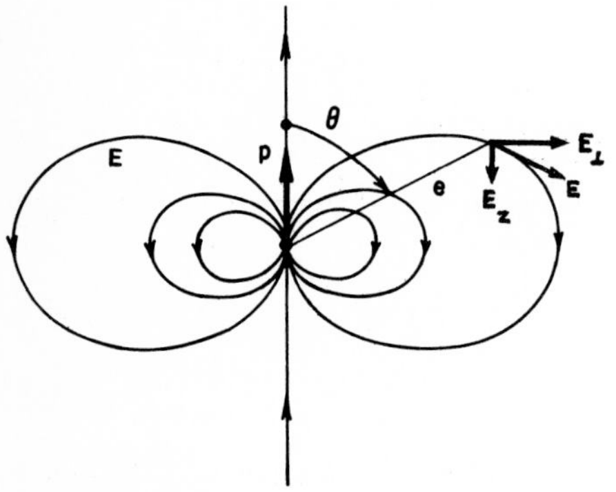

Fig. 6–4. The electric field of a dipole.

6–3 Remarks on vector equations

This is a good place to make a general remark about vector analysis. The fundamental proofs can be expressed by elegant equations in a general form, but in making various calculations and analyses it is always a good idea to choose the axes in some convenient way. Notice that when we were finding the potential of a dipole we chose the z-axis along the direction of the dipole, rather than at some arbitrary angle. This made the work much easier. But then we wrote the equations in vector form so that they would no longer depend on any particular coordinate system. After that, we are allowed to choose any coordinate system we wish, knowing that the relation is, in general, true. It clearly doesn't make any sense to bother with an arbitrary coordinate system at some complicated angle when you can choose a neat system for the particular problem—provided that the result can finally be expressed as a vector equation. So by all means take advantage of the fact that vector equations are independent of any coordinate system.

On the other hand, if you are trying to calculate the divergence of a vector, instead of just looking at $\nabla \cdot \boldsymbol{E}$ and wondering what it is, don't forget that it can always be spread out as

$$\frac{\partial E_x}{\partial x} + \frac{\partial E_y}{\partial y} + \frac{\partial E_z}{\partial z}.$$

If you can then work out the x-, y-, and z-components of the electric field and differentiate them, you will have the divergence. There often seems to be a feeling that there is something inelegant—some kind of defeat involved—in writing out the components; that somehow there ought always to be a way to do everything with the vector operators. There is often no advantage to it. The first time we encounter a particular kind of problem, it usually helps to write out the components to be sure we understand what is going on. There is nothing inelegant about putting numbers into equations, and nothing inelegant about substituting the derivatives for the fancy symbols. In fact, there is often a certain cleverness in doing just that. Of course when you publish a paper in a professional journal it will look better—and be more easily understood—if you can write everything in vector form. Besides, it saves print.

6–4 The dipole potential as a gradient

We would like to point out a rather amusing thing about the dipole formula, Eq. (6.13). The potential can also be written as

$$\phi = -\frac{1}{4\pi\epsilon_0}\,\boldsymbol{p}\cdot\nabla\left(\frac{1}{r}\right). \tag{6.16}$$

If you calculate the gradient of $1/r$, you get

$$\nabla\left(\frac{1}{r}\right) = -\frac{\boldsymbol{r}}{r^3} = -\frac{\boldsymbol{e}_r}{r^2},$$

and Eq. (6.16) is the same as Eq. (6.13).

How did we think of that? We just remembered that $\boldsymbol{e}_r/r^2$ appeared in the formula for the *field* of a point charge, and that the field was the gradient of a *potential* which has a $1/r$ dependence.

There is a *physical* reason for being able to write the dipole potential in the form of Eq. (6.16). Suppose we have a point charge q at the origin. The potential at the point P at (x, y, z) is

$$\phi_0 = \frac{q}{r}.$$

(Let's leave off the $1/4\pi\epsilon_0$ while we make these arguments; we can stick it in at the end.) Now if we move the charge $+q$ up a distance Δz, the potential at P will change a little, by, say, $\Delta\phi_+$. How much is $\Delta\phi_+$? Well, it is just the amount that the potential *would* change if we were to *leave* the charge at the origin and move P *downward* by the same distance Δz (Fig. 6–5). That is,

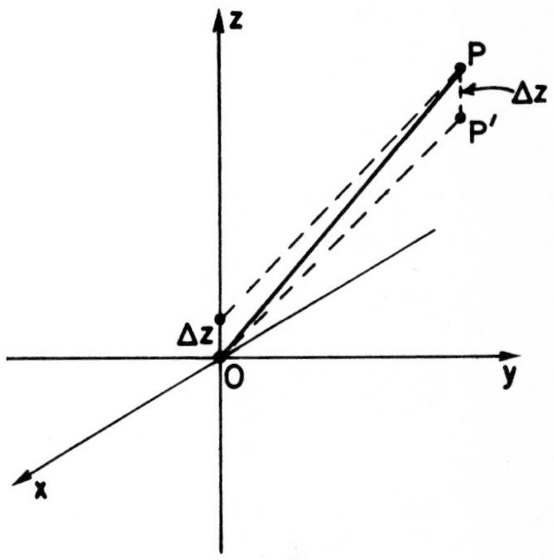

Fig. 6–5. The potential at P from a point charge at Δz above the origin is the same as the potential at P'(Δz below P) from the same charge at the origin.

$$\Delta\phi_+ = -\frac{\partial\phi_0}{\partial z}\,\Delta z,$$

where by Δz we mean the same as $d/2$. So, using $\phi = q/r$, we have that the potential from the positive charge is

$$\phi_+ = \frac{q}{r} - \frac{\partial}{\partial z}\left(\frac{q}{r}\right)\frac{d}{2}. \tag{6.17}$$

Applying the same reasoning for the potential from the negative charge, we can write

$$\phi_- = \frac{-q}{r} + \frac{\partial}{\partial z}\left(\frac{-q}{r}\right)\frac{d}{2}. \tag{6.18}$$

The total potential is the sum of (6.17) and (6.18):

$$\phi = \phi_+ + \phi_- = -\frac{\partial}{\partial z}\left(\frac{q}{r}\right)d \tag{6.19}$$

$$= -\frac{\partial}{\partial z}\left(\frac{1}{r}\right)qd.$$

For other orientations of the dipole, we could represent the displacement of the positive charge by the vector $\Delta\boldsymbol{r}_+$. We should then write Eq. (6.17) as

$$\Delta\phi_+ = -\nabla\phi_0\cdot\Delta\boldsymbol{r}_+,$$

where $\Delta\boldsymbol{r}$ is then to be replaced by $\boldsymbol{d}/2$. Completing the derivation as before, Eq. (6.19) would then become

$$\phi = -\nabla\left(\frac{1}{r}\right)\cdot q\boldsymbol{d}.$$

This is the same as Eq. (6.16), if we replace $q\boldsymbol{d} = \boldsymbol{p}$, and put back the $1/4\pi\epsilon_0$. Looking at it another way, we see that the dipole potential, Eq. (6.13), can be interpreted as

$$\phi = -\boldsymbol{p}\cdot\nabla\Phi_0, \tag{6.20}$$

where $\Phi_0 = 1/4\pi\epsilon_0 r$ is the potential of a *unit* point charge.

Although we can always find the potential of a known charge distribution by an integration, it is sometimes possible to save time by getting the answer with a clever trick. For example, one can often make use of the superposition principle. If we are given a charge distribution that can be made up of the sum of two distributions for which the potentials are already known, it is easy to find the desired potential by just adding the two known ones. One example of this is our derivation of (6.20), another is the following.

Suppose we have a spherical surface with a distribution of surface charge that varies as the cosine of the polar angle. The integration for this distribution is fairly messy. But, surprisingly, such a distribution can be analyzed by superposition. For imagine a sphere with a uniform *volume* density of positive charge, and another sphere with an equal uniform volume density of negative charge,

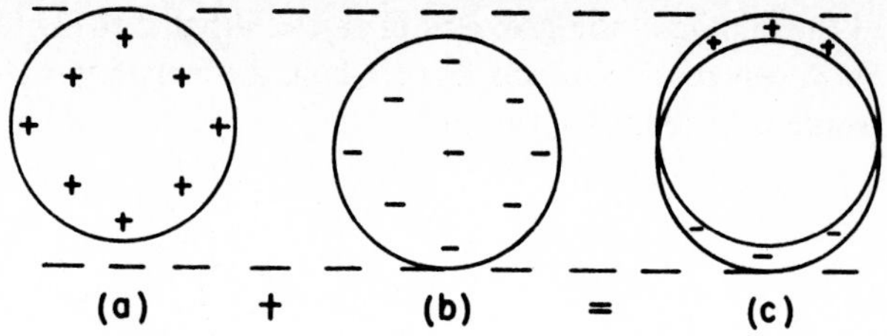

Fig. 6–6. Two uniformly charged spheres, superposed with a slight displacement, are equivalent to a nonuniform distribution of surface charge.

originally superposed to make a neutral—that is, uncharged—sphere. If the positive sphere is then displaced slightly with respect to the negative sphere, the body of the uncharged sphere would remain neutral, but a little positive charge will appear on one side, and some negative charge will appear on the opposite side, as illustrated in Fig. 6–6. If the relative displacement of the two spheres is small, the net charge is equivalent to a surface charge (on a spherical surface), and the surface charge density will be proportional to the cosine of the polar angle.

Now if we want the potential from this distribution, we do not need to do an integral. We know that the potential from each of the spheres of charge is—for points outside the sphere—the same as from a point charge. The two displaced spheres are like two point charges; the potential is just that of a dipole.

In this way you can show that a charge distribution on a sphere of radius a with a surface charge density

$$\sigma = \sigma_0 \cos\theta$$

produces a field outside the sphere which is just that of a dipole whose moment is

$$p = \frac{4\pi\sigma_0 a^3}{3}.$$

It can also be shown that inside the sphere the field is constant, with the value

$$E = \frac{\sigma_0}{3\epsilon_0}.$$

If θ is the angle from the positive z-axis, the electric field inside the sphere is in the *negative* z-direction. The example we have just considered is not as artificial as it may appear; we will encounter it again in the theory of dielectrics.

6–5 The dipole approximation for an arbitrary distribution

The dipole field appears in another circumstance both interesting and important. Suppose that we have an object that has a complicated distribution of charge—like the water molecule (Fig. 6–2)—and we are interested only in the fields far away. We will show that it is possible to find a relatively simple expression for the fields which is appropriate for distances large compared with the size of the object.

We can think of our object as an assembly of point charges q_i in a certain limited region, as shown in Fig. 6–7. (We can, later, replace q_i by $\rho\, dV$ if we wish.) Let each charge q_i be located at the displacement $\boldsymbol{d}_i$ from an origin chosen somewhere

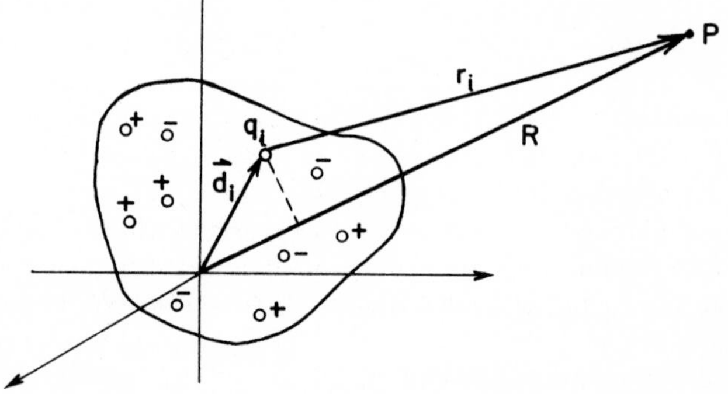

Fig. 6–7. Computation of the potential at a point P at a large distance from a set of charges.

in the middle of the group of charges. What is the potential at the point P, located at $\boldsymbol{R}$, where $\boldsymbol{R}$ is much larger than the maximum $\boldsymbol{d}_i$? The potential from the whole collection is given by

$$\phi = \frac{1}{4\pi\epsilon_0}\sum_i \frac{q_i}{r_i}, \tag{6.21}$$

where r_i is the distance from P to the charge q_i (the length of the vector $\boldsymbol{R} - \boldsymbol{d}_i$). Now if the distance from the charges to P, the point of observation, is enormous, each of the r_i's can be approximated by R. Each term becomes q_i/R, and we can take $1/R$ out as a factor in front of the summation. This gives us the simple result

$$\phi = \frac{1}{4\pi\epsilon_0}\frac{1}{R}\sum q_i = \frac{Q}{4\pi\epsilon_0 R}, \tag{6.22}$$

where Q is just the total charge of the whole object. Thus we find that for points far enough from any lump of charge, the lump looks like a point charge. The result is not too surprising.

But what if there are equal numbers of positive and negative charges? Then the total charge Q of the object is zero. This is not an unusual case; in fact, as we know, objects are usually neutral. The water molecule is neutral, but the charges are not all at one point, so if we are close enough we should be able to see some effects of the separate charges. We need a better approximation than (6.22) for the potential from an arbitrary distribution of charge in a neutral object. Equation (6.21) is still precise, but we can no longer just set $r_i = R$. We need a more accurate expression for r_i. If the point P is at a large distance, r_i will differ from R to an excellent approximation by the projection of $\boldsymbol{d}$ on $\boldsymbol{R}$, as can be seen from Fig. 6–7. (You should imagine that P is really farther away than is shown in the figure.) In other words, if $\boldsymbol{e}_r$ is the unit vector in the direction of $\boldsymbol{R}$, then our next approximation to r_i is

$$r_i \approx R - \boldsymbol{d}_i \cdot \boldsymbol{e}_r. \tag{6.23}$$

What we really want is $1/r_i$, which, since $d_i \ll R$, can be written to our approximation as

$$\frac{1}{r_i} \approx \frac{1}{R}\left(1 + \frac{\boldsymbol{d}_i \cdot \boldsymbol{e}_r}{R}\right). \tag{6.24}$$

Substituting this in (6.21), we get that the potential is

$$\phi = \frac{1}{4\pi\epsilon_0}\left(\frac{Q}{R} + \sum_i q_i \frac{\boldsymbol{d}_i \cdot \boldsymbol{e}_r}{R^2} + \cdots\right). \tag{6.25}$$

The three dots indicate the terms of higher order in d/R that we have neglected. These, as well as the ones we have already obtained, are successive terms in a Taylor expansion of $1/r_i$ about $1/R$ in powers of d_i/R.

The first term in (6.25) is what we got before; it drops out if the object is neutral. The second term depends on $1/R^2$, just as for a dipole. In fact, if we *define*

$$\boldsymbol{p} = \sum q_i \boldsymbol{d}_i \tag{6.26}$$

as a property of the charge distribution, the second term of the potential (6.25) is

$$\phi = \frac{1}{4\pi\epsilon_0}\frac{\boldsymbol{p} \cdot \boldsymbol{e}_r}{R^2}, \tag{6.27}$$

precisely a dipole potential. The quantity $\boldsymbol{p}$ is called the dipole moment of the distribution. It is a generalization of our earlier definition, and reduces to it for the special case of two point charges.

Our result is that, far enough away from *any* mess of charges that is as a whole neutral, the potential is a dipole potential. It decreases as $1/R^2$ and varies as $\cos\theta$—and its strength depends on the dipole moment of the distribution of charge. It is for these reasons that dipole fields are important, since the simple case of a pair of point charges is quite rare.

The water molecule, for example, has a rather strong dipole moment. The electric fields that result from this moment are responsible for some of the important properties of water. For many molecules, for example CO_2, the dipole moment vanishes because of the symmetry of the molecule. For them we should expand still more accurately, obtaining another term in the potential which decreases as $1/R^3$, and which is called a quadrupole potential. We will discuss such cases later.

6–6 The fields of charged conductors

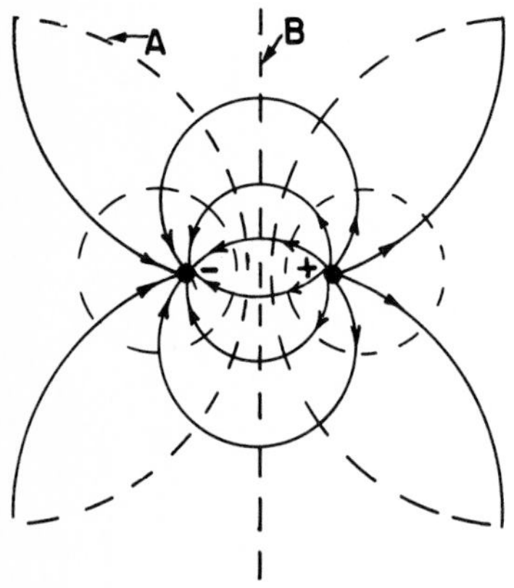

Fig. 6–8. The field lines and equipotentials for two point charges.

We have now finished with the examples we wish to cover of situations in which the charge distribution is known from the start. It has been a problem without serious complications, involving at most some integrations. We turn now to an entirely new kind of problem, the determination of the fields near charged conductors.

Suppose that we have a situation in which a total charge Q is placed on an arbitrary conductor. Now we will not be able to say exactly where the charges are. They will spread out in some way on the surface. How can we know how the charges have distributed themselves on the surface? They must distribute themselves so that the potential of the surface is constant. If the surface were not an equipotential, there would be an electric field inside the conductor, and the charges would keep moving until it became zero. The general problem of this kind can be solved in the following way. We guess at a distribution of charge and calculate the potential. If the potential turns out to be constant everywhere on the surface, the problem is finished. If the surface is not an equipotential, we have guessed the wrong distribution of charges, and should guess again—hopefully with an improved guess! This can go on forever, unless we are judicious about the successive guesses.

The question of how to guess at the distribution is mathematically difficult. Nature, of course, has time to do it; the charges push and pull until they all balance themselves. When we try to solve the problem, however, it takes us so long to make each trial that that method is very tedious. With an arbitrary group of conductors and charges the problem can be very complicated, and in general it cannot be solved without rather elaborate numerical methods. Such numerical computations, these days, are set up on a computing machine that will do the work for us, once we have told it how to proceed.

On the other hand, there are a lot of little practical cases where it would be nice to be able to find the answer by some more direct method—without having to write a program for a computer. Fortunately, there are a number of cases where the answer can be obtained by squeezing it out of Nature by some trick or other. The first trick we will describe involves making use of solutions we have already obtained for situations in which charges have specified locations.

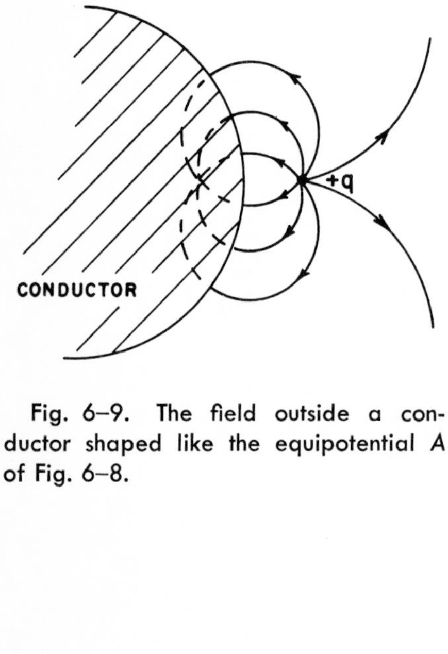

Fig. 6–9. The field outside a conductor shaped like the equipotential A of Fig. 6–8.

6–7 The method of images

We have solved, for example, the field of two point charges. Figure 6–8 shows some of the field lines and equipotential surfaces we obtained by the computations in Chapter 5. Now consider the equipotential surface marked A. Suppose we were to shape a thin sheet of metal so that it just fits this surface. If we place it right at the surface and adjust its potential to the proper value, no one would ever know it was there, because nothing would be changed.

But notice! We have really solved a *new* problem. We have a situation in which the surface of a curved conductor with a given potential is placed near a point charge. If the metal sheet we placed at the equipotential surface eventually closes on itself (or, in practice, if it goes far enough) we have the kind of situation considered in Section 5–10, in which our space is divided into two regions, one inside and one outside a closed conducting shell. We found there that the fields in the two regions are quite independent of each other. So we would have the same fields outside our curved conductor no matter what is inside. We can even fill up the whole inside with conducting material. We have found, therefore, the fields for the arrangement of Fig. 6–9. In the space outside the conductor the field is just like that of two point charges, as in Fig. 6–8. Inside the conductor, it is zero. Also—as it must be—the electric field just outside the conductor is normal to the surface.

Thus we can compute the fields in Fig. 6–9 by computing the field due to q and to an imaginary point charge $-q$ at a suitable point. The point charge we "imagine" existing behind the conducting surface is called an *image charge.*

In books you can find long lists of solutions for hyperbolic-shaped conductors and other complicated looking things, and you wonder how anyone ever solved these terrible shapes. They were solved backwards! Someone solved a simple problem with given charges. He then saw that some equipotential surface showed up in a new shape, and he wrote a paper in which he pointed out that the field outside that particular shape can be described in a certain way.

6–8 A point charge near a conducting plane

As the simplest application of the use of this method, let's make use of the plane equipotential surface B of Fig. 6–8. With it, we can solve the problem of a charge in front of a conducting sheet. We just cross out the left-hand half of the picture. The field lines for our solution are shown in Fig. 6–10. Notice that the plane, since it was halfway between the two charges, has zero potential. We have solved the problem of a positive charge next to a grounded conducting sheet.

We have now solved for the total field, but what about the *real* charges that are responsible for it? There are, in addition to our positive point charge, some induced negative charges on the conducting sheet that have been attracted by the positive charge (from large distances away). Now suppose that for some technical reason—or out of curiosity—you would like to know how the negative charges are distributed on the surface. You can find the surface charge density by using the result we worked out in Section 5–6 with Gauss' theorem. The normal com-

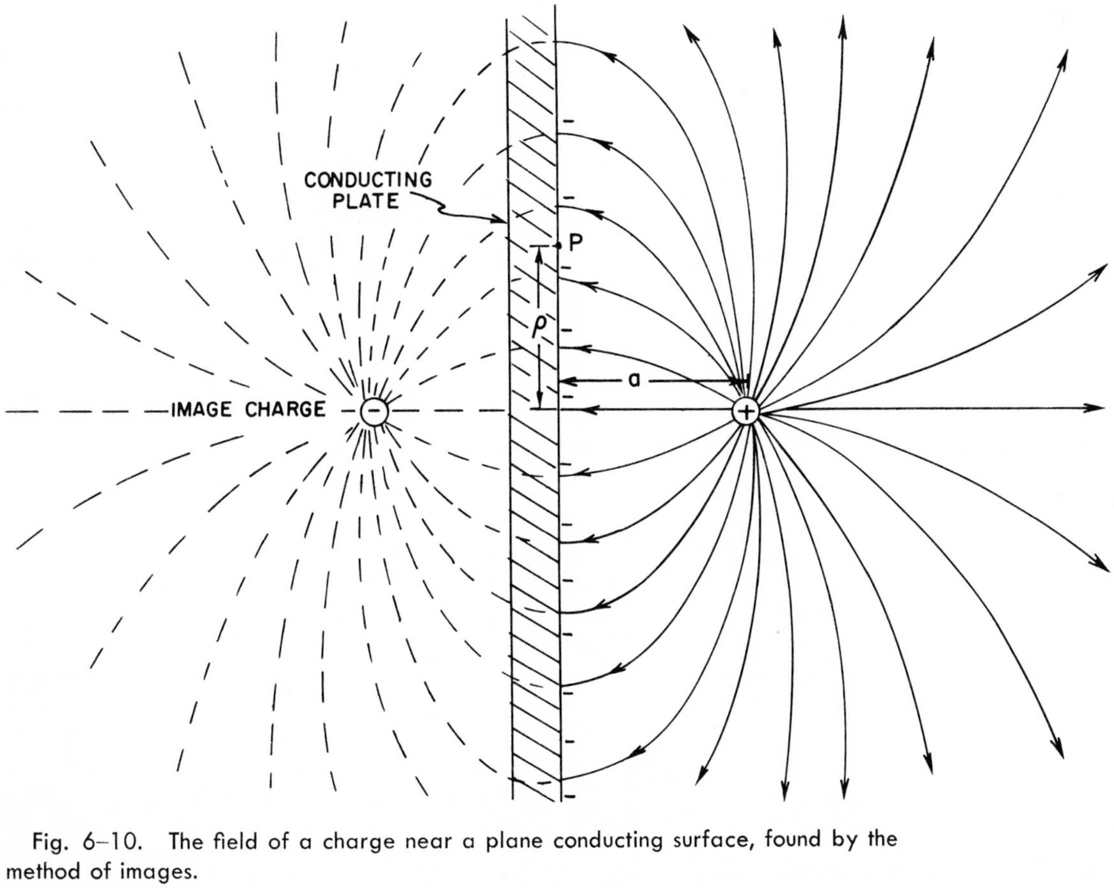

Fig. 6–10. The field of a charge near a plane conducting surface, found by the method of images.

ponent of the electric field just outside a conductor is equal to the density of surface charge σ divided by ϵ_0. We can obtain the density of charge at any point on the surface by working backwards from the normal component of the electric field at the surface. We know that, because we know the field everywhere.

Consider a point on the surface at the distance ρ from the point directly beneath the positive charge (Fig. 6–10). The electric field at this point is normal to the surface and is directed into it. The component normal to the surface of the field from the *positive* point charge is

$$E_{n+} = -\frac{1}{4\pi\epsilon_0}\frac{aq}{(a^2+\rho^2)^{3/2}}. \tag{6.28}$$

To this we must add the electric field produced by the negative image charge. That just doubles the normal component (and cancels all others), so the charge density σ at any point on the surface is

$$\sigma(\rho) = \epsilon_0 E(\rho) = -\frac{2aq}{4\pi(a^2+\rho^2)^{3/2}}. \tag{6.29}$$

An interesting check on our work is to integrate σ over the whole surface. We find that the total induced charge is $-q$, as it should be.

One further question: Is there a force on the point charge? Yes, because there is an attraction from the induced negative surface charge on the plate. Now that we know what the surface charges are (from Eq. (6.29)), we could compute the force on our positive point charge by an integral. But we also know that the force acting on the positive charge is exactly the same as it *would be* with the negative image charge instead of the plate, because the fields in the neighborhood are the same in both cases. The point charge feels a force toward the plate whose magnitude is

$$F = \frac{1}{4\pi\epsilon_0}\frac{q^2}{(2a)^2}. \tag{6.30}$$

We have found the force much more easily than by integrating over all the negative charges.

6–9 A point charge near a conducting sphere

What other surfaces besides a plane have a simple solution? The next most simple shape is a sphere. Let's find the fields around a metal sphere which has a point charge q near it, as shown in Fig. 6–11. Now we must look for a simple physical situation which gives a sphere for an equipotential surface. If we look around at problems people have already solved, we find that someone has noticed that the field of two *unequal* point charges has an equipotential that is a sphere. Aha! If we choose the location of an image charge—and pick the right amount of charge—maybe we can make the equipotential surface fit our sphere. Indeed, it can be done with the following prescription.

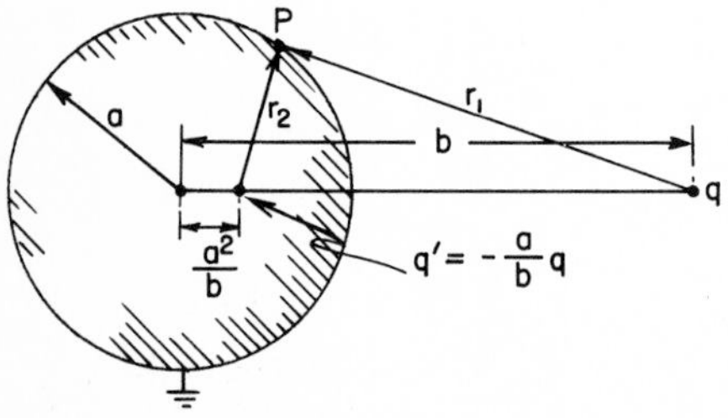

Fig. 6–11. The point charge q induces charges on a grounded conducting sphere whose fields are those of an image charge q' placed at the point shown.

Assume that you want the equipotential surface to be a sphere of radius a with its center at the distance b from the charge q. Put an image charge of strength $q' = -q(a/b)$ on the line from the charge to the center of the sphere, and at a distance a^2/b from the center. The sphere will be at zero potential.

The mathematical reason stems from the fact that a sphere is the locus of all points for which the distances from two points are in a constant ratio. Referring to Fig. 6–11, the potential at P from q and q' is proportional to

$$\frac{q}{r_1} + \frac{q'}{r_2}.$$

The potential will thus be zero at all points for which

$$\frac{q'}{r_2} = -\frac{q}{r_1} \quad \text{or} \quad \frac{r_2}{r_1} = -\frac{q'}{q}.$$

If we place q' at the distance a^2/b from the center, the ratio r_2/r_1 has the constant value a/b. Then if

$$\frac{q'}{q} = -\frac{a}{b}, \tag{6.31}$$

the sphere is an equipotential. Its potential is, in fact, zero.

What happens if we are interested in a sphere that is not at zero potential? That would be so only if its total charge happens accidentally to be q'. Of course if it is grounded, the charges induced on it would have to be just that. But what if it is insulated, and we have put no charge on it? Or if we know that the total charge Q has been put on it? Or just that it has a given potential *not* equal to zero? All these questions are easily answered. We can always add a point charge q'' at the center of the sphere. The sphere still remains an equipotential by superposition; only the magnitude of the potential will be changed.

If we have, for example, a conducting sphere which is initially uncharged and insulated from everything else, and we bring near to it the positive point charge q, the total charge of the sphere will remain zero. The solution is found by using an image charge q' as before, but, in addition, adding a charge q'' at the center of the sphere, choosing

$$q'' = -q' = \frac{a}{b}\,q. \tag{6.32}$$

The fields everywhere outside the sphere are given by the superposition of the fields of q, q', and q''. The problem is solved.

We can see now that there will be a force of attraction between the sphere and the point charge q. It is not zero even though there is no charge on the neutral sphere. Where does the attraction come from? When you bring a positive charge up to a conducting sphere, the positive charge attracts negative charges to the side closer to itself and leaves positive charges on the surface of the far side. The attraction by the negative charges exceeds the repulsion from the positive charges; there is a net attraction. We can find out how large the attraction is by computing the force on q in the field produced by q' and q''. The total force is the sum of the attractive force between q and a charge $q' = -(a/b)q$, at the distance $b - (a^2/b)$, and the repulsive force between q and a charge $q'' = +(a/b)q$ at the distance b.

Those who were entertained in childhood by the baking powder box which has on its label a picture of a baking powder box which has on its label a picture of a baking powder box which has . . . may be interested in the following problem. Two equal spheres, one with a total charge of $+Q$ and the other with a total charge of $-Q$, are placed at some distance from each other. What is the force between them? The problem can be solved with an infinite number of images. One first approximates each sphere by a charge at its center. These charges will have image charges in the other sphere. The image charges will have images, etc., etc., etc. The solution is like the picture on the box of baking powder—and it converges pretty fast.

6–10 Condensers; parallel plates

We take up now another kind of a problem involving conductors. Consider two large metal plates which are parallel to each other and separated by a distance small compared with their width. Let's suppose that equal and opposite charges have been put on the plates. The charges on each plate will be attracted by the charges on the other plate, and the charges will spread out uniformly on the inner surfaces of the plates. The plates will have surface charge densities $+\sigma$ and $-\sigma$, respectively, as in Fig. 6–12. From Chapter 5 we know that the field between the plates is σ/ϵ_0, and that the field outside the plates is zero. The plates will have different potentials ϕ_1 and ϕ_2. For convenience we will call the difference V; it is often called the "voltage":

$$\phi_1 - \phi_2 = V.$$

(You will find that sometimes people use V for the potential, but we have chosen to use ϕ.)

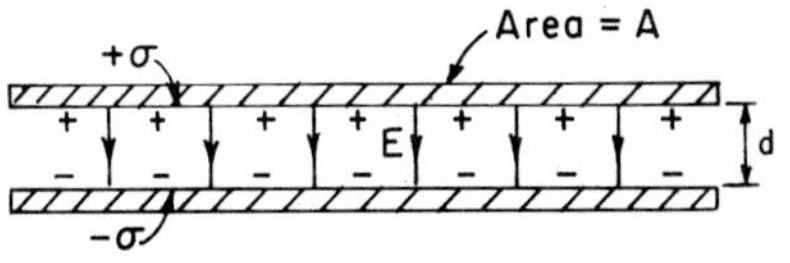

Fig. 6–12. A parallel-plate condenser.

The potential difference V is the work per unit charge required to carry a small charge from one plate to the other, so that

$$V = Ed = \frac{\sigma}{\epsilon_0} d = \frac{d}{\epsilon_0 A} Q, \tag{6.33}$$

where $\pm Q$ is the total charge on each plate, A is the area of the plates, and d is the separation.

We find that the voltage is proportional to the charge. Such a proportionality between V and Q is found for any two conductors in space if there is a plus charge on one and an equal minus charge on the other. The potential difference between them—that is, the voltage—will be proportional to the charge. (We are assuming that there are no other charges around.)

Why this proportionality? Just the superposition principle. Suppose we know the solution for one set of charges, and then we superimpose two such solutions. The charges are doubled, the fields are doubled, and the work done in carrying a unit charge from one point to the other is also doubled. Therefore the potential difference between any two points is proportional to the charges. In particular, the potential difference between the two conductors is proportional to the charges on them. Someone originally wrote the equation of proportionality the other way. That is, they wrote

$$Q = CV,$$

where C is a constant. This coefficient of proportionality is called the *capacity*, and such a system of two conductors is called a *condenser*.* For our parallel-plate condenser

$$C = \frac{\epsilon_0 A}{d} \quad \text{(parallel plates).} \tag{6.34}$$

This formula is not exact, because the field is not really uniform everywhere between the plates, as we assumed. The field does not just suddenly quit at the edges, but really is more as shown in Fig. 6–13. The total charge is not σA, as we have assumed—there is a little correction for the effects at the edges. To find out what the correction is, we will have to calculate the field more exactly and find out just what does happen at the edges. That is a complicated mathematical problem which can, however, be solved by techniques which we will not describe now. The result of such calculations is that the charge density rises somewhat near the edges of the plates. This means that the capacity of the plates is a little higher than we computed. [A very good approximation for the capacity is obtained if we use Eq. (6.34) but take for A the area one *would* get if the plates were extended artificially by a distance 3/8 of the separation between the plates.]

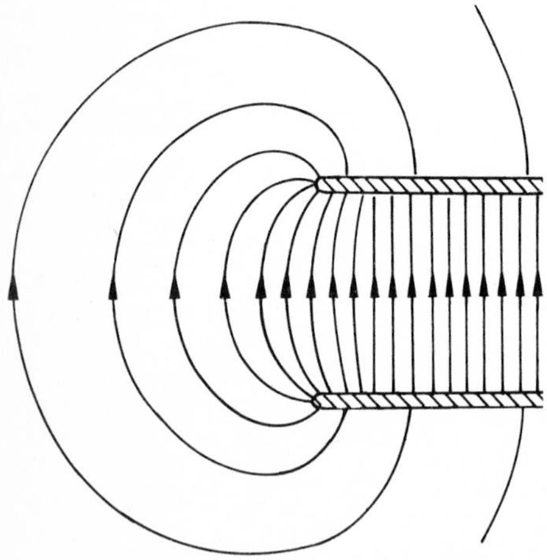
Fig. 6–13. The electric field near the edge of two parallel plates.

We have talked about the capacity for two conductors only. Sometimes people talk about the capacity of a single object. They say, for instance, that the capacity of a sphere of radius a is $4\pi\epsilon_0 a$. What they imagine is that the other terminal is another sphere of infinite radius—that when there is a charge $+Q$ on the sphere, the opposite charge, $-Q$, is on an infinite sphere. One can also speak of capacities when there are three or more conductors, a discussion we shall, however, defer.

Suppose that we wish to have a condenser with a very large capacity. We could get a large capacity by taking a very big area and a very small separation. We could put waxed paper between sheets of aluminum foil and roll it up. (If we seal it in plastic, we have a typical radio-type condenser.) What good is it? It is good for storing charge. If we try to store charge on a ball, for example, its potential rises rapidly as we charge it up. It may even get so high that the charge begins to escape into the air by way of sparks. But if we put the same charge on a condenser whose capacity is very large, the voltage developed across the condenser will be small.

---

* Some people think the words "capacitance" and "capacitor" should be used, instead of "capacity" and "condensor." We have decided to use the older terminology, because it is still more commonly heard in the physics laboratory—even if not in textbooks!

In many applications in electronic circuits, it is useful to have something which can absorb or deliver large quantities of charge without changing its potential much. A condenser (or "capacitor") does just that. There are also many applications in electronic instruments and in computers where a condenser is used to get a specified change in voltage in response to a particular change in charge. We have seen a similar application in Chapter 23, Vol. I, where we described the properties of resonant circuits.

From the definition of C, we see that its unit is one coul/volt. This unit is also called a *farad.* Looking at Eq. (6.34), we see that one can express the units of ϵ_0 as farad/meter, which is the unit most commonly used. Typical sizes of condensers run from one micro-microfarad (= 1 picofarad) to millifarads. Small condensers of a few picofarads are used in high-frequency tuned circuits, and capacities up to hundreds or thousands of microfarads are found in power-supply filters. A pair of plates one square centimeter in area with a one millimeter separation have a capacity of roughly one micro-microfarad.

$$\epsilon_0 \approx \frac{1}{36\pi \times 10^9} \text{ farad/meter}$$

6–11 High-voltage breakdown

We would like now to discuss qualitatively some of the characteristics of the fields around conductors. If we charge a conductor that is not a sphere, but one that has on it a point or a very sharp end, as, for example, the object sketched in Fig. 6–14, the field around the point is much higher than the field in the other regions. The reason is, qualitatively, that charges try to spread out as much as possible on the surface of a conductor, and the tip of a sharp point is as far away as it is possible to be from most of the surface. Some of the charges on the plate get pushed all the way to the tip. A relatively small *amount* of charge on the tip can still provide a large surface *density;* a high charge density means a high field just outside.

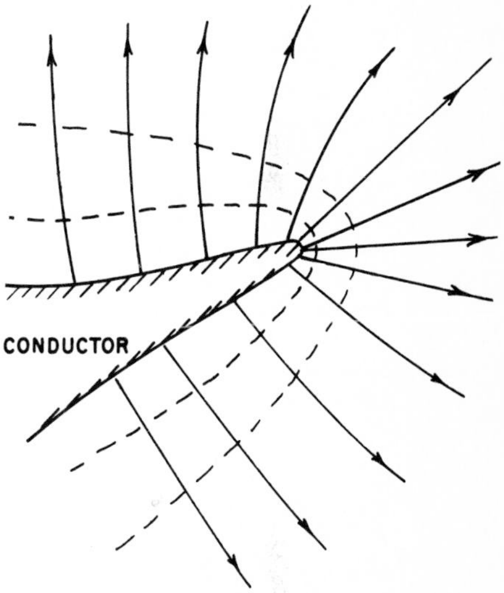

Fig. 6–14. The electric field near a sharp point on a conductor is very high.

One way to see that the field is highest at those places on a conductor where the radius of curvature is smallest is to consider the combination of a big sphere and a little sphere connected by a wire, as shown in Fig. 6–15. It is a somewhat idealized version of the conductor of Fig. 6–14. The wire will have little influence on the fields outside; it is there to keep the spheres at the same potential. Now, which ball has the biggest field at its surface? If the ball on the left has the radius a and carries a charge Q, its potential is about

$$\phi_1 = \frac{1}{4\pi\epsilon_0}\frac{Q}{a}.$$

(Of course the presence of one ball changes the charge distribution on the other, so that the charges are not really spherically symmetric on either. But if we are interested only in an estimate of the fields, we can use the potential of a spherical charge.) If the smaller ball, whose radius is b, carries the charge q, its potential is about

$$\phi_2 = \frac{1}{4\pi\epsilon_0}\frac{q}{b}.$$

But $\phi_1 = \phi_2$, so

$$\frac{Q}{a} = \frac{q}{b}.$$

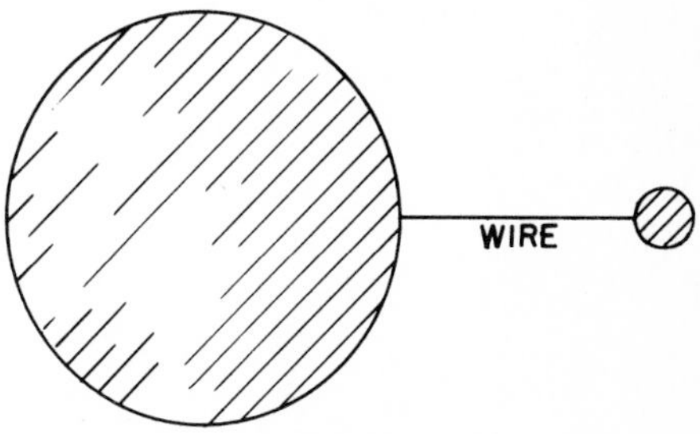

Fig. 6–15. The field of a pointed object can be approximated by that of two spheres at the same potential.

On the other hand, the field at the surface (see Eq. 5.8) is proportional to the surface charge density, which is like the total charge over the radius squared. We get that

$$\frac{E_a}{E_b} = \frac{Q/a^2}{q/b^2} = \frac{b}{a}. \tag{6.35}$$

Therefore the field is higher at the surface of the small sphere. The fields are in the inverse proportion of the radii.

This result is technically very important, because air will break down if the electric field is too great. What happens is that a loose charge (electron, or ion) somewhere in the air is accelerated by the field, and if the field is very great, the charge can pick up enough speed before it hits another atom to be able to knock an electron off that atom. As a result, more and more ions are produced. Their motion constitutes a discharge, or spark. If you want to charge an object to a high potential and not have it discharge itself by sparks in the air, you must be sure that the surface is smooth, so that there is no place where the field is abnormally large.

6–12 The field-emission microscope

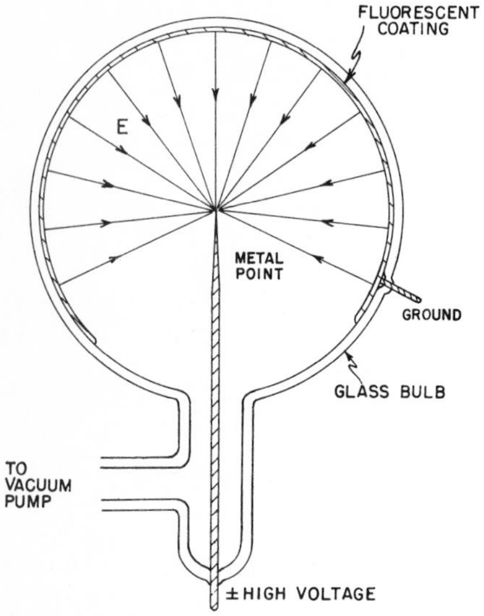

Fig. 6–16. Field-emission microscope.

There is an interesting application of the extremely high electric field which surrounds any sharp protuberance on a charged conductor. The *field-emission microscope* depends for its operation on the high fields produced at a sharp metal point.* It is built in the following way. A very fine needle, with a tip whose diameter is about 1000 angstroms, is placed at the center of an evacuated glass sphere (Fig. 6–16.) The inner surface of the sphere is coated with a thin conducting layer of fluorescent material, and a very high potential difference is applied between the fluorescent coating and the needle.

Let's first consider what happens when the needle is negative with respect to the fluorescent coating. The field lines are highly concentrated at the sharp point. The electric field can be as high as 40 million volts per centimeter. In such intense fields, electrons are pulled out of the surface of the needle and accelerated across the potential difference between the needle and the fluorescent layer. When they arrive there they cause light to be emitted, just as in a television picture tube.

The electrons which arrive at a given point on the fluorescent surface are, to an excellent approximation, those which leave the other end of the radial field line, because the electrons will travel along the field line passing from the point to the surface. Thus we see on the surface some kind of an image of the tip of the needle. More precisely, we see a picture of the *emissivity* of the surface of the needle—that is the ease with which electrons can leave the surface of the metal tip. If the resolution were high enough, one could hope to resolve the positions of the individual atoms on the tip of the needle. With electrons, this resolution is not possible for the following reasons. First, there is quantum-mechanical diffraction of the electron waves which blurs the image. Second, due to the internal motions of the electrons in the metal they have a small sideways initial velocity when they leave the needle, and this random transverse component of the velocity causes some smearing of the image. The combination of these two effects limits the resolution to 25 A or so.

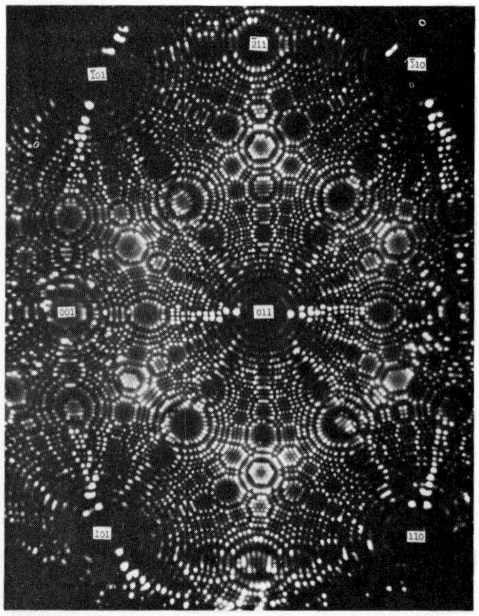

Fig. 6–17. Image produced by a field-emission microscope. [Courtesy of Erwin W. Mueller, Research Prof. of Physics, Pennsylvania State University.]

If, however, we reverse the polarity and introduce a small amount of helium gas into the bulb, much higher resolutions are possible. When a helium atom collides with the tip of the needle, the intense field there strips an electron off the helium atom, leaving it positively charged. The helium ion is then accelerated outward along a field line to the fluorescent screen. Since the helium ion is so much heavier than an electron, the quantum-mechanical wavelengths are much smaller. If the temperature is not too high, the effect of the thermal velocities is also smaller than in the electron case. With less smearing of the image a much sharper picture of the point is obtained. It has been possible to obtain magnifications up to 2,000,000 times with the positive ion field-emission microscope—a magnification ten times better than is obtained with the best electron microscope.

Figure 6–17 is an example of the results which were obtained with a field-ion microscope, using a tungsten needle. The center of a tungsten atom ionizes a helium atom at a slightly different rate than the spaces between the tungsten atoms. The pattern of spots on the fluorescent screen shows the arrangement of the *individual atoms* on the tungsten tip. The reason the spots appear in rings can be understood by visualizing a large box of balls packed in a rectangular array, representing the atoms in the metal. If you cut an approximately spherical section out of this box, you will see the ring pattern characteristic of the atomic structure. The field-ion microscope provided human beings with the means of seeing atoms for the first time. This is a remarkable achievement, considering the simplicity of the instrument.

* See E. W. Mueller: "The field-ion microscope," *Advances in Electronics and Electron Physics*, **13**, 83–179 (1960). Academic Press, New York.

7

The Electric Field in Various Circumstances (Continued)

7–1 Methods for finding the electrostatic field

This chapter is a continuation of our consideration of the characteristics of electric fields in various particular situations. We shall first describe some of the more elaborate methods for solving problems with conductors. It is not expected that these more advanced methods can be mastered at this time. Yet it may be of interest to have some idea about the kinds of problems that can be solved, using techniques that may be learned in more advanced courses. Then we take up two examples in which the charge distribution is neither fixed nor is carried by a conductor, but instead is determined by some other law of physics.

As we found in Chapter 6, the problem of the electrostatic field is fundamentally simple when the distribution of charges is specified; it requires only the evaluation of an integral. When there are conductors present, however, complications arise because the charge distribution on the conductors is not initially known; the charge must distribute itself on the surface of the conductor in such a way that the conductor is an equipotential. The solution of such problems is neither direct nor simple.

We have looked at an indirect method of solving such problems, in which we find the equipotentials for some specified charge distribution and replace one of them by a conducting surface. In this way we can build up a catalog of special solutions for conductors in the shapes of spheres, planes, etc. The use of images, described in Chapter 6, is an example of an indirect method. We shall describe another in this chapter.

If the problem to be solved does not belong to the class of problems for which we can construct solutions by the indirect method, we are forced to solve the problem by a more direct method. The mathematical problem of the direct method is the solution of Laplace's equation,

$$\nabla^2\phi = 0, \tag{7.1}$$

subject to the condition that ϕ is a suitable constant on certain boundaries—the surfaces of the conductors. Problems which involve the solution of a differential field equation subject to certain *boundary conditions* are called *boundary-value* problems. They have been the object of considerable mathematical study. In the case of conductors having complicated shapes, there are no general analytical methods. Even such a simple problem as that of a charged cylindrical metal can closed at both ends—a beer can—presents formidable mathematical difficulties. It can be solved only approximately, using numerical methods. The *only* general methods of solution are numerical.

There are a few problems for which Eq. (7.1) can be solved directly. For example, the problem of a charged conductor having the shape of an ellipsoid of revolution can be solved exactly in terms of known special functions. The solution for a thin disc can be obtained by letting the ellipsoid become infinitely oblate. In a similar manner, the solution for a needle can be obtained by letting the ellipsoid become infinitely prolate. However, it must be stressed that the only direct methods of general applicability are the numerical techniques.

Boundary-value problems can also be solved by measurements of a physical analog. Laplace's equation arises in many different physical situations: in steady-state heat flow, in irrotational fluid flow, in current flow in an extended medium, and in the deflection of an elastic membrane. It is frequently possible to set up a physical model which is analogous to an electrical problem which we wish to solve. By the measurement of a suitable analogous quantity on the model, the solution to the problem of interest can be determined. An example of the analog technique is the use of the electrolytic tank for the solution of two-dimensional problems in electrostatics. This works because the differential equation for the potential in a uniform conducting medium is the same as it is for a vacuum.

There are many physical situations in which the variations of the physical fields in one direction are zero, or can be neglected in comparison with the variations in the other two directions. Such problems are called two-dimensional; the field depends on two coordinates only. For example, if we place a long charged wire along the z-axis, then for points not too far from the wire the electric field depends on x and y, but not on z; the problem is two-dimensional. Since in a two-dimensional problem $\partial/\partial z = 0$, the equation for ϕ in free space is

$$\frac{\partial^2\phi}{\partial x^2} + \frac{\partial^2\phi}{\partial y^2} = 0. \tag{7.2}$$

Because the two-dimensional equation is comparatively simple, there is a wide range of conditions under which it can be solved analytically. There is, in fact, a very powerful indirect mathematical technique which depends on a theorem from the mathematics of functions of a complex variable, and which we will now describe.

7–2 Two-dimensional fields; functions of the complex variable

The complex variable $\mathfrak{z}$ is defined as

$$\mathfrak{z} = x + iy.$$

(Do not confuse $\mathfrak{z}$ with the z-coordinate, which we ignore in the following discussion because we assume there is no z-dependence of the fields.) Every point in x and y then corresponds to a complex number $\mathfrak{z}$. We can use $\mathfrak{z}$ as a single (complex) variable, and with it write the usual kinds of mathematical functions $F(\mathfrak{z})$. For example,

$$F(\mathfrak{z}) = \mathfrak{z}^2,$$

or

$$F(\mathfrak{z}) = 1/\mathfrak{z}^3,$$

or

$$F(\mathfrak{z}) = \mathfrak{z}\log\mathfrak{z},$$

and so forth.

Given any particular $F(\mathfrak{z})$ we can substitute $\mathfrak{z} = x + iy$, and we have a function of x and y—with real and imaginary parts. For example,

$$\mathfrak{z}^2 = (x + iy)^2 = x^2 - y^2 + 2ixy. \tag{7.3}$$

Any function $F(\mathfrak{z})$ can be written as a sum of a pure real part and a pure imaginary part, each part a function of x and y:

$$F(\mathfrak{z}) = U(x, y) + iV(x, y), \tag{7.4}$$

where $U(x, y)$ and $V(x, y)$ are real functions. Thus from any complex function $F(\mathfrak{z})$ two new functions $U(x, y)$ and $V(x, y)$ can be derived. For example, $F(\mathfrak{z}) = \mathfrak{z}^2$ gives us the two functions

$$U(x, y) = x^2 - y^2, \tag{7.5}$$

and

$$V(x, y) = 2xy. \tag{7.6}$$

Now we come to a miraculous mathematical theorem which is so delightful that we shall leave a proof of it for one of your courses in mathematics. (We should not reveal all the mysteries of mathematics, or that subject matter would become too dull.) It is this. For any "ordinary function" (mathematicians will define it better) the functions U and V *automatically* satisfy the relations

$$\frac{\partial U}{\partial x} = \frac{\partial V}{\partial y}, \tag{7.7}$$

$$\frac{\partial V}{\partial x} = -\frac{\partial U}{\partial y}. \tag{7.8}$$

It follows immediately that each of the functions U and V satisfy Laplace's equation:

$$\frac{\partial^2 U}{\partial x^2} + \frac{\partial^2 U}{\partial y^2} = 0, \tag{7.9}$$

$$\frac{\partial^2 V}{\partial x^2} + \frac{\partial^2 V}{\partial y^2} = 0. \tag{7.10}$$

These equations are clearly true for the functions of (7.5) and (7.6).

Thus, starting with any ordinary function, we can arrive at two functions $U(x, y)$ and $V(x, y)$, which are both solutions of Laplace's equation in two dimensions. Each function represents a possible electrostatic potential. We can pick *any* function $F(\mathfrak{z})$ and it should represent *some* electric field problem—in fact, *two* problems, because U and V *each* represent solutions. We can write down as many solutions as we wish—by just making up functions—then we just have to find the *problem* that goes with each solution. It may sound backwards, but it's a possible approach.

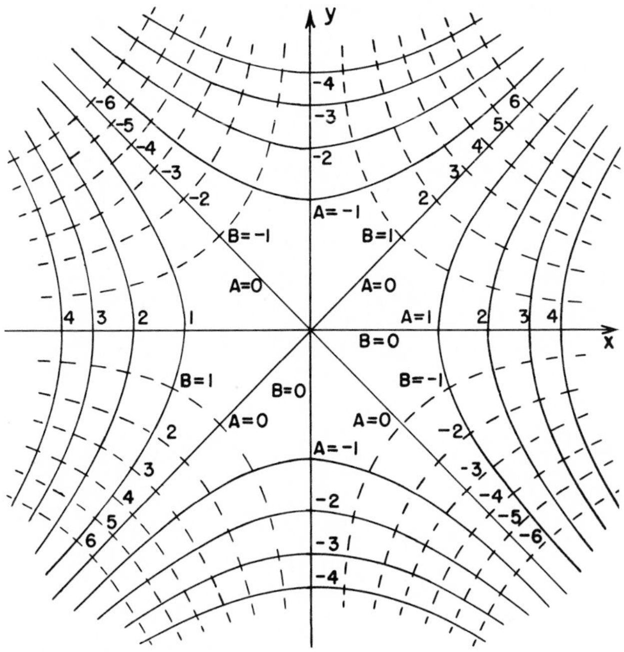

Fig. 7–1. Two sets of orthogonal curves which can represent equipotentials in a two-dimensional electrostatic field.

As an example, let's see what physics the function $F(\mathfrak{z}) = \mathfrak{z}^2$ gives us. From it we get the two potential functions of (7.5) and (7.6). To see what problem the function U belongs to, we solve for the equipotential surfaces by setting $U = A$, a constant:

$$x^2 - y^2 = A.$$

This is the equation of a rectangular hyperbola. For various values of A, we get the hyperbolas shown in Fig. 7–1. When $A = 0$, we get the special case of diagonal straight lines through the origin.

Such a set of equipotentials corresponds to several possible physical situations. First, it represents the fine details of the field near the point halfway between two

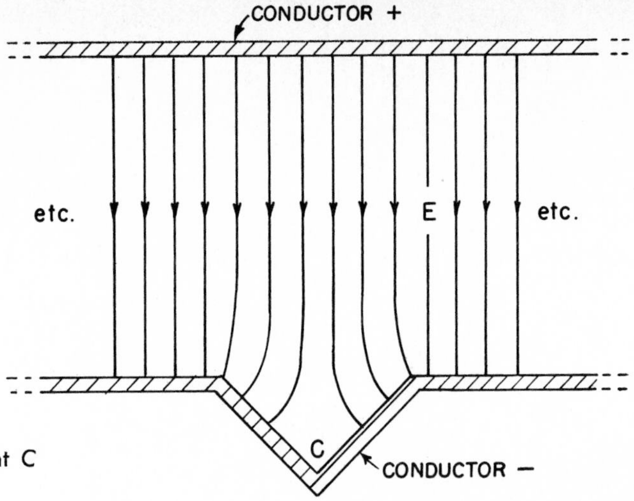

Fig. 7–2. The field near the point C is the same as that in Fig. 7–1.

equal point charges. Second, it represents the field at an inside right-angle corner of a conductor. If we have two electrodes shaped like those in Fig. 7–2, which are held at different potentials, the field near the corner marked C will look just like the field above the origin in Fig. 7–1. The solid lines are the equipotentials, and the broken lines at right angles correspond to lines of $\boldsymbol{E}$. Whereas at points or protuberances the electric field tends to be high, it tends to be *low* in dents or hollows.

The solution we have found also corresponds to that for a hyperbola-shaped electrode near a right-angle corner, or for two hyperbolas at suitable potentials. You will notice that the field of Fig. 7–1 has an interesting property. The x-component of the electric field, E_x, is given by

$$E_x = -\frac{\partial\phi}{\partial x} = -2x.$$

The electric field is proportional to the distance from the axis. This fact is used to make devices (called quadrupole lenses) that are useful for focusing particle beams (see Section 29–9). The desired field is usually obtained by using four hyperbola-shaped electrodes, as shown in Fig. 7–3. For the electric field lines in Fig. 7–3, we have simply copied from Fig. 7–1 the set of broken-line curves that represent $V =$ constant. We have a bonus! The curves for $V =$ constant are orthogonal to the ones for $U =$ constant because of the equations (7.7) and (7.8). Whenever we choose a function $F(\mathfrak{z})$, we get from U and V both the equipotentials and field lines. And you will remember that we have solved either of two problems, depending on which set of curves we call the equipotentials.

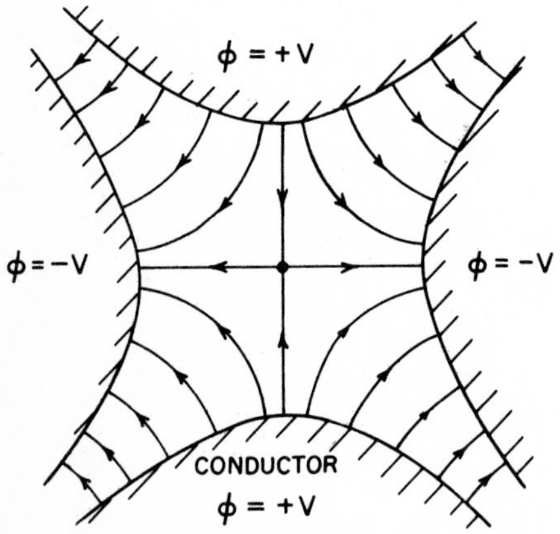

Fig. 7–3. The field in a quadrupole lens.

As a second example, consider the function

$$F(\mathfrak{z}) = \sqrt{\mathfrak{z}}. \tag{7.11}$$

If we write

$$\mathfrak{z} = x + iy = \rho e^{i\theta},$$

where

$$\rho = \sqrt{x^2 + y^2}$$

and

$$\tan\theta = y/x,$$

then

$$F(\mathfrak{z}) = \rho^{1/2}e^{i\theta/2}$$

$$= \rho^{1/2}\left(\cos\frac{\theta}{2} + i\sin\frac{\theta}{2}\right),$$

from which

$$F(\mathfrak{z}) = \left[\frac{(x^2 + y^2)^{1/2} + x}{2}\right]^{1/2} + i\left[\frac{(x^2 + y^2)^{1/2} - x}{2}\right]^{1/2}. \tag{7.12}$$

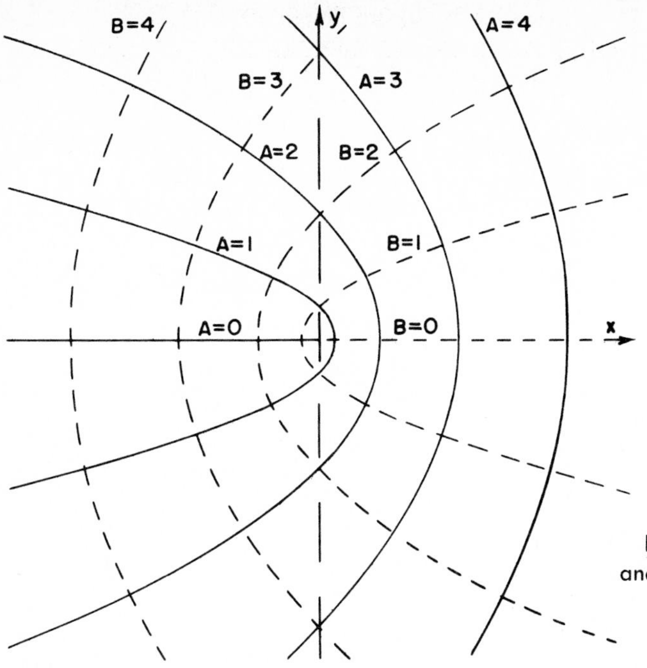

Fig. 7–4. Curves of constant $U(x, y)$ and $V(x, y)$ from Eq. (7.12).

The curves for $U(x, y) = A$ and $V(x, y) = B$, using U and V from Eq. (7.12), are plotted in Fig. 7–4. Again, there are many possible situations that could be described by these fields. One of the most interesting is the field near the edge of a thin plate. If the line $B = 0$—to the right of the y-axis—represents a thin charged plate, the field lines near it are given by the curves for various values of A. The physical situation is shown in Fig. 7–5.

Further examples are

$$F(\mathfrak{z}) = z^{3/2}, \tag{7.13}$$

which yields the field *outside* a rectangular corner

$$F(\mathfrak{z}) = \log \mathfrak{z}, \tag{7.14}$$

which yields the field for a line charge, and

$$F(\mathfrak{z}) = 1/\mathfrak{z}, \tag{7.15}$$

which gives the field for the two-dimensional analog of an electric dipole, i.e., two parallel line charges with opposite polarities, very close together.

We will not pursue this subject further in this course, but should emphasize that although the complex variable technique is often powerful, it is limited to two-dimensional problems; and also, it is an indirect method.

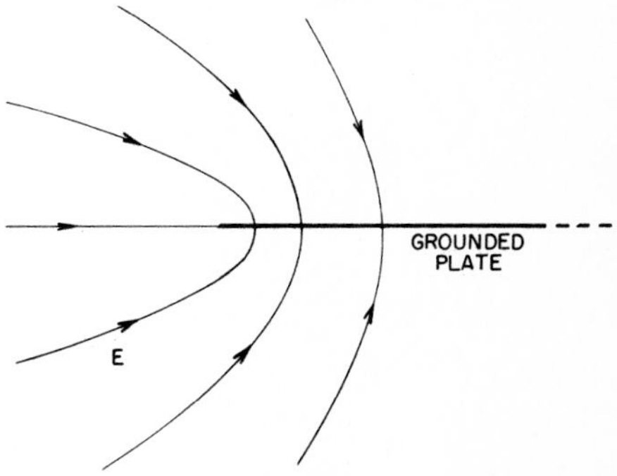

Fig. 7–5. The electric field near the edge of a thin grounded plate.

7–3 Plasma oscillations

We consider now some physical situations in which the field is determined neither by fixed charges nor by charges on conducting surfaces, but by a combination of two physical phenomena. In other words, the field will be governed simultaneously by two sets of equations: (1) the equations from electrostatics relating electric fields to charge distribution, and (2) an equation from another part of physics that determines the positions or motions of the charges in the presence of the field.

The first example that we will discuss is a dynamic one in which the motion of the charges is governed by Newton's laws. A simple example of such a situation occurs in a plasma, which is an ionized gas consisting of ions and free electrons distributed over a region in space. The ionosphere—an upper layer of the atmosphere—is an example of such a plasma. The ultraviolet rays from the sun knock electrons off the molecules of the air, creating free electrons and ions. In such a plasma the positive ions are very much heavier than the electrons, so we may neglect the ionic motion, in comparison to that of the electrons.

Let n_0 be the density of electrons in the undisturbed, equilibrium state. This must also be the density of positive ions, since the plasma is electrically neutral (when undisturbed). Now we suppose that the electrons are somehow moved from equilibrium and ask what happens. If the density of the electrons in one region is increased, they will repel each other and tend to return to their equilibrium positions. As the electrons move toward their original positions they pick up kinetic energy, and instead of coming to rest in their equilibrium configuration, they overshoot the mark. They will oscillate back and forth. The situation is similar to what occurs in sound waves, in which the restoring force is the gas pressure. In a plasma, the restoring force is the electrical force on the electrons.

To simplify the discussion, we will worry only about a situation in which the motions are all in one dimension, say x. Let us suppose that the electrons originally at x are, at the instant t, displaced from their equilibrium positions by a small amount $s(x, t)$. Since the electrons have been displaced, their density will, in general, be changed. The change in density is easily calculated. Referring to Fig. 7–6, the electrons initially contained between the two planes a and b have moved and are now contained between the planes a' and b'. The number of electrons that were between a and b is proportional to $n_0\Delta x$; the *same* number are now contained in the space whose width is $\Delta x + \Delta s$. The density has changed to

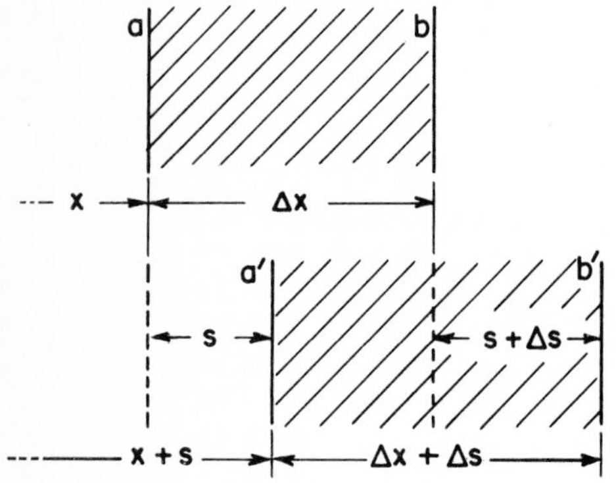

Fig. 7–6. Motion in a plasma wave. The electrons at the plane a move to a', and those at b move to b'.

$$n = \frac{n_0\Delta x}{\Delta x + \Delta s} = \frac{n_0}{1 + (\Delta s/\Delta x)}. \tag{7.16}$$

If the change in density is small, we can write [using the binomial expansion for $(1 + \epsilon)^{-1}$]

$$n = n_0\left(1 - \frac{\Delta s}{\Delta x}\right). \tag{7.17}$$

We assume that the positive ions do not move appreciably (because of the much larger inertia), so their density remains n_0. Each electron carries the charge $-q_e$, so the average charge density at any point is given by

$$\rho = -(n - n_0)q_e,$$

or

$$\rho = n_0 q_e \frac{ds}{dx} \tag{7.18}$$

(where we have written the differential form for $\Delta s/\Delta x$).

The charge density is related to the electric field by Maxwell's equations, in particular,

$$\nabla \cdot \boldsymbol{E} = \frac{\rho}{\epsilon_0}. \tag{7.19}$$

If the problem is indeed one-dimensional (and if there are no other fields but the one due to the displacements of the electrons), the electric field $\boldsymbol{E}$ has a single component E_x. Equation (7.19), together with (7.18), gives

$$\frac{\partial E_x}{\partial x} = \frac{n_0 q_e}{\epsilon_0}\frac{\partial s}{\partial x}. \tag{7.20}$$

Integrating Eq. (7.20) gives

$$E_x = \frac{n_0 q_e}{\epsilon_0} s + K. \tag{7.21}$$

Since $E_x = 0$ when $s = 0$, the integration constant K is zero.

The force on an electron in the displaced position is

$$F_x = -\frac{n_0 q_e^2}{\epsilon_0} s, \tag{7.22}$$

a restoring force proportional to the displacement s of the electron. This leads to a harmonic oscillation of the electrons. The equation of motion of a displaced electron is

$$m_e \frac{d^2 s}{dt^2} = -\frac{n_0 q_e^2}{\epsilon_0} s. \tag{7.23}$$

We find that s will vary harmonically. Its time variation will be as $\cos \omega t$, or—using the exponential notation of Vol. I—as

$$e^{i\omega_p t}. \tag{7.24}$$

The frequency of oscillation ω_p is determined from (7.23):

$$\omega_p^2 = \frac{n_0 q_e^2}{\epsilon_0 m_e}, \tag{7.25}$$

and is called the *plasma frequency*. It is a characteristic number of the plasma.

When dealing with electron charges many people prefer to express their answers in terms of a quantity e^2 defined by

$$e^2 = \frac{q_e^2}{4\pi\epsilon_0} = 2.3068 \times 10^{-28}\ \text{newton·meter}^2. \tag{7.26}$$

Using this convention, Eq. (7.25) becomes

$$\omega_p^2 = \frac{4\pi e^2 n_0}{m_e}, \tag{7.27}$$

which is the form you will find in most books.

Thus we have found that a disturbance of a plasma will set up free oscillations of the electrons about their equilibrium positions at the natural frequency ω_p, which is proportional to the square root of the density of the electrons. The plasma electrons behave like a resonant system, such as those we described in Chapter 23 of Vol. I.

This natural resonance of a plasma has some interesting effects. For example, if one tries to propagate a radiowave through the ionosphere, one finds that it can penetrate only if its frequency is higher than the plasma frequency. Otherwise the signal is reflected back. We must use high frequencies if we wish to communicate with a satellite in space. On the other hand, if we wish to communicate with a radio station beyond the horizon, we must use frequencies lower than the plasma frequency, so that the signal will be reflected back to the earth.

Another interesting example of plasma oscillations occurs in metals. In a metal we have a contained plasma of positive ions, and free electrons. The density n_0 is very high, so ω_p is also. But it should still be possible to observe the electron oscillations. Now, according to quantum mechanics, a harmonic oscillator with a natural frequency ω_p has energy levels which are separated by the the energy increment $\hbar\omega_p$. If, then, one shoots electrons through, say, an aluminum foil, and makes very careful measurements of the electron energies on the other side, one might expect to find that the electrons sometimes lose the energy $\hbar\omega_p$ to the plasma oscillations. This does indeed happen. It was first observed experimentally in 1936 that electrons with energies of a few hundred to a few thousand electron volts lost energy in jumps when scattering from or going through a thin metal foil. The effect was not understood until 1953 when Bohm and Pines* showed that the observations could be explained in terms of quantum excitations of the plasma oscillations in the metal.

---

* For some recent work and a bibliography see C. J. Powell and J. B. Swann, *Phys. Rev.* **115,** 869 (1959).

7–4 Colloidal particles in an electrolyte

We turn to another phenomenon in which the locations of charges is governed by a potential that arises in part from the same charges. The resulting effects influence in an important way the behavior of colloids. A colloid consists of a suspension in water of small charged particles which, though microscopic, from an atomic point of view are still very large. If the colloidal particles were not charged, they would tend to coagulate into large lumps; but because of their charge, they repel each other and remain in suspension.

Now if there is also some salt dissolved in the water, it will be dissociated into positive and negative ions. (Such a solution of ions is called an electrolyte.) The negative ions are attracted to the colloid particles (assuming their charge is positive) and the positive ions are repelled. We will determine how the ions which surround such a colloidal particle are distributed in space.

To keep the ideas simple, we will again solve only a one-dimensional case. If we think of a colloidal particle as a sphere having a very large radius—on an atomic scale!—we can then treat a small part of its surface as a plane. (Whenever one is trying to understand a new phenomenon it is a good idea to take a somewhat oversimplified model; then, having understood the problem with that model, one is better able to proceed to tackle the more exact calculation.)

We suppose that the distribution of ions generates a charge density $\rho(x)$, and an electrical potential ϕ, related by the electrostatic law $\nabla^2\phi = -\rho/\epsilon_0$ or, for fields that vary in only one dimension, by

$$\frac{d^2\phi}{dx^2} = -\frac{\rho}{\epsilon_0}. \tag{7.28}$$

Now supposing there were such a potential $\phi(x)$, how would the ions distribute themselves in it? This we can determine by the principles of statistical mechanics. Our problem then is to determine ϕ so that the resulting charge density from statistical mechanics *also* satisfies (7.28).

According to statistical mechanics (see Chapter 40, Vol. I), particles in thermal equilibrium in a force field are distributed in such a way that the density n of particles at the position x is given by

$$n(x) = n_0e^{-U(x)/kT}, \tag{7.29}$$

where $U(x)$ is the potential energy, k is Boltzmann's constant, and T is the absolute temperature.

We assume that the ions carry one electronic charge, positive or negative. At the distance x from the surface of a colloidal particle, a positive ion will have potential energy $q_e\phi(x)$, so that

$$U(x) = q_e\phi(x).$$

The density of positive ions, n_+, is then

$$n_+(x) = n_0e^{-q_e\phi(x)/kT}.$$

Similarly, the density of negative ions is

$$n_-(x) = n_0e^{+q_e\phi(x)/kT}.$$

The total charge density is

$$\rho = q_en_+ - q_en_-,$$

or

$$\rho = q_en_0(e^{-q_e\phi/kT} - e^{+q_e\phi/kT}). \tag{7.30}$$

Combining this with Eq. (7.28), we find that the potential ϕ must satisfy

$$\frac{d^2\phi}{dx^2} = -\frac{q_en_0}{\epsilon_0}(e^{-q_e\phi/kT} - e^{+q_e\phi/kT}). \tag{7.31}$$

This equation is readily solved in general [multiply both sides by $2(d\phi/dx)$, and integrate with respect to x], but to keep the problem as simple as possible, we will consider here only the limiting case in which the potentials are small or the temperature T is high. The case where ϕ is small corresponds to a dilute solution. For these cases the exponent is small, and we can approximate

$$e^{\pm q_e\phi/kT} = 1 \pm \frac{q_e\phi}{kT}. \tag{7.32}$$

Equation (7.31) then gives

$$\frac{d^2\phi}{dx^2} = +\frac{2n_0q_e^2}{\epsilon_0 kT}\,\phi(x). \tag{7.33}$$

Notice that this time the sign on the right is positive. The solutions for ϕ are not oscillatory, but exponential.

The general solution of Eq. (7.33) is

$$\phi = Ae^{-x/D} + Be^{+x/D}, \tag{7.34}$$

with

$$D^2 = \frac{\epsilon_0 kT}{2n_0q^2}. \tag{7.35}$$

The constants A and B must be determined from the conditions of the problem. In our case, B must be zero; otherwise the potential would go to infinity for large x. So we have that

$$\phi = Ae^{-x/D}, \tag{7.36}$$

in which A is the potential at $x = 0$, the surface of the colloidal particle.

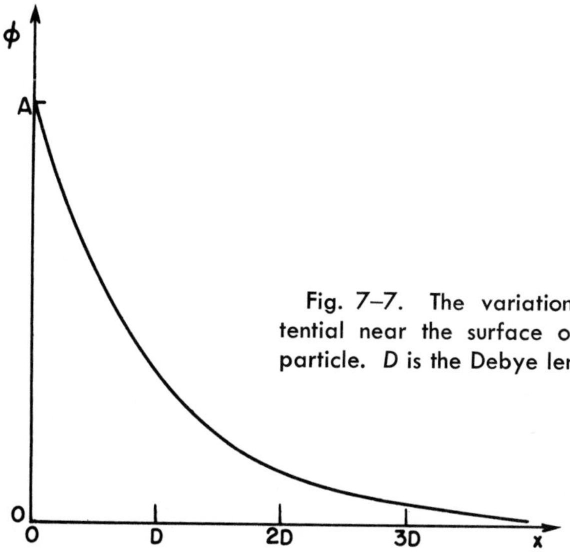

Fig. 7–7. The variation of the potential near the surface of a colloidal particle. D is the Debye length.

The potential decreases by a factor $1/e$ each time the distance increases by D, as shown in the graph of Fig. 7–7. The number D is called the *Debye length*, and is a measure of the thickness of the ion sheath that surrounds a large charged particle in an electrolyte. Equation (7.36) says that the sheath gets thinner with increasing concentration of the ions (n_0) or with decreasing temperature.

The constant A in Eq. (7.36) is easily obtained if we know the surface charge σ on the colloid particle. We know that

$$E_n = E_x(0) = \frac{\sigma}{\epsilon_0}. \tag{7.37}$$

But E is also the gradient of ϕ:

$$E_x(0) = -\left.\frac{\partial\phi}{\partial x}\right|_0 = +\frac{A}{D}, \tag{7.38}$$

from which we get

$$A = \frac{\sigma D}{\epsilon_0}. \tag{7.39}$$

Using this result in (7.36), we find (by taking $x = 0$) that the potential of the colloidal particle is

$$\phi(0) = \frac{\sigma D}{\epsilon_0}. \tag{7.40}$$

You will notice that this potential is the same as the potential difference across a condenser with a plate spacing D and a surface charge density σ.

We have said that the colloidal particles are kept apart by their electrical repulsion. But now we see that the field a little way from the surface of a particle is reduced by the ion sheath that collects around it. If the sheaths get thin enough, the particles have a good chance of knocking against each other. They will then stick, and the colloid will coagulate and precipitate out of the liquid. From our analysis, we understand why adding enough salt to a colloid should cause it to precipitate out. The process is called "salting out a colloid."

Another interesting example is the effect that a salt solution has on protein molecules. A protein molecule is a long, complicated, and flexible chain of amino acids. The molecule has various charges on it, and it sometimes happens that there is a net charge, say negative, which is distributed along the chain. Because of mutual repulsion of the negative charges, the protein chain is kept stretched out. Also, if there are other similar chain molecules present in the solution, they will be kept apart by the same repulsive effects. We can, therefore, have a suspension of chain molecules in a liquid. But if we add salt to the liquid we change the properties of the suspension. As salt is added to the solution, decreasing the Debye distance, the chain molecules can approach one another, and can also coil up. If enough salt is added to the solution, the chain molecules will precipitate out of the solution. There are many chemical effects of this kind that can be understood in terms of electrical forces.

7–5 The electrostatic field of a grid

As our last example, we would like to describe another interesting property of electric fields. It is one which is made use of in the design of electrical instruments, in the construction of vacuum tubes, and for other purposes. This is the character of the electric field near a grid of charged wires. To make the problem as simple as possible, let us consider an array of parallel wires lying in a plane, the wires being infinitely long and with a uniform spacing between them.

If we look at the field a large distance above the plane of the wires, we see a constant electric field, just as though the charge were uniformly spread over a plane. As we approach the grid of wires, the field begins to deviate from the uniform field we found at large distances from the grid. We would like to estimate how close to the grid we have to be in order to see appreciable variations in the potential. Figure 7–8 shows a rough sketch of the equipotentials at various distances from the grid. The closer we get to the grid, the larger the variations. As we travel parallel to the grid, we observe that the field fluctuates in a periodic manner.

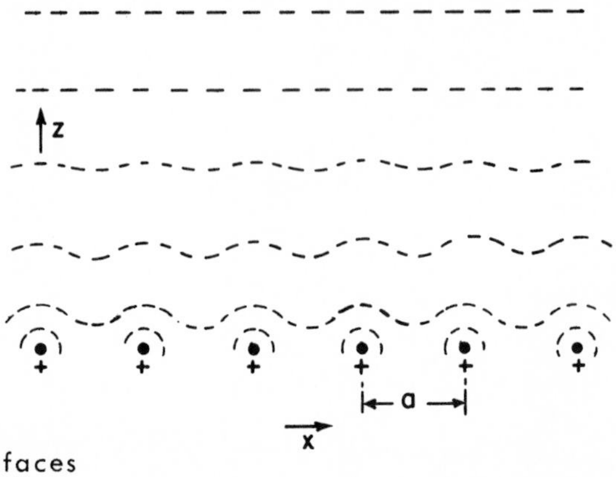

Fig. 7–8. Equipotential surfaces above a uniform grid of charged wires.

Now we have seen (Chapter 50, Vol. I) that any periodic quantity can be expressed as a sum of sine waves (Fourier's theorem). Let's see if we can find a suitable harmonic function that satisfies our field equations.

If the wires lie in the xy-plane and run parallel to the y-axis, then we might try terms like

$$\phi(x, z) = F_n(z) \cos \frac{2\pi n x}{a}, \tag{7.41}$$

where a is the spacing of the wires and n is the harmonic number. (We have assumed long wires, so there should be no variation with y.) A complete solution would be made up of a sum of such terms for $n = 1, 2, 3, \ldots$.

If this is to be a valid potential, it must satisfy Laplace's equation in the region above the wires (where there are no charges). That is,

$$\frac{\partial^2 \phi}{\partial x^2} + \frac{\partial^2 \phi}{\partial z^2} = 0.$$

Trying this equation on the ϕ in (7.41), we find that

$$-\frac{4\pi^2 n^2}{a^2} F_n(z) \cos \frac{2\pi n x}{a} + \frac{d^2 F_n}{dz^2} \cos \frac{2\pi n x}{a} = 0, \tag{7.42}$$

or that $F_n(z)$ must satisfy

$$\frac{d^2 F_n}{dz^2} = \frac{4\pi^2 n^2}{a^2} F_n. \tag{7.43}$$

So we must have

$$F_n = A_n e^{-z/z_0}, \tag{7.44}$$

where

$$z_0 = \frac{a}{2\pi n}. \tag{7.45}$$

We have found that if there is a Fourier component of the field of harmonic n, *that* component will decrease exponentially with a characteristic distance $z_0 = a/2\pi n$. For the first harmonic ($n = 1$), the amplitude falls by the factor $e^{-2\pi}$ (a large decrease) each time we increase z by one grid spacing a. The other harmonics fall off even more rapidly as we move away from the grid. We see that if we are only a few times the distance a away from the grid, the field is very nearly uniform, i.e., the oscillating terms are small. There would, of course, always remain the "zero harmonic" field

$$\phi_0 = -E_0 z$$

to give the uniform field at large z. For a complete solution, we would combine this term with a sum of terms like (7.41) with F_n from (7.44). The coefficients A_n would be adjusted so that the total sum would, when differentiated, give an electric field that would fit the charge density λ of the grid wires.

The method we have just developed can be used to explain why electrostatic shielding by means of a screen is often just as good as with a solid metal sheet. Except within a distance from the screen a few times the spacing of the screen wires, the fields inside a closed screen are zero. We see why copper screen—lighter and cheaper than copper sheet—is often used to shield sensitive electrical equipment from external disturbing fields.

8

Electrostatic Energy

8–1 The electrostatic energy of charges. A uniform sphere

Review: Chapter 4, Vol. I, *Conservation of Energy*
Chapters 13 and 14, Vol. I, *Work and Potential Energy*

In the study of mechanics, one of the most interesting and useful discoveries was the law of the conservation of energy. The expressions for the kinetic and potential energies of a mechanical system helped us to discover connections between the states of a system at two different times without having to look into the details of what was occurring in between. We wish now to consider the energy of electrostatic systems. In electricity also the principle of the conservation of energy will be useful for discovering a number of interesting things.

The law of the energy of interaction in electrostatics is very simple; we have, in fact, already discussed it. Suppose we have two charges q_1 and q_2 separated by the distance r_{12}. There is some energy in the system, because a certain amount of work was required to bring the charges together. We have already calculated the work done in bringing two charges together from a large distance. It is

$$\frac{q_1q_2}{4\pi\epsilon_0 r_{12}}. \tag{8.1}$$

We also know, from the principle of superposition, that if we have many charges present, the total force on any charge is the sum of the forces from the others. It follows, therefore, that the total energy of a system of a number of charges is the sum of terms due to the mutual interaction of each pair of charges. If q_i and q_j are any two of the charges and r_{ij} is the distance between them (Fig. 8–1), the energy of that particular pair is

$$\frac{q_iq_j}{4\pi\epsilon_0 r_{ij}}. \tag{8.2}$$

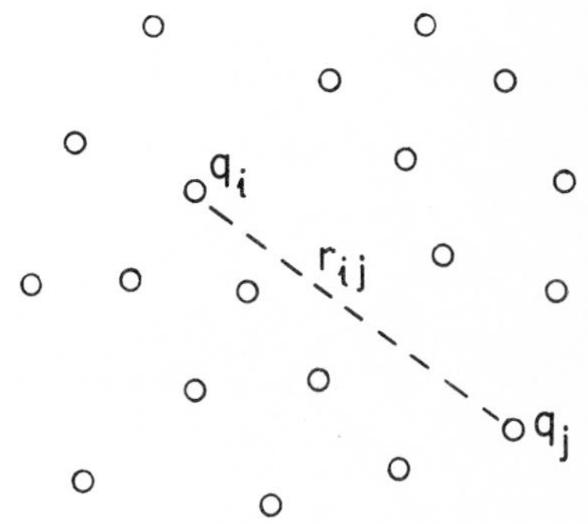

Fig. 8–1. The electrostatic energy of a system of particles is the sum of the electrostatic energy of each pair.

The total electrostatic energy U is the sum of the energies of all possible pairs of charges:

$$U = \sum_{\text{all pairs}} \frac{q_iq_j}{4\pi\epsilon_0 r_{ij}}. \tag{8.3}$$

If we have a distribution of charge specified by a charge density ρ, the sum of Eq. (8.3) is, of course, to be replaced by an integral.

We shall concern ourselves with two aspects of this energy. One is the *application* of the concept of energy to electrostatic problems; the other is the *evaluation* of the energy in different ways. Sometimes it is easier to compute the work done for some special case than to evaluate the sum in Eq. (8.3), or the corresponding integral. As an example, let us calculate the energy required to assemble a sphere of charge with a uniform charge density. The energy is just the work done in gathering the charges together from infinity.

Imagine that we assemble the sphere by building up a succession of thin spherical layers of infinitesimal thickness. At each stage of the process, we gather a small amount of charge and put it in a thin layer from r to $r + dr$. We continue the process until we arrive at the final radius a (Fig. 8–2). If Q_r is the charge of the sphere when it has been built up to the radius r, the work done in bringing a charge dQ to it is

$$dU = \frac{Q_r\,dQ}{4\pi\epsilon_0 r}. \tag{8.4}$$

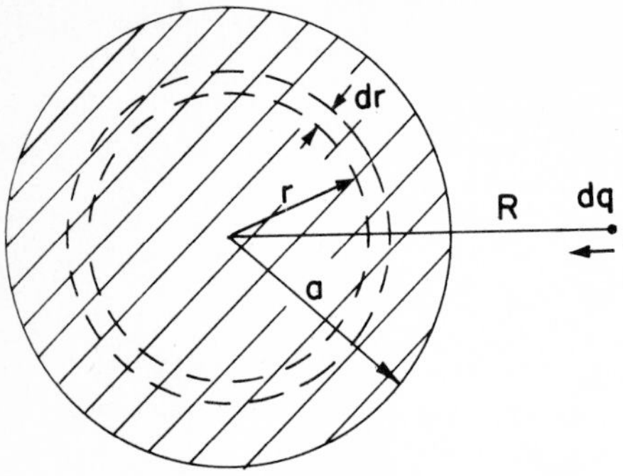

Fig. 8–2. The energy of a uniform sphere of charge can be computed by imagining that it is assembled from successive spherical shells.

If the density of charge in the sphere is ρ, the charge Q_r is

$$Q_r = \rho \cdot \frac{4}{3}\pi r^3,$$

and the charge dQ is

$$dQ = \rho \cdot 4\pi r^2\,dr.$$

Equation (8.4) becomes

$$dU = \frac{4\pi\rho^2 r^4\,dr}{3\epsilon_0}. \tag{8.5}$$

The total energy required to assemble the sphere is the integral of dU from $r = 0$ to $r = a$, or

$$U = \frac{4\pi\rho^2 a^5}{15\epsilon_0}. \tag{8.6}$$

Or if we wish to express the result in terms of the total charge Q of the sphere,

$$U = \frac{3}{5}\,\frac{Q^2}{4\pi\epsilon_0 a}. \tag{8.7}$$

The energy is proportional to the square of the total charge and inversely proportional to the radius. We can also interpret Eq. (8.7) as saying that the average of $(1/r_{ij})$ for all pairs of points in the sphere is $3/5a$.

8–2 The energy of a condenser. Forces on charged conductors

We consider now the energy required to charge a condenser. If the charge Q has been taken from one of the conductors of a condenser and placed on the other, the potential difference between them is

$$V = \frac{Q}{C}, \tag{8.8}$$

where C is the capacity of the condenser. How much work is done in charging the condenser? Proceeding as for the sphere, we imagine that the condenser has been charged by transferring charge from one plate to the other in small increments dQ. The work required to transfer the charge dQ is

$$dU = V\,dQ.$$

Taking V from Eq. (8.8), we write

$$dU = \frac{Q\,dQ}{C}.$$

Or integrating from zero charge to the final charge Q, we have

$$U = \frac{1}{2}\,\frac{Q^2}{C}. \tag{8.9}$$

This energy can also be written as

$$U = \tfrac{1}{2}CV^2. \tag{8.10}$$

Recalling that the capacity of a conducting sphere (relative to infinity) is

$$C_{\text{sphere}} = 4\pi\epsilon_0 a,$$

we can immediately get from Eq. (8.9) the energy of a charged sphere,

$$U = \frac{1}{2}\,\frac{Q^2}{4\pi\epsilon_0 a}. \tag{8.11}$$

This, of course, is also the energy of a thin *spherical shell* of total charge Q and is just 5/6 of the energy of a *uniformly charged* sphere, Eq. (8.7).

We now consider applications of the idea of electrostatic energy. Consider the following questions: What is the force between the plates of a condenser? Or what is the torque about some axis of a charged conductor in the presence of another with opposite charge? Such questions are easily answered by using our result Eq. (8.9) for electrostatic energy of a condenser, together with the principle of virtual work (Chapters 4, 13, and 14 of Vol. I).

Let's use this method for determining the force between the plates of a parallel-plate condenser. If we imagine that the spacing of the plates is increased by the small amount Δz, then the mechanical work done from the outside in moving the plates would be

$$\Delta W = F\,\Delta z, \tag{8.12}$$

where F is the force between the plates. This work must be equal to the change in the electrostatic energy of the condenser.

By Eq. (8.9), the energy of the condenser was originally

$$U = \frac{1}{2}\frac{Q^2}{C}.$$

The change in energy (if we do not let the charge change) is

$$\Delta U = \frac{1}{2}\,Q^2\,\Delta\left(\frac{1}{C}\right). \tag{8.13}$$

Equating (8.12) and (8.13), we have

$$F\,\Delta z = \frac{Q^2}{2}\,\Delta\left(\frac{1}{C}\right). \tag{8.14}$$

This can also be written as

$$F\,\Delta z = -\frac{Q^2}{2C^2}\,\Delta C. \tag{8.15}$$

The force, of course, results from the attraction of the charges on the plates, but we see that we do not have to worry in detail about how they are distributed; everything we need is taken care of in the capacity C.

It is easy to see how the idea is extended to conductors of any shape, and for other components of the force. In Eq. (8.14), we replace F by the component we are looking for, and we replace Δz by a small displacement in the corresponding direction. Or if we have an electrode with a pivot and we want to know the torque τ, we write the virtual work as

$$\Delta W = \tau\,\Delta\theta,$$

where $\Delta\theta$ is a small angular displacement. Of course, $\Delta(1/C)$ must be the change in $1/C$ which corresponds to $\Delta\theta$. We could, in this way, find the torque on the movable plates in a variable condenser of the type shown in Fig. 8–3.

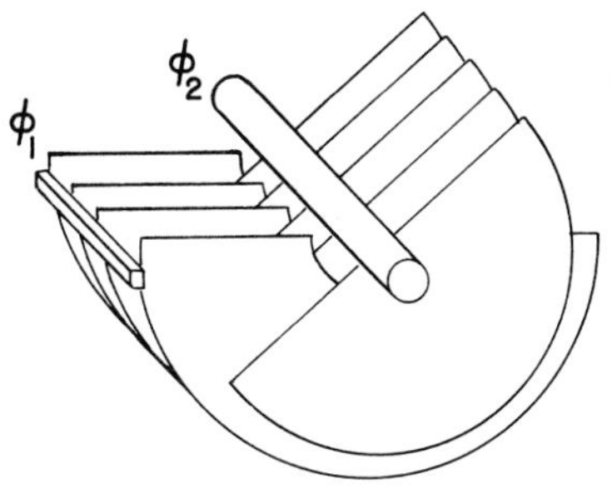

Fig. 8–3. What is the torque on a variable capacitor?

Returning to the special case of a parallel-plate condenser, we can use the formula we derived in Chapter 6 for the capacity:

$$\frac{1}{C} = \frac{d}{\epsilon_0 A}, \tag{8.16}$$

where A is the area of each plate. If we increase the separation by Δz,

$$\Delta\left(\frac{1}{C}\right) = \frac{\Delta z}{\epsilon_0 A}.$$

From Eq. (8.14) we get that the force between the plates is

$$F = \frac{Q^2}{2\epsilon_0 A}. \tag{8.17}$$

Let's look at Eq. (8.17) a little more closely and see if we can tell how the force arises. If for the charge on one plate we write

$$Q = \sigma A,$$

Eq. (8.17) can be rewritten as

$$F = \frac{1}{2}\,Q\,\frac{\sigma}{\epsilon_0}.$$

Or, since the electric field between the plates is

$$E_0 = \frac{\sigma}{\epsilon_0},$$

then

$$F = \tfrac{1}{2}QE_0. \tag{8.18}$$

One would immediately guess that the force acting on one plate is the charge Q on the plate times the field acting on the charge. But we have a surprising factor of one-half. The reason is that E_0 is not the field *at* the charges. If we imagine that the charge at the surface of the plate occupies a thin layer, as indicated in Fig. 8–4, the field will vary from zero at the inner boundary of the layer to E_0 in the space outside of the plate. The average field acting on the surface charges is $E_0/2$. That is why the factor one-half is in Eq. (8.18).

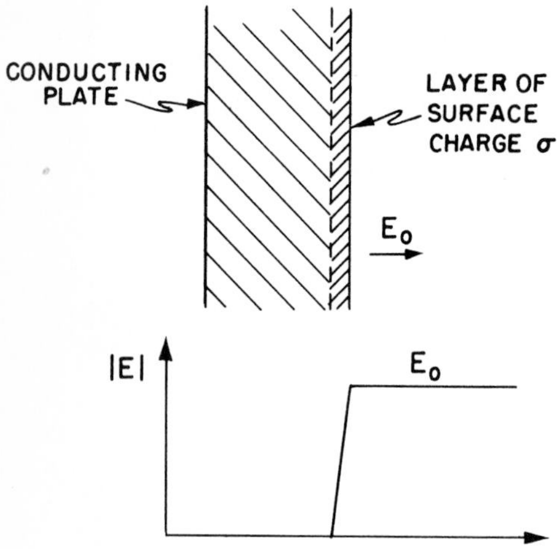

Fig. 8–4. The field at the surface of a conductor varies from zero to $E_0 = \sigma/\epsilon_0$, as one passes through the layer of surface charge.

You should notice that in computing the virtual work we have assumed that the charge on the condenser was constant—that it was not electrically connected to other objects, and so the total charge could not change.

Suppose we had imagined that the condenser was held at a constant potential difference as we made the virtual displacement. Then we should have taken

$$U = \tfrac{1}{2}CV^2$$

and in place of Eq. (8.15) we would have had

$$F\Delta z = \tfrac{1}{2}V^2\,\Delta C,$$

which gives a force equal in magnitude to the one in Eq. (8.15) (because $V = Q/C$), but with the opposite sign! Surely the force between the condenser plates doesn't reverse in sign as we disconnect it from its charging source. Also, we know that two plates with opposite electrical charges must attract. The principle of virtual work has been incorrectly applied in the second case—we have not taken into account the virtual work done on the charging source. That is, to keep the potential constant at V as the capacity changes, a charge $V\,\Delta C$ must be supplied by a source of charge. But this charge is supplied at a potential V, so the work done by the electrical system which keeps the potential constant is $V^2\,\Delta C$. The mechanical work $F\,\Delta z$ *plus* this electrical work $V^2\,\Delta C$ together make up the change in the total energy $\frac{1}{2}V^2\,\Delta C$ of the condenser. Therefore $F\,\Delta z$ is $-\frac{1}{2}V^2\,\Delta C$, as before.

8–3 The electrostatic energy of an ionic crystal

We now consider an application of the concept of electrostatic energy in atomic physics. We cannot easily measure the forces between atoms, but we are often interested in the energy differences between one atomic arrangement and another, as, for example, the energy of a chemical change. Since atomic forces are basically electrical, chemical energies are in large part just electrostatic energies.

Let's consider, for example, the electrostatic energy of an ionic lattice. An ionic crystal like NaCl consists of positive and negative ions which can be thought of as rigid spheres. They attract electrically until they begin to touch; then there is a repulsive force which goes up very rapidly if we try to push them closer together.

For our first approximation, therefore, we imagine a set of rigid spheres that represent the atoms in a salt crystal. The structure of the lattice has been determined by x-ray diffraction. It is a cubic lattice—like a three-dimensional checkerboard. Figure 8–5 shows a cross-sectional view. The spacing of the ions is 2.81 A ($= 2.81 \times 10^{-8}$ cm).

If our picture of this system is correct, we should be able to check it by asking the following question: How much energy will it take to pull all these ions apart—that is, to separate the crystal completely into ions? This energy should be equal to the heat of vaporization of NaCl plus the energy required to dissociate the molecules into ions. This total energy to separate NaCl to ions is determined experimentally to be 7.92 electron volts per molecule. Using the conversion

$$1 \text{ ev} = 1.602 \times 10^{-19} \text{ joule},$$

and Avogadro's number for the number of molecules in a mole,

$$N_0 = 6.02 \times 10^{23},$$

the energy of vaporization can also be given as

$$W = 7.64 \times 10^5 \text{ joules/mole}.$$

Physical chemists prefer for an energy unit the kilocalorie, which is 4190 joules; so that 1 ev per molecule is 23 kilocalories per mole. A chemist would then say that the dissociation energy of NaCl is

$$W = 183 \text{ kcal/mole}.$$

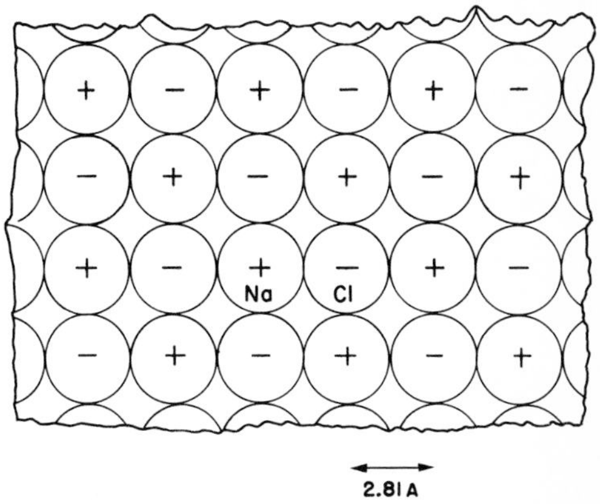

Fig. 8–5. Cross section of a salt crystal on an atomic scale. The checkerboard arrangement of Na and Cl ions is the same in the two cross sections perpendicular to the one shown. (See Vol. I, Fig. 1–7.)

Can we obtain this chemical energy theoretically by computing how much work it would take to pull apart the crystal? According to our theory, this work is the sum of the potential energies of all the pairs of ions. The easiest way to figure out this sum is to pick out a particular ion and compute its potential energy with each of the other ions. That will give us *twice* the energy per ion, because the energy belongs to the *pairs* of charges. If we want the energy to be associated with one particular ion, we should take half the sum. But we really want the energy *per molecule*, which contains two ions, so that the sum we compute will give directly the energy per molecule.

The energy of an ion with one of its nearest neighbors is e^2/a, where $e^2 = q_e^2/4\pi\epsilon_0$ and a is the center-to-center spacing between ions. (We are considering monovalent ions.) This energy is 5.12 ev, which we already see is going to give us a result of the correct order of magnitude. But it is still a long way from the infinite sum of terms we need.

Let's begin by summing all the terms from the ions along a straight line. Considering that the ion marked Na in Fig. 8–5 is our special ion, we shall consider first those ions on a horizontal line with it. There are two nearest Cl ions with negative charges, each at the distance a. Then there are two positive ions at the distance $2a$, etc. Calling the energy of this sum U_1, we write

$$\begin{aligned} U_1 &= \frac{e^2}{a}\left(-\frac{2}{1} + \frac{2}{2} - \frac{2}{3} + \frac{2}{4} + \cdots\right) \\ &= -\frac{2e^2}{a}\left(1 - \frac{1}{2} + \frac{1}{3} - \frac{1}{4} + \cdots\right). \end{aligned} \tag{8.19}$$

The series converges slowly, so it is difficult to evaluate numerically, but it is known to be equal to ln 2. So

$$U_1 = -\frac{2e^2}{a} \ln 2 = -1.386\,\frac{e^2}{a}. \tag{8.20}$$

Now consider the next adjacent line of ions above. The nearest is negative and at the distance a. Then there are two positives at the distance $\sqrt{2}\,a$. The next pair are at the distance $\sqrt{5}\,a$, the next at $\sqrt{10}\,a$, and so on. So for the whole line we get the series

$$\frac{e^2}{a}\left(-\frac{1}{1} + \frac{2}{\sqrt{2}} - \frac{2}{\sqrt{5}} + \frac{2}{\sqrt{10}} \cdots\right). \tag{8.21}$$

There are *four* such lines: above, below, in front, and in back. Then there are the four lines which are the nearest lines on diagonals, and on and on.

If you work patiently through for all the lines, and then take the sum, you find that the grand total is

$$U = -1.747 \frac{e^2}{a},$$

which is just somewhat more than what we obtained in (8.20) for the first line. Using $e^2/a = 5.12$ ev, we get

$$U = -8.94 \text{ ev}.$$

Our answer is about 10% above the experimentally observed energy. It shows that our idea that the whole lattice is held together by electrical Coulomb forces is fundamentally correct. This is the first time that we have obtained a specific property of a macroscopic substance from a knowledge of atomic physics. We will do much more later. The subject that tries to understand the behavior of bulk matter in terms of the laws of atomic behavior is called *solid-state physics.*

Now what about the error in our calculation? Why is it not exactly right? It is because of the repulsion between the ions at close distances. They are not perfectly rigid spheres, so when they are close together they are partly squashed. They are not very soft, so they squash only a little bit. Some energy, however, is used in deforming them, and when the ions are pulled apart this energy is released. The actual energy needed to pull the ions apart is a little less than the energy that we calculated; the repulsion helps in overcoming the electrostatic attraction.

Is there any way we can make an allowance for this contribution? We could if we knew the law of the repulsive force. We are not ready to analyze the details of this repulsive mechanism, but we can get some idea of its characteristics from some large-scale measurements. From a measurement of the *compressibility* of the whole crystal, it is possible to obtain a quantitative idea of the law of repulsion between the ions and therefore of its contribution to the energy. In this way it has been found that this contribution must be 1/9.4 of the contribution from the electrostatic attraction and, of course, of opposite sign. If we subtract this contribution from the pure electrostatic energy, we obtain 7.99 ev for the dissociation energy per molecule. It is much closer to the observed result of 7.92 ev, but still not in perfect agreement. There is one more thing we haven't taken into account: we have made no allowance for the kinetic energy of the crystal vibrations. If a correction is made for this effect, very good agreement with the experimental number is obtained. The ideas are then correct; the major contribution to the energy of a crystal like NaCl is electrostatic.

8–4 Electrostatic energy in nuclei

We will now take up another example of electrostatic energy in atomic physics, the electrical energy of atomic nuclei. Before we do this we will have to discuss some properties of the main forces (called nuclear forces) that hold the protons and neutrons together in a nucleus. In the early days of the discovery of nuclei—and of the neutrons and protons that make them up—it was hoped that the law of the strong, nonelectrical part of the force between, say, a proton and another proton would have some simple law, like the inverse square law of electricity. For once one had determined this law of force, and the corresponding ones between a proton and a neutron, and a neutron and a neutron, it would be possible to describe theoretically the complete behavior of these particles in nuclei. Therefore a big program was started for the study of the scattering of protons, in the hope of finding the law of force between them; but after thirty years of effort, nothing simple has emerged. A considerable knowledge of the force between proton and proton has been accumulated, but we find that the force is as complicated as it can possibly be.

What we mean by "as complicated as it can be" is that the force depends on as many things as it possibly can.

First, the force is not a simple function of the distance between the two protons. At large distances there is an attraction, but at closer distances there is a repulsion. The distance dependence is a complicated function, still imperfectly known.

Second, the force depends on the orientation of the protons' spin. The protons have a spin, and any two interacting protons may be spinning with their angular momenta in the same direction or in opposite directions. And the force is different when the spins are parallel from what it is when they are antiparallel, as in (a) and (b) of Fig. 8–6. The difference is quite large; it is not a small effect.

Third, the force is considerably different when the separation of the two protons is in the direction *parallel* to their spins, as in (c) and (d) of Fig. 8–6, than it is when the separation is in a direction *perpendicular* to the spins, as in (a) and (b).

Fourth, the force depends, as it does in magnetism, on the velocity of the protons, only much more strongly than in magnetism. And this velocity-dependent force is not a relativistic effect; it is strong even at speeds much less than the speed of light. Furthermore, this part of the force depends on other things besides the magnitude of the velocity. For instance, when a proton is moving near another proton, the force is different when the orbital motion has the same direction of rotation as the spin, as in (e) of Fig. 8–6, than when it has the opposite direction of rotation, as in (f). This is called the "spin orbit" part of the force.

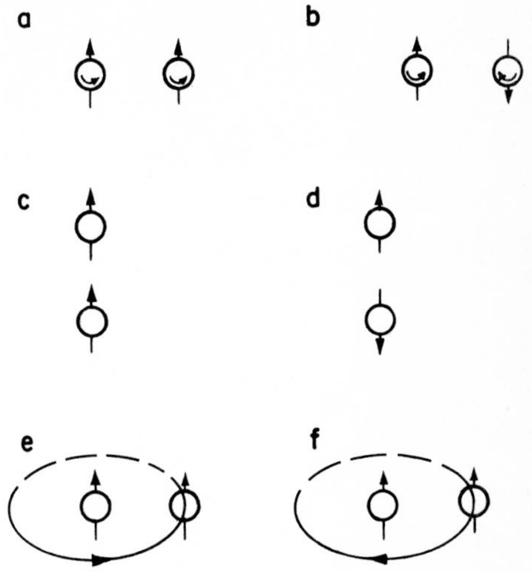

Fig. 8–6. The force between two protons depends on every possible parameter.

The force between a proton and a neutron and between a neutron and a neutron are also equally complicated. To this day we do not know the machinery behind these forces—that is to say, any simple way of understanding them.

There is, however, one important way in which the nucleon forces are *simpler* than they could be. That is that the *nuclear* force between two neutrons is the same as the force between a proton and a neutron, which is the same as the force between two protons! If, in any nuclear situation, we replace a proton by a neutron (or vice versa), the *nuclear interactions* are not changed. The "fundamental reason" for this equality is not known, but it is an example of an important principle that can be extended also to the interaction laws of other strongly interacting particles—such as the π-mesons and the "strange" particles.

This fact is nicely illustrated by the locations of the energy levels in similar nuclei. Consider a nucleus like B^{11} (boron-eleven), which is composed of five protons and six neutrons. In the nucleus the eleven particles interact with one another in a most complicated dance. Now, there is one configuration of all the possible interactions which has the lowest possible energy; this is the normal state of the nucleus, and is called the *ground state*. If the nucleus is disturbed (for example, by being struck by a high-energy proton or other particle) it can be put into any number of other configurations, called *excited states*, each of which will have a characteristic energy that is higher than that of the ground state. In nuclear physics research, such as is carried on with Van de Graaff generator (for example, in Caltech's Kellogg and Sloan Laboratories), the energies and other properties of these excited states are determined by experiment. The energies of the fifteen lowest known excited states of B^{11} are shown in a one-dimensional graph on the left half of Fig. 8–7. The lowest horizontal line represents the ground state. The first excited state has an energy 2.14 Mev higher than the ground state, the next an energy 4.46 Mev higher than the ground state, and so on. The study of nuclear physics attempts to find an explanation for this rather complicated pattern of energies; there is as yet, however, no complete general theory of such nuclear energy levels.

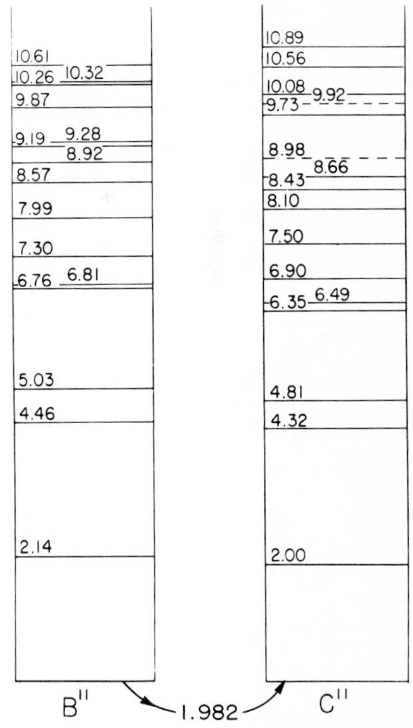

Fig. 8–7. The energy levels of B^{11} and C^{11} (energies in Mev). The ground state of C^{11} is 1.982 Mev higher than that of B^{11}.

If we replace one of the neutrons in B^{11} with a proton, we have the nucleus of an isotope of carbon, C^{11}. The energies of the lowest sixteen excited states of C^{11} have also been measured; they are shown in the right half of Fig. 8–7. (The broken lines indicate levels for which the experimental information is questionable.)

Looking at Fig. 8–7, we see a striking similarity between the pattern of the energy levels in the two nuclei. The first excited states are about 2 Mev above the ground states. There is a large gap of about 2.3 Mev to the second excited state, then a small jump of only 0.5 Mev to the third level. Again, between the fourth and fifth levels, a big jump; but between the fifth and sixth a tiny separation of the order of 0.1 Mev. And so on. After about the tenth level, the correspondence seems to become lost, but can still be seen if the levels are labeled with their other defining characteristics—for instance, their angular momentum and what they do to lose their extra energy.

The striking similarity of the pattern of the energy levels of B^{11} and C^{11} is surely not just a coincidence. It must reveal some physical law. It shows, in fact, that even in the complicated situation in a nucleus, replacing a neutron by a proton makes very little change. This can mean only that the neutron-neutron and proton-proton forces must be nearly identical. Only then would we expect the nuclear configurations with five protons and six neutrons to be the same as with six protons and five neutrons.

Notice that the properties of these two nuclei tell us nothing about the neutron-proton force; there are the same number of neutron-proton combinations in both nuclei. But if we compare two other nuclei, such as C^{14}, which has six protons and eight neutrons, with N^{14}, which has seven of each, we find a similar correspondence of energy levels. So we can conclude that the p-p, n-n, and p-n forces are identical in all their complexities. There is an unexpected principle in the laws of nuclear forces. Even though the force between each pair of nuclear particles is very complicated, the force between the three possible different pairs is the same.

But there are some small differences. The levels do not correspond exactly; also, the ground state of C^{11} has an absolute energy (its mass) which is higher than the ground state of B^{11} by 1.982 Mev. All the other levels are also higher in absolute energy by this same amount. So the forces are not exactly equal. But we know very well that the *complete* forces are not exactly equal; there is an *electrical* force between two protons because each has a positive charge, while between two neutrons there is no such electrical force. Can we perhaps explain the differences between B^{11} and C^{11} by the fact that the electrical interaction of the protons is different in the two cases? Perhaps even the remaining minor differences in the levels are caused by electrical effects? Since the nuclear forces are so much stronger than the electrical force, electrical effects would have only a small perturbing effect on the energies of the levels.

In order to check this idea, or rather to find out what the consequences of this idea are, we first consider the difference in the ground-state energies of the two nuclei. To take a very simple model, we suppose that the nuclei are spheres of radius r (to be determined), containing Z protons. If we consider that a nucleus is like a sphere with uniform charge density, we would expect the electrostatic energy (from Eq. 8.7) to be

$$U = \frac{3}{5}\frac{(Zq_e)^2}{4\pi\epsilon_0 r}, \tag{8.22}$$

where q_e is the elementary charge of the proton. Since Z is five for B^{11} and six for C^{11}, their electrostatic energies would be different.

With such a small number of protons, however, Eq. (8.22) is not quite correct. If we compute the electrical energy between all pairs of protons, considered as points which we assume to be nearly uniformly distributed throughout the sphere, we find that in Eq. (8.22) the quantity Z^2 should be replaced by $Z(Z - 1)$, so the energy is

$$U = \frac{3}{5}\frac{Z(Z-1)q_e^2}{4\pi\epsilon_0 r} = \frac{3}{5}\frac{Z(Z-1)e^2}{r}. \tag{8.23}$$

If we knew the nuclear radius r, we could use (8.23) to find the electrostatic energy difference between B^{11} and C^{11}. But let's do the opposite; let's instead use the observed energy difference to compute the radius, assuming that the energy difference is all electrostatic in origin.

That is, however, not quite right. The energy difference of 1.982 Mev between the ground states of B^{11} and C^{11} includes the rest energies—that is, the energy mc^2—of all the particles. In going from B^{11} to C^{11}, we replace a neutron by a proton, which has less mass. So part of the energy difference is the difference in the rest energies of a neutron and a proton, which is 0.784 Mev. The difference, to be accounted for by electrostatic energy, is thus more than 1.982 Mev; it is

$$1.982 + 0.784 = 2.786 \text{ Mev}.$$

Using this energy in Eq. (8.23), for the radius of either B^{11} or C^{11} we find

$$r = 3.12 \times 10^{-13} \text{ cm}. \tag{8.24}$$

Does this number have any meaning? To see whether it does, we should compare it with some other determination of the radius of these nuclei. For example, we can make another measurement of the radius of a nucleus by seeing how it scatters fast particles. From such measurements it has been found, in fact, that the *density* of matter in all nuclei is nearly the same, i.e., their volumes are proportional to the number of particles they contain. If we let A be the number of protons and neutrons in a nucleus (a number very nearly proportional to its mass), it is found that its radius is given by

$$r = A^{1/3} r_0, \tag{8.25}$$

where

$$r_0 = 1.2 \times 10^{-13} \text{ cm}. \tag{8.26}$$

From these measurements we find that the radius of a B^{11} (or a C^{11}) nucleus is expected to be

$$r = (1.2 \times 10^{-13})(11)^{1/3} = 2.7 \times 10^{-13} \text{ cm}.$$

Comparing this result with (8.24), we see that our assumptions that the energy difference between B^{11} and C^{11} is electrostatic is fairly good; the discrepancy is only about 15% (not bad for our first nuclear computation!).

The reason for the discrepancy is probably the following. According to the current understanding of nuclei, an even number of nuclear particles—in the case of B^{11}, five neutrons together with five protons—makes a kind of *core;* when one more particle is added to this core, it revolves around on the outside to make a new spherical nucleus, rather than being absorbed. If this is so, we should have taken a different electrostatic energy for the additional proton. We should have taken the excess energy of C^{11} over B^{11} to be just

$$\frac{Z_B q_e^2}{4\pi\epsilon_0 a},$$

which is the energy needed to add one more proton to the outside of the core. This number is just 5/6 of what Eq. (8.23) predicts, so the new prediction for the radius is 5/6 of (8.24), which is in much closer agreement with what is directly measured.

We can draw two conclusions from this agreement. One is that the electrical laws appear to be working at dimensions as small as 10^{-13} cm. The other is that we have verified the remarkable coincidence that the nonelectrical part of the forces between proton and proton, neutron and neutron, and proton and neutron are all equal.

8–5 Energy in the electrostatic field

We now consider other methods of calculating electrostatic energy. They can all be derived from the basic relation Eq. (8.3), the sum, over all pairs of charges, of the mutual energies of each charge-pair. First we wish to write an expression for the energy of a charge distribution. As usual, we consider that each volume element dV contains the element of charge $\rho\, dV$. Then Eq. (8.3) should be written

$$U = \frac{1}{2} \int\limits_{\substack{\text{all}\\\text{space}}} \frac{\rho(1)\rho(2)}{4\pi\epsilon_0 r_{12}}\, dV_1\, dV_2. \tag{8.27}$$

Notice the factor $\frac{1}{2}$, which is introduced because in the double integral over dV_1 and dV_2 we have counted all pairs of charge elements twice. (There is no convenient way of writing an integral that keeps track of the pairs so that each pair is counted only once.) Next we notice that the integral over dV_2 in (8.27) is just the potential at (1). That is,

$$\int \frac{\rho(2)}{4\pi\epsilon_0 r_{12}}\, dV_2 = \phi(1),$$

so that (8.27) can be written as

$$U = \frac{1}{2}\int \rho(1)\phi(1)\, dV_1.$$

Or, since the point (2) no longer appears, we can simply write

$$U = \frac{1}{2}\int \rho\phi\, dV. \tag{8.28}$$

This equation can be interpreted as follows. The potential energy of the charge $\rho\, dV$ is the product of this charge and the potential at the same point. The total energy is therefore the integral over $\phi\rho\, dV$. But there is again the factor $\frac{1}{2}$. It is still required because we are counting energies twice. The mutual energies of two charges is the charge of one times the potential at it due to the other. *Or*, it can be taken as the second charge times the potential at it from the first. Thus for two point charges we would write

$$U = q_1\phi(1) = q_1 \frac{q_2}{4\pi\epsilon_0 r_{12}}$$

or

$$U = q_2\phi(2) = q_2 \frac{q_1}{4\pi\epsilon_0 r_{12}}.$$

Notice that we could also write

$$U = \tfrac{1}{2}[q_1\phi(1) + q_2\phi(2)]. \tag{8.29}$$

The integral in (8.28) corresponds to the sum of both terms in the brackets of (8.29). That is why we need the factor $\frac{1}{2}$.

An interesting question is: Where is the electrostatic energy located? One might also ask: Who cares? What is the meaning of such a question? If there is a pair of interacting charges, the combination has a certain energy. Do we need to say that the energy is located at one of the charges or the other, or at both, or in between? These questions may not make sense because we really know only that the total energy is conserved. The idea that the energy is located *somewhere* is not necessary.

Yet suppose that it *did* make sense to say, in general, that energy is located at a certain place, as it does for heat energy. We might then *extend* our principle of the conservation of energy with the idea that if the energy in a given volume changes, we should be able to account for the change by the flow of energy into or out of that volume. You realize that our early statement of the principle of the conservation of energy is still perfectly all right if some energy disappears at one place and appears somewhere else far away without anything passing (that is, without any special phenomena occurring) in the space between. We are, therefore, now discussing an extension of the idea of the conservation of energy. We might call it a principle of the *local* conservation of energy. Such a principle would say that the energy in any given volume changes only by the amount that flows into or out of the volume. It is indeed possible that energy is conserved locally in such a way. If it is, we would have a much more detailed law than the simple statement of the conservation of total energy. It does turn out that in nature *energy is conserved locally*. We can find formulas for where the energy is located and how it travels from place to place.

There is also a *physical* reason why it is imperative that we be able to say where energy is located. According to the theory of gravitation, all mass is a source of gravitational attraction. We also know, by $E = mc^2$, that mass and energy are equivalent. All energy is, therefore, a source of gravitational force. If we could not locate the energy, we could not locate all the mass. We would not be able to say where the sources of the gravitational field are located. The theory of gravitation would be incomplete.

If we restrict ourselves to electrostatics there is really no way to tell where the energy is located. The complete Maxwell equations of electrodynamics give us much more information (although even then the answer is, strictly speaking, not unique.) We will therefore discuss this question in detail again in a later chapter. We will give you now only the result for the particular case of electrostatics. The energy is located in space, where the electric field is. This seems reasonable because we know that when charges are accelerated they radiate electric fields. We would like to say that when light or radiowaves travel from one point to another, they carry their energy with them. But there are no charges in the waves. So we would like to locate the energy where the electromagnetic field is and not at the charges from which it came. We thus describe the energy, not in terms of the charges, but in terms of the fields they produce. We can, in fact, show that Eq. (8.28) is *numerically* equal to

$$U = \frac{\epsilon_0}{2}\int \boldsymbol{E}\cdot\boldsymbol{E}\,dV. \tag{8.30}$$

We can then interpret this formula as saying that when an electric field is present, there is located in space an energy whose *density* (energy per unit volume) is

$$u = \frac{\epsilon_0}{2}\,\boldsymbol{E}\cdot\boldsymbol{E} = \frac{\epsilon_0 E^2}{2}\cdot \tag{8.31}$$

This idea is illustrated in Fig. 8–8.

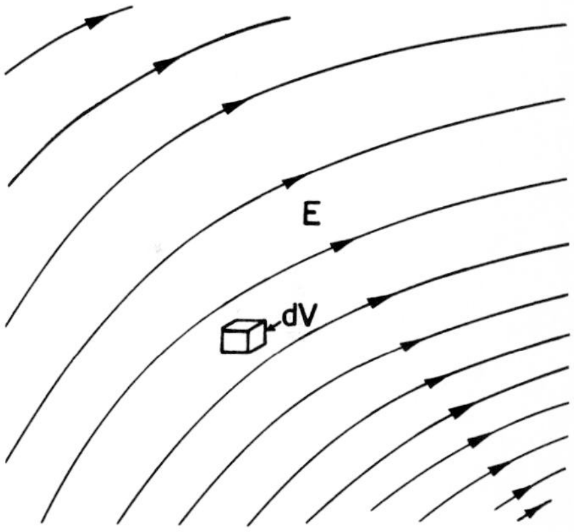

Fig. 8–8. Each volume element $dV = dx\,dy\,dz$ in an electric field contains the energy $(\epsilon_0/2)E^2\,dV$.

To show that Eq. (8.30) is consistent with our laws of electrostatics, we begin by introducing into Eq. (8.28) the relation between ρ and ϕ that we obtained in Chapter 6:

$$\rho = -\epsilon_0\,\nabla^2\phi.$$

We get

$$U = -\frac{\epsilon_0}{2}\int \phi\,\nabla^2\phi\,dV. \tag{8.32}$$

Writing out the components of the integrand, we see that

$$\begin{aligned}
\phi\,\nabla^2\phi &= \phi\left(\frac{\partial^2\phi}{\partial x^2}+\frac{\partial^2\phi}{\partial y^2}+\frac{\partial^2\phi}{\partial z^2}\right)\\
&= \frac{\partial}{\partial x}\left(\phi\,\frac{\partial\phi}{\partial x}\right)-\left(\frac{\partial\phi}{\partial x}\right)^2+\frac{\partial}{\partial y}\left(\phi\,\frac{\partial\phi}{\partial y}\right)-\left(\frac{\partial\phi}{\partial y}\right)^2+\frac{\partial}{\partial z}\left(\phi\,\frac{\partial\phi}{\partial z}\right)-\left(\frac{\partial\phi}{\partial z}\right)^2\\
&= \boldsymbol{\nabla}\cdot(\phi\,\boldsymbol{\nabla}\phi)-(\boldsymbol{\nabla}\phi)\cdot(\boldsymbol{\nabla}\phi).
\end{aligned} \tag{8.33}$$

Our energy integral is then

$$U = \frac{\epsilon_0}{2}\int(\boldsymbol{\nabla}\phi)\cdot(\boldsymbol{\nabla}\phi)\,dV - \frac{\epsilon_0}{2}\int\boldsymbol{\nabla}\cdot(\phi\,\boldsymbol{\nabla}\phi)\,dV.$$

We can use Gauss' theorem to change the second integral into a surface integral:

$$\int_{\text{vol.}}\boldsymbol{\nabla}\cdot(\phi\,\boldsymbol{\nabla}\phi)\,dV = \int_{\text{surface}}(\phi\,\boldsymbol{\nabla}\phi)\cdot\boldsymbol{n}\,da. \tag{8.34}$$

We evaluate the surface integral in the case that the surface goes to infinity (so the volume integrals become integrals over all space), supposing that all the charges are located within some finite distance. The simple way to proceed is to take a spherical surface of enormous radius R whose center is at the origin of coordinates. We know that when we are very far away from all charges, ϕ varies as $1/R$ and $\boldsymbol{\nabla}\phi$ as $1/R^2$. (Both will decrease even faster with R if there the net charge in the distribution is zero.) Since the surface area of the large sphere increases as R^2, we see that the surface integral falls off as $(1/R)(1/R^2)R^2 = (1/R)$ as the radius of the sphere increases. So if we include all space in our integration $(R \to \infty)$, the surface integral goes to zero and we have that

$$U = \frac{\epsilon_0}{2} \int\limits_{\substack{\text{all}\\\text{space}}} (\nabla\phi) \cdot (\nabla\phi)\, dV = \frac{\epsilon_0}{2} \int\limits_{\substack{\text{all}\\\text{space}}} \boldsymbol{E} \cdot \boldsymbol{E}\, dV. \tag{8.35}$$

We see that it is possible for us to represent the energy of any charge distribution as being the integral over an energy density located in the field.

8–6 The energy of a point charge

Our new relation, Eq. (8.35), says that even a single point charge q will have some electrostatic energy. In this case, the electric field is given by

$$E = \frac{q}{4\pi\epsilon_0 r^2}\cdot$$

So the energy density at the distance r from the charge is

$$\frac{\epsilon_0 E^2}{2} = \frac{q^2}{32\pi^2\epsilon_0 r^4}\cdot$$

We can take for an element of volume a spherical shell of thickness dr and area $4\pi r^2$. The total energy is

$$U = \int\limits_{r=0}^{\infty} \frac{q^2}{8\pi\epsilon_0 r^2}\, dr = -\frac{q^2}{8\pi\epsilon_0}\frac{1}{r}\bigg|_{r=0}^{r=\infty}\cdot \tag{8.36}$$

Now the limit at $r = \infty$ gives no difficulty. But for a point charge we are supposed to integrate down to $r = 0$, which gives an infinite integral. Equation (8.35) says that there is an infinite amount of energy in the field of a point charge, although we began with the idea that there was energy only *between* point charges. In our original energy formula for a collection of point charges (Eq. 8.3), we did not include any interaction energy of a charge with itself. What has happened is that when we went over to a continuous distribution of charge in Eq. (8.27), we counted the energy of interaction of every *infinitesimal* charge with all other infinitesimal charges. The same account is included in Eq. (8.35), so when we apply it to a *finite* point charge, we are including the energy it would take to assemble that charge from infinitesimal parts. You will notice, in fact, that we would also get the result in Eq. (8.36) if we used our expression (8.11) for the energy of a charged sphere and let the radius tend toward zero.

We must conclude that the idea of locating the energy in the field is inconsistent with the assumption of the existence of point charges. One way out of the difficulty would be to say that elementary charges, such as an electron, are not points but are really small distributions of charge. Alternatively, we could say that there is something wrong in our theory of electricity at very small distances, or with the idea of the local conservation of energy. There are difficulties with either point of view. These difficulties have never been overcome; they exist to this day. Sometime later, when we have discussed some additional ideas, such as the momentum in an electromagnetic field, we will give a more complete account of these fundamental difficulties in our understanding of nature.

9

Electricity in the Atmosphere

Reference: Chalmers, J. Alan, *Atmospheric Electricity*, Pergamon Press, London (1957).

9–1 The electric potential gradient of the amosphere

On an ordinary day over flat desert country, or over the sea, as one goes upward from the surface of the ground the electric potential increases by about 100 volts per meter. Thus there is a vertical electric field $\boldsymbol{E}$ of 100 volts/m in the air. The sign of the field corresponds to a negative charge on the earth's surface. This means that outdoors the potential at the height of your nose is 200 volts higher than the potential at your feet! You might ask: "Why don't we just stick a pair of electrodes out in the air one meter apart and use the 100 volts to power our electric lights?" Or you might wonder: "If there is *really* a potential difference of 200 volts between my nose and my feet, why is it I don't get a shock when I go out into the street?"

We will answer the second question first. Your body is a relatively good conductor. If you are in contact with the ground, you and the ground will tend to make one equipotential surface. Ordinarily, the equipotentials are parallel to the surface, as shown in Fig. 9–1(a), but when you are there, the equipotentials are distorted, and the field looks somewhat as shown in Fig. 9–1(b). So you still have very nearly zero potential difference between your head and your feet. There are charges that come from the earth to your head, changing the field. Some of them may be discharged by ions collected from the air, but the current of these is very small because air is a poor conductor.

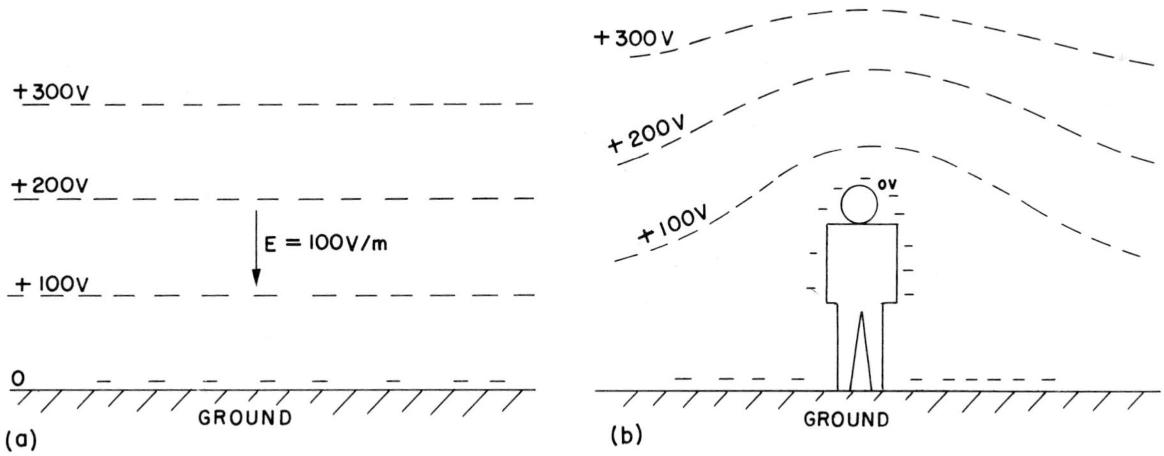

Fig. 9–1. (a) The potential distribution above the earth. (b) The potential distribution near a man in an open flat place.

How can we measure such a field if the field is changed by putting something there? There are several ways. One way is to place an insulated conductor at some distance above the ground and leave it there until it is at the same potential as the air. If we leave it long enough, the very small conductivity in the air will let the charges leak off (or onto) the conductor until it comes to the potential at its level. Then we can bring it back to the ground, and measure the shift of its potential as we do so. A faster way is to let the conductor be a bucket of water with a small leak. As the water drops out, it carries away any excess charges and the bucket will approach the same potential as the air. (The charges, as you know, reside on the surface, and as the drops come off "pieces of surface" break off.) We can measure the potential of the bucket with an electrometer.

There is another way to directly measure the potential *gradient*. Since there is an electric field, there is a surface charge on the earth ($\sigma = \epsilon_0 E$). If we place a flat metal plate at the earth's surface and ground it, negative charges appear on it (Fig. 9–2a). If this plate is now covered by another grounded conducting cover B, the charges will appear on the cover, and there will be no charges on the original plate A. If we measure the charge that flows from plate A to the ground (by, say, a galvanometer in the grounding wire) as we cover it, we can find the surface charge density that was there, and therefore also find the electric field.

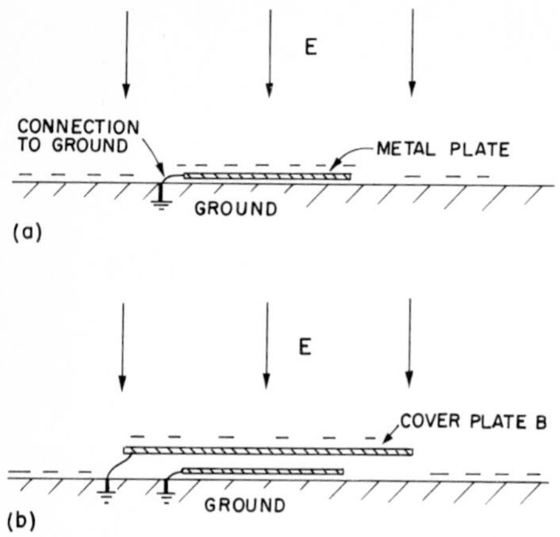

Fig. 9–2. (a) A grounded metal plate will have the same surface charge as the earth. (b) If the plate is covered with a grounded conductor it will have no surface charge.

Having suggested how we can measure the electric field in the atmosphere, we now continue our description of it. Measurements show, first of all, that the field continues to exist, but gets weaker, as one goes up to high altitudes. By about 50 kilometers, the field is very small, so most of the potential change (the integral of E) is at lower altitudes. The total potential difference from the surface of the earth to the top of the atmosphere is about 400,000 volts.

9–2 Electric currents in the atmosphere

Another thing that can be measured, in addition to the potential gradient, is the current in the atmosphere. The current density is small—about 10 micromicroamperes crosses each square meter parallel to the earth. The air is evidently not a perfect insulator, and because of this conductivity, a small current—caused by the electric field we have just been describing—passes from the sky down to the earth.

Why does the atmosphere have conductivity? Here and there among the air molecules there is an ion—a molecule of oxygen, say, which has acquired an extra electron, or perhaps lost one. These ions do not stay as single molecules; because of their electric field they usually accumulate a few other molecules around them. Each ion then becomes a little lump which, along with other lumps, drifts in the field—moving slowly upward or downward—making the observed current. Where do the *ions* come from? It was first guessed that the ions were produced by the radioactivity of the earth. (It was known that the radiation from radioactive materials would make air conducting by ionizing the air molecules.) Particles like β-rays coming out of the atomic nuclei are moving so fast that they tear electrons from the atoms, leaving ions behind. This would imply, of course, that if we were to go to higher altitudes, we should find less ionization, because the radioactivity is all in the dirt on the ground—in the traces of radium, uranium, potassium, etc.

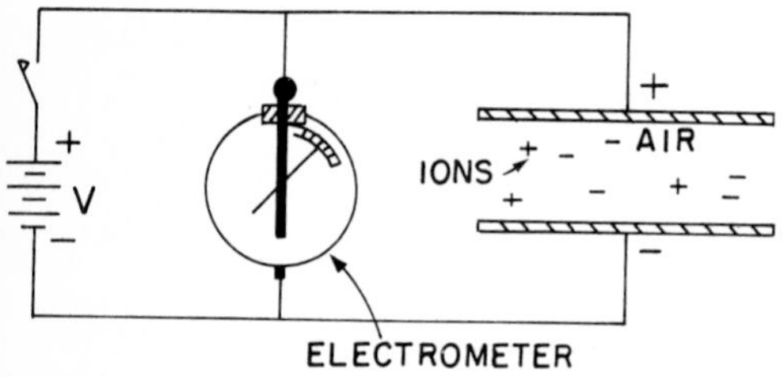

Fig. 9–3. Measuring the conductivity of air due to the motion of ions.

To test this theory, some physicists carried an experiment up in balloons to measure the ionization of the air (Hess, in 1912) and discovered that the opposite was true—the ionization per unit volume *increased* with altitude! (The apparatus was like that of Fig. 9–3. The two plates were charged periodically to the potential V. Due to the conductivity of the air, the plates slowly discharged; the rate of discharge was measured with the electrometer.) This was a most mysterious result—the most dramatic finding in the entire history of atmospheric electricity. It was so dramatic, in fact, that it required a branching off of an entirely new subject—cosmic rays. Atmospheric electricity itself remained less dramatic. Ionization was evidently being produced by something from outside the earth; the investigation of this source led to the discovery of the cosmic rays. We will not discuss the subject of cosmic rays now, except to say that they maintain the supply of ions. Although the ions are being swept away all the time, new ones are being created by the cosmic-ray particles coming from the outside.

To be precise, we must say that besides the ions made of molecules, there are also other kinds of ions. Tiny pieces of dirt, like extremely fine bits of dust, float in the air and become charged. They are sometimes called "nuclei." For example, when a wave breaks in the sea, little bits of spray are thrown into the air. When one of these drops evaporates, it leaves an infinitesimal crystal of NaCl floating in the air. These tiny crystals can then pick up charges and become ions; they are called "large ions."

The small ions—those formed by cosmic rays—are the most mobile. Because they are so small, they move rapidly through the air—with a speed of about 1 cm/sec in a field of 100 volts/meter, or 1 volt/cm. The much bigger and heavier ions move much more slowly. It turns out that if there are many "nuclei," they will pick up the charges from the small ions. Then, since the "large ions" move so slowly in a field, the total conductivity is reduced. The conductivity of air, therefore, is quite variable, since it is very sensitive to the amount of "dirt" there is in it. There is much more of such dirt over land—where the winds can blow up dust or where man throws all kinds of pollution into the air—than there is over water. It is not surprising that from day to day, from moment to moment, from place to place, the conductivity near the earth's surface varies enormously. The voltage gradient observed at any particular place on the earth's surface also varies greatly because roughly the same current flows down from high altitudes in different places, and the varying conductivity near the earth results in a varying voltage gradient.

The conductivity of the air due to the drifting of ions also increases rapidly with altitude—for two reasons. First of all, the ionization from cosmic rays increases with altitude. Secondly, as the density of air goes down, the mean free path of the ions increases, so that they can travel farther in the electric field before they have a collision—resulting in a rapid increase of conductivity as one goes up.

Although the electric current-density in the air is only a few micromicroamperes per square meter, there are very many square meters on the earth's surface. The total electric current reaching the earth's surface at any time is very nearly constant at 1800 amperes. This current, of course, is "positive"—it carries plus charges to the earth. So we have a voltage supply of 400,000 volts with a current of 1800 amperes—a power of 700 megawatts!

With such a large current coming down, the negative charge on the earth should soon be discharged. In fact, it should take only about half an hour to discharge the entire earth. But the atmospheric electric field has already lasted more than a half-hour since its discovery. How is it maintained? What maintains the voltage? And between what and the earth? There are many questions.

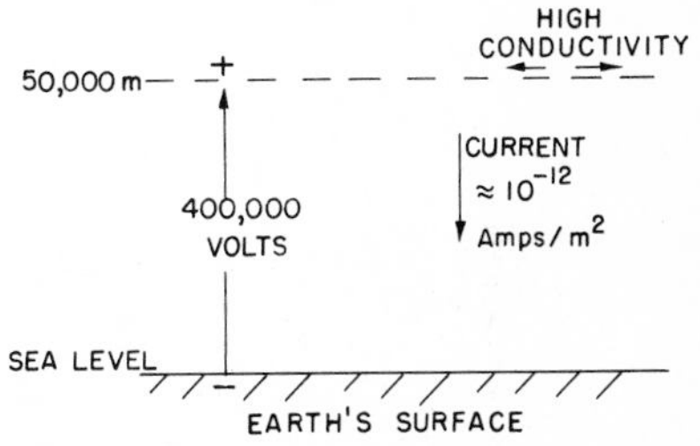

Fig. 9–4. Typical electrical conditions in a clear atmosphere.

The earth is negative, and the potential in the air is positive. If you go high enough, the conductivity is so great that horizontally there is no more chance for voltage variations. The air, for the scale of times that we are talking about, becomes effectively a conductor. This occurs at a height in the neighborhood of 50 kilometers. This is not as high as what is called the "ionosphere," in which there are very large numbers of ions produced by photoelectricity from the sun. Nevertheless, for our discussions of atmospheric electricity, the air becomes sufficiently conductive at about 50 kilometers that we can imagine that there is practically a perfect conducting surface at this height, from which the currents come down. Our picture of the situation is shown in Fig. 9–4. The problem is: How is the positive charge maintained there? How is it pumped back? Because if it comes down to the earth, it has to be pumped back somehow. That was one of the greatest puzzles of atmospheric electricity for quite a while.

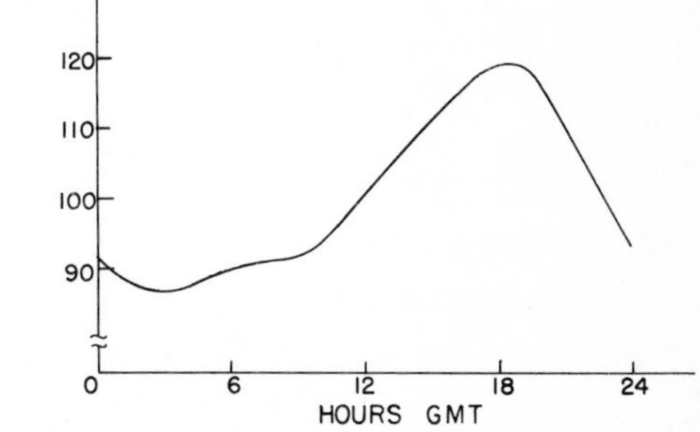

Fig. 9–5. The average daily variation of the atmospheric potential gradient on a clear day over the oceans; referred to Greenwich time.

Each piece of information we can get should give a clue or, at least, tell you something about it. Here is an interesting phenomenon: If we measure the current (which is more stable than the potential gradient) over the sea, for instance, or in careful conditions, and average very carefully so that we get rid of the irregularities, we discover that there is still a daily variation. The average of many measurements over the oceans has a variation with time roughly as shown in Fig. 9–5. The current varies by about ± 15 percent, and it is largest at 7:00 P.M. in London. The strange part of the thing is that no matter *where* you measure the current—in the Atlantic Ocean, the Pacific Ocean, or the Arctic Ocean—it is at its peak value when the clocks in *London* say 7:00 P.M.! All over the world the current is at its maximum at 7:00 P.M. London time and it is at a minimum at 4:00 A.M. London time. In other words, it depends upon the absolute time on the earth, *not* upon the local time at the place of observation. In one respect this is not mysterious; it checks with our idea that there is a very high conductivity laterally at the top, because that makes it impossible for the voltage difference from the ground to the top to vary locally. Any potential variations should be worldwide, as indeed they are. What we now know, therefore, is that the voltage at the "top" surface is dropping and rising by 15 percent with the absolute time on the earth.

9–3 Origin of the atmospheric currents

We must next talk about the source of the large negative currents which must be flowing from the "top" to the surface of the earth to keep charging it up negatively. Where are the batteries that do this? The "battery" is shown in Fig. 9–6. It is the thunderstorm and its lightning. It turns out that the bolts of lightning do not "discharge" the potential we have been talking about (as you might at first guess). Lightning storms carry *negative* charges to the earth. When a lightning bolt strikes, nine times out of ten it brings down negative charges to the earth in large amounts. It is the thunderstorms throughout the world that are charging the earth with an average of 1800 amperes, which is then being discharged through regions of fair weather.

There are about 40,000 thunderstorms per day all over the earth, and we can think of them as batteries pumping the electricity to the upper layer and maintaining the voltage difference. Then take into account the geography of the earth—there are thunderstorms in the afternoon in Brazil, tropical thunderstorms in Africa, and so forth. People have made estimates of how much lightning is striking world-wide at any time, and perhaps needless to say, their estimates more or less agree with the voltage difference measurements: the total amount of thunderstorm activity is highest on the whole earth at about 7:00 P.M. in London. However, the thunderstorm estimates are very difficult to make and were made only *after* it was known that the variation should have occurred. These things are very difficult because we don't have enough observations on the seas and over all parts of the world to know the number of thunderstorms accurately. But those people who think they "do it right" obtain the result that there are about 100 lightning flashes per second world-wide with a peak in the activity at 7:00 P.M. Greenwich Mean Time.

Fig. 9–6. The mechanism that generates the atmospheric electric field. [Photo by William L. Widmayer.]

In order to understand how these batteries work, we will look at a thunderstorm in detail. What is going on inside a thunderstorm? We will describe this insofar as it is known. As we get into this marvelous phenomenon of real nature—instead of the idealized spheres of perfect conductors inside of other spheres that we can solve so neatly—we discover that we don't know very much. Yet it is really quite exciting. Anyone who has been in a thunderstorm has enjoyed it, or has been frightened, or at least has had some emotion. And in those places in nature where we get an emotion, we find that there is generally a corresponding complexity and mystery about it. It is not going to be possible to describe exactly how a thunderstorm works, because we do not yet know very much. But we will try to describe a little bit about what happens.

9–4 Thunderstorms

In the first place, an ordinary thunderstorm is made up of a number of "cells" fairly close together, but almost independent of each other. So it is best to analyze one cell at a time. By a "cell" we mean a region with a limit area in the horizontal direction in which all of the basic processes occur. Usually there are several cells side by side, and in each one about the same thing is happening, although perhaps with a different timing. Figure 9–7 indicates in an idealized fashion what such a cell looks like in the early stage of the thunderstorm. It turns out that in a certain place in the air, under certain conditions which we shall describe, there is a general rising of the air, with higher and higher velocities near the top. As the warm, moist air at the bottom rises, it cools and condenses. In the figure the little crosses indicate snow and the dots indicate rain, but because the updraft currents are great enough and the drops are small enough, the snow and rain do not come down at this stage. This is the beginning stage, and not the real thunderstorm yet—in the sense that we don't have anything happening at the ground. At the same time that the warm air rises, there is an entrainment of air from the sides—an important point which was neglected for many years. Thus it is not just the air from below which is rising, but also a certain amount of other air from the sides.

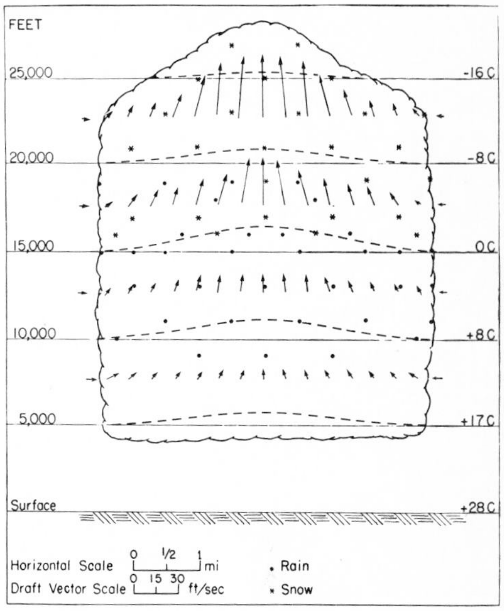

Fig. 9–7. A thunderstorm cell in the early stages of development. [From U.S. Department of Commerce Weather Bureau Report, June 1949.]

Why does the air rise like this? As you know, when you go up in altitude the air is colder. The *ground* is heated by the sun, and the re-radiation of heat to the sky comes from water vapor high in the atmosphere; so at high altitudes the air is cold—very cold—whereas lower down it is warm. You may say, "Then it's very simple. Warm air is lighter than cold; therefore the combination is mechanically unstable and the warm air rises." Of course, if the temperature is different at different heights, the air *is* unstable *thermodynamically*. Left to itself infinitely long, the air would all come to the same temperature. But it is not left to itself; the sun is always shining (during the day). So the problem is indeed not one of thermodynamic equilibrium, but of *mechanical* equilibrium. Suppose we plot—as in Fig. 9–8—the temperature of the air against height above the ground. In ordinary circumstances we would get a decrease along a curve like the one labeled (a); as the height goes up, the temperature goes down. How can the atmosphere be stable? Why doesn't the hot air below simply rise up into the cold air? The answer is this: if the air were to go up, its pressure would go down, and if we consider a particular parcel of air going up, it would be expanding adiabatically. (There would be no heat coming in or out because in the large dimensions considered here, there isn't time for much heat flow.) Thus the parcel of air would cool as it rises. Such an adiabatic process would give a temperature-height relationship like curve (b) in Fig. 9–8. Any air which rose from below would be *colder* than the environment it goes into. Thus there is no reason for the hot air below to rise; if it were to rise, it would cool to a lower temperature than the air already there, would be heavier than the air there, and would just want to come down again. On a good, bright day with very little humidity there is a certain rate at which the temperature in the atmosphere falls, and this rate is, in general, lower than the "maximum stable gradient," which is represented by curve (b). The air is in stable mechanical equilibrium.

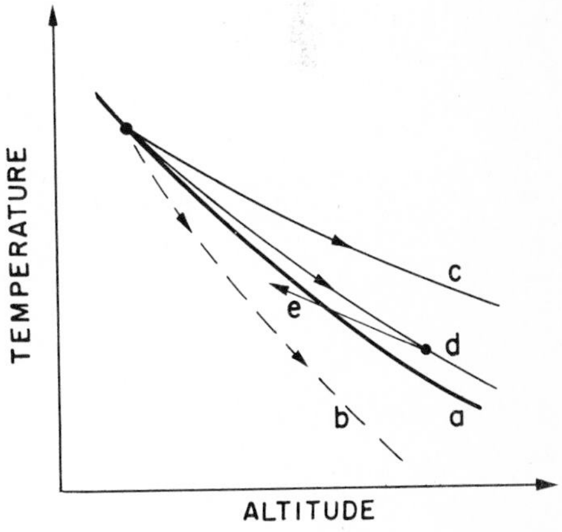

Fig. 9–8. Atmospheric temperature. (a) Static atmosphere; (b) adiabatic cooling of dry air; (c) adiabatic cooling of wet air; (d) wet air with some mixing of ambient air.

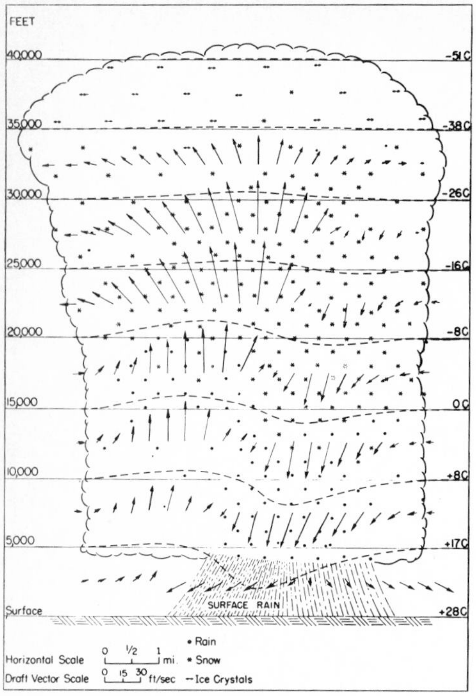

Fig. 9–9. A mature thunderstorm cell. [From U.S. Department of Commerce Weather Bureau Report, June 1949.]

On the other hand, if we think of a parcel of air that contains a lot of water vapor being carried up into the air, its adiabatic cooling curve will be different. As it expands and cools, the water vapor in it will condense, and the condensing water will liberate heat. Moist air, therefore, does not cool nearly as much as dry air does. So if air that is wetter than the average starts to rise, its temperature will follow a curve like (c) in Fig. 9–8. It will cool off somewhat, but will still be warmer than the surrounding air at the same level. If we have a region of warm moist air and something starts it rising, it will always find itself lighter and warmer than the air around it and will continue to rise until it gets to enormous heights. This is the machinery that makes the air in the thunderstorm cell rise.

For many years the thunderstorm cell was explained simply in this manner. But then measurements showed that the temperature of the cloud at different heights was not nearly as high as indicated by curve (c). The reason is that as the moist air "bubble" goes up, it entrains air from the environment and is cooled off by it. The temperature-versus-height curve looks more like curve (d), which is much closer to the original curve (a) than to curve (c).

After the convection just described gets under way, the cross section of a thunderstorm cell looks like Fig. 9–9. We have what is called a "mature" thunderstorm. There is a very rapid updraft which, in this stage, goes up to about 10,000 to 15,000 meters—sometimes even much higher. The thunderheads, with their condensation, climb way up out of the general cloud bank, carried by an updraft that is usually about 60 miles an hour. As the water vapor is carried up and condenses, it forms tiny drops which are rapidly cooled to temperatures below zero degrees. They should freeze, but do not freeze immediately—they are "supercooled." Water and other liquids will usually cool well below their freezing points before crystallizing if there are no "nuclei" present to start the crystallization process. Only if there is some small piece of material present, like a tiny crystal of NaCl, will the water drop freeze into a little piece of ice. Then the equilibrium is such that the water drops evaporate and the ice crystals grow. Thus at a certain point there is a rapid disappearance of the water and a rapid buildup of ice. Also, there may be direct collisions between the water drops and the ice—collisions in which the supercooled water becomes attached to the ice crystals, which causes it to suddenly crystallize. So at a certain point in the cloud expansion there is a rapid accumulation of large ice particles.

When the ice particles are heavy enough, they begin to fall through the rising air—they get too heavy to be supported any longer in the updraft. As they come down, they draw a little air with them and start a downdraft. And surprisingly enough, it is easy to see that once the downdraft is started, it will maintain itself. The air now drives itself down!

Notice that the curve (d) in Fig. 9–8 for the actual distribution of temperature in the cloud is not as steep as curve (c), which applies to wet air. So if we have wet air falling, its temperature will drop with the slope of curve (c) and will go *below* the temperature of the environment if it gets down far enough, as indicated by curve (e) in the figure. The moment it does that, it is denser than the environment and continues to fall rapidly. You say, "That is perpetual motion. First, you argue that the air should rise, and when you have it up there, you argue equally well that the air should fall." But it isn't perpetual motion. When the situation is unstable and the warm air should rise, then clearly something has to replace the warm air. It is equally true that cold air coming down would energetically replace the warm air, but you realize that what is coming down is *not* the original air. The early arguments, that had a particular cloud without entrainment going up and then coming down, had some kind of a puzzle. They needed the rain to maintain the downdraft—an argument which is hard to believe. As soon as you realize that there is a lot of original air mixed in with the rising air, the thermodynamic argument shows that there can be a descent of the cold air which was originally at some great height. This explains the picture of the active thunderstorm sketched in Fig. 9–9.

As the air comes down, rain begins to come out of the bottom of the thunderstorm. In addition, the relatively cold air spreads out when it arrives at the earth's surface. So just before the rain comes there is a certain little cold wind that gives us a forewarning of the coming storm. In the storm itself there are rapid and irregular gusts of air, there is an enormous turbulence in the cloud, and so on. But basically we have an updraft, then a downdraft—in general, a very complicated process.

The moment at which precipitation starts is the same moment that the large downdraft begins and is the same moment, in fact, when the electrical phenomena arise. Before we describe lightning, however, we can finish the story by looking at what happens to the thunderstorm cell after about one-half an hour to an hour. The cell looks as shown in Fig. 9–10. The updraft stops because there is no longer enough warm air to maintain it. The downward precipitation continues for a while, the last little bits of water come out, and things get quieter and quieter—although there are small ice crystals left way up in the air. Because the winds at very great altitude are in different directions, the top of the cloud usually spreads into an anvil shape. The cell comes to the end of its life.

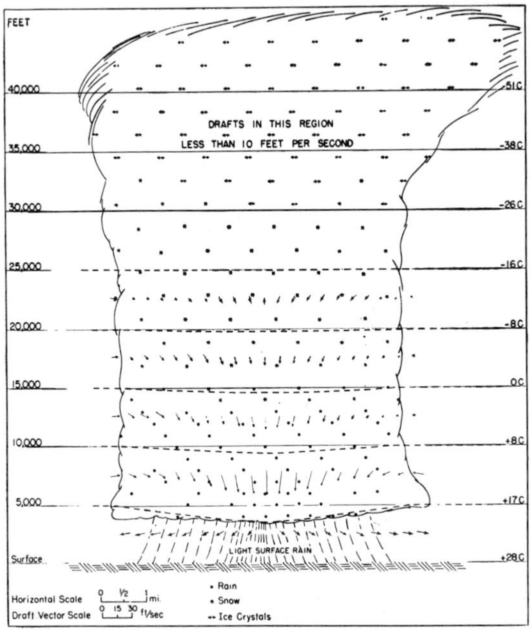

Fig. 9–10. The late phase of a thunderstorm cell. [From U.S. Department of Commerce Weather Bureau Report, June 1949.]

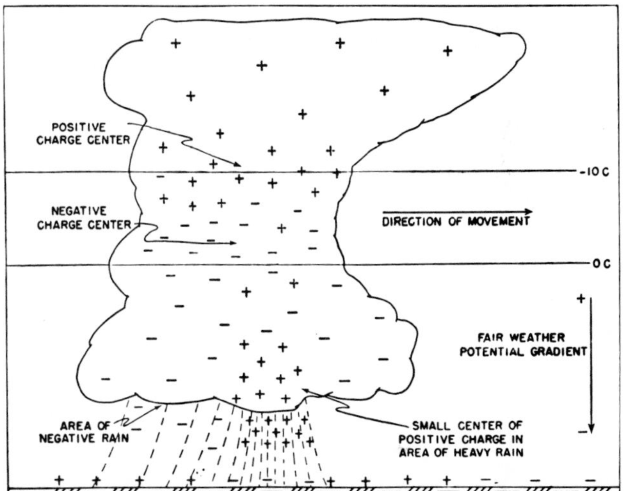

Fig. 9–11. The distribution of electrical charges in a mature thunderstorm cell. [From U.S. Department of Commerce Weather Bureau Report, June 1949.]

9–5 The mechanism of charge separation

We want now to discuss the most important aspect for our purposes—the development of the electrical charges. Experiments of various kinds—including flying airplanes through thunderstorms (the pilots who do this are brave men!)—tell us that the charge distribution in a thunderstorm cell is something like that shown in Fig. 9–11. The top of the thunderstorm has a positive charge, and the bottom a negative one—except for a small local region of positive charge in the bottom of the cloud, which has caused everybody a lot of worry. No one seems to know why it is there, how important it is—whether it is a secondary effect of the positive rain coming down, or whether it is an essential part of the machinery. Things would be much simpler if it weren't there. Anyway, the predominantly negative charge at the bottom and the positive charge at the top have the correct sign for the battery needed to drive the earth negative. The positive charges are 6 or 7 kilometers up in the air, where the temperature is about -20°C, whereas the negative charges are 3 or 4 kilometers high, where the temperature is between zero and -10°C.

The charge at the bottom of the cloud is large enough to produce potential differences of 20, or 30, or even 100 million volts between the cloud and the earth—much bigger than the 0.4 million volts from the "sky" to the ground in a clear atmosphere. These large voltages break down the air and create giant arc discharges. When the breakdown occurs the negative charges at the bottom of the thunderstorm are carried down to the earth in the lightning strokes.

Now we will describe in some detail the character of the lightning. First of all, there are large voltage differences around, so that the air breaks down. There are lightning strokes between one piece of a cloud and another piece of a cloud, or between one cloud and another cloud, or between a cloud and the earth. In each of the independent discharge flashes—the kind of lightning strokes you see—there are approximately 20 or 30 coulombs of charge brought down. One question is: How long does it take for the cloud to regenerate the 20 or 30 coulombs which are taken away by the lightning bolt? This can be seen by measuring, far from a cloud, the electric field produced by the cloud's dipole moment. In such measurements you see a sudden decrease in the field when the lightning strikes, and then an exponential return to the previous value with a time constant which is slightly different for different cases but which is in the neighborhood of 5 seconds. It takes a thunderstorm only 5 seconds after each lightning stroke to build its charge up again. That doesn't necessarily mean that another stroke is going to occur in exactly 5 seconds every time, because, of course, the geometry is changed, and so on. The strokes occur more or less irregularly, but the important point is that it takes about 5 seconds to recreate the original condition. Thus there are approximately 4 amperes of current in the generating machine of the thunderstorm. This means that any model made to explain how this storm generates its electricity must be one with plenty of juice—it must be a big, rapidly operating device.

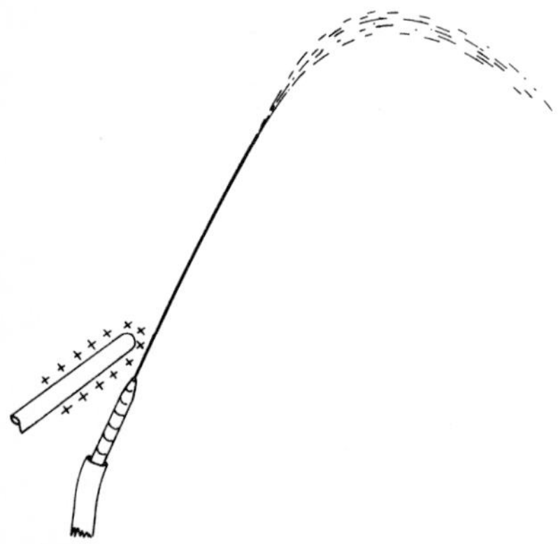

Fig. 9–12. A jet of water with an electric field near the nozzle.

Before we go further we shall consider something which is almost certainly completely irrelevant, but nevertheless interesting, because it does show the effect of an electric field on water drops. We say that it may be irrelevant because it relates to an experiment one can do in the laboratory with a stream of water to show the rather strong effects of the electric field on drops of water. In a thunderstorm there is no stream of water; there is a cloud of condensing ice and drops of water. So the question of the mechanisms at work in a thunderstorm is probably not at all related to what you can see in the simple experiment we will describe. If you take a small nozzle connected to a water faucet and direct it upward at a steep angle, as in Fig. 9–12, the water will come out in a fine stream that eventually breaks up into a spray of fine drops. If you now put an electric field across the stream at the nozzle (by bringing up a charged rod, for example), the form of the stream will change. With a weak electric field you will find that the stream breaks up into a smaller number of large-sized drops. But if you apply a stronger field, the stream breaks up into many, many fine drops—smaller than before.* With a weak electric field there is a tendency to inhibit the breakup of the stream into drops. With a stronger field, however, there is an increase in the tendency to separate into drops.

The explanation of these effects is probably the following. If we have the stream of water coming out of the nozzle and we put a small electric field across it one side of the water gets slightly positive and the other side gets slightly negative. Then, when the stream breaks, the drops on one side may be positive, and those on the other side may be negative. They will attract each other and will have a tendency to stick together more than they would have before—the stream doesn't break up as much. On the other hand, if the field is stronger, the charge in each one of the drops gets much larger, and there is a tendency for the charge *itself* to help break up the drops through their own repulsion. Each drop will break into many smaller ones, each carrying a charge, so that they are all repelled, and spread out so rapidly. So as we increase the field, the stream becomes more finely separated. The only point we wish to make is that in certain circumstances electric fields can have considerable influence on the drops. The exact machinery by which something happens in a thunderstorm is not at all known, and is not at all necessarily related to what we have just described. We have included it just so that

* A handy way to observe the sizes of the drops is to let the stream fall on a large thin metal plate. The larger drops make a louder noise.

you will appreciate the complexities that could come into play. In fact, nobody has a theory applicable to clouds based on that idea.

We would like to describe two theories which have been invented to account for the separation of the charges in a thunderstorm. All the theories involve the idea that there should be some charge on the precipitation particles and a different charge in the air. Then by the movement of the precipitation particles—the water or the ice—through the air there is a separation of electric charge. The only question is: How does the charging of the drops begin? One of the older theories is called the "breaking-drop" theory. Somebody discovered that if you have a drop of water that breaks into two pieces in a windstream, there is positive charge on the water and negative charge in the air. This breaking-drop theory has several disadvantages, among which the most serious is that the *sign* is wrong. Second, in the large number of temperate-zone thunderstorms which do exhibit lightning, the precipitation effects at high altitudes are in ice, *not* in water.

From what we have just said, we note that if we could imagine some way for the charge to be different at the top and bottom of a drop and if we could also see some reason why drops in a high-speed airstream would break up into unequal pieces—a large one in the front and a smaller one in the back because of the motion through the air or something—we would have a theory. (Different from any known theory!) Then the small drops would not fall through the air as fast as the big ones, because of the air resistance, and we would get a charge separation. You see, it is possible to concoct all kinds of possibilities.

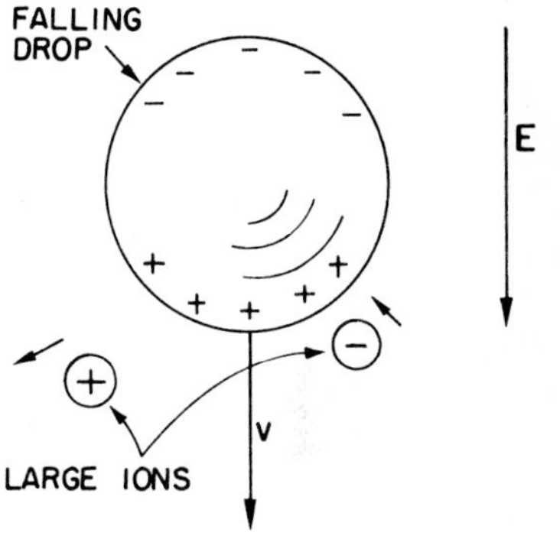

Fig. 9–13. C. T. R. Wilson's theory of charge separation in a thundercloud.

One of the more ingenious theories, which is more satisfactory in many respects than the breaking-drop theory, is due to C. T. R. Wilson. We will describe it, as Wilson did, with reference to water drops, although the same phenomenon would also work with ice. Suppose we have a water drop that is falling in the electric field of about 100 volts per meter toward the negatively charged earth. The drop will have an induced dipole moment—with the bottom of the drop positive and the top of the drop negative, as drawn in Fig. 9–13. Now there are in the air the "nuclei" that we mentioned earlier—the large slow-moving ions. (The fast ions do not have an important effect here.) Suppose that as a drop comes down, it approaches a large ion. If the ion is positive, it is repelled by the positive bottom of the drop and is pushed away. So it does not become attached to the drop. If the ion were to approach from the top, however, it might attach to the negative, top side. But since the drop is falling through the air, there is an air drift relative to it, going upwards, which carries the ions away if their motion through the air is slow enough. Thus the positive ions cannot attach at the top either. This would apply, you see, only to the large, slow-moving ions. The positive ions of this type will not attach themselves either to the front or the back of a falling drop. On the other hand, as the large, slow, *negative* ions are approached by a drop, they will be attracted and will be caught. The drop will acquire negative charge—the sign of the charge having been determined by the original potential difference on the entire earth—and we get the right sign. Negative charge will be brought down to the bottom part of the cloud by the drops, and the positively charged ions which are left behind will be blown to the top of the cloud by the various updraft currents. The theory looks pretty good, and it at least gives the right sign. Also it doesn't depend on having liquid drops. We will see, when we learn about polarization in a dielectric, that pieces of ice will do the same thing. They also will develop positive and negative charges on their extremities when they are in an electric field.

There are, however, some problems even with this theory. First of all, the total charge involved in a thunderstorm is very high. After a short time, the supply of large ions would get used up. So Wilson and others have had to propose that there are additional sources of the large ions. Once the charge separation starts, very large electric fields are developed, and in these large fields there may be places where the air will become ionized. If there is a highly charged point, or any small object like a drop, it may concentrate the field enough to make a "brush discharge." When there is a strong enough electric field—let us say it is positive—electrons will fall into the field and will pick up a lot of speed between collisions. Their speed will be such that in hitting another atom they will tear electrons off at that atom, leaving positive charges behind. These new electrons also pick up speed and collide with more electrons. So a kind of chain reaction or avalanche occurs, and there is a rapid accumulation of ions. The positive charges are left near their original positions, so the net effect is to distribute the positive charge on the point into a region around the point. Then, of course, there is no longer a strong field, and the process stops. This is the character of a brush discharge. It is possible that the fields may become strong enough in the cloud to produce a little bit of brush discharge; there may also be other mechanisms, once the thing is started, to produce a large amount of ionization. But nobody knows exactly how it works. So the fundamental origin of lightning is really not thoroughly understood. We know it comes from the thunderstorms. (And we know, of course, that thunder comes from the lightning—from the thermal energy released by the bolt.)

At least we can understand, in part, the origin of atmospheric electricity. Due to the air currents, ions, and water drops on ice particles in a thunderstorm, positive and negative charges are separated. The positive charges are carried upward to the top of the cloud (see Fig. 9–11), and the negative charges are dumped into the ground in lightning strokes. The positive charges leave the top of the cloud, enter the high-altitude layers of more highly conducting air, and spread throughout the earth. In regions of clear weather, the positive charges in this layer are slowly conducted to the earth by the ions in the air—ions formed by cosmic rays, by the sea, and by man's activities. The atmosphere is a busy electrical machine!

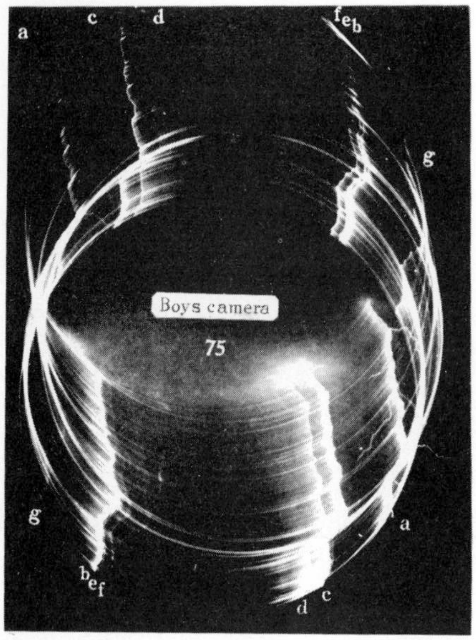

Fig. 9–14. Photograph of a lightning flash taken with a "Boys" camera. [From Schonland, Malan, and Collens, *Proc. Roy. Soc. London*, Vol. 152 (1935).]

9–6 Lightning

The first evidence of what happens in a lightning stroke was obtained in photographs taken with a camera held by hand and moved back and forth with the shutter open—while pointed toward a place where lightning was expected. The first photographs obtained this way showed clearly that lightning strokes are usually multiple discharges along the same path. Later, the "Boys" camera, which has *two* lenses mounted 180° apart on a rapidly rotating disc, was developed. The image made by each lens moves across the film—the picture is spread out in time. If, for instance, the stroke repeats, there will be two images side by side. By comparing the images of the two lenses, it is possible to work out the details of the time sequence of the flashes. Figure 9–14 shows a photograph taken with a "Boys" camera.

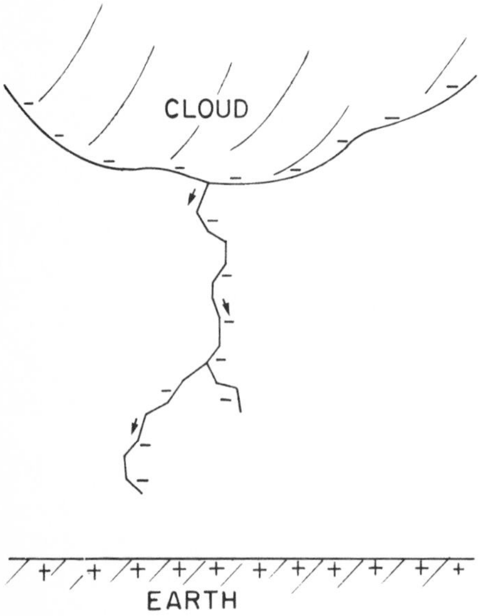

Fig. 9–15. The formation of the "step leader."

We will now describe the lightning. Again, we don't understand exactly how it works. We will give a qualitative description of what it *looks* like, but we won't go into any details of *why* it does what it appears to do. We will describe only the ordinary case of the cloud with a negative bottom over flat country. Its potential is much more negative than the earth underneath, so negative electrons will be accelerated toward the earth. What happens is the following. It all starts with a thing called a "step leader," which is not as bright as the stroke of lightning. On the photographs one can see a little bright spot at the beginning that starts from the cloud and moves downward very rapidly—at a sixth of the speed of light! It goes only about 50 meters and stops. It pauses for about 50 microseconds, and then takes another step. It pauses again and then goes another step, and so on. It moves in a series of steps toward the ground, along a path like that shown in Fig. 9–15. In the leader there are negative charges from the cloud; the whole column is full of negative charge. Also, the air is becoming ionized by the rapidly moving charges that produce the leader, so the air becomes a conductor along the path traced out. The moment the leader touches the ground, we have a conducting "wire" that runs all the way up to the cloud and is full of negative charge. Now, at last, the negative charge of the cloud can simply escape and run out. The electrons at the bottom of the leader are the first ones to realize this; they dump out, leaving positive charge behind that attracts more negative charge from higher up in the leader, which in its turn pours out, etc. So finally all the negative charge in a part of the cloud runs out along the column in a rapid and energetic way. So the lightning stroke you *see* runs *upwards* from the ground, as indicated in Fig. 9–16. In fact, this main stroke—by far the brightest part—is called the *return stroke*. It is what produces the very bright light, and the heat, which by causing a rapid expansion of the air makes the thunder clap.

The current in a lightning stroke is about 10,000 amperes at its peak, and it carries down about 20 coulombs.

But we are still not finished. After a time of, perhaps, a few hundredths of a second, when the return stroke has disappeared, another leader comes down. But this time there are no pauses. It is called a "dark leader" this time, and it goes all the way down—from top to bottom in one swoop. It goes full steam on exactly the old track, because there is enough debris there to make it the easiest route. The new leader is again full of negative charge. The moment it touches the ground—zing!—there is a return stroke going straight up along the path. So you see the lightning strike again, and again, and again. Sometimes it strikes only once or twice, sometimes five or ten times—once as many as 42 times on the same track was seen—but always in rapid succession.

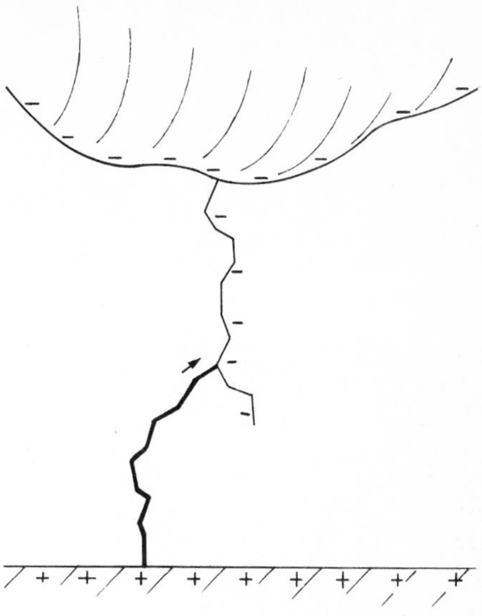

Fig. 9–16. The return lightning stroke runs back up the path made by the leader.

Sometimes things get even more complicated. For instance, after one of its pauses the leader may develop a branch by sending out *two* steps—both toward the ground but in somewhat different directions, as shown in Fig. 9–15. What happens then depends on whether one branch reaches the ground definitely before the other. If that does happen, the bright return stroke (of negative charge dumping into the ground) works its way *up* along the branch that touches the ground, and when it reaches and passes the branching point on its way up to the cloud, a bright stroke appears to go *down* the other branch. Why? Because negative charge is dumping out and that is what lights up the bolt. This charge begins to move at the top of the secondary branch, emptying successive, longer pieces of the branch, so the bright lightning bolt appears to work its way down that branch, at the same time as it works up toward the cloud. If, however, one of these extra leader branches happens to have reached the ground almost simultaneously with the original leader, it can sometimes happen that the *dark* leader of the second stroke will take the second branch. Then you will see the first main flash in one place and the second flash in another place. It is a variant of the original idea.

Also, our description is oversimplified for the region very near the ground. When the step leader gets to within a hundred meters or so from the ground, there is evidence that a discharge rises from the ground to meet it. Presumably, the field gets big enough for a brush-type discharge to occur. If, for instance, there is a sharp object, like a building with a point at the top, then as the leader comes down nearby the fields are so large that a discharge starts from the sharp point and reaches up to the leader. The lightning tends to strike such a point.

It has apparently been known for a long time that high objects are struck by lightning. There is a quotation of Artabanis, the advisor to Xerxes, giving his master advice on a contemplated attack on the Greeks—during Xerxes' campaign to bring the entire known world under the control of the Persians. Artabanis said, "See how God with his lightning always smites the bigger animals and will not suffer them to wax insolent, while these of a lesser bulk chafe him not. How likewise his bolts fall ever on the highest houses and tallest trees." And then he explains the reason: "So, plainly, doth he love to bring down everything that exalts itself."

Do you think—now that you know a true account of lightning striking tall trees—that you have a greater wisdom in advising kings on military matters than did Artabanis 2300 years ago? Do not exalt yourself. You could only do it less poetically.

10

Dielectrics

10–1 The dielectric constant

Here we begin to discuss another of the peculiar properties of matter under the influence of the electric field. In an earlier chapter we considered the behavior of *conductors*, in which the charges move freely in response to an electric field to such points that there is no field left inside a conductor. Now we will discuss *insulators*, materials which do not conduct electricity. One might at first believe that there should be no effect whatsoever. However, using a simple electroscope and a parallel-plate capacitor, Faraday discovered that this was not so. His experiments showed that the capacitance of such a capacitor is *increased* when an insulator is put between the plates. If the insulator completely fills the space between the plates, the capacitance is increased by a factor κ which depends only on the nature of the insulating material. Insulating materials are also called *dielectrics;* the factor κ is then a property of the dielectric, and is called the *dielectric constant.* The dielectric constant of a vacuum is, of course, unity.

Our problem now is to explain why there is any electrical effect if the insulators are indeed insulators and do not conduct electricity. We begin with the experimental fact that the capacitance is increased and try to reason out what might be going on. Consider a parallel-plate capacitor with some charges on the surfaces of the conductors, let us say negative charge on the top plate and positive charge on the bottom plate. Suppose that the spacing between the plates is d and the area of each plate is A. As we have proved earlier, the capacitance is

$$C = \frac{\epsilon_0 A}{d}, \tag{10.1}$$

and the charge and voltage on the capacitor are related by

$$Q = CV. \tag{10.2}$$

Now the experimental fact is that if we put a piece of insulating material like lucite or glass between the plates, we find that the capacitance is larger. That means, of course, that the voltage is lower for the same charge. But the voltage difference is the integral of the electric field across the capacitor; so we must conclude that inside the capacitor, the electric field is reduced even though the charges on the plates remain unchanged.

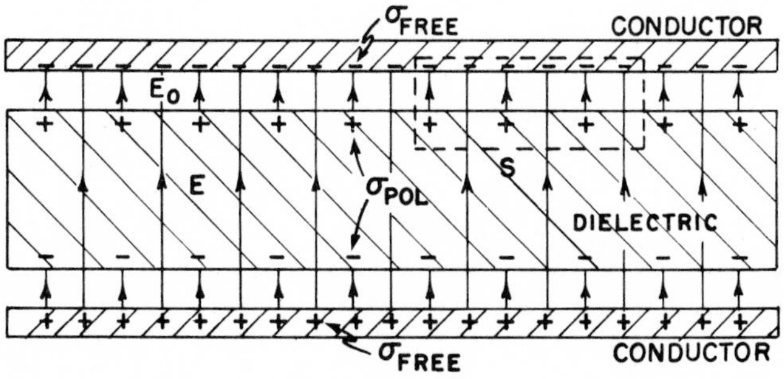

Fig. 10–1. A parallel-plate capacitor with a dielectric. The lines of E are shown.

Now how can that be? We have a law due to Gauss that tells us that the flux of the electric field is directly related to the enclosed charge. Consider the gaussian surface S shown by broken lines in Fig. 10–1. Since the electric field is reduced with the dielectric present, we conclude that the net charge inside the surface must be lower than it would be without the material. There is only one possible conclusion, and that is that there must be positive charges on the surface of the dielectric. Since the field is reduced but is not zero, we would expect this positive charge to be smaller than the negative charge on the conductor. So the phenomena can be explained if we could understand in some way that when a dielectric material is placed in an electric field there is positive charge induced on one surface and negative charge induced on the other.

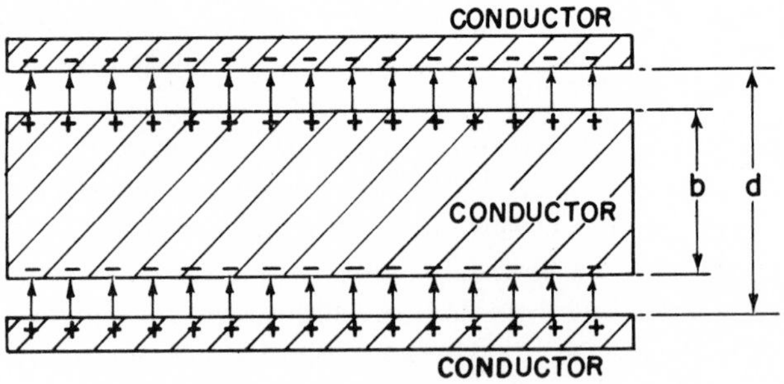

Fig. 10–2. If we put a conducting plate in the gap of a parallel-plate condenser, the induced charges reduce the field in the conductor to zero.

We would expect that to happen for a conductor. For example, suppose that we had a capacitor with a plate spacing d, and we put between the plates a neutral conductor whose thickness is b, as in Fig. 10–2. The electric field induces a positive charge on the upper surface and a negative charge on the lower surface, so there is no field inside the conductor. The field in the rest of the space is the same as it was without the conductor, because it is the surface density of charge divided by ϵ_0; but the distance over which we have to integrate to get the voltage (the potential difference) is reduced. The voltage is

$$V = \frac{\sigma}{\epsilon_0}(d - b).$$

The resulting equation for the capacitance is like Eq. (10.1), with $(d - b)$ substituted for d:

$$C = \frac{\epsilon_0 A}{d[1 - (b/d)]}. \tag{10.3}$$

The capacitance is increased by a factor which depends upon (b/d), the proportion of the volume which is occupied by the conductor.

This gives us an obvious model for what happens with dielectrics—that inside the material there are many little sheets of conducting material. The trouble with such a model is that it has a specific axis, the normal to the sheets, whereas most dielectrics have no such axis. However, this difficulty can be eliminated if we assume that all insulating materials contain small conducting spheres separated from each other by insulation, as shown in Fig. 10–3. The phenomenon of the dielectric constant is explained by the effect of the charges which would be induced on each sphere. This is one of the earliest physical models of dielectrics used to explain the phenomenon that Faraday observed. More specifically, it was assumed that each of the atoms of a material was a perfect conductor, but insulated from the others. The dielectric constant κ would depend on the proportion of space which was occupied by the conducting spheres. This is not, however, the model that is used today.

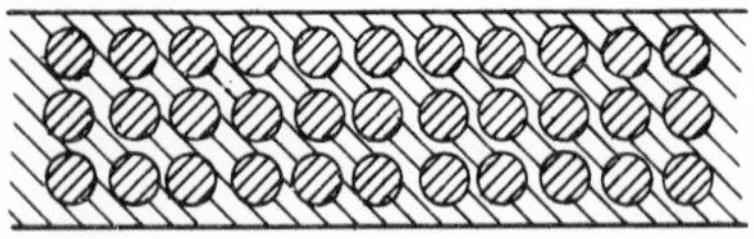

Fig. 10–3. A model of a dielectric: small conducting spheres embedded in an idealized insulator.

10–2 The polarization vector *P*

If we follow the above analysis further, we discover that the idea of regions of perfect conductivity and insulation is not essential. Each of the small spheres acts like a dipole, the moment of which is induced by the external field. The only thing that is essential to the understanding of dielectrics is that there are many little dipoles induced in the material. Whether the dipoles are induced because there are tiny conducting spheres or for any other reason is irrelevant.

Why should a field induce a dipole moment in an atom if the atom is not a conducting sphere? This subject will be discussed in much greater detail in the next chapter, which will be about the inner workings of dielectric materials. However, we give here one example to illustrate a possible mechanism. An atom has a positive charge on the nucleus, which is surrounded by negative electrons. In an electric field, the nucleus will be attracted in one direction and the electrons in the other. The orbits or wave patterns of the electrons (or whatever picture is used in quantum mechanics) will be distorted to some extent, as shown in Fig. 10–4; the center of gravity of the negative charge will be displaced and will no longer coincide with the positive charge of the nucleus. We have already discussed such distributions of charge. If we look from a distance, such a neutral configuration is equivalent, to a first approximation, to a little dipole.

It seems reasonable that if the field is not too enormous, the amount of induced dipole moment will be proportional to the field. That is, a small field will displace the charges a little bit and a larger field will displace them further—and in proportion to the field—unless the displacement gets too large. For the remainder of this chapter, it will be supposed that the dipole moment is exactly proportional to the field.

We will now assume that in each atom there are charges q separated by a distance $\boldsymbol{\delta}$, so that $q\boldsymbol{\delta}$ is the dipole moment per atom. (We use $\boldsymbol{\delta}$ because we are already using d for the plate separation.) If there are N atoms per unit volume, there will be a *dipole moment per unit volume* equal to $Nq\boldsymbol{\delta}$. This dipole moment per unit volume will be represented by a vector, $\boldsymbol{P}$. Needless to say, it is in the direction of the individual dipole moments, i.e., in the direction of the charge separation $\boldsymbol{\delta}$:

$$\boldsymbol{P} = Nq\boldsymbol{\delta}. \tag{10.4}$$

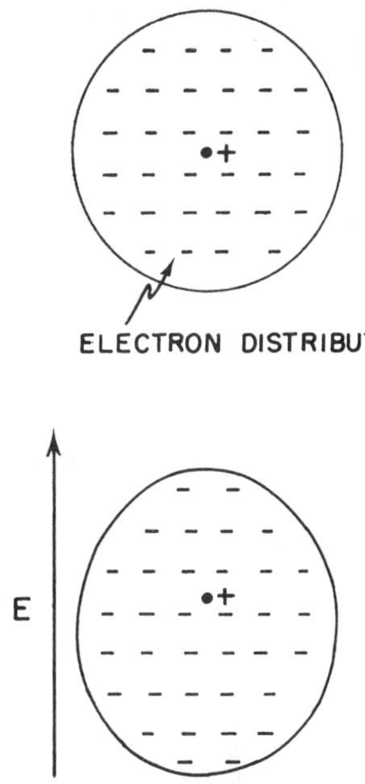

Fig. 10–4. An atom in an electric field has its distribution of electrons displaced with respect to the nucleus.

In general, $\boldsymbol{P}$ will vary from place to place in the dielectric. However, at any point in the material, $\boldsymbol{P}$ is proportional to the electric field $\boldsymbol{E}$. The constant of proportionality, which depends on the ease with which the electron are displaced, will depend on the kinds of atoms in the material.

What actually determines how this constant of proportionality behaves, how accurately it is constant for very large fields, and what is going on inside different materials, we will discuss at a later time. For the present, we will simply suppose that there exists a mechanism by which a dipole moment is induced which is proportional to the electric field.

10–3 Polarization charges

Now let us see what this model gives for the theory of a condenser with a dielectric. First consider a sheet of material in which there is a certain dipole moment per unit volume. Will there be on the average any charge density produced by this? Not if $\boldsymbol{P}$ is uniform. If the positive and negative charges being displaced relative to each other have the same average density, the fact that they are displaced does not produce any net charge inside the volume. On the other hand, if $\boldsymbol{P}$ were larger at one place and smaller at another, that would mean that more charge would be moved into some region than away from it; we would then expect to get a volume density of charge. For the parallel-plate condenser, we suppose that $\boldsymbol{P}$ is uniform, so we need to look only at what happens at the surfaces. At one surface the negative charges, the electrons, have effectively moved out a distance δ; at the other surface they have moved in, leaving some positive charge effectively out a distance δ. As shown in Fig. 10–5, we will have a surface density of charge, which will be called the surface *polarization charge*.

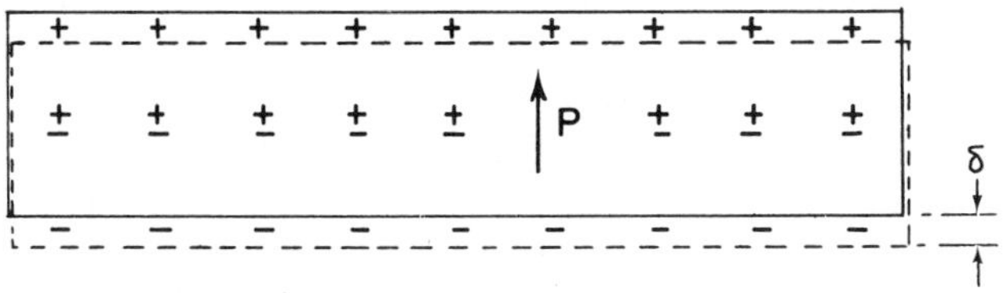

Fig. 10–5. A dielectric slab in a uniform field. The positive charges displaced the distance δ with respect to the negatives.

This charge can be calculated as follows. If A is the area of the plate, the number of electrons that appear at the surface is the product of A and N, the number per unit volume, and the displacement δ, which we assume here is perpendicular to the surface. The total charge is obtained by multiplying by the electronic charge q_e. To get the surface density of the polarization charge induced on the surface, we divide by A. The magnitude of the surface charge density is

$$\sigma_{\text{pol}} = Nq_e\,\delta.$$

But this is just equal to the magnitude P of the polarization vector $\boldsymbol{P}$, Eq. (10.4):

$$\sigma_{\text{pol}} = P. \tag{10.5}$$

The surface density of charge is equal to the polarization inside the material. The surface charge is, of course, positive on one surface and negative on the other.

Now let us assume that our slab is the dielectric of a parallel-plate capacitor. The *plates* of the capacitor also have a surface charge, which we will call σ_{free}, because they can move "freely" anywhere on the conductor. This is, of course, the charge that we put on when we charged the capacitor. It should be emphasized that σ_{pol} exists only because of σ_{free}. If σ_{free} is removed by discharging the capacitor, then σ_{pol} will disappear, not by going out on the discharging wire, but by moving back into the material—by the relaxation of the polarization inside the material.

We can now apply Gauss' law to the gaussian surface S in Fig. 10–1. The electric field E in the dielectric is equal to the *total* surface charge density divided by ϵ_0. It is clear that σ_{pol} and σ_{free} have opposite signs, so

$$E = \frac{\sigma_{\text{free}} - \sigma_{\text{pol}}}{\epsilon_0}. \tag{10.6}$$

Note that the field E_0 between the metal plate and the surface of the dielectric is higher than the field E; it corresponds to σ_{free} alone. But here we are concerned with the field inside the dielectric which, if the dielectric nearly fills the gap, is the field over nearly the whole volume. Using Eq. (10.5), we can write

$$E = \frac{\sigma_{\text{free}} - P}{\epsilon_0}. \tag{10.7}$$

This equation doesn't tell us what the electric field is unless we know what P is. Here, however, we are assuming that P depends on E—in fact, that it is proportional to E. This proportionality is usually written as

$$\boldsymbol{P} = \chi\epsilon_0\boldsymbol{E}. \tag{10.8}$$

The constant χ (Greek "khi") is called the *electric susceptibility* of the dielectric.

Then Eq. (10.7) becomes

$$E = \frac{\sigma_{\text{free}}}{\epsilon_0}\,\frac{1}{(1+\chi)}, \tag{10.9}$$

which gives us the factor $1/(1+\chi)$ by which the field is reduced.

The voltage between the plates is the integral of the electric field. Since the field is uniform, the integral is just the product of E and the plate separation d. We have that

$$V = Ed = \frac{\sigma_{\text{free}}d}{\epsilon_0(1+\chi)}.$$

The total charge on the capacitor is $\sigma_{\text{free}}A$, so that the capacitance defined by (10.2) becomes

$$C = \frac{\epsilon_0 A(1+\chi)}{d} = \frac{\kappa\epsilon_0 A}{d}. \tag{10.10}$$

We have explained the observed facts. When a parallel-plate capacitor is filled with a dielectric, the capacitance is increased by the factor

$$\kappa = 1 + \chi, \tag{10.11}$$

which is a property of the material. Our explanation, of course, is not complete until we have explained—as we will do later—how the atomic polarization comes about.

Let's now consider something a little bit more complicated—the situation in which the polarization $\boldsymbol{P}$ is not everywhere the same. As mentioned earlier, if the polarization is not constant, we would expect in general to find a charge density in the volume, because more charge might come into one side of a small volume element than leaves it on the other. How can we find out how much charge is gained or lost from a small volume?

First let's compute how much charge moves across any imaginary surface when the material is polarized. The amount of charge that goes across a surface is just P times the surface area if the polarization is *normal* to the surface. Of course, if the polarization is *tangential* to the surface, no charge moves across it.

Following the same arguments we have already used, it is easy to see that the charge moved across any surface element is proportional to the *component* of $\boldsymbol{P}$ *perpendicular* to the surface. Compare Fig. 10–6 with Fig. 10–5. We see that Eq. (10.5) should, in the general case, be written

$$\sigma_{\text{pol}} = \boldsymbol{P} \cdot \boldsymbol{n}. \tag{10.12}$$

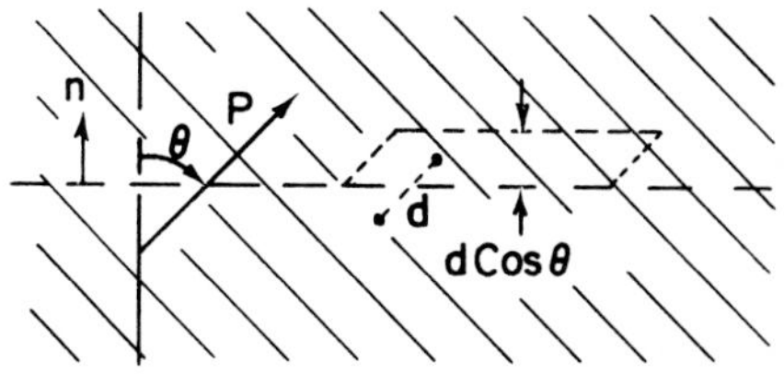

Fig. 10–6. The charge moved across an element of an imaginary surface in a dielectric is proportional to the component of **P** normal to the surface.

If we are thinking of an imagined surface element *inside* the dielectric, Eq. (10.12) gives the charge moved across the surface but doesn't result in a net surface charge, because there are equal and opposite contributions from the dielectric on the two sides of the surface.

The displacements of the charges can, however, result in a *volume* charge density. The total charge displaced *out* of any volume V by the polarization is the integral of the outward normal component of $\boldsymbol{P}$ over the surface S that bounds the volume (see Fig. 10–7). An equal excess charge of the opposite sign is left behind. Denoting the net charge inside V by ΔQ_{pol} we write

$$\Delta Q_{\text{pol}} = -\int_S \boldsymbol{P} \cdot \boldsymbol{n}\, da. \tag{10.13}$$

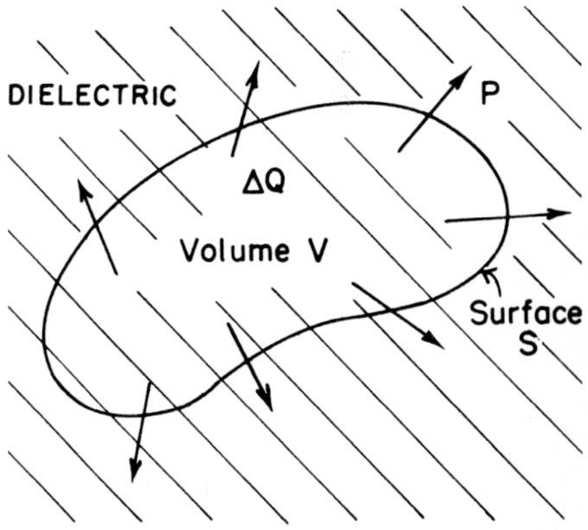

Fig. 10–7. A nonuniform polarization **P** can result in a net charge in the body of a dielectric.

We can attribute ΔQ_{pol} to a volume distribution of charge with the density ρ_{pol}, and so

$$\Delta Q_{\text{pol}} = \int_V \rho_{\text{pol}}\, dV. \tag{10.14}$$

Combining the two equations yields

$$\int_V \rho_{\text{pol}}\, dV = -\int_S \boldsymbol{P} \cdot \boldsymbol{n}\, da. \tag{10.15}$$

We have a kind of Gauss' theorem that relates the charge density from polarized materials to the polarization vector $\boldsymbol{P}$. We can see that it agrees with the result we got for the surface polarization charge or the dielectric in a parallel-plate capacitor. Using Eq. (10.15) with the gaussian surface of Fig. 10–1, the surface integral gives $P\,\Delta A$, and the charge inside is $\sigma_{\text{pol}}\,\Delta A$, so we get again that $\sigma = P$.

Just as we did for Gauss' law of electrostatics, we can convert Eq. (10.15) to a differential form—using Gauss' mathematical theorem:

$$\int_S \boldsymbol{P} \cdot \boldsymbol{n}\, da = \int_V \boldsymbol{\nabla} \cdot \boldsymbol{P}\, dV.$$

We get

$$\rho_{\text{pol}} = -\boldsymbol{\nabla} \cdot \boldsymbol{P}. \tag{10.16}$$

If there is a nonuniform polarization, its divergence gives the net density of charge appearing in the material. We emphasize that this is a perfectly *real* charge density; we call it "polarization charge" only to remind ourselves how it got there.

10–4 The electrostatic equations with dielectrics

Now let's combine the above result with our theory of electrostatics. The fundamental equation is

$$\nabla \cdot \boldsymbol{E} = \frac{\rho}{\epsilon_0}. \tag{10.17}$$

The ρ here is the density of *all* electric charges. Since it is not easy to keep track of the polarization charges, it is convenient to separate ρ into two parts. Again we call ρ_{pol} the charges due to nonuniform polarizations, and call ρ_{free} all the rest. Usually ρ_{free} is the charge we put on conductors, or at known places in space. Equation (10.17) then becomes

$$\nabla \cdot \boldsymbol{E} = \frac{\rho_{\text{free}} + \rho_{\text{pol}}}{\epsilon_0} = \frac{\rho_{\text{free}} - \nabla \cdot \boldsymbol{P}}{\epsilon_0},$$

or

$$\nabla \cdot \left(\boldsymbol{E} + \frac{\boldsymbol{P}}{\epsilon_0}\right) = \frac{\rho_{\text{free}}}{\epsilon_0}. \tag{10.18}$$

Of course, the equation for the curl of E is unchanged:

$$\nabla \times \boldsymbol{E} = 0. \tag{10.19}$$

Taking $\boldsymbol{P}$ from Eq. (10.8), we get the simpler equation

$$\nabla \cdot [(1 + \chi)\boldsymbol{E}] = \nabla \cdot (\kappa \boldsymbol{E}) = \frac{\rho_{\text{free}}}{\epsilon_0}. \tag{10.20}$$

These are the equations of electrostatics when there are dielectrics. They don't, of course, say anything new, but they are in a form which is more convenient for computation in cases where ρ_{free} is known and the polarization $\boldsymbol{P}$ is proportional to $\boldsymbol{E}$.

Notice that we have not taken the dielectric "constant," κ, out of the divergence. That is because it may not be the same everywhere. If it has everywhere the same value, it can be factored out and the equations are just those of electrostatics with the charge density ρ_{free} divided by κ. In the form we have given, the equations apply to the general case where different dielectrics may be in different places in the field. Then the equations may be quite difficult to solve.

There is a matter of some historical importance which should be mentioned here. In the early days of electricity, the atomic mechanism of polarization was not known and the existence of ρ_{pol} was not appreciated. The charge ρ_{free} was considered to be the entire charge density. In order to write Maxwell's equations in a simple form, a new vector $\boldsymbol{D}$ was defined to be equal to a linear combination of $\boldsymbol{E}$ and $\boldsymbol{P}$:

$$\boldsymbol{D} = \epsilon_0 \boldsymbol{E} + \boldsymbol{P}. \tag{10.21}$$

As a result, Eqs. (10.18) and (10.19) were written in an apparently very simple form:

$$\nabla \cdot \boldsymbol{D} = \rho_{\text{free}}, \qquad \nabla \times \boldsymbol{E} = 0. \tag{10.22}$$

Can one solve these? Only if a third equation is given for the relationship between $\boldsymbol{D}$ and $\boldsymbol{E}$. When Eq. (10.8) holds, this relationship is

$$\boldsymbol{D} = \epsilon_0(1 + \chi)\boldsymbol{E} = \kappa\epsilon_0\boldsymbol{E}. \tag{10.23}$$

This equation was usually written

$$\boldsymbol{D} = \epsilon \boldsymbol{E}, \tag{10.24}$$

where ϵ is still another constant for describing the dielectric property of materials. It is called the "permittivity." (Now you see why we have ϵ_0 in our equations, it is the "permittivity of empty space.") Evidently,

$$\epsilon = \kappa\epsilon_0 = (1 + \chi)\epsilon_0. \tag{10.25}$$

Today we look upon these matters from another point of view, namely, that we have simpler equations in a vacuum, and if we exhibit in every case all the charges, whatever their origin, the equations are always correct. If we separate some of the charges away for convenience, or because we do not want to discuss what is going on in detail, then we can, if we wish, write our equations in any other form that may be convenient.

One more point should be emphasized. An equation like $\boldsymbol{D} = \epsilon \boldsymbol{E}$ is an attempt to describe a property of matter. But matter is extremely complicated, and such an equation is in fact not correct. For instance, if $\boldsymbol{E}$ gets too large, then $\boldsymbol{D}$ is no longer proportional to $\boldsymbol{E}$. For some substances, the proportionality breaks down even with relatively small fields. Also, the "constant" of proportionality may depend on how fast $\boldsymbol{E}$ changes with time. Therefore this kind of equation is a kind of approximation, like Hooke's law. It cannot be a deep and fundamental equation. On the other hand, our fundamental equations for $\boldsymbol{E}$, (10.17) and (10.19), represent our deepest and most complete understanding of electrostatics.

10–5 Fields and forces with dielectrics

We will now prove some rather general theorems for electrostatics in situations where dielectrics are present. We have seen that the capacitance of a parallel-plate capacitor is increased by a definite factor if it is filled with a dielectric. We can show that this is true for a capacitor of *any* shape, provided the entire region in the neighborhood of the two conductors is filled with a uniform linear dielectric. Without the dielectric, the equations to be solved are

$$\nabla \cdot \boldsymbol{E}_0 = \frac{\rho_{\text{free}}}{\epsilon_0} \quad \text{and} \quad \nabla \times \boldsymbol{E}_0 = 0.$$

With the dielectric present, the first of these equations is modified; we have instead the equations

$$\nabla \cdot (\kappa \boldsymbol{E}) = \frac{\rho_{\text{free}}}{\epsilon_0} \quad \text{and} \quad \nabla \times \boldsymbol{E} = 0. \tag{10.26}$$

Now since we are taking κ to be everywhere the same, the last two equations can be written as

$$\nabla \cdot (\kappa \boldsymbol{E}) = \frac{\rho_{\text{free}}}{\epsilon_0} \quad \text{and} \quad \nabla \times (\kappa \boldsymbol{E}) = 0. \tag{10.27}$$

We therefore have the same equations for $\kappa \boldsymbol{E}$ as for $\boldsymbol{E}_0$, so they have the solution $\kappa \boldsymbol{E} = \boldsymbol{E}_0$. In other words, the field is everywhere smaller, by the factor $1/\kappa$, than in the case without the dielectric. Since the voltage difference is a line integral of the field, the voltage is reduced by this same factor. Since the charge on the electrodes of the capacitor has been taken the same in both cases, Eq. (10.2) tells us that the capacitance, in the case of an everywhere uniform dielectric, is increased by the factor κ.

Let us now ask what the *force* would be between two charged conductors in a dielectric. We consider a liquid dielectric that is homogeneous everywhere. We have seen earlier that one way to obtain the force is to differentiate the energy with respect to the appropriate distance. If the conductors have equal and opposite charges, the energy $U = Q^2/2C$, where C is their capacitance. Using the principle of virtual work, any component is given by a differentiation; for example,

$$F_x = -\frac{\partial U}{\partial x} = -\frac{Q^2}{2}\frac{\partial}{\partial x}\left(\frac{1}{C}\right). \tag{10.28}$$

Since the dielectric increases the capacity by a factor κ, all forces will be *reduced* by this same factor.

One point should be emphasized. What we have said is true only if the dielectric is a liquid. Any motion of conductors that are embedded in solid dielectric changes the mechanical stress conditions of the dielectric and alters its electrical properties, as well as causing some mechanical energy change in the dielectric. Moving the conductors in a liquid does not change the liquid. The liquid moves to a new place but its electrical characteristics are not changed.

Many older books on electricity start with the "fundamental" law that the force between two charges is

$$F = \frac{q_1 q_2}{4\pi\epsilon_0 \kappa r^2}, \tag{10.29}$$

a point of view which is thoroughly unsatisfactory. For one thing, it is not true in general; it is true only for a world filled with a liquid. Secondly, it depends on the fact that κ is a constant, which is only approximately true for most real materials. It is much better to start with Coulomb's law for charges in a *vacuum*, which is always right (for stationary charges).

What does happen in a solid? This is a very difficult problem which has not been solved, because it is, in a sense, indeterminate. If you put charges inside a dielectric solid, there are many kinds of pressures and strains. You cannot deal with virtual work without including also the mechanical energy required to compress the solid, and it is a difficult matter, generally speaking, to make a unique distinction between the electrical forces and the mechanical forces due to the solid material itself. Fortunately, no one ever really needs to know the answer to the question proposed. He may sometimes want to know how much strain there is going to be in a solid, and that can be worked out. But it is much more complicated than the simple result we got for liquids.

A surprisingly complicated problem in the theory of dielectrics is the following: Why does a charged object pick up little pieces of dielectric? If you comb your hair on a dry day, the comb readily picks up small scraps of paper. If you thought casually about it, you probably assumed the comb had one charge on it and the paper had the opposite charge on it. But the paper is initially electrically neutral. It hasn't any net charge, but it is attracted anyway. It is true that sometimes the paper will come up to the comb and then fly away, repelled immediately after it touches the comb. The reason is, of course, that when the paper touches the comb, it picks up some negative charges and then the like charges repel. But that doesn't answer the original question. Why did the paper come toward the comb in the first place?

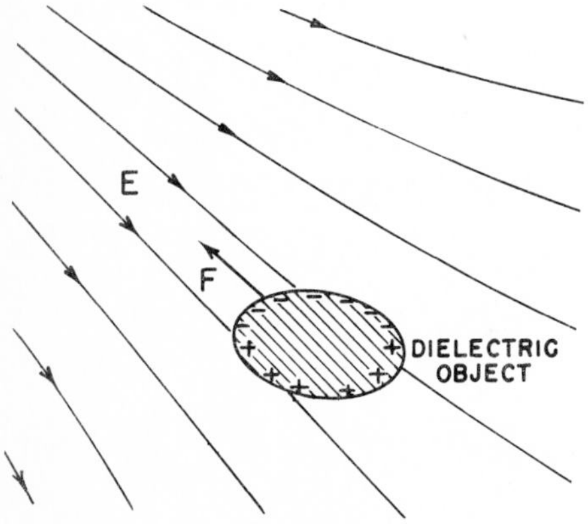

Fig. 10–8. A dielectric object in a nonuniform field feels a force toward regions of higher field strength.

The answer has to do with the polarization of a dielectric when it is placed in an electric field. There are polarization charges of both signs, which are attracted and repelled by the comb. There is a net attraction, however, because the field nearer the comb is stronger than the field farther away—the comb is not an infinite sheet. Its charge is localized. A neutral piece of paper will not be attracted to either plate inside the parallel plates of a capacitor. The variation of the field is an essential part of the attraction mechanism.

As illustrated in Fig. 10–8, a dielectric is always drawn from a region of weak field toward a region of stronger field. In fact, one can prove that for small objects the force is proportional to the gradient of the *square* of the electric field. Why does it depend on the square of the field? Because the induced polarization charges are proportional to the fields, and for given charges the forces are proportional to the field. However, as we have just indicated, there will be a *net* force only if the square of the field is changing from point to point. So the force is proportional to the gradient of the square of the field. The constant of proportionality involves, among other things, the dielectric constant of the object, and it also depends upon the size and shape of the object.

There is a related problem in which the force on a dielectric can be worked out quite accurately. If we have a parallel-plate capacitor with a dielectric slab only partially inserted, as shown in Fig. 10–9, there will be a force driving the sheet in. A detailed examination of the force is quite complicated; it is related to nonuniformities in the field near the edges of the dielectric and the plates. However, if we do not look at the details, but merely use the principle of conservation of energy, we can easily calculate the force. We can find the force from the formula we de-

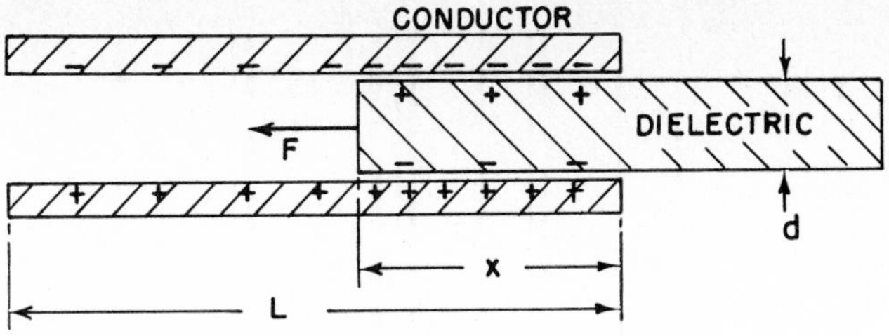

Fig. 10–9. The force on a dielectric sheet in a parallel-plate capacitor can be computed by applying the principle of energy conservation.

rived earlier. Equation (10.28) is equivalent to

$$F_x = -\frac{\partial U}{\partial x} = +\frac{V^2}{2}\frac{\partial C}{\partial x}. \tag{10.30}$$

We need only find out how the capacitance varies with the position of the dielectric slab.

Let's suppose that the total length of the plates is L, that the width of the plates is W, that the plate separation and dielectric thickness are d, and that the distance to which the dielectric has been inserted is x. The capacitance is the ratio of the total free charge on the plates to the voltage between the plates. We have seen above that for a given voltage V the surface charge density of free charge is $\kappa\epsilon_0 V/d$. So the total charge on the plates is

$$Q = \frac{\kappa\epsilon_0 V}{d}\,xW + \frac{\epsilon_0 V}{d}\,(L - x)W,$$

from which we get the capacitance:

$$C = \frac{\epsilon_0 W}{d}\,(\kappa x + L - x). \tag{10.31}$$

Using (10.30), we have

$$F_x = \frac{V^2}{2}\frac{\epsilon_0 W}{d}\,(\kappa - 1). \tag{10.32}$$

Now this equation is not particularly useful for anything unless you happen to need to know the force in such circumstances. We only wished to show that the theory of energy can often be used to avoid enormous complications in determining the forces on dielectric materials—as there would be in the present case.

Our discussion of the theory of dielectrics has dealt only with electrical phenomena, accepting the fact that the material has a polarization which is proportional to the electric field. Why there is such a proportionality is perhaps of greater interest to physics. Once we understand the origin of the dielectric constants from an atomic point of view, we can use electrical measurements of the dielectric constants in varying circumstances to obtain detailed information about atomic or molecular structure. This aspect will be treated in part in the next chapter.

11

Inside Dielectrics

Review: Chapter 31, Vol. I, *The Origin of the Refractive Index*
Chapter 40, Vol. I, *The Principles of Statistical Mechanics*

11–1 Molecular dipoles

In this chapter we are going to discuss why it is that materials are dielectric. We said in the last chapter that we could understand the properties of electrical systems with dielectrics once we appreciated that when an electric field is applied to a dielectric it induces a dipole moment in the atoms. Specifically, if the electric field E induces an average dipole moment per unit volume P, then κ, the dielectric constant, is given by

$$\kappa - 1 = \frac{P}{\epsilon_0 E}. \tag{11.1}$$

We have already discussed how this equation is applied; now we have to discuss the mechanism by which polarization arises when there is an electric field inside a material. We begin with the simplest possible example—the polarization of gases. But even gases already have complications: there are two types. The molecules of some gases, like oxygen, which has a symmetric pair of atoms in each molecule, have no inherent dipole moment. But the molecules of others, like water vapor (which has a nonsymmetric arrangement of hydrogen and oxygen atoms) carry a permanent electric dipole moment. As we pointed out in Chapters 6 and 7, there is in the water vapor molecule an average plus charge on the hydrogen atoms and a negative charge on the oxygen. Since the center of gravity of the negative charge and the center of gravity of the positive charge do not coincide, the total charge distribution of the molecule has a dipole moment. Such a molecule is called a *polar* molecule. In oxygen, because of the symmetry of the molecule, the centers of gravity of the positive and negative charges are the same, so it is a *nonpolar* molecule. It does, however, become a dipole when placed in an electric field. The forms of the two types of molecules are sketched in Fig. 11–1.

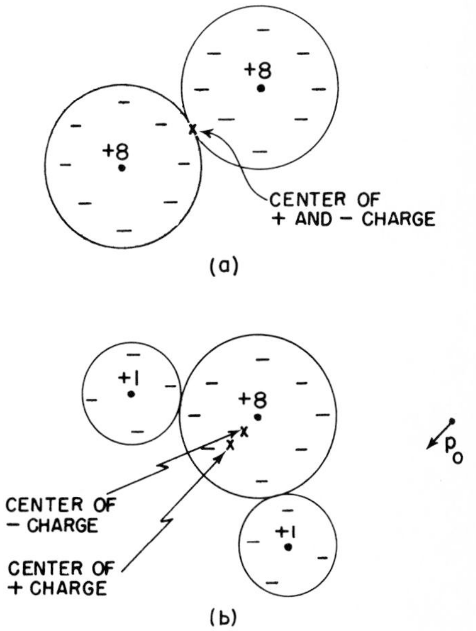

Fig. 11–1. (a) An oxygen molecule with zero dipole moment. (b) The water molecule has a permanent dipole moment $\mathbf{p}_0$.

11–2 Electronic polarization

We will first discuss the polarization of nonpolar molecules. We can start with the simplest case of a monatomic gas (for instance, helium). When an atom of such a gas is in an electric field, the electrons are pulled one way by the field while the nucleus is pulled the other way, as shown in Fig. 10–4. Although the atoms are very stiff with respect to the electrical forces we can apply experimentally, there is a slight net displacement of the centers of charge, and a dipole moment is induced. For small fields, the amount of displacement, and so also the dipole moment, is proportional to the electric field. The displacement of the electron distribution which produces this kind of induced dipole moment is called *electronic polarization.*

We have already discussed the influence of an electric field on an atom in Chapter 31 of Vol. I, when we were dealing with the theory of the index of refraction. If you think about it for a moment, you will see that what we must do now is exactly the same as we did then. But now we need worry only about fields that do not vary with time, while the index of refraction depended on time-varying fields.

In Chapter 31 of Vol. I we supposed that when an atom is placed in an oscillating electric field the center of charge of the electrons obeys the equation

$$m\frac{d^2x}{dt^2} + m\omega_0^2x = q_eE. \tag{11.2}$$

The first term is the electron mass times its acceleration and the second is a restoring force, while the right-hand side is the force from the outside electric field. If the electric field varies with the frequency ω, Eq. (11.2) has the solution

$$x = \frac{q_e E}{m(\omega_0^2 - \omega^2)}, \tag{11.3}$$

which has a resonance at $\omega = \omega_0$. When we previously found this solution, we interpreted it as saying that ω_0 was the frequency at which light (in the optical region or in the ultraviolet, depending on the atom) was absorbed. For our purposes, however, we are interested oı ly in the case of constant fields, i.e., for $\omega = 0$, so we can disregard the acceleration term in (11.2), and we find that the displacement is

$$x = \frac{q_e E}{m\omega_0^2}. \tag{11.4}$$

From this we see that the dipole moment p of a single atom is

$$p = q_e x = \frac{q_e^2 E}{m\omega_0^2}. \tag{11.5}$$

In this theory the dipole moment p is indeed proportional to the electric field.

People usually write

$$\boldsymbol{p} = \alpha\epsilon_0 \boldsymbol{E}. \tag{11.6}$$

(Again the ϵ_0 is put in for historical reasons.) The constant α is called the polarizability of the atom, and has the dimensions L^3. It is a measure of how easy it is to induce a moment in an atom with an electric field. Comparing (11.5) and (11.6), our simple theory says that

$$\alpha = \frac{q_e^2}{\epsilon_0 m\omega_0^2} = \frac{4\pi e^2}{m\omega_0^2}. \tag{11.7}$$

If there are N atoms in a unit volume, the polarization P—the dipole moment per unit volume—is given by

$$\boldsymbol{P} = N\boldsymbol{p} = N\alpha\epsilon_0 \boldsymbol{E}. \tag{11.8}$$

Putting (11.1) and (11.8) together, we get

$$\kappa - 1 = \frac{P}{\epsilon_0 E} = N\alpha \tag{11.9}$$

or, using (11.7),

$$\kappa - 1 = \frac{4\pi N e^2}{m\omega_0^2}. \tag{11.10}$$

From Eq. (11.9) we would predict that the dielectric constant κ of different gases should depend on the density of the gas and on the frequency ω_0 of its optical absorption.

Our formula is, of course, only a very rough approximation, because in Eq. (11.2) we have taken a model which ignores the complications of quantum mechanics. For example, we have assumed that an atom has only one resonant frequency, when it really has many. To calculate properly the polarizability α of atoms we must use the complete quantum-mechanical theory, but the classical ideas above give us a reasonable estimate.

Let's see if we can get the right order of magnitude for the dielectric constant of some substance. Suppose we try hydrogen. We have once estimated (Chapter 38, Vol. I) that the energy needed to ionize the hydrogen atom should be approximately

$$E \approx \frac{1}{2}\frac{me^4}{\hbar^2}. \tag{11.11}$$

For an estimate of the natural frequency ω_0, we can set this energy equal to $\hbar\omega_0$—the energy of an atomic oscillator whose natural frequency is ω_0. We get

$$\omega_0 \approx \frac{1}{2}\frac{me^4}{\hbar^3}.$$

If we now use this value of ω_0 in Eq. (11.7), we find for the electronic polarizability

$$\alpha \approx 16\pi\left[\frac{\hbar^2}{me^2}\right]^3. \tag{11.12}$$

The quantity $(\hbar^2/me^2)$ is the radius of the ground-state orbit of a Bohr atom (see Chapter 38, Vol. I) and equals 0.528 angstroms. In a gas at standard pressure and temperature (1 atmosphere, 0°C) there are 2.69×10^{19} atoms/cm^3, so Eq. (11.9) gives us

$$\kappa = 1 + (2.69 \times 10^{19})16\pi(0.528 \times 10^{-8})^3 = 1.00020. \tag{11.13}$$

The dielectric constant for hydrogen gas is measured to be

$$\kappa_{\text{exp}} = 1.00026.$$

We see that our theory is about right. We should not expect any better, because the measurements were, of course, made with normal hydrogen gas, which has diatomic molecules, not single atoms. We should not be surprised if the polarization of the atoms in a molecule is not quite the same as that of the separate atoms. The molecular effect, however, is not really that large. An exact quantum-mechanical calculation of α for hydrogen atoms gives a result about 12% higher than (11.12) (the 16π is changed to 18π), and therefore predicts a dielectric constant somewhat closer to the observed one. In any case, it is clear that our model of a dielectric is fairly good.

Another check on our theory is to try Eq. (11.12) on atoms which have a higher frequency of excitation. For instance, it takes about 24.5 volts to pull the electron off helium, compared with the 13.5 volts required to ionize hydrogen. We would, therefore, expect that the absorption frequency ω_0 for helium would be about twice as big as for hydrogen and that α would be one-quarter as large. We expect that

$$\kappa_{\text{helium}} \approx 1.000050.$$

Experimentally,

$$\kappa_{\text{helium}} = 1.000068,$$

so you see that our rough estimates are coming out on the right track. So we have understood the dielectric constant of nonpolar gas, but only qualitatively, because we have not yet used a correct atomic theory of the motions of the atomic electrons.

11–3 Polar molecules; orientation polarization

Next we will consider a molecule which carries a permanent dipole moment p_0—such as a water molecule. With no electric field, the individual dipoles point in random directions, so the net moment per unit volume is zero. But when an electric field is applied, two things happen: First, there is an extra dipole moment induced because of the forces on the electrons; this part gives just the same kind of electronic polarizability we found for a nonpolar molecule. For very accurate work, this effect should, of course, be included, but we will neglect it for the moment. (It can always be added in at the end.) Second, the electric field tends to line up the individual dipoles to produce a net moment per unit volume. If all the dipoles in a gas were to line up, there would be a very large polarization, but that does not happen. At ordinary temperatures and electric fields the collisions of the molecules in their thermal motion keep them from lining up very much. But there is some net alignment, and so some polarization (see Fig. 11–2). The polarization that does occur can be computed by the methods of statistical mechanics we described in Chapter 40 of Vol. I.

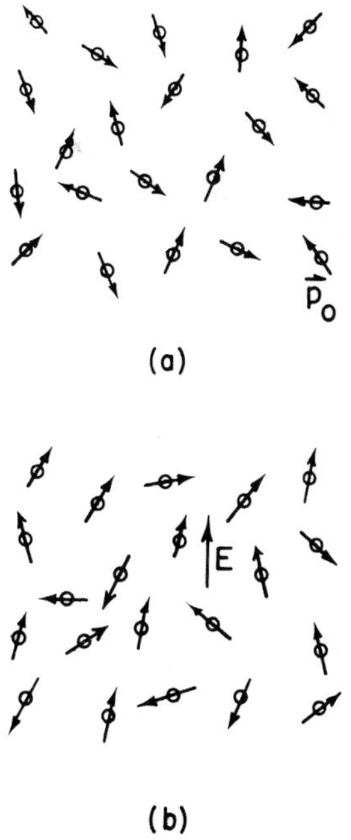

Fig. 11–2. (a) In a gas of polar molecules, the individual moments are oriented at random; the average moment in a small volume is zero. (b) When there is an electric field, there is some average alignment of the molecules.

To use this method we need to know the energy of a dipole in an electric field. Consider a dipole of moment $\boldsymbol{p}_0$ in an electric field, as shown in Fig. 11–3. The energy of the positive charge is $q\phi(1)$, and the energy of the negative charge is $-q\phi(2)$. Thus the energy of the dipole is

$$U = q\phi(1) - q\phi(2) = q\boldsymbol{d}\cdot\nabla\phi,$$

or

$$U = -\boldsymbol{p}_0\cdot\boldsymbol{E} = -p_0 E\cos\theta, \tag{11.14}$$

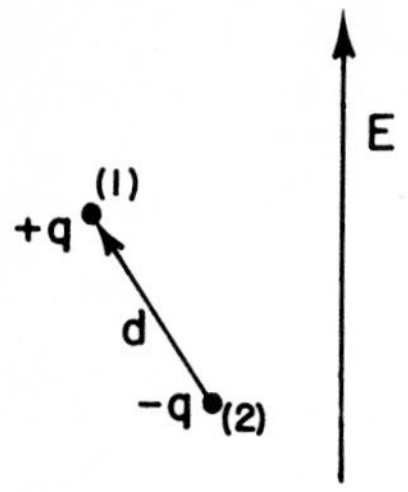

Fig. 11–3. The energy of a dipole $\boldsymbol{p}_0$ in the field $\boldsymbol{E}$ is $-\boldsymbol{p}_0\cdot\boldsymbol{E}$.

where θ is the angle between $\boldsymbol{p}_0$ and $\boldsymbol{E}$. As we would expect, the energy is lower when the dipoles are lined up with the field.

We now find out how much lining up occurs by using the methods of statistical mechanics. We found in Chapter 40 of Vol. I that in a state of thermal equilibrium, the relative number of molecules with the potential energy U is proportional to

$$e^{-U/kT}, \tag{11.15}$$

where $U(x, y, z)$ is the potential energy as a function of position. The same arguments would say that using Eq. (11.14) for the potential energy as a function of *angle*, the number of molecules at θ *per unit solid angle* is proportional to $e^{-U/kT}$.

Letting $n(\theta)$ be the number of molecules per unit solid angle at θ, we have

$$n(\theta) = n_0 e^{+p_0E\cos\theta/kT}. \tag{11.16}$$

For normal temperatures and fields, the exponent is small, so we can approximate by expanding the exponential:

$$n(\theta) = n_0\left(1 + \frac{p_0E\cos\theta}{kT}\right). \tag{11.17}$$

We can find n_0 if we integrate (11.17) over all angles; the result should be just N, the total number of molecules per unit volume. The average value of $\cos\theta$ over all angles is zero, so the integral is just n_0 times the total solid angle 4π. We get

$$n_0 = \frac{N}{4\pi}. \tag{11.18}$$

We see from (11.17) that there will be more molecules oriented along the field ($\cos\theta = 1$) than against the field ($\cos\theta = -1$). So in any small volume containing many molecules there will be a net dipole moment per unit volume—that is, a polarization P. To calculate P, we want the vector sum of all the molecular moments in a unit volume. Since we know that the result is going to be in the direction of $\boldsymbol{E}$, we will just sum the components in that direction (the components at right angles to $\boldsymbol{E}$ will sum to zero):

$$P = \sum_{\substack{\text{unit}\\ \text{volume}}} p_0\cos\theta_i.$$

We can evaluate the sum by integrating over the angular distribution. The solid angle at θ is $2\pi\sin\theta\,d\theta$, so

$$P = \int_0^\pi n(\theta)p_0\cos\theta\,2\pi\sin\theta\,d\theta. \tag{11.19}$$

Substituting for $n(\theta)$ from (11.17), we have

$$P = -\frac{N}{2}\int_0^\pi\left(1 + \frac{p_0E}{kT}\cos\theta\right)p_0\cos\theta\,d(\cos\theta),$$

which is easily integrated to give

$$P = \frac{Np_0^2E}{3kT}. \tag{11.20}$$

The polarization is proportional to the field E, so there will be normal dielectric behavior. Also, as we expect, the polarization depends inversely on the temperature, because at higher temperatures there is more disalignment by collisions. This $1/T$ dependence is called Curie's law. The permanent moment p_0 appears squared for the following reason: In a given electric field, the aligning force depends upon p_0, and the mean moment that is produced by the lining up is again proportional to p_0. The average induced moment is proportional to p_0^2.

We should now try to see how well Eq. (11.20) agrees with experiment. Let's look at the case of steam. Since we don't know what p_0 is, we cannot compute P directly, but Eq. (11.20) does predict that $\kappa - 1$ should vary inversely as the temperature, and this we should check.

From (11.20) we get

$$\kappa - 1 = \frac{P}{\epsilon_0 E} = \frac{Np_0^2}{3\epsilon_0 kT}, \tag{11.21}$$

so $\kappa - 1$ should vary in direct proportion to the density N, and inversely as the absolute temperature. The dielectric constant has been measured at several different pressures and temperatures, chosen such that the number of molecules in a unit volume remained fixed.* [Notice that if the measurements had all been taken at constant pressure, the number of molecules per unit volume would decrease linearly with increasing temperature and $\kappa - 1$ would vary as T^{-2} instead of as T^{-1}.] In Fig. 11–4 we plot the experimental observations for $\kappa - 1$ as a function of $1/T$. The dependence predicted by (11.21) is followed quite well.

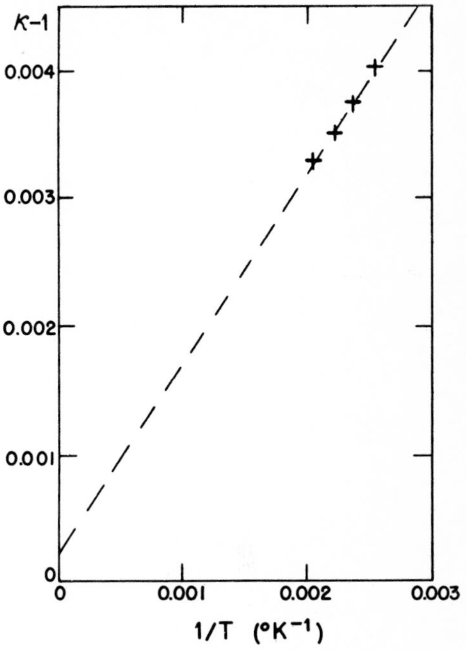

Fig. 11–4. Experimental measurements of the dielectric constant of water vapor at various temperatures.

There is another characteristic of the dielectric constant of polar molecules—its variation with the frequency of the applied field. Due to the moment of inertia of the molecules, it takes a certain amount of time for the heavy molecules to turn toward the direction of the field. So if we apply frequencies in the high microwave region or above, the polar contribution to the dielectric constant begins to fall away because the molecules cannot follow. In contrast to this, the electronic polarizability still remains the same up to optical frequencies, because of the smaller inertia in the electrons.

11–4 Electric fields in cavities of a dielectric

We now turn to an interesting but complicated question—the problem of the dielectric constant in dense materials. Suppose that we take liquid helium or liquid argon or some other nonpolar material. We still expect electronic polarization. But in a dense material, $\boldsymbol{P}$ can be large, so the field on an individual atom will be influenced by the polarization of the atoms in its close neighborhood. The question is, what electric field acts on the individual atom?

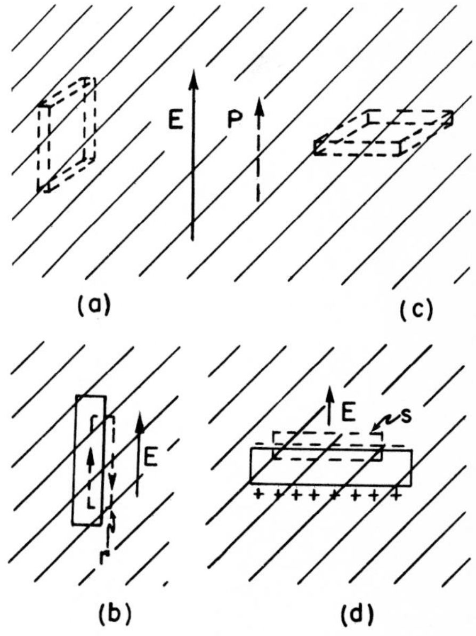

Fig. 11–5. The field in a slot cut in a dielectric depends on the shape and orientation of the slot.

Imagine that the liquid is put between the plates of a condenser. If the plates are charged they will produce an electric field in the liquid. But there are also charges in the individual atoms, and the total field $\boldsymbol{E}$ is the sum of both of these effects. This true electric field varies very, very rapidly from point to point in the liquid. It is very high inside the atoms—particularly right next to the nucleus—and relatively small between the atoms. The potential difference between the plates is the line integral of this total field. If we ignore all the fine-grained variations, we can think of an *average* electric field E, which is just V/d. (This is the field we were using in the last chapter.) We should think of this field as the average over a space containing many atoms.

Now you might think that an "average" atom in an "average" location would feel this average field. But it is not that simple, as we can show by considering what happens if we imagine different-shaped holes in a dielectric. For instance, suppose that we cut a slot in a polarized dielectric, with the slot oriented parallel to the field, as shown in part (a) of Fig. 11–5. Since we know that $\boldsymbol{\nabla} \times \boldsymbol{E} = 0$, the line integral of $\boldsymbol{E}$ around the curve, Γ, which goes as shown in (b) of the figure, should

* Sänger, Steiger, and Gächter, *Helvetica Physica Acta* **5**, 200 (1932).

be zero. The field inside the slot must give a contribution which just cancels the part from the field outside. Therefore the field E_0 actually found in the center of a long thin slot is equal to E, the average electric field found in the dielectric.

Now consider another slot whose large sides are perpendicular to E, as shown in part (c) of Fig. 11–5. In this case, the field E_0 in the slot is not the same as E because polarization charges appear on the surfaces. If we apply Gauss' law to a surface S drawn as in (d) of the figure, we find that the field E_0 *in the slot* is given by

$$E_0 = E + \frac{P}{\epsilon_0}, \tag{11.22}$$

where E is again the electric field in the dielectric. (The gaussian surface contains the surface polarization charge $\sigma_{\text{pol}} = P$.) We mentioned in Chapter 10 that $\epsilon_0 E + P$ is often called D, so $\epsilon_0 E_0 = D_0$ is equal to D in the dielectric.

Earlier in the history of physics, when it was supposed to be very important to define every quantity by direct experiment, people were delighted to discover that they could define what they meant by E and D in a dielectric without having to crawl around between the atoms. The average field $\boldsymbol{E}$ is numerically equal to the field $\boldsymbol{E}_0$ that would be measured in a slot cut parallel to the field. And the field $\boldsymbol{D}$ could be measured by finding E_0 in a slot cut normal to the field. But nobody ever measures them that way anyway, so it was just one of those philosophical things.

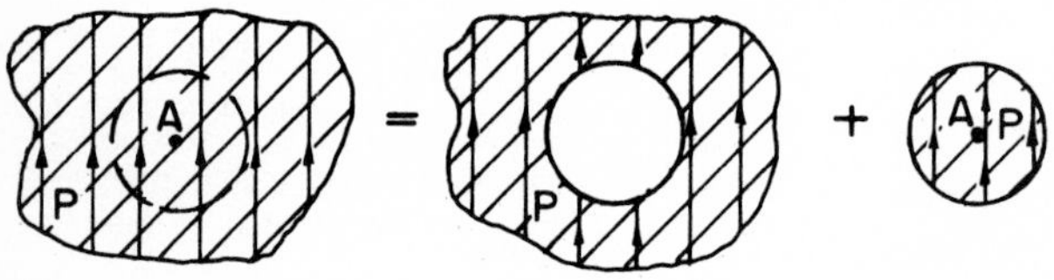

Fig. 11–6. The field at any point A in a dielectric can be considered as the sum of the field in a spherical hole plus the field due to a spherical plug.

For most liquids which are not too complicated in structure, we could expect that an atom finds itself, on the average, surrounded by the other atoms in what would be a good approximation to a *spherical hole*. And so we should ask: "What would be the field in a spherical hole?" We can find out by noticing that if we imagine carving out a spherical hole in a uniformly polarized material, we are just removing a sphere of polarized material. (We must imagine that the polarization is "frozen in" before we cut out the hole.) By superposition, however, the fields inside the dielectric, before the sphere was removed, is the sum of the fields from all charges outside the spherical volume plus the fields from the charges within the polarized sphere. That is, if we call E the field in the uniform dielectric, we can write

$$E = E_{\text{hole}} + E_{\text{plug}}, \tag{11.23}$$

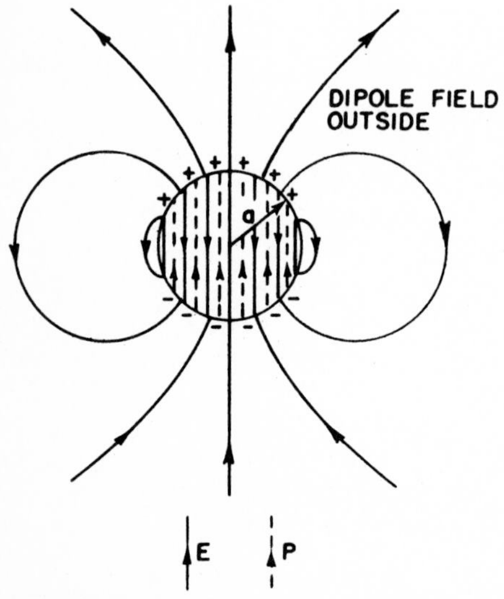

Fig. 11–7. The electric field of a uniformly polarized sphere.

where E_{hole} is the field in the hole and E_{plug} is the field inside a sphere which is uniformly polarized (see Fig. 11–6). The fields due to a uniformly polarized sphere are shown in Fig. 11–7. The electric field inside the sphere is uniform, and its value is

$$E_{\text{plug}} = -\frac{P}{3\epsilon_0}. \tag{11.24}$$

Using (11.23), we get

$$E_{\text{hole}} = E + \frac{P}{3\epsilon_0}. \tag{11.25}$$

The field in a spherical cavity is greater than the average field by the amount $P/3\epsilon_0$. (The spherical hole gives a field 1/3 of the way between a slot parallel to the field and a slot perpendicular to the field.)

11–5 The dielectric constant of liquids; the Clausius-Mossotti equation

In a liquid we expect that the field which will polarize an individual atom is more like E_{hole} than just E. If we use the E_{hole} of (11.25) for the polarizing field in

Eq. (11.6), then Eq. (11.8) becomes

$$P = N\alpha\epsilon_0\left(E + \frac{P}{3\epsilon_0}\right), \tag{11.26}$$

or

$$P = \frac{N\alpha}{1 - (N\alpha/3)}\,\epsilon_0 E. \tag{11.27}$$

Remembering that $\kappa - 1$ is just $P/\epsilon_0 E$, we have

$$\kappa - 1 = \frac{N\alpha}{1 - (N\alpha/3)}, \tag{11.28}$$

which gives us the dielectric constant of a liquid in terms of α, the atomic polarizability. This is called the *Clausius-Mossotti* equation.

Whenever $N\alpha$ is very small, as it is for a gas (because the density N is small), then the term $N\alpha/3$ can be neglected compared with 1, and we get our old result, Eq. (11.9), that

$$\kappa - 1 = N\alpha. \tag{11.29}$$

Let's compare Eq. (11.28) with some experimental results. It is first necessary to look at gases for which, using the measurement of κ, we can find α from Eq. (11.29). For instance, for carbon disulfide at zero degrees centigrade the dielectric constant is 1.0029, so $N\alpha$ is 0.0029. Now the density of the gas is easily worked out and the density of the liquid can be found in handbooks. At 20°C, the density of liquid CS_2 is 381 times higher than the density of the gas at 0°C. This means that N is 381 times higher in the liquid than it is in the gas so, that—if we make the approximation that the basic atomic polarizability of the carbon disulfide doesn't change when it is condensed into a liquid—$N\alpha$ in the liquid is equal to 381 times 0.0029, or 1.11. Notice that the $N\alpha/3$ term amounts to almost 0.4, so it is quite significant. With these numbers we predict a dielectric constant of 2.76, which agrees reasonably well with the observed value of 2.64.

In Table 11–1 we give some experimental data on various materials (taken from the *Handbook of Chemistry and Physics*), together with the dielectric constants calculated from Eq. (11.28) in the way just described. The agreement between observation and theory is even better for argon and oxygen than for CS_2—and not so good for carbon tetrachloride. On the whole, the results show that Eq. (11.28) works very well.

Table 11–1

Computation of the dielectric constants of liquids from the dielectric constant of the gas.

	Gas			Liquid				
Substance	κ (exp)	$N\alpha$	Density	Density	Ratio*	$N\alpha$	κ (predict)	κ (exp)
CS_2	1.0029	0.0029	0.00339	1.293	381	1.11	2.76	2.64
O_2	1.000523	0.000523	0.00143	1.19	832	0.435	1.509	1.507
CCl_4	1.0030	0.0030	0.00489	1.59	325	0.977	2.45	2.24
A	1.000545	0.000545	0.00178	1.44	810	0.441	1.517	1.54

* Ratio = density of liquid/density of gas.

Our derivation of Eq. (11.28) is valid only for *electronic* polarization in liquids. It is not right for a polar molecule like H_2O. If we go through the same calculations for water, we get 13.2 for $N\alpha$, which means that the dielectric constant for the liquid is *negative*, while the observed value of κ is 80. The problem has to do with the correct treatment of the permanent dipoles, and Onsager has pointed out the right way to go. We do not have the time to treat the case now, but if you are interested it is discussed in Kittel's book, *Introduction to Solid State Physics*.

11–6 Solid dielectrics

Now we turn to the solids. The first interesting fact about solids is that there can be a permanent polarization built in—which exists even without applying an electric field. An example occurs with a material like wax, which contains long molecules having a permanent dipole moment. If you melt some wax and put a strong electric field on it when it is a liquid, so that the dipole moments get partly lined up, they will stay that way when the liquid freezes. The solid material will have a permanent polarization which remains when the field is removed. Such a solid is called an *electret*.

An electret has permanent polarization charges on its surface. It is the electrical analog of a magnet. It is not as useful, though, because free charges from the air are attracted to its surfaces, eventually cancelling the polarization charges. The electret is "discharged" and there are no visible external fields.

A permanent internal polarization P is also found occurring naturally in some crystalline substances. In such crystals, each unit cell of the lattice has an identical permanent dipole moment, as drawn in Fig. 11–8. All the dipoles point in the same direction, even with no applied electric field. Many complicated crystals have, in fact, such a polarization; we do not normally notice it because the external fields are discharged, just as for the electrets.

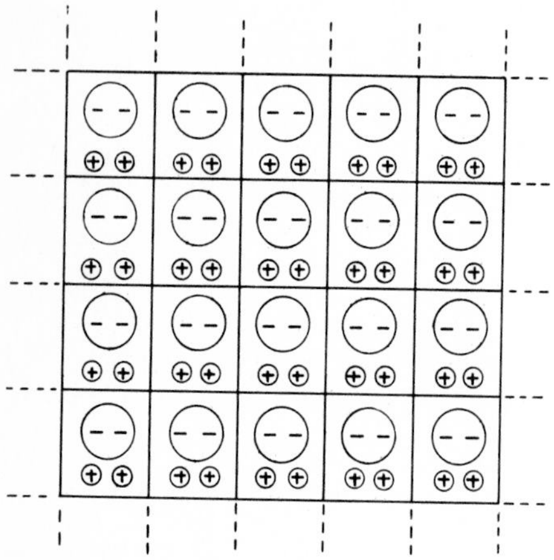

Fig. 11–8. A complex crystal lattice can have a permanent intrinsic polarization P.

If these internal dipole moments of a crystal are changed, however, external fields appear because there is not time for stray charges to gather and cancel the polarization charges. If the dielectric is in a condenser, free charges will be induced on the electrodes. For example, the moments can change when a dielectric is heated, because of thermal expansion. The effect is called *pyroelectricity*. Similarly, if we change the stresses in a crystal—for instance, if we bend it—again the moment may change a little bit, and a small electrical effect, called *piezoelectricity*, can be detected.

For crystals that do not have a permanent moment, one can work out a theory of the dielectric constant that involves the electronic polarizability of the atoms. It goes much the same as for liquids. Some crystals also have rotatable dipoles inside, and the rotation of these dipoles will also contribute to κ. In ionic crystals such as NaCl there is also *ionic polarizability*. The crystal consists of a checkerboard of positive and negative ions, and in an electric field the positive ions are pulled one way and the negatives the other; there is a net relative motion of the plus and minus charges, and so a volume polarization. We could estimate the magnitude of the ionic polarizability from our knowledge of the stiffness of salt crystals, but we will not go into that subject here.

11–7 Ferroelectricity; $BaTiO_3$

We want to describe now one special class of crystals which have, just by accident almost, a built-in permanent moment. The situation is so marginal that if we increase the temperature a little bit they lose the permanent moment completely. On the other hand, if they are nearly cubic crystals, so that their moments can be turned in different directions, we can detect a large change in the moment when an applied electric field is changed. All the moments flip over and we get a large effect. Substances which have this kind of permanent moment are called *ferroelectric*, after the corresponding ferromagnetic effects which were first discovered in iron.

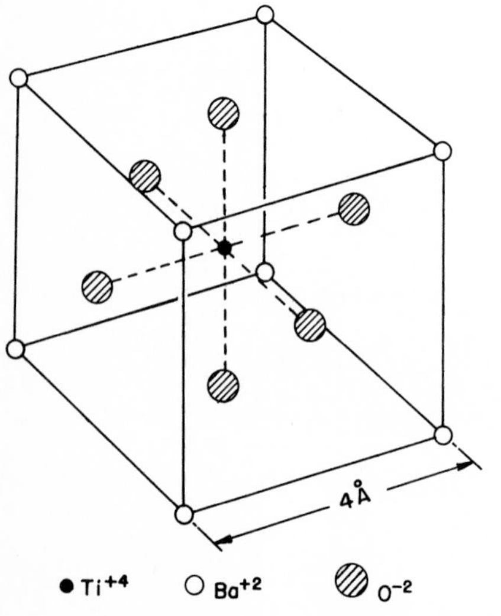

Fig. 11–9. The unit cell of $BaTiO_3$. The atoms really fill up most of the space; for clarity, only the positions of their centers are shown.

We would like to explain how ferroelectricity works by describing a particular example of a ferroelectric material. There are several ways in which the ferroelectric property can originate; but we will take up only one mysterious case—that of barium titanate, $BaTiO_3$. This material has a crystal lattice whose basic cell is sketched in Fig. 11–9. It turns out that above a certain temperature, specifically 118°C, barium titanate is an ordinary dielectric with an enormous dielectric constant. Below this temperature, however, it suddenly takes on a permanent moment.

In working out the polarization of solid material, we must first find what are the local fields in each unit cell. We must include the fields from the polarization itself, just as we did for the case of a liquid. But a crystal is not a homogeneous liquid, so we cannot use for the local field what we would get in a spherical hole. If you work it out for a crystal, you find that the factor 1/3 in Eq. (11.24) becomes slightly different, but not far from 1/3. (For a simple cubic crystal, it is just 1/3.) We will, therefore, assume for our preliminary discussion that the factor is 1/3 for $BaTiO_3$.

Now when we wrote Eq. (11.28) you may have wondered what would happen if $N\alpha$ became greater than 3. It appears as though κ would become negative. But that surely cannot be right. Let's see what should happen if we were gradually to increase α in a particular crystal. As α gets larger, the polarization gets bigger, making a bigger local field. But a bigger local field will polarize each atom more, raising the local fields still more. If the "give" of the atoms is enough, the process keeps going; there is a kind of feedback that causes the polarization to increase without limit—assuming that the polarization of each atom increases in proportion to the field. The "runaway" condition occurs when $N\alpha = 3$. The polarization does not become infinite, of course, because the proportionality between the induced moment and the electric field breaks down at high fields, so that our formulas are no longer correct. What happens is that the lattice gets "locked in" with a high, self-generated, internal polarization.

In the case of $BaTiO_3$, there is, in addition to an electronic polarization, also a rather large ionic polarization, presumed to be due to titanium ions which can move a little within the cubic lattice. The lattice resists large motions, so after the titanium has gone a little way, it jams up and stops. But the crystal cell is then left with a permanent dipole moment.

In most crystals, this is really the situation for all temperatures that can be reached. The very interesting thing about barium titanate is that there is such a delicate condition that if $N\alpha$ is decreased just a little bit it comes unstuck. Since N decreases with increasing temperature—because of thermal expansion—we can vary $N\alpha$ by varying the temperature. Below the critical temperature it is just barely stuck, so it is easy—by applying an external field—to shift the polarization and have it lock in a different direction.

Let's see if we can analyze what happens in more detail. We call T_c the critical temperature at which $N\alpha$ is exactly 3. As the temperature increases, N goes down a little bit because of the expansion of the lattice. Since the expansion is small, we can say that near the critical temperature

$$N\alpha = 3 - \beta(T - T_c), \tag{11.30}$$

where β is a small constant, of the same order of magnitude as the thermal expansion coefficient, or about 10^{-5} to 10^{-6} per degree C. Now if we substitute this relation into Eq. (11.28), we get that

$$\kappa - 1 = \frac{3 - \beta(T - T_c)}{\beta(T - T_c)/3}.$$

Since we have assumed that $\beta(T - T_c)$ is small compared with one, we can approximate this formula by

$$\kappa - 1 = \frac{9}{\beta(T - T_c)}. \tag{11.31}$$

This relation is right, of course, only for $T > T_c$. We see that just above the critical temperature κ is enormous. Because $N\alpha$ is so close to 3, there is a tremendous magnification effect, and the dielectric constant can easily be as high as 50,000 to 100,000. It is also very sensitive to temperature. For increases in temperature, the dielectric constant goes down inversely as the temperature, but, unlike the case of a dipolar gas, for which $\kappa - 1$ goes inversely as the *absolute* temperature, for ferroelectrics it varies inversely as the difference between the absolute temperature and the critical temperature (this law is called the Curie-Weiss law).

When we lower the temperature to the critical temperature, what happens? If we imagine a lattice of unit cells like that in Fig. 11–9, we see that it is possible to pick out chains of ions along vertical lines. One of them consists of alternating oxygen and titanium ions. There are other lines made up of either barium or oxygen ions, but the spacing along these lines is greater. We make a simple model to imitate this situation by imagining, as shown in Fig. 11–10(a), a series of chains of ions. Along what we call the main chain, the separation of the ions is a, which is *half* the lattice constant; the lateral distance between identical chains is $2a$. There are less-dense chains in between which we will ignore for the moment. To make the analysis a little easier, we will also suppose that all the ions on the main chain are identical. (It is not a serious simplification because all the important effects will still appear. This is one of the tricks of theoretical physics. One does a different problem because it is easier to figure out the first time—then when one understands how the thing works, it is time to put in all the complications.)

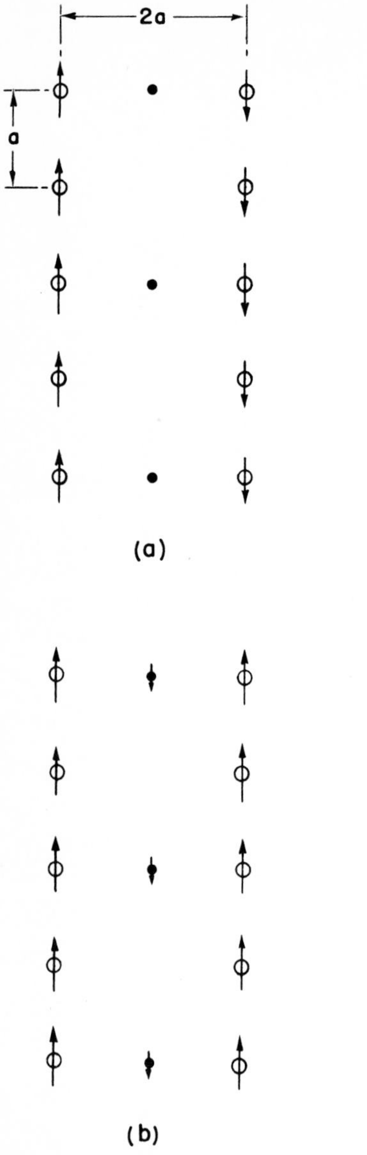

Fig. 11–10. Models of a ferroelectric: (a) corresponds to an antiferroelectric, and (b) to a normal ferroelectric.

Now let's try to find out what would happen with our model. We suppose that the dipole moment of each atom is p and we wish to calculate the field at one of the atoms of the chain. We must find the sum of the fields from all the other atoms. We will first calculate the field from the dipoles in only one vertical chain; we will talk about the other chains later. The field at the distance r from a dipole in a direction along its axis is given by

$$E = \frac{1}{4\pi\epsilon_0}\frac{2p}{r^3}. \tag{11.32}$$

At any given atom, the dipoles at equal distances above and below it give fields in the same direction, so for the whole chain we get

$$E_{\text{chain}} = \frac{p}{4\pi\epsilon_0}\frac{2}{a^3}\cdot\left(2 + \frac{2}{8} + \frac{2}{27} + \frac{2}{64} + \cdots\right) = \frac{p}{\epsilon_0}\frac{0.383}{a^3}. \tag{11.33}$$

It is not too hard to show that if our model were like a completely cubic crystal—that is, if the next identical lines were only the distance a away—the number 0.383 would be changed to 1/3. In other words, if the next lines were at the distance a they would contribute only -0.050 unit to our sum. However, the next main chain we are considering is at the distance $2a$ and, as you remember from Chapter 7, the field from a periodic structure dies off exponentially with distance. Therefore these lines contribute much less than -0.050 and we can just ignore all the other chains.

It is necessary now to find out what polarizability α is needed to make the runaway process work. Suppose that the induced moment p of each atom of the chain is proportional to the field on it, as in Eq. (11.6). We get the polarizing field on the atom from E_{chain}, using Eq. (11.32). So we have the two equations

$$p = \alpha\epsilon_0 E_{\text{chain}}$$

and

$$E_{\text{chain}} = \frac{0.383}{a^3}\frac{p}{\epsilon_0}.$$

There are two solutions: E and p both zero, or

$$\alpha = \frac{a^3}{0.383},$$

with E and p both finite. Thus if α is as large as $a^3/0.383$, a permanent polarization sustained by its own field will set in. This critical equality must be reached for barium titanate at just the temperature T_c. (Notice that if α were larger than the critical value for small fields, it would decrease at larger fields and at equilibrium the same equality we have found would hold.)

For $BaTiO_3$, the spacing a is 2×10^{-8} cm, so we must expect that $\alpha = 21.8 \times 10^{-24}\ \text{cm}^3$. We can compare this with the known polarizabilities of the individual atoms. For oxygen, $\alpha = 30.2 \times 10^{-24}\ \text{cm}^3$; we're on the right track! But for titanium, $\alpha = 2.4 \times 10^{-24}\ \text{cm}^3$; rather small. To use our model we should probably take the average. (We could work out the chain again for alternating atoms, but the result would be about the same.) So $\alpha(\text{average}) = 16.3 \times 10^{-24}$, which is not high enough to give a permanent polarization.

But wait a moment! We have so far only added up the electronic polarizabilities. There is also some ionic polarization due to the motion of the titanium ion. All we need is an ionic polarizability of $9.2 \times 10^{-24}\ \text{cm}^3$. (A more precise computation using alternating atoms shows that actually 11.9×10^{-24} is needed.) To understand the properties of $BaTiO_3$, we have to assume that such an ionic polarizability exists.

Why the titanium ion in barium titanate should have that much ionic polarizability is not known. Furthermore, why, at a lower temperature, it polarizes along the cube diagonal and the face diagonal equally well is not clear. If we figure out the actual size of the spheres in Fig. 11–9, and ask whether the titanium is a little bit loose in the box formed by its neighboring oxygen atoms—which is what you would hope, so that it could be easily shifted—you find quite the contrary. It fits very tightly. The *barium* atoms are slightly loose, but if you let them be the ones that move, it doesn't work out. So you see that the subject is really not one-hundred percent clear; there are still mysteries we would like to understand.

Returning to our simple model of Fig. 11–10(a), we see that the field from one chain would tend to polarize the neighboring chain in the *opposite* direction, which means that although each chain would be locked, there would be no net permanent moment per unit volume! (Although there would be no external electric effects, there are still certain thermodynamic effects one could observe.) Such systems exist, and are called antiferroelectric. So what we have explained is really an antiferroelectric. Barium titanate, however, is really like the arrangement in Fig. 11–10(b). The oxygen-titanium chains are all polarized in the same direction because there are intermediate chains of atoms in between. Although the atoms in these chains are not very polarizable, or very dense, they will be somewhat polarized, in the direction antiparallel to the oxygen-titanium chains. The small fields produced at the next oxygen-titanium chain will get it started parallel to the first. So $BaTiO_3$ is really ferroelectric, and it is because of the atoms in between. You may be wondering: "But what about the direct effect between the two O-Ti chains?" Remember, though, the direct effect dies off exponentially with the separation; the effect of the chain of *strong* dipoles at $2a$ can be less than the effect of a chain of weak ones at the distance a.

This completes our rather detailed report on our present understanding of the dielectric constants of gases, of liquids, and of solids.

12

Electrostatic Analogs

12–1 The same equations have the same solutions

The total amount of information which has been acquired about the physical world since the beginning of scientific progress is enormous, and it seems almost impossible that any one person could know a reasonable fraction of it. But it is actually quite possible for a physicist to retain a broad knowledge of the physical world rather than to become a specialist in some narrow area. The reasons for this are threefold: First, there are great principles which apply to all the different kinds of phenomena—such as the principles of the conservation of energy and of angular momentum. A thorough understanding of such principles gives an understanding of a great deal all at once. Second, there is the fact that many complicated phenomena, such as the behavior of solids under compression, really basically depend on electrical and quantum-mechanical forces, so that if one understands the fundamental laws of electricity and quantum mechanics, there is at least some possibility of understanding many of the phenomena that occur in complex situations. Finally, there is a most remarkable coincidence: *The equations for many different physical situations have exactly the same appearance.* Of course, the symbols may be different—one letter is substituted for another—but the mathematical form of the equations is the same. This means that having studied one subject, we immediately have a great deal of direct and precise knowledge about the solutions of the equations of another.

We are now finished with the subject of electrostatics, and will soon go on to study magnetism and electrodynamics. But before doing so, we would like to show that while learning electrostatics we have simultaneously learned about a large number of other subjects. We will find that the equations of electrostatics appear in several other places in physics. By a direct translation of the solutions (of course the same mathematical equations must have the same solutions) it is possible to solve problems in other fields with the same ease—or with the same difficulty—as in electrostatics.

The equations of electrostatics, we know, are

$$\nabla \cdot (\kappa \boldsymbol{E}) = \frac{\rho_{\text{free}}}{\epsilon_0}, \tag{12.1}$$

$$\nabla \times \boldsymbol{E} = 0. \tag{12.2}$$

(We take the equations of electrostatics with dielectrics so as to have the most general situation.) The same physics can be expressed in another mathematical form:

$$\boldsymbol{E} = -\nabla\phi, \tag{12.3}$$

$$\nabla \cdot (\kappa\, \nabla\phi) = -\frac{\rho_{\text{free}}}{\epsilon_0}. \tag{12.4}$$

Now the point is that there are many physics problems whose mathematical equations have the same form. There is a potential (ϕ) whose gradient multiplied by a scalar function (κ) has a divergence equal to another scalar function ($-\rho/\epsilon_0$).

Whatever we know about electrostatics can immediately be carried over into that other subject, and *vice versa.* (It works both ways, of course—if the other subject has some particular characteristics that are known, then we can apply that knowledge to the corresponding electrostatic problem.) We want to consider a series of examples from different subjects that produce equations of this form.

12–2 The flow of heat; a point source near an infinite plane boundary

We have discussed one example earlier (Section 3–4)—the flow of heat. Imagine a block of material, which need not be homogeneous but may consist of different materials at different places, in which the temperature varies from point to point. As a consequence of these temperature variations there is a flow of heat, which can be represented by the vector $\boldsymbol{h}$. It represents the amount of heat energy which flows per unit time through a unit area perpendicular to the flow. The divergence of $\boldsymbol{h}$ represents the rate per unit volume at which heat is leaving a region:

$$\nabla \cdot \boldsymbol{h} = \text{rate of heat out per unit volume.}$$

(We could, of course, write the equation in integral form—just as we did in electrostatics with Gauss' law—which would say that the flux through a surface is equal to the rate of change of heat energy inside the material. We will not bother to translate the equations back and forth between the differential and the integral forms, because it goes exactly the same as in electrostatics.)

The rate at which heat is generated or absorbed at various places depends, of course, on the problem. Suppose, for example, that there is a source of heat inside the material (perhaps a radioactive source, or a resistor heated by an electrical current). Let us call s the heat energy produced per unit volume per second by this source. There may also be losses (or gains) of thermal energy to other internal energies in the volume. If u is the internal energy per unit volume, $-du/dt$ will also be a "source" of heat energy. We have, then,

$$\nabla \cdot \boldsymbol{h} = s - \frac{du}{dt}. \tag{12.5}$$

We are not going to discuss just now the complete equation in which things change with time, because we are making an analogy to electrostatics, where nothing depends on the time. We will consider only *steady heat-flow* problems, in which constant sources have produced an equilibrium state. In these cases,

$$\nabla \cdot \boldsymbol{h} = s. \tag{12.6}$$

It is, of course, necessary to have another equation, which describes how the heat flows at various places. In many materials the heat current is approximately proportional to the rate of change of the temperature with position: the larger the temperature difference, the more the heat current. As we have seen, the *vector* heat current is proportional to the temperature gradient. The constant of proportionality K, a property of the material, is called the *thermal conductivity.*

$$\boldsymbol{h} = -K\,\nabla T. \tag{12.7}$$

If the properties of the material vary from place to place, then $K = K(x, y, z)$, a function of position. [Equation (12.7) is not as fundamental as (12.5), which expresses the conservation of heat energy, since the former depends upon a special property of the substance.] If now we substitute Eq. (12.7) into Eq. (12.6) we have

$$\nabla \cdot (K\,\nabla T) = -s, \tag{12.8}$$

which has exactly the same form as (12.4). *Steady heat-flow problems and electrostatic problems are the same.* The heat flow vector $\boldsymbol{h}$ corresponds to $\boldsymbol{E}$, and the temperature T corresponds to ϕ. We have already noticed that a point heat source produces a temperature field which varies as $1/r$ and a heat flow which varies as $1/r^2$. This is nothing more than a translation of the statements from electrostatics that a point charge generates a potential which varies as $1/r$ and an electric field which varies as $1/r^2$. We can, in general, solve static heat problems as easily as we can solve electrostatic problems.

Consider a simple example. Suppose that we have a cylinder of radius a at the temperature T_1, maintained by the generation of heat in the cylinder. (It could be, for example, a wire carrying a current, or a pipe with steam condensing inside.)

The cylinder is covered with a concentric sheath of insulating material which has a conductivity K. Say the outside radius of the insulation is b and the outside is kept at temperature T_2 (Fig. 12–1a). We want to find out at what rate heat will be lost by the wire, or steampipe, or whatever it is in the center. Let the total amount of heat lost from a length L of the pipe be called G—which is what we are trying to find.

How can we solve this problem? We have the differential equations, but since these are the same as those of electrostatics, we have really already solved the mathematical problem. The analogous problem is that of a conductor of radius a at the potential ϕ_1, separated from another conductor of radius b at the potential ϕ_2, with a concentric layer of dielectric material in between, as drawn in Fig. 12–1(b). Now since the heat flow $\boldsymbol{h}$ corresponds to the electric field $\boldsymbol{E}$, the quantity G that we want to find corresponds to the flux of the electric field from a unit length (in other words, to the electric charge per unit length over ϵ_0). We have solved the electrostatic problem by using Gauss' law. We follow the same procedure for our heat-flow problem.

From the symmetry of the situation, we know that h depends only on the distance from the center. So we enclose the pipe in a gaussian cylinder of length L and radius r. From Gauss' law, we know that the heat flow h multiplied by the area $2\pi rL$ of the surface must be equal to the total amount of heat generated inside, which is what we are calling G:

$$2\pi rLh = G \qquad \text{or} \qquad h = \frac{G}{2\pi rL}. \tag{12.9}$$

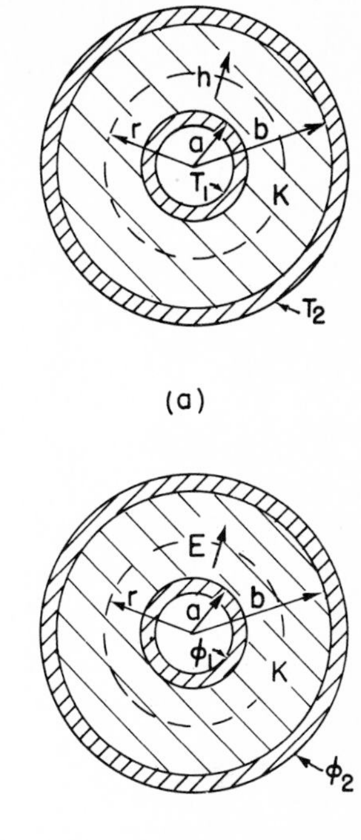

Fig. 12–1. (a) Heat flow in a cylindrical geometry. (b) The corresponding electrical problem.

The heat flow is proportional to the temperature gradient:

$$\boldsymbol{h} = -K\,\boldsymbol{\nabla} T,$$

or, in this case, the magnitude of $\boldsymbol{h}$ is

$$h = -K\frac{dT}{dr}.$$

This, together with (12.9), gives

$$\frac{dT}{dr} = -\frac{G}{2\pi KLr}. \tag{12.10}$$

Integrating from $r = a$ to $r = b$, we get

$$T_2 - T_1 = -\frac{G}{2\pi KL}\ln\frac{b}{a}. \tag{12.11}$$

Solving for G, we find

$$G = \frac{2\pi KL(T_1 - T_2)}{\ln(b/a)}. \tag{12.12}$$

This result corresponds exactly to the result for the charge on a cylindrical condenser:

$$Q = \frac{2\pi\epsilon_0 L(\phi_1 - \phi_2)}{\ln(b/a)}.$$

The problems are the same, and they have the same solutions. From our knowledge of electrostatics, we also know how much heat is lost by an insulated pipe.

Let's consider another example of heat flow. Suppose we wish to know the heat flow in the neighborhood of a point source of heat located a little way beneath the surface of the earth, or near the surface of a large metal block. The localized heat source might be an atomic bomb that was set off underground, leaving an intense source of heat, or it might correspond to a small radioactive source inside a block of iron—there are numerous possibilities.

We will treat the idealized problem of a point heat source of strength G at the distance a beneath the surface of an infinite block of uniform material whose thermal conductivity is K. And we will neglect the thermal conductivity of the air outside the material. We want to determine the distribution of the temperature on the surface of the block. How hot is it right above the source and at various places on the surface of the block?

How shall we solve it? It is like an electrostatic problem with two materials with different dielectric coefficients κ on opposite sides of a plane boundary. Aha! Perhaps it is the analog of a point charge near the boundary between a dielectric and a conductor, or something similar. Let's see what the situation is near the surface. The physical condition is that the normal component of $\boldsymbol{h}$ on the surface is *zero*, since we have assumed there is no heat flow out of the block. We should ask: In what electrostatic problem do we have the condition that the normal component of the electric field $\boldsymbol{E}$ (which is the analog of $\boldsymbol{h}$) is *zero* at a surface? There is none!

That is one of the things that we have to watch out for. For physical reasons, there may be certain restrictions in the kinds of mathematical conditions which arise in any one subject. So if we have analyzed the differential equation only for certain limited cases, we may have missed some kinds of solutions that can occur in other physical situations. For example, there is no material with a dielectric constant of zero, whereas a vacuum does have zero thermal conductivity. So there is no electrostatic analogy for a perfect heat insulator. We can, however, still use the same *methods*. We can try to *imagine* what would happen if the dielectric constant *were* zero. (Of course, the dielectric constant is never zero in any real situation. But we might have a case in which there is a material with a very *high* dielectric constant, so that we could neglect the dielectric constant of the air outside.)

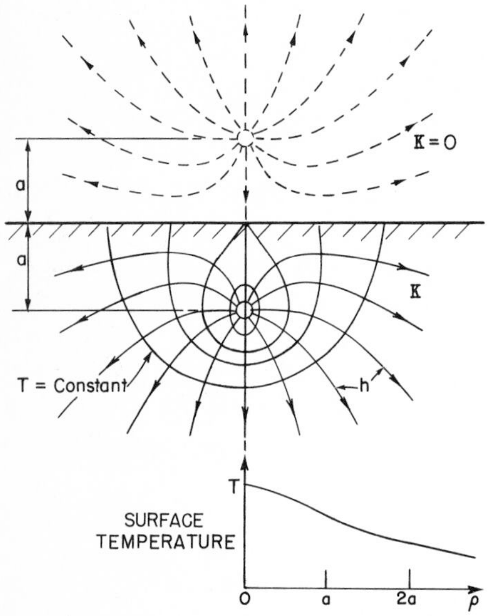

Fig. 12–2. The heat flow and isothermals near a point heat source at the distance a below the surface of a good thermal conductor. An image source is shown outside the material.

How shall we find an electric field that has *no* component perpendicular to the surface? That is, one which is always *tangent* at the surface? You will notice that our problem is opposite to the one of a point charge near a plane conductor. There we wanted the field to be *perpendicular* to the surface, because the conductor was all at the same potential. In the electrical problem, we invented a solution by imagining a point charge behind the conducting plate. We can use the same idea again. We try to pick an "image source" that will automatically make the normal component of the field zero at the surface. The solution is shown in Fig. 12–2. An image source of *the same sign* and the same strength placed at the distance a above the surface will cause the field to be always horizontal at the surface. The normal components of the two sources cancel out.

Thus our heat flow problem is solved. The temperature everywhere is the same, by direct analogy, as the potential due to two equal point charges! The temperature T at the distance r from a single point source G in an infinite medium is

$$T = \frac{G}{4\pi K r}. \tag{12.13}$$

(This, of course, is just the analog of $\phi = q/4\pi\epsilon_0 r$.) The temperature for a point source, together with its image source, is

$$T = \frac{G}{4\pi K r_1} + \frac{G}{4\pi K r_2}. \tag{12.14}$$

This formula gives us the temperature everywhere in the block. Several isothermal surfaces are shown in Fig. 12–2. Also shown are lines of $\boldsymbol{h}$, which can be obtained from $\boldsymbol{h} = -K\,\nabla T$.

We originally asked for the temperature distribution on the surface. For a point on the surface at the distance ρ from the axis, $r_1 = r_2 = \sqrt{\rho^2 + a^2}$, so

$$T(\text{surface}) = \frac{1}{4\pi K}\frac{2G}{\sqrt{\rho^2 + a^2}}. \tag{12.15}$$

This function is also shown in the figure. The temperature is, naturally, higher right above the source than it is farther away. This is the kind of problem that geophysicists often need to solve. We now see that it is the same kind of thing we have already been solving for electricity.

12–3 The stretched membrane

Now let us consider a completely different physical situation which, nevertheless, gives the same equations again. Consider a thin rubber sheet—a membrane—which has been stretched over a large horizontal frame (like a drumhead). Suppose now that the membrane is pushed up in one place and down in another; as shown in Fig. 12–3. Can we describe the shape of the surface? We will show how the problem can be solved when the deflections of the membrane are not too large.

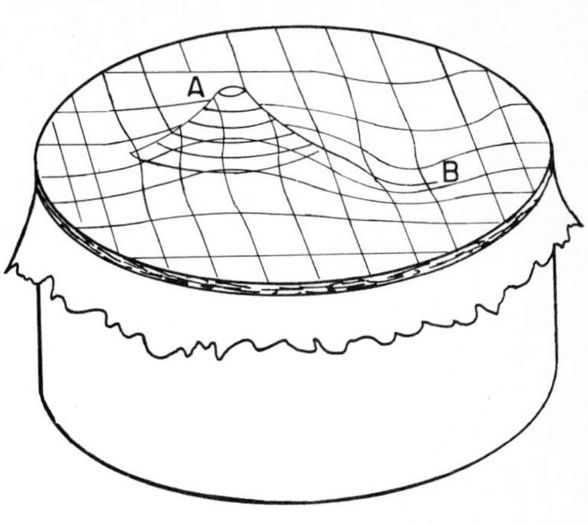

Fig. 12–3. A thin rubber sheet stretched over a cylindrical frame (like a drumhead). If the sheet is pushed up at *A* and down at *B*, what is the shape of the surface?

There are forces in the sheet because it is stretched. If we were to make a small cut anywhere, the two sides of the cut would pull apart (see Fig. 12–4). So there is a *surface tension* in the sheet, analogous to the one-dimensional tension in a stretched string. We define the magnitude of the surface tension τ as the force *per unit length* which will just hold together the two sides of a cut such as one of those shown in Fig. 12–4.

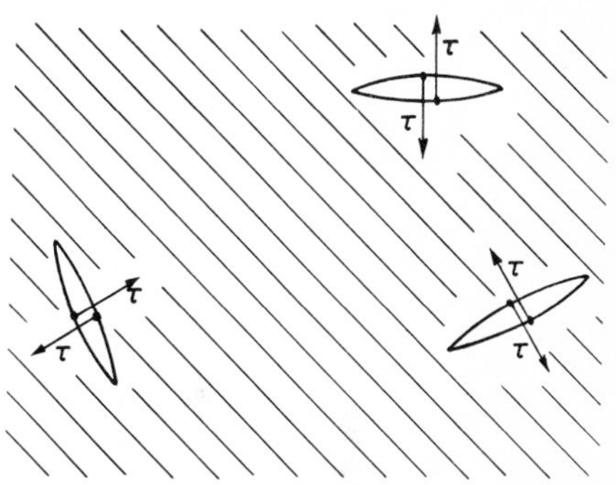

Fig. 12–4. The surface tension τ of a stretched rubber sheet is the force per unit length across a line.

Suppose now that we look at a vertical cross section of the membrane. It will appear as a curve, like the one in Fig. 12–5. Let u be the vertical displacement of the membrane from its normal position, and x and y the coordinates in the horizontal plane. (The cross section shown is parallel to the x-axis.)

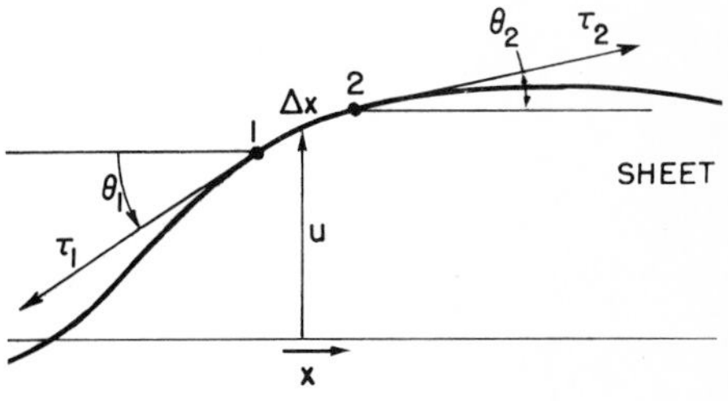

Fig. 12–5. Cross section of the deflected sheet.

Consider a little piece of the surface of length Δx and width Δy. There will be forces on the piece from the surface tension along each edge. The force along edge 1 of the figure will be $\tau_1\,\Delta y$, directed tangent to the surface—that is, at the angle θ_1 from the horizontal. Along edge 2, the force will be $\tau_2\,\Delta y$ at the angle θ_2. (There will be similar forces on the other two edges of the piece, but we will forget them for the moment.) The net *upward* force on the piece from edges 1 and 2 is

$$\Delta F = \tau_2\,\Delta y \sin\theta_2 - \tau_1\,\Delta y\sin\theta_1.$$

We will limit our considerations to small distortions of the membrane, i.e., to *small slopes:* we can then replace $\sin\theta$ by $\tan\theta$, which can be written as $\partial u/\partial x$. The force is then

$$\Delta F = \left[\tau_2\left(\frac{\partial u}{\partial x}\right)_2 - \tau_1\left(\frac{\partial u}{\partial x}\right)_1\right]\Delta y.$$

The quantity in brackets can be equally well written (for small Δx) as

$$\frac{\partial}{\partial x}\left(\tau\,\frac{\partial u}{\partial x}\right)\Delta x;$$

then

$$\Delta F = \frac{\partial}{\partial x}\left(\tau\,\frac{\partial u}{\partial x}\right)\Delta x\,\Delta y.$$

There will be another contribution to ΔF from the forces on the other two edges; the total is evidently

$$\Delta F = \left[\frac{\partial}{\partial x}\left(\tau\,\frac{\partial u}{\partial x}\right) + \frac{\partial}{\partial y}\left(\tau\,\frac{\partial u}{\partial y}\right)\right]\Delta x\,\Delta y. \tag{12.16}$$

The distortions of the diaphragm are caused by external forces. Let's let f represent the *upward* force *per unit area* on the sheet (a kind of "pressure") *from the external forces*. When the membrane is in equilibrium (the *static* case), this force must be balanced by the internal force we have just computed, Eq. (12.16). That is

$$f = -\frac{\Delta F}{\Delta x\,\Delta y}\cdot$$

Equation (12.16) can then be written

$$f = -\nabla\cdot(\tau\,\nabla u), \tag{12.17}$$

where by ∇ we now mean, of course, the two-dimensional gradient operator $(\partial/\partial x, \partial/\partial y)$. We have the differential equation that relates $u(x, y)$ to the applied forces $f(x, y)$ and the surface tension $\tau(x, y)$, which may, in general, vary from place to place in the sheet. (The distortions of a three-dimensional elastic body are also governed by similar equations, but we will stick to two-dimensions.) We will worry only about the case in which the tension τ is constant throughout the sheet. We can then write for Eq. (12.17),

$$\nabla^2 u = -\frac{f}{\tau}. \tag{12.18}$$

We have another equation that is the same as for electrostatics!—only this time, limited to two-dimensions. The displacement u corresponds to ϕ, and f/τ corresponds to ρ/ϵ_0. So all the work we have done for infinite plane charged sheets, or long parallel wires, or charged cylinders is directly applicable to the stretched membrane.

Suppose we push the membrane at some points up to a definite *height*—that is, we fix the value of u at some places. That is the analog of having a definite *potential* at the corresponding places in an electrical situation. So, for instance, we may make a positive "potential" by pushing up on the membrane with an object having the cross-sectional shape of the corresponding cylindrical conductor. For example, if we push the sheet up with a round rod, the surface will take on the shape shown in Fig. 12–6. The height u is the same as the electrostatic potential ϕ of a charged cylindrical rod. It falls off as $\ln(1/r)$. (The *slope*, which corresponds to the electric field E, drops off as $1/r$.)

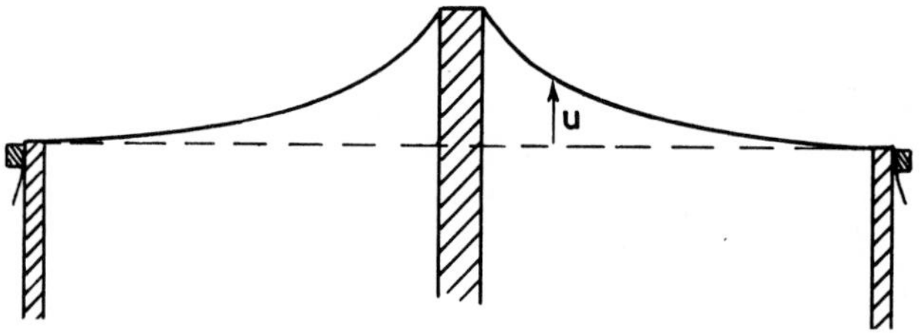

Fig. 12–6. Cross section of a stretched rubber sheet pushed up by a round rod. The function $u(x, y)$ is the same as the electric potential $\phi(x, y)$ near a very long charged rod.

The stretched rubber sheet has often been used as a way of solving complicated *electrical* problems experimentally. The analogy is used backwards! Various rods and bars are pushed against the sheet to heights that correspond to the potentials of a set of electrodes. Measurements of the height then give the electrical potential for the electrical situation. The analogy has been carried even further. If little balls are placed on the membrane, their motion corresponds approximately to the motion of electrons in the corresponding electric field. One can actually *watch* the "electrons" move on their trajectories. This method was used to design the complicated geometry of many photomultiplier tubes (such as the ones used for scintillation counters, and the one used for controlling the headlight beams on Cadillacs). The method is still used, but the accuracy is limited. For the most accurate work, it is better to determine the fields by numerical methods, using the large electronic computing machines.

12–4 The diffusion of neutrons; a uniform spherical source in a homogeneous medium

We take another example that gives the same kind of equation, this time having to do with diffusion. In Chapter 43 of Vol. I we considered the diffusion of ions in a single gas, and of one gas through another. This time, let's take a different example—the diffusion of neutrons in a material like graphite. We choose to speak of graphite (a pure form of carbon) because carbon doesn't absorb slow neutrons. In it the neutrons are free to wander around. They travel in a straight line for several centimeters, on the average, before being scattered by a nucleus and deflected into a new direction. So if we have a large block—many meters on a side—the neutrons initially at one place will diffuse to other places. We want to find a description of their average behavior—that is, their *average flow*.

Let $N(x, y, z)\,\Delta V$ be the number of neutrons in the element of volume ΔV at the point (x, y, z). Because of their motion, some neutrons will be leaving ΔV, and others will be coming in. If there are more neutrons in one region than in a nearby region, more neutrons will go from the first region to the second than come back; there will be a net flow. Following the arguments of Chapter 43 in Vol. I, we describe the flow by a flow vector $\boldsymbol{J}$. Its x-component J_x is the *net* number of neutrons that pass in unit time a unit area perpendicular to the x-direction. We found that

$$J_x = -D\,\frac{\partial N}{\partial x}, \tag{12.19}$$

where the diffusion constant D is given in terms of the mean velocity v, and the mean-free-path l between scatterings is given by

$$D = \frac{1}{3}\,lv.$$

The vector equation for $\boldsymbol{J}$ is

$$\boldsymbol{J} = -D\,\boldsymbol{\nabla} N. \tag{12.20}$$

The rate at which neutrons flow across any surface element da is $\boldsymbol{J}\cdot\boldsymbol{n}\,da$ (where, as usual, $\boldsymbol{n}$ is the unit normal). The net flow *out of a volume element* is then (following the usual gaussian argument) $\boldsymbol{\nabla}\cdot\boldsymbol{J}\,dV$. This flow would result in a decrease with time of the number in ΔV unless neutrons are being created in ΔV (by some nuclear process). If there are sources in the volume that generate S neutrons per unit time in a unit volume, then the net flow out of ΔV will be equal to $(S - \partial N/\partial t)\,\Delta V$. We have then that

$$\boldsymbol{\nabla}\cdot\boldsymbol{J} = S - \frac{\partial N}{\partial t}. \tag{12.21}$$

Combining (12.21) with (12.20), we get the *neutron diffusion equation*

$$\boldsymbol{\nabla}\cdot(-D\,\boldsymbol{\nabla} N) = S - \frac{\partial N}{\partial t}. \tag{12.22}$$

In the static case—where $\partial N/\partial t = 0$—we have Eq. (12.4) all over again! We can use our knowledge of electrostatics to solve problems about the diffusion of neutrons. So let's solve a problem. (You may wonder: *Why* do a problem if we have already done all the problems in electrostatics? We can do it *faster* this time because we *have* done the electrostatic problems!)

Suppose we have a block of material in which neutrons are being generated—say by uranium fission—uniformly throughout a spherical region of radius a (Fig. 12–7). We would like to know: What is the density of neutrons everywhere? How uniform is the density of neutrons in the region where they are being generated? What is the ratio of the neutron density at the center to the neutron density at the surface of the source region? Finding the answers is easy. The source density S_0 replaces the charge density ρ, so our problem is the same as the problem of a sphere of uniform charge density. Finding N is just like finding the potential ϕ. We have already worked out the fields inside and outside of a uniformly charged sphere; we can integrate them to get the potential. Outside, the potential is $Q/4\pi\epsilon_0 r$, with the total charge Q given by $4\pi a^3\rho/3$. So

$$\phi_{\text{outside}} = \frac{\rho a^3}{3\epsilon_0 r}. \tag{12.23}$$

For points inside, the field is due only to the charge $Q(r)$ inside the sphere of radius r, $Q(r) = 4\pi r^3\rho/3$, so

$$E = \frac{\rho r}{3\epsilon_0}. \tag{12.24}$$

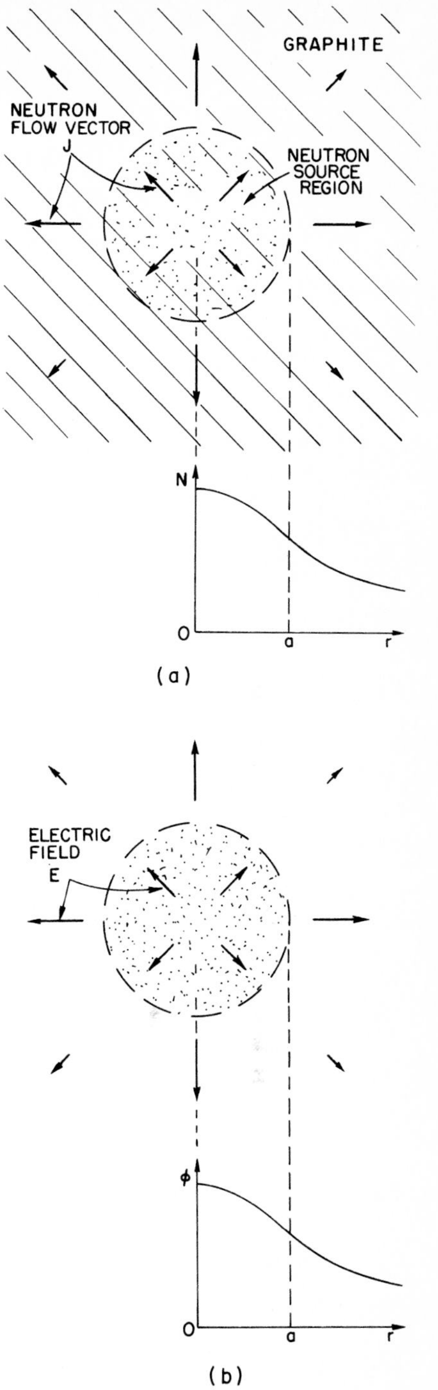

Fig. 12–7. (a) Neutrons are produced uniformly throughout a sphere of radius a in a large graphite block and diffuse outward. The neutron density N is found as a function of r, the distance from the center of the source. (b) The analogous electrostatic situation: a uniform sphere of charge, where N corresponds to ϕ and $\boldsymbol{J}$ corresponds to $\boldsymbol{E}$.

The field increases linearly with r. Integrating E to get ϕ, we have

$$\phi_{\text{inside}} = -\frac{\rho r^2}{6\epsilon_0} + \text{a constant.}$$

At the radius a, ϕ_{inside} must be the same as ϕ_{outside}, so the constant must be $\rho a^2/2\epsilon_0$. (We are assuming that ϕ is zero at large distances from the source, which will correspond to N being zero for the neutrons.) Therefore,

$$\phi_{\text{inside}} = \frac{\rho}{3\epsilon_0}\left(\frac{3a^2}{2} - \frac{r^2}{2}\right). \tag{12.25}$$

We know immediately the neutron density in our other problem. The answer is

$$N_{\text{outside}} = \frac{Sa^3}{3Dr}, \tag{12.26}$$

and

$$N_{\text{inside}} = \frac{S}{3D}\left(\frac{3a^2}{2} - \frac{r^2}{2}\right). \tag{12.27}$$

N is shown as a function of r in Fig. 12–7.

Now what is the ratio of density at the center to that at the edge? At the center ($r = 0$), it is proportional to $3a^2/2$. At the edge ($r = a$) it is proportional to $2a^2/2$, so the ratio of densities is $3/2$. A uniform source doesn't produce a uniform density of neutrons. You see, our knowledge of electrostatics gives us a good start on the physics of nuclear reactors.

There are many physical circumstances in which diffusion plays a big part. The motion of ions through a liquid, or of electrons through a semiconductor, obeys the same equation. We find again and again the same equations.

12–5 Irrotational fluid flow; the flow past a sphere

Let's now consider an example which is not really a very good one, because the equations we will use will not really represent the subject with complete generality but only in an artificial idealized situation. We take up the problem of *water flow*. In the case of the stretched sheet, our equations were an approximation which was correct only for *small deflections*. For our consideration of water flow, we will not make that kind of an approximation; we must make restrictions that do not apply at all to real water. We treat only the case of the steady flow of an *incompressible, nonviscous, circulation-free* liquid. Then we represent the flow by giving the velocity $v(\boldsymbol{r})$ as a function of position $\boldsymbol{r}$. If the motion is steady (the only case for which there is an electrostatic analog) $\boldsymbol{v}$ is independent of time. If ρ is the density of the fluid, then $\rho\boldsymbol{v}$ is the amount of mass which passes per unit time through a unit area. By the conservation of matter, the divergence of $\rho\boldsymbol{v}$ will be, in general, the time rate of change of the mass of the material per unit volume. We will assume that there are no processes for the continuous creation or destruction of matter. The conservation of matter then requires that $\nabla \cdot \rho\boldsymbol{v} = 0$. (It should, in general, be equal to $-\partial\rho/\partial t$, but since our fluid is incompressible, ρ cannot change.) Since ρ is everywhere the same, we can factor it out, and our equation is simply

$$\nabla \cdot \boldsymbol{v} = 0.$$

Good! We have electrostatics again (with no charges); it's just like $\nabla \cdot \boldsymbol{E} = 0$. Not so! Electrostatics is *not* simply $\nabla \cdot \boldsymbol{E} = 0$. It is a *pair* of equations. One equation does not tell us enough; we need still an additional equation. To match electrostatics, we should have also that the *curl* of $\boldsymbol{v}$ is zero. But that is not generally true for real liquids. Most liquids will ordinarily develop some circulation. So we are restricted to the situation in which there is no circulation of the fluid. Such flow is often called *irrotational*. Anyway, if we make all our assumptions, we can *imagine* a case of fluid flow that is analogous to electrostatics. So we take

$$\nabla \cdot \boldsymbol{v} = 0 \tag{12.28}$$

and

$$\nabla \times \boldsymbol{v} = 0. \tag{12.29}$$

We want to emphasize that the number of circumstances in which liquid flow follows these equations is far from the great majority, but there are a few. They must be cases in which we can neglect surface tension, compressibility, and viscosity, and in which we can assume that the flow is irrotational. These assumptions are valid so rarely for real water that the mathematician John von Neumann said that people who analyze Eqs. (12.28) and (12.29) are studying "dry water"! (We take up the problem of fluid flow in more detail in Chapters 40 and 41.)

Because $\nabla \times \boldsymbol{v} = 0$, the velocity of "dry water" can be written as the gradient of some potential:

$$\boldsymbol{v} = -\nabla\psi. \tag{12.30}$$

What is the physical meaning of ψ? There isn't any very useful meaning. The velocity can be written as the gradient of a potential simply because the flow is irrotational. And by analogy with electrostatics, ψ is called the *velocity potential*, but it is not related to a potential energy in the way that ϕ is. Since the divergence of $\boldsymbol{v}$ is zero, we have

$$\nabla \cdot (\nabla\psi) = \nabla^2\psi = 0. \tag{12.31}$$

The velocity potential ψ obeys the same differential equation as the electrostatic potential in free space ($\rho = 0$).

Let's pick a problem in irrotational flow and see whether we can solve it by the methods we have learned. Consider the problem of a spherical ball falling through a liquid. If it is going too slowly, the viscous forces, which we are disregarding, will be important. If it is going too fast, little whirlpools (turbulence) will appear in its wake and there will be some circulation of the water. But if the ball is going neither too fast nor too slow, it is more or less true that the water flow will fit our assumptions, and we can describe the motion of the water by our simple equations.

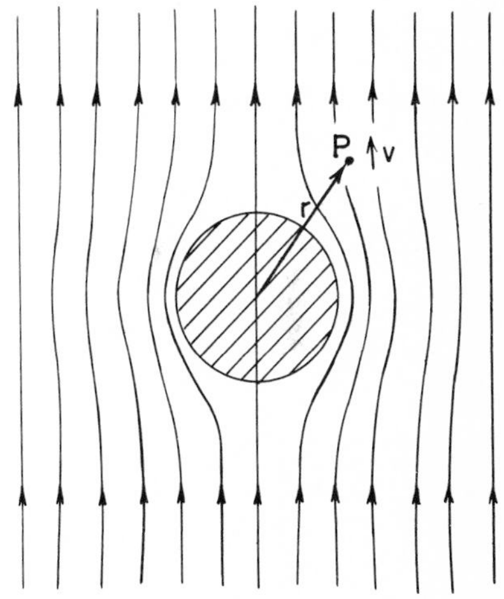

Fig. 12–8. The velocity field of irrotational fluid flow past a sphere.

It is convenient to describe what happens in a frame of reference *fixed in the sphere*. In this frame we are asking the question: How does water flow past a sphere at rest when the flow at large distances is uniform? That is, when, far from the sphere, the flow is everywhere the same. The flow near the sphere will be as shown by the streamlines drawn in Fig. 12–8. These lines, always parallel to $\boldsymbol{v}$, correspond to lines of electric field. We want to get a quantative description for the velocity field, i.e., an expression for the velocity at any point P.

We can find the velocity from the gradient of ψ, so we first work out the potential. We want a potential that satisfies Eq. (12.31) everywhere, and which also satisfies two restrictions: (1) there is no flow in the spherical region inside the surface of the ball, and (2) the flow is constant at large distances. To satisfy (1), the component of $\boldsymbol{v}$ normal to the surface of the sphere must be zero. That means that $\partial\psi/\partial r$ is zero at $r = a$. To satisfy (2), we must have $\partial\psi/\partial z = v_0$ at all points where $r \gg a$. Strictly speaking, there is no electrostatic case which corresponds exactly to our problem. It really corresponds to putting a sphere of dielectric constant *zero* in a uniform electric field. If we had worked out the solution to the problem of a sphere of a dielectric constant κ in a uniform field, then by putting $\kappa = 0$ we would immediately have the solution to this problem.

We have not actually worked out this particular electrostatic problem in detail, but let's do it now. (We could work directly on the fluid problem with $\boldsymbol{v}$ and ψ, but we will use $\boldsymbol{E}$ and ϕ because we are so used to them.)

The problem is: Find a solution of $\nabla^2\phi = 0$ such that $\boldsymbol{E} = -\nabla\phi$ is a constant, say $\boldsymbol{E}_0$, for large r, and such that the radial component of $\boldsymbol{E}$ is equal to zero at $r = a$. That is,

$$\left.\frac{\partial\phi}{\partial r}\right|_{r=a} = 0. \tag{12.32}$$

Our problem involves a new kind of boundary condition, not one for which ϕ is a constant on a surface, but for which $\partial\phi/\partial r$ is a constant. That is a little different. It is not easy to get the answer immediately. First of all, without the sphere, ϕ would be $-E_0z$. Then $\boldsymbol{E}$ would be in the z-direction and have the constant magnitude E_0, everywhere. Now we have analyzed the case of a dielectric sphere which has a uniform polarization inside it, and we found that the field inside such a polarized sphere is a uniform field, and that outside it is the same as the field of a point dipole located at the center. So let's guess that the solution we want is a superposition of a uniform field plus the field of a dipole. The potential of a dipole (Chapter 6) is $pz/4\pi\epsilon_0r^3$. Thus we assume that

$$\phi = -E_0z + \frac{pz}{4\pi\epsilon_0r^3}. \tag{12.33}$$

Since the dipole field falls off as $1/r^3$, at large distances we have just the field E_0. Our guess will automatically satisfy condition (2) above. But what do we take for the dipole strength p? To find out, we may use the other condition on ϕ, Eq. (12.32). We must differentiate ϕ with respect to r, but of course we must do so at a constant angle θ, so it is more convenient if we first express ϕ in terms of r and θ, rather than of z and r. Since $z = r\cos\theta$, we get

$$\phi = -E_0r\cos\theta + \frac{p\cos\theta}{4\pi\epsilon_0r^2}. \tag{12.34}$$

The radial component of $\boldsymbol{E}$ is

$$-\frac{\partial\phi}{\partial r} = +E_0\cos\theta + \frac{p\cos\theta}{2\pi\epsilon_0r^3}. \tag{12.35}$$

This must be zero at $r = a$ for all θ. This will be true if

$$p = -2\pi\epsilon_0a^3E_0. \tag{12.36}$$

Note carefully that if both terms in Eq. (12.35) had not had the same θ-dependence, it would not have been possible to choose p so that (12.35) turned out to be zero at $r = a$ for all angles. The fact that it works out means that we have guessed wisely in writing Eq. (12.33). Of course, when we made the guess we were looking ahead; we knew that we would need another term that (a) satisfied $\nabla^2\phi = 0$ (any real field would do that), (b) dependent on $\cos\theta$, and (c) fell to zero at large r. The dipole field is the only one that does all three.

Using (12.36), our potential is

$$\phi = -E_0\cos\theta\left(r + \frac{a^3}{2r^2}\right). \tag{12.37}$$

The solution of the fluid flow problem can be written simply as

$$\psi = -v_0\cos\theta\left(r + \frac{a^3}{2r^2}\right). \tag{12.38}$$

It is straightforward to find $\boldsymbol{v}$ from this potential. We will not pursue the matter further.

12–6 Illumination; the uniform lighting of a plane

In this section we turn to a completely different physical problem—we want to illustrate the great variety of possibilities. This time we will do something that leads to the same kind of *integral* that we found in electrostatics. (If we have a mathematical problem which gives us a certain integral, then we know something about the properties of that integral if it is the same integral that we had to do for another problem.) We take our example from illumination engineering. Suppose there is a light source at the distance a above a plane surface. What is the illumination of the surface? That is, what is the radiant energy per unit time arriving at a unit area of the surface? (See Fig. 12–9.) We suppose that the source is spherically

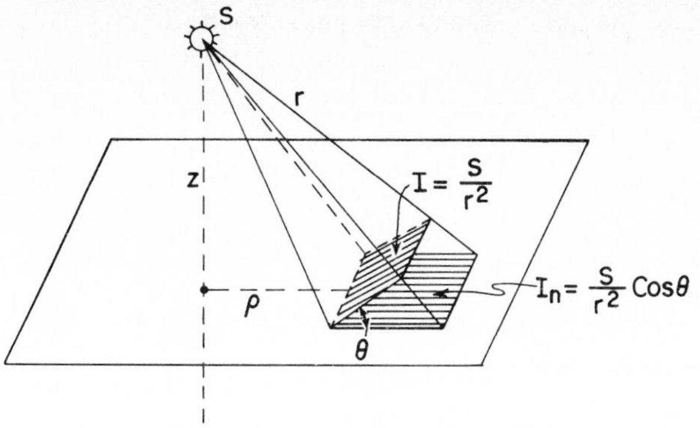

Fig. 12–9. The illumination I_n of a surface is the radiant energy per unit time arriving at a unit area of the surface.

symmetric, so that light is radiated equally in all directions. Then the amount of radiant energy which passes through a unit area *at right angles* to a light flow varies inversely as the square of the distance. It is evident that the intensity of the light in the direction normal to the flow is given by the same kind of formula as for the electric field from a point source. If the light rays meet the surface at an angle θ to the normal, then I, the energy arriving *per unit area* of the surface, is only $\cos\theta$ as great, because the same energy goes onto an area larger by $1/\cos\theta$. If we call the strength of our light source S, then I_n, the illumination of a surface, is

$$I_n = \frac{S}{r^2}\,\boldsymbol{e}_r\cdot\boldsymbol{n}, \tag{12.39}$$

where $\boldsymbol{e}_r$ is the unit vector from the source and $\boldsymbol{n}$ is the unit normal to the surface. The illumination I_n corresponds to the normal component of the electric field from a point charge of strength $4\pi\epsilon_0 S$. Knowing that, we see that for any distribution of light sources, we can find the answer by solving the corresponding electrostatic problem. We calculate the vertical component of electric field on the plane due to a distribution of charge in the same way as for that of the light sources.*

Consider the following example. We wish for some special experimental situation to arrange that the top surface of a table will have a very uniform illumination. We have available long tubular fluorescent lights which radiate uniformly along their lengths. We can illuminate the table by placing the fluorescent tubes in a regular array on the ceiling, which is at the height z above the table. What is the widest spacing b from tube to tube that we should use if we want the surface illumination to be uniform to, say, within one part in a thousand? *Answer;* (1) Find the electric field from a grid of wires with the spacing b, each charged uniformly; (2) compute the vertical component of the electric field; (3) find out what b must be so that the ripples of the field are not more than one part in a thousand.

In Chapter 7 we saw that the electric field of a grid of charged wires could be represented as a sum of terms, each one of which gave a sinusoidal variation of the field with a period of b/n, where n is an integer. The amplitude of any one of these terms is given by Eq. (7.44):

$$F_n = A_n e^{-2\pi nz/b}.$$

We need consider only $n = 1$, so long as we only want the field at points not too close to the grid. For a complete solution, we would still need to determine the coefficients A_n, which we have not yet done (although it is a straightforward calculation). Since we need only A_1, we can estimate that its magnitude is roughly the same as that of the average field. The exponential factor would then give us directly the *relative* amplitude of the variations. If we want this factor to be 10^{-3}, we find that b must be $0.91z$. If we make the spacing of the fluorescent tubes 3/4

* Since we are talking about *incoherent* sources whose *intensities* always add linearly, the analogous electric charges will always have the same sign. Also, our analogy applies only to the light energy arriving at the top of an opaque surface, so we must include in our integral only the sources which shine on the surface (and, naturally, not sources located below the surface!).

of the distance to the ceiling, the exponential factor is then $1/4000$, and we have a safety factor of 4, so we are fairly sure that we will have the illumination constant to one part in a thousand. (An exact calculation shows that A_1 is really twice the average field, so the exact answer is $b = 0.8z$.) It is somewhat surprising that for such a uniform illumination the allowed separation of the tubes comes out so large.

12–7 The "underlying unity" of nature

In this chapter, we wished to show that in learning electrostatics you have learned at the same time how to handle many subjects in physics, and that by keeping this in mind, it is possible to learn almost all of physics in a limited number of years.

However, a question surely suggests itself at the end of such a discussion: *Why are the equations from different phenomena so similar*? We might say: "It is the underlying unity of nature." But what does that mean? What *could* such a statement mean? It could mean simply that the equations are similar for different phenomena; but then, of course, we have given no explanation. The "underlying unity" might mean that everything is made out of the same stuff, and therefore obeys the same equations. That sounds like a good explanation, but let us think. The electrostatic potential, the diffusion of neutrons, heat flow—are we really dealing with the same stuff? Can we really imagine that the electrostatic potential is *physically* identical to the temperature, or to the density of particles? Certainly ϕ is not *exactly the same* as the thermal energy of particles. The displacement of a membrane is certainly *not* like a temperature. Why, then, is there "an underlying unity"?

A closer look at the physics of the various subjects shows, in fact, that the equations are not really identical. The equation we found for neutron diffusion is only an approximation that is good when the distance over which we are looking is large compared with the mean free path. If we look more closely, we would see the individual neutrons running around. Certainly the motion of an individual neutron is a completely different thing from the smooth variation we get from solving the differential equation. The differential equation is an approximation, because we assume that the neutrons are smoothly distributed in *space*.

Is it possible that *this* is the clue? That the thing which is common to all the phenomena is the *space*, the framework into which the physics is put? As long as things are reasonably smooth in space, then the important things that will be involved will be the rates of change of quantities with position in space. That is why we always get an equation with a gradient. The derivatives *must* appear in the form of a gradient or a divergence; because the laws of physics are *independent of direction*, they must be expressible in vector form. The equations of electrostatics are the simplest vector equations that one can get which involve only the spatial derivatives of quantities. Any other *simple* problem—or simplification of a complicated problem—must look like electrostatics. What is common to all our problems is that they involve *space* and that we have *imitated* what is actually a complicated phenomenon by a simple differential equation.

That leads us to another interesting question. Is the same statement perhaps also true for the *electrostatic* equations? Are they also correct only as a smoothed-out imitation of a really much more complicated microscopic world? Could it be that the real world consists of little X-ons which can be seen only at *very* tiny distances? And that in our measurements we are always observing on such a large scale that we can't see these little X-ons, and that is why we get the differential equations?

Our currently most complete theory of electrodynamics does indeed have its difficulties at very short distances. So it is possible, in principle, that these equations are smoothed-out versions of something. They appear to be correct at distances down to about 10^{-14} cm, but then they begin to look wrong. It is possible that there is some as yet undiscovered underlying "machinery," and that the details of an underlying complexity are hidden in the smooth-looking equations—as is so in the "smooth" diffusion of neutrons. But no one has yet formulated a successful theory that works that way.

Strangely enough, it turns out (for reasons that we do not at all understand) that the combination of relativity and quantum mechanics as we know them seems to *forbid* the invention of an equation that is fundamentally different from Eq. (12.4), and which does not at the same time lead to some kind of contradiction. Not simply a disagreement with experiment, but an *internal contradiction.* As, for example, the prediction that the sum of the probabilities of all possible occurrences is not equal to unity, or that energies may sometimes come out as complex numbers, or some other such idiocy. No one has yet made up a theory of electricity for which $\nabla^2\phi = -\rho/\epsilon_0$ is understood as a smoothed-out approximation to a mechanism underneath, and which does not lead ultimately to some kind of an absurdity. But, it must be added, it is also true that the assumption that $\nabla^2\phi = -\rho/\epsilon_0$ is valid for all distances, no matter how small, leads to absurdities of its own (the electrical energy of an electron is infinite)—absurdities from which no one yet knows an escape.

13

Magnetostatics

Review: Chapter 15, Vol. I: *The Special Theory of Relativity*

13–1 The magnetic field

The force on an electric charge depends not only on where it is, but also on how fast it is moving. Every point in space is characterized by two vector quantities which determine the force on any charge. First, there is the *electric force*, which gives a force component independent of the motion of the charge. We describe it by the electric field, $\boldsymbol{E}$. Second, there is an additional force component, called the *magnetic force*, which depends on the velocity of the charge. This magnetic force has a strange directional character. At any particular point in space, both the *direction* of the force and its *magnitude* depend on the direction of motion of the particle: at every instant the force is always at right angles to the velocity vector; also, at any particular point, the force is always at right angles to a *fixed direction in space* (see Fig. 13–1); and finally, the magnitude of the force is proportional to the *component* of the velocity at right angles to this unique direction. It is possible to describe all of this behavior by defining the magnetic field vector $\boldsymbol{B}$, which specifies both the unique direction in space and the constant of proportionality with the velocity, and to write the magnetic force as $q\boldsymbol{v} \times \boldsymbol{B}$. The total electromagnetic force on a charge can, then, be written as

$$\boldsymbol{F} = q(\boldsymbol{E} + \boldsymbol{v} \times \boldsymbol{B}). \tag{13.1}$$

This is called the *Lorentz force*.

The magnetic force is easily demonstrated by bringing a bar magnet close to a cathode-ray tube. The deflection of the electron beam shows that the presence of the magnet results in forces on the electrons transverse to their direction of motion, as we described in Chapter 12 of Vol. I.

The unit of magnetic field $\boldsymbol{B}$ is evidently one newton·second per coulomb·meter. The same unit is also one volt·second per meter^2. It is also called one *weber per square meter.*

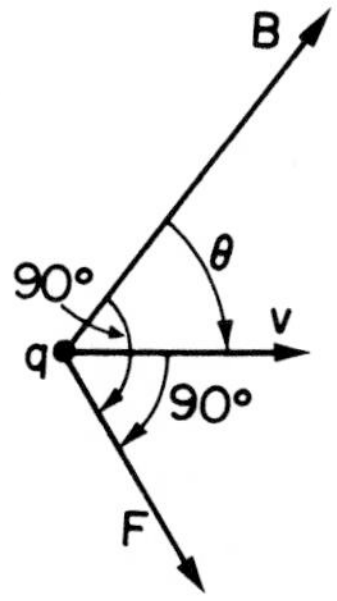

Fig. 13–1. The velocity-dependent component of the force on a moving charge is at right angles to **v** and to the direction of **B**. It is also proportional to the component of **v** at right angles to **B**, that is, to $v \sin \theta$.

13–2 Electric current; the conservation of charge

We consider first how we can understand the magnetic forces on wires carrying electric currents. In order to do this, we define what is meant by the current density. Electric currents are electrons or other charges in motion with a net drift or flow. We can represent the charge flow by a vector which gives the amount of charge passing per unit area and per unit time through a surface element at right angles to the flow (just as we did for the case of heat flow). We call this the *current density* and represent it by the vector $\boldsymbol{j}$. It is directed along the motion of the charges. If we take a small area ΔS at a given place in the material, the amount of charge flowing across that area in a unit time is

$$\boldsymbol{j} \cdot \boldsymbol{n}\, \Delta S, \tag{13.2}$$

where $\boldsymbol{n}$ is the unit vector normal to ΔS.

The current density is related to the average flow velocity of the charges. Suppose that we have a distribution of charges whose average motion is a drift with the velocity $\boldsymbol{v}$. As this distribution passes over a surface element ΔS, the charge Δq passing through the surface element in a time Δt is equal to the charge contained in a parallelepiped whose base is ΔS and whose height is $v\,\Delta t$, as shown in Fig. 13–2. The volume of the parallelepiped is the projection of ΔS at right angles to $\boldsymbol{v}$ times

$v\,\Delta t$, which when multiplied by the charge density ρ will give Δq. Thus

$$\Delta q = \rho \boldsymbol{v} \cdot \boldsymbol{n}\, \Delta S\, \Delta t.$$

The charge per unit time is then $\rho \boldsymbol{v} \cdot \boldsymbol{n}\, \Delta S$, from which we get

$$\boldsymbol{j} = \rho \boldsymbol{v}. \tag{13.3}$$

If the charge distribution consists of individual charges, say electrons, each with the charge q and moving with the mean velocity $\boldsymbol{v}$, then the current density is

$$\boldsymbol{j} = Nq\boldsymbol{v}, \tag{13.4}$$

where N is the number of charges per unit volume.

The total charge passing per unit time through any surface S is called the *electric current*, I. It is equal to the integral of the normal component of the flow through all of the elements of the surface:

$$I = \int_S \boldsymbol{j} \cdot \boldsymbol{n}\, dS \tag{13.5}$$

(see Fig. 13–3).

The current I out of a closed surface S represents the rate at which charge leaves the volume V enclosed by S. One of the basic laws of physics is that *electric charge is indestructible;* it is never lost or created. Electric charges can move from place to place but never appear from nowhere. We say that *charge is conserved.* If there is a net current out of a closed surface, the amount of charge inside must decrease by the corresponding amount (Fig. 13–4). We can, therefore, write the law of the conservation of charge as

$$\int_{\substack{\text{any closed}\\ \text{surface}}} \boldsymbol{j} \cdot \boldsymbol{n}\, dS = -\frac{d}{dt}(Q_{\text{inside}}). \tag{13.6}$$

The charge inside can be written as a volume integral of the charge density:

$$Q_{\text{inside}} = \int_{\substack{V\\ \text{inside } S}} \rho\, dV. \tag{13.7}$$

If we apply (13.6) to a small volume ΔV, we know that the left-hand integral is $\boldsymbol{\nabla} \cdot \boldsymbol{j}\, \Delta V$. The charge inside is $\rho\, \Delta V$, so the conservation of charge can also be written as

$$\boldsymbol{\nabla} \cdot \boldsymbol{j} = -\frac{\partial \rho}{\partial t} \tag{13.8}$$

(Gauss' mathematics once again!).

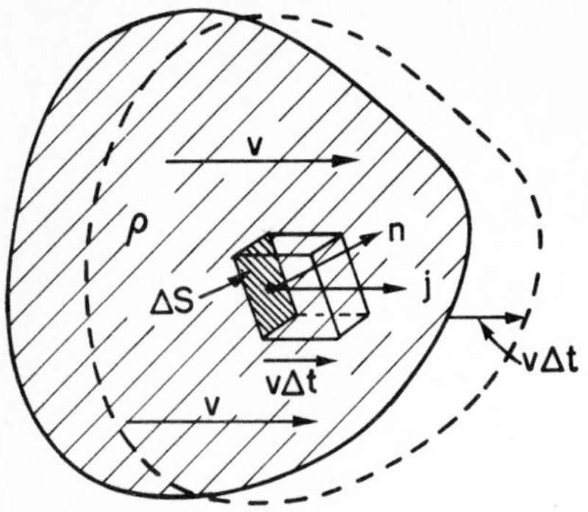

Fig. 13–2. If a charge distribution of density ρ moves with the velocity $\boldsymbol{v}$, the charge per unit time through ΔS is $\rho \boldsymbol{v} \cdot \boldsymbol{n}\, \Delta S$.

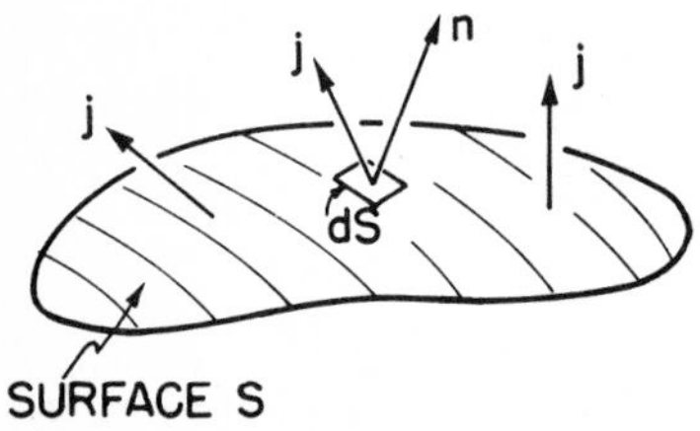

Fig. 13–3. The current I through the surface S is $\int \boldsymbol{j} \cdot \boldsymbol{n}\, dS$.

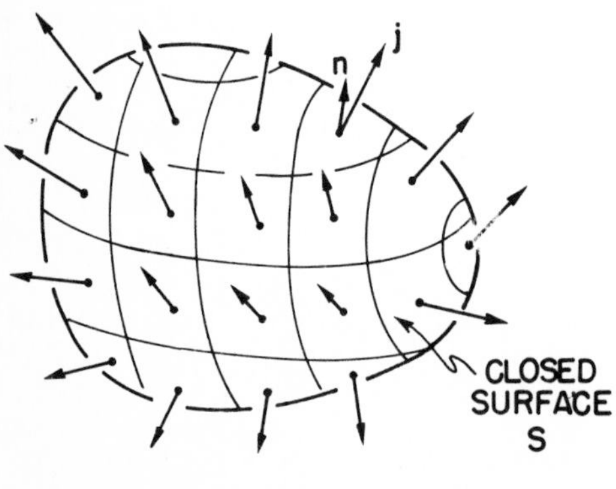

Fig. 13–4. The integral of $\boldsymbol{j} \cdot \boldsymbol{n}$ over a closed surface is the rate of change of the total charge Q inside.

13–3 The magnetic force on a current

Now we are ready to find the force on a current-carrying wire in a magnetic field. The current consists of charged particles moving with the velocity $\boldsymbol{v}$ along the wire. Each charge feels a transverse force

$$\boldsymbol{F} = q\boldsymbol{v} \times \boldsymbol{B}$$

(Fig. 13–5a). If there are N such charges per unit volume, the number in a small volume ΔV of the wire is $N\,\Delta V$. The total magnetic force $\Delta \boldsymbol{F}$ on the volume ΔV is the sum of the forces on the individual charges, that is,

$$\Delta \boldsymbol{F} = (N\,\Delta V)(q\boldsymbol{v} \times \boldsymbol{B}).$$

But $Nq\boldsymbol{v}$ is just $\boldsymbol{j}$, so

$$\Delta \boldsymbol{F} = \boldsymbol{j} \times \boldsymbol{B}\,\Delta V \tag{13.9}$$

(Fig. 13–5b). The force per unit volume is $\boldsymbol{j} \times \boldsymbol{B}$.

If the current is uniform across a wire whose cross-sectional area is A, we may take as the volume element a cylinder with the base area A and the length ΔL. Then

$$\Delta F = j \times BA\,\Delta L. \tag{13.10}$$

Now we can call jA the vector current I in the wire. (Its magnitude is the electric current in the wire, and its direction is along the wire.) Then

$$\Delta F = I \times B\,\Delta L. \tag{13.11}$$

The force per unit length on a wire is $I \times B$.

This equation gives the important result that the magnetic force on a wire, due to the movement of charges in it, depends only on the total current, and not on the amount of charge carried by each particle—or even its sign! The magnetic force on a wire near a magnet is easily shown by observing its deflection when a current is turned on, as was described in Chapter 1 (see Fig. 1–6).

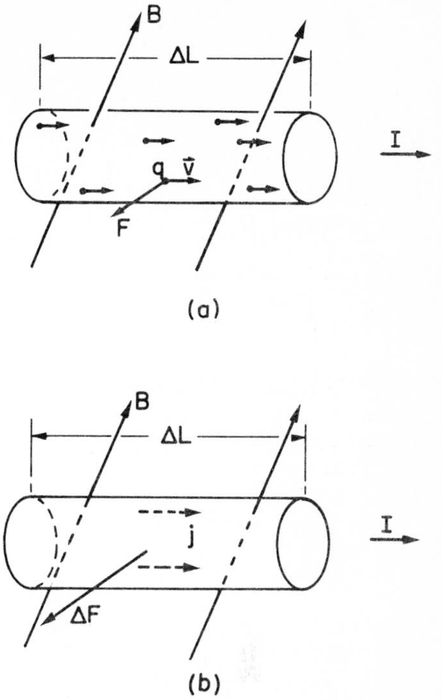

Fig. 13–5. The magnetic force on a current-carrying wire is the sum of the forces on the individual moving charges.

13–4 The magnetic field of steady currents; Ampere's law

We have seen that there is a force on a wire in the presence of a magnetic field, produced, say, by a magnet. From the principle that action equals reaction we might expect that there should be a force on the source of the magnetic field, i.e., on the magnet, when there is a current through the wire.* There are indeed such forces, as is seen by the deflection of a compass needle near a current-carrying wire. Now we know that magnets feel forces from other magnets, so that means that when there is a current in a wire, the wire itself generates a magnetic field. Moving charges, then, *produce* a magnetic field. We would like now to try to discover the laws that determine how such magnetic fields are created. The question is: Given a current, what magnetic field does it make? The answer to this question was determined experimentally by three critical experiments and a brilliant theoretical argument given by Ampere. We will pass over this interesting historical development and simply say that a large number of experiments have demonstrated the validity of Maxwell's equations. We take them as our starting point. If we drop the terms involving time derivatives in these equations we get the equations of *magnetostatics:*

$$\nabla \cdot B = 0 \tag{13.12}$$

and

$$c^2\nabla \times B = \frac{j}{\epsilon_0}. \tag{13.13}$$

These equations are valid only if all electric charge densities are constant and all currents are steady, so that the electric and magnetic fields are not changing with time—all of the fields are "static."

We may remark that it is rather dangerous to think that there is such a thing as a static magnetic situation, because there must be currents in order to get a magnetic field at all—and currents can come only from moving charges. "Magnetostatics" is, therefore, an approximation. It refers to a special kind of dynamic situation with *large numbers* of charges in motion, which we can approximate by a *steady* flow of charge. Only then can we speak of a current density j which does not change with time. The subject should more accurately be called the study of steady currents. Assuming that all fields are steady, we drop all terms in $\partial E/\partial t$ and $\partial B/\partial t$ from the complete Maxwell equations, Eqs. (2.41), and obtain the two equations (13.12) and (13.13) above. Also notice that since the divergence of the curl of any vector is necessarily zero, Eq. (13.13) requires that $\nabla \cdot j = 0$. This is true, by Eq. (13.8), only if $\partial\rho/\partial t$ is zero. But that must be so if E is not changing with time, so our assumptions are consistent.

---

* We will see later, however, that such assumptions are *not* generally correct for electromagnetic forces!

The requirement that $\nabla \cdot \boldsymbol{j} = 0$ means that we may only have charges which flow in paths that close back on themselves. They may, for instance, flow in wires that form complete loops—called circuits. The circuits may, of course, contain generators or batteries that keep the charges flowing. But they may not include condensers which are charging or discharging. (We will, of course, extend the theory later to include dynamic fields, but we want first to take the simpler case of steady currents.)

Now let us look at Eqs. (13.12) and (13.13) to see what they mean. The first one says that the divergence of $\boldsymbol{B}$ is zero. Comparing it to the analogous equation in electrostatics, which says that $\nabla \cdot \boldsymbol{E} = \rho/\epsilon_0$, we can conclude that there is no magnetic analog of an electric charge. There are *no magnetic charges* from which lines of $\boldsymbol{B}$ can emerge. If we think in terms of "lines" of the vector field $\boldsymbol{B}$, they can never start and they never stop. Then where do they come from? Magnetic fields "appear" *in the presence of* currents; they have a *curl* proportional to the current density. Wherever there are currents, there are lines of magnetic field making loops around the currents. Since lines of $\boldsymbol{B}$ do not begin or end, they will often close back on themselves, making closed loops. But there can also be complicated situations in which the lines are not simple closed loops. But whatever they do, they never diverge from points. No magnetic charges have ever been discovered, so $\nabla \cdot \boldsymbol{B} = 0$. This much is true not only for magnetostatics, it is *always* true—even for dynamic fields.

The connection between the $\boldsymbol{B}$ field and currents is contained in Eq. (13.13). Here we have a new kind of situation which is quite different from electrostatics, where we had $\nabla \times \boldsymbol{E} = 0$. That equation meant that the line integral of $\boldsymbol{E}$ around any closed path is zero:

$$\oint_{\text{loop}} \boldsymbol{E} \cdot d\boldsymbol{s} = 0.$$

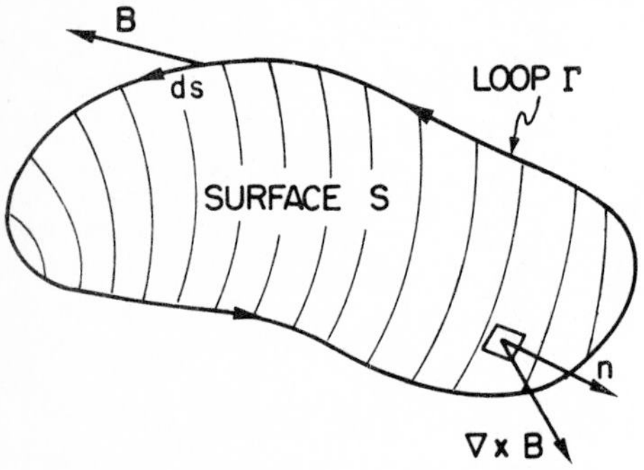

Fig. 13–6. The line integral of the tangential component of $\boldsymbol{B}$ is equal to the surface integral of the normal component of $\nabla \times \boldsymbol{B}$.

We got that result from Stokes' theorem, which says that the integral around any closed path of *any* vector field is equal to the surface integral of the normal component of the curl of the vector (taken over any surface which has the closed loop as its periphery). Applying the same theorem to the magnetic field vector and using the symbols shown in Fig. 13–6, we get

$$\oint_{\Gamma} \boldsymbol{B} \cdot d\boldsymbol{s} = \int_S (\nabla \times \boldsymbol{B}) \cdot \boldsymbol{n}\, dS. \tag{13.14}$$

Taking the curl of $\boldsymbol{B}$ from Eq. (13.13), we have

$$\oint_{\Gamma} \boldsymbol{B} \cdot d\boldsymbol{s} = \frac{1}{\epsilon_0 c^2} \int_S \boldsymbol{j} \cdot \boldsymbol{n}\, dS. \tag{13.15}$$

The integral over $\boldsymbol{j}$, according to (13.5), is the total current I through the surface S. Since for steady currents the current through S is independent of the shape of S, so long as it is bounded by the curve Γ, one usually speaks of "the current through the loop Γ." We have, then, a general law: the circulation of $\boldsymbol{B}$ around any closed curve is equal to the current I through the loop, divided by $\epsilon_0 c^2$:

$$\oint_{\Gamma} \boldsymbol{B} \cdot d\boldsymbol{s} = \frac{I_{\text{through } \Gamma}}{\epsilon_0 c^2}. \tag{13.16}$$

This law—called *Ampere's law*—plays the same role in magnetostatics that Gauss' law played in electrostatics. Ampere's law alone does not determine $\boldsymbol{B}$ from currents; we must, in general, also use $\nabla \cdot \boldsymbol{B} = 0$. But, as we will see in the next section, it can be used to find the field in special circumstances which have certain simple symmetries.

13–5 The magnetic field of a straight wire and of a solenoid; atomic currents

We can illustrate the use of Ampere's law by finding the magnetic field near a wire. We ask: What is the field outside a long straight wire with a cylindrical cross section? We will assume something which may not be at all evident, but which is nevertheless true: that the field lines of $\boldsymbol{B}$ go around the wire in closed circles. If we make this assumption, then Ampere's law, Eq. (13.16), tells us how strong the field is. From the symmetry of the problem, $\boldsymbol{B}$ has the same magnitude at all points on a circle concentric with the wire (see Fig. 13–7). We can then do the line integral of $\boldsymbol{B}\cdot d\boldsymbol{s}$ quite easily; it is just the magnitude of $\boldsymbol{B}$ times the circumference. If r is the radius of the circle, then

$$\oint \boldsymbol{B}\cdot d\boldsymbol{s} = B\cdot 2\pi r.$$

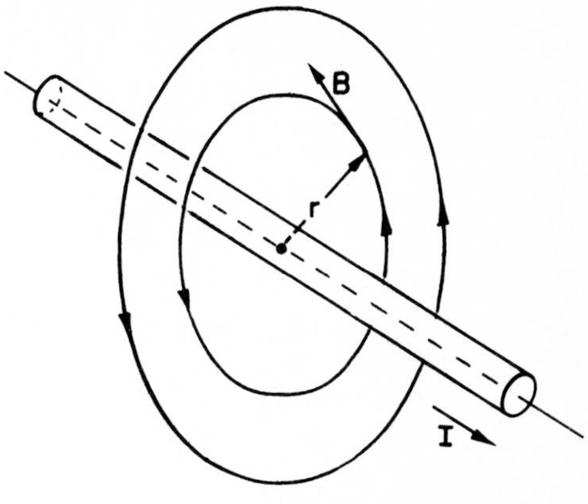

Fig. 13–7. The magnetic field outside of a long wire carrying the current I.

The total current through the loop is merely the current I in the wire, so

$$B\cdot 2\pi r = \frac{I}{\epsilon_0 c^2},$$

or

$$B = \frac{1}{4\pi\epsilon_0 c^2}\,\frac{2I}{r}. \tag{13.17}$$

The strength of the magnetic field drops off inversely as r, the distance from the axis of the wire. We can, if we wish, write Eq. (13.17) in vector form. Remembering that $\boldsymbol{B}$ is at right angles both to $\boldsymbol{I}$ and to $\boldsymbol{r}$, we have

$$\boldsymbol{B} = \frac{1}{4\pi\epsilon_0 c^2}\,\frac{2\boldsymbol{I}\times\boldsymbol{e}_r}{r}. \tag{13.18}$$

We have separated out the factor $1/4\pi\epsilon_0 c^2$, because it appears often. It is worth remembering that it is exactly 10^{-7} (in the mks system), since an equation like (13.17) is used to *define* the unit of current, the ampere. At one meter from a current of one ampere the magnetic field is 2×10^{-7} webers per square meter.

Since a current produces a magnetic field, it will exert a force on a nearby wire which is also carrying a current. In Chapter 1 we described a simple demonstration of the forces between two current-carrying wires. If the wires are parallel, each is at right angles to the $\boldsymbol{B}$ field of the other; the wires should then be pushed either toward or away from each other. When currents are in the same direction, the wires attract; when the currents are moving in opposite directions, the wires repel.

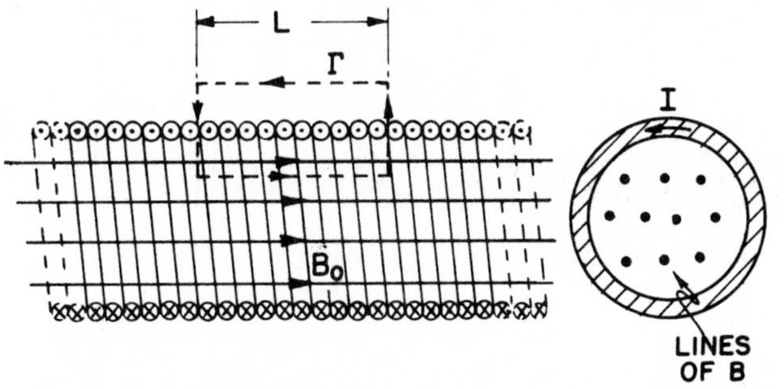

Fig. 13–8. The magnetic field of a long solenoid.

Let's take another example that can be analyzed by Ampere's law if we add some knowledge about the field. Suppose we have a long coil of wire wound in a tight spiral, as shown by the cross sections in Fig. 13–8. Such a coil is called a *solenoid.* We observe experimentally that when a solenoid is very long compared with its diameter, the field outside is very small compared with the field inside. Using just that fact, together with Ampere's law, we can find the size of the field inside.

Since the field *stays* inside (and has zero divergence), its lines must go along parallel to the axis, as shown in Fig. 13–8. That being the case, we can use Ampere's law with the rectangular "curve" Γ shown in the figure. This loop goes the distance

L inside the solenoid, where the field is, say, $\boldsymbol{B}_0$, then goes at right angles to the field, and returns along the outside, where the field is negligible. The line integral of $\boldsymbol{B}$ for this curve is just B_0L, and it must be $1/\epsilon_0c^2$ times the total current through Γ, which is NI if there are N turns of the solenoid in the length L. We have

$$B_0L = \frac{NI}{\epsilon_0c^2}.$$

Or, letting n be the number of turns *per unit length* of the solenoid (that is, $n = N/L$), we get

$$B_0 = \frac{nI}{\epsilon_0c^2}. \tag{13.19}$$

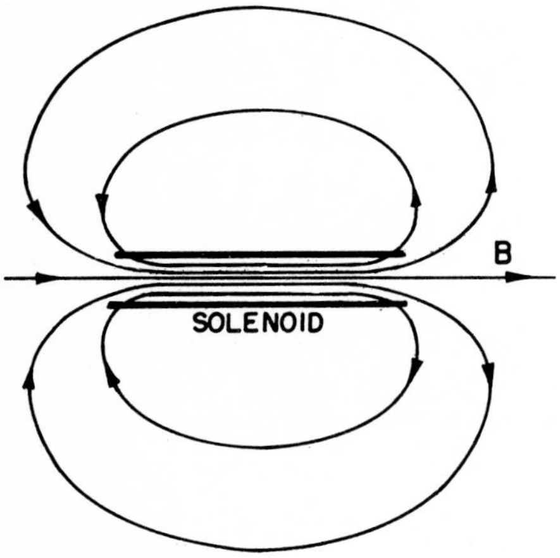

Fig. 13–9. The magnetic field outside of a solenoid.

What happens to the lines of $\boldsymbol{B}$ when they get to the end of the solenoid? Presumably, they spread out in some way and return to enter the solenoid at the other end, as sketched in Fig. 13–9. Such a field is just what is observed outside of a bar magnet. But what *is* a magnet anyway? Our equations say that $\boldsymbol{B}$ comes from the presence of currents. Yet we know that ordinary bars of iron (no batteries or generators) also produce magnetic fields. You might expect that there should be some other terms on the right-hand side of (13.12) or (13.13) to represent "the density of magnetic iron" or some such quantity. But there is no such term. Our theory says that the magnetic effects of iron come from some internal currents which are already taken care of by the $\boldsymbol{j}$ term.

Matter is very complex when looked at from a fundamental point of view—as we saw when we tried to understand dielectrics. In order not to interrupt our present discussion, we will wait until later to deal in detail with the interior mechanisms of magnetic materials like iron. You will have to accept, for the moment, that all magnetism is produced from currents, and that in a permanent magnet there are permanent internal currents. In the case of iron, these currents come from electrons spinning around their own axes. Every electron has such a spin, which corresponds to a tiny circulating current. Of course, one electron doesn't produce much magnetic field, but in an ordinary piece of matter there are billions and billions of electrons. Normally these spin and point every which way, so that there is no net effect. The miracle is that in a very few substances, like iron, a large fraction of the electrons spin with their axes in the same direction—for iron, two electrons of each atom take part in this cooperative motion. In a bar magnet there are large numbers of electrons all spinning in the same direction and, as we will see, their total effect is equivalent to a current circulating on the surface of the bar. (This is quite analogous to what we found for dielectrics—that a uniformly polarized dielectric is equivalent to a distribution of charges on its surface.) It is, therefore, no accident that a bar magnet is equivalent to a solenoid.

13–6 The relativity of magnetic and electric fields

When we said that the magnetic force on a charge was proportional to its velocity, you may have wondered: "What velocity? With respect to which reference frame?" It is, in fact, clear from the definition of $\boldsymbol{B}$ given at the beginning of this chapter that what this vector is will depend on what we choose as a reference frame for our specification of the velocity of charges. But we have said nothing about which is the proper frame for specifying the magnetic field.

It turns out that *any* inertial frame will do. We will also see that magnetism and electricity are not independent things—that they should always be taken together as *one* complete electromagnetic field. Although in the static case Maxwell's equations separate into two distinct pairs, one pair for electricity and one pair for magnetism, with no apparent connection between the two fields, nevertheless, in nature itself there is a very intimate relationship between them that arises from the principle of relativity. Historically, the principle of relativity was discovered after Maxwell's equations. It was, in fact, the study of electricity and magnetism which led ultimately to Einstein's discovery of his principle of relativity. But let's see

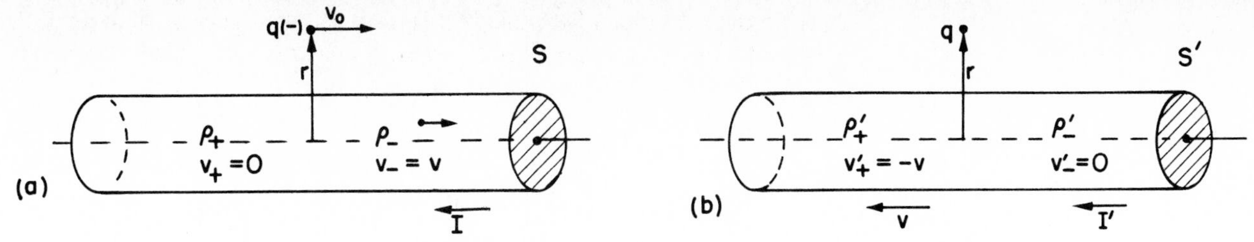

Fig. 13–10. The interaction of a current-carrying wire and a particle with the charge q as seen in two frames. In frame S (part a), the wire is at rest; in frame S' (part b), the charge is at rest.

what our knowledge of relativity would tell us about magnetic forces if we assume that the relativity principle is applicable—as it is—to electromagnetism.

Suppose we think about what happens when a negative charge moves with velocity v_0 parallel to a current-carrying wire, as in Fig. 13–10. We will try to understand what goes on in two reference frames: one fixed with respect to the wire, as in part (a) of the figure, and one fixed with respect to the particle, as in part (b). We will call the first frame S and the second S'.

In the S-frame, there is clearly a magnetic force on the particle. The force is directed toward the wire, so if the charge were moving freely we would see it curve in toward the wire. But in the S'-frame there can be no magnetic force on the particle, because its velocity is zero. Does it, therefore, stay where it is? Would we see different things happening in the two systems? The principle of relativity would say that in S' we should also see the particle move closer to the wire. We must try to understand why that would happen.

We return to our atomic description of a wire carrying a current. In a normal conductor, like copper, the electric currents come from the motion of some of the negative electrons—called the conduction electrons—while the positive nuclear charges and the remainder of the electrons stay fixed in the body of the material. We let the density of the conduction electrons be ρ_- and their velocity in S be $\boldsymbol{v}$. The density of the charges at rest in S is ρ_+, which must be equal to the negative of ρ_-, since we are considering an uncharged wire. There is thus no electric field outside the wire, and the force on the moving particle is just

$$\boldsymbol{F} = q\boldsymbol{v}_0 \times \boldsymbol{B}.$$

Using the result we found in Eq. (13.18) for the magnetic field at the distance r from the axis of a wire, we conclude that the force on the particle is directed toward the wire and has the magnitude

$$F = \frac{1}{4\pi\epsilon_0 c^2}\cdot\frac{2Iqv_0}{r}.$$

Using Eqs. (13.4) and (13.5), the current I can be written as $\rho_- vA$, where A is the area of a cross section of the wire. Then

$$F = \frac{1}{4\pi\epsilon_0 c^2}\cdot\frac{2q\rho_- Avv_0}{r}. \tag{13.20}$$

We could continue to treat the general case of arbitrary velocities for v and v_0, but it will be just as good to look at the special case in which the velocity v_0 of the particle is the same as the velocity v of the conduction electrons. So we write $v_0 = v$, and Eq. (13.20) becomes

$$F = \frac{q}{2\pi\epsilon_0}\frac{\rho_- A}{r}\frac{v^2}{c^2}. \tag{13.21}$$

Now we turn our attention to what happens in S', in which the particle is at rest and the wire is running past (toward the left in the figure) with the speed v. The positive charges moving with the wire will make some magnetic field B' at the particle. But the particle is now at *rest*, so there is no *magnetic* force on it! If there is any force on the particle, it must come from an electric field. It must be that the moving wire has produced an electric field. But it can do that only if it appears *charged*—it must be that a neutral wire with a current appears to be charged when set in motion.

We must look into this. We must try to compute the charge density in the wire in S' from what we know about it in S. One might, at first, think they are the same; but we know that lengths are changed between S and S' (see Chapter 15, Vol. I), so volumes will change also. Since the charge *densities* depend on the volume occupied by charges, the densities will change, too.

Before we can decide about the charge *densities* in S', we must know what happens to the electric *charge* of a bunch of electrons when the charges are moving. We know that the apparent mass of a particle changes by $1/\sqrt{1 - v^2/c^2}$. Does its charge do something similar? No! *Charges* are always the *same*, moving or not. Otherwise we would not always observe that the total charge is conserved.

Suppose that we take a block of material, say a conductor, which is initially uncharged. Now we heat it up. Because the electrons have a different mass than the protons, the velocities of the electrons and of the protons will change by different amounts. If the charge of a particle depended on the speed of the particle carrying it, in the heated block the charge of the electrons and protons would no longer balance. A block would become charged when heated. As we have seen earlier, a very small fractional change in the charge of all the electrons in a block would give rise to enormous electric fields. No such effect has ever been observed.

Also, we can point out that the mean speed of the electrons in matter depends on its chemical composition. If the charge on an electron changed with speed, the net charge in a piece of material would be changed in a chemical reaction. Again, a straightforward calculation shows that even a very small dependence of charge on speed would give enormous fields from the simplest chemical reactions. No such effect is observed, and we conclude that the electric charge of a single particle is independent of its state of motion.

So the charge q on a particle is an invariant scalar quantity, independent of the frame of reference. That means that in any frame the charge density of a distribution of electrons is just proportional to the number of electrons per unit volume. We need only worry about the fact that the volume *can* change because of the relativistic contraction of distances.

We now apply these ideas to our moving wire. If we take a length L_0 of the wire, in which there is a charge density ρ_0 of *stationary* charges, it will contain the total charge $Q = \rho_0 L_0 A_0$. If the same charges are observed in a different frame to be moving with velocity v, they will all be found in a piece of the material with the *shorter* length

$$L = L_0\sqrt{1 - v^2/c^2}, \tag{13.22}$$

but with the same area A_0 (since dimensions transverse to the motion are unchanged). See Fig. 13–11.

If we call ρ the density of charges in the frame in which they are moving, the total charge Q will be $\rho L A_0$. This must also be equal to $\rho_0 L_0 A$, because charge is the same in any system, so that $\rho L = \rho_0 L_0$ or, from (13.22),

$$\rho = \frac{\rho_0}{\sqrt{1 - v^2/c^2}}\cdot \tag{13.23}$$

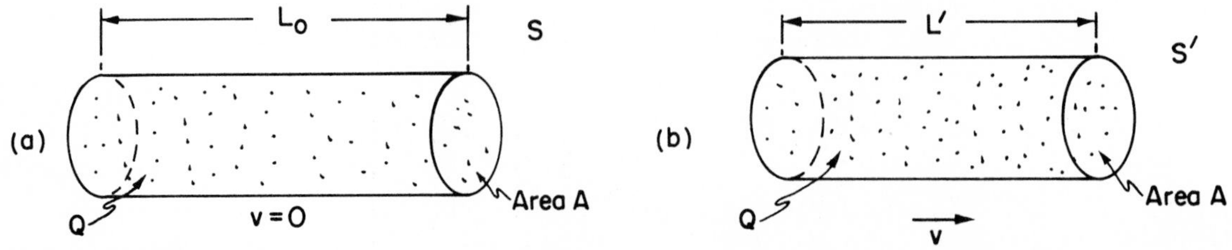

Fig. 13–11. If a distribution of charged particles at rest has the charge density ρ_0, the same charges will have the density $\rho = \rho_0/\sqrt{1 - v^2/c^2}$ when seen from a frame with the relative velocity v.

The charge *density* of a moving *distribution* of charges varies in the same way as the relativistic mass of a particle.

We now use this general result for the positive charge density ρ_+ of our wire. These charges are at rest in frame S. In S', however, where the wire moves with the speed v, the positive charge density becomes

$$\rho'_+ = \frac{\rho_+}{\sqrt{1 - v^2/c^2}}. \tag{13.24}$$

The *negative* charges are at rest in S'. So they have their "rest density" ρ_0 in this frame. In Eq. (13.23) $\rho_0 = \rho'_-$, because they have the density ρ'_- when the *wire* is at rest, i.e., in frame S, where the speed of the negative charges is v. For the conduction electrons, we then have that

$$\rho_- = \frac{\rho'_-}{\sqrt{1 - v^2/c^2}}, \tag{13.25}$$

or

$$\rho'_- = \rho_-\sqrt{1 - v^2/c^2}. \tag{13.26}$$

Now we can see why there are electric fields in S'—because in this frame the wire has the net charge density ρ' given by

$$\rho' = \rho'_+ + \rho'_-.$$

Using (13.24) and (13.26), we have

$$\rho' = \frac{\rho_+}{\sqrt{1 - v^2/c^2}} + \rho_-\sqrt{1 - v^2/c^2}.$$

Since the stationary wire is neutral, $\rho_- = -\rho_+$, and we have

$$\rho' = \rho_+ \frac{v^2/c^2}{\sqrt{1 - v^2/c^2}}. \tag{13.27}$$

Our moving wire is positively charged and will produce an electric field E' at the external stationary particle. We have already solved the electrostatic problem of a uniformly charged cylinder. The electric field at the distance r from the axis of the cylinder is

$$E' = \frac{\rho' A}{2\pi\epsilon_0 r} = \frac{\rho_+ A v^2/c^2}{2\pi\epsilon_0 r\sqrt{1 - v^2/c^2}}. \tag{13.28}$$

The force on the negatively charged particle is toward the wire. We have, at least, a force in the same direction from the two points of view; the electric force in S' has the same direction as the magnetic force in S.

The magnitude of the force in S' is

$$F' = \frac{q}{2\pi\epsilon_0}\frac{\rho_+ A}{r}\frac{v^2/c^2}{\sqrt{1 - v^2/c^2}}. \tag{13.29}$$

Comparing this result for F' with our result for F in Eq. (13.21), we see that the magnitudes of the forces are almost identical from the two points of view. In fact,

$$F' = \frac{F}{\sqrt{1 - v^2/c^2}}, \tag{13.30}$$

so for the small velocities we have been considering, the two forces are equal. We can say that for low velocities, at least, we understand that magnetism and electricity are just "two ways of looking at the same thing."

But things are even better than that. If we take into account the fact that *forces* also transform when we go from one system to the other, we find that the two ways of looking at what happens do indeed give the same *physical* result for any velocity.

One way of seeing this is to ask a question like: What transverse momentum will the particle have after the force has acted for a little while? We know from Chapter 16 of Vol. I that the transverse momentum of a particle should be the same in both the S- and S'-frames. Calling the transverse coordinate y, we want to compare Δp_y and $\Delta p'_y$. Using the relativistically correct equation of motion, $\boldsymbol{F} = d\boldsymbol{p}/dt$, we expect that after the time Δt our particle will have a transverse momentum Δp_y in the S-system given by

$$\Delta p_y = F\,\Delta t. \tag{13.31}$$

In the S'-system, the transverse momentum will be

$$\Delta p'_y = F'\,\Delta t'. \tag{13.32}$$

We must, of course, compare Δp_y and $\Delta p'_y$ for corresponding time intervals Δt and $\Delta t'$. We have seen in Chapter 15 of Vol. I that the time intervals referred to a *moving* particle appear to be *longer* than those in the rest system of the particle. Since our particle is initially at rest in S', we expect, for small Δt, that

$$\Delta t = \frac{\Delta t'}{\sqrt{1 - v^2/c^2}}, \tag{13.33}$$

and everything comes out O.K. From (13.31) and (13.32),

$$\frac{\Delta p'_y}{\Delta p_y} = \frac{F'\,\Delta t'}{F\,\Delta t},$$

which is just $= 1$ if we combine (13.30) and (13.33).

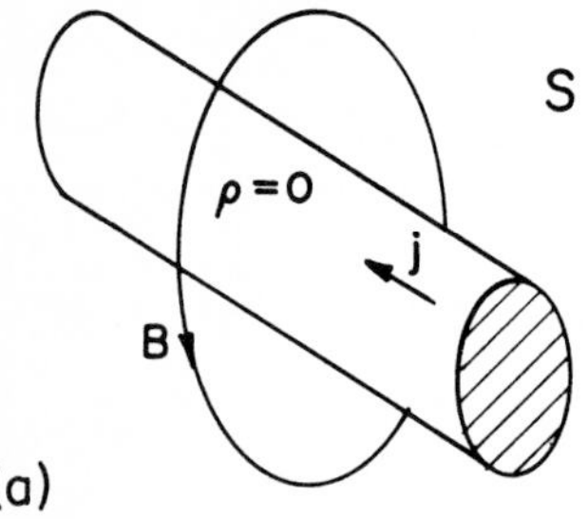

(a)

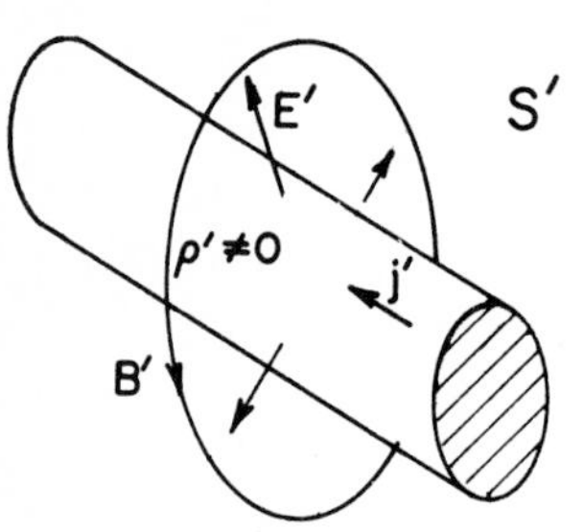

(b)

Fig. 13–12. In frame S the charge density is zero and the current density is $\boldsymbol{j}$. There is only a magnetic field. In S', there is a charge density ρ', and a different current density $\boldsymbol{j}'$. The magnetic field $\boldsymbol{B}'$ is different and there is an electric field $\boldsymbol{E}'$.

We have found that we get the same physical result whether we analyze the motion of a particle moving along a wire in a coordinate system at rest with respect to the wire, or in a system at rest with respect to the particle. In the first instance, the force was purely "magnetic," in the second, it was purely "electric." The two points of view are illustrated in Fig. 13–12 (although there is still a magnetic field B' in the second frame, it produces no forces on the stationary particle).

If we had chosen still another coordinate system, we would have found a different mixture of $\boldsymbol{E}$ and $\boldsymbol{B}$ fields. Electric and magnetic forces are part of *one* physical phenomenon—the electromagnetic interactions of particles. The separation of this interaction into electric and magnetic parts depends very much on the reference frame chosen for the description. But a complete electromagnetic description is invariant; electricity and magnetism taken together are consistent with Einstein's relativity.

Since electric and magnetic fields appear in different mixtures if we change our frame of reference, we must be careful about how we look at the fields $\boldsymbol{E}$ and $\boldsymbol{B}$. For instance, if we think of "lines" of $\boldsymbol{E}$ or $\boldsymbol{B}$, we must not attach too much reality to them. The lines may disappear if we try to observe them from a different coordinate system. For example, in system S' there are electric field lines, which we do *not* find "moving past us with velocity v in system S." In system S there are no electric field lines at all! Therefore it makes no sense to say something like: When I move a magnet, it takes its field with it, so the lines of $\boldsymbol{B}$ are also moved. There is no way to make sense, in general, out of the idea of "the speed of a moving field line." The fields are our way of describing what goes on at a point in space. In particular, $\boldsymbol{E}$ and $\boldsymbol{B}$ tell us about the forces that will act on a moving particle. The question "What is the force on a charge from a *moving* magnetic field?" doesn't mean anything precise. The force is given by the values of $\boldsymbol{E}$ and $\boldsymbol{B}$ at the charge, and the formula (13.1) is not to be altered if the *source* of $\boldsymbol{E}$ or $\boldsymbol{B}$ is moving (it is the values of $\boldsymbol{E}$ and $\boldsymbol{B}$ that will be altered by the motion). Our mathematical description deals only with the fields as a function of x, y, z, and t *with respect to some inertial frame*.

We will later be speaking of "a *wave* of electric and magnetic fields travelling through space," as, for instance, a light wave. But that is like speaking of a *wave* travelling on a string. We don't then mean that some part of the *string* is moving in the direction of the wave, we mean that the *displacement* of the string appears first at one place and later at another. Similarly, in an electromagnetic wave, the *wave* travels, but the magnitude of the fields *change*. So in the future when we—or someone else—speaks of a "moving" field, you should think of it as just a handy, short way of describing a changing field in some circumstances.

13–7 The transformation of currents and charges

You may have worried about the simplification we made above when we took the same velocity v for the particle and for the conduction electrons in the wire. We could go back and carry through the analysis again for two different velocities, but it is easier to simply notice that charge and current density are the components of a four-vector (see Chapter 17, Vol. I).

We have seen that if ρ_0 is the density of the charges in their rest frame, then in a frame in which they have the velocity $\boldsymbol{v}$, the density is

$$\rho = \frac{\rho_0}{\sqrt{1 - v^2/c^2}}.$$

In that frame their current density is

$$\boldsymbol{j} = \rho\boldsymbol{v} = \frac{\rho_0\boldsymbol{v}}{\sqrt{1 - v^2/c^2}}. \tag{13.34}$$

Now we know that the energy U and momentum $\boldsymbol{p}$ of a particle moving with velocity $\boldsymbol{v}$ are given by

$$U = \frac{m_0c^2}{\sqrt{1 - v^2/c^2}}, \qquad \boldsymbol{p} = \frac{m_0\boldsymbol{v}}{\sqrt{1 - v^2/c^2}},$$

where m_0 is its rest mass. We also know that U and $\boldsymbol{p}$ form a relativistic four-vector. Since ρ and $\boldsymbol{j}$ depend on the velocity $\boldsymbol{v}$ exactly as do U and $\boldsymbol{p}$, we can conclude that ρ and $\boldsymbol{j}$ are *also* the components of a relativistic four-vector. This property is the key to a general analysis of the field of a wire moving with any velocity, which we would need if we want to do the problem again with the velocity $\boldsymbol{v}_0$ of the particle different from the velocity of the conduction electrons.

If we wish to transform ρ and $\boldsymbol{j}$ to a coordinate system moving with a velocity u in the x-direction, we know that they transform just like t and (x, y, z), so that we have (see Chapter 15, Vol. I)

$$\begin{aligned}
x' &= \frac{x - ut}{\sqrt{1 - u^2/c^2}}, & j'_x &= \frac{j_x - u\rho}{\sqrt{1 - u^2/c^2}},\\
y' &= y, & j'_y &= j_y,\\
z' &= z, & j'_z &= j_z,\\
t' &= \frac{t - ux/c^2}{\sqrt{1 - u^2/c^2}}, & \rho' &= \frac{\rho - uj_x/c^2}{\sqrt{1 - u^2/c^2}}.
\end{aligned} \tag{13.35}$$

With these equations we can relate charges and currents in one frame to those in another. Taking the charges and currents in either frame, we can solve the electromagnetic problem in that frame by using our Maxwell equations. The result we obtain *for the motions of particles* will be the same no matter which frame we choose. We will return at a later time to the relativistic transformations of the electromagnetic fields.

13–8 Superposition; the right-hand rule

We will conclude this chapter by making two further points regarding the subject of magnetostatics. First, our basic equations for the magnetic field,

$$\boldsymbol{\nabla}\cdot\boldsymbol{B} = 0, \qquad \boldsymbol{\nabla}\times\boldsymbol{B} = \boldsymbol{j}/c^2\epsilon_0,$$

are linear in $\boldsymbol{B}$ and $\boldsymbol{j}$. That means that the principle of superposition also applies to magnetic fields. The field produced by two different steady currents is the sum of the individual fields from each current acting alone. Our second remark concerns the right-hand rules which we have encountered (such as the right-hand rule for the magnetic field produced by a current). We have also observed that the magnetization of an iron magnet is to be understood from the spin of the electrons in the material. The direction of the magnetic field of a spinning electron is related to its spin axis by the same right-hand rule. Because $\boldsymbol{B}$ is determined by a "handed" rule—involving either a cross product or a curl—it is called an *axial* vector. (Vectors whose direction in space does not depend on a reference to a right or left hand are called *polar* vectors. Displacement, velocity, force, and $\boldsymbol{E}$, for example, are polar vectors.)

Physically observable quantities in electromagnetism are *not,* however, right- (or left-) handed. Electromagnetic interactions are symmetrical under reflection (see Chapter 52, Vol. I). Whenever magnetic forces between two sets of currents are computed, the result is invariant with respect to a change in the hand convention. Our equations lead, independently of the right-hand convention, to the end result that parallel currents attract, or that currents in opposite directions repel. (Try working out the force using "left-hand rules.") An attraction or repulsion is a polar vector. This happens because in describing any complete interaction, we use the right-hand rule twice—once to find $\boldsymbol{B}$ from currents, again to find the force this $\boldsymbol{B}$ produces on a second current. Using the right-hand rule twice is the same as using the left-hand rule twice. If we were to change our conventions to a left-hand system all our $\boldsymbol{B}$ fields would be reversed, but all forces—or, what is perhaps more relevant, the observed accelerations of objects—would be unchanged.

Although physicists have recently found to their surprise that *all* the laws of nature are not always invariant for mirror reflections, the laws of electromagnetism do have such a basic symmetry.

14

The Magnetic Field in Various Situations

14–1 The vector potential

In this chapter we continue our discussion of magnetic fields associated with steady currents—the subject of magnetostatics. The magnetic field is related to electric currents by our basic equations

$$\nabla \cdot \boldsymbol{B} = 0, \tag{14.1}$$

$$c^2\nabla \times \boldsymbol{B} = \frac{\boldsymbol{j}}{\epsilon_0}. \tag{14.2}$$

We want now to solve these equations mathematically in a *general* way, that is, without requiring any special symmetry or intuitive guessing. In electrostatics, we found that there was a straightforward procedure for finding the field when the positions of all electric charges are known: One simply works out the scalar potential ϕ by taking an integral over the charges—as in Eq. (4.25). Then if one wants the electric field, it is obtained from the derivatives of ϕ. We will now show that there is a corresponding procedure for finding the magnetic field $\boldsymbol{B}$ if we know the current density $\boldsymbol{j}$ of all moving charges.

In electrostatics we saw that (because the curl of $\boldsymbol{E}$ was always zero) it was possible to represent $\boldsymbol{E}$ as the gradient of a scalar field ϕ. Now the curl of $\boldsymbol{B}$ is *not* always zero, so it is not possible, in general, to represent it as a gradient. However, the *divergence* of $\boldsymbol{B}$ *is* always zero, and this means that we can always represent $\boldsymbol{B}$ as the *curl* of another vector field. For, as we saw in Section 2–8, the divergence of a curl is always zero. Thus we can always relate $\boldsymbol{B}$ to a field we will call $\boldsymbol{A}$ by

$$\boldsymbol{B} = \nabla \times \boldsymbol{A}. \tag{14.3}$$

Or, by writing out the components,

$$\begin{aligned} B_x &= (\nabla \times \boldsymbol{A})_x = \frac{\partial A_z}{\partial y} - \frac{\partial A_y}{\partial z}, \\ B_y &= (\nabla \times \boldsymbol{A})_y = \frac{\partial A_x}{\partial z} - \frac{\partial A_z}{\partial x}, \\ B_z &= (\nabla \times \boldsymbol{A})_z = \frac{\partial A_y}{\partial x} - \frac{\partial A_x}{\partial y}. \end{aligned} \tag{14.4}$$

Writing $\boldsymbol{B} = \nabla \times \boldsymbol{A}$ guarantees that Eq. (14.1) is satisfied, since, necessarily,

$$\nabla \cdot \boldsymbol{B} = \nabla \cdot (\nabla \times \boldsymbol{A}) = 0.$$

The field $\boldsymbol{A}$ is called the *vector potential.*

You will remember that the scalar potential ϕ was not completely specified by its definition. If we have found ϕ for some problem, we can always find another potential ϕ' that is equally good by adding a constant:

$$\phi' = \phi + C.$$

The new potential ϕ' gives the same electric fields, since the gradient ∇C is zero; ϕ' and ϕ represent the same physics.

Similarly, we can have different vector potentials $\boldsymbol{A}$ which give the same magnetic fields. Again, because $\boldsymbol{B}$ is obtained from $\boldsymbol{A}$ by differentiation, adding a constant to $\boldsymbol{A}$ doesn't change anything physical. But there is even more latitude for $\boldsymbol{A}$. We can add to $\boldsymbol{A}$ any field which is the gradient of some scalar field, without changing the physics. We can show this as follows. Suppose we have an $\boldsymbol{A}$ that gives correctly the magnetic field $\boldsymbol{B}$ for some real situation, and ask in what circumstances some other new vector potential $\boldsymbol{A}'$ will give the *same* field $\boldsymbol{B}$ if substituted into (14.3). Then A and A' must have the same curl:

$$\boldsymbol{B} = \nabla \times \boldsymbol{A}' = \nabla \times \boldsymbol{A}.$$

Therefore

$$\nabla \times \boldsymbol{A}' - \nabla \times \boldsymbol{A} = \nabla \times (\boldsymbol{A}' - \boldsymbol{A}) = 0.$$

But if the curl of a vector is zero it must be the gradient of some scalar field, say ψ, so $\boldsymbol{A}' - \boldsymbol{A} = \nabla\psi$. That means that if $\boldsymbol{A}$ is a satisfactory vector potential for a problem then, for any ψ at all,

$$\boldsymbol{A}' = \boldsymbol{A} + \nabla\psi \tag{14.5}$$

will be an equally satisfactory vector potential, leading to the same field $\boldsymbol{B}$.

It is usually convenient to take some of the "latitude" out of $\boldsymbol{A}$ by arbitrarily placing some other condition on it (in much the same way that we found it convenient—often—to choose to make the potential ϕ zero at large distances). We can, for instance, restrict $\boldsymbol{A}$ by choosing arbitrarily what the divergence of $\boldsymbol{A}$ must be. We can always do that without affecting $\boldsymbol{B}$. This is because although $\boldsymbol{A}'$ and $\boldsymbol{A}$ have the same curl, and give the same B, they do not need to have the same divergence. In fact, $\nabla \cdot \boldsymbol{A}' = \nabla \cdot \boldsymbol{A} + \nabla^2\psi$, and by a suitable choice of ψ we can make $\nabla \cdot \boldsymbol{A}'$ anything we wish.

What should we choose for $\nabla \cdot \boldsymbol{A}$? The choice should be made to get the greatest mathematical convenience and will depend on the problem we are doing. For *magnetostatics*, we will make the simple choice

$$\nabla \cdot \boldsymbol{A} = 0. \tag{14.6}$$

(Later, when we take up electrodynamics, we will change our choice.) Our complete definition* of $\boldsymbol{A}$ is then, for the moment, $\nabla \times \boldsymbol{A} = \boldsymbol{B}$ and $\nabla \cdot \boldsymbol{A} = 0$.

To get some experience with the vector potential, let's look first at what it is for a uniform magnetic field $\boldsymbol{B}_0$. Taking our z-axis in the direction of $\boldsymbol{B}_0$, we must have

$$\begin{aligned} B_x &= \frac{\partial A_z}{\partial y} - \frac{\partial A_y}{\partial z} = 0,\\ B_y &= \frac{\partial A_x}{\partial z} - \frac{\partial A_z}{\partial x} = 0,\\ B_z &= \frac{\partial A_y}{\partial x} - \frac{\partial A_x}{\partial y} = B_0. \end{aligned} \tag{14.7}$$

By inspection, we see that one *possible* solution of these equations is

$$A_y = xB_0, \qquad A_x = 0, \qquad A_z = 0.$$

Or we could equally well take

$$A_x = -yB_0, \qquad A_y = 0, \qquad A_z = 0.$$

Still another solution is a linear combination of the two:

$$A_x = -\tfrac{1}{2}yB_0, \qquad A_y = \tfrac{1}{2}xB_0, \qquad A_z = 0. \tag{14.8}$$

---

* Our definition still does not uniquely determine A. For a *unique* specification we would also have to say something about how the field A behaves on some boundary, or at large distances. It is sometimes convenient, for example, to choose a field which goes to zero at large distances.

It is clear that for any particular field $\boldsymbol{B}$, the vector potential $\boldsymbol{A}$ is not unique; there are many possibilities.

The third solution, Eq. (14.8), has some interesting properties. Since the x-component is proportional to $-y$ and the y-component is proportional to $+x$, $\boldsymbol{A}$ must be at right angles to the vector from the z-axis, which we will call $\boldsymbol{r}'$ (the "prime" is to remind us that it is *not* the vector displacement from the origin). Also, the magnitude of $\boldsymbol{A}$ is proportional to $\sqrt{x^2+y^2}$ and, hence, to r'. So $\boldsymbol{A}$ can be simply written (for our uniform field) as

$$\boldsymbol{A} = \tfrac{1}{2}\boldsymbol{B} \times \boldsymbol{r}'. \tag{14.9}$$

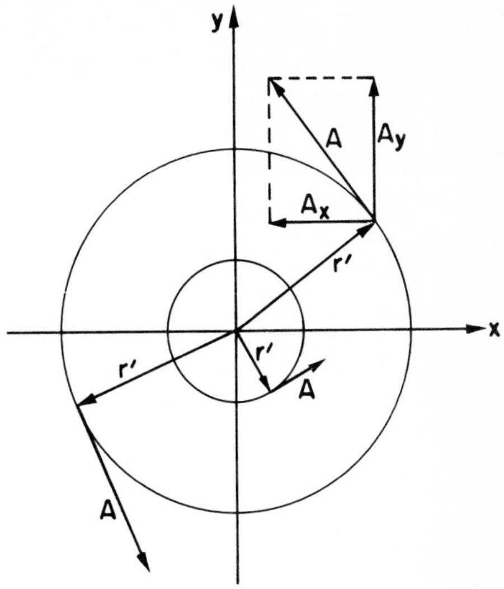

Fig. 14–1. A uniform magnetic field **B** in the z-direction corresponds to a vector potential **A** that rotates about the z-axis, with the magnitude $A = Br'/2$ (r' is the displacement from the z-axis).

The vector potential $\boldsymbol{A}$ has the magnitude $Br'/2$ and rotates about the z-axis as shown in Fig. 14–1. If, for example, the $\boldsymbol{B}$ field is the axial field inside a solenoid, then the vector potential circulates in the same sense as do the currents of the solenoid.

The vector potential for a uniform field can be obtained in another way. The circulation of $\boldsymbol{A}$ on any closed loop Γ can be related to the surface integral of $\nabla \times \boldsymbol{A}$ by Stokes' theorem, Eq. (3.38):

$$\oint_{\Gamma} \boldsymbol{A} \cdot d\boldsymbol{s} = \int_{\text{inside } \Gamma} (\nabla \times \boldsymbol{A}) \cdot \boldsymbol{n}\, da. \tag{14.10}$$

But the integral on the right is equal to the flux of $\boldsymbol{B}$ through the loop, so

$$\oint_{\Gamma} \boldsymbol{A} \cdot d\boldsymbol{s} = \int_{\text{inside } \Gamma} \boldsymbol{B} \cdot \boldsymbol{n}\, da. \tag{14.11}$$

So the circulation of $\boldsymbol{A}$ around *any* loop is equal to the flux of $\boldsymbol{B}$ through the loop. If we take a circular loop, of radius r' in a plane perpendicular to a uniform field $\boldsymbol{B}$, the flux is just

$$\pi r'^2 B.$$

If we choose our origin on an axis of symmetry, so that we can take $\boldsymbol{A}$ as circumferential and a function only of r', the circulation will be

$$\oint \boldsymbol{A} \cdot d\boldsymbol{s} = 2\pi r' A = \pi r'^2 B.$$

We get, as before,

$$A = \frac{Br'}{2}.$$

In the example we have just given, we have calculated the vector potential from the magnetic field, which is opposite to what one normally does. In complicated problems it is usually easier to solve for the vector potential, and then determine the magnetic field from it. We will now show how this can be done.

14–2 The vector potential of known currents

Since $\boldsymbol{B}$ is determined by currents, so also is $\boldsymbol{A}$. We want now to find $\boldsymbol{A}$ in terms of the currents. We start with our basic equation (14.2):

$$c^2 \nabla \times \boldsymbol{B} = \frac{\boldsymbol{j}}{\epsilon_0},$$

which means, of course, that

$$c^2 \nabla \times (\nabla \times \boldsymbol{A}) = \frac{\boldsymbol{j}}{\epsilon_0}. \tag{14.12}$$

This equation is for magnetostatics what the equation

$$\nabla \cdot \nabla \phi = -\frac{\rho}{\epsilon_0} \tag{14.13}$$

was for electrostatics.

Our equation (14.12) for the vector potential looks even more like that for ϕ if we rewrite $\nabla \times (\nabla \times A)$ using the vector identity Eq. (2.58):

$$\nabla \times (\nabla \times A) = \nabla(\nabla \cdot A) - \nabla^2 A. \tag{14.14}$$

Since we have chosen to make $\nabla \cdot A = 0$ (and now you see why), Eq. (14.12) becomes

$$\nabla^2 A = -\frac{j}{\epsilon_0 c^2}. \tag{14.15}$$

This vector equation means, of course, three equations:

$$\nabla^2 A_x = -\frac{j_x}{\epsilon_0 c^2}, \qquad \nabla^2 A_y = -\frac{j_y}{\epsilon_0 c^2}, \qquad \nabla^2 A_z = -\frac{j_z}{\epsilon_0 c^2}. \tag{14.16}$$

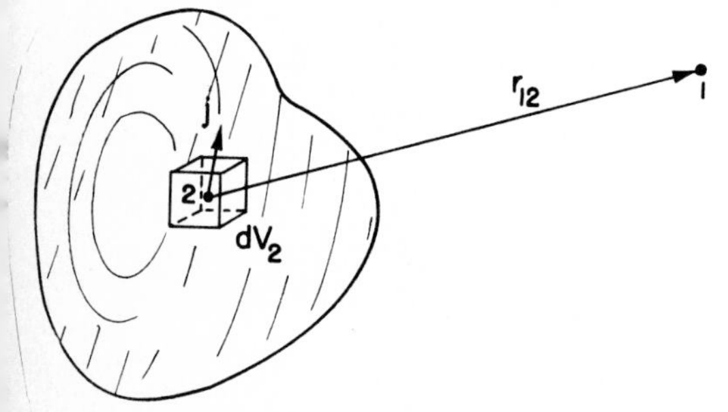

Fig. 14–2. The vector potential A at point 1 is given by an integral over the current elements $j\,dV$ at all points 2.

And each of these equations is *mathematically identical* to

$$\nabla^2 \phi = -\frac{\rho}{\epsilon_0}. \tag{14.17}$$

All we have learned about solving for potentials when ρ is known can be used for solving for each component of A when j is known!

We have seen in Chapter 4 that a general solution for the electrostatic equation (14.17) is

$$\phi(1) = \frac{1}{4\pi\epsilon_0}\int \frac{\rho(2)\,dV_2}{r_{12}}.$$

So we know immediately that a general solution for A_x is

$$A_x(1) = \frac{1}{4\pi\epsilon_0 c^2}\int \frac{j_x(2)\,dV_2}{r_{12}}, \tag{14.18}$$

and similarly for A_y and A_z. (Figure 14–2 will remind you of our conventions for r_{12} and dV_2.) We can combine the three solutions in the vector form

$$A(1) = \frac{1}{4\pi\epsilon_0 c^2}\int \frac{j(2)\,dV_2}{r_{12}}. \tag{14.19}$$

(You can verify if you wish, by direct differentiation of components, that this integral for A satisfies $\nabla \cdot A = 0$ so long as $\nabla \cdot j = 0$, which, as we saw, must happen for steady currents.)

We have, then, a general method for finding the magnetic field of steady currents. The principle is: the x-component of vector potential arising from a current density j is the same as the electric potential ϕ that would be produced by a charge density ρ equal to j_x/c^2—and similarly for the y- and z-components. (This principle works only with components in fixed directions. The "radial" component of A does not come in the same way from the "radial" component of j, for example.) So from the vector current density j, we can find A using Eq. (14.19)—that is, we find each component of A by solving three imaginary electrostatic problems for the charge distributions $\rho_1 = j_x/c^2$, $\rho_2 = j_y/c^2$, and $\rho_3 = j_z/c^2$. Then we get B by taking various derivatives of A to obtain $\nabla \times A$. It's a little more complicated than electrostatics, but the same idea. We will now illustrate the theory by solving for the vector potential in a few special cases.

14–3 A straight wire

For our first example, we will again find the field of a straight wire—which we solved in the last chapter by using Eq. (14.2) and some arguments of symmetry. We take a long straight wire of radius a, carrying the steady current I. Unlike the charge on a conductor in the electrostatic case, a steady current in a wire is uniformly distributed throughout the cross section of the wire. If we choose our coordinates as shown in Fig. 14–3, the current density vector $\boldsymbol{j}$ has only a z-component. Its magnitude is

$$j_z = \frac{I}{\pi a^2} \tag{14.20}$$

inside the wire, and zero outside.

Since j_x and j_y are both zero, we have immediately

$$A_x = 0, \qquad A_y = 0.$$

To get A_z we can use our solution for the electrostatic potential ϕ of a wire with a uniform charge density $\rho = j_z/c^2$. For points outside an infinite charged cylinder, the electrostatic potential is

$$\phi = -\frac{\lambda}{2\pi\epsilon_0} \ln r',$$

where $r' = \sqrt{x^2 + y^2}$ and λ is the charge per unit length, $\pi a^2 \rho$. So A_z must be

$$A_z = -\frac{\pi a^2 j_z}{2\pi\epsilon_0 c^2} \ln r'$$

for points outside a long wire carrying a uniform current. Since $\pi a^2 j_z = I$, we can also write

$$A_z = -\frac{I}{2\pi\epsilon_0 c^2} \ln r'. \tag{14.21}$$

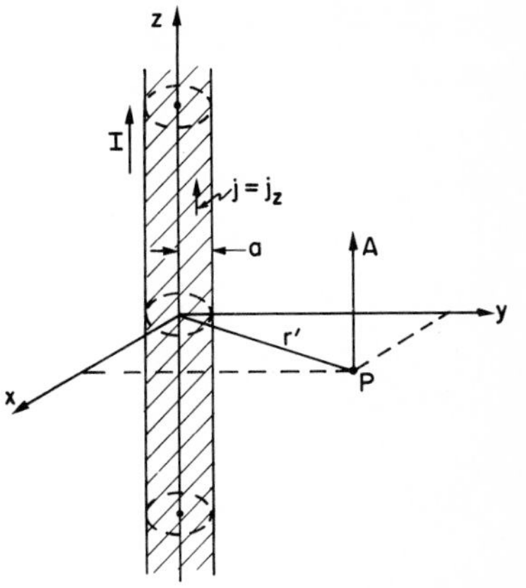

Fig. 14–3. A long cylindrical wire along the z-axis with a uniform current density $\boldsymbol{j}$.

Now we can find $\boldsymbol{B}$ from (14.4). There are only two of the six derivatives that are not zero. We get

$$B_x = -\frac{I}{2\pi\epsilon_0 c^2} \frac{\partial}{\partial y} \ln r' = -\frac{I}{2\pi\epsilon_0 c^2} \frac{y}{r'^2}, \tag{14.22}$$

$$B_y = \frac{I}{2\pi\epsilon_0 c^2} \frac{\partial}{\partial x} \ln r' = \frac{I}{2\pi\epsilon_0 c^2} \frac{x}{r'^2}, \tag{14.23}$$

$$B_z = 0.$$

We get the same result as before: $\boldsymbol{B}$ circles around the wire, and has the magnitude

$$B = \frac{1}{4\pi\epsilon_0 c^2} \frac{2I}{r'}. \tag{14.24}$$

14–4 A long solenoid

Next, we consider again the infinitely long solenoid with a circumferential current on the surface of nI per unit length. (We imagine there are n turns of wire per unit length, carrying the current I, and we neglect the slight pitch of the winding.)

Just as we have defined a "surface charge density" σ, we define here a "surface current density" $\boldsymbol{J}$ equal to the current per unit length on the surface of the solenoid (which is, of course, just the average $\boldsymbol{j}$ times the thickness of the thin winding). The magnitude of $\boldsymbol{J}$ is, here, nI. This surface current (see Fig. 14–4) has the components.

$$J_x = -J \sin\phi, \qquad J_y = J\cos\phi, \qquad J_z = 0.$$

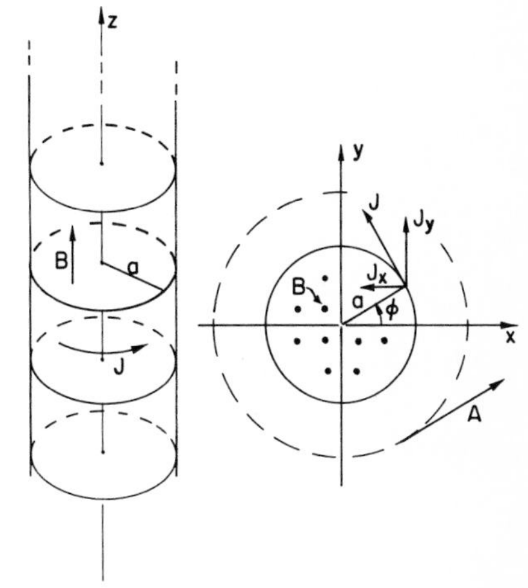

Fig. 14–4. A long solenoid with a surface current density $\boldsymbol{J}$.

Now we must find $\boldsymbol{A}$ for such a current distribution.

First, we wish to find A_x for points outside the solenoid. The result is the same as the electrostatic potential outside a cylinder with a surface charge

$$\sigma = \sigma_0 \sin\phi,$$

with $\sigma_0 = J/c^2$. We have not solved such a charge distribution, but we have done something similar. This charge distribution is equivalent to two *solid* cylinders of charge, one positive and one negative, with a slight relative displacement of their axes in the y-direction. The potential of such a pair of cylinders is proportional to the derivative with respect to y of the potential of a single uniformly charged cylinder. We could work out the constant of proportionality, but let's not worry about it for the moment.

The potential of a cylinder of charge is proportional to $\ln r'$; the potential of the pair is then

$$\phi \propto \frac{\partial \ln r'}{\partial y} = \frac{y}{r'^2}.$$

So we know that

$$A_x = -K\frac{y}{r'^2}, \tag{14.25}$$

where K is some constant. Following the same argument, we would find

$$A_y = K\frac{x}{r'^2}. \tag{14.26}$$

Although we said before that there was no *magnetic* field outside a solenoid, we find now that there *is* an A-field which circulates around the z-axis, as in Fig. 14–4. The question is: Is its curl zero?

Clearly, B_x and B_y are zero, and

$$B_z = \frac{\partial}{\partial x}\left(K\frac{x}{r'^2}\right) - \frac{\partial}{\partial y}\left(-K\frac{y}{r'^2}\right)$$

$$= K\left(\frac{1}{r'^2} - \frac{2x^2}{r'^4} + \frac{1}{r'^2} - \frac{2y^2}{r'^4}\right) = 0.$$

So the magnetic field outside a very long solenoid is indeed zero, even though the vector potential is not.

We can check our result against something else we know: The circulation of the vector potential around the solenoid should be equal to the flux of $\boldsymbol{B}$ inside the coil (Eq. 14.11). The circulation is $A \cdot 2\pi r'$ or, since $A = K/r'$, the circulation is $2\pi K$. Notice that it is independent of r'. That is just as it should be if there is no $\boldsymbol{B}$ outside, because the flux is just the magnitude of $\boldsymbol{B}$ *inside* the solenoid times πa^2. It is the same for all circles of radius $r' > a$. We have found in the last chapter that the field inside is $nI/\epsilon_0 c^2$, so we can determine the constant K:

$$2\pi K = \pi a^2 \frac{nI}{\epsilon_0 c^2},$$

or

$$K = \frac{nIa^2}{2\epsilon_0 c^2}.$$

So the vector potential *outside* has the magnitude

$$A = \frac{nIa^2}{2\epsilon_0 c^2}\frac{1}{r'}, \tag{14.27}$$

and is always perpendicular to the vector $\boldsymbol{r}'$.

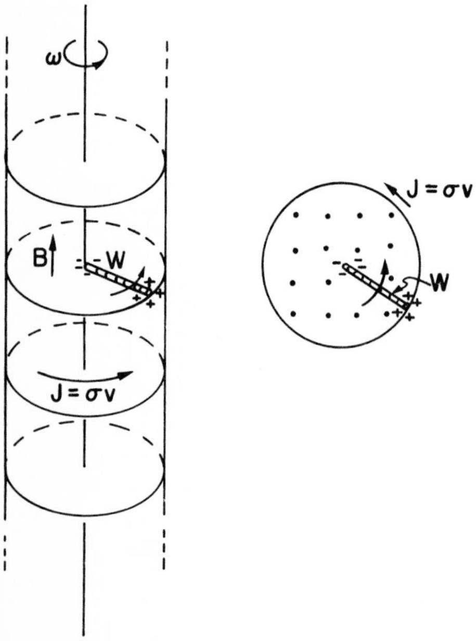

Fig. 14–5. A rotating charged cylinder produces a magnetic field inside. A short radial wire rotating with the cylinder has charges induced on its ends.

We have been thinking of a solenoidal coil of wire, but we would produce the same fields if we rotated a long cylinder with an electrostatic charge on the surface. If we have a thin cylindrical shell of radius a with a surface charge σ, rotating the cylinder makes a surface current $J = \sigma v$, where $v = a\omega$ is the velocity of the surface charge. There will then be a magnetic field $B = \sigma a\omega/\epsilon_0 c^2$ inside the cylinder.

Now we can raise an interesting question. Suppose we put a short piece of wire W perpendicular to the axis of the cylinder, extending from the axis out to the surface, and fastened to the cylinder so that it rotates with it, as in Fig. 14–5. This wire is moving in a magnetic field, so the $\boldsymbol{v} \times \boldsymbol{B}$ forces will cause the ends of the wire to be charged (they will charge up until the $\boldsymbol{E}$-field from the charges just balances the $\boldsymbol{v} \times \boldsymbol{B}$ force). If the cylinder has a positive charge, the end of the wire at the axis will have a negative charge. By measuring the charge on the end of the wire, we could measure the speed of rotation of the system. We would have an "angular-velocity meter"!

But are you wondering: "What if I put myself in the frame of reference of the rotating cylinder? Then there is just a charged cylinder at rest, and I know that the electrostatic equations say there will be *no* electric fields inside, so there will be no force pushing charges to the center. So something must be wrong." But there is nothing wrong. There is no "relativity of rotation." A rotating system is *not* an inertial frame, and the laws of physics are different. We must be sure to use equations of electromagnetism only with respect to inertial coordinate systems.

It would be nice if we could measure the absolute rotation of the earth with such a charged cylinder, but unfortunately the effect is much too small to observe even with the most delicate instruments now available.

14–5 The field of a small loop; the magnetic dipole

Let's use the vector-potential method to find the magnetic field of a small loop of current. As usual, by "small" we mean simply that we are interested in the fields only at distances large compared with the size of the loop. It will turn out that any small loop is a "magnetic dipole." That is, it produces a *magnetic* field like the electric field from an electric dipole.

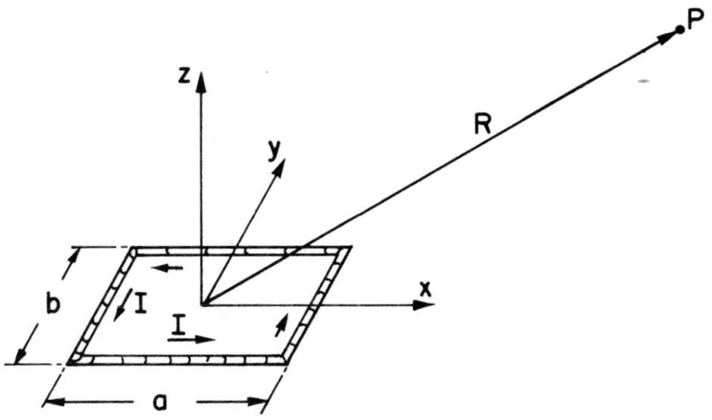

Fig. 14–6. A rectangular loop of wire with the current I. What is the magnetic field at P? ($R \gg a$, or b.)

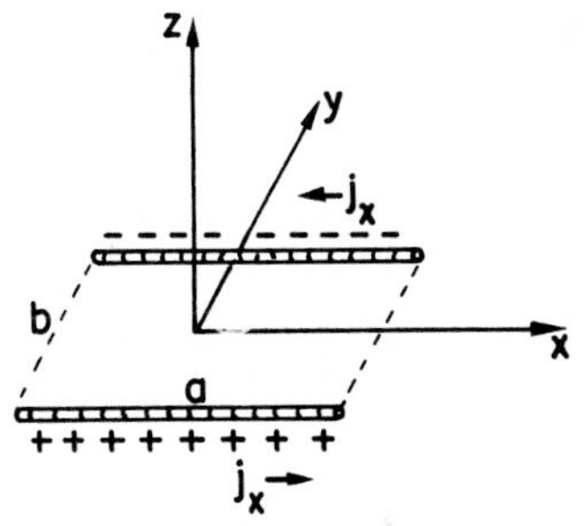

Fig. 14–7. The distribution of j_x in the current loop of Fig. 14–6.

We take first a rectangular loop, and choose our coordinates as shown in Fig. 14–6. There are no currents in the z-direction, so A_z is zero. There are currents in the x-direction on the two sides of length a. In each leg, the current density (and current) is uniform. So the solution for A_x is just like the electrostatic potential from two charged rods (see Fig. 14–7). Since the rods have opposite charges, their electric potential at large distances would be just the dipole potential (Section 6–5). At the point P in Fig. 14–6, the potential would be

$$\phi = \frac{1}{4\pi\epsilon_0}\frac{\boldsymbol{p}\cdot\boldsymbol{e}_R}{R^2}, \tag{14.28}$$

where $\boldsymbol{p}$ is the dipole moment of the charge distribution. The dipole moment, in this case, is the total charge on one rod times the separation between them:

$$p = \lambda ab. \tag{14.29}$$

The dipole moment points in the negative y-direction, so the cosine of the angle between $\boldsymbol{R}$ and $\boldsymbol{p}$ is $-y/R$ (where y is the coordinate of P). So we have

$$\phi = -\frac{1}{4\pi\epsilon_0}\frac{\lambda ab}{R^2}\frac{y}{R}.$$

We get A_x simply by replacing λ by I/c^2:

$$A_x = -\frac{Iab}{4\pi\epsilon_0 c^2}\frac{y}{R^3}. \tag{14.30}$$

By the same reasoning,

$$A_y = \frac{Iab}{4\pi\epsilon_0c^2}\,\frac{x}{R^3}. \tag{14.31}$$

Again, A_y is proportional to x and A_x is proportional to $-y$, so the vector potential (at large distances) goes in circles around the z-axis, circulating in the same sense as I in the loop, as shown in Fig. 14–8.

The strength of $\boldsymbol{A}$ is proportional to Iab, which is the current times the area of the loop. This product is called the *magnetic dipole moment* (or, often, just "magnetic moment") of the loop. We represent it by μ:

$$\mu = Iab. \tag{14.32}$$

The vector potential of a small plane loop of *any* shape (circle, triangle, etc.) is also given by Eqs. (14.30) and (14.31) provided we replace Iab by

$$\mu = I\cdot(\text{area of loop}). \tag{14.33}$$

We leave the proof of this to you.

We can put our equation in vector form if we define the direction of the vector $\boldsymbol{\mu}$ to be the normal to the plane of the loop, with a positive sense given by the right-hand rule (Fig. 14–8). Then we can write

$$\boldsymbol{A} = \frac{1}{4\pi\epsilon_0c^2}\,\frac{\boldsymbol{\mu}\times\boldsymbol{R}}{R^3} = \frac{1}{4\pi\epsilon_0c^2}\,\frac{\boldsymbol{\mu}\times\boldsymbol{e}_R}{R^2}. \tag{14.34}$$

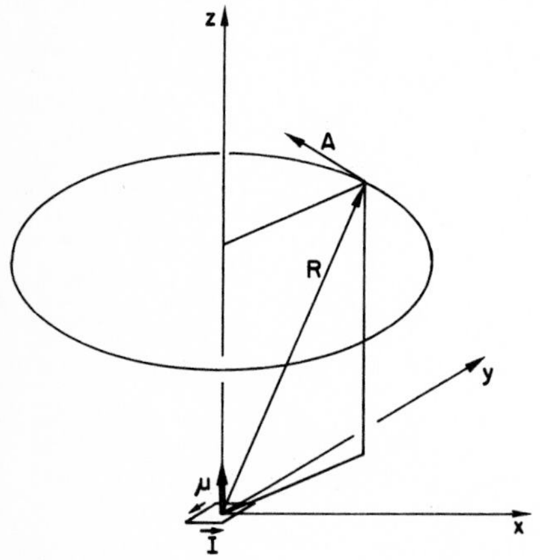

Fig. 14–8. The vector potential of a small current loop at the origin (in the xy-plane); a magnetic dipole field.

We have still to find $\boldsymbol{B}$. Using (14.33) and (14.34), together with (14.4), we get

$$B_x = -\frac{\partial}{\partial z}\,\frac{\mu}{4\pi\epsilon_0c^2}\,\frac{x}{R^3} = \cdots\frac{3xz}{R^5} \tag{14.35}$$

(where by . . . we mean $\mu/4\pi\epsilon_0c^2$),

$$B_y = \frac{\partial}{\partial z}\left(-\cdots\frac{y}{R^3}\right) = \cdots\frac{3yz}{R^5},$$

$$B_z = \frac{\partial}{\partial x}\left(\cdots\frac{x}{R^3}\right) - \frac{\partial}{\partial y}\left(-\cdots\frac{y}{R^3}\right) \tag{14.36}$$

$$= -\cdots\left(\frac{1}{r^3} - \frac{3z^2}{r^5}\right).$$

The components of the $\boldsymbol{B}$-field behave exactly like those of the $\boldsymbol{E}$-field for a dipole oriented along the z-axis. (See Eqs. (6.14) and (6.15); also Fig. 6–5.) That's why we call the loop a magnetic dipole. The word "dipole" is slightly misleading when applied to a magnetic field because there are *no* magnetic "poles" that correspond to electric charges. The magnetic "dipole field" is not produced by two "charges," but by an elementary current loop.

It is curious, though, that starting with completely different laws, $\nabla\cdot\boldsymbol{E} = \rho/\epsilon_0$ and $\nabla\times\boldsymbol{B} = \boldsymbol{j}/\epsilon_0c^2$, we can end up with the same kind of a field. Why should that be? It is because the dipole fields appear only when we are far away from all charges or currents. So through most of the relevant space the equations for $\boldsymbol{E}$ and $\boldsymbol{B}$ are identical: both have zero divergence and zero curl. So they give the same solutions. However, the *sources* whose configuration we summarize by the dipole moments are physically quite different—in one case, it's a circulating current; in the other, a pair of charges, one above and one below the plane of the loop for the corresponding field.

14–6 The vector potential of a circuit

We are often interested in the magnetic fields produced by circuits of wire in which the diameter of the wire is very small compared with the dimensions of the whole system. In such cases, we can simplify the equations for the magnetic field.

For a thin wire we can write our volume element as

$$dV = S\,ds,$$

where S is the cross-sectional area of the wire and ds is the element of distance along the wire. In fact, since the vector $\boldsymbol{ds}$ is in the same direction as $\boldsymbol{j}$, as shown in Fig. 14–9 (and we can assume that $\boldsymbol{j}$ is constant across any given cross section), we can write a vector equation:

$$\boldsymbol{j}\,dV = jS\,\boldsymbol{ds}. \tag{14.37}$$

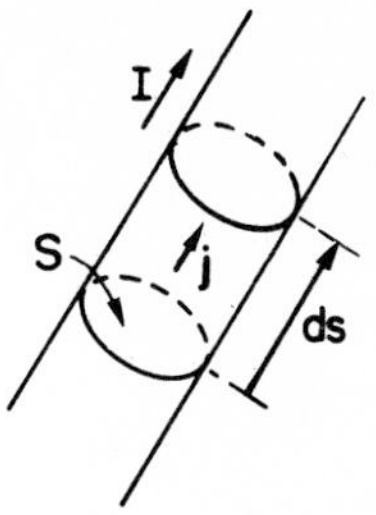

Fig. 14–9. For a fine wire $\boldsymbol{j}\,dV$ is the same as $I\,\boldsymbol{ds}$.

But jS is just what we call the current I in a wire, so our integral for the vector potential (14.19) becomes

$$\boldsymbol{A}(1) = \frac{1}{4\pi\epsilon_0 c^2}\int \frac{I\,\boldsymbol{ds}_2}{r_{12}} \tag{14.38}$$

(see Fig. 14–10). (We assume that I is the same throughout the circuit. If there are several branches with different currents, we should, of course, use the appropriate I for each branch.)

Again, we can find the fields from (14.38) either by integrating directly or by solving the corresponding electrostatic problems.

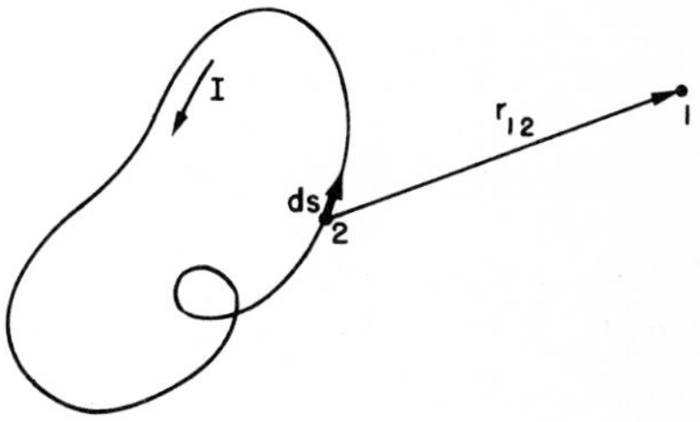

Fig. 14–10. The magnetic field of a wire can be obtained from an integral around the circuit.

14–7 The law of Biot and Savart

In studying electrostatics we found that the electric field of a known charge distribution could be obtained directly with an integral (Eq. 4–16):

$$\boldsymbol{E}(1) = \frac{1}{4\pi\epsilon_0}\int \frac{\rho(2)\boldsymbol{e}_{12}\,dV_2}{r_{12}^2}.$$

As we have seen, it is usually more work to evaluate this integral—there are really three integrals, one for each component—than to do the integral for the potential and take its gradient.

There is a similar integral which relates the magnetic field to the currents. We already have an integral for $\boldsymbol{A}$, Eq. (14.19); we can get an integral for $\boldsymbol{B}$ by taking the curl of both sides:

$$\boldsymbol{B}(1) = \nabla\times\boldsymbol{A}(1) = \nabla\times\left[\frac{1}{4\pi\epsilon_0 c^2}\int\frac{\boldsymbol{j}(2)\,dV_2}{r_{12}}\right]. \tag{14.39}$$

Now we must be careful: The curl operator means taking the derivatives of $\boldsymbol{A}(1)$, that is, it operates only on the coordinates (x_1, y_1, z_1). We can move the $\nabla\times$ operator inside the integral sign if we remember that it operates only on variables with the subscript 1, which of course, appear only in

$$r_{12} = [(x_1 - x_2)^2 + (y_1 - y_2)^2 + (z_1 - z_2)^2]^{1/2}. \tag{14.40}$$

We have, for the x-component of $\boldsymbol{B}$,

$$\begin{aligned} B_x &= \frac{\partial A_z}{\partial y_1} - \frac{\partial A_y}{\partial z_1} \\ &= \frac{1}{4\pi\epsilon_0 c^2}\int\left[j_z\frac{\partial}{\partial y_1}\left(\frac{1}{r_{12}}\right) - j_y\frac{\partial}{\partial z_1}\left(\frac{1}{r_{12}}\right)\right]dV_2 \\ &= -\frac{1}{4\pi\epsilon_0 c^2}\int\left[j_z\frac{y_1-y_2}{r_{12}^3} - j_y\frac{z_1-z_2}{r_{12}^3}\right]dV_2. \end{aligned} \tag{14.41}$$

The quantity in brackets is just the negative of the x-component of

$$\frac{\boldsymbol{j}\times\boldsymbol{r}_{12}}{r_{12}^3} = \frac{\boldsymbol{j}\times\boldsymbol{e}_{12}}{r_{12}^2}.$$

Corresponding results will be found for the other components, so we have

$$\boldsymbol{B}(1) = \frac{1}{4\pi\epsilon_0 c^2} \int \frac{\boldsymbol{j}(2) \times \boldsymbol{e}_{12}}{r_{12}^2}\, dV_2. \tag{14.42}$$

The integral gives $\boldsymbol{B}$ directly in terms of the known currents. The geometry involved is the same as that shown in Fig. 14–2.

If the currents exist only in circuits of small wires we can, as in the last section, immediately do the integral across the wire, replacing $\boldsymbol{j}\,dV$ by $I\,d\boldsymbol{s}$, where $d\boldsymbol{s}$ is an element of length of the wire. Then, using the symbols in Fig. 14–10,

$$\boldsymbol{B}(1) = -\frac{1}{4\pi\epsilon_0 c^2} \int \frac{I\boldsymbol{e}_{12} \times d\boldsymbol{s}_2}{r_{12}^2}. \tag{14.43}$$

(The minus sign appears because we have reversed the order of the cross product.) This equation for $\boldsymbol{B}$ is called the *Biot-Savart law*, after its discoverers. It gives a formula for obtaining directly the magnetic field produced by wires carrying currents.

You may wonder: "What is the advantage of the vector potential if we can find $\boldsymbol{B}$ directly with a vector integral? After all, $\boldsymbol{A}$ also involves three integrals!" Because of the cross product, the integrals for $\boldsymbol{B}$ are usually more complicated, as is evident from Eq. (14.41). Also, since the integrals for $\boldsymbol{A}$ are like those of electrostatics, we may already know them. Finally, we will see that in more advanced theoretical matters (in relativity, in advanced formulations of the laws of mechanics, like the principle of least action to be discussed later, and in quantum mechanics) the vector potential plays an important role.

15

The Vector Potential

15–1 The forces on a current loop; energy of a dipole

In the last chapter we studied the magnetic field produced by a small rectangular current loop. We found that it is a dipole field, with the dipole moment given by

$$\mu = IA, \tag{15.1}$$

where I is the current and A is the area of the loop. The direction of the moment is normal to the plane of the loop, so we can also write

$$\boldsymbol{\mu} = IA\mathbf{n},$$

where $\mathbf{n}$ is the unit normal to the area A.

A current loop—or magnetic dipole—not only produces magnetic fields, but will also experience forces when placed in the magnetic field of other currents. We will look first at the forces on a rectangular loop in a uniform magnetic field. Let the z-axis be along the direction of the field, and the plane of the loop be placed through the y-axis, making the angle θ with the xy-plane as in Fig. 15–1. Then the magnetic moment of the loop—which is normal to its plane—will make the angle θ with the magnetic field.

Since the currents are opposite on opposite sides of the loop, the forces are also opposite, so there is no net force on the loop (when the field is uniform). Because of forces on the two sides marked 1 and 2 in the figure, however, there is a torque which tends to rotate the loop about the y-axis. The magnitude of these forces F_1 and F_2 is

$$F_1 = F_2 = IBb.$$

Their moment arm is

$$a \sin\theta,$$

so the torque is

$$\tau = Iab\, B \sin\theta,$$

or, since Iab is the magnetic moment of the loop,

$$\tau = \mu B \sin\theta.$$

The torque can be written in vector notation:

$$\boldsymbol{\tau} = \boldsymbol{\mu} \times \boldsymbol{B}. \tag{15.2}$$

Although we have only shown that the torque is given by Eq. (15.2) in one rather special case, the result is right for a small loop of any shape, as we will see. You will remember that we found the same kind of relation for the torque on an electric dipole:

$$\boldsymbol{\tau} = \boldsymbol{p} \times \boldsymbol{E}.$$

We now ask about the mechanical energy of our current loop. Since there is a torque, the energy evidently depends on the orientation. The principle of virtual work says that the torque is the rate of change of energy with angle, so we can write

$$dU = -\tau\, d\theta.$$

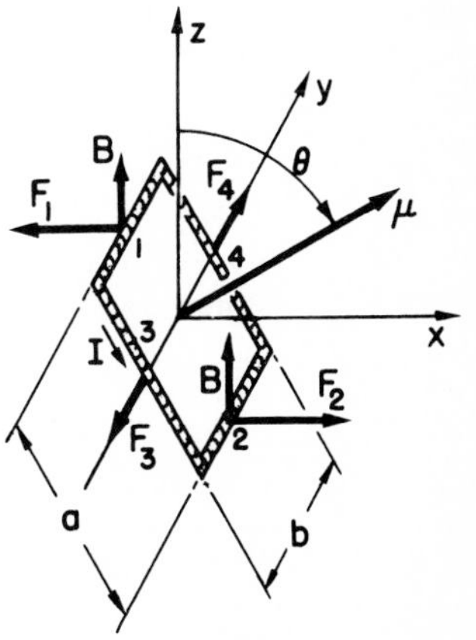

Fig. 15–1. A rectangular loop carrying the current I sits in a uniform field $\mathbf{B}$ (in the z-direction). The torque on the loop is $\boldsymbol{\tau} = \boldsymbol{\mu} \times \mathbf{B}$, where the magnetic moment $\mu = Iab$.

Setting $\tau = -\mu B \sin\theta$, and integrating, we can write for the energy

$$U = -\mu B \cos\theta + \text{a constant}. \tag{15.3}$$

(The sign is negative because the torque tries to line up the moment with the field; the energy is lowest when μ and B are parallel.)

For reasons which we will discuss later, this energy is *not* the total energy of a current loop. (We have, for one thing, not taken into account the energy required to maintain the current in the loop.) We will, therefore, call this energy U_{mech}, to remind us that it is only part of the energy. Also, since we are leaving out some of the energy anyway, we can set the constant of integration equal to zero in Eq. (15.3). So we rewrite the equation:

$$U_{\text{mech}} = -\boldsymbol{\mu}\cdot\boldsymbol{B}. \tag{15.4}$$

Again, this corresponds to our result for an electric dipole:

$$U = -\boldsymbol{p}\cdot\boldsymbol{E}. \tag{15.5}$$

Now the electrostatic energy U in Eq. (15.5) is the true energy, but U_{mech} in (15.4) is not the real energy. It *can*, however, be used in computing forces, by the principle of virtual work, supposing that the current in the loop—or at least μ—is kept constant.

We can show for our rectangular loop that U_{mech} also corresponds to the mechanical work done in bringing the loop into the field. The total force on the loop is zero only in a uniform field; in a nonuniform field there *are* net forces on a current loop. In putting the loop into a region with a field, we must have gone through places where the field was not uniform, and so work was done. To make the calculation simple, we shall imagine that the loop is brought into the field with its moment pointing along the field. (It can be rotated to its final position after it is in place.)

Imagine that we want to move the loop in the x-direction—toward a region of stronger field—and that the loop is oriented as shown in Fig. 15–2. We start somewhere where the field is zero and integrate the force times the distance as we bring the loop into the field.

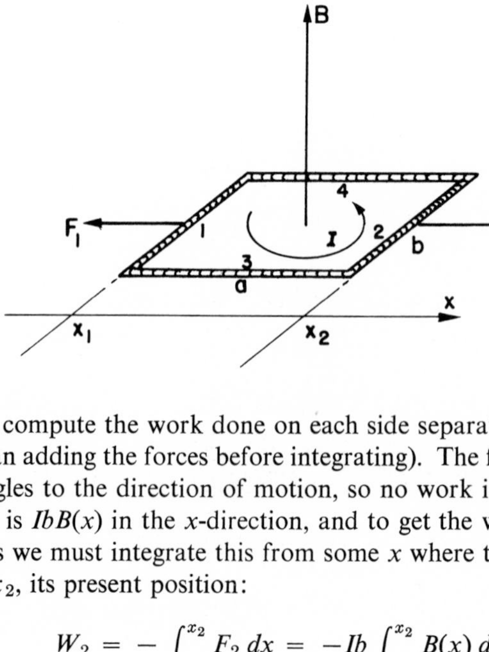

Fig. 15–2. A loop is carried along the x-direction through the field *B*, at right angles to x.

First, let's compute the work done on each side separately and then take the sum (rather than adding the forces before integrating). The forces on sides 3 and 4 are at right angles to the direction of motion, so no work is done on them. The force on side 2 is $IbB(x)$ in the x-direction, and to get the work done against the magnetic forces we must integrate this from some x where the field is zero, say at $x = -\infty$, to x_2, its present position:

$$W_2 = -\int_{-\infty}^{x_2} F_2\,dx = -Ib\int_{-\infty}^{x_2} B(x)\,dx. \tag{15.6}$$

Similarly, the work done against the forces on side 1 is

$$W_1 = -\int_{-\infty}^{x_1} F_1\,dx = Ib\int_{-\infty}^{x_1} B(x)\,dx. \tag{15.7}$$

To find each integral, we need to know how $B(x)$ depends on x. But notice that side 1 follows along right behind side 2, so that its integral includes most of the work done on side 2. In fact, the sum of (15.6) and (15.7) is just

$$W = -Ib \int_{x_1}^{x_2} B(x)\, dx. \tag{15.8}$$

But if we are in a region where B is nearly the same on both sides 1 and 2, we can write the integral as

$$\int_{x_1}^{x_2} B(x)\, dx = (x_2 - x_1)B = aB,$$

where B is the field at the center of the loop. The total mechanical energy we have put in is

$$U_{\text{mech}} = W = -Iab\, B = -\mu B. \tag{15.9}$$

The result agrees with the energy we took for Eq. (15.4).

We would, of course, have gotten the same result if we had added the forces on the loop before integrating to find the work. If we let B_1 be the field at side 1 and B_2 be the field at side 2, then the total force in the x-direction is

$$F_x = Ib(B_2 - B_1).$$

If the loop is "small," that is, if B_2 and B_1 are not too different, we can write

$$B_2 = B_1 + \frac{\partial B}{\partial x}\,\Delta x = B_1 + \frac{\partial B}{\partial x}\, a.$$

So the force is

$$F_x = Iab\, \frac{\partial B}{\partial x}\cdot \tag{15.10}$$

The total work done on the loop by *external* forces is

$$-\int_{-\infty}^{x} F_x\, dx = -Iab \int \frac{\partial B}{\partial x}\, dx = -IabB,$$

which is again just $-\mu B$. Only now we see why it is that the *force* on a small current loop is proportional to the derivative of the magnetic field, as we would expect from

$$F_x\, \Delta x = -\Delta U_{\text{mech}} = -\Delta(-\boldsymbol{\mu}\cdot\boldsymbol{B}). \tag{15.11}$$

Our result, then, is that even though $U_{\text{mech}} = -\boldsymbol{\mu}\cdot\boldsymbol{B}$ may not include all the energy of a system—it is a fake kind of energy—it can still be used with the principle of virtual work to find the forces on steady current loops.

15–2 Mechanical and electrical energies

We want now to show why the energy U_{mech} discussed in the previous section is not the correct energy associated with steady currents—that it does not keep track of the total energy in the world. We have, indeed, emphasized that it can be used like the energy, for computing forces from the principle of virtual work, *provided* that the current in the loop (and all *other* currents) do not change. Let's see why all this works.

Imagine that the loop in Fig. 15–2 is moving in the $+x$-direction and take the z-axis in the direction of $\boldsymbol{B}$. The conduction electrons in side 2 will experience a force along the wire, in the y-direction. But because of their flow—as an electric current—there is a component of their motion in the same direction as the force. Each electron is, therefore, having work done on it at the rate $F_y v_y$, where v_y is the component of the electron velocity along the wire. We will call this work done on the electrons *electrical* work. Now it turns out that if the loop is moving in a *uniform* field, the total electrical work is zero, since positive work is done on some parts of the loop and an equal amount of negative work is done on other parts.

But this is not true if the circuit is moving in a nonuniform field—then there *will* be a net amount of work done on the electrons. In general, this work would tend to change the flow of the electrons, but if the current is being held constant, energy must be absorbed or delivered by the battery or other source that is keeping the current steady. This energy was not included when we computed U_{mech} in Eq. (15.9), because our computations included only the mechanical forces on the body of the wire.

You may be thinking: But the force on the electrons depends on how *fast* the wire is moved; perhaps if the wire is moved slowly enough this electrical energy can be neglected. It is true that the *rate* at which the electrical energy is delivered is proportional to the speed of the wire, but the *total* energy delivered is proportional also to the *time* that this rate goes on. So the total electrical energy is proportional to the velocity times the time, which is just the distance moved. For a given distance moved in a field the same amount of electrical work is done.

Let's consider a segment of wire of unit length carrying the current I and moving in a direction perpendicular to itself and to a magnetic field $\boldsymbol{B}$ with the speed v_{wire}. Because of the current the electrons will have a drift velocity v_{drift} along the wire. The component of the magnetic force on each electron in the direction of the drift is $q_e v_{\text{wire}} B$. So the rate at which electrical work is being done is $F v_{\text{drift}} = (q_e v_{\text{wire}} B) v_{\text{drift}}$. If there are N conduction electrons in the unit length of the wire, the total rate at which electrical work is being done is

$$\frac{dU_{\text{elect}}}{dt} = N q_e v_{\text{wire}} B v_{\text{drift}}.$$

But $N q_e v_{\text{drift}} = I$, the current in the wire, so

$$\frac{dU_{\text{elect}}}{dt} = I v_{\text{wire}} B.$$

Now since the current is held constant, the forces on the conduction electrons do not cause them to accelerate; the electrical energy is not going into the electrons but into the source that is keeping the current constant.

But notice that the force on the *wire* is IB, so IBv_{wire} is also the rate of *mechanical work* done on the wire, $dU_{\text{mech}}/dt = IBv_{\text{wire}}$. We conclude that the mechanical work done on the wire is just equal to the electrical work done on the current source, so the energy of the loop *is a constant!*

This is not a coincidence, but a consequence of the law we already know. The total force on each charge in the wire is

$$\boldsymbol{F} = q(\boldsymbol{E} + \boldsymbol{v} \times \boldsymbol{B}).$$

The rate at which work is done is

$$\boldsymbol{v} \cdot \boldsymbol{F} = q[\boldsymbol{v} \cdot \boldsymbol{E} + \boldsymbol{v} \cdot (\boldsymbol{v} \times \boldsymbol{B})]. \tag{15.12}$$

If there are no electric fields we have only the second term, which is always zero. We shall see later that *changing* magnetic fields produce electric fields, so our reasoning applies only to moving wires in steady magnetic fields.

How is it then that the principle of virtual work gives the right answer? Because we *still* have not taken into account the *total* energy of the world. We have not included the energy of the currents that are *producing* the magnetic field we start out with.

Suppose we imagine a complete system such as that drawn in Fig. 15–3(a), in which we are moving our loop with the current I_1 into the magnetic field $\boldsymbol{B}_1$ produced by the current I_2 in a coil. Now the current I_1 in the loop will also be producing some magnetic field $\boldsymbol{B}_2$ at the coil. If the loop is moving, the field $\boldsymbol{B}_2$ will be changing. As we shall see in the next chapter, a changing magnetic field generates an $\boldsymbol{E}$-field; and this $\boldsymbol{E}$-field will do work on the charges in the coil. This energy must also be included in our balance sheet of the total energy.

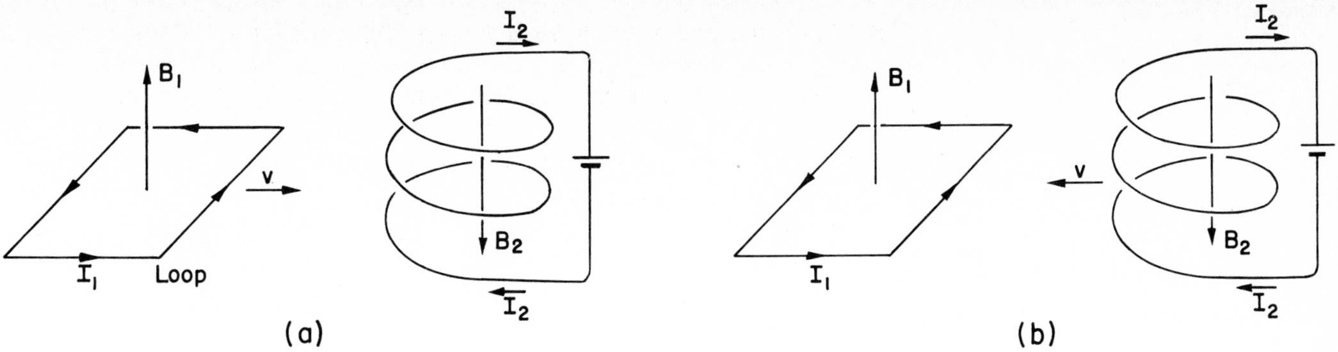

Fig. 15–3. Finding the energy of a small loop in a magnetic field.

We could wait until the next chapter to find out about this new energy term, but we can also see what it will be if we use the principle of relativity in the following way. When we are moving the loop toward the stationary coil we know that its electrical energy is just equal and opposite to the mechanical work done. So

$$U_{\text{mech}} + U_{\text{elect}}(\text{loop}) = 0.$$

Suppose now we look at what is happening from a different point of view, in which the loop is at rest, and the coil is moved toward it. The coil is then moving into the field produced by the loop. The same arguments would give that

$$U_{\text{mech}} + U_{\text{elect}}(\text{coil}) = 0.$$

The mechanical energy is the same in the two cases because it comes from the force between the two circuits.

The sum of the two equations gives

$$2U_{\text{mech}} + U_{\text{elect}}(\text{loop}) + U_{\text{elect}}(\text{coil}) = 0.$$

The total energy of the whole system is, of course, the sum of the two electrical energies plus the mechanical energy taken only *once*. So we have

$$U_{\text{total}} = U_{\text{elect}}(\text{loop}) + U_{\text{elect}}(\text{coil}) + U_{\text{mech}} = -U_{\text{mech}}. \tag{15.13}$$

The total energy of the world is really the *negative* of U_{mech}. If we want the true energy of a magnetic dipole, for example, we should write

$$U_{\text{total}} = +\boldsymbol{\mu}\cdot\boldsymbol{B}.$$

It is only if we make the condition that all currents are constant that we can use only a part of the energy, U_{mech} (which is always the negative of the true energy), to find the mechanical forces. In a more general problem, we must be careful to include all energies.

We have seen an analogous situation in electrostatics. We showed that the energy of a capacitor is equal to $Q^2/2C$. When we use the principle of virtual work to find the force between the plates of the capacitor, the change in energy is equal to $Q^2/2$ times the change in $1/C$. That is,

$$\Delta U = \frac{Q^2}{2}\,\Delta\left(\frac{1}{C}\right) = -\frac{Q^2}{2}\frac{\Delta C}{C^2}. \tag{15.14}$$

Now suppose that we were to calculate the work done in moving two conductors subject to the different condition that the voltage between them is held constant. Then we can get the right answers for force from the principle of virtual work if we do something artificial. Since $Q = CV$, the real energy is $\frac{1}{2}CV^2$. But if we define an artificial energy equal to $-\frac{1}{2}CV^2$, then the principle of virtual work can be used to get forces by setting the change in the artificial energy equal to the mechanical work, provided that we insist that the voltage V be held constant. Then

$$\Delta U_{\text{mech}} = \Delta\left(-\frac{CV^2}{2}\right) = -\frac{V^2}{2}\,\Delta C, \tag{15.15}$$

which is the same as Eq. (15.14). We get the correct result even though we are neglecting the work done by the electrical system to keep the voltage constant. Again, this electrical energy is just twice as big as the mechanical energy and of the opposite sign.

Thus if we calculate artificially, disregarding the fact that the source of the potential has to do work to maintain the voltages constant, we get the right answer. It is exactly analogous to the situation in magnetostatics.

15–3 The energy of steady currents

We can now use our knowledge that $U_{\text{total}} = -U_{\text{mech}}$ to find the true energy of steady currents in magnetic fields. We can begin with the true energy of a small current loop. Calling U_{total} just U, we write

$$U = \boldsymbol{\mu}\cdot\boldsymbol{B}. \tag{15.16}$$

Although we calculated this energy for a plane rectangular loop, the same result holds for a small plane loop of any shape.

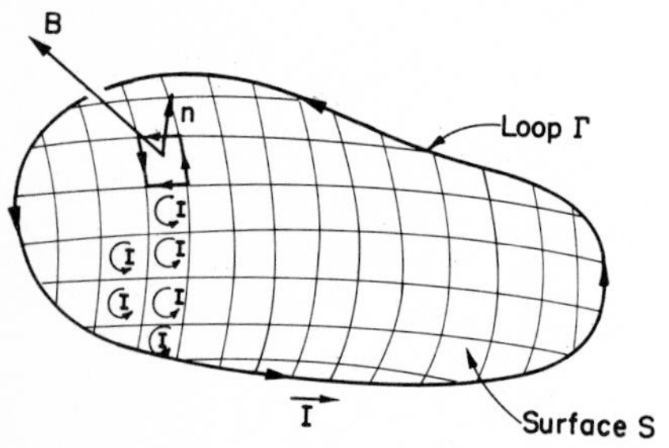

Fig. 15–4. The energy of a large loop in a magnetic field can be considered as the sum of energies of smaller loops.

We can find the energy of a circuit of any shape by imagining that it is made up of small current loops. Say we have a wire in the shape of the loop Γ of Fig. 15–4. We fill in this curve with the surface S, and on the surface mark out a large number of small loops, each of which can be considered plane. If we let the current I circulate around *each* of the little loops, the net result will be the same as a current around Γ, since the currents will cancel on all lines internal to Γ. Physically, the system of little currents is indistinguishable from the original circuit. The energy must also be the same, and so is just the sum of the energies of the little loops.

If the area of each little loop is Δa, its energy is $I\,\Delta a B_n$, where B_n is the component normal to Δa. The total energy is

$$U = \sum IB_n\,\Delta a.$$

Going to the limit of infinitesimal loops, the sum becomes an integral, and

$$U = I\int B_n\,da = I\int \boldsymbol{B}\cdot\boldsymbol{n}\,da, \tag{15.17}$$

where $\boldsymbol{n}$ is the unit normal to da.

If we set $\boldsymbol{B} = \boldsymbol{\nabla}\times\boldsymbol{A}$, we can connect the surface integral to a line integral, using Stokes' theorem,

$$I\int_S (\boldsymbol{\nabla}\times\boldsymbol{A})\cdot\boldsymbol{n}\,da = I\oint_\Gamma \boldsymbol{A}\cdot d\boldsymbol{s}, \tag{15.18}$$

where $d\boldsymbol{s}$ is the line element along Γ. So we have the energy for a circuit of any shape:

$$U = I\oint_{\text{circuit}} \boldsymbol{A}\cdot d\boldsymbol{s}. \tag{15.19}$$

In this expression $\boldsymbol{A}$ refers, of course, to the vector potential due to those currents (other than the I in the wire) which produce the field $\boldsymbol{B}$ at the wire.

Now any distribution of steady currents can be imagined to be made up of filaments that run parallel to the lines of current flow. For each pair of such circuits, the energy is given by (15.19), where the integral is taken around one circuit, using the vector potential $\boldsymbol{A}$ from the other circuit. For the total energy we want the sum of all such pairs. If, instead of keeping track of the pairs, we take the complete sum over all the filaments, we would be counting the energy twice (we saw a similar effect in electrostatics), so the total energy can be written

$$U = \tfrac{1}{2}\int \boldsymbol{j}\cdot\boldsymbol{A}\,dV. \tag{15.20}$$

This formula corresponds to the result we found for the electrostatic energy:

$$U = \tfrac{1}{2}\int \rho\phi\, dV. \tag{15.21}$$

So we may if we wish think of $\boldsymbol{A}$ as a kind of potential energy for currents in magnetostatics. Unfortunately, this idea is not too useful, because it is true only for static fields. In fact, neither of the equations (15.20) and (15.21) gives the correct energy when the fields change with time.

15–4 *B* versus *A*

In this section we would like to discuss the following questions: Is the vector potential merely a device which is useful in making calculations—as the scalar potential is useful in electrostatics—or is the vector potential a "real" field? Isn't the magnetic field the "real" field, because it is responsible for the force on a moving particle? First we should say that the phrase "a real field" is not very meaningful. For one thing, you probably don't feel that the magnetic field is very "real" anyway, because even the whole idea of a field is a rather abstract thing. You cannot put out your hand and feel the magnetic field. Furthermore, the value of the magnetic field is not very definite; by choosing a suitable moving coordinate system, for instance, you can make a magnetic field at a given point disappear.

What we mean here by a "real" field is this: a real field is a mathematical function we use for avoiding the idea of action at a distance. If we have a charged particle at the position P, it is affected by other charges located at some distance from P. One way to describe the interaction is to say that the other charges make some "condition"—whatever it may be—in the environment at P. If we know that condition, which we describe by giving the electric and magnetic fields, then we can determine completely the behavior of the particle—with no further reference to how those conditions came about.

In other words, if those other charges were altered in some way, but the conditions at P that are described by the electric and magnetic field at P remain the same, then the motion of the charge will also be the same. A "real" field is then a set of numbers we specify in such a way that what happens *at a point* depends only on the numbers *at that point*. We do not need to know any more about what's going on at other places. It is in this sense that we will discuss whether the vector potential is a "real" field.

You may be wondering about the fact that the vector potential is not unique—that it can be changed by adding the gradient of any scalar with no change at all in the forces on particles. That has not, however, anything to do with the question of reality in the sense that we are talking about. For instance, the magnetic field is in a sense altered by a relativity change (as are also $\boldsymbol{E}$ and $\boldsymbol{A}$). But we are not worried about what happens if the field *can* be changed in this way. That doesn't really make any difference; that has nothing to do with the question of whether the vector potential is a proper "real" field for describing magnetic effects, or whether it is just a useful mathematical tool.

We should also make some remarks on the usefulness of the vector potential $\boldsymbol{A}$. We have seen that it can be used in a formal procedure for calculating the magnetic fields of known currents, just as ϕ can be used to find electric fields. In electrostatics we saw that ϕ was given by the scalar integral

$$\phi(1) = \frac{1}{4\pi\epsilon_0}\int \frac{\rho(2)}{r_{12}}\, dV_2. \tag{15.22}$$

From this ϕ, we get the three components of $\boldsymbol{E}$ by three differential operations. This procedure is usually easier to handle than evaluating the three integrals in the vector formula

$$\boldsymbol{E}(1) = \frac{1}{4\pi\epsilon_0}\int \frac{\rho(2)\boldsymbol{e}_{12}}{r_{12}^2}\, dV_2. \tag{15.23}$$

First, there are three integrals; and second, each integral is in general somewhat more difficult.

The advantages are much less clear for magnetostatics. The integral for $\boldsymbol{A}$ is already a vector integral:

$$\boldsymbol{A}(1) = \frac{1}{4\pi\epsilon_0 c^2} \int \frac{\boldsymbol{j}(2)\,dV_2}{r_{12}}, \tag{15.24}$$

which is, of course, three integrals. Also, when we take the curl of $\boldsymbol{A}$ to get $\boldsymbol{B}$, we have six derivatives to do and combine by pairs. It is not immediately obvious whether in most problems this procedure is really any easier than computing $\boldsymbol{B}$ directly from

$$\boldsymbol{B}(1) = \frac{1}{4\pi\epsilon_0 c^2} \int \frac{\boldsymbol{j}(2) \times \boldsymbol{e}_{12}}{r_{12}^2}\,dV_2. \tag{15.25}$$

Using the vector potential is often more difficult for simple problems for the following reason. Suppose we are interested only in the magnetic field $\boldsymbol{B}$ at one point, and that the problem has some nice symmetry—say we want the field at a point on the axis of a ring of current. Because of the symmetry, we can easily get $\boldsymbol{B}$ by doing the integral of Eq. (15.25). If, however, we were to find $\boldsymbol{A}$ first, we would have to compute $\boldsymbol{B}$ from *derivatives* of $\boldsymbol{A}$, so we must know what $\boldsymbol{A}$ is at all points in the *neighborhood* of the point of interest. And most of these points are off the axis of symmetry, so the integral for $\boldsymbol{A}$ gets complicated. In the ring problem, for example, we would need to use elliptic integrals. In such problems, $\boldsymbol{A}$ is clearly not very useful. It is true that in many complex problems it is easier to work with $\boldsymbol{A}$, but it would be hard to argue that this ease of technique would justify making you learn about one more vector field.

We have introduced $\boldsymbol{A}$ because it *does* have an important physical significance. Not only is it related to the energies of currents, as we saw in the last section, but it is also a "real" physical field in the sense that we described above. In classical mechanics it is clear that we can write the force on a particle as

$$\boldsymbol{F} = q(\boldsymbol{E} + \boldsymbol{v} \times \boldsymbol{B}), \tag{15.26}$$

so that, given the forces, everything about the motion is determined. In any region where $\boldsymbol{B} = 0$ even if $\boldsymbol{A}$ is not zero, such as outside a solenoid, there is no discernible effect of $\boldsymbol{A}$. Therefore for a long time it was believed that $\boldsymbol{A}$ was not a "real" field. It turns out, however, that there are phenomena involving quantum mechanics which show that the field $\boldsymbol{A}$ is in fact a "real" field in the sense we have defined it. In the next section we will show you how that works.

15–5 The vector potential and quantum mechanics

There are many changes in what concepts are important when we go from classical to quantum mechanics. We have already discussed some of them in Vol. I. In particular, the force concept gradually fades away, while the concepts of energy and momentum become of paramount importance. You remember that instead of particle motions, one deals with probability amplitudes which vary in space and time. In these amplitudes there are wavelengths related to momenta, and frequencies related to energies. The momenta and energies, which determine the phases of wave functions, are therefore the important quantities in quantum mechanics. Instead of forces, we deal with the way interactions change the wavelength of the waves. The idea of a force becomes quite secondary—if it is there at all. When people talk about nuclear forces, for example, what they usually analyze and work with are the energies of interaction of two nucleons, and not the force between them. Nobody ever differentiates the energy to find out what the force looks like. In this section we want to describe how the vector and scalar potentials enter into quantum mechanics. It is, in fact, just because momentum and energy play a central role in quantum mechanics that $\boldsymbol{A}$ and ϕ provide the most direct way of introducing electromagnetic effects into quantum descriptions.

We must review a little how quantum mechanics works. We will consider again the imaginary experiment described in Chapter 37 of Vol. I, in which elec-

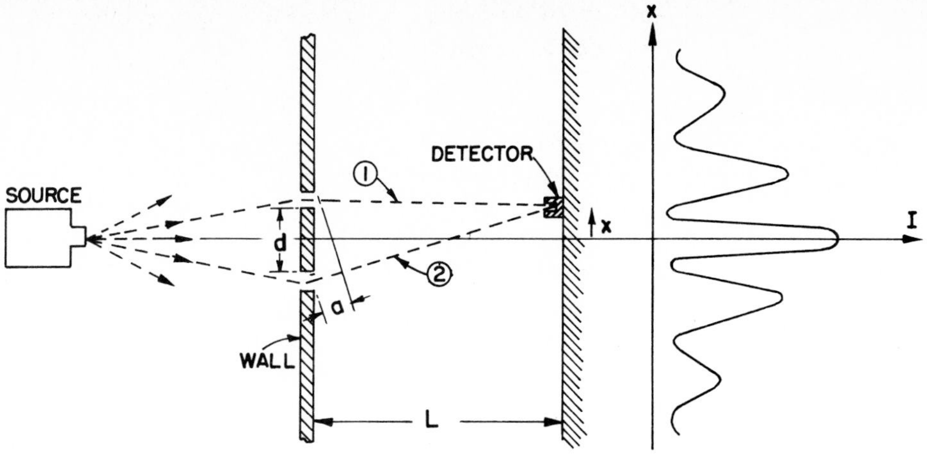

Fig. 15–5. An interference experiment with electrons (see also Chapter 37 of Vol. I).

trons are diffracted by two slits. The arrangement is shown again in Fig. 15–5. Electrons, all of nearly the same energy, leave the source and travel toward a wall with two narrow slits. Beyond the wall is a "backstop" with a movable detector. The detector measures the rate, which we call I, at which electrons arrive at a small region of the backstop at the distance x from the axis of symmetry. The rate is proportional to the probability that an individual electron that leaves the source will reach that region of the backstop. This probability has the complicated-looking distribution shown in the figure, which we understand as due to the interference of two amplitudes, one from each slit. The interference of the two amplitudes depends on their phase difference. That is, if the amplitudes are $C_1e^{i\Phi_1}$ and $C_2e^{i\Phi_2}$, the phase difference $\delta = \Phi_1 - \Phi_2$ determines their interference pattern [see Eq. (29.12) in Vol. I]. If the distance between the screen and the slits is L, and if the difference in the path lengths for electrons going through the two slits is a, as shown in the figure, then the phase difference of the two waves is given by

$$\delta = \frac{a}{\lambdabar}. \tag{15.27}$$

As usual, we let $\lambdabar = \lambda/2\pi$, where λ is the wavelength of the space variation of the probability amplitude. For simplicity, we will consider only values of x much less than L; then we can set

$$a = \frac{x}{L}\,d$$

and

$$\delta = \frac{x}{L}\,\frac{d}{\lambdabar}. \tag{15.28}$$

When x is zero, δ is zero; the waves are in phase, and the probability has a maximum. When δ is π, the waves are out of phase, they interfere destructively, and the probability is a minimum. So we get the wavy function for the electron intensity.

Now we would like to state the law that for quantum mechanics replaces the force law $\boldsymbol{F} = q\boldsymbol{v} \times \boldsymbol{B}$. It will be the law that determines the behavior of quantum-mechanical particles in an electromagnetic field. Since what happens is determined by amplitudes, the law must tell us how the magnetic influences affect the amplitudes; we are no longer dealing with the acceleration of a particle. The law is the following: the phase of the amplitude to arrive via any trajectory is changed by the presence of a magnetic field by an amount equal to the integral of the vector potential along the whole trajectory times the charge of the particle over Planck's constant. That is,

$$\text{Magnetic change in phase} = \frac{q}{\hbar}\int_{\text{trajectory}} \boldsymbol{A}\cdot d\boldsymbol{s}. \tag{15.29}$$

If there were no magnetic field there would be a certain phase of arrival. If there is a magnetic field anywhere, the phase of the arriving wave is increased by the integral in Eq. (15.29).

Although we will not need to use it for our present discussion, we mention that the effect of an electrostatic field is to produce a phase change given by the *negative* of the *time* integral of the scalar potential ϕ:

$$\text{Electric change in phase} = -\frac{q}{\hbar}\int \phi\, dt.$$

These two expressions are correct not only for static fields, but together give the correct result for *any* electromagnetic field, static or dynamic. This is the law that replaces $\boldsymbol{F} = q(\boldsymbol{E} + \boldsymbol{v} \times \boldsymbol{B})$. We want now, however, to consider only a static magnetic field.

Suppose that there is a magnetic field present in the two-slit experiment. We want to ask for the phase of arrival at the screen of the two waves whose paths pass through the two slits. Their interference determines where the maxima in the probability will be. We may call Φ_1 the phase of the wave along trajectory (1). If $\Phi_1(B = 0)$ is the phase without the magnetic field, then when the field is turned on the phase will be

$$\Phi_1 = \Phi_1(B = 0) + \frac{q}{\hbar}\int_{(1)} \boldsymbol{A} \cdot d\boldsymbol{s}. \tag{15.30}$$

Similarly, the phase for trajectory (2) is

$$\Phi_2 = \Phi_2(B = 0) + \frac{q}{\hbar}\int_{(2)} \boldsymbol{A} \cdot d\boldsymbol{s}. \tag{15.31}$$

The interference of the waves at the detector depends on the phase difference

$$\delta = \Phi_1(B = 0) - \Phi_2(B = 0) + \frac{q}{\hbar}\int_{(1)} \boldsymbol{A} \cdot d\boldsymbol{s} - \frac{q}{\hbar}\int_{(2)} \boldsymbol{A} \cdot d\boldsymbol{s}. \tag{15.32}$$

The no-field difference we will call $\delta(B = 0)$; it is just the phase difference we have calculated above in Eq. (15.28). Also, we notice that the two integrals can be written as *one* integral that goes forward along (1) and back along (2); we call this the closed path (1–2). So we have

$$\delta = \delta(B = 0) + \frac{q}{\hbar}\oint_{(1-2)} \boldsymbol{A} \cdot d\boldsymbol{s}. \tag{15.33}$$

This equation tells us how the electron motion is changed by the magnetic field; with it we can find the new positions of the intensity maxima and minima at the backstop.

Before we do that, however, we want to raise the following interesting and important point. You remember that the vector potential function has some arbitrariness. Two different vector potential functions $\boldsymbol{A}$ and $\boldsymbol{A}'$ whose difference is the gradient of some scalar function $\nabla\psi$, both represent the same magnetic field, since the curl of a gradient is zero. They give, therefore, the same classical force $q\boldsymbol{v} \times \boldsymbol{B}$. If in quantum mechanics the effects depend on the vector potential, *which* of the many possible $\boldsymbol{A}$-functions is correct?

The answer is that the same arbitrariness in $\boldsymbol{A}$ continues to exist for quantum mechanics. If in Eq. (15.33) we change $\boldsymbol{A}$ to $\boldsymbol{A}' = \boldsymbol{A} + \nabla\psi$, the integral on $\boldsymbol{A}$ becomes

$$\oint_{(1-2)} \boldsymbol{A}' \cdot d\boldsymbol{s} = \oint_{(1-2)} \boldsymbol{A} \cdot d\boldsymbol{s} + \oint_{(1-2)} \nabla\psi \cdot d\boldsymbol{s}.$$

The integral of $\nabla\psi$ is around the *closed* path (1–2), but the integral of the tangential component of a gradient on a closed path is always zero, by Stokes' theorem. Therefore both $\boldsymbol{A}$ and $\boldsymbol{A}'$ give the same phase differences and the same quantum-mechanical interference effects. In both classical and quantum theory it is only the curl of $\boldsymbol{A}$ that matters; any choice of the function of $\boldsymbol{A}$ which has the correct curl gives the correct physics.

The same conclusion is evident if we use the results of Section 14–1. There we found that the line integral of $\boldsymbol{A}$ around a closed path is the flux of $\boldsymbol{B}$ through the path, which here is the flux between paths (1) and (2). Equation (15.33) can, if we wish, be written as

$$\delta = \delta(B = 0) + \frac{q}{\hbar}\,[\text{flux of } \boldsymbol{B} \text{ between (1) and (2)}], \tag{15.34}$$

where by the flux of $\boldsymbol{B}$ we mean, as usual, the surface integral of the normal component of $\boldsymbol{B}$. The result depends only on $\boldsymbol{B}$, and therefore only on the curl of $\boldsymbol{A}$.

Now because we can write the result in terms of $\boldsymbol{B}$ as well as in terms of $\boldsymbol{A}$, you might be inclined to think that the $\boldsymbol{B}$ holds its own as a "real" field and that the $\boldsymbol{A}$ can still be thought of as an artificial construction. But the definition of "real" field that we originally proposed was based on the idea that a "real" field would not act on a particle from a distance. We can, however, give an example in which $\boldsymbol{B}$ is zero—or at least arbitrarily small—at any place where there is some chance to find the particles, so that it is not possible to think of it acting *directly* on them.

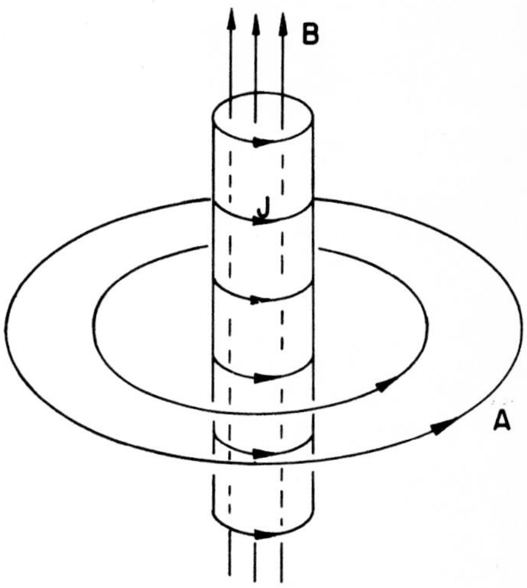

Fig. 15–6. The magnetic field and vector potential of a long solenoid.

You remember that for a long solenoid carrying an electric current there is a $\boldsymbol{B}$-field inside but none outside, while there is lots of $\boldsymbol{A}$ circulating around outside, as shown in Fig. 15–6. If we arrange a situation in which electrons are to be found only *outside* of the solenoid—only where there is $\boldsymbol{A}$—there will still be an influence on the motion, according to Eq. (15.33). Classically, that is impossible. Classically, the force depends only on $\boldsymbol{B}$; in order to know that the solenoid is carrying current, the particle must go through it. But quantum-mechanically you can find out that there is a magnetic field inside the solenoid by going *around* it—without ever going close to it!

Suppose that we put a very long solenoid of small diameter just behind the wall and between the two slits, as shown in Fig. 15–7. The diameter of the solenoid is to be much smaller than the distance d between the two slits. In these circumstances, the diffraction of the electrons at the slit gives no appreciable probability that the electrons will get near the solenoid. What will be the effect on our interference experiment?

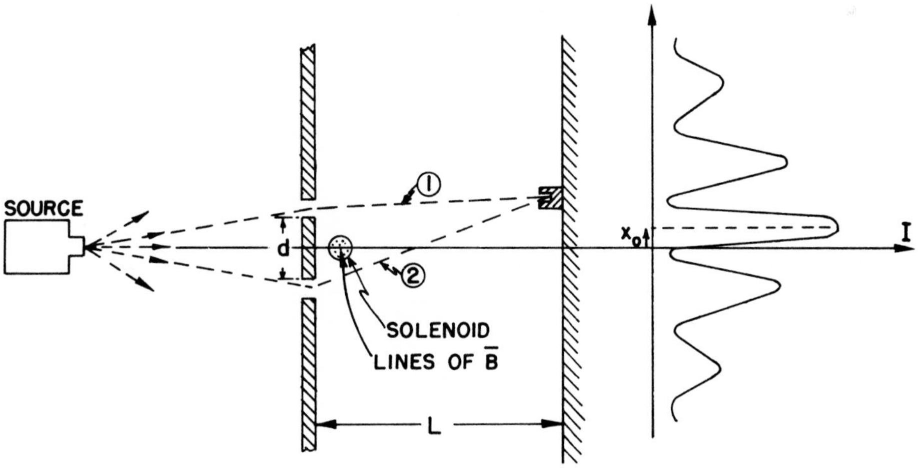

Fig. 15–7. A magnetic field can influence the motion of electrons even though it exists only in regions where there is an arbitrarily small probability of finding the electrons.

We compare the situation with and without a current through the solenoid. If we have no current, we have no $\boldsymbol{B}$ or $\boldsymbol{A}$ and we get the original pattern of electron intensity at the backstop. If we turn the current on in the solenoid and build up a magnetic field $\boldsymbol{B}$ inside, then there is an $\boldsymbol{A}$ outside. There is a shift in the phase difference proportional to the circulation of $\boldsymbol{A}$ outside the solenoid, which will mean that the pattern of maxima and minima is shifted to a new position. In fact, since the flux of $\boldsymbol{B}$ inside is a constant for any pair of paths, so also is the circulation of $\boldsymbol{A}$. For every arrival point there is the same phase change; this corresponds to shifting the entire pattern in x by a constant amount, say x_0, that we can easily calculate. The maximum intensity will occur where the phase difference between the two waves is zero. Using Eq. (15.32) or Eq. (15.33) for δ and Eq. (15.28) for $\delta(B = 0)$, we have

$$x_0 = -\frac{L}{d}\,\lambda\!\!\!{}^{-}\,\frac{q}{\hbar}\oint_{(1-2)} \boldsymbol{A}\cdot d\boldsymbol{s}, \tag{15.35}$$

or

$$x_0 = -\frac{L}{d}\,\lambda\!\!\!{}^{-}\,\frac{q}{\hbar}\,[\text{flux of } \boldsymbol{B} \text{ between (1) and (2)}]. \tag{15.36}$$

The pattern with the solenoid in place should appear* as shown in Fig. 15–7. At least, that is the prediction of quantum mechanics.

Precisely this experiment has recently been done. It is a very, very difficult experiment. Because the wavelength of the electrons is so small, the apparatus must be on a tiny scale to observe the interference. The slits must be very close together, and that means that one needs an exceedingly small solenoid. It turns out that in certain circumstances, iron crystals will grow in the form of very long, microscopically thin filaments called whiskers. When these iron whiskers are magnetized they are like a tiny solenoid, and there is no field outside except near the ends. The electron interference experiment was done with such a whisker between two slits, and the predicted displacement in the pattern of electrons was observed.

In our sense then, the $\boldsymbol{A}$-field is "real." You may say: "But there *was* a magnetic field." There was, but remember our original idea—that a field is "real" if it is what must be specified *at the position* of the particle in order to get the motion. The $\boldsymbol{B}$-field in the whisker acts at a distance. If we want to describe its influence not as action-at-a-distance, we must use the vector potential.

This subject has an interesting history. The theory we have described was known from the beginning of quantum mechanics in 1926. The fact that the vector potential appears in the wave equation of quantum mechanics (called the Schrödinger equation) was obvious from the day it was written. That it cannot be replaced by the magnetic field in any easy way was observed by one man after the other who tried to do so. This is also clear from our example of electrons moving in a region where there is no field and being affected nevertheless. But because in classical mechanics $\boldsymbol{A}$ did not appear to have any direct importance and, furthermore, because it could be changed by adding a gradient, people repeatedly said that the vector potential had no direct physical significance—that only the magnetic and electric fields are "right" even in quantum mechanics. It seems strange in retrospect that no one thought of discussing this experiment until 1956, when Bohm and Aharanov first suggested it and made the whole question crystal clear. The implication was there all the time, but no one paid attention to it. Thus many people were rather shocked when the matter was brought up. That's why someone thought it would be worth while to do the experiment to see that it really was right, even though quantum mechanics, which had been believed for so many years, gave an unequivocal answer. It is interesting that something like this can be around for thirty years but, because of certain prejudices of what is and is not significant, continues to be ignored.

Now we wish to continue in our analysis a little further. We will show the connection between the quantum-mechanical formula and the classical formula—to show why it turns out that if we look at things on a large enough scale it will look as though the particles are acted on by a force equal to $q\boldsymbol{v} \times$ the curl of $\boldsymbol{A}$. To get classical mechanics from quantum mechanics, we need to consider cases in which all the wavelengths are very small compared with distances over which external conditions, like fields, vary appreciably. We shall not prove the result in great generality, but only in a very simple example, to show how it works. Again we consider the same slit experiment. But instead of putting all the magnetic field in a very tiny region between the slits, we imagine a magnetic field that extends

---

* If the field $\boldsymbol{B}$ comes out of the plane of the figure, the flux as we have defined it is negative and x_0 is positive.

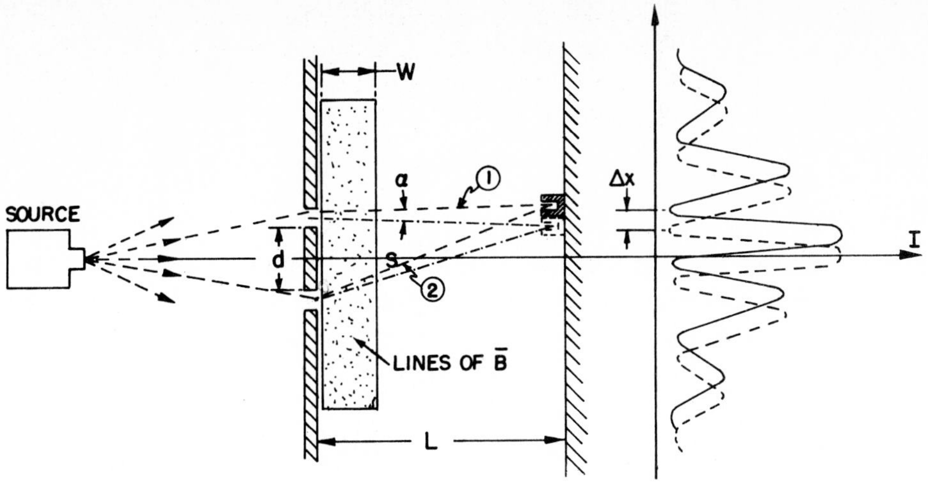

Fig. 15–8. The shift of the interference pattern due to a strip of magnetic field.

over a larger region behind the slits, as shown in Fig. 15–8. We will take the idealized case where we have a magnetic field which is uniform in a narrow strip of width w, considered small as compared with L. (That can easily be arranged; the backstop can be put as far out as we want.) In order to calculate the shift in phase, we must take the two integrals of $\boldsymbol{A}$ along the two trajectories (1) and (2). They differ, as we have seen, merely by the flux of $\boldsymbol{B}$ between the paths. To our approximation, the flux is Bwd. The phase difference for the two paths is then

$$\delta = \delta(B = 0) + \frac{q}{\hbar} Bwd. \tag{15.37}$$

We note that, to our approximation, the phase shift is independent of the angle. So again the effect will be to shift the whole pattern upward by an amount Δx. Using Eq. (15.28),

$$\Delta x = \frac{L\lambda\!\!\!^{-}}{d}\,\Delta\delta = \frac{L\lambda\!\!\!^{-}}{d}\,[\delta - \delta(B = 0)].$$

Using (15.37) for $\delta - \delta(B = 0)$,

$$\Delta x = L\lambda\!\!\!^{-}\,\frac{q}{\hbar}\,Bw. \tag{15.38}$$

Such a shift is equivalent to deflecting all the trajectories by the small angle α (see Fig. 15–8), where

$$\alpha = \frac{\Delta x}{L} = \frac{\lambda\!\!\!^{-}}{\hbar}\,qBw. \tag{15.39}$$

Now classically we would also expect a thin strip of magnetic field to deflect all trajectories through some small angle, say α', as shown in Fig. 15–9(a). As the electrons go through the magnetic field, they feel a transverse force $q\boldsymbol{v} \times \boldsymbol{B}$ which lasts for a time w/v. The change in their transverse momentum is just equal to this impulse, so

$$\Delta p_x = qwB. \tag{15.40}$$

The angular deflection [Fig. 15–9(b)] is equal to the ratio of this transverse momentum to the total momentum p. We get that

$$\alpha' = \frac{\Delta p_x}{p} = \frac{qwB}{p}. \tag{15.41}$$

We can compare this result with Eq. (15.39), which gives the same quantity computed quantum-mechanically. But the connection between classical mechanics and quantum mechanics is this: A particle of momentum p corresponds to a quan-

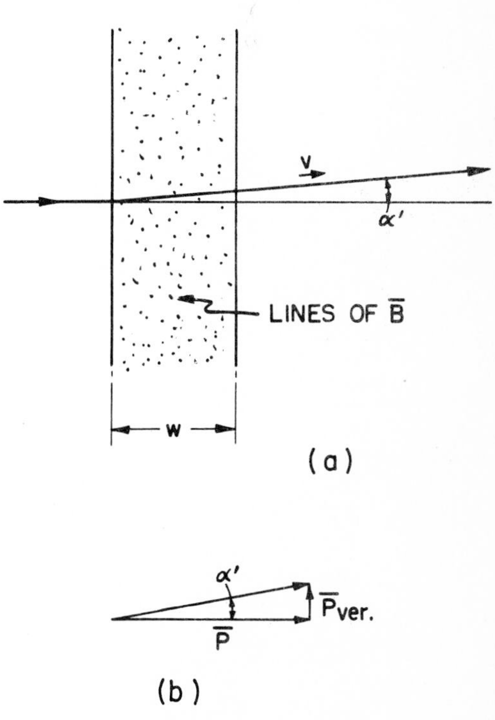

Fig. 15–9. Deflection of a particle due to passage through a strip of magnetic field.

tum amplitude varying with the wavelength $\lambda\!\!\!^- = \hbar/p$. With this equality, α and α' are identical; the classical and quantum calculations give the same result.

From the analysis we see how it is that the vector potential which appears in quantum mechanics in an explicit form produces a classical force which depends only on its derivatives. In quantum mechanics what matters is the interference between nearby paths; it always turns out that the effects depend only on how much the field A *changes* from point to point, and therefore only on the derivatives of A and not on the value itself. Nevertheless, the vector potential A (together with the scalar potential ϕ that goes with it) appears to give the most direct description of the physics. This becomes more and more apparent the more deeply we go into the quantum theory. In the general theory of quantum electrodynamics, one takes the vector and scalar potentials as the fundamental quantities in a set of equations that replace the Maxwell equations: $\boldsymbol{E}$ and $\boldsymbol{B}$ are slowly disappearing from the modern expression of physical laws; they are being replaced by $\boldsymbol{A}$ and ϕ.

15–6 What is true for statics is false for dynamics

We are now at the end of our exploration of the subject of static fields. Already in this chapter we have come perilously close to having to worry about what happens when fields change with time. We were barely able to avoid it in our treatment of magnetic energy by taking refuge in a relativistic argument. Even so, our treatment of the energy problem was somewhat artificial and perhaps even mysterious, because we ignored the fact that moving coils must, in fact, produce changing fields. It is now time to take up the treatment of time-varying fields—the subject of electrodynamics. We will do so in the next chapter. First, however, we would like to emphasize a few points.

Although we began this course with a presentation of the complete and correct equations of electromagnetism, we immediately began to study some incomplete pieces—because that was easier. There is a great advantage in starting with the simpler theory of static fields, and proceeding only later to the more complicated theory which includes dynamic fields. There is less new material to learn all at once, and there is time for you to develop your intellectual muscles in preparation for the bigger task.

But there is the danger in this process that before we get to see the complete story, the incomplete truths learned on the way may become ingrained and taken as the whole truth—that what is true and what is only sometimes true will become confused. So we give in Table 15–1 a summary of the important formulas we have covered, separating those which are true in general from those which are true for statics, but false for dynamics. This summary also shows, in part, where we are going, since as we treat dynamics we will be developing in detail what we must just state here without proof.

It may be useful to make a few remarks about the table. First, you should notice that the equations we started with are the *true* equations—we have not misled you there. The electromagnetic force (often called the *Lorentz force*) $\boldsymbol{F} = q(\boldsymbol{E} + \boldsymbol{v} \times \boldsymbol{B})$ is *true*. It is only Coulomb's law that is false, to be used only for statics. The four Maxwell equations for $\boldsymbol{E}$ and $\boldsymbol{B}$ are also true. The equations we took for statics are false, of course, because we left off all terms with time derivatives.

Gauss' law, $\nabla \cdot \boldsymbol{E} = \rho/\epsilon_0$, remains, but the curl of $\boldsymbol{E}$ is *not* zero in general. So $\boldsymbol{E}$ cannot always be equated to the gradient of a scalar—the electrostatic potential. We will see that a scalar potential still remains, but it is a time-varying quantity that must be used together with vector potentials for a complete description of the electric field. The equations governing this new scalar potential are, necessarily, also new.

We must also give up the idea that $\boldsymbol{E}$ is zero in conductors. When the fields are changing, the charges in conductors do not, in general, have time to rearrange themselves to make the field zero. They are set in motion, but never reach equilibrium. The only general statement is: electric fields in conductors produce cur-

Table 15–1

FALSE IN GENERAL (true only for statics)	TRUE ALWAYS
$F = \dfrac{1}{4\pi\epsilon_0}\dfrac{q_1q_2}{r^2}$ (Coulomb's law)	$\boldsymbol{F} = q(\boldsymbol{E} + \boldsymbol{v} \times \boldsymbol{B})$ (Lorentz force) → $\nabla \cdot \boldsymbol{E} = \dfrac{\rho}{\epsilon_0}$ (Gauss' law)
$\nabla \times \boldsymbol{E} = 0$ $\boldsymbol{E} = -\nabla\phi$ $\boldsymbol{E}(1) = \dfrac{1}{4\pi\epsilon_0}\displaystyle\int \dfrac{\rho(2)\boldsymbol{e}_{12}}{r_{12}^2}\,dV_2$ For conductors, $\boldsymbol{E} = 0$, $\phi =$ constant. $Q = CV$	→ $\nabla \times \boldsymbol{E} = -\dfrac{\partial \boldsymbol{B}}{\partial t}$ (Faraday's law) $\boldsymbol{E} = -\nabla\phi - \dfrac{\partial \boldsymbol{A}}{\partial t}$ In a conductor, $\boldsymbol{E}$ makes currents.
$c^2\nabla \times \boldsymbol{B} = \dfrac{\boldsymbol{j}}{\epsilon_0}$ (Ampere's law) $\boldsymbol{B}(1) = \dfrac{1}{4\pi\epsilon_0 c^2}\displaystyle\int \dfrac{\boldsymbol{j}(2) \times \boldsymbol{e}_{12}}{r_{12}^2}\,dV_2$	→ $\nabla \cdot \boldsymbol{B} = 0$ (No magnetic charges) $\boldsymbol{B} = \nabla \times \boldsymbol{A}$ → $c^2\nabla \times \boldsymbol{B} = \dfrac{\boldsymbol{j}}{\epsilon_0} + \dfrac{\partial \boldsymbol{E}}{\partial t}$
$\nabla^2\phi = -\dfrac{\rho}{\epsilon_0}$ (Poisson's equation) $\nabla^2\boldsymbol{A} = -\dfrac{\boldsymbol{j}}{\epsilon_0 c^2}$ with $\nabla \cdot \boldsymbol{A} = 0$	$\nabla^2\phi - \dfrac{1}{c^2}\dfrac{\partial^2\phi}{\partial t^2} = -\dfrac{\rho}{\epsilon_0}$ and $\nabla^2\boldsymbol{A} - \dfrac{1}{c^2}\dfrac{\partial^2\boldsymbol{A}}{\partial t^2} = -\dfrac{\boldsymbol{j}}{\epsilon_0 c^2}$ with $c^2\nabla \cdot \boldsymbol{A} + \dfrac{\partial\phi}{\partial t} = 0$
$\phi(1) = \dfrac{1}{4\pi\epsilon_0}\displaystyle\int \dfrac{\rho(2)}{r_{12}}\,dV_2$ $\boldsymbol{A}(1) = \dfrac{1}{4\pi\epsilon_0 c^2}\displaystyle\int \dfrac{\boldsymbol{j}(2)}{r_{12}}\,dV_2$	$\phi(1, t) = \dfrac{1}{4\pi\epsilon_0}\displaystyle\int \dfrac{\rho(2, t')}{r_{12}}\,dV_2$ and $\boldsymbol{A}(1, t) \pm \dfrac{1}{4\pi\epsilon_0 c^2}\displaystyle\int \dfrac{\boldsymbol{j}(2, t')}{r_{12}}\,dV_2$ with $t' = t - \dfrac{r_{12}}{c}$
$U = \frac{1}{2}\displaystyle\int \rho\phi\,dV + \frac{1}{2}\int \boldsymbol{j} \cdot \boldsymbol{A}\,dV$	$U = \displaystyle\int \left(\dfrac{\epsilon_0}{2}\boldsymbol{E} \cdot \boldsymbol{E} + \dfrac{\epsilon_0 c^2}{2}\boldsymbol{B} \cdot \boldsymbol{B}\right) dV$

The equations marked by an arrow (→) are Maxwell's equations.

rents. So in varying fields a conductor is *not* an equipotential. It also follows that the idea of a capacitance is no longer precise.

Since there are no magnetic charges, the divergence of $\boldsymbol{B}$ is *always* zero. So $\boldsymbol{B}$ can always be equated to $\nabla \times \boldsymbol{A}$. (Everything doesn't change!) But the generation of $\boldsymbol{B}$ is not only from currents: $\nabla \times \boldsymbol{B}$ is proportional to the current density *plus* a new term $\partial \boldsymbol{E}/\partial t$. This means that $\boldsymbol{A}$ is related to currents by a new equation. It is also related to ϕ. If we make use of our freedom to choose $\nabla \cdot \boldsymbol{A}$ for our own convenience, the equations for $\boldsymbol{A}$ or ϕ can be arranged to take on a simple and elegant form. We therefore make the condition that $c^2\nabla \cdot \boldsymbol{A} = -\partial\phi/\partial t$, and the differential equations for $\boldsymbol{A}$ or ϕ appear as shown in the table.

The potentials $\boldsymbol{A}$ and ϕ can still be found by integrals over the currents and charges, but not the *same* integrals as for statics. Most wonderfully, though, the true integrals are like the static ones, with only a small and physically appealing modification. When we do the integrals to find the potentials at some point, say point (1) in Fig. 15–10, we must use the values of $\boldsymbol{j}$ and ρ at the point (2) *at an earlier time* $t' = t - r_{12}/c$. As you would expect, the influences propagate from point (2) to point (1) at the speed c. With this small change, one can solve for the fields of varying currents and charges, because once we have $\boldsymbol{A}$ and ϕ, we get $\boldsymbol{B}$ from $\nabla \times \boldsymbol{A}$, as before, and $\boldsymbol{E}$ from $-\nabla\phi - \partial\boldsymbol{A}/\partial t$.

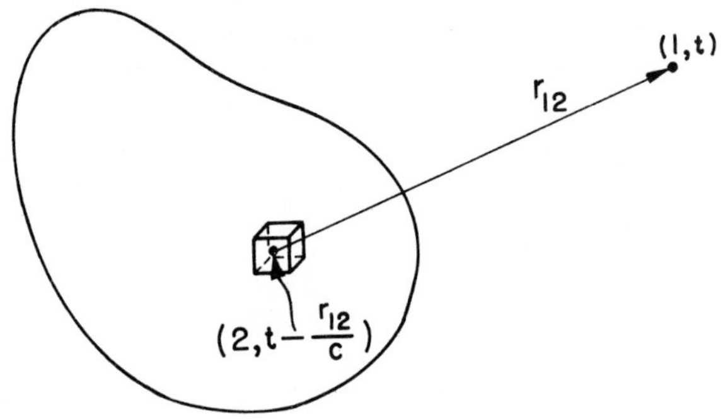

Fig. 15–10. The potentials at point (1) and at the time t are given by summing the contributions from each element of the source at the roving point (2), using the currents and charges which were present at the earlier time $t - r_{12}/c$.

Finally, you will notice that some results—for example, that the energy density in an electric field is $\epsilon_0 E^2/2$—are true for electrodynamics as well as for statics. You should not be misled into thinking that this is at all "natural." The validity of any formula derived in the static case must be demonstrated over again for the dynamic case. A contrary example is the expression for the electrostatic energy in terms of a volume integral of $\rho\phi$. This result is true *only* for statics.

We will consider all these matters in more detail in due time, but it will perhaps be useful to keep in mind this summary, so you will know what you can forget, and what you should remember as always true.

16

Induced Currents

16–1 Motors and generators

The discovery in 1820 that there was a close connection between electricity and magnetism was very exciting—until then, the two subjects had been considered as quite independent. The first discovery was that currents in wires make magnetic fields; then, in the same year, it was found that wires carrying current in a magnetic field have forces on them.

One of the excitements whenever there is a mechanical force is the possibility of using it in an engine to do work. Almost immediately after their discovery, people started to design electric motors using the forces on current-carrying wires. The principle of the electromagnetic motor is shown in bare outline in Fig. 16–1. A permanent magnet—usually with some pieces of soft iron—is used to produce a magnetic field in two slots. Across each slot there is a north and south pole, as shown. A rectangular coil of copper is placed with one side in each slot. When a current passes through the coil, it flows in opposite directions in the two slots, so the forces are also opposite, producing a torque on the coil about the axis shown. If the coil is mounted on a shaft so that it can turn, it can be coupled to pulleys or gears and can do work.

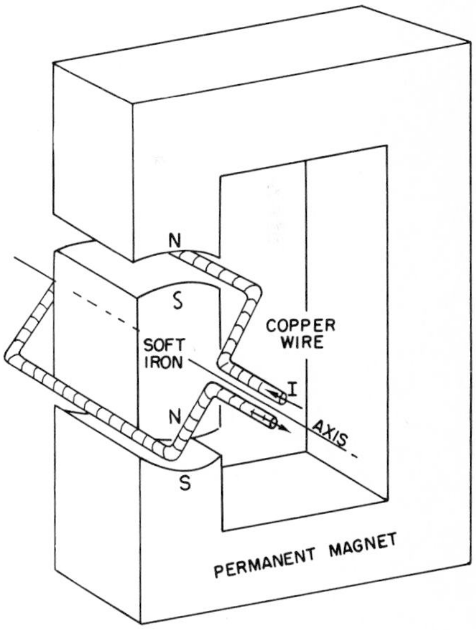

Fig. 16–1. Schematic outline of a simple electromagnetic motor.

The same idea can be used for making a sensitive instrument for electrical measurements. Thus the moment the force law was discovered the precision of electrical measurements was greatly increased. First, the torque of such a motor can be made much greater for a given current by making the current go around many turns instead of just one. Then the coil can be mounted so that it turns with very little torque—either by supporting its shaft on very delicate jewel bearings or by hanging the coil on a very fine wire or a quartz fiber. Then an exceedingly small current will make the coil turn, and for small angles the amount of rotation will be proportional to the current. The rotation can be measured by gluing a pointer to the coil or, for the most delicate instruments, by attaching a small mirror to the coil and looking at the shift of the image of a scale. Such instruments are called galvanometers. Voltmeters and ammeters work on the same principle.

The same ideas can be applied on a large scale to make large motors for providing mechanical power. The coil can be made to go around and around by arranging that the connections to the coil are reversed each half-turn by contacts mounted on the shaft. Then the torque is always in the same direction. Small dc motors are made just this way. Larger motors, dc or ac, are often made by replacing the permanent magnet by an electromagnet, energized from the electrical power source.

With the realization that electric currents make magnetic fields, people immediately suggested that, somehow or other, magnets might also make electric fields. Various experiments were tried. For example, two wires were placed parallel to each other and a current was passed through one of them in the hope of finding a current in the other. The thought was that the magnetic field might in some way drag the electrons along in the second wire, giving some such law as "likes prefer to move alike." With the largest available current and the most sensitive galvanometer to detect any current, the result was negative. Large magnets next to wires also produced no observed effects. Finally, Faraday discovered in 1840 the essential feature that had been missed—that electric effects exist only when there is something *changing*. If one of a pair of wires has a *changing* current, a current is induced in the other, or if a magnet is *moved* near an electric circuit, there is a current. We say that currents are *induced*. This was the induction effect discovered by Faraday. It transformed the rather dull subject of static fields into a very exciting dynamic subject with an enormous range of wonderful phenomena. This chapter is devoted to a qualitative description of some of them. As we will see, one can quickly get into fairly complicated situations that are hard to analyze quantitatively in all their details. But never mind, our main purpose in this chapter is first to acquaint you with the phenomena involved. We will take up the detailed analysis later.

We can easily understand one feature of magnetic induction from what we already know, although it was not known in Faraday's time. It comes from the $\boldsymbol{v} \times \boldsymbol{B}$ force on a moving charge that is proportional to its velocity in a magnetic field. Suppose that we have a wire which passes near a magnet, as shown in Fig. 16–2, and that we connect the ends of the wire to a galvanometer. If we move the wire across the end of the magnet the galvanometer pointer moves.

The magnet produces some vertical magnetic field, and when we push the wire across the field, the electrons in the wire feel a *sideways* force—at right angles to the field and to the motion. The force pushes the electrons along the wire. But why does this move the galvanometer, which is so far from the force? Because when the electrons which feel the magnetic force try to move, they push—by electric repulsion—the electrons a little farther down the wire; they, in turn, repel the electrons a little farther on, and so on for a long distance. An amazing thing.

It was so amazing to Gauss and Weber—who first built a galvanometer—that they tried to see how far the forces in the wire would go. They strung a wire all the way across their city. Mr. Gauss, at one end, connected the wires to a battery (batteries were known before generators) and Mr. Weber watched the galvanometer move. They had a way of signaling long distances—it was the beginning of the telegraph! Of course, this has nothing directly to do with induction—it has to do with the way wires carry currents, whether the currents are pushed by induction or not.

Now suppose in the setup of Fig. 16–2 we leave the wire alone and move the magnet. We still see an effect on the galvanometer. As Faraday discovered, moving the magnet under the wire—one way—has the same effect as moving the wire over the magnet—the other way. But when the magnet is moved, we no longer have any $\boldsymbol{v} \times \boldsymbol{B}$ force on the electrons in the wire. This is the new effect that Faraday found. Today, we might hope to understand it from a relativity argument.

We already understand that the magnetic field of a magnet comes from its internal currents. So we expect to observe the same effect if instead of a magnet in Fig. 16–2 we use a coil of wire in which there is a current. If we move the wire past the coil there will be a current through the galvanometer, or also if we move the coil past the wire. But there is now a more exciting thing: If we change the magnetic field of the coil *not* by moving it, but by *changing its current*, there is again an effect in the galvanometer. For example, if we have a loop of wire near a coil, as shown in Fig. 16–3, and if we keep both of them stationary but switch off the current, there is a pulse of current through the galvanometer. When we switch the coil on again, the galvanometer kicks in the other direction.

Whenever the galvanometer in a situation such as the one shown in Fig. 16–2, or in Fig. 16–3, has a current, there is a net push on the electrons in the wire in one direction along the wire. There may be pushes in different directions at different places, but there is more push in one direction than another. What counts is the push integrated around the complete circuit. We call this net integrated push *the electromotive force* (abbreviated emf) in the circuit. More precisely, the emf is defined as the tangential force per unit charge in the wire integrated over length, once around the complete circuit. Faraday's complete discovery was that emf's can be generated in a wire in three different ways: by moving the wire, by moving a magnet near the wire, or by changing a current in a nearby wire.

Let's consider the simple machine of Fig. 16–1 again, only now, instead of putting a current through the wire to make it turn, let's turn the loop by an external force, for example by hand or by a waterwheel. When the coil rotates, its wires are moving in the magnetic field and we will find an emf in the circuit of the coil. The motor becomes a generator.

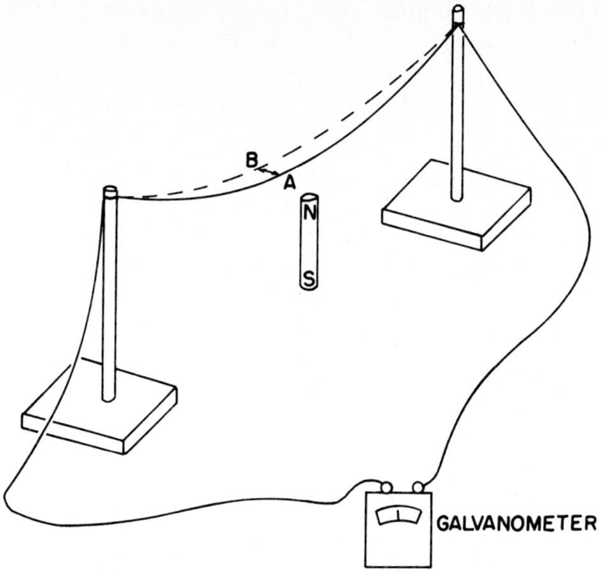

Fig. 16–2. Moving a wire through a magnetic field produces a current, as shown by the galvanometer.

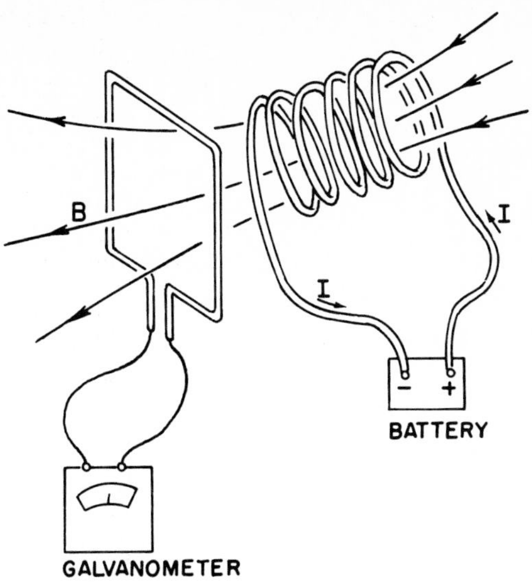

Fig. 16–3. A coil with current produces a current in a second coil if the first coil is moved or if its current is changed.

The coil of the generator has an induced emf from its motion. The amount of the emf is given by a simple rule discovered by Faraday. (We will just state the rule now and wait until later to examine it in detail.) The rule is that when the magnetic flux that passes through the loop (this flux is the normal component of $\boldsymbol{B}$ integrated over the area of the loop) is changing with time, the emf is equal to the rate of change of the flux. We will refer to this as "the flux rule." You see that when the coil of Fig. 16–1 is rotated, the flux through it changes. At the start some flux goes through one way; then when the coil has rotated 180° the same flux goes through the other way. If we continuously rotate the coil the flux is first positive, then negative, then positive, and so on. The rate of change of the flux must alternate also. So there is an alternating emf in the coil. If we connect the two ends of the coil to outside wires through some sliding contacts—called slip-rings—(just so the wires won't get twisted) we have an alternating-current generator.

Or we can also arrange, by means of some sliding contacts, that after every one-half rotation, the connection between the coil ends and the outside wires is reversed, so that when the emf reverses, so do the connections. Then the pulses of emf will always push currents in the same direction through the external circuit. We have what is called a direct-current generator.

The machine of Fig. 16–1 is either a motor or a generator. The reciprocity between motors and generators is nicely shown by using two identical dc "motors" of the permanent magnet kind, with their coils connected by two copper wires. When the shaft of one is turned mechanically, it becomes a generator and drives the other as a motor. If the shaft of the second is turned, it becomes the generator and drives the first as a motor. So here is an interesting example of a new kind of equivalence of nature: motor and generator are equivalent. The quantitative equivalence is, in fact, not completely accidental. It is related to the law of conservation of energy.

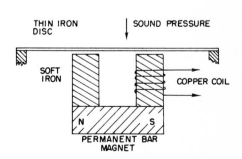

Fig. 16–4. A telephone transmitter or receiver.

Another example of a device that can operate either to generate emf's or to respond to emf's is the receiver of a standard telephone—that is, an "earphone." The original telephone of Bell consisted of two such "earphones" connected by two long wires. The basic principle is shown in Fig. 16–4. A permanent magnet produces a magnetic field in two "yokes" of soft iron and in a thin diaphragm that is moved by sound pressure. When the diaphragm moves, it changes the amount of magnetic field in the yokes. Therefore a coil of wire wound around one of the yokes will have the flux through it changed when a sound wave hits the diaphragm.

So there is an emf in the coil. If the ends of the coil are connected to a circuit, a current which is an electrical representation of the sound is set up.

If the ends of the coil of Fig. 16–4 are connected by two wires to another identical gadget, varying currents will flow in the second coil. These currents will produce a varying magnetic field and will make a varying attraction on the iron diaphragm. The diaphragm will wiggle and make sound waves approximately similar to the ones that moved the original diaphragm. With a few bits of iron and copper the human voice is transmitted over wires!

(The modern home telephone uses a receiver like the one described but uses an improved invention to get a more powerful transmitter. It is the "carbon-button microphone," that uses sound pressure to vary the electric current from a battery.)

16–2 Transformers and inductances

One of the most interesting features of Faraday's discoveries is not that an emf exists in a moving coil—which we can understand in terms of the magnetic force $q\boldsymbol{v} \times \boldsymbol{B}$—but that a changing current in one coil makes an emf in a second coil. And quite surprisingly the amount of emf induced in the second coil is given by the same "flux rule": that the emf is equal to the rate of change of the magnetic flux through the coil. Suppose that we take two coils, each wound around separate bundles of iron sheets (these help to make stronger magnetic fields), as shown in Fig. 16–5. Now we connect one of the coils—coil (a)—to an alternating-current generator. The continually changing current produces a continuously varying magnetic field. This varying field generates an alternating emf in the second coil—coil (b). This emf can, for example, produce enough power to light an electric bulb.

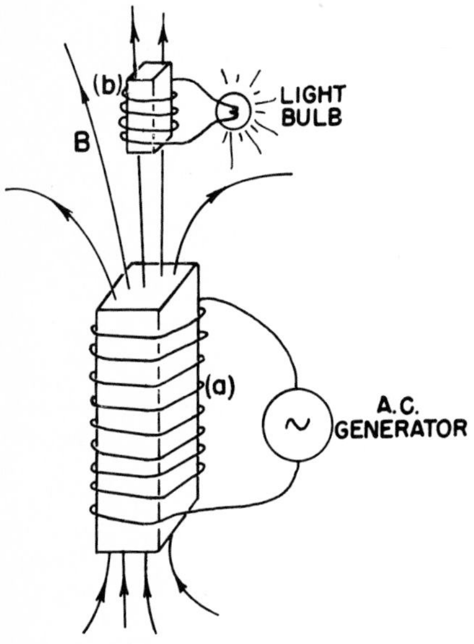

Fig. 16–5. Two coils, wrapped around bundles of iron sheets, allow a generator to light a bulb with no direct connection.

The emf alternates in coil (b) at a frequency which is, of course, the same as the frequency of the original generator. But the current in coil (b) can be larger or smaller than the current in coil (a). The current in coil (b) depends on the emf induced in it and on the resistance and inductance of the rest of its circuit. The emf can be less than that of the generator if, say, there is little flux change. Or the emf in coil (b) can be made much larger than that in the generator by winding coil (b) with many turns, since in a given magnetic field the flux through the coil is then greater. (Or if you prefer to look at it another way, the emf is the same in each turn, and since the total emf is the sum of the emf's of the separate turns, many turns in series produce a large emf.)

Such a combination of two coils—usually with an arrangement of iron sheets to guide the magnetic fields—is called a *transformer*. It can "transform" one emf (also called a "voltage") to another.

There are also induction effects in a single coil. For instance, in the setup in Fig. 16–5 there is a changing flux not only through coil (b), which lights the bulb, but also through coil (a). The varying current in coil (a) produces a varying magnetic field inside itself and the flux of this field is continually changing, so there is a *self-induced* emf in coil (a). There is an emf acting on any current when it is building up a magnetic field—or, in general, when its field is changing in any way. The effect is called *self-inductance*.

When we gave "the flux rule" that the emf is equal to the rate of change of the flux linkage, we didn't specify the direction of the emf. There is a simple rule, called Lenz's rule, for figuring out which way the emf goes: the emf *tries to oppose* any flux change. That is, the direction of an induced emf is always such that if a current were to flow in the direction of the emf, it would produce a flux of $\boldsymbol{B}$ that opposes the change in $\boldsymbol{B}$ that produces the emf. Lenz's rule can be used to find the direction of the emf in the generator of Fig. 16–1, or in the transformer winding of Fig. 16–3.

In particular, if there is a changing current in a single coil (or in any wire) there is a "back" emf in the circuit. This emf acts on the charges flowing in coil (a) of Fig. 16–5 to oppose the change in magnetic field, and so in the direction to oppose the change in current. It tries to keep the current constant; it is opposite to the current when the current is increasing, and it is in the direction of the current

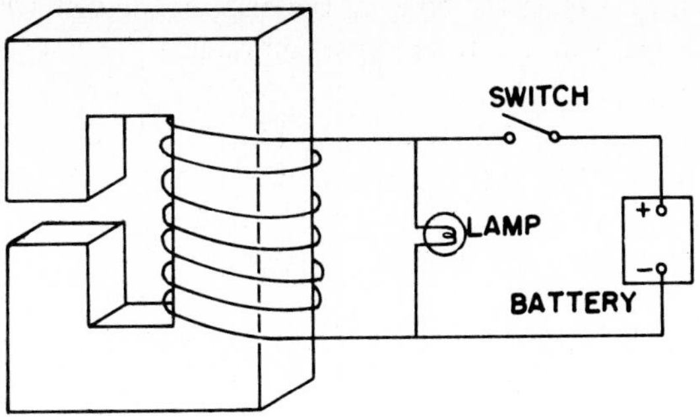

Fig. 16–6. Circuit connections for an electromagnet. The lamp allows the passage of current when the switch is opened, preventing the appearance of excessive emf's.

when it is decreasing. A current in a self-inductance has "inertia," because the inductive effects try to keep the flow constant, just as mechanical inertia tries to keep the velocity of an object constant.

Any large electromagnet will have a large self-inductance. Suppose that a battery is connected to the coil of a large electromagnet, as in Fig. 16–6, and that a strong magnetic field has been built up. (The current reaches a steady value determined by the battery voltage and the resistance of the wire in the coil.) But now suppose that we try to disconnect the battery by opening the switch. If we really opened the circuit, the current would go to zero rapidly, and in doing so it would generate an enormous emf. In most cases this emf would be large enough to develop an arc across the opening contacts of the switch. The high voltage that appears might also damage the insulation of the coil—or you, if you are the person who opens the switch! For these reasons, electromagnets are usually connected in a circuit like the one shown in Fig. 16–6. When the switch is opened, the current does not change rapidly but remains steady, flowing instead through the lamp, being driven by the emf from the self-inductance of the coil.

16–3 Forces on induced currents

You have probably seen the dramatic demonstration of Lenz's rule made with the gadget shown in Fig. 16–7. It is an electromagnet, just like coil (a) of Fig. 16–5. An aluminum ring is placed on the end of the magnet. When the coil is connected to an alternating-current generator by closing the switch, the ring flies into the air. The force comes, of course, from the induced currents in the ring. The fact that the ring flies away shows that the currents in it oppose the change of the field through it. When the magnet is making a north pole at its top, the induced current in the ring is making a downward-point north pole. The ring and the coil are repelled just like two magnets with like poles opposite. If a thin radial cut is made in the ring the force disappears, showing that it does indeed come from the currents in the ring.

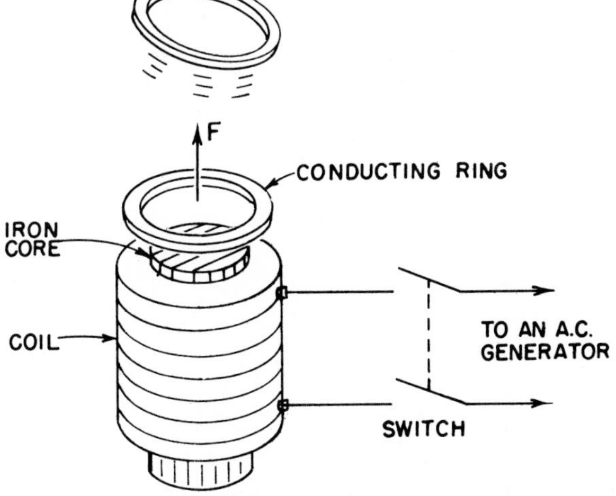

Fig. 16–7. A conducting ring is strongly repelled by an electromagnet with a varying current.

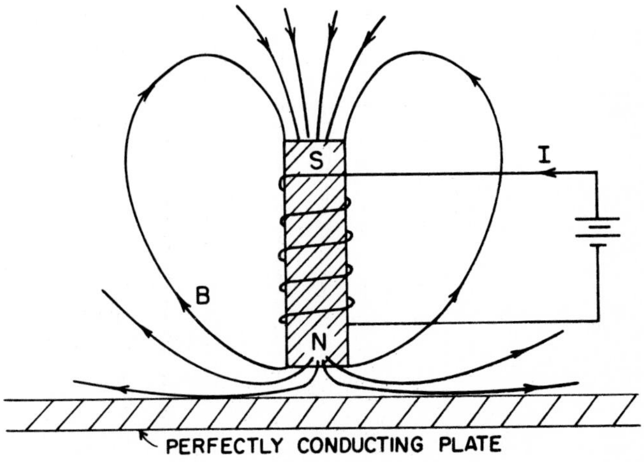

Fig. 16–8. An electromagnet near a perfectly conducting plate.

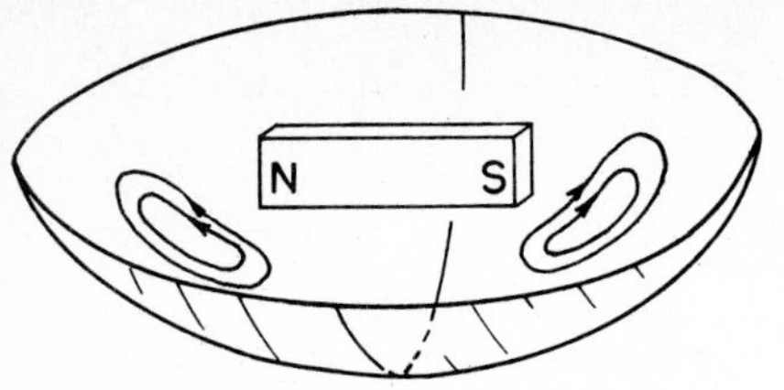

Fig. 16–9. A bar magnet is suspended above a superconducting bowl, by the repulsion of eddy currents.

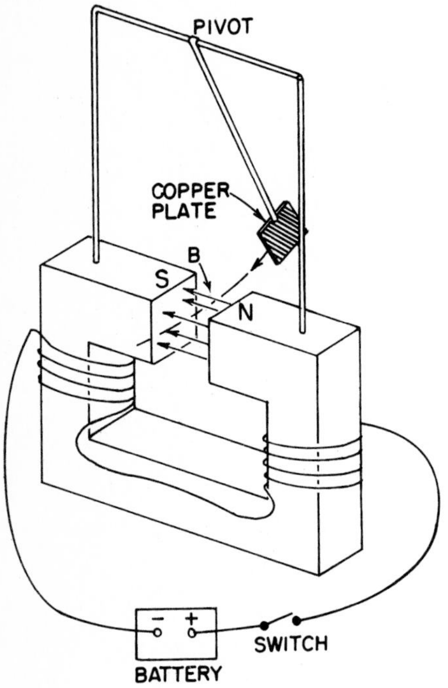

Fig. 16–10. The braking of the pendulum shows the forces due to eddy currents.

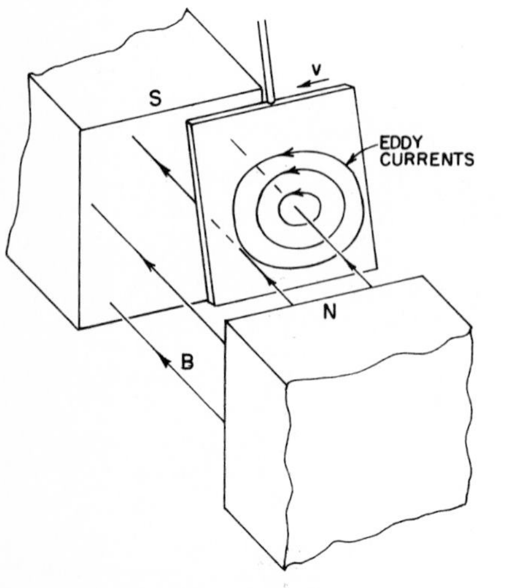

Fig. 16–11. The eddy currents in the copper pendulum.

If, instead of the ring, we place a disc of aluminum or copper across the end of the electromagnet of Fig. 16–7, it is also repelled; induced currents circulate in the material of the disc, and again produce a repulsion.

An interesting effect, similar in origin, occurs with a sheet of a perfect conductor. In a "perfect conductor" there is no resistance whatever to the current. So if currents are generated in it, they can keep going forever. In fact, the *slightest* emf would generate an arbitrarily large current—which really means that there can be no emf's at all. Any attempt to make a magnetic flux go through such a sheet generates currents that create opposite $\boldsymbol{B}$ fields—all with infinitesimal emf's, so with no flux entering.

If we have a sheet of a perfect conductor and put an electromagnet next to it, when we turn on the current in the magnet, currents called eddy currents appear in the sheet, so that no magnetic flux enters. The field lines would look as shown in Fig. 16–8. The same thing happens, of course, if we bring a bar magnet near a perfect conductor. Since the eddy currents are creating opposing fields, the magnets are repelled from the conductor. This makes it possible to suspend a bar magnet in air above a sheet of perfect conductor shaped like a dish, as shown in Fig. 16–9. The magnet is suspended by the repulsion of the induced eddy currents in the perfect conductor. There are no perfect conductors at ordinary temperatures, but some materials become perfect conductors at low enough temperatures. For instance, below 3.8°K tin conducts perfectly. It is called a superconductor.

If the conductor in Fig. 16–8 is not quite perfect there will be some resistance to flow of the eddy currents. The currents will tend to die out and the magnet will slowly settle down. The eddy currents in an imperfect conductor need an emf to keep them going, and to have an emf the flux must keep changing. The flux of the magnetic field gradually penetrates the conductor.

In a normal conductor, there are not only repulsive forces from eddy currents, but there can also be sidewise forces. For instance, if we move a magnet sideways along a conducting surface the eddy currents produce a force of drag, because the induced currents are opposing the changing of the location of flux. Such forces are proportional to the velocity and are like a kind of viscous force.

These effects show up nicely in the apparatus shown in Fig. 16–10. A square sheet of copper is suspended on the end of a rod to make a pendulum. The copper swings back and forth between the poles of an electromagnet. When the magnet is turned on, the pendulum motion is suddenly arrested. As the metal plate enters the gap of the magnet, there is a current induced in the plate which acts to oppose the change in flux through the plate. If the sheet were a perfect conductor, the currents would be so great that they would push the plate out again—it would bounce back. With a copper plate there is some resistance in the plate, so the currents at first bring the plate almost to a dead stop as it starts to enter the field. Then, as the currents die down, the plate slowly settles to rest in the magnetic field.

The nature of the eddy currents in the copper pendulum is shown in Fig. 16–11. The strength and geometry of the currents are quite sensitive to the shape of the plate. If, for instance, the copper plate is replaced by one which has several narrow slots cut in it, as shown in Fig. 16–12, the eddy-current effects are drastically reduced. The pendulum swings through the magnetic field with only a small retarding force. The reason is that the currents in each section of the copper have less flux to drive them, so the effects of the resistance of each loop are greater. The currents are smaller and the drag is less. The viscous character of the force is seen even more clearly if a sheet of copper is placed between the poles of the magnet of Fig. 16–10 and then released. It doesn't fall; it just sinks slowly downward. The eddy currents exert a strong resistance to the motion—just like the viscous drag in honey.

If, instead of dragging a conductor past a magnet, we try to rotate it in a magnetic field, there will be a resistive torque from the same effects. Alternatively, if we rotate a magnet—end over end—near a conducting plate or ring, the ring is dragged around; currents in the ring will create a torque that tends to rotate the ring with the magnet.

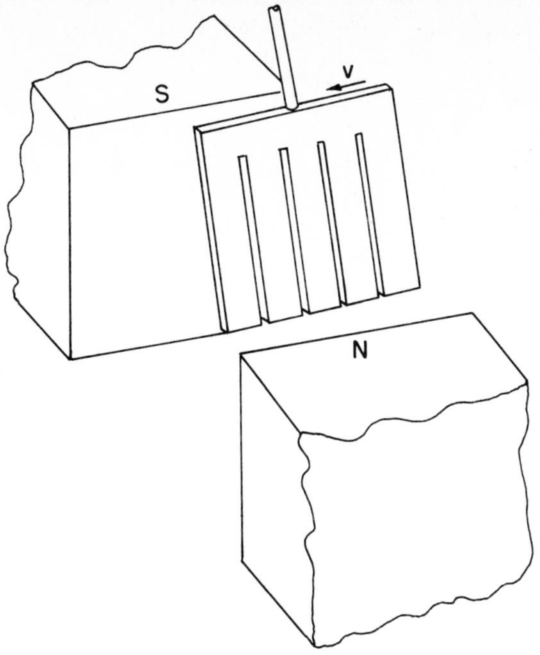

Fig. 16–12. Eddy-current effects are drastically reduced by cutting slots in the plate.

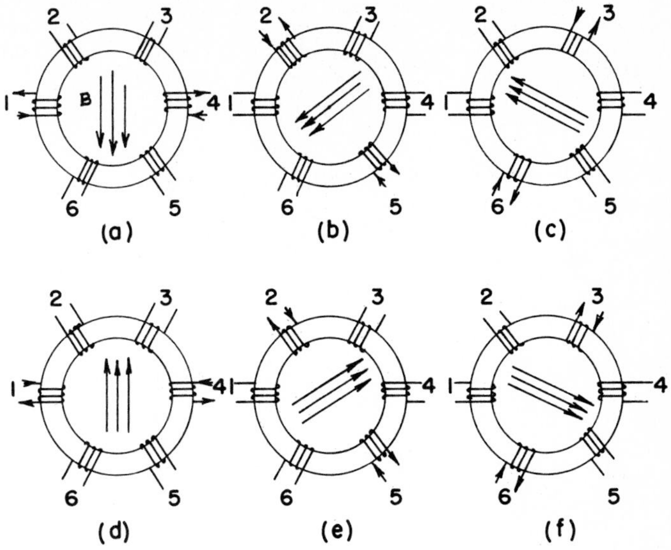

Fig. 16–13. Making a rotating magnetic field.

A field just like that of a rotating magnet can be made with an arrangement of coils such as is shown in Fig. 16–13. We take a torus of iron (that is, a ring of iron like a doughnut) and wind six coils on it. If we put a current, as shown in part (a), through windings (1) and (4), there will be a magnetic field in the direction shown in the figure. If we now switch the current to windings (2) and (5), the magnetic field will be in a new direction, as shown in part (b) of the figure. Continuing the process, we get the sequence of fields shown in the rest of the figure. If the process is done smoothly, we have a "rotating" magnetic field. We can easily get the required sequence of currents by connecting the coils to a three-phase power line, which provides just such a sequence of currents. "Three-phase power" is made in a generator using the principle of Fig. 16–1, except that there are *three* loops fastened together on the same shaft in a symmetrical way—that is, with an angle of 120° from one loop to the next. When the coils are rotated as a unit, the emf is a maximum in one, then in the next, and so on in a regular sequence. There are many practical advantages of three-phase power. One of them is the possibility of making a rotating magnetic field. The torque produced on a conductor by such a rotating field is easily shown by standing a metal ring on an insulating table just above the torus, as shown in Fig. 16–14. The rotating field causes the ring to spin about a vertical axis. The basic elements seen here are quite the same as those at play in a large commercial three-phase induction motor.

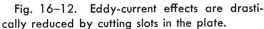

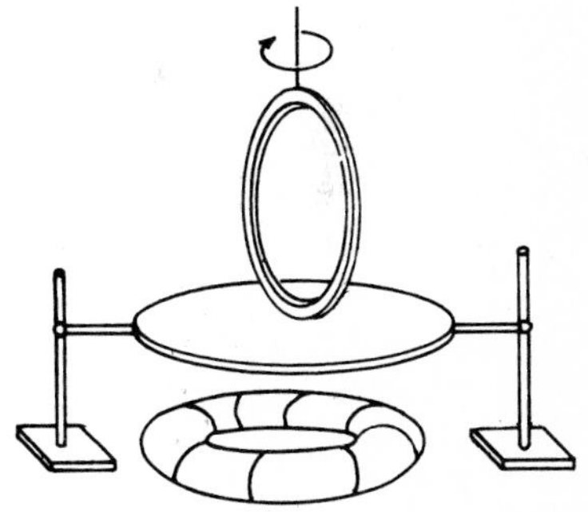

Fig. 16–14. The rotating field of Fig. 16–13 can be used to provide torque on a conducting ring.

Another form of induction motor is shown in Fig. 16–15. The arrangement shown is not suitable for a practical high-efficiency motor but will illustrate the principle. The electromagnet M, consisting of a bundle of laminated iron sheets wound with a solenoidal coil, is powered with alternating current from a generator. The magnet produces a varying flux of $\boldsymbol{B}$ through the aluminum disc. If we have just these two components, as shown in part (a) of the figure, we do not yet have a motor. There are eddy currents in the disc, but they are symmetric and there is no torque. (There will be some heating of the disc due to the induced currents.) If we now cover only one-half of the magnet pole with an aluminum plate, as shown in part (b) of the figure, the disc begins to rotate, and we have a motor. The operation depends on *two* eddy-current effects. First, the eddy currents in the aluminum plate oppose the change of flux through it, so the magnetic field above the plate always lags the field above that half of the pole which is not covered. This so-called "shaded-pole" effect produces a field which in the "shaded" region varies

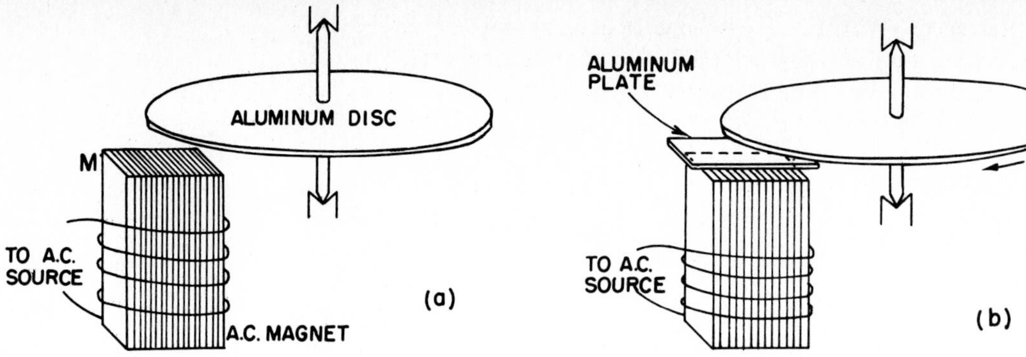

Fig. 16–15. A simple example of a shaded-pole induction motor.

much like that in the "unshaded" region except that it is delayed a constant amount in time. The whole effect is as if there were a magnet only half as wide which is continually being moved from the unshaded region toward the shaded one. Then the varying fields interact with the eddy currents in the disc to produce the torque on it.

16–4 Electrical technology

When Faraday first made public his remarkable discovery that a changing magnetic flux produces an emf, he was asked (as anyone is asked when he discovers a new fact of nature), "What is the use of it?" All he had found was the oddity that a tiny current was produced when he moved a wire near a magnet. Of what possible "use" could that be? His answer was: "What is the use of a newborn baby?"

Yet think of the tremendous practical applications his discovery has led to. What we have been describing are not just toys but examples chosen in most cases to represent the principle of some practical machine. For instance, the rotating ring in the turning field is an induction motor. There are, of course, some differences between it and a practical induction motor. The ring has a very small torque; it can be stopped with your hand. For a good motor, things have to be put together more intimately: there shouldn't be so much "wasted" magnetic field out in the air. First, the field is concentrated by using iron. We have not discussed how iron does that, but iron can make the magnetic field tens of thousands of times stronger than copper coils alone could do. Second, the gaps between the pieces of iron are made small; to do that, some iron is even built into the rotating ring. Everything is arranged so as to get the greatest forces and the greatest efficiency—that is, conversion of electrical power to mechanical power—until the "ring" can no longer be held still by your hand.

This problem of closing the gaps and making the thing work in the most practical way is *engineering*. It requires serious study of design problems, although there are no new basic principles from which the forces are obtained. But there is a long way to go from the basic principles to a practical and economic design. Yet it is just such careful engineering design that has made possible such a tremendous thing as Boulder Dam and all that goes with it.

What is Boulder Dam? A huge river is stopped by a concrete wall. But what a wall it is! Shaped with a perfect curve that is very carefully worked out so that the least possible amount of concrete will hold back a whole river. It thickens at the bottom in that wonderful shape that the artists like but that the engineers can appreciate because they know that such thickening is related to the increase of pressure with the depth of the water. But we are getting away from electricity.

Then the water of the river is diverted into a huge pipe. That's a nice engineering accomplishment in itself. The pipe feeds the water into a "waterwheel"—a huge turbine—and makes wheels turn. (Another engineering feat.) But why turn wheels? They are coupled to an exquisitely intricate mess of copper and iron, all twisted and interwoven. With two parts—one that turns and one that doesn't. All a complex intermixture of a few materials, mostly iron and copper but also some paper and shellac for insulation. A revolving monster thing. A generator. Somewhere out of the mess of copper and iron come a few special pieces of copper. The dam, the turbine, the iron, the copper, all put there to make something special happen to a few bars of copper—an emf. Then the copper bars go a little way and circle for several times around another piece of iron in a transformer; then their job is done.

But around that same piece of iron curls another cable of copper which has no direct connection whatsoever to the bars from the generator; they have just been influenced because they passed near it—to get their emf. The transformer converts the power from the relatively low voltages required for the efficient design of the generator to the very high voltages that are best for efficient transmission of electrical energy over long cables.

And everything must be enormously efficient—there can be no waste, no loss. Why? The power for a metropolis is going through. If a small fraction were lost—one or two percent—think of the energy left behind! If one percent of the power were left in the transformer, that energy would need to be taken out somehow. If it appeared as heat, it would quickly melt the whole thing. There is, of course, some small inefficiency, but all that is required are a few pumps which circulate some oil through a radiator to keep the transformer from heating up.

Out of the Boulder Dam come a few dozen rods of copper—long, long, long rods of copper perhaps the thickness of your wrist that go for hundreds of miles in all directions. Small rods of copper carrying the power of a giant river. Then the rods are split to make more rods . . . then to more transformers . . . sometimes to great generators which recreate the current in another form . . . sometimes to engines turning for big industrial purposes . . . to more transformers . . . then more splitting and spreading . . . until finally the river is spread throughout the whole city—turning motors, making heat, making light, working gadgetry. The miracle of hot lights from cold water over 600 miles away—all done with specially arranged pieces of copper and iron. Large motors for rolling steel, or tiny motors for a dentist's drill. Thousands of little wheels, turning in response to the turning of the big wheel at Boulder Dam. Stop the big wheel, and all the wheels stop; the lights go out. They really are connected.

Yet there is more. The same phenomena that take the tremendous power of the river and spread it through the countryside, until a few drops of the river are running the dentist's drill, come again into the building of extremely fine instruments . . . for the detection of incredibly small amounts of current . . . for the transmission of voices, music, and pictures . . . for computers . . . for automatic machines of fantastic precision.

All this is possible because of carefully designed arrangements of copper and iron—efficiently created magnetic fields . . . blocks of rotating iron six feet in diameter whirling with clearances of 1/16 of an inch . . . careful proportions of copper for the optimum efficiency . . . strange shapes all serving a purpose, like the curve of the dam.

If some future archaeologist uncovers Boulder Dam, we may guess that he would admire the beauty of its curves. But also the explorers from some great future civilizations will look at the generators and transformers and say: "Notice that every iron piece has a beautifully efficient shape. Think of the thought that has gone into every piece of copper!"

This is the power of engineering and the careful design of our electrical technology. There has been created in the generator something which exists nowhere else in nature. It is true that there are forces of induction in other places. Certainly in some places around the sun and stars there are effects of electromagnetic induction. Perhaps also (though it's not certain) the magnetic field of the earth is maintained by an analog of an electric generator that operates on circulating currents in the interior of the earth. But nowhere have there been pieces put together with moving parts to generate electrical power as is done in the generator—with great efficiency and regularity.

You may think that designing electric generators is no longer an interesting subject, that it is a dead subject because they are all designed. Almost perfect generators or motors can be taken from a shelf. Even if this were true, we can admire the wonderful accomplishment of a problem solved to near perfection. But there remain as many unfinished problems. Even generators and transformers are returning as problems. It is likely that the whole field of low temperatures and superconductors will soon be applied to the problem of electric power distribution. With a radically new factor in the problem, new optimum designs will have to be created. Power networks of the future may have little resemblance to those of today.

You can see that there is an endless number of applications and problems that one could take up while studying the laws of induction. The study of the design of electrical machinery is a life work in itself. We cannot go very far in that direction, but we should be aware of the fact that when we have discovered the law of induction, we have suddenly connected our theory to an enormous practical development. We must, however, leave that subject to the engineers and applied scientists who are interested in working out the details of particular applications. Physics only supplies the base—the basic principles that apply, no matter what. (We have not yet completed the base, because we have yet to consider in detail the properties of iron and of copper. Physics has something to say about these as we will see a little later.)

Modern electrical technology began with Faraday's discoveries. The useless baby developed into a prodigy and changed the face of the earth in ways its proud father could never have imagined.

17

The Laws of Induction

17–1 The physics of induction

In the last chapter we described many phenomena which show that the effects of induction are quite complicated and interesting. Now we want to discuss the fundamental principles which govern these effects. We have already defined the emf in a conducting circuit as the total accumulated force on the charges throughout the length of the loop. More specifically, it is the tangential component of the force per unit charge, integrated along the wire once around the circuit. This quantity is equal, therefore, to the total work done on a single charge that travels once around the circuit.

We have also given the "flux rule," which says that the emf is equal to the rate at which the magnetic flux through such a conducting circuit is changing. Let's see if we can understand why that might be. First, we'll consider a case in which the flux changes because a circuit is moved in a steady field.

In Fig. 17–1 we show a simple loop of wire whose dimensions can be changed. The loop has two parts, a fixed U-shaped part (a) and a movable crossbar (b) that can slide along the two legs of the U. There is always a complete circuit, but its area is variable. Suppose we now place the loop in a uniform magnetic field with the plane of the U perpendicular to the field. According to the rule, when the crossbar is moved there should be in the loop an emf that is proportional to the rate of change of the flux through the loop. This emf will cause a current in the loop. We will assume that there is enough resistance in the wire that the currents are small. Then we can neglect any magnetic field from this current.

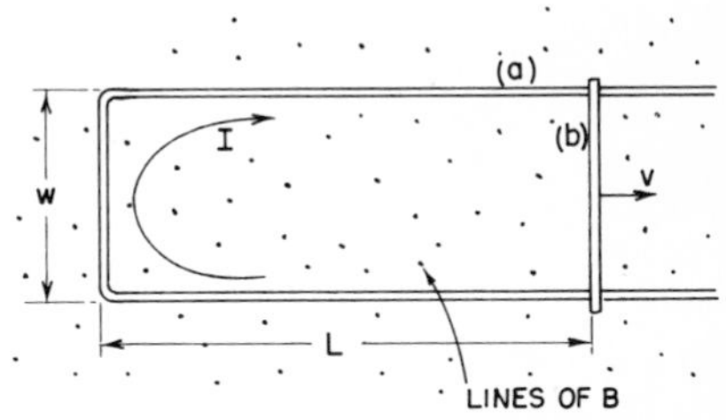

Fig. 17–1. An emf is induced in a loop if the flux is changed by varying the area of the circuit.

The flux through the loop is wLB, so the "flux rule" would give for the emf—which we write as $\mathcal{E}$—

$$\mathcal{E} = wB\frac{dL}{dt} = wBv,$$

where v is the speed of translation of the crossbar.

Now we should be able to understand this result from the magnetic $\boldsymbol{v} \times \boldsymbol{B}$ forces on the charges in the moving crossbar. These charges will feel a force, tangential to the wire, equal to vB per unit charge. It is constant along the length w of the crossbar and zero elsewhere, so the integral is

$$\mathcal{E} = wvB,$$

which is the same result we got from the rate of change of the flux.

The argument just given can be extended to any case where there is a fixed magnetic field and the wires are moved. One can prove, in general, that for any circuit whose parts move in a fixed magnetic field the emf is the time derivative of the flux, regardless of the shape of the circuit.

On the other hand, what happens if the loop is stationary and the magnetic field is changed? We cannot deduce the answer to this question from the same argument. It was Faraday's discovery—from experiment—that the "flux rule" is still correct no matter why the flux changes. The force on electric charges is given in complete generality by $\boldsymbol{F} = q(\boldsymbol{E} + \boldsymbol{v} \times \boldsymbol{B})$; there are no new special "forces due to changing magnetic fields." Any forces on charges at rest in a stationary wire come from the $\boldsymbol{E}$ term. Faraday's observations led to the discovery that electric and magnetic fields are related by a new law: in a region where the magnetic field is changing with time, electric fields are generated. It is this electric field which drives the electrons around the wire—and so is responsible for the emf in a stationary circuit when there is a changing magnetic flux.

The general law for the electric field associated with a changing magnetic field is

$$\nabla \times \boldsymbol{E} = -\frac{\partial \boldsymbol{B}}{\partial t}. \tag{17.1}$$

We will call this Faraday's law. It was discovered by Faraday but was first written in differential form by Maxwell, as one of his equations. Let's see how this equation gives the "flux rule" for circuits.

Using Stokes' theorem, this law can be written in integral form as

$$\oint_{\Gamma} \boldsymbol{E} \cdot d\boldsymbol{s} = \int_S (\nabla \times \boldsymbol{E}) \cdot \boldsymbol{n}\, da = -\int_S \frac{\partial \boldsymbol{B}}{\partial t} \cdot \boldsymbol{n}\, da, \tag{17.2}$$

where, as usual, Γ is any closed curve and S is any surface bounded by it. Here, remember, Γ is a *mathematical* curve fixed in space, and S is a fixed surface. Then the time derivative can be taken outside the integral and we have

$$\oint_{\Gamma} \boldsymbol{E} \cdot d\boldsymbol{s} = -\frac{\partial}{\partial t} \int_S \boldsymbol{B} \cdot \boldsymbol{n}\, da$$

$$= -\frac{\partial}{\partial t} \text{(flux through } S\text{)}. \tag{17.3}$$

Applying this relation to a curve Γ that follows a *fixed* circuit of conductor, we get the "flux rule" once again. The integral on the left is the emf, and that on the right is the negative rate of change of the flux linked by the circuit. So Eq. (17.1) applied to a fixed circuit is equivalent to the "flux rule."

So the "flux rule"—that the emf in a circuit is equal to the rate of change of the magnetic flux through the circuit—applies whether the flux changes because the field changes or because the circuit moves (or both). The two possibilities—"circuit moves" or "field changes"—are not distinguished in the statement of the rule. Yet in our explanation of the rule we have used two completely distinct laws for the two cases—$\boldsymbol{v} \times \boldsymbol{B}$ for "circuit moves" and $\nabla \times \boldsymbol{E} = -\partial \boldsymbol{B}/\partial t$ for "field changes."

We know of no other place in physics where such a simple and accurate general principle requires for its real understanding an analysis in terms of *two different phenomena*. Usually such a beautiful generalization is found to stem from a single deep underlying principle. Nevertheless, in this case there does not appear to be any such profound implication. We have to understand the "rule" as the combined effects of two quite separate phenomena.

We must look at the "flux rule" in the following way. In general, the force per unit charge is $\boldsymbol{F}/q = \boldsymbol{E} + \boldsymbol{v} \times \boldsymbol{B}$. In moving wires there is the force from the second term. Also, there is an $\boldsymbol{E}$-field if there is somewhere a changing magnetic field. They are independent effects, but the emf around the loop of wire is always equal to the rate of change of magnetic flux through it.

17–2 Exceptions to the "flux rule"

We will now give some examples, due in part to Faraday, which show the importance of keeping clearly in mind the distinction between the two effects responsible for induced emf's. Our examples involve situations to which the "flux rule" cannot be applied—either because there is no wire at all or because the *path* taken by induced currents moves about within an extended volume of a conductor.

We begin by making an important point: The part of the emf that comes from the $\boldsymbol{E}$-field does not depend on the existence of a physical wire (as does the $\boldsymbol{v} \times \boldsymbol{B}$ part). The $\boldsymbol{E}$-field can exist in free space, and its line integral around any imaginary line fixed in space is the rate of change of the flux of $\boldsymbol{B}$ through that line. (Note that this is quite unlike the $\boldsymbol{E}$-field produced by static charges, for in that case the line integral of $\boldsymbol{E}$ around a closed loop is always zero.)

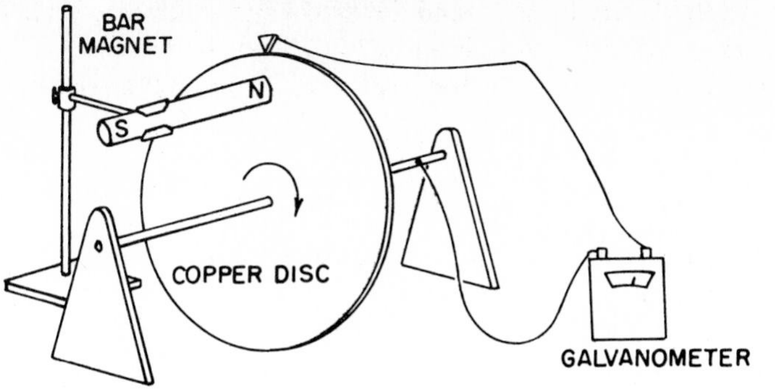

Fig. 17–2. When the disc rotates there is an emf from $\boldsymbol{v} \times \boldsymbol{B}$, but with no change in the linked flux.

Now we will describe a situation in which the flux through a circuit does not change, but there is nevertheless an emf. Figure 17–2 shows a conducting disc which can be rotated on a fixed axis in the presence of a magnetic field. One contact is made to the shaft and another rubs on the outer periphery of the disc. A circuit is completed through a galvanometer. As the disc rotates, the "circuit," in the sense of the place in space where the currents are, is always the same. But the part of the "circuit" in the disc is in material which is moving. Although the flux through the "circuit" is constant, there is still an emf, as can be observed by the deflection of the galvanometer. Clearly, here is a case where the $\boldsymbol{v} \times \boldsymbol{B}$ force in the moving disc gives rise to an emf which cannot be equated to a change of flux.

Now we consider, as an opposite example, a somewhat unusual situation in which the flux through a "circuit" (again in the sense of the place where the current is) changes but where there is *no* emf. Imagine two metal plates with slightly curved edges, as shown in Fig. 17–3, placed in a uniform magnetic field perpendicular to their surfaces. Each plate is connected to one of the terminals of a galvanometer, as shown. The plates make contact at one point P, so there is a complete circuit. If the plates are now rocked through a small angle, the point of contact will move to P'. If we imagine the "circuit" to be completed through the plates on the dotted line shown in the figure, the magnetic flux through this circuit changes by a large amount as the plates are rocked back and forth. Yet the rocking can be done with small motions, so that $\boldsymbol{v} \times \boldsymbol{B}$ is very small and there is practically no emf. The "flux rule" does not work in this case. It must be applied to circuits in which the *material* of the circuit remains the same. When the material of the circuit is changing, we must return to the basic laws. The *correct* physics is always given by the two basic laws

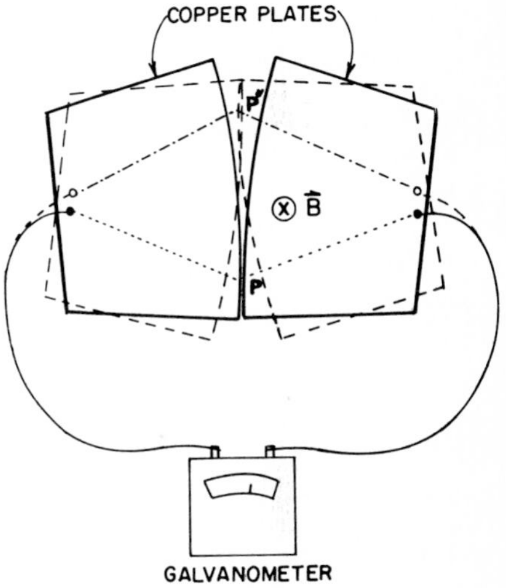

Fig. 17–3. When the plates are rocked in a uniform magnetic field, there can be a large change in the flux linkage without the generation of an emf.

$$\boldsymbol{F} = q(\boldsymbol{E} + \boldsymbol{v} \times \boldsymbol{B}),$$

$$\nabla \times \boldsymbol{E} = -\frac{\partial \boldsymbol{B}}{\partial t}.$$

17–3 Particle acceleration by an induced electric field; the betatron

We have said that the electromotive force generated by a changing magnetic field can exist even without conductors; that is, there can be magnetic induction without wires. We may still imagine an electromotive force around an arbitrary mathematical curve in space. It is defined as the tangential component of $\boldsymbol{E}$ integrated around the curve. Faraday's law says that this line integral is equal to the rate of change of the magnetic flux through the closed curve, Eq. (17.3).

As an example of the effect of such an induced electric field, we want now to consider the motion of an electron in a changing magnetic field. We imagine a magnetic field which, everywhere on a plane, points in a vertical direction, as shown in Fig. 17–4. The magnetic field is produced by an electromagnet, but we will not worry about the details. For our example we will imagine that the magnetic field is symmetric about some axis, i.e., that the strength of the magnetic field will depend only on the distance from the axis. The magnetic field is also varying with time. We now imagine an electron that is moving in this field on a path that is a circle of constant radius with its center at the axis of the field. (We will see later

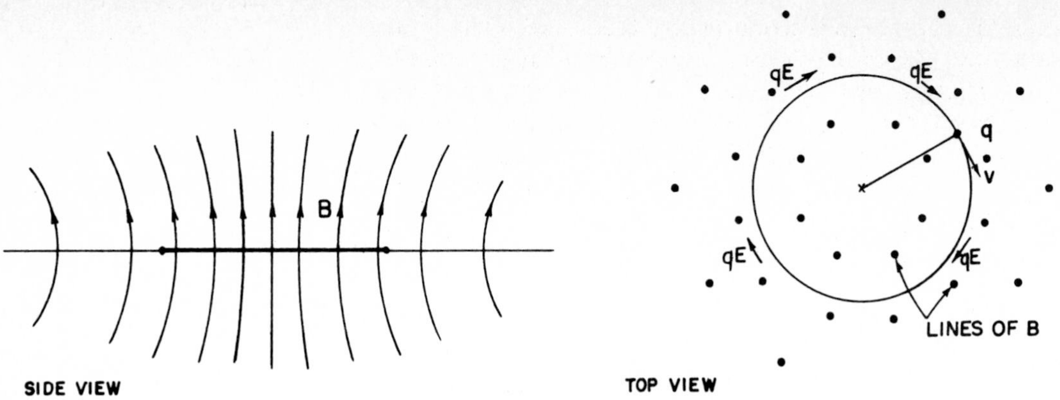

Fig. 17–4. An electron accelerating in an axially symmetric, time-varying magnetic field.

how this motion can be arranged.) Because of the changing magnetic field, there will be an electric field $\boldsymbol{E}$ tangential to the electron's orbit which will drive it around the circle. Because of the symmetry, this electric field will have the same value everywhere on the circle. If the electron's orbit has the radius r, the line integral of $\boldsymbol{E}$ around the orbit is equal to the rate of change of the magnetic flux through the circle. The line integral of $\boldsymbol{E}$ is just its magnitude times the circumference of the circle, $2\pi r$. The magnetic flux must, in general, be obtained from an integral. For the moment, we let B_{av} represent the average magnetic field in the interior of the circle; then the flux is this average magnetic field times the area of the circle. We will have

$$2\pi r E = \frac{\partial}{\partial t}(B_{\text{av}} \cdot \pi r^2).$$

Since we are assuming r is constant, E is proportional to the time derivative of the average field:

$$E = \frac{r}{2}\frac{dB_{\text{av}}}{dt}. \tag{17.4}$$

The electron will feel the electric force $q\boldsymbol{E}$ and will be accelerated by it. Remembering that the relativistically correct equation of motion is that the rate of change of the momentum is proportional to the force, we have

$$qE = \frac{dp}{dt}. \tag{17.5}$$

For the circular orbit we have assumed, the electric force on the electron is always in the direction of its motion, so its total momentum will be increasing at the rate given by Eq. (17.5). Combining Eqs. (17.5) and (17.4), we may relate the rate of change of momentum to the change of the average magnetic field:

$$\frac{dp}{dt} = \frac{qr}{2}\frac{dB_{\text{av}}}{dt}. \tag{17.6}$$

Integrating with respect to t, we find for the electron's momentum

$$p = p_0 + \frac{qr}{2}\Delta B_{\text{av}}, \tag{17.7}$$

where p_0 is the momentum with which the electrons start out, and ΔB_{av} is the subsequent change in B_{av}. The operation of a *betatron*—a machine for accelerating electrons to high energies—is based on this idea.

To see how the betatron operates in detail, we must now examine how the electron can be constrained to move on a circle. We have discussed in Chapter 11 of Vol. I the principle involved. If we arrange that there is a magnetic field $\boldsymbol{B}$ at the orbit of the electron, there will be a transverse force $q\boldsymbol{v} \times \boldsymbol{B}$ which, for a suitably chosen $\boldsymbol{B}$, can cause the electron to keep moving on its assumed orbit. In the betatron this transverse force causes the electron to move in a circular orbit of constant radius. We can find out what the magnetic field at the orbit must be by using again the relativistic equation of motion, but this time, for the transverse component of the force. In the betatron (see Fig. 17–4), $\boldsymbol{B}$ is at right angles to $\boldsymbol{v}$, so the transverse force is qvB. Thus the force is equal to the rate of change of the transverse component p_t of the momentum:

$$qvB = \frac{dp_t}{dt}. \tag{17.8}$$

When a particle is moving in a *circle*, the rate of change of its transverse momentum is equal to the magnitude of the total momentum times ω, the angular velocity of rotation (following the arguments of Chapter 11, Vol. I):

$$\frac{dp_t}{dt} = \omega p, \tag{17.9}$$

where, since the motion is circular,

$$\omega = \frac{v}{r}. \tag{17.10}$$

Setting the magnetic force equal to the transverse acceleration, we have

$$qvB_{\text{orbit}} = p\,\frac{v}{r}, \tag{17.11}$$

where B_{orbit} is the field at the radius r.

As the betatron operates, the momentum of the electron grows in proportion to B_{av}, according to Eq. (17.7), and if the electron is to continue to move in its proper circle, Eq. (17.11) must continue to hold as the momentum of the electron increases. The value of B_{orbit} must increase in proportion to the momentum p. Comparing Eq. (17.11) with Eq. (17.7), which determines p, we see that the following relation must hold between B_{av}, the average magnetic field *inside* the orbit at the radius r, and the magnetic field B_{orbit} at the orbit:

$$\Delta B_{\text{av}} = 2\,\Delta B_{\text{orbit}}. \tag{17.12}$$

The correct operation of a betatron requires that the average magnetic field inside the orbit increase at twice the rate of the magnetic field at the orbit itself. In these circumstances, as the energy of the particle is increased by the induced electric field the magnetic field at the orbit increases at just the rate required to keep the particle moving in a circle.

The betatron is used to accelerate electrons to energies of tens of millions of volts, or even to hundreds of millions of volts. However, it becomes impractical for the acceleration of electrons to energies much higher than a few hundred million volts for several reasons. One of them is the practical difficulty of attaining the required high average value for the magnetic field inside the orbit. Another is that Eq. (17.6) is no longer correct at very high energies because it does not include the loss of energy from the particle due to its radiation of electromagnetic energy (the so-called synchrotron radiation discussed in Chapter 36, Vol. I). For these reasons, the acceleration of electrons to the highest energies—to many billions of electron volts—is accomplished by means of a different kind of machine, called a *synchrotron*.

17–4 A paradox

We would now like to describe for you an apparent paradox. A paradox is a situation which gives one answer when analyzed one way, and a different answer when analyzed another way, so that we are left in somewhat of a quandary as to actually what should happen. Of course, in physics there are never any real paradoxes because there is only one correct answer; at least we believe that nature will act in only one way (and that is the *right way*, naturally). So in physics a paradox is only a confusion in our own understanding. Here is our paradox.

Imagine that we construct a device like that shown in Fig. 17–5. There is a thin, circular plastic disc supported on a concentric shaft with excellent bearings, so that it is quite free to rotate. On the disc is a coil of wire in the form of a short solenoid concentric with the axis of rotation. This solenoid carries a steady current I provided by a small battery, also mounted on the disc. Near the edge of the disc and spaced uniformly around its circumference are a number of small metal spheres insulated from each other and from the solenoid by the plastic material of the disc. Each of these small conducting spheres is charged with the same electrostatic charge Q. Everything is quite stationary, and the disc is at rest. Suppose now that by some accident—or by prearrangement—the current in the solenoid is interrupted, without, however, any intervention from the outside. So long as the current continued, there was a magnetic flux through the solenoid more or less parallel to the axis of the disc. When the current is interrupted, this flux must go to zero. There will, therefore, be an electric field induced which will circulate around in circles centered at the axis. The charged spheres on the perimeter of the disc will all experience an electric field tangential to the perimeter of the disc. This electric force is in the same sense for all the charges and so will result in a net torque on the disc. From these arguments we would expect that as the current in the solenoid disappears, the disc would begin to rotate. If we knew the moment of inertia of the disc, the current in the solenoid, and the charges on the small spheres, we could compute the resulting angular velocity.

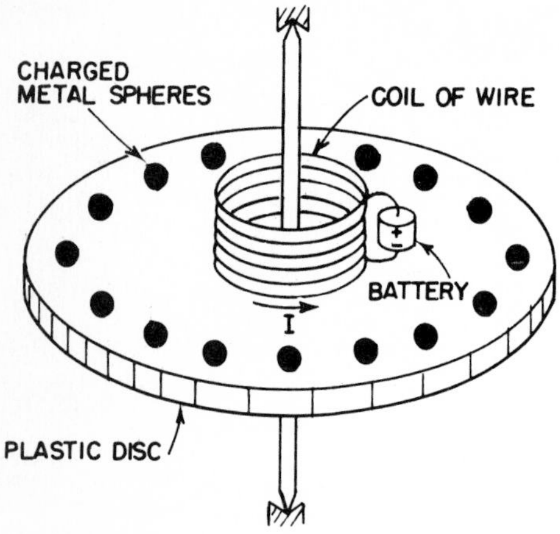

Fig. 17–5. Will the disc rotate if the current I is stopped?

But we could also make a different argument. Using the principle of the conservation of angular momentum, we could say that the angular momentum of the disc with all its equipment is initially zero, and so the angular momentum of the assembly should remain zero. There should be no rotation when the current is stopped. Which argument is correct? Will the disc rotate or will it not? We will leave this question for you to think about.

We should warn you that the correct answer does not depend on any nonessential feature, such as the asymmetric position of a battery, for example. In fact, you can imagine an ideal situation such as the following: The solenoid is made of superconducting wire through which there is a current. After the disc has been carefully placed at rest, the temperature of the solenoid is allowed to rise slowly. When the temperature of the wire reaches the transition temperature between superconductivity and normal conductivity, the current in the solenoid will be brought to zero by the resistance of the wire. The flux will, as before, fall to zero, and there will be an electric field around the axis. We should also warn you that the solution is not easy, nor is it a trick. When you figure it out, you will have discovered an important principle of electromagnetism.

17–5 Alternating-current generator

In the remainder of this chapter we apply the principles of Section 17–1 to analyze a number of the phenomena discussed in Chapter 16. We first look in more detail at the alternating-current generator. Such a generator consists basically of a coil of wire rotating in a uniform magnetic field. The same result can also be achieved by a fixed coil in a magnetic field whose direction rotates in the manner described in the last chapter. We will consider only the former case. Suppose we have a circular coil of wire which can be turned on an axis along one of its diameters. Let this coil be located in a uniform magnetic field perpendicular to the axis of rotation, as in Fig. 17–6. We also imagine that the two ends of the coil are brought to external connections through some kind of sliding contacts.

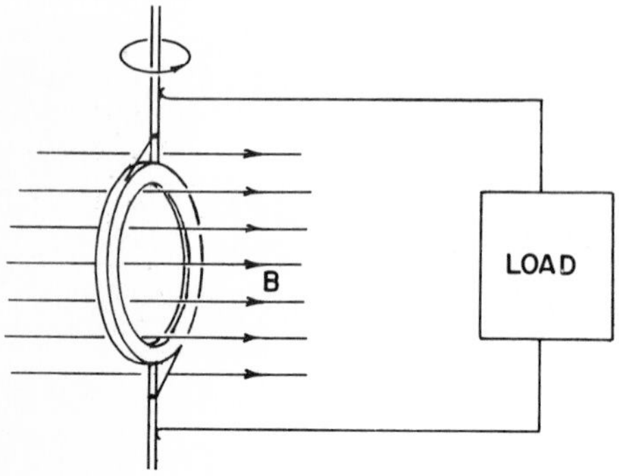

Fig. 17–6. A coil of wire rotating in a uniform magnetic field—the basic idea of the ac generator.

Due to the rotation of the coil, the magnetic flux through it will be changing. The circuit of the coil will therefore have an emf in it. Let S be the area of the coil and θ the angle between the magnetic field and the normal to the plane of the coil.*

* Now that we are using the letter A for the vector potential, we prefer to let S stand for a Surface area.

The flux through the coil is then

$$BS\cos\theta. \tag{17.13}$$

If the coil is rotating at the uniform angular velocity ω, θ varies with time as $\theta = \omega t$.

Each turn of the coil will have an emf equal to the rate of change of this flux. If the coil has N turns of wire the total emf will be N times larger, so

$$\mathcal{E} = -N\frac{d}{dt}(BS\cos\omega t) = NBS\omega\sin\omega t. \tag{17.14}$$

If we bring the wires from the generator to a point some distance from the rotating coil, where the magnetic field is zero, or at least is not varying with time, the curl of $\boldsymbol{E}$ in this region will be zero and we can define an electric potential. In fact, if there is no current being drawn from the generator, the potential difference V between the two wires will be equal to the emf in the rotating coil. That is,

$$V = NBS\omega\sin\omega t = V_0\sin\omega t.$$

The potential difference between the wires varies as $\sin\omega t$. Such a varying potential difference is called an alternating voltage.

Since there is an electric field between the wires, they must be electrically charged. It is clear that the emf of the generator has pushed some excess charges out to the wire until the electric field from them is strong enough to exactly counterbalance the induction force. Seen from outside the generator, the two wires appear as though they had been electrostatically charged to the potential difference V, and as though the charge was being changed with time to give an alternating potential difference. There is also another difference from an electrostatic situation. If we connect the generator to an external circuit that permits passage of a current, we find that the emf does not permit the wires to be discharged but continues to provide charge to the wires as current is drawn from them, attempting to keep the wires always at the same potential difference. If, in fact, the generator is connected in a circuit whose total resistance is R, the current through the circuit will be proportional to the emf of the generator and inversely proportional to R. Since the emf has a sinusoidal time variation, so also does the current. There is an alternating current

$$I = \frac{\mathcal{E}}{R} = \frac{V_0}{R}\sin\omega t.$$

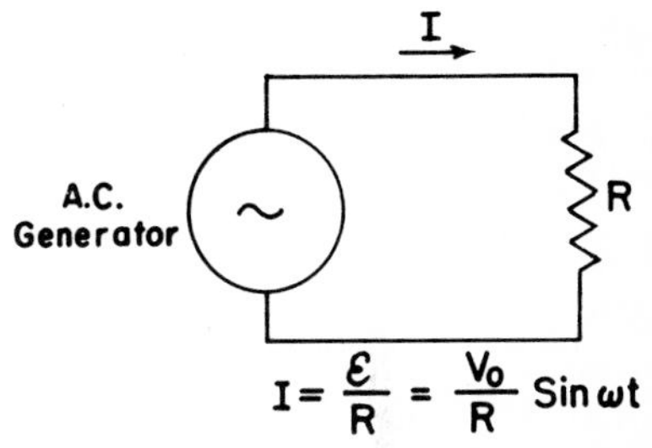

Fig. 17–7. A circuit with an ac generator and a resistance.

The schematic diagram of such a circuit is shown in Fig. 17–7.

We can also see that the emf determines how much energy is supplied by the generator. Each charge in the wire is receiving energy at the rate $\boldsymbol{F}\cdot\boldsymbol{v}$, where $\boldsymbol{F}$ is the force on the charge and $\boldsymbol{v}$ is its velocity. Now let the number of moving charges per unit length of the wire be n; then the power being delivered into any element ds of the wire is

$$\boldsymbol{F}\cdot\boldsymbol{v}n\,ds.$$

For a wire, $\boldsymbol{v}$ is always along $\boldsymbol{ds}$, so we can rewrite the power as

$$nv\boldsymbol{F}\cdot\boldsymbol{ds}.$$

The total power being delivered to the complete circuit is the integral of this expression around the complete loop:

$$\text{Power} = \oint nv\boldsymbol{F}\cdot\boldsymbol{ds}. \tag{17.15}$$

Now remember that qnv is the current I, and that the emf is defined as the integral of F/q around the circuit. We get the result

$$\text{Power from a generator} = \mathcal{E}I. \tag{17.16}$$

When there is a current in the coil of the generator, there will also be mechanical forces on it. In fact, we know that the torque on the coil is proportional to its magnetic moment, to the magnetic field strength B, and to the sine of the angle between. The magnetic moment is the current in the coil times its area. Therefore the torque is

$$\tau = NISB \sin \theta. \tag{17.17}$$

The rate at which mechanical work must be done to keep the coil rotating is the angular velocity ω times the torque:

$$\frac{dW}{dt} = \omega\tau = \omega NISB \sin \theta. \tag{17.18}$$

Comparing this equation with Eq. (17.14), we see that the rate of mechanical work required to rotate the coil against the magnetic forces is just equal to $\mathcal{E}I$, the rate at which electrical energy is delivered by the emf of the generator. All of the mechanical energy used up in the generator appears as electrical energy in the circuit.

As another example of the currents and forces due to an induced emf, let's analyze what happens in the setup described in Section 12, and shown in Fig. 17–1. There are two parallel wires and a sliding crossbar located in a uniform magnetic field perpendicular to the plane of the parallel wires. Now let's assume that the "bottom" of the U (the left side in the figure) is made of wires of high resistance, while the two side wires are made of a good conductor like copper—then we don't need to worry about the change of the circuit resistance as the crossbar is moved. As before, the emf in the circuit is

$$\mathcal{E} = vBw. \tag{17.19}$$

The current in the circuit is proportional to this emf and inversely proportional to the resistance of the circuit:

$$I = \frac{\mathcal{E}}{R} = \frac{vBw}{R} \cdot \tag{17.20}$$

Because of this current there will be a magnetic force on the crossbar that is proportional to its length, to the current in it, and to the magnetic field, such that

$$F = BIw. \tag{17.21}$$

Taking I from Eq. (17.20), we have for the force

$$F = \frac{B^2w^2}{R}\,v. \tag{17.22}$$

We see that the force is proportional to the velocity of the crossbar. The direction of the force, as you can easily see, is opposite to its velocity. Such a "velocity-proportional" force, which is like the force of viscosity, is found whenever induced currents are produced by moving conductors in a magnetic field. The examples of eddy currents we gave in the last chapter also produced forces on the conductors proportional to the velocity of the conductor, even though such situations, in general, give a complicated distribution of currents which is difficult to analyze.

It is often convenient in the design of mechanical systems to have damping forces which are proportional to the velocity. Eddy-current forces provide one of the most convenient ways of getting such a velocity-dependent force. An example of the application of such a force is found in the conventional domestic wattmeter. In the wattmeter there is a thin aluminum disc that rotates between the poles of a permanent magnet. This disc is driven by a small electric motor whose torque is proportional to the power being consumed in the electrical circuit of the house. Because of the eddy-current forces in the disc, there is a resistive force proportional to the velocity. In equilibrium, the velocity is therefore proportional to the rate of consumption of electrical energy. By means of a counter attached to the rotating disc, a record is kept of the number of revolutions it makes. This count is an indication of the total energy consumption, i.e., the number of watthours used.

We may also point out that Eq. (17.22) shows that the force from induced currents—that is, any eddy-current force—is inversely proportional to the resistance. The force will be larger, the better the conductivity of the material. The reason, of course, is that an emf produces more current if the resistance is low, and the stronger currents represent greater mechanical forces.

We can also see from our formulas how mechanical energy is converted into electrical energy. As before, the electrical energy supplied to the resistance of the circuit is the product $\mathcal{E}I$. The rate at which work is done in moving the conducting crossbar is the force on the bar times its velocity. Using Eq. (17.21) for the force, the rate of doing work is

$$\frac{dW}{dt} = \frac{v^2B^2w^2}{R}.$$

We see that this is indeed equal to the product $\mathcal{E}I$ we would get from Eqs. (17.19) and (17.20). Again the mechanical work appears as electrical energy.

17–6 Mutual inductance

We now want to consider a situation in which there are fixed coils of wire but changing magnetic fields. When we described the production of magnetic fields by currents, we considered only the case of steady currents. But so long as the currents are changed slowly, the magnetic field will at each instant be nearly the same as the magnetic field of a steady current. We will assume in the discussion of this section that the currents are always varying sufficiently slowly that this is true.

In Fig. 17–8 is shown an arrangement of two coils which demonstrates the basic effects responsible for the operation of a transformer. Coil 1 consists of a conducting wire wound in the form of a long solenoid. Around this coil—and insulated from it—is wound coil 2, consisting of a few turns of wire. If now a current is passed through coil 1, we know that a magnetic field will appear inside it. This magnetic field also passes through coil 2. As the current in coil 1 is varied, the magnetic flux will also vary, and there will be an induced emf in coil 2. We will now calculate this induced emf.

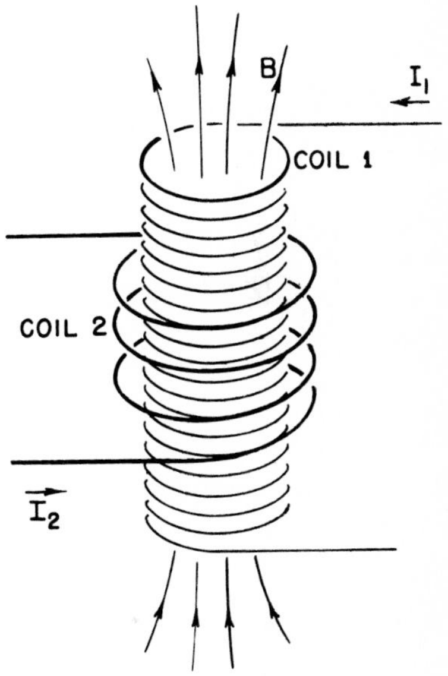

Fig. 17–8. A current in coil 1 produces a magnetic field through coil 2.

We have seen in Section 13–5 that the magnetic field inside a long solenoid is uniform and has the magnitude

$$B = \frac{1}{\epsilon_0 c^2}\frac{N_1 I_1}{l}, \tag{17.23}$$

where N_1 is the number of turns in coil 1, I_1 is the current through it, and l is its length. Let's say that the cross-sectional area of coil 1 is S; then the flux of $\boldsymbol{B}$ is its magnitude times S. If coil 2 has N_2 turns, this flux links the coil N_2 times. Therefore the emf in coil 2 is given by

$$\mathcal{E}_2 = -N_2 S\frac{dB}{dt}. \tag{17.24}$$

The only quantity in Eq. (17.23) which varies with time is I_1. The emf is therefore given by

$$\mathcal{E}_2 = -\frac{N_1N_2S}{\epsilon_0c^2l}\frac{dI_1}{dt}. \tag{17.25}$$

We see that the emf in coil 2 is proportional to the rate of change of the current in coil 1. The constant of proportionality, which is basically a geometric factor of the two coils, is called the *mutual inductance*, and is usually designated $\mathfrak{M}_{21}$. Equation (17.25) is then written

$$\mathcal{E}_2 = \mathfrak{M}_{21}\frac{dI_1}{dt}. \tag{17.26}$$

Suppose now that we were to pass a current through coil 2 and ask about the emf in coil 1. We would compute the magnetic field, which is everywhere proportional to the current I_2. The flux linkage through coil 1 would depend on the geometry, but would be proportional to the current I_2. The emf in coil 1 would, therefore, again be proportional to dI_2/dt: We can write

$$\mathcal{E}_1 = \mathfrak{M}_{12} \frac{dI_2}{dt}. \tag{17.27}$$

The computation of $\mathfrak{M}_{12}$ would be more difficult than the computation we have just done for $\mathfrak{M}_{21}$. We will not carry through that computation now, because we will show later in this chapter that $\mathfrak{M}_{12}$ is necessarily equal to $\mathfrak{M}_{21}$.

Since for *any* coil its field is proportional to its current, the same kind of result would be obtained for any two coils of wire. The equations (17.26) and (17.27) would have the same form; only the constants $\mathfrak{M}_{21}$ and $\mathfrak{M}_{12}$ would be different. Their values would depend on the shapes of the coils and their relative positions.

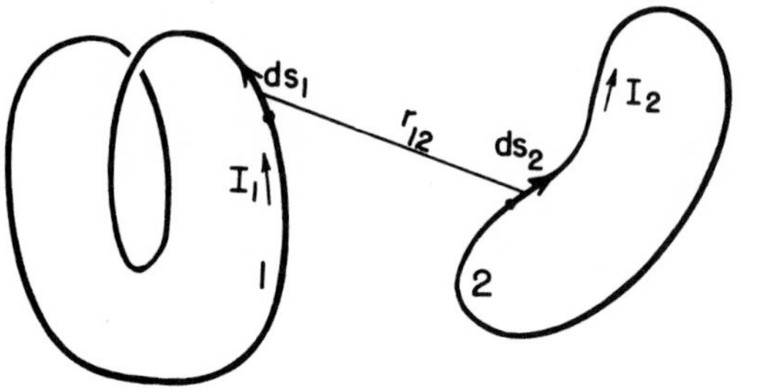

Fig. 17–9. Any two coils have a mutual inductance $\mathfrak{M}$ proportional to the integral of $d\mathbf{s}_1 \cdot d\mathbf{s}_2/r_{12}$.

Suppose that we wish to find the mutual inductance between any two arbitrary coils—for example, those shown in Fig. 17–9. We know that the general expression for the emf in coil 1 can be written as

$$\mathcal{E}_1 = -\frac{d}{dt}\int_{(1)} \boldsymbol{B} \cdot \boldsymbol{n}\, da,$$

where $\boldsymbol{B}$ is the magnetic field and the integral is to be taken over a surface bounded by circuit 1. We have seen in Section 14–1 that such a surface integral of $\boldsymbol{B}$ can be related to a line integral of the vector potential. In particular,

$$\int_{(1)} \boldsymbol{B} \cdot \boldsymbol{n}\, da = \oint_{(1)} \boldsymbol{A} \cdot d\boldsymbol{s}_1,$$

where $\boldsymbol{A}$ represents the vector potential and $d\boldsymbol{s}_1$ is an element of circuit 1. The line integral is to be taken around circuit 1. The emf in coil 1 can therefore be written as

$$\mathcal{E}_1 = -\frac{d}{dt}\oint_{(1)} \boldsymbol{A} \cdot d\boldsymbol{s}_1. \tag{17.28}$$

Now let's assume that the vector potential at circuit 1 comes from currents in circuit 2. Then it can be written as a line integral around circuit 2:

$$\boldsymbol{A} = \frac{1}{4\pi\epsilon_0 c^2}\oint_{(2)} \frac{I_2\, d\boldsymbol{s}_2}{r_{12}}, \tag{17.29}$$

where I_2 is the current in circuit 2, and r_{12} is the distance from the element of the circuit $d\boldsymbol{s}_2$ to the point on circuit 1 at which we are evaluating the vector potential. (See Fig. 17–9.) Combining Eqs. (17.28) and (17.29), we can express the emf in circuit 1 as a double line integral:

$$\mathcal{E}_1 = -\frac{1}{4\pi\epsilon_0 c^2}\frac{d}{dt}\oint_{(1)}\oint_{(2)} \frac{I_2\, d\boldsymbol{s}_2}{r_{12}} \cdot d\boldsymbol{s}_1.$$

In this equation the integrals are all taken with respect to stationary circuits. The only variable quantity is the current I_2, which does not depend on the variables of integration. We may therefore take it out of the integrals. The emf can then be written as

$$\mathcal{E}_1 = \mathfrak{M}_{12}\frac{dI_2}{dt},$$

where the coefficient $\mathfrak{M}_{12}$ is

$$\mathfrak{M}_{12} = -\frac{1}{4\pi\epsilon_0 c^2}\oint_{(1)}\oint_{(2)}\frac{d\boldsymbol{s}_2\cdot d\boldsymbol{s}_1}{r_{12}}. \tag{17.30}$$

We see from this integral that $\mathfrak{M}_{12}$ depends only on the circuit geometry. It depends on a kind of average separation of the two circuits, with the average weighted most for parallel segments of the two coils. Our equation can be used for calculating the mutual inductance of any two circuits of arbitrary shape. Also, it shows that the integral for $\mathfrak{M}_{12}$ is identical to the integral for $\mathfrak{M}_{21}$. We have therefore shown that the two coefficients are identical. For a system with only two coils, the coefficients $\mathfrak{M}_{12}$ and $\mathfrak{M}_{21}$ are often represented by the symbol $\mathfrak{M}$ without subscripts, called simply the *mutual inductance*:

$$\mathfrak{M}_{12} = \mathfrak{M}_{21} = \mathfrak{M}.$$

17–7 Self-inductance

In discussing the induced electromotive forces in the two coils of Figs. 17–8 or 17–9, we have considered only the case in which there was a current in one coil or the other. If there are currents in the two coils simultaneously, the magnetic flux linking either coil will be the sum of the two fluxes which would exist separately, because the law of superposition applies for magnetic fields. The emf in either coil will therefore be proportional not only to the change of the current in the other coil, but also to the change in the current of the coil itself. Thus the total emf in coil 2 should be written*

$$\mathcal{E}_2 = \mathfrak{M}_{21}\frac{dI_1}{dt} + \mathfrak{M}_{22}\frac{dI_2}{dt}. \tag{17.31}$$

Similarly, the emf in coil 1 will depend not only on the changing current in coil 2, but also on the changing current in itself:

$$\mathcal{E}_1 = \mathfrak{M}_{12}\frac{dI_2}{dt} + \mathfrak{M}_{11}\frac{dI_1}{dt}. \tag{17.32}$$

The coefficients $\mathfrak{M}_{22}$ and $\mathfrak{M}_{11}$ are always negative numbers. It is usual to write

$$\mathfrak{M}_{11} = -\mathcal{L}_1, \qquad \mathfrak{M}_{22} = -\mathcal{L}_2, \tag{17.33}$$

where $\mathcal{L}_1$ and $\mathcal{L}_2$ are called the *self-inductances* of the two coils.

The self-induced emf will, of course, exist even if we have only one coil. Any coil by itself will have a self-inductance $\mathcal{L}$. The emf will be proportional to the rate of change of the current in it. For a single coil, it is usual to adopt the convention that the emf and the current are considered positive if they are in the same direction. With this convention, we may write for the emf of a single coil

$$\mathcal{E} = -\mathcal{L}\frac{dI}{dt}. \tag{17.34}$$

The negative sign indicates that the emf opposes the change in current—it is often called a "back emf."

Since any coil has a self-inductance which opposes the change in current, the current in the coil has a kind of inertia. In fact, if we wish to change the current in

---

* The sign of $\mathfrak{M}_{12}$ and $\mathfrak{M}_{21}$ in Eqs. (17.31) and (17.32) depends on the arbitrary choices for the sense of a positive current in the two coils.

a coil we must overcome this inertia by connecting the coil to some external voltage source such as a battery or a generator, as shown in the schematic diagram of Fig. 17–10(a). In such a circuit, the current I depends on the voltage $\mathcal{V}$ according to the relation

$$\mathcal{V} = \mathcal{L}\,\frac{dI}{dt}. \tag{17.35}$$

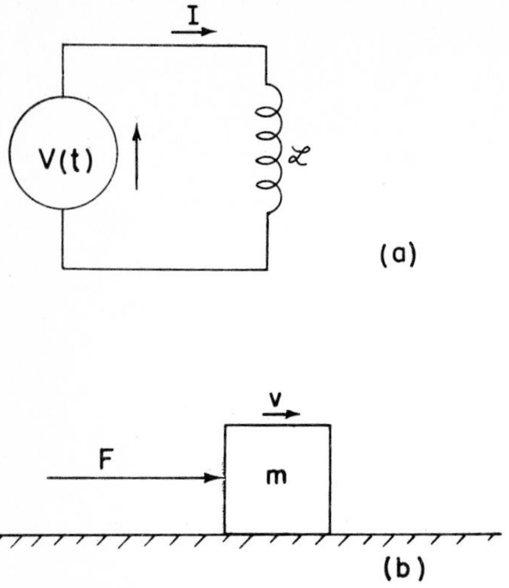

Fig. 17–10 (a) A circuit with a voltage source and an inductance. (b) An analogous mechanical system.

This equation has the same form as Newton's law of motion for a particle in one dimension. We can therefore study it by the principle that "the same equations have the same solutions." Thus, if we make the externally applied voltage $\mathcal{V}$ correspond to an externally applied force F, and the current I in a coil correspond to the velocity v of a particle, the inductance $\mathcal{L}$ of the coil corresponds to the mass m of the particle.* See Fig. 17–10(b). We can make the following table of corresponding quantities.

Particle	*Coil*
F (force)	$\mathcal{V}$ (potential difference)
v (velocity)	I (current)
x (displacement)	q (charge)
$F = m\,\dfrac{dv}{dt}$	$\mathcal{V} = \mathcal{L}\,\dfrac{dI}{dt}$
mv (momentum)	$\mathcal{L}I$
$\frac{1}{2}mv^2$ (kinetic energy)	$\frac{1}{2}\mathcal{L}I^2$ (magnetic energy)

17–8 Inductance and magnetic energy

Continuing with the analogy of the preceding section, we would expect that corresponding to the mechanical momentum $p = mv$, whose rate of change is the applied force, there should be an analogous quantity equal to $\mathcal{L}I$, whose rate of change is $\mathcal{V}$. We have no right, of course, to say that $\mathcal{L}I$ is the real momentum of the circuit; in fact, it isn't. The whole circuit may be standing still and have no momentum. It is only that $\mathcal{L}I$ is analogous to the momentum mv in the sense of satisfying corresponding equations. In the same way, to the kinetic energy $\frac{1}{2}mv^2$, there corresponds an analogous quantity $\frac{1}{2}\mathcal{L}I^2$. But there we have a surprise. This $\frac{1}{2}\mathcal{L}I^2$ is really the energy in the electrical case also. This is because the rate of doing work on the inductance is $\mathcal{V}I$, and in the mechanical system it is Fv, the corresponding quantity. Therefore, in the case of the energy, the quantities not only correspond mathematically, but also have the same physical meaning as well.

We may see this in more detail as follows. As we found in Eq. (17.16), the rate of electrical work by induced forces is the product of the electromotive force and the current:

$$\frac{dW}{dt} = \mathcal{E}I.$$

Replacing $\mathcal{E}$ by its expression in terms of the current from Eq. (17.34), we have

$$\frac{dW}{dt} = -\mathcal{L}I\,\frac{dI}{dt}. \tag{17.36}$$

Integrating this equation, we find that the energy required from an external source to overcome the emf in the self-inductance while building up the current† (which must equal the energy stored, U) is

$$-W = U = \tfrac{1}{2}\mathcal{L}I^2. \tag{17.37}$$

Therefore the energy stored in an inductance is $\frac{1}{2}\mathcal{L}I^2$.

* This is, incidentally, *not* the *only* way a correspondence can be set up between mechanical and electrical quantities.

† We are neglecting any energy loss to heat from the current in the resistance of the coil. Such losses require additional energy from the source but do not change the energy which goes into the inductance.

Applying the same arguments to a pair of coils such as those in Figs. 17–8 or 17–9, we can show that the total electrical energy of the system is given by

$$U = \tfrac{1}{2}\mathcal{L}_1 I_1^2 + \tfrac{1}{2}\mathcal{L}_2 I_2^2 + \mathfrak{M} I_1 I_2. \tag{17.38}$$

For, starting with $I = 0$ in both coils, we could first turn on the current I_1 in coil 1, with $I_2 = 0$. The work done is just $\frac{1}{2}\mathcal{L}_1 I_1^2$. But now, on turning up I_2, we not only do the work $\frac{1}{2}\mathcal{L}_2 I_2^2$ against the emf in circuit 2, but also an additional amount $\mathfrak{M} I_1 I_2$, which is the integral of the emf $[\mathfrak{M}(dI_2/dt)]$ in circuit 1 times the now *constant* current I_1 in that circuit.

Suppose we now wish to find the force between any two coils carrying the currents I_1 and I_2. We might at first expect that we could use the principle of virtual work, by taking the change in the energy of Eq. (17.38). We must remember, of course, that as we change the relative positions of the coils the only quantity which varies is the mutual inductance $\mathfrak{M}$. We might then write the equation of virtual work as

$$-F\,\Delta x = \Delta U = I_1 I_2\,\Delta\mathfrak{M} \text{ (wrong)}.$$

But this equation is wrong because, as we have seen earlier, it includes only the change in the energy of the two coils and not the change in the energy of the sources which are maintaining the currents I_1 and I_2 at their constant values. We can now understand that these sources must supply energy against the induced emf's in the coils as they are moved. If we wish to apply the principle of virtual work correctly, we must also include these energies. As we have seen, however, we may take a short cut and use the principle of virtual work by remembering that the total energy is the negative of what we have called U_{mech}, the "mechanical energy." We can therefore write for the force

$$-F\,\Delta x = \Delta U_{\text{mech}} = -\Delta U. \tag{17.39}$$

The force between two coils is then given by

$$F\,\Delta x = I_1 I_2\,\Delta\mathfrak{M}.$$

Equation (17.38) for the energy of a system of two coils can be used to show that an interesting inequality exists between mutual inductance $\mathfrak{M}$ and the self-inductances $\mathcal{L}_1$ and $\mathcal{L}_2$ of the two coils. It is clear that the energy of two coils must be positive. If we begin with zero currents in the coils and increase these currents to some values, we have been adding energy to the system. If not, the currents would spontaneously increase with release of energy to the rest of the world—an unlikely thing to happen! Now our energy equation, Eq. (17.38), can equally well be written in the following form:

$$U = \frac{1}{2}\,\mathcal{L}_1\left(I_1 + \frac{\mathfrak{M}}{\mathcal{L}_1}\,I_2\right)^2 + \frac{1}{2}\left(\mathcal{L}_2 - \frac{\mathfrak{M}^2}{\mathcal{L}_1}\right) I_2^2. \tag{17.40}$$

That is just an algebraic transformation. This quantity must always be positive for any values of I_1 and I_2. In particular, it must be positive if I_2 should happen to have the special value

$$I_2 = -\frac{\mathcal{L}_1}{\mathfrak{M}}\,I_1. \tag{17.41}$$

But with this current for I_2, the first term in Eq. (17.40) is zero. If the energy is to be positive, the last term in (17.40) must be greater than zero. We have the requirement that

$$\mathcal{L}_1\mathcal{L}_2 > \mathfrak{M}^2.$$

We have thus proved the general result that the magnitude of the mutual inductance $\mathfrak{M}$ of any two coils is necessarily less than or equal to the geometric mean of the two self-inductances. ($\mathfrak{M}$ itself may be positive or negative, depending on the sign conventions for the currents I_1 and I_2.)

$$|\mathfrak{M}| < \sqrt{\mathfrak{L}_1\mathfrak{L}_2}. \tag{17.42}$$

The relation between $\mathfrak{M}$ and the self-inductances is usually written as

$$\mathfrak{M} = k\sqrt{\mathfrak{L}_1\mathfrak{L}_2}. \tag{17.43}$$

The constant k is called the coefficient of coupling. If most of the flux from one coil links the other coil, the coefficient of coupling is near one; we say the coils are "tightly coupled." If the coils are far apart or otherwise arranged so that there is very little mutual flux linkage, the coefficient of coupling is near zero and the mutual inductance is very small.

For calculating the mutual inductance of two coils, we have given in Eq. (17.30) a formula which is a double line integral around the two circuits. We might think that the same formula could be used to get the self-inductance of a single coil by carrying out both line integrals around the same coil. This, however, will not work, because in integrating around the two coils, the denominator r_{12} of the integrand will go to zero when the two line elements are at the same point. The self-inductance obtained from this formula is infinite. The reason is that this formula is an approximation that is valid only when the cross sections of the wires of the two circuits are small compared with the distance from one circuit to the other. Clearly, this approximation doesn't hold for a single coil. It is, in fact, true that the inductance of a single coil tends logarithmically to infinity as the diameter of its wire is made smaller and smaller.

We must, then, look for a different way of calculating the self-inductance of a single coil. It is necessary to take into account the distribution of the currents within the wires because the size of the wire is an important parameter. We should therefore ask not what is the inductance of a "circuit," but what is the inductance of a *distribution* of conductors. Perhaps the easiest way to find this inductance is to make use of the magnetic energy. We found earlier, in Section 15–3, an expression for the magnetic energy of a distribution of stationary currents:

$$U = \tfrac{1}{2}\int \boldsymbol{j}\cdot\boldsymbol{A}\,dV. \tag{17.44}$$

If we know the distribution of current density $\boldsymbol{j}$, we can compute the vector potential $\boldsymbol{A}$ and then evaluate the integral of Eq. (17.44) to get the energy. This energy is equal to the magnetic energy of the self-inductance, $\tfrac{1}{2}\mathfrak{L}I^2$. Equating the two gives us a formula for the inductance:

$$\mathfrak{L} = \frac{1}{I^2}\int \boldsymbol{j}\cdot\boldsymbol{A}\,dV. \tag{17.45}$$

We expect, of course, that the inductance is a number depending only on the geometry of the circuit and not on the current I in the circuit. The formula of Eq. (17.45) will indeed give such a result, because the integral in this equation is proportional to the square of the current—the current appears once through $\boldsymbol{j}$ and again through the vector potential $\boldsymbol{A}$. The integral divided by I^2 will depend on the geometry of the circuit but not on the current I.

Equation (17.44) for the energy of a current distribution can be put in a quite different form which is sometimes more convenient for calculation. Also, as we will see later, it is a form that is important because it is more generally valid. In the energy equation, Eq. (17.44), both $\boldsymbol{A}$ and $\boldsymbol{j}$ can be related to $\boldsymbol{B}$, so we can hope to express the energy in terms of the magnetic field—just as we were able to relate the electrostatic energy to the electric field. We begin by replacing $\boldsymbol{j}$ by $\epsilon_0 c^2\boldsymbol{\nabla}\times\boldsymbol{B}$. We cannot replace $\boldsymbol{A}$ so easily, since $\boldsymbol{B} = \boldsymbol{\nabla}\times\boldsymbol{A}$ cannot be reversed to give $\boldsymbol{A}$ in terms of $\boldsymbol{B}$. Anyway, we can write

$$U = \frac{\epsilon_0 c^2}{2}\int (\boldsymbol{\nabla}\times\boldsymbol{B})\cdot\boldsymbol{A}\,dV. \tag{17.46}$$

The interesting thing is that—with some restrictions—this integral can be written as

$$U = \frac{\epsilon_0 c^2}{2} \int \boldsymbol{B} \cdot (\boldsymbol{\nabla} \times \boldsymbol{A})\, dV. \tag{17.47}$$

To see this, we write out in detail a typical term. Suppose that we take the term $(\boldsymbol{\nabla} \times \boldsymbol{B})_z A_z$ which occurs in the integral of Eq. (17.46). Writing out the components, we get

$$\int \left(\frac{\partial B_y}{\partial x} - \frac{\partial B_x}{\partial y}\right) A_z\, dx\, dy\, dz.$$

(There are, of course, two more integrals of the same kind.) We now integrate the first term with respect to x—integrating by parts. That is, we can say

$$\int \frac{\partial B_y}{\partial x} A_z\, dx = B_y A_z - \int B_y \frac{\partial A_z}{\partial x}\, dx.$$

Now suppose that our system—meaning the sources and fields—is finite, so that as we go to large distances all fields go to zero. Then if the integrals are carried out over all space, evaluating the term $B_y A_z$ at the limits will give zero. We have left only the term with $B_y(\partial A_z/\partial x)$, which is evidently one part of $B_y(\boldsymbol{\nabla} \times \boldsymbol{A})_y$ and, therefore, of $\boldsymbol{B} \cdot (\boldsymbol{\nabla} \times \boldsymbol{A})$. If you work out the other five terms, you will see that Eq. (17.47) is indeed equivalent to Eq. (17.46).

But now we can replace $(\boldsymbol{\nabla} \times \boldsymbol{A})$ by $\boldsymbol{B}$, to get

$$U = \frac{\epsilon_0 c^2}{2} \int \boldsymbol{B} \cdot \boldsymbol{B}\, dV. \tag{17.48}$$

We have expressed the energy of a magnetostatic situation in terms of the magnetic field only. The expression corresponds closely to the formula we found for the electrostatic energy:

$$U = \frac{\epsilon_0}{2} \int \boldsymbol{E} \cdot \boldsymbol{E}\, dV. \tag{17.49}$$

One reason for emphasizing these two energy formulas is that sometimes they are more convenient to use. More important, it turns out that for dynamic fields (when $\boldsymbol{E}$ and $\boldsymbol{B}$ are changing with time) the two expressions (17.48) and (17.49) remain true, whereas the other formulas we have given for electric or magnetic energies are no longer correct—they hold only for static fields.

If we know the magnetic field $\boldsymbol{B}$ of a single coil, we can find the self-inductance by equating the energy expression (17.48) to $\frac{1}{2}\mathfrak{L}I^2$. Let's see how this works by finding the self-inductance of a long solenoid. We have seen earlier that the magnetic field inside a solenoid is uniform and $\boldsymbol{B}$ outside is zero. The magnitude of the field inside is $B = nI/\epsilon_0 c^2$, where n is the number of turns per unit length in the winding and I is the current. If the radius of the coil is r and its length is L (we take L very long, so that we can neglect end effects, i.e., $L \gg r$), the volume inside is $\pi r^2 L$. The magnetic energy is therefore

$$U = \frac{\epsilon_0 c^2}{2} B^2 \cdot (\text{Vol}) = \frac{n^2 I^2}{2\epsilon_0 c^2}\, \pi r^2 L,$$

which is equal to $\frac{1}{2}\mathfrak{L}I^2$. Or,

$$\mathfrak{L} = \frac{\pi r^2 n^2}{\epsilon_0 c^2} L. \tag{17.50}$$

18

The Maxwell Equations

18–1 Maxwell's equations

In this chapter we come back to the complete set of the four Maxwell equations that we took as our starting point in Chapter 1. Until now, we have been studying Maxwell's equations in bits and pieces; it is time to add one final piece, and to put them all together. We will then have the complete and correct story for electromagnetic fields that may be changing with time in any way. Anything said in this chapter that contradicts something said earlier is true and what was said earlier is false—because what was said earlier applied to such special situations as, for instance, steady currents or fixed charges. Although we have been very careful to point out the restrictions whenever we wrote an equation, it is easy to forget all of the qualifications and to learn too well the wrong equations. Now we are ready to give the whole truth, with no qualifications (or almost none).

The complete Maxwell equations are written in Table 18–1, in words as well as in mathematical symbols. The fact that the words are equivalent to the equations should by this time be familiar—you should be able to translate back and forth from one form to the other.

The first equation—that the divergence of $\boldsymbol{E}$ is the charge density over ϵ_0—is true in general. In dynamic as well as in static fields, Gauss' law is always valid. The flux of $\boldsymbol{E}$ through any closed surface is proportional to the charge inside. The third equation is the corresponding general law for magnetic fields. Since there are no magnetic charges, the flux of $\boldsymbol{B}$ through any closed surface is always zero. The second equation, that the curl of $\boldsymbol{E}$ is $-\partial\boldsymbol{B}/\partial t$, is Faraday's law and was discussed in the last two chapters. It also is generally true. The last equation has something new. We have seen before only the part of it which holds for steady currents. In that case we said that the curl of $\boldsymbol{B}$ is $\boldsymbol{j}/\epsilon_0 c^2$, but the correct general equation has a new part that was discovered by Maxwell.

Until Maxwell's work, the known laws of electricity and magnetism were those we have studied in Chapters 3 through 17. In particular, the equation for the magnetic field of steady currents was known only as

$$\boldsymbol{\nabla} \times \boldsymbol{B} = \frac{\boldsymbol{j}}{\epsilon_0 c^2}. \tag{18.1}$$

Maxwell began by considering these known laws and expressing them as differential equations, as we have done here. (Although the $\boldsymbol{\nabla}$ notation was not yet invented, it is mainly due to Maxwell that the importance of the combinations of derivatives, which we today call the curl and the divergence, first became apparent.) He then noticed that there was something strange about Eq. (18.1). If one takes the divergence of this equation, the left-hand side will be zero, because the divergence of a curl is always zero. So this equation requires that the divergence of $\boldsymbol{j}$ also be zero. But if the divergence of $\boldsymbol{j}$ is zero, then the total flux of current out of any closed surface is also zero.

The flux of current from a closed surface is the decrease of the charge inside the surface. This certainly cannot in general be zero because we know that the charges can be moved from one place to another. The equation

$$\boldsymbol{\nabla} \cdot \boldsymbol{j} = -\frac{\partial\rho}{\partial t} \tag{18.2}$$

has, in fact, been almost our definition of $\boldsymbol{j}$. This equation expresses the very funda-

Table 18–1 Classical Physics

Maxwell's equations	
I. $\nabla \cdot \boldsymbol{E} = \frac{\rho}{\epsilon_0}$	(Flux of $\boldsymbol{E}$ through a closed surface) = (Charge inside)/ϵ_0
II. $\nabla \times \boldsymbol{E} = -\frac{\partial \boldsymbol{B}}{\partial t}$	(Line integral of $\boldsymbol{E}$ around a loop) $= -\frac{d}{dt}$ (Flux of $\boldsymbol{B}$ through the loop)
III. $\nabla \cdot \boldsymbol{B} = 0$	(Flux of $\boldsymbol{B}$ through a closed surface) = 0
IV. $c^2\nabla \times \boldsymbol{B} = \frac{\boldsymbol{j}}{\epsilon_0} + \frac{\partial \boldsymbol{E}}{\partial t}$	c^2(Integral of $\boldsymbol{B}$ around a loop) = (Current through the loop)/ϵ_0 $+\frac{\partial}{\partial t}$ (Flux of $\boldsymbol{E}$ through the loop)
[**Conservation of charge**	
$\nabla \cdot \boldsymbol{j} = -\frac{\partial \rho}{\partial t}$	(Flux of current through a closed surface) $= -\frac{\partial}{\partial t}$ (Charge inside)]
Force law	
$\boldsymbol{F} = q(\boldsymbol{E} + \boldsymbol{v} \times \boldsymbol{B})$	
Law of motion	
$\frac{d}{dt}(\boldsymbol{p}) = \boldsymbol{F}$, where $\boldsymbol{p} = \frac{m\boldsymbol{v}}{\sqrt{1 - v^2/c^2}}$	(Newton's law, with Einstein's modification)
Gravitation	
$\boldsymbol{F} = -G\frac{m_1 m_2}{r^2}\boldsymbol{e}_r$	

mental law that electric charge is conserved—any flow of charge must come from some supply. Maxwell appreciated this difficulty and proposed that it could be avoided by adding the term $\partial \boldsymbol{E}/\partial t$ to the right-hand side of Eq. (18.1); he then got the fourth equation in Table 18–1:

$$\text{IV.} \qquad c^2\nabla \times \boldsymbol{B} = \frac{\boldsymbol{j}}{\epsilon_0} + \frac{\partial \boldsymbol{E}}{\partial t}.$$

It was not yet customary in Maxwell's time to think in terms of abstract fields. Maxwell discussed his ideas in terms of a model in which the vacuum was like an elastic solid. He also tried to explain the meaning of his new equation in terms of the mechanical model. There was much reluctance to accept his theory, first because of the model, and second because there was at first no experimental justification. Today, we understand better that what counts are the equations themselves and not the model used to get them. We may only question whether the equations are true or false. This is answered by doing experiments, and untold numbers of experiments have confirmed Maxwell's equations. If we take away the scaffolding he used to build it, we find that Maxwell's beautiful edifice stands on its own. He brought together all of the laws of electricity and magnetism and made one complete and beautiful theory.

Let us show that the extra term is just what is required to straighten out the difficulty Maxwell discovered. Taking the divergence of his equation (IV in Table 18–1), we must have that the divergence of the right-hand side is zero:

$$\nabla \cdot \frac{\boldsymbol{j}}{\epsilon_0} + \nabla \cdot \frac{\partial \boldsymbol{E}}{\partial t} = 0. \tag{18.3}$$

In the second term, the order of the derivatives with respect to coordinates and time can be reversed, so the equation can be rewritten as

$$\nabla \cdot \boldsymbol{j} + \epsilon_0 \frac{\partial}{\partial t} \nabla \cdot \boldsymbol{E} = 0. \tag{18.4}$$

But the first of Maxwell's equations says that the divergence of $\boldsymbol{E}$ is ρ/ϵ_0. Inserting this equality in Eq. (18.4), we get back Eq. (18.2), which we know is true. Conversely, if we accept Maxwell's equations—and we do because no one has ever found an experiment that disagrees with them—we must conclude that charge is always conserved.

The laws of physics have no answer to the question: "What happens if a charge is suddenly created at this point—what electromagnetic effects are produced?" No answer can be given because our equations say it doesn't happen. If it *were* to happen, we would need new laws, but we cannot say what they would be. We have not had the chance to observe how a world without charge conservation behaves. According to our equations, if you suddenly place a charge at some point, you had to carry it there from somewhere else. In that case, we can say what would happen.

When we added a new term to the equation for the curl of $\boldsymbol{E}$, we found that a whole new class of phenomena was described. We shall see that Maxwell's little addition to the equation for $\nabla \times \boldsymbol{B}$ also has far-reaching consequences. We can touch on only a few of them in this chapter.

18–2 How the new term works

As our first example we consider what happens with a spherically symmetric radial distribution of current. Suppose we imagine a little sphere with radioactive material on it. This radioactive material is squirting out some charged particles. (Or we could imagine a large block of jello with a small hole in the center into which some charge had been injected with a hypodermic needle and from which the charge is slowly leaking out.) In either case we would have a current that is everywhere radially outward. We will assume that it has the same magnitude in all directions.

Let the total charge inside any radius r be $Q(r)$. If the radial current density at the same radius is $j(r)$, then Eq. (18.2) requires that Q decreases at the rate

$$\frac{\partial Q(r)}{\partial t} = -4\pi r^2 j(r). \tag{18.5}$$

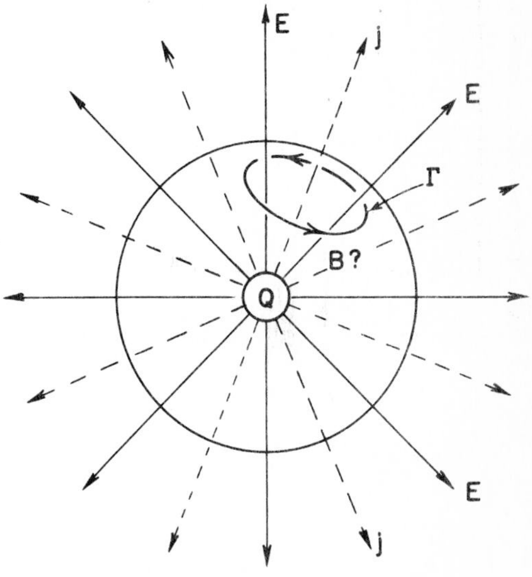

Fig. 18–1. What is the magnetic field of a spherically symmetric current?

We now ask about the magnetic field produced by the currents in this situation. Suppose we draw some loop Γ on a sphere of radius r, as shown in Fig. 18–1. There is some current through this loop, so we might expect to find a magnetic field circulating in the direction shown.

But we are already in difficulty. How can the $\boldsymbol{B}$ have any particular direction on the sphere? A different choice of Γ would allow us to conclude that its direction is exactly opposite to that shown. So how *can* there be any circulation of $\boldsymbol{B}$ around the currents?

We are saved by Maxwell's equation. The circulation of $\boldsymbol{B}$ depends not only on the total *current* through Γ but also on the rate of change with time of the *electric flux* through it. It must be that these two parts just cancel. Let's see if that works out.

The electric field at the radius r must be $Q(r)/4\pi\epsilon_0 r^2$—so long as the charge is symmetrically distributed, as we assume. It is radial, and its rate of change is then

$$\frac{\partial E}{\partial t} = \frac{1}{4\pi\epsilon_0 r^2} \frac{\partial Q}{\partial t}. \tag{18.6}$$

Comparing this with Eq. (18.5), we see that at any radius

$$\frac{\partial E}{\partial t} = -\frac{j}{\epsilon_0}. \tag{18.7}$$

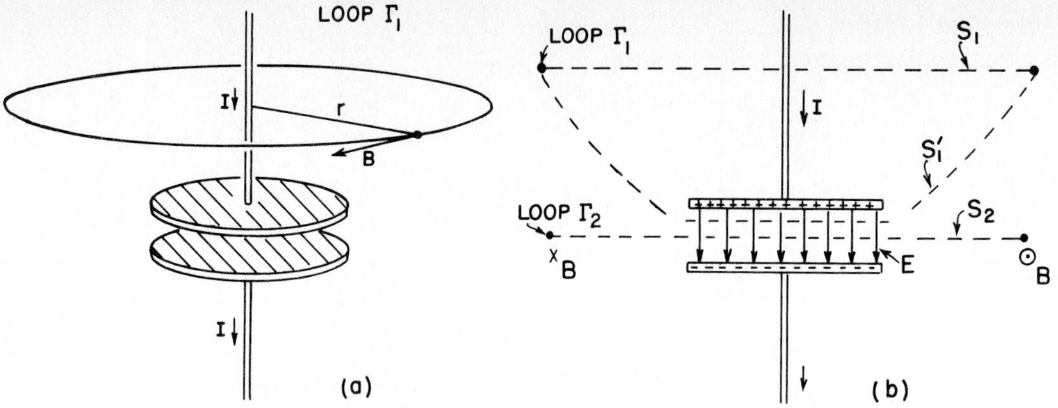

Fig. 18–2. The magnetic field near a charging capacitor.

In Eq. IV the two source terms cancel and the curl of $\boldsymbol{B}$ is always zero. There is no magnetic field in our example.

As our second example, we consider the magnetic field of a wire used to charge a parallel-plate condenser (see Fig. 18–2). If the charge Q on the plates is changing with time (but not too fast), the current in the wires is equal to dQ/dt. We would expect that this current will produce a magnetic field that encircles the wire. Surely, the current close to the wire must produce the normal magnetic field—it cannot depend on where the current is going.

Suppose we take a loop Γ_1 which is a circle with radius r, as shown in part (a) of the figure. The line integral of the magnetic field should be equal to the current I divided by $\epsilon_0 c^2$. We have

$$2\pi r B = \frac{I}{\epsilon_0 c^2}. \tag{18.8}$$

This is what we would get for a steady current, but it is also correct with Maxwell's addition, because if we consider the plane surface S inside the circle, there are no electric fields on it (assuming the wire to be a very good conductor). The surface integral of $\partial \boldsymbol{E}/\partial t$ is zero.

Suppose, however, that we now slowly move the curve Γ downward. We get always the same result until we draw even with the plates of the condenser. Then the current I goes to zero. Does the magnetic field disappear? That would be quite strange. Let's see what Maxwell's equation says for the curve Γ_2, which is a circle of radius r whose plane passes between the condenser plates [Fig. 18–2(b)]. The line integral of $\boldsymbol{B}$ around Γ_2 is $2\pi r B$. This must equal the time derivative of the flux of $\boldsymbol{E}$ through the plane circular surface S_2. This flux of $\boldsymbol{E}$, we know from Gauss' law, must be equal to $1/\epsilon_0$ times the charge Q on one of the condenser plates. We have

$$c^2\, 2\pi r B = \frac{d}{dt}\left(\frac{Q}{\epsilon_0}\right). \tag{18.9}$$

That is very convenient. It is the same result we found in Eq. (18.8). Integrating over the changing electric field gives the same magnetic field as does integrating over the current in the wire. Of course, that is just what Maxwell's equation says. It is easy to see that this must always be so by applying our same arguments to the two surfaces S_1 and S_1' that are bounded by the same circle Γ_1 in Fig. 18–2(b). Through S_1 there is the current I, but no electric flux. Through S_1' there is no current, but an electric flux changing at the rate I/ϵ_0. The same $\boldsymbol{B}$ is obtained if we use Eq. IV with either surface.

From our discussion so far of Maxwell's new term, you may have the impression that it doesn't add much—that it just fixes up the equations to agree with what we already expect. It is true that if we just consider Eq. IV *by itself*, nothing particularly new comes out. The words "*by itself*" are, however, all-important. Maxwell's small change in Eq. IV, when *combined with the other* equations, does indeed produce much that is new and important. Before we take up these matters, however, we want to speak more about Table 18–1.

18–3 All of classical physics

In Table 18–1 we have all that was known of fundamental *classical* physics, that is, the physics that was known by 1905. Here it all is, in one table. With these equations we can understand the complete realm of classical physics.

First we have the Maxwell equations—written in both the expanded form and the short mathematical form. Then there is the conservation of charge, which is even written in parentheses, because the moment we have the complete Maxwell equations, we can deduce from them the conservation of charge. So the table is even a little redundant. Next, we have written the force law, because having all the electric and magnetic fields doesn't tell us anything until we know what they do to charges. Knowing $\boldsymbol{E}$ and $\boldsymbol{B}$, however, we can find the force on an object with the charge q moving with velocity $\boldsymbol{v}$. Finally, having the force doesn't tell us anything until we know what happens when a force pushes on something; we need the law of motion, which is that the force is equal to the rate of change of the momentum. (Remember? We had that in Volume I.) We even include relativity effects by writing the momentum as $p = m_0v/\sqrt{1 - v^2/c^2}$.

If we really want to be complete, we should add one more law—Newton's law of gravitation—so we put that at the end.

Therefore in one small table we have all the fundamental laws of classical physics—even with room to write them out in words and with some redundancy. This is a great moment. We have climbed a great peak. We are on the top of K-2—we are nearly ready for Mount Everest, which is quantum mechanics. We have climbed the peak of a "Great Divide," and now we can go down the other side.

We have mainly been trying to learn how to understand the equations. Now that we have the whole thing put together, we are going to study what the equations mean—what new things they say that we haven't already seen. We've been working hard to get up to this point. It has been a great effort, but now we are going to have nice coasting downhill as we see all the consequences of our accomplishment.

18–4 A travelling field

Now for the new consequences. They come from putting together all of Maxwell's equations. First, let's see what would happen in a circumstance which we pick to be particularly simple. By assuming that all the quantities vary only in one coordinate, we will have a one-dimensional problem. The situation is shown in Fig. 18–3. We have a sheet of charge located on the yz-plane. The sheet is first at rest, then instantaneously given a velocity u in the y-direction, and kept moving with this constant velocity. You might worry about having such an "infinite" acceleration, but it doesn't really matter; just imagine that the velocity is brought to u very quickly. So we have suddenly a surface current J (J is the current per unit

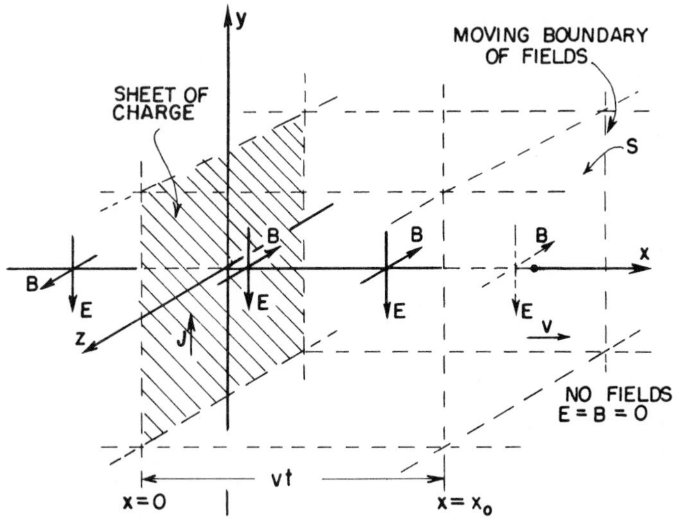

Fig. 18–3. An infinite sheet of charge is suddenly set into motion parallel to itself. There are magnetic and electric fields that propagate out from the sheet at a constant speed.

width in the z-direction). To keep the problem simple, we suppose that there is also a stationary sheet of charge of opposite sign superposed on the yz-plane, so that there are no electrostatic effects. Also, although in the figure we show only what is happening in a finite region, we imagine that the sheet extends to infinity in $\pm y$ and $\pm z$. In other words, we have a situation where there is no current, and then suddenly there is a uniform sheet of current. What will happen?

Well, when there is a sheet of current in the plus y-direction, there is, as we know, a magnetic field generated which will be in the minus z-direction for $x > 0$ and in the opposite direction for $x < 0$. We could find the magnitude of $\boldsymbol{B}$ by using the fact that the line integral of the magnetic field will be equal to the current over $\epsilon_0 c^2$. We would get that $B = J/2\epsilon_0 c^2$ (since the current I in a strip of width w is Jw and the line integral of $\boldsymbol{B}$ is $2Bw$).

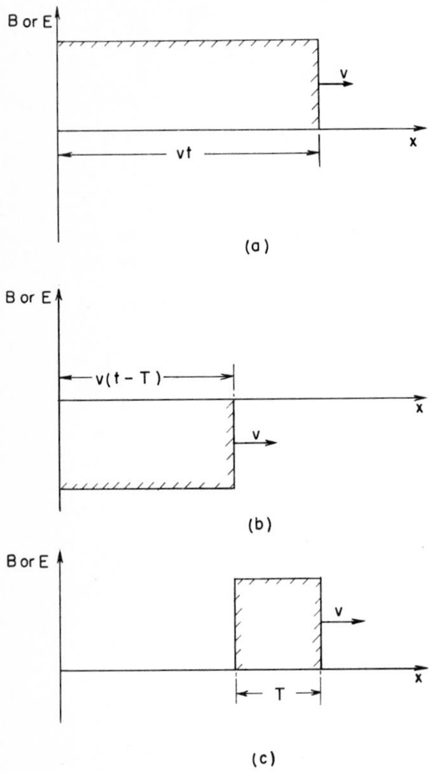

Fig. 18–4. (a) The magnitude of **B** (or **E**) as a function of x at the time t after the charge sheet is set in motion. (b) The fields for a charge sheet set in motion, toward negative y at $t = T$. (c) The sum of (a) and (b).

This gives us the field next to the sheet—for small x—but since we are imagining an infinite sheet, we would expect the same argument to give the magnetic field farther out for larger values of x. However, that would mean that the moment we turn on the current, the magnetic field is suddenly changed from zero to a finite value everywhere. But wait! If the magnetic field is suddenly changed, it will produce tremendous electrical effects. (If it changes in *any* way, there are electrical effects.) So because we moved the sheet of charge, we make a changing magnetic field, and therefore electric fields must be generated. If there are electric fields generated, they had to start from zero and change to something else. There will be some $\partial \boldsymbol{E}/\partial t$ that will make a contribution, together with the current J, to the production of the magnetic field. So through the various equations there is a big intermixing, and we have to try to solve for all the fields at once.

By looking at the Maxwell equations alone, it is not easy to see directly how to get the solution. So we will first show you what the answer is and then verify that it does indeed satisfy the equations. The answer is the following: The field $\boldsymbol{B}$ that we computed is, in fact, generated right next to the current sheet (for small x). It must be so, because if we make a tiny loop around the sheet, there is no room for any electric flux to go through it. But the field $\boldsymbol{B}$ out farther—for larger x—is, at first, zero. It stays zero for awhile, and then suddenly turns on. In short, we turn on the current and the magnetic field immediately next to it turns on to a constant value $\boldsymbol{B}$; then the turning on of $\boldsymbol{B}$ spreads out from the source region. After a certain time, there is a uniform magnetic field everywhere out to some value x, and then zero beyond. Because of the symmetry, it spreads in both the plus and minus x-directions.

The $\boldsymbol{E}$-field does the same thing. Before $t = 0$ (when we turn on the current), the field is zero everywhere. Then after the time t, both $\boldsymbol{E}$ and $\boldsymbol{B}$ are uniform out to the distance $x = vt$, and zero beyond. The fields make their way forward like a tidal wave, with a front moving at a uniform velocity which turns out to be c, but for a while we will just call it v. A graph of the magnitude of $\boldsymbol{E}$ or $\boldsymbol{B}$ versus x, as they appear at the time t, is shown in Fig. 18–4(a). Looking again at Fig. 18–3, at the time t, the region between $x = \pm vt$ is "filled" with the fields, but they have not yet reached beyond. We emphasize again that we are assuming that the current sheet and, therefore the fields $\boldsymbol{E}$ and $\boldsymbol{B}$, extend infinitely far in both the y- and z-directions. (We cannot draw an infinite sheet, so we have shown only what happens in a finite area.)

We want now to analyze quantitatively what is happening. To do that, we want to look at two cross-sectional views, a top view looking down along the y-axis, as shown in Fig. 18–5, and a side view looking back along the z-axis, as shown in Fig. 18–6. Suppose we start with the side view. We see the charged sheet moving up; the magnetic field points into the page for $+x$, and out of the page for $-x$, and the electric field is downward everywhere—out to $x = \pm vt$.

Let's see if these fields are consistent with Maxwell's equations. Let's first draw one of those loops that we use to calculate a line integral, say the rectangle Γ_2 shown in Fig. 18–6. You notice that one side of the rectangle is in the region where there are fields, but one side is in the region the fields have still not reached. There is some magnetic flux through this loop. If it is changing, there should be an emf around it. If the wavefront is moving, we will have a changing magnetic

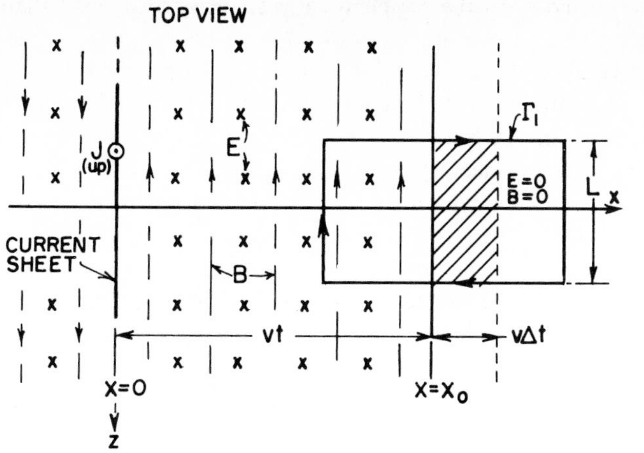

Fig. 18–5. Top view of Fig. 18–3.

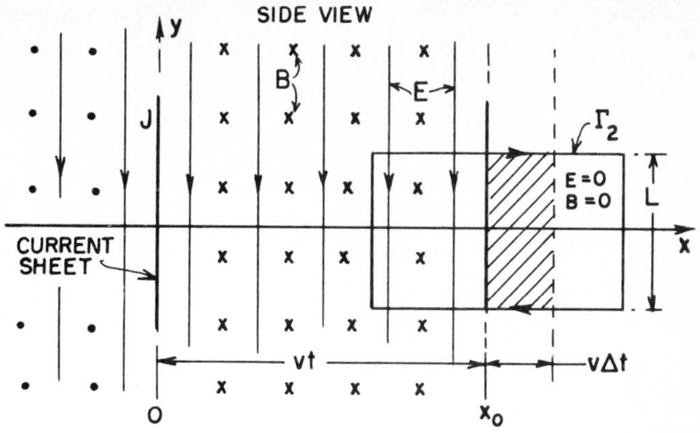

Fig. 18–6. Side view of Fig. 18–3.

flux, because the area in which $\boldsymbol{B}$ exists is progressively increasing at the velocity v. The flux inside Γ_2 is B times the part of the area inside Γ_2 which has a magnetic field. The rate of change of the flux, since the magnitude of $\boldsymbol{B}$ is constant, is the magnitude times the rate of change of the area. The rate of change of the area is easy. If the width of the rectangle Γ_2 is L, the area in which $\boldsymbol{B}$ exists changes by $Lv\,\Delta t$ in the time Δt. (See Fig. 18–6.) The rate of change of flux is then BLv. According to Faraday's law, this should equal the line integral of $\boldsymbol{E}$ around Γ_2, which is just EL. We have the equation

$$E = vB. \tag{18.10}$$

So if the ratio of E to B is v, the fields we have assumed will satisfy Faraday's equation.

But that is not the only equation; we have the other equation relating $\boldsymbol{E}$ and $\boldsymbol{B}$:

$$c^2\nabla \times \boldsymbol{B} = \frac{\boldsymbol{j}}{\epsilon_0} + \frac{\partial \boldsymbol{E}}{\partial t}. \tag{18.11}$$

To apply this equation, we look at the top view in Fig. 18–5. We have seen that this equation will give us the value of B next to the current sheet. Also, for any loop drawn outside the sheet but behind the wavefront, there is no curl of $\boldsymbol{B}$ nor any $\boldsymbol{j}$ or changing $\boldsymbol{E}$, so the equation is correct there. Now let's look at what happens for the curve Γ_1 that intersects the wavefront, as shown in Fig. 18–5. Here there are no currents, so Eq. (18.11) can be written—in integral form—as

$$c^2 \oint_{\Gamma_1} \boldsymbol{B}\cdot d\boldsymbol{s} = \frac{d}{dt}\int_{\text{inside }\Gamma_1} \boldsymbol{E}\cdot\boldsymbol{n}\,da. \tag{18.12}$$

The line integral of $\boldsymbol{B}$ is just B times L. The rate of change of the flux of $\boldsymbol{E}$ is due only to the advancing wavefront. The area inside Γ_1, where $\boldsymbol{E}$ is not zero, is increasing at the rate vL. The right-hand side of Eq. (18.12) is then vLE. That equation becomes

$$c^2B = Ev. \tag{18.13}$$

We have a solution in which we have a constant $\boldsymbol{B}$ and a constant $\boldsymbol{E}$ behind the front, both at right angles to the direction in which the front is moving and at right angles to each other. Maxwell's equations specify the ratio of E to B. From Eqs. (18.10) and (18.13),

$$E = vB, \qquad \text{and} \qquad E = \frac{c^2}{v}B.$$

But one moment! We have found *two different* conditions on the ratio E/B. Can such a field as we describe really exist? There is, of course, only one velocity v for which both of these equations can hold, namely $v = c$. The wavefront must travel with the velocity c. We have an example in which the electrical influence from a current propagates at a certain finite velocity c.

Now let's ask what happens if we suddenly stop the motion of the charged sheet after it has been on for a short time T. We can see what will happen by the principle of superposition. We had a current that was zero and then was suddenly turned on. We know the solution for that case. Now we are going to add another set of fields. We take another charged sheet and suddenly start it moving, in the opposite direction with the same speed, only at the time T after we started the first current. The total current of the two added together is first zero, then on for a time T, then off again—because the two currents cancel. We have a square "pulse" of current.

The new negative current produces the same fields as the positive one, only with all the signs reversed and, of course, delayed in time by T. A wavefront again travels out at the velocity c. At the time t it has reached the distance $x = \pm c(t - T)$, as shown in Fig. 18–4(b). So we have two "blocks" of field marching out at the speed c, as in parts (a) and (b) of Fig. 18–4. The combined fields are as shown in part (c) of the figure. The fields are zero for $x > ct$, they are constant (with the values we found above) between $x = c(t - T)$ and $x = ct$, and again zero for $x < c(t - T)$.

In short, we have a little piece of field—a block of thickness cT—which has left the current sheet and is travelling through space all by itself. The fields have "taken off"; they are propagating freely through space, no longer connected in any way with the source. The caterpillar has turned into a butterfly!

How can this bundle of electric and magnetic fields maintain itself? The answer is: by the combined effects of the Faraday law, $\nabla \times \boldsymbol{E} = -\partial \boldsymbol{B}/\partial t$, and the new term of Maxwell, $c^2\nabla \times \boldsymbol{B} = \partial \boldsymbol{E}/\partial t$. They cannot help maintaining themselves. Suppose the magnetic field were to disappear. There would be a changing magnetic field which would produce an electric field. If this electric field tries to go away, the changing electric field would create a magnetic field back again. So by a perpetual interplay—by the swishing back and forth from one field to the other—they must go on forever. It is impossible for them to disappear.* They maintain themselves in a kind of a dance—one making the other, the second making the first—propagating onward through space.

18–5 The speed of light

We have a wave which leaves the material source and goes outward at the velocity c, which is the speed of light. But let's go back a moment. From a historical point of view, it wasn't known that the coefficient c in Maxwell's equations was also the speed of light propagation. There was just a constant in the equations. We have called it c from the beginning, because we knew what it would turn out to be. We didn't think it would be sensible to make you learn the formulas with a different constant and then go back to substitute c wherever it belonged. From the point of view of electricity and magnetism, however, we just start out with two constants, ϵ_0 and c^2, that appear in the equations of electrostatics and magnetostatics:

$$\nabla \cdot \boldsymbol{E} = \frac{\rho}{\epsilon_0} \tag{18.14}$$

and

$$\nabla \times \boldsymbol{B} = \frac{\boldsymbol{j}}{\epsilon_0 c^2}\cdot \tag{18.15}$$

If we take any *arbitrary* definition of a unit of charge, we can determine experimentally the constant ϵ_0 required in Eq. (18.14)—say by measuring the force between two unit charges at rest, using Coulomb's law. We must also determine experimentally the constant $\epsilon_0 c^2$ that appears in Eq. (18.15), which we can do, say, by measuring the force between two unit currents. (A unit current means one unit of charge per second.) The ratio of these two experimental constants is c^2—just another "electromagnetic constant."

---

* Well, not quite. They can be "absorbed" if they get to a region where there are charges. By which we mean that other fields can be produced somewhere which superpose on these fields and "cancel" them by destructive interference (see Chapter 31, Vol. I).

Notice now that this constant c^2 is the same no matter what we choose for our unit of charge. If we put twice as much "charge"—say twice as many proton charges—in our "unit" of charge, ϵ_0 would need to be one-fourth as large. When we pass two of these "unit" currents through two wires, there will be twice as much "charge" per second in each wire, so the force between two wires is four times larger. The constant $\epsilon_0 c^2$ must be reduced by one-fourth. But the ratio $\epsilon_0 c^2/\epsilon_0$ is unchanged.

So just by experiments with charges and currents we find a number c^2 which turns out to be the square of the velocity of propagation of electromagnetic influences. From static measurements—by measuring the forces between two unit charges and between two unit currents—we find that $c = 3.00 \times 10^8$ meters/sec. When Maxwell first made this calculation with his equations, he said that bundles of electric and magnetic fields should be propagated at this speed. He also remarked on the mysterious coincidence that this was the same as the speed of light. "We can scarcely avoid the inference," said Maxwell, "that light consists in the transverse undulations of the same medium which is the cause of electric and magnetic phenomena."

Maxwell had made one of the great unifications of physics. Before his time, there was light, and there was electricity and magnetism. The latter two had been unified by the experimental work of Faraday, Oersted, and Ampere. Then, all of a sudden, light was no longer "something else," but was only electricity and magnetism in this new form—little pieces of electric and magnetic fields which propagate through space on their own.

We have called your attention to some characteristics of this special solution, which turn out to be true, however, for *any* electromagnetic wave: that the magnetic field is perpendicular to the direction of motion of the wavefront; that the electric field is likewise perpendicular to the direction of motion of the wavefront; and that the two vectors $\boldsymbol{E}$ and $\boldsymbol{B}$ are perpendicular to each other. Furthermore, the magnitude of the electric field E is equal to c times the magnitude of the magnetic field B. These three facts—that the two fields are transverse to the direction of propagation, that $\boldsymbol{B}$ is perpendicular to $\boldsymbol{E}$, and that $E = cB$—are generally true for any electromagnetic wave. Our special case is a good one—it shows all the main features of electromagnetic waves.

18–6 Solving Maxwell's equations; the potentials and the wave equation

Now we would like to do something mathematical; we want to write Maxwell's equations in a simpler form. You may consider that we are complicating them, but if you will be patient a little bit, they will suddenly come out simpler. Although by this time you are thoroughly used to each of the Maxwell equations, there are many pieces that must all be put together. That's what we want to do.

We begin with $\nabla \cdot \boldsymbol{B} = 0$—the simplest of the equations. We know that it implies that $\boldsymbol{B}$ is the curl of something. So, if we write

$$\boldsymbol{B} = \nabla \times \boldsymbol{A}, \tag{18.16}$$

we have already solved one of Maxwell's equations. (Incidentally, you appreciate that it remains true that another vector $\boldsymbol{A}'$ would be just as good if $\boldsymbol{A}' = \boldsymbol{A} + \nabla\psi$—where ψ is any scalar field—because the curl of $\nabla\psi$ is zero, and $\boldsymbol{B}$ is still the same. We have talked about that before.)

We take next the Faraday law, $\nabla \times \boldsymbol{E} = -\partial\boldsymbol{B}/\partial t$, because it doesn't involve any currents or charges. If we write $\boldsymbol{B}$ as $\nabla \times \boldsymbol{A}$ and differentiate with respect to t, we can write Faraday's law in the form

$$\nabla \times \boldsymbol{E} = -\frac{\partial}{\partial t}\nabla \times \boldsymbol{A}.$$

Since we can differentiate either with respect to time or to space first, we can also write this equation as

$$\nabla \times \left(\boldsymbol{E} + \frac{\partial \boldsymbol{A}}{\partial t}\right) = 0. \tag{18.17}$$

We see that $\boldsymbol{E} + \partial \boldsymbol{A}/\partial t$ is a vector whose curl is equal to zero. Therefore that vector is the gradient of something. When we worked on electrostatics, we had $\boldsymbol{\nabla} \times \boldsymbol{E} = 0$, and then we decided that $\boldsymbol{E}$ itself was the gradient of something. We took it to be the gradient of $-\phi$ (the minus for technical convenience). We do the same thing for $\boldsymbol{E} + \partial \boldsymbol{A}/\partial t$; we set

$$\boldsymbol{E} + \frac{\partial \boldsymbol{A}}{\partial t} = -\boldsymbol{\nabla}\phi. \tag{18.18}$$

We use the same symbol ϕ so that, in the electrostatic case where nothing changes with time and the $\partial \boldsymbol{A}/\partial t$ term disappears, $\boldsymbol{E}$ will be our old $-\boldsymbol{\nabla}\phi$. So Faraday's equation can be put in the form

$$\boldsymbol{E} = -\boldsymbol{\nabla}\phi - \frac{\partial \boldsymbol{A}}{\partial t}. \tag{18.19}$$

We have solved two of Maxwell's equations already, and we have found that to describe the electromagnetic fields $\boldsymbol{E}$ and $\boldsymbol{B}$, we need four potential functions: a scalar potential ϕ and a vector potential $\boldsymbol{A}$, which is, of course, three functions.

Now that $\boldsymbol{A}$ determines part of $\boldsymbol{E}$, as well as $\boldsymbol{B}$, what happens when we change $\boldsymbol{A}$ to $\boldsymbol{A}' = \boldsymbol{A} + \boldsymbol{\nabla}\psi$? In general, $\boldsymbol{E}$ would change if we didn't take some special precaution. We can, however, still allow $\boldsymbol{A}$ to be changed in this way without affecting the fields $\boldsymbol{E}$ and $\boldsymbol{B}$—that is, without changing the physics—if we always change $\boldsymbol{A}$ and ϕ *together* by the rules

$$\boldsymbol{A}' = \boldsymbol{A} + \boldsymbol{\nabla}\psi, \qquad \phi' = \phi - \frac{\partial \psi}{\partial t}. \tag{18.20}$$

Then neither $\boldsymbol{B}$ nor $\boldsymbol{E}$, obtained from Eq. (18.19), is changed.

Previously, we chose to make $\boldsymbol{\nabla} \cdot \boldsymbol{A} = 0$, to make the equations of statics somewhat simpler. We are not going to do that now; we are going to make a different choice. But we'll wait a bit before saying what the choice is, because later it will be clear *why* the choice is made.

Now we return to the two remaining Maxwell equations which will give us relations between the potentials and the sources ρ and $\boldsymbol{j}$. Once we can determine $\boldsymbol{A}$ and ϕ from the currents and charges, we can always get $\boldsymbol{E}$ and $\boldsymbol{B}$ from Eqs. (18.16) and (18.19), so we will have another form of Maxwell's equations.

We begin by substituting Eq. (18.19) into $\boldsymbol{\nabla} \cdot \boldsymbol{E} = \rho/\epsilon_0$; we get

$$\boldsymbol{\nabla} \cdot \left(-\boldsymbol{\nabla}\phi - \frac{\partial \boldsymbol{A}}{\partial t}\right) = \frac{\rho}{\epsilon_0},$$

which we can write also as

$$-\nabla^2\phi - \frac{\partial}{\partial t}\,\boldsymbol{\nabla} \cdot \boldsymbol{A} = \frac{\rho}{\epsilon_0}. \tag{18.21}$$

This is one equation relating ϕ and $\boldsymbol{A}$ to the sources.

Our final equation will be the most complicated. We start by rewriting the fourth Maxwell equation as

$$c^2\boldsymbol{\nabla} \times \boldsymbol{B} - \frac{\partial \boldsymbol{E}}{\partial t} = \frac{\boldsymbol{j}}{\epsilon_0},$$

and then substitute for $\boldsymbol{B}$ and $\boldsymbol{E}$ in terms of the potentials, using Eqs. (18.16) and (18.19):

$$c^2\boldsymbol{\nabla} \times (\boldsymbol{\nabla} \times \boldsymbol{A}) - \frac{\partial}{\partial t}\left(-\boldsymbol{\nabla}\phi - \frac{\partial \boldsymbol{A}}{\partial t}\right) = \frac{\boldsymbol{j}}{\epsilon_0}.$$

The first term can be rewritten using the algebraic identity: $\boldsymbol{\nabla} \times (\boldsymbol{\nabla} \times \boldsymbol{A}) = \boldsymbol{\nabla}(\boldsymbol{\nabla} \cdot \boldsymbol{A}) - \boldsymbol{\nabla}^2\boldsymbol{A}$; we get

$$-c^2\nabla^2\boldsymbol{A} + c^2\boldsymbol{\nabla}(\boldsymbol{\nabla} \cdot \boldsymbol{A}) + \frac{\partial}{\partial t}\,\boldsymbol{\nabla}\phi + \frac{\partial^2 \boldsymbol{A}}{\partial t^2} = \frac{\boldsymbol{j}}{\epsilon_0}. \tag{18.22}$$

It's not very simple!

Fortunately, we can now make use of our freedom to choose arbitrarily the divergence of $\boldsymbol{A}$. What we are going to do is to use our choice to fix things so that the equations for $\boldsymbol{A}$ and for ϕ are separated but have the same form. We can do this by taking*

$$\nabla \cdot \boldsymbol{A} = -\frac{1}{c^2}\frac{\partial \phi}{\partial t}. \tag{18.23}$$

When we do that, the two middle terms in $\boldsymbol{A}$ and ϕ in Eq. (18.22) cancel, and that equation becomes much simpler:

$$\nabla^2 \boldsymbol{A} - \frac{1}{c^2}\frac{\partial^2 \boldsymbol{A}}{\partial t^2} = -\frac{\boldsymbol{j}}{\epsilon_0 c^2}. \tag{18.24}$$

And our equation for ϕ—Eq. (18.21)—takes on the same form:

$$\nabla^2 \phi - \frac{1}{c^2}\frac{\partial^2 \phi}{\partial t^2} = -\frac{\rho}{\epsilon_0}. \tag{18.25}$$

What a beautiful set of equations! They are beautiful, first, because they are nicely separated—with the charge density, goes ϕ; with the current, goes $\boldsymbol{A}$. Furthermore, although the left side looks a little funny—a Laplacian together with a $(\partial/\partial t)^2$—when we unfold it we see

$$\frac{\partial^2 \phi}{\partial x^2} + \frac{\partial^2 \phi}{\partial y^2} + \frac{\partial^2 \phi}{\partial z^2} - \frac{1}{c^2}\frac{\partial^2 \phi}{\partial t^2} = -\frac{\rho}{\epsilon_0}. \tag{18.26}$$

It has a nice symmetry in x, y, z, t—the $-1/c^2$ is necessary because, of course, time and space *are* different; they have different units.

Maxwell's equations have led us to a new kind of equation for the potentials ϕ and $\boldsymbol{A}$ but to the same mathematical form for all four functions ϕ, A_x, A_y, and A_z. Once we learn how to solve these equations, we can get $\boldsymbol{B}$ and $\boldsymbol{E}$ from $\nabla \times \boldsymbol{A}$ and $-\nabla\phi - \partial \boldsymbol{A}/\partial t$. We have another form of the electromagnetic laws exactly equivalent to Maxwell's equations, and in many situations they are much simpler to handle.

We have, in fact, already solved an equation much like Eq. (18.26). When we studied sound in Chapter 47 of Vol. I, we had an equation of the form

$$\frac{\partial^2 \phi}{\partial x^2} = \frac{1}{c^2}\frac{\partial^2 \phi}{\partial t^2},$$

and we saw that it described the propagation of waves in the x-direction at the speed c. Equation (18.26) is the corresponding wave equation for three dimensions. So in regions where there are no longer any charges and currents, the solution of these equations is *not* that ϕ and $\boldsymbol{A}$ are zero. (Although that is indeed one possible solution.) There are solutions in which there is some set of ϕ and $\boldsymbol{A}$ which are changing in time but always moving out at the speed c. The fields travel onward through free space, as in our example at the beginning of the chapter.

With Maxwell's new term in Eq. IV, we have been able to write the field equations in terms of $\boldsymbol{A}$ and ϕ in a form that is simple and that makes immediately apparent that there are electromagnetic waves. For many practical purposes, it will still be convenient to use the original equations in terms of $\boldsymbol{E}$ and $\boldsymbol{B}$. But they are on the other side of the mountain we have already climbed. Now we are ready to cross over to the other side of the peak. Things will look different—we are ready for some new and beautiful views.

---

* Choosing the $\nabla \cdot \boldsymbol{A}$ is called "choosing a gauge." Changing $\boldsymbol{A}$ by adding $\nabla\psi$ is called a "gauge transformation." Equation (18.23) is called "the Lorentz gauge."

19

The Principle of Least Action

A special lecture—almost verbatim*

"When I was in high school, my physics teacher—whose name was Mr. Bader—called me down one day after physics class and said, 'You look bored; I want to tell you something interesting.' Then he told me something which I found absolutely fascinating, and have, since then, always found fascinating. Every time the subject comes up, I work on it. In fact, when I began to prepare this lecture I found myself making more analyses on the thing. Instead of worrying about the lecture, I got involved in a new problem. The subject is this—the principle of least action.

"Mr. Bader told me the following: Suppose you have a particle (in a gravitational field, for instance) which starts somewhere and moves to some other point by free motion—you throw it, and it goes up and comes down.

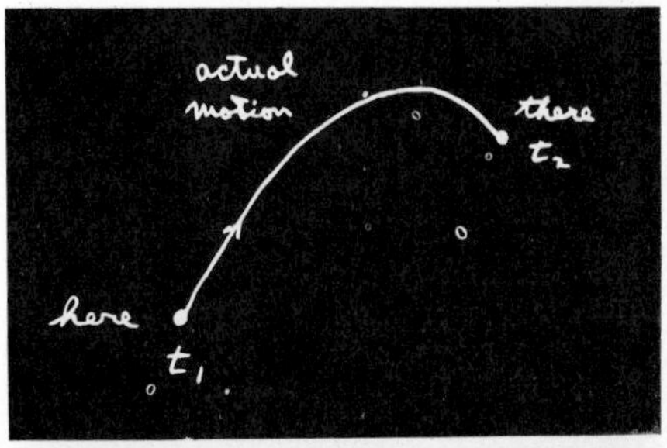

It goes from the original place to the final place in a certain amount of time. Now, you try a different motion. Suppose that to get from here to there, it went like this

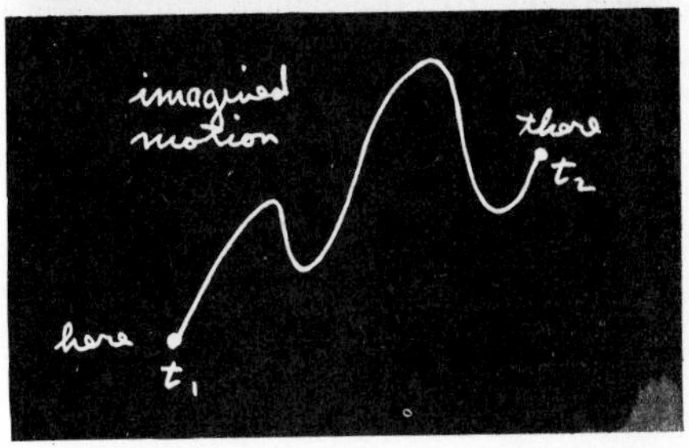

but got there in just the same amount of time. Then he said this: If you calculate the kinetic energy at every moment on the path, take away the potential energy, and integrate it over the time during the whole path, you'll find that the number you'll get is *bigger* than that for the actual motion.

---

* Later chapters do not depend on the material of this special lecture—which is intended to be for "entertainment."

"In other words, the laws of Newton could be stated not in the form $F = ma$ but in the form: the average kinetic energy less the average potential energy is as little as possible for the path of an object going from one point to another.

"Let me illustrate a little bit better what it means. If you take the case of the gravitational field, then if the particle has the path $x(t)$ (let's just take one dimension for a moment; we take a trajectory that goes up and down and not sideways), where x is the height above the ground, the kinetic energy is $\frac{1}{2}m\,(dx/dt)^2$, and the potential energy at any time is mgx. Now I take the kinetic energy minus the potential energy at every moment along the path and integrate that with respect to time from the initial time to the final time. Let's suppose that at the original time t_1 we started at some height and at the end of the time t_2 we are definitely ending at some other place.

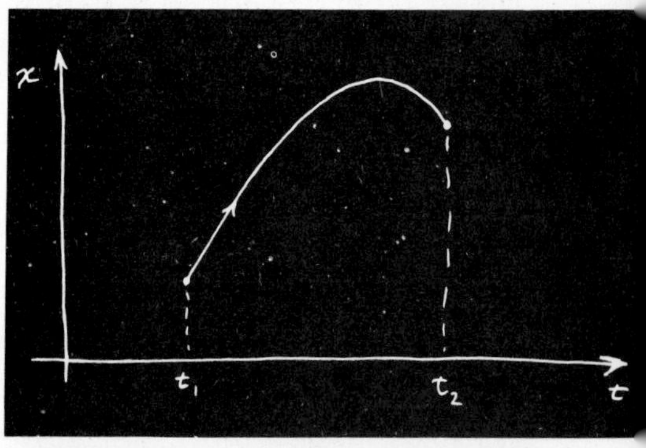

"Then the integral is

$$\int_{t_1}^{t_2}\left[\frac{1}{2}\,m\left(\frac{dx}{dt}\right)^2 - mgx\right]dt.$$

The actual motion is some kind of a curve—it's a parabola if we plot against the time—and gives a certain value for the integral. But we could *imagine* some other motion that went very high and came up and down in some peculiar way.

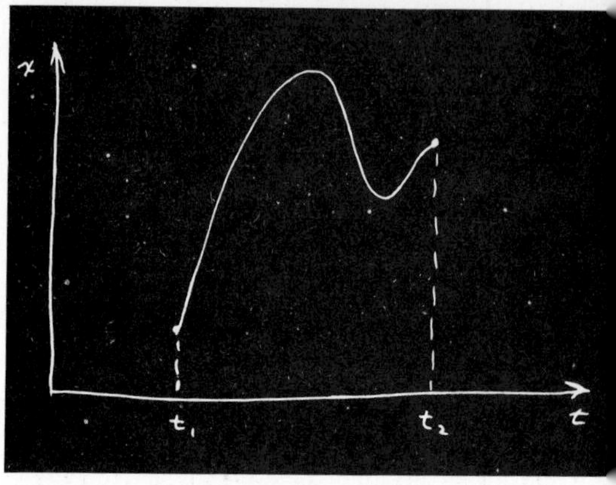

We can calculate the kinetic energy minus the potential energy and integrate for such a path . . . or for any other path we want. The miracle is that the true path is the one for which that integral is least.

"Let's try it out. First, suppose we take the case of a free particle for which there is no potential energy at all. Then the rule says that in going from one point to another in a given amount of time, the kinetic energy integral is least, so it must go at a uniform speed. (We know that's the right answer—to go at a uniform speed.) Why is that? Because if the particle were to go any other way, the velocities would be sometimes higher and sometimes lower than the average. The average velocity is the same for every case because it has to get from 'here' to 'there' in a given amount of time.

"As an example, say your job is to start from home and get to school in a given length of time with the car. You can do it several ways: You can accelerate like mad at the beginning and slow down with the brakes near the end, or you can go at a uniform speed, or you can go backwards for a while and then go forward, and so on. The thing is that the average speed has got to be, of course, the total distance that you have gone over the time. But if you do anything but go at a uniform speed, then sometimes you are going too fast and sometimes you are going too slow. Now the mean *square* of something that deviates around an average, as you know, is always greater than the square of the mean; so the kinetic energy integral would always be higher if you wobbled your velocity than if you went at a uniform velocity. So we see that the integral is a minimum if the velocity is a constant (when there are no forces). The correct path is like this.

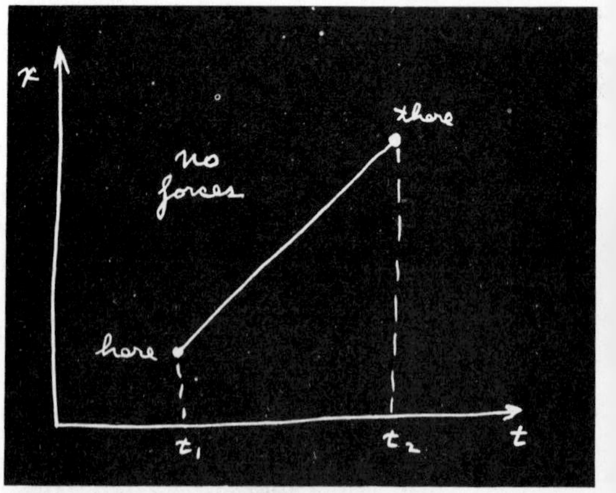

"Now, an object thrown up in a gravitational field does rise faster first and then slow down. That is because there is also the potential energy, and we must have the least *difference* of kinetic and potential energy on the average. Because the potential energy rises as we go up in space, we will get a lower *difference* if we can get as soon as possible up to where there is a high potential energy. Then we can take that potential away from the kinetic energy and get a lower average. So it is better to take a path which goes up and gets a lot of negative stuff from the potential energy.

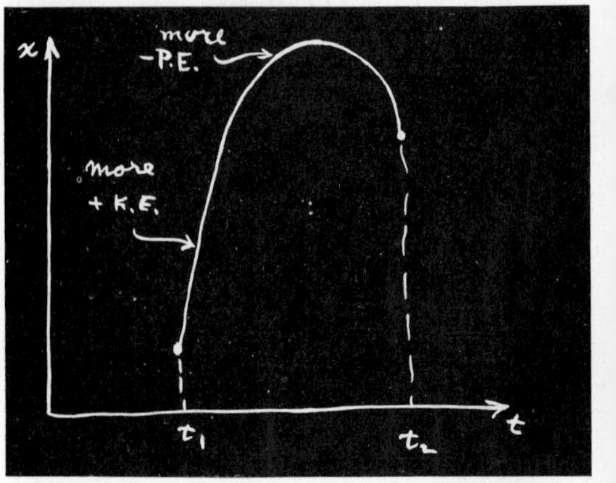

"On the other hand, you can't go up too fast, or too far, because you will then have too much kinetic energy involved—you have to go very fast to get way up and come down again in the fixed amount of time available. So you don't want to go too far up, but you want to go up some. So it turns out that the solution is some kind of balance between trying to get more potential energy with the least amount of extra kinetic energy—trying to get the difference, kinetic minus the potential, as small as possible.

"That is all my teacher told me, because he was a very good teacher and knew when to stop talking. But I don't know when to stop talking. So instead of leaving it as an interesting remark, I am going to horrify and disgust you with the complexities of life by proving that it is so. The kind of mathematical problem we will have is very difficult and a new kind. We have a certain quantity which is called the *action*, S. It is the kinetic energy, minus the potential energy, integrated over time.

$$\text{Action} = S = \int_{t_1}^{t_2} (\text{KE} - \text{PE})\, dt.$$

Remember that the PE and KE are both functions of time. For each different possible path you get a different number for this action. Our mathematical problem is to find out for what curve that number is the least.

"You say—Oh, that's just the ordinary calculus of maxima and minima. You calculate the action and just differentiate to find the minimum.

"But watch out. Ordinarily we just have a function of some variable, and we have to find the value of that *variable* where the function is least or most. For instance, we have a rod which has been heated in the middle and the heat is spread around. For each point on the rod we have a temperature, and we must find the point at which that temperature is largest. But now for *each path in space* we have a number—quite a different thing—and we have to find the *path in space* for which the number is the minimum. That is a completely different branch of mathematics. It is not the ordinary calculus. In fact, it is called the *calculus of variations*.

"There are many problems in this kind of mathematics. For example, the circle is usually defined as the locus of all points at a constant distance from a fixed point, but another way of defining a circle is this: a circle is that curve *of given length* which encloses the biggest area. Any other curve encloses less area for a given perimeter than the circle does. So if we give the problem: find that curve which encloses the greatest area for a given perimeter, we would have a problem of the calculus of variations—a different kind of calculus than you're used to.

"So we make the calculation for the path of an object. Here is the way we are going to do it. The idea is that we imagine that there is a true path and that any other curve we draw is a false path, so that if we calculate the action for the false path we will get a value that is bigger than if we calculate the action for the true path.

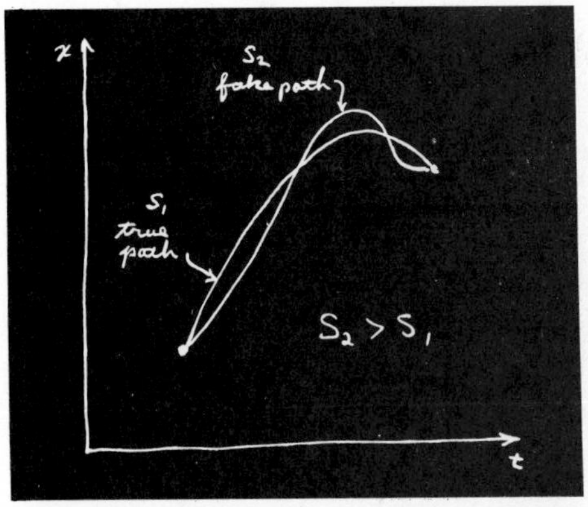

"Problem: Find the true path. Where is it? One way, of course, is to calculate the action for millions and millions of paths and look at which one is lowest. When you find the lowest one, that's the true path.

"That's a possible way. But we can do it better than that. When we have a quantity which has a minimum—for instance, in an ordinary function like the temperature—one of the properties of the minimum is that if we go away from the minimum in the *first* order, the deviation of the function from its minimum value is only *second* order. At any place else on the curve, if we move a small distance the value of the function changes also in the first order. But at a minimum, a tiny motion away makes, in the first approximation, no difference.

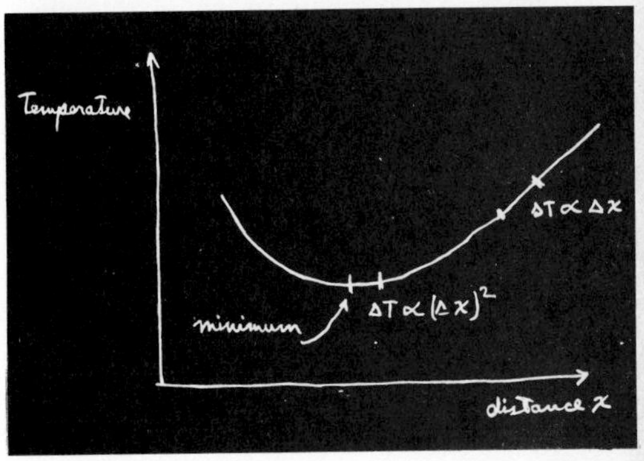

"That is what we are going to use to calculate the true path. If we have the true path, a curve which differs only a little bit from it will, in the first approximation, make no difference in the action. Any difference will be in the second approximation, if we really have a minimum.

"That is easy to prove. If there is a change in the first order when I deviate the curve a certain way, there is a change in the action that is *proportional* to the deviation. The change presumably makes the action greater; otherwise we haven't got a minimum. But then if the change is *proportional* to the deviation, reversing the sign of the deviation will make the action less. We would get the action to increase one way and to decrease the other way. The only way that it could really be a minimum is that in the *first* approximation it doesn't make any change, that the changes are proportional to the square of the deviations from the true path.

"So we work it this way: We call $\underline{x(t)}$ (with an underline) the true path—the one we are trying to find. We take some trial path $x(t)$ that differs from the true path by a small amount which we will call $\eta(t)$ (eta of t).

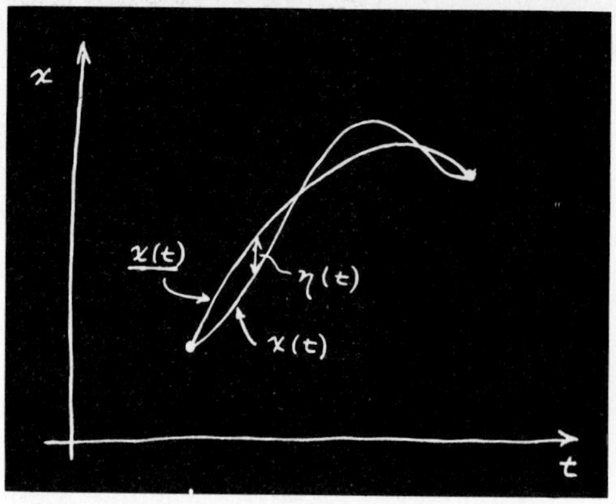

"Now the idea is that if we calculate the action S for the path $x(t)$, then the difference between that S and the action that we calculated for the path $\underline{x(t)}$—to simplify the writing we can call it $\underline{S}$—the difference of $\underline{S}$ and S must be zero in the first-order approximation of small η. It can differ in the second order, but in the first order the difference must be zero.

"And that must be true for any η at all. Well, not quite. The method doesn't mean anything unless you consider paths which all begin and end at the same two points—each path begins at a certain point at t_1 and ends at a certain other point at t_2, and those points and times are kept fixed. So the deviations in our η have to be zero at each end, $\eta(t_1) = 0$ and $\eta(t_2) = 0$. With that condition, we have specified our mathematical problem.

"If you didn't know any calculus, you might do the same kind of thing to find the minimum of an ordinary function $f(x)$. You could discuss what happens if you take $f(x)$ and add a small amount h to x and argue that the correction to $f(x)$ in the first order in h must be zero at the minimum. You would substitute $x + h$ for x and expand out to the first order in h . . . just as we are going to do with η.

"The idea is then that we substitute $x(t) = \underline{x(t)} + \eta(t)$ in the formula for the action:

$$S = \int \left[\frac{m}{2}\left(\frac{dx}{dt}\right)^2 - V(x)\right] dt,$$

where I call the potential energy $V(x)$. The derivative dx/dt is, of course, the derivative of $\underline{x(t)}$ plus the derivative of $\eta(t)$, so for the action I get this expression:

$$S = \int_{t_1}^{t_2} \left[\frac{m}{2}\left(\frac{d\underline{x}}{dt} + \frac{d\eta}{dt}\right)^2 - V(\underline{x} + \eta)\right] dt.$$

"Now I must write this out in more detail. For the squared term I get

$$\left(\frac{d\underline{x}}{dt}\right)^2 + 2\,\frac{d\underline{x}}{dt}\,\frac{d\eta}{dt} + \left(\frac{d\eta}{dt}\right)^2.$$

But wait. I'm not worrying about higher than the first order, so I will take all the terms which involve η^2 and higher powers and put them in a little box called 'second and higher order.' From this term I get only second order, but there will be more from something else. So the kinetic energy part is

$$\frac{m}{2}\left(\frac{d\underline{x}}{dt}\right)^2 + m\,\frac{d\underline{x}}{dt}\,\frac{d\eta}{dt} + \text{(second and higher order)}.$$

"Now we need the potential V at $\underline{x} + \eta$. I consider η small, so I can write $V(x)$ as a Taylor series. It is approximately $V(\underline{x})$; in the next approximation (from the ordinary nature of derivatives) the correction is η times the rate of change of V with respect to x, and so on:

$$V(\underline{x} + \eta) = V(\underline{x}) + \eta V'(\underline{x}) + \frac{\eta^2}{2}\,V''(\underline{x}) + \cdots$$

I have written V' for the derivative of V with respect to x in order to save writing. The term in η^2 and the ones beyond fall into the 'second and higher order' category and we don't have to worry about them. Putting it all together,

$$S = \int_{t_1}^{t_2} \left[\frac{m}{2}\left(\frac{d\underline{x}}{dt}\right)^2 - V(\underline{x}) + m\,\frac{d\underline{x}}{dt}\,\frac{d\eta}{dt} - \eta V'(\underline{x}) + \text{(second and higher order)}\right] dt\,.$$

Now if we look carefully at the thing, we see that the first two terms which I have arranged here correspond to the action $\underline{S}$ that I would have calculated with the true path $\underline{x}$. The thing I want to concentrate on is the change in S—the difference between the S and the $\underline{S}$ that we would get for the right path. This difference we will write as δS, called the variation in S. Leaving out the 'second and higher order' terms, I have for δS

$$\delta S = \int_{t_1}^{t_2} \left[m \frac{d\underline{x}}{dt} \frac{d\eta}{dt} - \eta V'(\underline{x}) \right] dt.$$

"Now the problem is this: Here is a certain integral. I don't know what the $\underline{x}$ is yet, but I do know that *no matter what* η is, this integral must be zero. Well, you think, the only way that that can happen is that what multiplies η must be zero. But what about the first term with $d\eta/dt$? Well, after all, if η can be anything at all, its derivative is anything also, so you conclude that the coefficient of $d\eta/dt$ must also be zero. That isn't quite right. It isn't quite right because there is a connection between η and its derivative; they are not absolutely independent, because $\eta(t)$ must be zero at both t_1 and t_2.

"The method of solving all problems in the calculus of variations always uses the same general principle. You make the shift in the thing you want to vary (as we did by adding η); you look at the first-order terms; *then* you always arrange things in such a form that you get an integral of the form 'some kind of stuff times the shift (η),' but with no other derivatives (no $d\eta/dt$). It must be rearranged so it is always 'something' times η. You will see the great value of that in a minute. (There are formulas that tell you how to do this in some cases without actually calculating, but they are not general enough to be worth bothering about; the best way is to calculate it out this way.)

"How can I rearrange the term in $d\eta/dt$ to make it have an η? I can do that by integrating by parts. It turns out that the whole trick of the calculus of variations consists of writing down the variation of S and then integrating by parts so that the derivatives of η disappear. It is always the same in every problem in which derivatives appear.

"You remember the general principle for integrating by parts. If you have any function f times $d\eta/dt$ integrated with respect to t, you write down the derivative of ηf:

$$\frac{d}{dt}(\eta f) = \eta \frac{df}{dt} + f \frac{d\eta}{dt}.$$

The integral you want is over the last term, so

$$\int f \frac{d\eta}{dt}\, dt = \eta f - \int \eta \frac{df}{dt}\, dt.$$

"In our formula for δS, the function f is m times $d\underline{x}/dt$; therefore, I have the following formula for δS.

$$\delta S = m \frac{d\underline{x}}{dt} \eta(t) \bigg|_{t_1}^{t_2} - \int_{t_1}^{t_2} \frac{d}{dt}\left(m \frac{d\underline{x}}{dt} \right) \eta(t)\, dt - \int_{t_1}^{t_2} V'(\underline{x})\, \eta(t)\, dt.$$

The first term must be evaluated at the two limits t_1 and t_2. Then I must have the integral from the rest of the integration by parts. The last term is brought down without change.

"Now comes something which always happens—the integrated part disappears. (In fact, if the integrated part does not disappear, you restate the principle, adding conditions to make sure it does!) We have already said that η must be zero at both ends of the path, because the principle is that the action is a minimum provided that the varied curve begins and ends at the chosen points. The condition is that

$\eta(t_1) = 0$, and $\eta(t_2) = 0$. So the integrated term is zero. We collect the other terms together and obtain this:

$$\delta S = \int_{t_1}^{t_2} \left[-m \frac{d^2\underline{x}}{dt^2} - V'(\underline{x}) \right] \eta(t)\, dt.$$

The variation in S is now the way we wanted it—there is the stuff in brackets, say F, all multiplied by $\eta(t)$ and integrated from t_1 to t_2.

"We have that an integral of something or other times $\eta(t)$ is always zero:

$$\int F(t)\, \eta(t)\, dt = 0.$$

I have some function of t; I multiply it by $\eta(t)$; and I integrate it from one end to the other. And no matter what the η is, I get zero. That means that the function $F(t)$ is zero. That's obvious, but anyway I'll show you one kind of proof.

"Suppose that for $\eta(t)$ I took something which was zero for all t except right near one particular value. It stays zero until it gets to this t, →

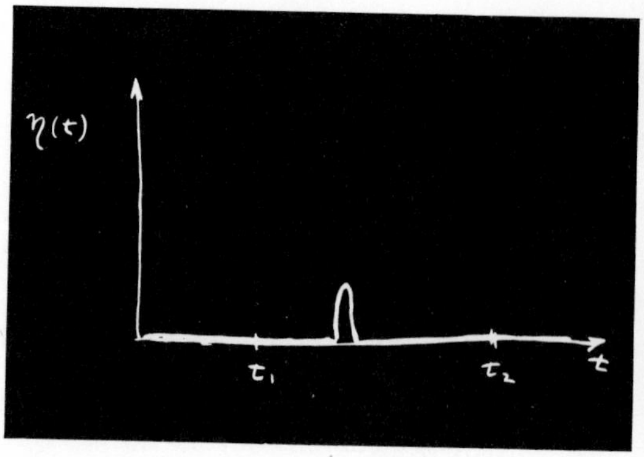

then it blips up for a moment and blips right back down. When we do the integral of this η times any function F, the only place that you get anything other than zero was where $\eta(t)$ was blipping, and then you get the value of F at that place times the integral over the blip. The integral over the blip alone isn't zero, but when multiplied by F it has to be; so the function F has to be zero where the blip was. But the blip was anywhere I wanted to put it, so F must be zero everywhere.

"We see that if our integral is zero for any η, then the coefficient of η must be zero. The action integral will be a minimum for the path that satisfies this complicated differential equation:

$$\left[-m \frac{d^2\underline{x}}{dt^2} - V'(\underline{x}) \right] = 0.$$

It's not really so complicated; you have seen it before. It is just $F = ma$. The first term is the mass times acceleration, and the second is the derivative of the potential energy, which is the force.

"So, for a conservative system at least, we have demonstrated that the principle of least action gives the right answer; it says that the path that has the minimum action is the one satisfying Newton's law.

"One remark: I did not prove it was a *minimum*—maybe it's a maximum. In fact, it doesn't really have to be a minimum. It is quite analogous to what we found for the 'principle of least time' which we discussed in optics. There also, we said at first it was 'least' time. It turned out, however, that there were situations in which it wasn't the *least* time. The fundamental principle was that for any *first-order variation* away from the optical path, the *change* in time was zero; it is the same story. What we really mean by 'least' is that the first-order change in the value of S, when you change the path, is zero. It is not necessarily a 'minimum.'

"Next, I remark on some generalizations. In the first place, the thing can be done in three dimensions. Instead of just x, I would have x, y, and z as functions of t; the action is more complicated. For three-dimensional motion, you have to use the complete kinetic energy—$(m/2)$ times the whole velocity squared. That is,

$$\text{KE} = \frac{m}{2} \left[\left(\frac{dx}{dt}\right)^2 + \left(\frac{dy}{dt}\right)^2 + \left(\frac{dz}{dt}\right)^2 \right].$$

Also, the potential energy is a function of x, y, and z. And what about the path? The path is some general curve in space, which is not so easily drawn, but the idea is the same. And what about the η? Well, η can have three components. You could shift the paths in x, or in y, or in z—or you could shift in all three directions simultaneously. So η would be a vector. This doesn't really complicate things too much, though. Since only the *first-order* variation has to be zero, we can do the calculation by three successive shifts. We can shift η only in the x-direction and say that coefficient must be zero. We get one equation. Then we shift it in the y-direction and get another. And in the z-direction and get another. Or, of course, in any order that you want. Anyway, you get three equations. And, of course, Newton's law is really three equations in the three dimensions—one for each component. I think that you can practically see that it is bound to work, but we will leave you to show for yourself that it will work for three dimensions. Incidentally, you could use any coordinate system you want, polar or otherwise, and get Newton's laws appropriate to that system right off by seeing what happens if you have the shift η in radius, or in angle, etc.

"Similarly, the method can be generalized to any number of particles. If you have, say, two particles with a force between them, so that there is a mutual potential energy, then you just add the kinetic energy of both particles and take the potential energy of the mutual interaction. And what do you vary? You vary the paths of *both* particles. Then, for two particles moving in three dimensions, there are six equations. You can vary the position of particle 1 in the x-direction, in the y-direction, and in the z-direction, and similarly for particle 2; so there are six equations. And that's as it should be. There are the three equations that determine the acceleration of particle 1 in terms of the force on it and three for the acceleration of particle 2, from the force on it. You follow the same game through, and you get Newton's law in three dimensions for any number of particles.

"I have been saying that we get Newton's law. That is not quite true, because Newton's law includes nonconservative forces like friction. Newton said that ma is equal to any F. But the principle of least action only works for *conservative* systems—where all forces can be gotten from a potential function. You know, however, that on a microscopic level—on the deepest level of physics—there are no nonconservative forces. Nonconservative forces, like friction, appear only because we neglect microscopic complications—there are just too many particles to analyze. But the *fundamental* laws *can* be put in the form of a principle of least action.

"Let me generalize still further. Suppose we ask what happens if the particle moves relativistically. We did not get the right relativistic equation of motion; $F = ma$ is only right nonrelativistically. The question is: Is there a corresponding principle of least action for the relativistic case? There is. The formula in the case of relativity is the following:

$$S = -m_0c^2 \int_{t_1}^{t_2} \sqrt{1 - v^2/c^2}\, dt - q \int_{t_1}^{t_2} [\phi(x, y, z, t) - \boldsymbol{v} \cdot \boldsymbol{A}(x, y, z, t)]\, dt.$$

The first part of the action integral is the rest mass m_0 times c^2 times the integral of a function of velocity, $\sqrt{1 - v^2/c^2}$. Then instead of just the potential energy, we have an integral over the scalar potential ϕ and over $\boldsymbol{v}$ times the vector potential $\boldsymbol{A}$. Of course, we are then including only electromagnetic forces. All electric and magnetic fields are given in terms of ϕ and $\boldsymbol{A}$. This action function gives the complete theory of relativistic motion of a single particle in an electromagnetic field.

"Of course, wherever I have written $\boldsymbol{v}$, you understand that before you try to figure anything out, you must substitute dx/dt for v_x and so on for the other components. Also, you put the point along the path at time t, $x(t)$, $y(t)$, $z(t)$ where I wrote simply x, y, z. Properly, it is only after you have made those replacements for the $\boldsymbol{v}$'s that you have the formula for the action for a relativistic particle. I will leave to the more ingenious of you the problem to demonstrate that this action formula does, in fact, give the correct equations of motion for relativity. May I suggest you do it first without the $\boldsymbol{A}$, that is, for no magnetic field? Then you should get the components of the equation of motion, $d\boldsymbol{p}/dt = -q\, \nabla\phi$, where, you remember, $\boldsymbol{p} = m\boldsymbol{v}/\sqrt{1 - v^2/c^2}$.

"It is much more difficult to include also the case with a vector potential. The variations get much more complicated. But in the end, the force term does come out equal to $q(\boldsymbol{E} + \boldsymbol{v} \times \boldsymbol{B})$, as it should. But I will leave that for you to play with.

"I would like to emphasize that in the general case, for instance in the relativistic formula, the action integrand no longer has the form of the kinetic energy minus the potential energy. That's only true in the nonrelativistic approximation. For example, the term $m_0c^2\sqrt{1 - v^2/c^2}$ is not what we have called the kinetic energy. The question of what the action should be for any particular case must be determined by some kind of trial and error. It is just the same problem as determining what are the laws of motion in the first place. You just have to fiddle around with the equations that you know and see if you can get them into the form of the principle of least action.

"One other point on terminology. The function that is integrated over time to get the action S is called the *Lagrangian*, $\mathcal{L}$, which is a function only of the velocities and positions of particles. So the principle of least action is also written

$$S = \int_{t_1}^{t_2} \mathcal{L}(x_i, v_i)\,dt,$$

where by x_i and v_i are meant all the components of the positions and velocities. So if you hear someone talking about the 'Lagrangian,' you know they are talking about the function that is used to find S. For relativistic motion in an electromagnetic field

$$\mathcal{L} = -m_0c^2\sqrt{1 - v^2/c^2} - q(\phi + \boldsymbol{v}\cdot\boldsymbol{A}).$$

"Also, I should say that S is not really called the 'action' by the most precise and pedantic people. It is called 'Hamilton's first principal function.' Now I hate to give a lecture on 'the-principle-of-least-Hamilton's-first-principal-function.' So I call it 'the action.' Also, more and more people are calling it the action. You see, historically something else which is not quite as useful was called the action, but I think it's more sensible to change to a newer definition. So now you too will call the new function the action, and pretty soon everybody will call it by that simple name.

"Now I want to say some things on this subject which are similar to the discussions I gave about the principle of least time. There is quite a difference in the characteristic of a law which says a certain integral from one place to another is a minimum—which tells something about the whole path—and of a law which says that as you go along, there is a force that makes it accelerate. The second way tells how you inch your way along the path, and the other is a grand statement about the whole path. In the case of light, we talked about the connection of these two. Now, I would like to explain why it is true that there are differential laws when there is a least action principle of this kind. The reason is the following: Consider the actual path in space and time. As before, let's take only one dimension, so we can plot the graph of x as a function of t. Along the true path, S is a minimum. Let's suppose that we have the true path and that it goes through some point a in space and time, and also through another nearby point b.

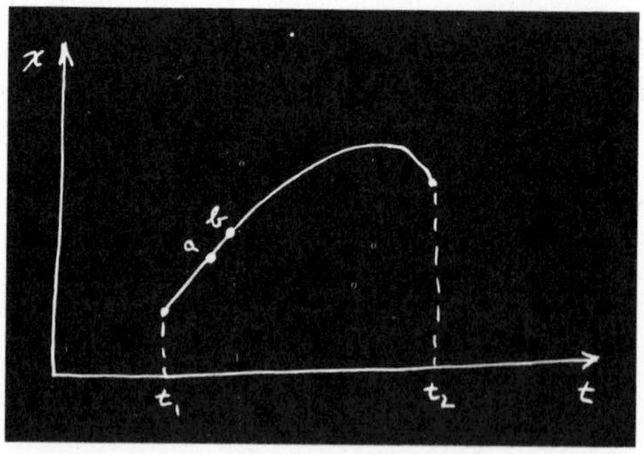

Now if the entire integral from t_1 to t_2 is a minimum, it is also necessary that the integral along the little section from a to b is also a minimum. It can't be that the part from a to b is a little bit more. Otherwise you could just fiddle with just that piece of the path and make the whole integral a little lower.

"So every subsection of the path must also be a minimum. And this is true no matter how short the subsection. Therefore, the principle that the whole path gives a minimum can be stated also by saying that an infinitesimal section of path also has a curve such that it has a minimum action. Now if we take a short enough section of path—between two points a and b very close together—how the potential varies from one place to another far away is not the important thing, because you are staying almost in the same place over the whole little piece of the path. The only thing that you have to discuss is the first-order change in the potential. The answer can only depend on the derivative of the potential and not on the potential everywhere. So the statement about the gross property of the whole path becomes a statement of what happens for a short section of the path—a differential statement. And this differential statement only involves the derivatives of the potential, that is, the force at a point. That's the qualitative explanation of the relation between the gross law and the differential law.

"In the case of light we also discussed the question: How does the particle find the right path? From the differential point of view, it is easy to understand. Every moment it gets an acceleration and knows only what to do at that instant. But all your instincts on cause and effect go haywire when you say that the particle decides to take the path that is going to give the minimum action. Does it 'smell' the neighboring paths to find out whether or not they have more action? In the case of light, when we put blocks in the way so that the photons could not test all the paths, we found that they couldn't figure out which way to go, and we had the phenomenon of diffraction.

"Is the same thing true in mechanics? Is it true that the particle doesn't just 'take the right path' but that it looks at all the other possible trajectories? And if by having things in the way, we don't let it look, that we will get an analog of diffraction? The miracle of it all is, of course, that it does just that. That's what the laws of quantum mechanics say. So our principle of least action is incompletely stated. It isn't that a particle takes the path of least action but that it smells all the paths in the neighborhood and chooses the one that has the least action by a method analogous to the one by which light chose the shortest time. You remember that the way light chose the shortest time was this: If it went on a path that took a different amount of time, it would arrive at a different phase. And the total amplitude at some point is the sum of contributions of amplitude for all the different ways the light can arrive. All the paths that give wildly different phases don't add up to anything. But if you can find a whole sequence of paths which have phases almost all the same, then the little contributions will add up and you get a reasonable total amplitude to arrive. The important path becomes the one for which there are many nearby paths which give the same phase.

"It is just exactly the same thing for quantum mechanics. The complete quantum mechanics (for the nonrelativistic case and neglecting electron spin) works as follows: The probability that a particle starting at point 1 at the time t_1 will arrive at point 2 at the time t_2 is the square of a probability amplitude. The total amplitude can be written as the sum of the amplitudes for each possible path—for each way of arrival. For every $x(t)$ that we could have—for every possible imaginary trajectory—we have to calculate an amplitude. Then we add them all together. What do we take for the amplitude for each path? Our action integral tells us what the amplitude for a single path ought to be. The amplitude is proportional to some constant times $e^{iS/\hbar}$, where S is the action for that path. That is, if we represent the phase of the amplitude by a complex number, the phase angle is $S/\hbar$. The action S has dimensions of energy times time, and Planck's constant $\hbar$ has the same dimensions. It is the constant that determines when quantum mechanics is important.

"Here is how it works: Suppose that for all paths, S is very large compared to $\hbar$. One path contributes a certain amplitude. For a nearby path, the phase is quite different, because with an enormous S even a small change in S means a completely different phase—because $\hbar$ is so tiny. So nearby paths will normally cancel their effects out in taking the sum—except for one region, and that is when a path and a nearby path all give the same phase in the first approximation (more precisely, the same action within $\hbar$). Only those paths will be the important ones. So in the limiting case in which Planck's constant $\hbar$ goes to zero, the correct quantum-mechanical laws can be summarized by simply saying: 'Forget about all these probability amplitudes. The particle does go on a special path, namely, that one for which S does not vary in the first approximation.' That's the relation between the principle of least action and quantum mechanics. The fact that quantum mechanics can be formulated in this way was discovered in 1942 by a student of that same teacher, Bader, I spoke of at the beginning of this lecture. [Quantum mechanics was originally formulated by giving a differential equation for the amplitude (Schrödinger) and also by some other matrix mathematics (Heisenberg).]

"Now I want to talk about other minimum principles in physics. There are many very interesting ones. I will not try to list them all now but will only describe one more. Later on, when we come to a physical phenomenon which has a nice minimum principle, I will tell about it then. I want now to show that we can describe electrostatics, not by giving a differential equation for the field, but by saying that a certain integral is a maximum or a minimum. First, let's take the case where the charge density is known everywhere, and the problem is to find the potential ϕ everywhere in space. You know that the answer should be

$$\nabla^2\phi = -\rho/\epsilon_0.$$

But another way of stating the same thing is this: Calculate the integral U^*, where

$$U^* = \frac{\epsilon_0}{2}\int (\nabla\phi)^2\,dV - \int \rho\phi\,dV,$$

which is a volume integral to be taken over all space. This thing is a minimum for the correct potential distribution $\phi(x, y, z)$.

"We can show that the two statements about electrostatics are equivalent. Let's suppose that we pick any function ϕ. We want to show that when we take for ϕ the correct potential $\underline{\phi}$, plus a small deviation f, then in the first order, the change in U^* is zero. So we write

$$\phi = \underline{\phi} + f.$$

The $\underline{\phi}$ is what we are looking for, but we are making a variation of it to find what it has to be so that the variation of U^* is zero to first order. For the first part of U^*, we need

$$(\nabla\phi)^2 = (\nabla\underline{\phi})^2 + 2\,\nabla\underline{\phi}\cdot\nabla f + (\nabla f)^2.$$

The only first-order term that will vary is

$$2\,\nabla\underline{\phi}\cdot\nabla f.$$

In the second term of the quantity U^*, the integrand is

$$\rho\phi = \rho\underline{\phi} + \rho f,$$

whose variable part is ρf. So, keeping only the variable parts, we need the integral

$$\Delta U^* = \int (\epsilon_0\nabla\underline{\phi}\cdot\nabla f - \rho f)\,dV.$$

"Now, following the old general rule, we have to get the darn thing all clear of derivatives of f. Let's look at what the derivatives are. The dot product is

$$\frac{\partial\underline{\phi}}{\partial x}\frac{\partial f}{\partial x} + \frac{\partial\underline{\phi}}{\partial y}\frac{\partial f}{\partial y} + \frac{\partial\underline{\phi}}{\partial z}\frac{\partial f}{\partial z},$$

which we have to integrate with respect to x, to y, and to z. Now here is the trick: to get rid of $\partial f/\partial x$ we integrate by parts with respect to x. That will carry the derivative over onto the $\underline{\phi}$. It's the same general idea we used to get rid of derivatives with respect to t. We use the equality

$$\int \frac{\partial\underline{\phi}}{\partial x}\frac{\partial f}{\partial x}\,dx = f\frac{\partial\underline{\phi}}{\partial x} - \int f\frac{\partial^2\underline{\phi}}{\partial x^2}\,dx.$$

The integrated term is zero, since we have to make f zero at infinity. (That corresponds to making η zero at t_1 and t_2. So our principle should be more accurately stated: U^* is less for the true ϕ than for any other $\phi(x, y, z)$ having the same values at infinity.) Then we do the same thing for y and z. So our integral ΔU^* is

$$\Delta U^* = \int (-\epsilon_0\nabla^2\underline{\phi} - \rho)f\,dV.$$

In order for this variation to be zero for any f, no matter what, the coefficient of f must be zero and, therefore,

$$\nabla^2 \underline{\phi} = -\rho/\epsilon_0.$$

We get back our old equation. So our 'minimum' proposition is correct.

"We can generalize our proposition if we do our algebra in a little different way. Let's go back and do our integration by parts without taking components. We start by looking at the following equality:

$$\nabla \cdot (f\,\nabla\underline{\phi}) = \nabla f \cdot \nabla\underline{\phi} + f\,\nabla^2\underline{\phi}.$$

If I differentiate out the left-hand side, I can show that it is just equal to the right-hand side. Now we can use this equation to integrate by parts. In our integral ΔU^*, we replace $-\nabla\underline{\phi}\cdot\nabla f$ by $f\,\nabla^2\underline{\phi} - \nabla\cdot(f\,\nabla\underline{\phi})$, which gets integrated over volume. The divergence term integrated over volume can be replaced by a surface integral:

$$\int \nabla\cdot(f\,\nabla\underline{\phi})\,dV = \int f\,\nabla\underline{\phi}\cdot\boldsymbol{n}\,da.$$

Since we are integrating over all space, the surface over which we are integrating is at infinity. There, f is zero and we get the same answer as before.

"Only now we see how to solve a problem when we *don't* know where all the charges are. Suppose that we have conductors with charges spread out on them in some way. We can still use our minimum principle if the potentials of all the conductors are fixed. We carry out the integral for U^* only in the space outside of all conductors. Then, since we can't vary $\underline{\phi}$ on the conductor, f is zero on all those surfaces, and the surface integral

$$\int f\,\nabla\underline{\phi}\cdot\boldsymbol{n}\,da$$

is still zero. The remaining volume integral

$$\Delta U^* = \int (-\epsilon_0\,\nabla^2\underline{\phi} - \rho\underline{\phi})f\,dV$$

is only to be carried out in the spaces between conductors. Of course, we get Poisson's equation again,

$$\nabla^2\underline{\phi} = -\rho/\epsilon_0.$$

So we have shown that our original integral U^* is also a minimum if we evaluate it over the space outside of conductors all at fixed potentials (that is, such that any trial $\phi(x, y, z)$ must equal the given potential of the conductors when x, y, z is a point on the surface of a conductor).

"There is an interesting case when the only charges are on conductors. Then

$$U^* = \frac{\epsilon_0}{2}\int (\nabla\phi)^2\,dV.$$

Our minimum principle says that in the case where there are conductors set at certain given potentials, the potential between them adjusts itself so that integral U^* is least. What is this integral? The term $\nabla\phi$ is the electric field, so the integral is the electrostatic energy. The true field is the one, of all those coming from the gradient of a potential, with the minimum total energy.

"I would like to use this result to calculate something particular to show you that these things are really quite practical. Suppose I take two conductors in the form of a cylindrical condenser.

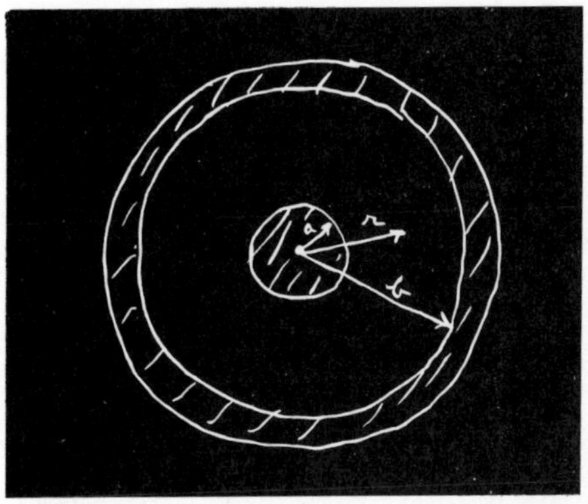

The inside conductor has the potential V, and the outside is at the potential zero. Let the radius of the inside conductor be a and that of the outside, b. Now we can suppose *any* distribution of potential between the two. If we use the *correct* $\underline{\phi}$, and calculate $\epsilon_0/2\int(\nabla\underline{\phi})^2\,dV$, it should be the energy of the system, $\frac{1}{2}CV^2$.

So we can also calculate C by our principle. But if we use a wrong distribution of potential and try to calculate the capacity C by this method, we will get a capacity that is too big, since V is specified. Any assumed potential ϕ that is not the exactly correct one will give a fake C that is larger than the correct value. But if my false ϕ is any rough approximation, the C will be a good approximation, because the error in C is second order in the error in ϕ.

"Suppose I don't know the capacity of a cylindrical condenser. I can use this principle to find it. I just guess at the potential function ϕ until I get the lowest C. Suppose, for instance, I pick a potential that corresponds to a constant field. (You know, of course, that the field isn't really constant here; it varies as $1/r$.) A field which is constant means a potential which goes linearly with distance. To fit the conditions at the two conductors, it must be

$$\phi = V\left(1 - \frac{r - a}{b - a}\right).$$

This function is V at $r = a$, zero at $r = b$, and in between has a constant slope equal to $-V/(b - a)$. So what one does to find the integral U^* is multiply the square of this gradient by $\epsilon_0/2$ and integrate over all volume. Let's do this calculation for a cylinder of unit length. A volume element at the radius r is $2\pi r\,dr$. Doing the integral, I find that my first try at the capacity gives

$$\frac{1}{2}\,CV^2(\text{first try}) = \frac{\epsilon_0}{2}\int_a^b \frac{V^2}{(b - a)^2}\,2\pi r\,dr.$$

The integral is easy; it is just

$$\pi V^2\left(\frac{b + a}{b - a}\right).$$

So I have a formula for the capacity which is not the true one but is an approximate job:

$$\frac{C}{2\pi\epsilon_0} = \frac{b + a}{2(b - a)}.$$

It is, naturally, different from the correct answer $C = 2\pi\epsilon_0/\ln(b/a)$, but it's not too bad. Let's compare it with the right answer for several values of b/a. I have computed out the answers in this table:

$\frac{b}{a}$	$\frac{C_{\text{true}}}{2\pi\epsilon_0}$	$\frac{C\text{ (first approx.)}}{2\pi\epsilon_0}$
2	1.4423	1.500
4	0.721	0.833
10	0.434	0.612
100	0.267	0.51
1.5	2.4662	2.50
1.1	10.492070	10.500000

Even when b/a is as big as 2—which gives a pretty big variation in the field compared with a linearly varying field—I get a pretty fair approximation. The answer is, of course, a little too high, as expected. The thing gets much worse if you have a tiny wire inside a big cylinder. Then the field has enormous variations and if you represent it by a constant, you're not doing very well. With $b/a = 100$, we're off by nearly a factor of two. Things are much better for small b/a. To take the opposite extreme, when the conductors are not very far apart—say $b/a = 1.1$—then the constant field is a pretty good approximation, and we get the correct value for C to within a tenth of a percent.

"Now I would like to tell you how to improve such a calculation. (Of course, you *know* the right answer for the cylinder, but the method is the same for some other odd shapes, where you may not know the right answer.) The next step is to try a better approximation to the unknown true ϕ. For example, we might try a constant plus an exponential ϕ, etc. But how do you know when you have a better approximation unless you know the true ϕ? Answer: You calculate C; the lowest C is the value nearest the truth. Let us try this idea out. Suppose that the potential is not linear but say quadratic in r—that the electric field is not constant but linear. The most *general* quadratic form that fits $\phi = 0$ at $r = b$ and $\phi = V$ at $r = a$ is

$$\phi = V\left[1 + \alpha\left(\frac{r - a}{b - a}\right) - (1 + \alpha)\left(\frac{r - a}{b - a}\right)^2\right],$$

where α is any constant number. This formula is a little more complicated. It involves a quadratic term in the potential as well as a linear term. It is very easy to get the field out of it. The field is just

$$E = -\frac{d\phi}{dr} = -\frac{\alpha V}{b - a} + 2(1 + \alpha)\frac{(r - a)V}{(b - a)^2}.$$

Now we have to square this and integrate over volume. But wait a moment. What should I take for α? I can take a parabola for the ϕ; but what parabola? Here's what I do: Calculate the capacity with *an arbitrary* α. What I get is

$$\frac{C}{2\pi\epsilon_0} = \frac{a}{b - a}\left[\frac{b}{a}\left(\frac{\alpha^2}{6} + \frac{2\alpha}{3} + 1\right) + \frac{1}{6}\alpha^2 + \frac{1}{3}\right].$$

It looks a little complicated, but it comes out of integrating the square of the field. Now I can pick my α. I know that the truth lies lower than anything that I am going to calculate, so whatever I put in for α is going to give me an answer too big. But if I keep playing with α and get the lowest possible value I can, that lowest value is nearer to the truth than any other value. So what I do next is to pick the α that gives the minimum value for C. Working it out by ordinary calculus, I get that the minimum C occurs for $\alpha = -2b/(b + a)$. Substituting that value into the formula, I obtain for the minimum capacity

$$\frac{C}{2\pi\epsilon_0} = \frac{b^2 + 4ab + a^2}{3(b^2 - a^2)}.$$

"I've worked out what this formula gives for C for various values of b/a. I call these numbers C(quadratic). Here is a table that compares C(quadratic) with the true C.

$\dfrac{b}{a}$	$\dfrac{C_{\text{true}}}{2\pi\epsilon_0}$	$\dfrac{C(\text{quadratic})}{2\pi\epsilon_0}$
2	1.4423	1.444
4	0.721	0.733
10	0.434	0.475
100	0.267	0.346
1.5	2.4662	2.4667
1.1	10.492070	10.492065

"For example, when the ratio of the radii is 2 to 1, I have 1.444, which is a very good approximation to the true answer, 1.4423. Even for larger b/a, it stays pretty good—it is much, much better than the first approximation. It is even fairly good—only off by 10 percent—when b/a is 10 to 1. But when it gets to be 100 to 1—well, things begin to go wild. I get that C is 0.346 instead of 0.267. On the other hand, for a ratio of radii of 1.5, the answer is excellent; and for a b/a of 1.1, the answer comes out 10.492065 instead of 10.492070. Where the answer should be good, it is very, very good.

"I have given these examples, first, to show the theoretical value of the principles of minimum action and minimum principles in general and, second, to show their practical utility—not just to calculate a capacity when we already know the answer. For any other shape, you can guess an approximate field with some unknown parameters like α and adjust them to get a minimum. You will get excellent numerical results for otherwise intractable problems."

A note added after the lecture

"I should like to add something that I didn't have time for in the lecture. (I always seem to prepare more than I have time to tell about.) As I mentioned earlier, I got interested in a problem while working on this lecture. I want to tell you what that problem is. Among the minimum principles that I could mention, I noticed that most of them sprang in one way or another from the least action principle of mechanics and electrodynamics. But there is also a class that does not. As an example, if currents are made to go through a piece of material obeying Ohm's law, the currents distribute themselves inside the piece so that the rate at which heat is generated is as little as possible. Also we can say (if things are kept isothermal) that the rate at which energy is generated is a minimum. Now, this principle also holds, according to classical theory, in determining even the distribution of velocities of the electrons inside a metal which is carrying a current. The distribution of velocities is not exactly the equilibrium distribution [Chapter 40, Vol. I; Eq. (40.6)] because they are drifting sideways The new distribution can be found from the principle that it is the distribution for a given current for which the entropy developed per second by collisions is as small as possible. The true description of the electrons' behavior ought to be by quantum mechanics, however. The question is: Does the same principle of minimum entropy generation also hold when the situation is described quantum-mechanically? I haven't found out yet.

"The question is interesting academically, of course. Such principles are fascinating, and it is always worth while to try to see how general they are. But also from a more practical point of view, I *want* to know. I, with some colleagues, have published a paper in which we calculated by quantum mechanics approximately the electrical resistance felt by an electron moving through an ionic crystal like NaCl. [Feynman, Hellworth, Iddings, and Platzman, "Mobility of Slow Electrons in a Polar Crystal," *Phys Rev.* **127,** 1004 (1962).] But if a minimum principle existed, we could use it to make the results much more accurate, just as the minimum principle for the capacity of a condenser permitted us to get such accuracy for that capacity even though we had only a rough knowledge of the electric field."

20

Solutions of Maxwell's Equations in Free Space

References: Chapter 47, Vol. I: *Sound: The Wave Equation*
Chapter 28, Vol. I: *Electromagnetic Radiation*

20–1 Waves in free space; plane waves

In Chapter 18 we had reached the point where we had the Maxwell equations in complete form. All there is to know about the classical theory of the electric and magnetic fields can be found in the four equations:

$$\text{I.}\quad \nabla\cdot\boldsymbol{E} = \frac{\rho}{\epsilon_0} \qquad\qquad \text{II.}\quad \nabla\times\boldsymbol{E} = -\frac{\partial\boldsymbol{B}}{\partial t}$$

$$\text{III.}\quad \nabla\cdot\boldsymbol{B} = 0 \qquad\qquad \text{IV.}\quad c^2\nabla\times\boldsymbol{B} = \frac{\boldsymbol{j}}{\epsilon_0} + \frac{\partial\boldsymbol{E}}{\partial t} \tag{20.1}$$

When we put all these equations together, a remarkable new phenomenon occurs: fields generated by moving charges can leave the sources and travel alone through space. We considered a special example in which an infinite current sheet is suddenly turned on. After the current has been on for the time t, there are uniform electric and magnetic fields extending out the distance ct from the source. Suppose that the current sheet lies in the yz-plane with a surface current density J going toward positive y. The electric field will have only a y-component, and the magnetic field, only a z-component. The magnitude of the field components is given by

$$E_y = cB_z = -\frac{J}{2\epsilon_0 c}, \tag{20.2}$$

for positive values of x less than ct. For larger x the fields are zero. There are, of course, similar fields extending the same distance from the current sheet in the negative x-direction. In Fig. 20–1 we show a graph of the magnitude of the fields as a function of x at the instant t. As time goes on, the "wavefront" at ct moves outward in x at the constant velocity c.

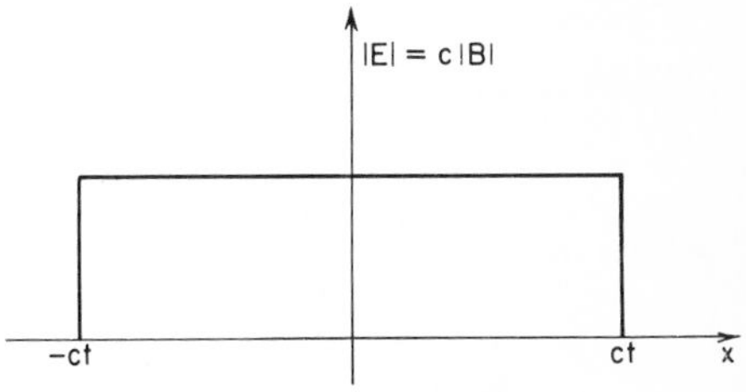

Fig. 20–1. The electric and magnetic field as a function of x at the time t after the current sheet is turned on.

Now consider the following sequence of events. We turn on a current of unit strength for a while, then suddenly increase the current strength to three units, and hold it constant at this value. What do the fields look like then? We can see what the fields will look like in the following way. First, we imagine a current of unit strength that is turned on at $t = 0$ and left constant forever. The fields for positive x are then given by the graph in part (a) of Fig. 20–2. Next, we ask what would happen if we turn on a steady current of two units at the time t_1.

The fields in this case will be twice as high as before, but will extend out in x only the distance $c(t - t_1)$, as shown in part (b) of the figure. When we add these two solutions, using the principle of superposition, we find that the sum of the two sources is a current of one unit for the time from zero to t_1 and a current of three units for times greater than t_1. At the time t the fields will vary with x as shown in part (c) of Fig. 20–2.

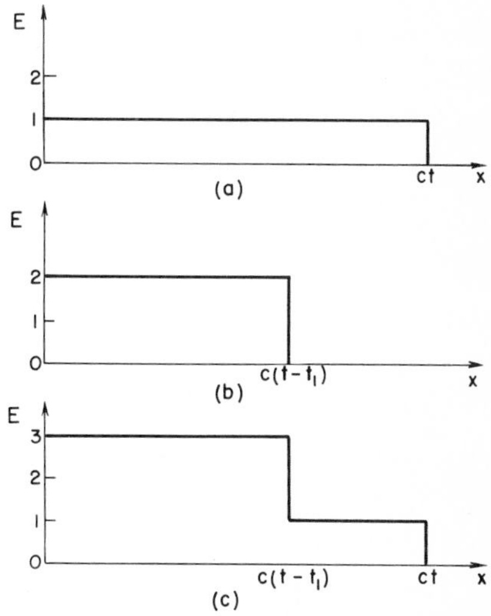

Fig. 20–2. The electric field of a current sheet. (a) One unit of current turned on at $t = 0$; (b) Two units of current turned on at $t = t_1$; (c) Superposition of (a) and (b).

Now let's take a more complicated problem. Consider a current which is turned on to one unit for a while, then turned up to three units, and later turned off to zero. What are the fields for such a current? We can find the solution in the same way—by adding the solutions of three separate problems. First, we find the fields for a step current of unit strength. (We have solved that problem already.) Next, we find the fields produced by a step current of two units. Finally, we solve for the fields of a step current of *minus* three units. When we add the three solutions, we will have a current which is one unit strong from $t = 0$ to some later time, say t_1, then three units strong until a still later time t_2, and then turned off—that

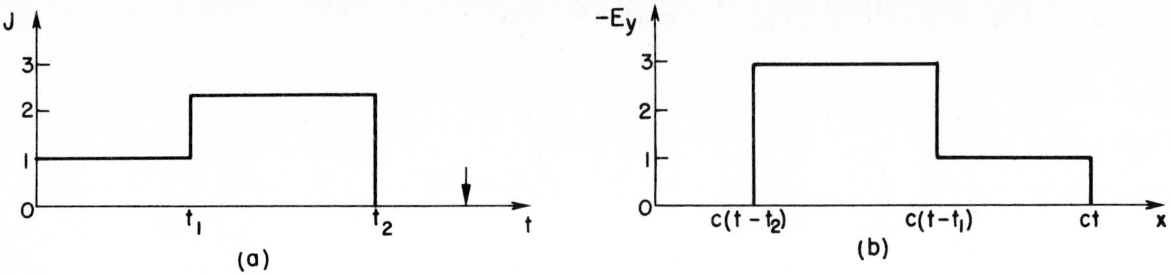

Fig. 20–3. If the current source strength varies as shown in (a), then at the time t shown by the arrow the electric field as a function of x is as shown in (b).

is, to zero. A graph of the current as a function of time is shown in Fig. 20–3(a). When we add the three solutions for the electric field, we find that its variation with x, at a given instant t, is as shown in Fig. 20–3(b). The field is an exact representation of the current. The field distribution in space is a nice graph of the current variation with time—only drawn backwards. As time goes on the whole picture moves outward at the speed c, so there is a little blob of field, travelling toward positive x, which contains a completely detailed memory of the history of all the current variations. If we were to stand miles away, we could tell from the variation of the electric or magnetic field exactly how the current had varied at the source.

You will also notice that long after all activity at the source has completely stopped and all charges and currents are zero, the block of field continues to travel through space. We have a distribution of electric and magnetic fields that exist independently of any charges or currents. That is the new effect that comes from the complete set of Maxwell's equations. If we want, we can give a complete mathematical representation of the analysis we have just done by writing that the electric field at a given place and a given time is proportional to the current at the source, only not at the *same* time, but at the *earlier* time $t - x/c$. We can write

$$E_y(t) = -\frac{J(t - x/c)}{2\epsilon_0 c}. \tag{20.3}$$

We have, believe it or not, already derived this same equation from another point of view in Vol. I, when we were dealing with the theory of the index of refraction. Then, we had to figure out what fields were produced by a thin layer of oscillating dipoles in a sheet of dielectric material with the dipoles set in motion by the electric field of an incoming electromagnetic wave. Our problem was to calculate the combined fields of the original wave and the waves radiated by the oscillating dipoles. How could we have calculated the fields generated by moving charges when we didn't have Maxwell's equations? At that time we took as our starting point (without any derivation) a formula for the radiation fields produced at large distances from an accelerating point charge. If you will look in Chapter 31 of Vol. I, you will see that Eq. (31.10) there is just the same as the Eq. (20.3) that we have just written down. Although our earlier derivation was correct only at large distances from the source, we see now that the same result continues to be correct even right up to the source.

We want now to look in a general way at the behavior of electric and magnetic fields in empty space far away from the sources, i.e., from the currents and charges. Very near the sources—near enough so that during the delay in transmission, the source has not had time to change much—the fields are very much the same as we have found in what we called the electrostatic or magnetostatic cases. If we go out to distances large enough so that the delays become important, however, the nature of the fields can be radically different from the solutions we have found. In a sense, the fields begin to take on a character of their own when they have gone a long way from all the sources. So we can begin by discussing the behavior of the fields in a region where there are no currents or charges.

Suppose we ask: What kind of fields can there be in regions where ρ and $\boldsymbol{j}$ are both zero? In Chapter 18 we saw that the physics of Maxwell's equations could also be expressed in terms of differential equations for the scalar and vector potentials:

$$\nabla^2\phi - \frac{1}{c^2}\frac{\partial^2\phi}{\partial t^2} = -\frac{\rho}{\epsilon_0}, \tag{20.4}$$

$$\nabla^2\boldsymbol{A} = \frac{1}{c^2}\frac{\partial^2\boldsymbol{A}}{\partial t^2} = -\frac{\boldsymbol{j}}{\epsilon_0 c^2}. \tag{20.5}$$

If ρ and $\boldsymbol{j}$ are zero, these equations take on the simpler form

$$\nabla^2\phi - \frac{1}{c^2}\frac{\partial^2\phi}{\partial t^2} = 0, \tag{20.6}$$

$$\nabla^2\boldsymbol{A} - \frac{1}{c^2}\frac{\partial^2\boldsymbol{A}}{\partial t^2} = 0. \tag{20.7}$$

Thus in free space the scalar potential ϕ and each component of the vector potential $\boldsymbol{A}$ all satisfy the same mathematical equation. Suppose we let ψ (psi) stand for any one of the four quantities ϕ, A_x, A_y, A_z; then we want to investigate the general solutions of the following equation:

$$\nabla^2\psi - \frac{1}{c^2}\frac{\partial^2\psi}{\partial t^2} = 0. \tag{20.8}$$

This equation is called the three-dimensional wave equation—three-dimensional, because the function ψ may depend in general on x, y, and z, and we need to worry about variations in all three coordinates. This is made clear if we write out explicitly the three terms of the Laplacian operator:

$$\frac{\partial^2\psi}{\partial x^2} + \frac{\partial^2\psi}{\partial y^2} + \frac{\partial^2\psi}{\partial z^2} - \frac{1}{c^2}\frac{\partial^2\psi}{\partial t^2} = 0. \tag{20.9}$$

In free space, the electric fields $\boldsymbol{E}$ and $\boldsymbol{B}$ also satisfy the wave equation. For example, since $\boldsymbol{B} = \boldsymbol{\nabla}\times\boldsymbol{A}$, we can get a differential equation for $\boldsymbol{B}$ by taking the curl of Eq. (20.7). Since the Laplacian is a scalar operator, the order of the Laplacian and curl operations can be interchanged:

$$\boldsymbol{\nabla}\times(\nabla^2\boldsymbol{A}) = \nabla^2(\boldsymbol{\nabla}\times\boldsymbol{A}) = \nabla^2\boldsymbol{B}.$$

Similarly, the order of the operations curl and $\partial/\partial t$ can be interchanged:

$$\boldsymbol{\nabla}\times\frac{1}{c^2}\frac{\partial^2\boldsymbol{A}}{\partial t^2} = \frac{1}{c^2}\frac{\partial^2}{\partial t^2}(\boldsymbol{\nabla}\times\boldsymbol{A}) = \frac{1}{c^2}\frac{\partial^2\boldsymbol{B}}{\partial t^2}.$$

Using these results, we get the following differential equation for $\boldsymbol{B}$:

$$\nabla^2\boldsymbol{B} - \frac{1}{c^2}\frac{\partial^2\boldsymbol{B}}{\partial t^2} = 0. \tag{20.10}$$

So each component of the magnetic field $\boldsymbol{B}$ satisfies the three-dimensional wave equation. Similarly, using the fact that $\boldsymbol{E} = -\boldsymbol{\nabla}\phi - d\boldsymbol{A}/dt$, it follows that the electric field $\boldsymbol{E}$ in free space also satisfies the three-dimensional wave equation:

$$\nabla^2\boldsymbol{E} - \frac{1}{c^2}\frac{\partial^2\boldsymbol{E}}{\partial t^2} = 0. \tag{20.11}$$

All of our electromagnetic fields satisfy the same wave equation, Eq. (20.8). We might well ask: What is the most general solution to this equation? However, rather than tackling that difficult question right away, we will look first at what can be said in general about those solutions in which nothing varies in y and z. (Always do an easy case first so that you can see what is going to happen, and then you can go to the more complicated cases.) Let's suppose that the magnitudes of the fields depend only upon x—that there are no *variations* of the fields with y and z. We are, of course, considering plane waves again. We should expect to get results something like those in the previous section. In fact, we will find precisely the same answers. You may ask: "Why do it all over again?" It is important to do it again, first, because we did not show that the waves we found were the most general solutions for plane waves, and second, because we found the fields only from a very particular kind of current source. We would like to ask now: What is the most general kind of one-dimensional wave there can be in free space? We cannot find that by seeing what happens for this or that particular source, but must work with greater generality. Also we are going to work this time with differential equations instead of with integral forms. Although we will get the same results, it is a way of practicing back and forth to show that it doesn't make any difference which way you go. You should know how to do things every which way, because when you get a hard problem, you will often find that only one of the various ways is tractable.

We could consider directly the solution of the wave equation for some electromagnetic quantity. Instead, we want to start right from the beginning with Maxwell's equations in free space so that you can see their close relationship to the electromagnetic waves. So we start with the equations in (20.1), setting the charges and currents equal to zero. They become

$$\begin{aligned} &\text{I.} \quad \nabla \cdot \boldsymbol{E} = 0 \\ &\text{II.} \quad \nabla \times \boldsymbol{E} = -\frac{\partial \boldsymbol{B}}{\partial t} \\ &\text{III.} \quad \nabla \cdot \boldsymbol{B} = 0 \\ &\text{IV.} \quad c^2 \nabla \times \boldsymbol{B} = \frac{\partial \boldsymbol{E}}{\partial t} \end{aligned} \tag{20.12}$$

We write the first equation out in components:

$$\nabla \cdot \boldsymbol{E} = \frac{\partial E_x}{\partial x} + \frac{\partial E_y}{\partial y} + \frac{\partial E_z}{\partial z} = 0. \tag{20.13}$$

We are assuming that there are no variations with y and z, so the last two terms are zero. This equation then tells us that

$$\frac{\partial E_x}{\partial x} = 0. \tag{20.14}$$

Its solution is that E_x, the component of the electric field in the x-direction, is a constant in space. If you look at IV in (20.12), supposing no $\boldsymbol{B}$-variation in y and z either, you can see that E_x is also constant in time. Such a field could be the steady DC field from some charged condenser plates a long distance away. We are not interested now in such an uninteresting static field; we are at the moment interested only in dynamically varying fields. For *dynamic* fields, $E_x = 0$.

We have then the important result that for the propagation of plane waves in any direction, *the electric field must be at right angles to the direction of propagation.* It can, of course, still vary in a complicated way with the coordinate x.

The transverse $\boldsymbol{E}$-field can always be resolved into two components, say the y-component and the z-component. So let's first work out a case in which the electric field has only one transverse component. We'll take first an electric field that is always in the y-direction, with zero z-component. Evidently, if we solve this problem we can also solve for the case where the electric field is always in the z-direction. The general solution can always be expressed as the superposition of two such fields.

How easy our equations now get. The only component of the electric field that is not zero is E_y, and all derivatives—except those with respect to x—are zero. The rest of Maxwell's equations then become quite simple.

Let's look next at the second of Maxwell's equations [II of Eq. (20.12)]. Writing out the components of the curl $\boldsymbol{E}$, we have

$$(\nabla \times \boldsymbol{E})_x = \frac{\partial E_z}{\partial y} - \frac{\partial E_y}{\partial z} = 0,$$

$$(\nabla \times \boldsymbol{E})_y = \frac{\partial E_x}{\partial z} - \frac{\partial E_z}{\partial x} = 0,$$

$$(\nabla \times \boldsymbol{E})_z = \frac{\partial E_y}{\partial x} - \frac{\partial E_x}{\partial y} = \frac{\partial E_y}{\partial x}.$$

The x-component of $\nabla \times \boldsymbol{E}$ is zero because the derivatives with respect to y and z are zero. The y-component is also zero; the first term is zero because the derivative with respect to z is zero, and the second term is zero because E_z is zero. The only components of the curl of $\boldsymbol{E}$ that is not zero is the z-component, which is equal to $\partial E_y/\partial x$. Setting the three components of $\nabla \times \boldsymbol{E}$ equal to the corresponding components of $-\partial \boldsymbol{B}/\partial t$, we can conclude the following:

$$\frac{\partial B_x}{\partial t} = 0, \qquad \frac{\partial B_y}{\partial t} = 0. \tag{20.15}$$

$$\frac{\partial B_z}{\partial t} = -\frac{\partial E_y}{\partial x}. \tag{20.16}$$

Since the x-component of the magnetic field and the y-component of the magnetic field both have zero time derivatives, these two components are just constant fields and correspond to the magnetostatic solutions we found earlier. Somebody may have left some permanent magnets near where the waves are propagating. We will ignore these constant fields and set B_x and B_y equal to zero.

Incidentally, we would already have concluded that the x-component of $\boldsymbol{B}$ should be zero for a different reason. Since the divergence of $\boldsymbol{B}$ is zero (from the third Maxwell equation), applying the same arguments we used above for the electric field, we would conclude that the longitudinal component of the magnetic field can have no variation with x. Since we are ignoring such uniform fields in our wave solutions, we would have set B_x equal to zero. In plane electromagnetic waves the $\boldsymbol{B}$-field, as well as the $\boldsymbol{E}$-field, must be directed at right angles to the direction of propagation.

Equation (20.16) gives us the additional proposition that if the electric field has only a y-component, the magnetic field will have only a z-component. So $\boldsymbol{E}$ and $\boldsymbol{B}$ *are at right angles* to each other. This is exactly what happened in the special wave we have already considered.

We are now ready to use the last of Maxwell's equations for free space [IV of Eq. (20.12)]. Writing out the components, we have

$$\begin{aligned}
c^2(\nabla \times \boldsymbol{B})_x &= c^2\frac{\partial B_z}{\partial y} - c^2\frac{\partial B_y}{\partial z} = \frac{\partial E_x}{\partial t},\\
c^2(\nabla \times \boldsymbol{B})_y &= c^2\frac{\partial B_x}{\partial z} - c^2\frac{\partial B_z}{\partial x} = \frac{\partial E_y}{\partial t},\\
c^2(\nabla \times \boldsymbol{B})_z &= c^2\frac{\partial B_y}{\partial x} - c^2\frac{\partial B_x}{\partial y} = \frac{\partial E_z}{\partial t}.
\end{aligned} \tag{20.17}$$

Of the six derivatives of the components of $\boldsymbol{B}$, only the term $\partial B_z/\partial x$ is not equal to zero. So the three equations give us simply

$$-c^2\frac{\partial B_z}{\partial x} = \frac{\partial E_y}{\partial t}. \tag{20.18}$$

The result of all our work is that only one component each of the electric and magnetic fields is not zero, and that these components must satisfy Eqs. (20.16) and (20.18). The two equations can be combined into one if we differentiate the first with respect to x and the second with respect to t; the left-hand sides of the two equations will then be the same (except for the factor c^2). So we find that E_y satisfies the equation

$$\frac{\partial^2 E_y}{\partial x^2} - \frac{1}{c^2}\frac{\partial^2 E_y}{\partial t^2} = 0. \tag{20.19}$$

We have seen the same differential equation before, when we studied the propagation of sound. It is the wave equation for one-dimensional waves.

You should note that in the process of our derivation we have found something *more* than is contained in Eq. (20.11). Maxwell's equations have given us the further information that electromagnetic waves have field components only at right angles to the direction of the wave propagation.

Let's review what we know about the solutions of the one-dimensional wave equation. If any quantity ψ satisfies the one-dimensional wave equation

$$\frac{\partial^2 \psi}{\partial x^2} - \frac{1}{c^2}\frac{\partial^2 \psi}{\partial t^2} = 0, \tag{20.20}$$

then one possible solution is a function $\psi(x, t)$ of the form

$$\psi(x, t) = f(x - ct), \tag{20.21}$$

that is, some function of the *single* variable $(x - ct)$. The function $f(x - ct)$ represents a "rigid" pattern in x which travels toward positive x at the speed c (see Fig. 20–4). For example, if the function f has a maximum when its argument is zero, then for $t = 0$ the maximum of ψ will occur at $x = 0$. At some later time, say $t = 10$, ψ will have its maximum at $x = 10c$. As time goes on, the maximum moves toward positive x at the speed c.

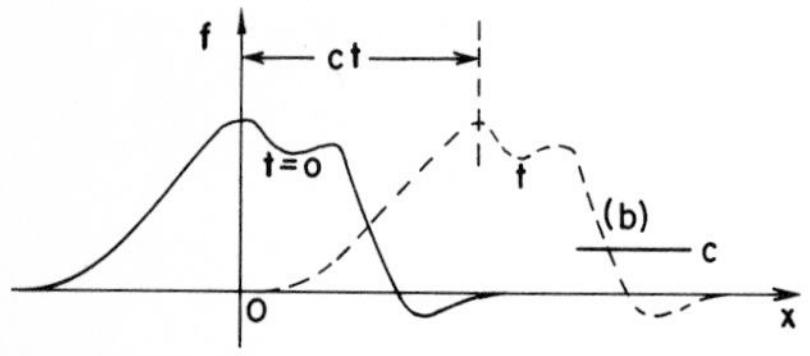

Fig. 20–4. The function $f(x - ct)$ represents a constant "shape" that travels toward positive x with the speed c.

Sometimes it is more convenient to say that a solution of the one-dimensional wave equation is a function of $(t - x/c)$. However, this is saying the same thing, because any function of $(t - x/c)$ is also a function of $(x - ct)$:

$$F(t - x/c) = F\left[-\frac{x - ct}{c}\right] = f(x - ct).$$

Let's show that $f(x - ct)$ is indeed a solution of the wave equation. Since it is a function of only one variable—the variable $(x - ct)$—we will let f' represent the derivative of f with respect to its variable and f'' represent the second derivative of f. Differentiating Eq. (20.21) with respect to x, we have

$$\frac{\partial \psi}{\partial x} = f'(x - ct),$$

since the derivative of $(x - ct)$ with respect to x is 1. The second derivative of ψ with respect to x is clearly

$$\frac{\partial^2 \psi}{\partial x^2} = f''(x - ct). \tag{20.22}$$

Taking derivatives of ψ with respect to t, we find

$$\frac{\partial \psi}{\partial t} = f'(x - ct)(-c),$$

$$\frac{\partial^2 \psi}{\partial t^2} = +c^2 f''(x - ct). \tag{20.23}$$

We see that ψ does indeed satisfy the one-dimensional wave equation.

You may be wondering: "If I have the wave equation, how do I know that I should take $f(x - ct)$ as a solution? I don't like this backward method. Isn't there some *forward* way to find the solution?" Well, one good forward way is to know the solution. It is possible to "cook up" an apparently forward mathematical argument, expecially because we know what the solution is supposed to be, but with an equation as simple as this we don't have to play games. Soon you will get so that when you see Eq. (20.20), you nearly simultaneously see

$\psi = f(x - ct)$ as a solution. (Just as now when you see the integral of $x^2\,dx$, you know right away that the answer is $x^3/3$.)

Actually you should also see a little more. Not only is any function of $(x - ct)$ a solution, but any function of $(x + ct)$ is also a solution. Since the wave equation contains only c^2, changing the sign of c makes no difference. In fact, the *most general* solution of the one-dimensional wave equation is the sum of two arbitrary functions, one of $(x - ct)$ and the other of $(x + ct)$:

$$\psi = f(x - ct) + g(x + ct). \tag{20.24}$$

The first term represents a wave travelling toward positive x, and the second term an arbitrary wave travelling toward negative x. The general solution is the superposition of two such waves both existing at the same time.

We will leave the following amusing question for you to think about. Take a function ψ of the following form:

$$\psi = \cos kx \cos kct.$$

This equation isn't in the form of a function of $(x - ct)$ or of $(x + ct)$. Yet you can easily show that this function is a solution of the wave equation by direct substitution into Eq. (20.20). How can we then say that the general solution is of the form of Eq. (20.24)?

Applying our conclusions about the solution of the wave equation to the y-component of the electric field, E_y, we conclude that E_y can vary with x in any arbitrary fashion. However, the fields which do exist can always be considered as the sum of two patterns. One wave is sailing through space in one direction with speed c, with an associated magnetic field perpendicular to the electric field; another wave is travelling in the opposite direction with the same speed. Such waves correspond to the electromagnetic waves that we know about—light, radiowaves, infrared radiation, ultraviolet radiation, x-rays, and so on. We have already discussed the radiation of light in great detail in Vol. I. Since everything we learned there applies to any electromagnetic wave, we don't need to consider in great detail here the behavior of these waves.

We should perhaps make a few further remarks on the question of the polarization of the electromagnetic waves. In our solution we chose to consider the special case in which the electric field has only a y-component. There is clearly another solution for waves travelling in the plus or minus x-direction, with an electric field which has only a z-component. Since Maxwell's equations are linear, the general solution for one-dimensional waves propagating in the x-direction is the sum of waves of E_y and waves of E_z. This general solution is summarized in the following equations:

$$\begin{aligned}
\boldsymbol{E} &= (0, E_y, E_z)\\
E_y &= f(x - ct) + g(x + ct)\\
E_z &= F(x - ct) + G(x + ct)\\
\boldsymbol{B} &= (0, B_y, B_z)\\
cB_z &= f(x-ct) - g(x + ct)\\
cB_y &= -F(x - ct) + G(x + ct).
\end{aligned} \tag{20.25}$$

Such electromagnetic waves have an $\boldsymbol{E}$-vector whose direction is not constant but which gyrates around in some arbitrary way in the yz-plane. At every point the magnetic field is always perpendicular to the electric field and to the direction of propagation.

If there are only waves travelling in one direction, say the positive x-direction, there is a simple rule which tells the relative orientation of the electric and magnetic fields. The rule is that the cross product $\boldsymbol{E} \times \boldsymbol{B}$—which is, of course, a vector at right angles to both $\boldsymbol{E}$ and $\boldsymbol{B}$—points in the direction in which the wave is travelling. If $\boldsymbol{E}$ is rotated into $\boldsymbol{B}$ by a right-hand screw, the screw points in the direction of the wave velocity. (We shall see later that the vector $\boldsymbol{E} \times \boldsymbol{B}$ has a special physical significance: it is a vector which describes the flow of energy in an electromagnetic field.)

20–2 Three-dimensional waves

We want now to turn to the subject of three-dimensional waves. We have already seen that the vector $\boldsymbol{E}$ satisfies the wave equation. It is also easy to arrive at the same conclusion by arguing directly from Maxwell's equations. Suppose we start with the equation

$$\nabla \times \boldsymbol{E} = -\frac{\partial \boldsymbol{B}}{\partial t}$$

and take the curl of both sides:

$$\nabla \times (\nabla \times \boldsymbol{E}) = -\frac{\partial}{\partial t}(\nabla \times \boldsymbol{B}). \tag{20.26}$$

You will remember that the curl of the curl of any vector can be written as the sum of two terms, one involving the divergence and the other the Laplacian,

$$\nabla \times (\nabla \times \boldsymbol{E}) = \nabla(\nabla \cdot \boldsymbol{E}) - \nabla^2 \boldsymbol{E}.$$

In free space, however, the divergence of $\boldsymbol{E}$ is zero, so only the Laplacian term remains. Also, from the fourth of Maxwell's equations in free space [Eq. (20.12)] the time derivative of $c^2\, \nabla \times \boldsymbol{B}$ is the second derivative of $\boldsymbol{E}$ with respect to t:

$$c^2 \frac{\partial}{\partial t}(\nabla \times \boldsymbol{B}) = \frac{\partial^2 \boldsymbol{E}}{\partial t^2}.$$

Equation (20.26) then becomes

$$\nabla^2 \boldsymbol{E} = \frac{1}{c^2}\frac{\partial^2 \boldsymbol{E}}{\partial t^2},$$

which is the three-dimensional wave equation. Written out in all its glory, this equation is, of course,

$$\frac{\partial^2 \boldsymbol{E}}{\partial x^2} + \frac{\partial^2 \boldsymbol{E}}{\partial y^2} + \frac{\partial^2 \boldsymbol{E}}{\partial z^2} - \frac{1}{c^2}\frac{\partial^2 \boldsymbol{E}}{\partial t^2} = 0. \tag{20.27}$$

How shall we find the general wave solution? The answer is that all the solutions of the three-dimensional wave equation can be represented as a superposition of the one-dimensional solutions we have already found. We obtained the equation for waves which move in the x-direction by supposing that the field did not depend on y and z. Obviously, there are other solutions in which the fields do not depend on x and z, representing waves going in the y-direction. Then there are solutions which do not depend on x and y, representing waves travelling in the z-direction. Or in general, since we have written our equations in vector form, the three-dimensional wave equation can have solutions which are plane waves moving in any direction at all. Again, since the equations are linear, we may have simultaneously as many plane waves as we wish, travelling in as many different directions. Thus the most general solution of the three-dimensional wave equation is a superposition of all sorts of plane waves moving in all sorts of directions.

Try to imagine what the electric and magnetic fields look like at present in the space in this lecture room. First of all, there is a steady magnetic field; it comes from the currents in the interior of the earth—that is, the earth's steady magnetic field. Then there are some irregular, nearly static electric fields produced perhaps by electric charges generated by friction as various people move about in their chairs and rub their coat sleeves against the chair arms. Then there are other magnetic fields produced by oscillating currents in the electrical wiring—fields which vary at a frequency of 60 cycles per second, in synchronism with the generator at Boulder Dam. But more interesting are the electric and magnetic fields varying at much higher frequencies. For instance, as light travels from window to floor and wall to wall, there are little wiggles of the electric and magnetic fields moving along at 186,000 miles per second. Then there are also infrared waves travelling from the warm foreheads to the cold blackboard. And we have forgotten the ultraviolet light, the x-rays, and the radiowaves travelling through the room.

Flying across the room are electromagnetic waves which carry music of a jazz band. There are waves modulated by a series of impulses representing pictures of events going on in other parts of the world, or of imaginary aspirins dissolving in imaginary stomachs. To demonstrate the reality of these waves it is only necessary to turn on electronic equipment that converts these waves into pictures and sounds.

If we go into further detail to analyze even the smallest wiggles, there are tiny electromagnetic waves that have come into the room from enormous distances. There are now tiny oscillations of the electric field, whose crests are separated by a distance of one foot, that have come from millions of miles away, transmitted to the earth from the Mariner II space craft which has just passed Venus. Its signals carry summaries of information it has picked up about the planets (information obtained from electromagnetic waves that travelled from the planet to the space craft).

There are very tiny wiggles of the electric and magnetic fields that are waves which originated billions of light years away—from galaxies in the remotest corners of the universe. That this is true has been found by "filling the room with wires"—by building antennas as large as this room. Such radiowaves have been detected from places in space beyond the range of the greatest optical telescopes. Even they, the optical telescopes, are simply gatherers of electromagnetic waves. What we call the stars are only inferences, inferences drawn from the only physical reality we have yet gotten from them—from a careful study of the unendingly complex undulations of the electric and magnetic fields reaching us on earth.

There is, of course, more: the fields produced by lightning miles away, the fields of the charged cosmic ray particles as they zip through the room, and more, and more. What a complicated thing is the electric field in the space around you! Yet it always satisfies the three-dimensional wave equation.

20–3 Scientific imagination

I have asked you to imagine these electric and magnetic fields. What do you do? Do you know how? How do *I* imagine the electric and magnetic field? What do *I* actually see? What are the demands of scientific imagination? Is it any different from trying to imagine that the room is full of invisible angels? No, it is not like imagining invisible angels. It requires a much higher degree of imagination to understand the electromagnetic field than to understand invisible angels. Why? Because to make invisible angels understandable, all I have to do is to alter their properties *a little bit*—I make them slightly visible, and then I can see the shapes of their wings, and bodies, and halos. Once I succeed in imagining a visible angel, the abstraction required—which is to take almost invisible angels and imagine them completely invisible—is relatively easy. So you say, "Professor, please give me an approximate description of the electromagnetic waves, even though it may be slightly inaccurate, so that I too can see them as well as I can see almost invisible angels. Then I will modify the picture to the necessary abstraction."

I'm sorry I can't do that for you. I don't know how. I have no picture of this electromagnetic field that is in any sense accurate. I have known about the electromagnetic field a long time—I was in the same position 25 years ago that you are now, and I have had 25 years more of experience thinking about these wiggling waves. When I start describing the magnetic field moving through space, I speak of the $\boldsymbol{E}$- and $\boldsymbol{B}$ fields and wave my arms and you may imagine that I can see them.

I'll tell you what I see. I see some kind of vague shadowy, wiggling lines—here and there is an E and B written on them somehow, and perhaps some of the lines have arrows on them—an arrow here or there which disappears when I look too closely at it. When I talk about the fields swishing through space, I have a terrible confusion between the symbols I use to describe the objects and the objects themselves. I cannot really make a picture that is even nearly like the true waves. So if you have some difficulty in making such a picture, you should not be worried that your difficulty is unusual.

Our science makes terrific demands on the imagination. The degree of imagination that is required is much more extreme than that required for some of the ancient ideas. The modern ideas are much harder to imagine. We use a lot of tools, though. We use mathematical equations and rules, and make a lot of pictures. What I realize now is that when I talk about the electromagnetic field in space, I see some kind of a superposition of all of the diagrams which I've ever seen drawn about them. I don't see little bundles of field lines running about because it worries me that if I ran at a different speed the bundles would disappear. I don't even always see the electric and magnetic fields because sometimes I think I should have made a picture with the vector potential and the scalar potential, for those were perhaps the more physically significant things that were wiggling.

Perhaps the only hope, you say, is to take a mathematical view. Now what is a mathematical view? From a mathematical view, there is an electric field vector and a magnetic field vector at every point in space; that is, there are six numbers associated with every point. Can you imagine six numbers associated with each point in space? That's too hard. Can you imagine even *one* number associated with every point? I cannot! I can imagine such a thing as the temperature at every point in space. That seems to be understandable. There is a hotness and coldness that varies from place to place. But I honestly do not understand the idea of a *number* at every point.

So perhaps we should put the question: Can we represent the electric field by something more like a temperature, say like the displacement of a piece of jello? Suppose that we were to begin by imagining that the world was filled with thin jello and that the fields represented some distortion—say a stretching or twisting—of the jello. Then we could visualize the field. After we "see" what it is like we could abstract the jello away. For many years that's what people tried to do. Maxwell, Ampere, Faraday, and others tried to understand electromagnetism this way. (Sometimes they called the abstract jello "ether.") But it turned out that the attempt to imagine the electromagnetic field in that way was really standing in the way of progress. We are unfortunately limited to abstractions, to using instruments to detect the field, to using mathematical symbols to describe the field, etc. But nevertheless, in some sense the fields are real, because after we are all finished fiddling around with mathematical equations—with or without making pictures and drawings or trying to visualize the thing—we can still make the instruments detect the signals from Mariner II and find out about galaxies a billion miles away, and so on.

The whole question of imagination in science is often misunderstood by people in other disciplines. They try to test our imagination in the following way. They say, "Here is a picture of some people in a situation. What do you imagine will happen next?" When we say, "I can't imagine," they may think we have a weak imagination. They overlook the fact that whatever we are *allowed* to imagine in science must be *consistent with everything else we know:* that the electric fields and the waves we talk about are not just some happy thoughts which we are free to make as we wish, but ideas which must be consistent with all the laws of physics we know. We can't allow ourselves to seriously imagine things which are obviously in contradiction to the known laws of nature. And so our kind of imagination is quite a difficult game. One has to have the imagination to think of something that has never been seen before, never been heard of before. At the same time the thoughts are restricted in a strait jacket, so to speak, limited by the conditions that come from our knowledge of the way nature really is. The problem of creating something which is new, but which is consistent with everything which has been seen before, is one of extreme difficulty.

While I'm on this subject I want to talk about whether it will ever be possible to imagine *beauty* that we can't *see*. It is an interesting question. When we look at a rainbow, it looks beautiful to us. Everybody says, "Ooh, a rainbow." (You see how scientific I am. I am afraid to say something is beautiful unless I have an experimental way of defining it.) But how would we describe a rainbow if we were blind? We *are* blind when we measure the infrared reflection coefficient of sodium chloride, or when we talk about the frequency of the waves that are coming from some galaxy that we can't see—we make a diagram, we make a plot. For instance, for the rainbow, such a plot would be the intensity of radiation vs. wavelength measured with a spectrophotometer for each direction in the sky. Generally, such measurements would give a curve that was rather flat. Then some day, someone would discover that for certain conditions of the weather, and at certain angles in the sky, the spectrum of intensity as a function of wavelength would behave strangely; it would have a bump. As the angle of the instrument was varied only a little bit, the maximum of the bump would move from one wavelength to another. Then one day the physical review of the blind men might publish a technical article with the title "The Intensity of Radiation as a Function of Angle under Certain Conditions of the Weather." In this article there might appear a graph such as the one in Fig. 20–5. The author would perhaps remark that at the larger angles there was more radiation at long wavelengths, whereas for the smaller angles the maximum in the radiation came at shorter wavelengths. (From our point of view, we would say that the light at 40° is predominantly green and the light at 42° is predominantly red.)

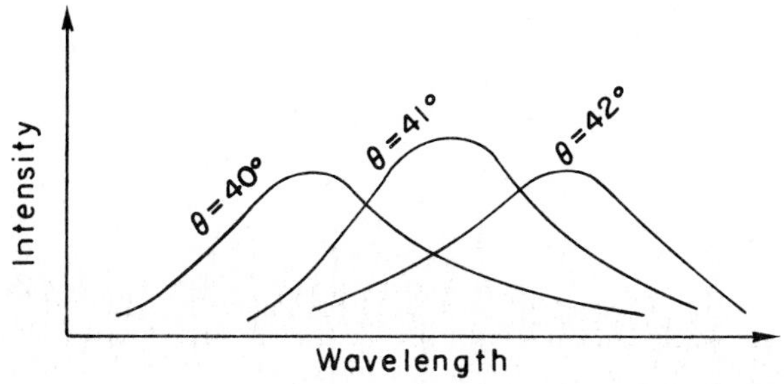

Fig. 20–5. The intensity of electromagnetic waves as a function of wavelength for three angles (measured from the direction opposite the sun), observed only with certain meteorological conditions.

Now do we find the graph of Fig. 20–5 beautiful? It contains much more detail than we apprehend when we look at a rainbow, because our eyes cannot see the exact details in the shape of a spectrum. The eye, however, finds the rainbow beautiful. Do we have enough imagination to see in the spectral curves the same beauty we see when we look directly at the rainbow? I don't know.

But suppose I have a graph of the reflection coefficient of a sodium chloride crystal as a function of wavelength in the infrared, and also as a function of angle. I would have a representation of how it would look to my eyes if they could see in the infrared—perhaps some glowing, shiny "green," mixed with reflections from the surface in a "metallic red." That would be a beautiful thing, but I don't know whether I can ever look at a graph of the reflection coefficient of NaCl measured with some instrument and say that it has the same beauty.

On the other hand, even if we cannot see beauty in particular measured results, we *can* already claim to see a certain beauty in the equations which describe general physical laws. For example, in the wave equation (20.9), there's something nice about the regularity of the appearance of the x, the y, the z, and the t. And this nice symmetry in appearance of the x, y, z, and t suggests to the mind still a greater beauty which has to do with the four dimensions, the possibility that space has four-dimensional symmetry, the possibility of analyzing that and the developments of the special theory of relativity. So there is plenty of intellectual beauty associated with the equations.

20–4 Spherical waves

We have seen that there are solutions of the wave equation which correspond to plane waves, and that any electromagnetic wave can be described as a superposition of many plane waves. In certain special cases, however, it is more convenient to describe the wave field in a different mathematical form. We would like to discuss now the theory of spherical waves—waves which correspond to spherical surfaces that are spreading out from some center. When you drop a stone into a lake, the ripples spread out in circular waves on the surface—they are two-dimensional waves. A spherical wave is a similar thing except that it spreads out in three dimensions.

Before we start describing spherical waves, we need a little mathematics. Suppose we have a function that depends only on the radial distance r from a certain origin—in other words, a function that is spherically symmetric. Let's call the function $\psi(r)$, where by r we mean

$$r = \sqrt{x^2 + y^2 + z^2},$$

the radial distance from the origin. In order to find out what functions $\psi(r)$ satisfy the wave equation, we will need an expression for the Laplacian of ψ. So we want to find the sum of the second derivatives of ψ with respect to x, y, and z. We will use the notation that $\psi'(r)$ represents the derivative of ψ with respect to r and $\psi''(r)$ represents the second derivative of ψ with respect to r.

First, we find the derivatives with respect to x. The first derivative is

$$\frac{\partial\psi(r)}{\partial x} = \psi'(r)\,\frac{\partial r}{\partial x}.$$

The second derivative of ψ with respect to x is

$$\frac{\partial^2\psi}{\partial x^2} = \psi''\left(\frac{\partial r}{\partial x}\right)^2 + \psi'\,\frac{\partial^2 r}{\partial x^2}.$$

We can evaluate the partial derivatives of r with respect to x from

$$\frac{\partial r}{\partial x} = \frac{x}{r}, \qquad \frac{\partial^2 r}{\partial x^2} = \frac{1}{r}\left(1 - \frac{x^2}{r^2}\right).$$

So the second derivative of ψ with respect to x is

$$\frac{\partial^2\psi}{\partial x^2} = \frac{x^2}{r^2}\,\psi'' + \frac{1}{r}\left(1 - \frac{x^2}{r^2}\right)\psi'. \tag{20.28}$$

Likewise,

$$\frac{\partial^2\psi}{\partial y^2} = \frac{y^2}{r^2}\,\psi'' + \frac{1}{r}\left(1 - \frac{y^2}{r^2}\right)\psi', \tag{20.29}$$

$$\frac{\partial^2\psi}{\partial z^2} = \frac{z^2}{r^2}\,\psi'' + \frac{1}{r}\left(1 - \frac{z^2}{r^2}\right)\psi'. \tag{20.30}$$

The Laplacian is the sum of these three derivatives. Remembering that $x^2 + y^2 + z^2 = r^2$, we get

$$\nabla^2\psi(r) = \psi''(r) + \frac{2}{r}\,\psi'(r). \tag{20.31}$$

It is often more convenient to write this equation in the following form:

$$\nabla^2\psi = \frac{1}{r}\,\frac{d^2}{dr^2}\,(r\psi). \tag{20.32}$$

If you carry out the differentiation indicated in Eq. (20.32), you will see that the right-hand side is the same as in Eq. (20.31).

If we wish to consider spherically symmetric fields which can propagate as spherical waves, our field quantity must be a function of both r and t. Suppose we ask, then, what functions $\psi(r, t)$ are solutions of the three-dimensional wave equation

$$\nabla^2\psi(r, t) - \frac{1}{c^2}\frac{\partial^2}{\partial t^2}\psi(r, t) = 0. \tag{20.33}$$

Since $\psi(r, t)$ depends only on the spatial coordinates through r, we can use the equation for the Laplacian we found above, Eq. (20.32). To be precise, however, since ψ is also a function of t, we should write the derivatives with respect to r as partial derivatives. Then the wave equation becomes

$$\frac{1}{r}\frac{\partial^2}{\partial r^2}(r\psi) - \frac{1}{c^2}\frac{\partial^2}{\partial t^2}\psi = 0.$$

We must now solve this equation, which appears to be much more complicated than the plane wave case. But notice that if we multiply this equation by r, we get

$$\frac{\partial^2}{\partial r^2}(r\psi) - \frac{1}{c^2}\frac{\partial^2}{\partial t^2}(r\psi) = 0. \tag{20.34}$$

This equation tells us that the function $r\psi$ satisfies the one-dimensional wave equation in the variable r. Using the general principle which we have emphasized so often, that the same equations always have the same solutions, we know that if $r\psi$ is a function only of $(r - ct)$ then it will be a solution of Eq. (20.34). So we know that spherical waves must have the form

$$r\psi(r, t) = f(r - ct).$$

Or, as we have seen before, we can equally well say that $r\psi$ can have the form

$$r\psi = f(t - r/c).$$

Dividing by r, we find that the field quantity ψ (whatever it may be) has the following form:

$$\psi = \frac{f(t - r/c)}{r}. \tag{20.35}$$

Such a function represents a general spherical wave travelling outward from the origin at the speed c. If we forget about the r in the denominator for a moment, the amplitude of the wave as a function of the distance from the origin at a given time has a certain shape that travels outward at the speed c. The factor r in the denominator, however, says that the amplitude of the wave decreases in proportion to $1/r$ as the wave propagates. In other words, unlike a plane wave in which the amplitude remains constant as the wave runs along, in a spherical wave the amplitude steadily decreases, as shown in Fig. 20–6. This effect is easy to understand from a simple physical argument.

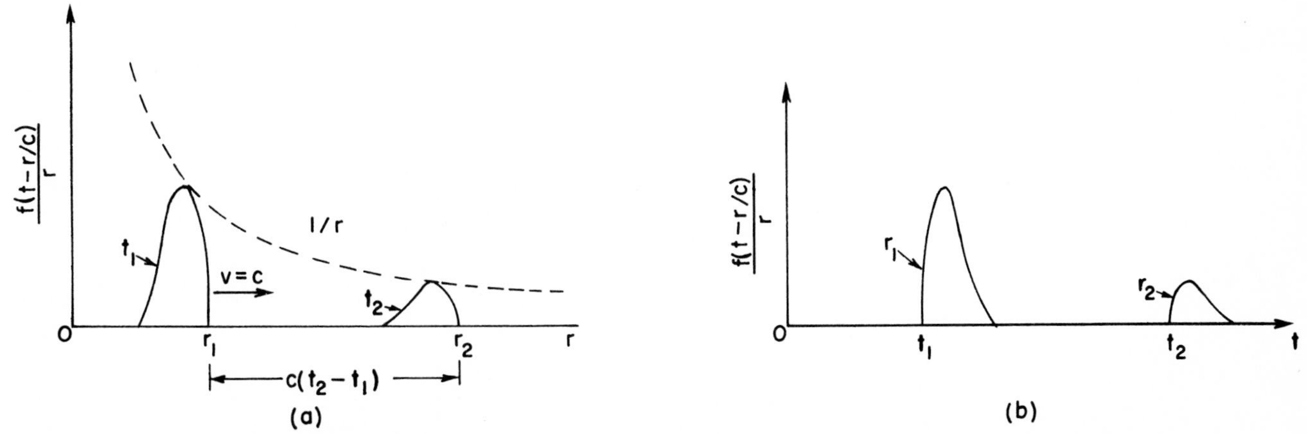

Fig. 20–6. A spherical wave $\psi = f(t - r/c)/r$. (a) ψ as a function of r for $t = t_1$ and the same wave for the later time t_2. (b) ψ as a function of t for $r = r_1$ and the same wave seen at r_2.

We know that the energy density in a wave depends on the square of the wave amplitude. As the wave spreads, its energy is spread over larger and larger areas proportional to the radial distance squared. If the total energy is conserved, the energy density must fall as $1/r^2$, and the amplitude of the wave must decrease as $1/r$. So Eq. (20.35) is the "reasonable" form for a spherical wave.

We have disregarded the second possible solution to the one-dimensional wave equation:

$$r\psi = g(t + r/c),$$

or

$$\psi = \frac{g(t + r/c)}{r}.$$

This also represents a spherical wave, but one which travels *inward* from large r toward the origin.

We are now going to make a special assumption. We say, without any demonstration whatever, that the waves generated by a source are only the waves which go *outward*. Since we know that waves are caused by the motion of charges, we want to think that the waves proceed outward from the charges. It would be rather strange to imagine that before charges were set in motion, a spherical wave started out from infinity and arrived at the charges just at the time they began to move. That is a possible solution, but experience shows that when charges are accelerated the waves travel outward from the charges. Although Maxwell's equations would allow either possibility, we will put in an *additional fact*—based on experience—that only the outgoing wave solution makes "physical sense."

We should remark, however, that there is an interesting consequence to this additional assumption: we are removing the symmetry with respect to time that exists in Maxwell's equations. The original equations for $\boldsymbol{E}$ and $\boldsymbol{B}$, and also the wave equations we derived from them, have the property that if we change the sign of t, the equation is unchanged. These equations say that for every solution corresponding to a wave going in one direction there is an equally valid solution for a wave travelling in the opposite direction. Our statement that we will consider only the outgoing spherical waves is an important additional assumption. (A formulation of electrodynamics in which this additional assumption is avoided has been carefully studied. Surprisingly, in many circumstances it does *not* lead to physically absurd conclusions, but it would take us too far astray to discuss these ideas just now. We will talk about them a little more in Chapter 28.)

We must mention another important point. In our solution for an outgoing wave, Eq. (20.35), the function ψ is infinite at the origin. That is somewhat peculiar. We would like to have a wave solution which is smooth everywhere. Our solution must represent physically a situation in which there is some source at the origin. In other words, we have inadvertently made a mistake. We have not solved the free wave equation (20.33) *everywhere;* we have solved Eq. (20.33) with zero on the right everywhere, except at the origin. Our mistake crept in because some of the steps in our derivation are not "legal" when $r = 0$.

Let's show that it is easy to make the same kind of mistake in an electrostatic problem. Suppose we want a solution of the equation for an electrostatic potential in free space, $\nabla^2\phi = 0$. The Laplacian is equal to zero, because we are assuming that there are no charges anywhere. But what about a spherically symmetric solution to this equation—that is, some function ϕ that depends only on r. Using the formula of Eq. (20.32) for the Laplacian, we have

$$\frac{1}{r}\frac{d^2}{dr^2}(r\phi) = 0.$$

Multiplying this equation by r, we have an equation which is readily integrated:

$$\frac{d^2}{dr^2}(r\phi) = 0.$$

If we integrate once with respect to r, we find that the first derivative of $r\phi$ is a constant, which we may call a:

$$\frac{d}{dr}(r\phi) = a.$$

Integrating again, we find that $r\phi$ is of the form

$$r\phi = ar + b,$$

where b is another constant of integration. So we have found that the following ϕ is a solution for the electrostatic potential in free space:

$$\phi = a + \frac{b}{r}\cdot$$

Something is evidently wrong. In the region where there are no electric charges, we know the solution for the electrostatic potential: the potential is everywhere a constant. That corresponds to the first term in our solution. But we also have the second term, which says that there is a contribution to the potential that varies as one over the distance from the origin. We know, however, that such a potential corresponds to a point charge at the origin. So, although we thought we were solving for the potential in free space, our solution also gives the field for a point source at the origin. Do you see the similarity between what happened now and what happened when we solved for a spherically symmetric solution to the wave equation? If there were really no charges or currents at the origin, there would not be spherical outgoing waves. The spherical waves must, of course, be produced by sources at the origin. In the next chapter we will investigate the connection between the outgoing electromagnetic waves and the currents and voltages which produce them.

21

Solutions of Maxwell's Equations with Currents and Charges

Review: Chapter 28, Vol. I, *Electromagnetic Radiation*
Chapter 31, Vol. I, *The Origin of the Refractive Index*
Chapter 36, Vol. I, *Relativistic Effects in Radiation*

21–1 Light and electromagnetic waves

We saw in the last chapter that among their solutions, Maxwell's equations have waves of electricity and magnetism. These waves correspond to the phenomena of radio, light, x-rays, and so on, depending on the wavelength. We have already studied light in great detail in Vol. I. In this chapter we want to tie together the two subjects—we want to show that Maxwell's equations can indeed form the base for our earlier treatment of the phenomena of light.

When we studied light, we began by writing down an equation for the electric field produced by a charge which moves in any arbitrary way. That equation was

$$\boldsymbol{E} = \frac{q}{4\pi\epsilon_0}\left[\frac{\boldsymbol{e}_{r'}}{r'^2} + \frac{r'}{c}\frac{d}{dt}\left(\frac{\boldsymbol{e}_{r'}}{r'^2}\right) + \frac{1}{c^2}\frac{d^2}{dt^2}\boldsymbol{e}_{r'}\right], \tag{21.1}$$

$$c\boldsymbol{B} = \boldsymbol{e}_{r'} \times \boldsymbol{E}.$$

[See Eq. (28.3), Vol. I.]

If a charge moves in an arbitrary way, the electric field we would find *now* at some point depends only on the position and motion of the charge not now, but at an *earlier* time—at an instant which is earlier by the time it would take light, going at the speed c, to travel the distance r' from the charge to the field point. In other words, if we want the electric field at point (1) at the time t, we must calculate the location (2′) of the charge and its motion at the time $(t - r'/c)$, where r' is the distance to the point (1) from the position of the charge (2′) at the time $(t - r'/c)$. The prime is to remind you that r' is the so-called "retarded distance" from the point (2′) to the point (1), and not the actual distance between point (2), the position of the charge at the time t, and the field point (1) (see Fig. 21–1). Note that we are using a different convention now for the *direction* of the unit vector $\boldsymbol{e}_r$. In Chapters 28 and 36 of Vol. I it was convenient to take $\boldsymbol{r}$ (and hence $\boldsymbol{e}_r$) pointing *toward* the source. Now we are following the definition we took for Coulomb's law, in which $\boldsymbol{r}$ is directed *from* the charge, at (2), *toward* the field point at (1). The only difference, of course, is that our new $\boldsymbol{r}$ (and $\boldsymbol{e}_r$) are the negatives of the old ones.

We have also seen that if the velocity v of a charge is always much less than c, and if we consider only points at large distances from the charge, so that only the last term of Eq. (21.1) is important, the fields can also be written as

$$\boldsymbol{E} = -\frac{q}{4\pi\epsilon_0 c^2 r'}\begin{bmatrix}\text{acceleration of the charge at } (t - r'/c)\\ \text{projected at right angles to } r'\end{bmatrix}, \tag{21.1′}$$

and

$$c\boldsymbol{B} = \boldsymbol{e}_{r'} \times \boldsymbol{E}.$$

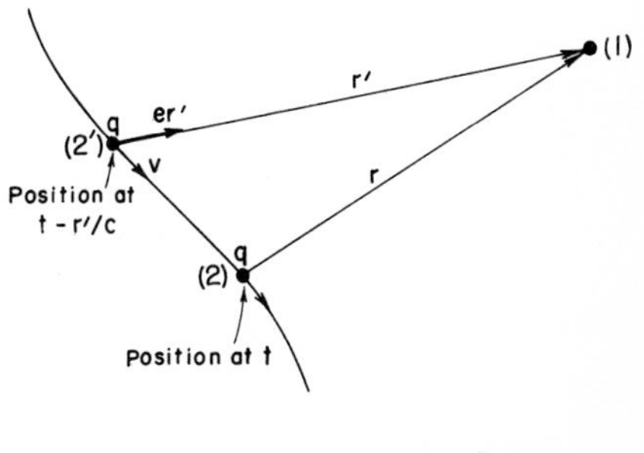

Fig. 21–1. The fields at (1) at the time t depend on the position (2′) occupied by the charge q at the time $(t - r'/c)$.

Let's look at what the complete equation, Eq. (21.1), says in a little more detail. The vector $\boldsymbol{e}_{r'}$ is the unit vector to point (1) from the retarded position (2′). The first term, then, is what we would expect for the Coulomb field of the charge at its retarded position—we may call this "the retarded Coulomb field." The electric field depends inversely on the square of the distance and is directed away from the retarded position of the charge (that is, in the direction of $\boldsymbol{e}_{r'}$).

But that is only the first term. The other terms tell us that the laws of electricity do *not* say that all the fields are the same as the static ones, but just retarded (which is what people sometimes like to say). To the "retarded Coulomb field" we must add the other two terms. The second term says that there is a "correction" to the retarded Coulomb field which is the *rate of change* of the retarded Coulomb field multiplied by r'/c, the retardation delay. In a way of speaking, this term tends to *compensate* for the retardation in the first term. The first *two* terms correspond to computing the "retarded Coulomb field" and then extrapolating it toward the future by the amount r'/c, that is, *right up to the time t*! The extrapolation is linear, as if we were to assume that the "retarded Coulomb field" would continue to change at the rate computed for the charge at the point (2′). If the field is changing slowly, the effect of the retardation is almost completely removed by the correction term, and the two terms together give us an electric field that is the "instantaneous Coulomb field"—that is, the Coulomb field of the charge at the point (2)—to a very good approximation.

Finally, there is a third term in Eq. (21.1) which is the second derivative of the unit vector $\boldsymbol{e}_{r'}$. For our study of the phenomena of light, we made use of the fact that far away from the charge the first two terms went inversely as the square of the distance and, for large distances, became very weak in comparison to the last term, which decreases as $1/r$. So we concentrated entirely on the last term, and we showed that it is (again, for large distances) proportional to the component of the acceleration of the charge at right angles to the line of sight. (Also, for most of our work in Vol. I, we took the case in which the charges were moving nonrelativistically. We considered the relativistic effects in only one chapter, Chapter 36.)

Now we should try to connect the two things together. We have the Maxwell equations, and we have Eq. (21.1) for the field of a point charge. We should certainly ask whether they are equivalent. If we can deduce Eq. (21.1) from Maxwell's equations, we will really understand the connection between light and electromagnetism. To make this connection is the main purpose of this chapter.

It turns out that we won't quite make it—that the mathematical details get too complicated for us to carry through in all their gory details. But we will come close enough so that you should easily see how the connection could be made. The missing pieces will only be in the mathematical details. Some of you may find the mathematics in this chapter rather complicated, and you may not wish to follow the argument very closely. We think it is important, however, to make the connection between what you have learned earlier and what you are learning now, or at least to indicate how such a connection can be made. You will notice, if you look over the earlier chapters, that whenever we have taken a statement as a starting point for a discussion, we have carefully explained whether it is a new "assumption" that is a "basic law," or whether it can ultimately be deduced from some other laws. We owe it to you in the spirit of these lectures to make the connection between light and Maxwell's equations. If it gets difficult in places, well, that's life—there is no other way.

21–2 Spherical waves from a point source

In Chapter 18 we found that Maxwell's equations could be solved by letting

$$\boldsymbol{E} = -\boldsymbol{\nabla}\phi - \frac{\partial \boldsymbol{A}}{\partial t} \tag{21.2}$$

and

$$\boldsymbol{B} = \boldsymbol{\nabla} \times \boldsymbol{A}, \tag{21.3}$$

where ϕ and $\boldsymbol{A}$ must then be solutions of the equations

$$\nabla^2\phi - \frac{1}{c^2}\frac{\partial^2\phi}{\partial t^2} = -\frac{\rho}{\epsilon_0} \tag{21.4}$$

and

$$\nabla^2\boldsymbol{A} - \frac{1}{c^2}\frac{\partial^2\boldsymbol{A}}{\partial t^2} = -\frac{\boldsymbol{j}}{\epsilon_0 c^2}, \tag{21.5}$$

and must also satisfy the condition that

$$\boldsymbol{\nabla}\cdot\boldsymbol{A} = -\frac{1}{c^2}\frac{\partial\phi}{\partial t}. \tag{21.6}$$

Now we will find the solution of Eqs. (21.4) and (21.5). To do that we have to find the solution ψ of the equation

$$\nabla^2\psi - \frac{1}{c^2}\frac{\partial^2\psi}{\partial t^2} = -s, \tag{21.7}$$

where s, which we call the source, is known. Of course, s corresponds to ρ/ϵ_0 and ψ to ϕ for Eq. (21.4), or s is j_x/ϵ_0c^2 if ψ is A_x, etc., but we want to solve Eq. (21.7) as a mathematical problem no matter what ψ and s are physically.

In places where ρ and $\boldsymbol{j}$ are zero—in what we have called "free" space—the potentials ϕ and A, and the fields $\boldsymbol{E}$ and $\boldsymbol{B}$, all satisfy the three-dimensional wave equation without sources, whose mathematical form is

$$\nabla^2\psi - \frac{1}{c^2}\frac{\partial^2\psi}{\partial t^2} = 0. \tag{21.8}$$

In Chapter 20 we saw that solutions of this equation can represent waves of various kinds: plane waves in the x-direction, $\psi = f(t - x/c)$; plane waves in the y- or z-direction, or in any other direction; or spherical waves of the form

$$\psi(x, y, z, t) = \frac{f(t - r/c)}{r}. \tag{21.9}$$

(The solutions can be written in still other ways, for example cylindrical waves that spread out from an axis.)

We also remarked that, physically, Eq. (21.9) does not represent a wave in free space—that there must be charges at the origin to get the outgoing wave started. In other words, Eq. (21.9) is a solution of Eq. (21.8) everywhere except right near $r = 0$, where it must be a solution of the complete equation (21.7), including some sources. Let's see how that works. What kind of a source s in Eq. (21.7) would give rise to a wave like Eq. (21.9)?

Suppose we have the spherical wave of Eq. (21.9) and look at what is happening for very small r. Then the retardation $-r/c$ in $f(t - r/c)$ can be neglected—provided f is a smooth function—and ψ becomes

$$\psi = \frac{f(t)}{r} \qquad (r \to 0). \tag{21.10}$$

So ψ is just like a Coulomb field for a charge at the origin that varies with time. That is, if we had a little lump of charge, limited to a very small region near the origin, with a density ρ, we know that

$$\phi = \frac{Q/4\pi\epsilon_0}{r},$$

where $Q = \int\rho\,dV$. Now we know that such a ϕ satisfies the equation

$$\nabla^2\phi = -\frac{\rho}{\epsilon_0}.$$

Following the same mathematics, we would say that the ψ of Eq. (21.10) satisfies

$$\nabla^2\psi = -s \qquad (r \to 0), \tag{21.11}$$

where s is related to f by

$$f = \frac{S}{4\pi},$$

with

$$S = \int s\,dV.$$

The only difference is that in the general case, s, and therefore S, can be a function of time.

Now the important thing is that if ψ satisfies Eq. (21.11) for small r, it also satisfies Eq. (21.7). As we go very close to the origin, the $1/r$ dependence of ψ

causes the space derivatives to become very large. But the time derivatives keep their same values. [They are just the time derivatives of $f(t)$.] So as r goes to zero, the term $\partial^2\psi/\partial t^2$ in Eq. (21.7) can be neglected in comparison with $\nabla^2\psi$, and Eq. (21.7) becomes equivalent to Eq. (21.11).

To summarize, then, if the source function $s(t)$ of Eq. (21.7) is localized at the origin and has the total strength

$$S(t) = \int s(t)\,dV, \tag{21.12}$$

the solution of Eq. (21.7) is

$$\psi(x, y, z, t) = \frac{1}{4\pi}\,\frac{S(t - r/c)}{r}. \tag{21.13}$$

The only effect of the term $\partial^2\psi/\partial t^2$ in Eq. (21.7) is to introduce the retardation $(t - r/c)$ in the Coulomb-like potential.

21–3 The general solution of Maxwell's equations

We have found the solution of Eq. (21.7) for a "point" source. The next question is: What is the solution for a spread-out source? That's easy; we can think of any source $s(x, y, z, t)$ as made up of the sum of many "point" sources, one for each volume element dV, and each with the source strength $s(x, y, z, t)\,dV$. Since Eq. (21.7) is linear, the resultant field is the superposition of the fields from all of such source elements.

Using the results of the preceding section [Eq. (21.13)] we know that the field $d\psi$ at the point (x_1, y_1, z_1)—or (1) for short—at the time t, from a source element $s\,dV$ at the point (x_2, y_2, z_2)—or (2) for short—is given by

$$d\psi(1, t) = \frac{s(2, t - r_{12}/c)\,dV_2}{4\pi r_{12}},$$

where r_{12} is the distance from (2) to (1). Adding the contributions from all the pieces of the source means, of course, doing an integral over all regions where $s \neq 0$; so we have

$$\psi(1, t) = \int \frac{s(2, t - r_{12}/c)}{4\pi r_{12}}\,dV_2. \tag{21.14}$$

That is, the field at (1) at the time t is the sum of all the spherical waves which leave the source elements at (2) at the times $(t - r_{12}/c)$. This is the solution of our wave equation for any set of sources.

We see now how to obtain a general solution for Maxwell's equations. If for ψ we mean the scalar potential ϕ, the source function s becomes ρ/ϵ_0. Or we can let ψ represent any one of the three components of the vector potential $\boldsymbol{A}$, replacing s by the corresponding component of $\boldsymbol{j}/\epsilon_0 c^2$. Thus, if we know the charge density $\rho(x, y, z, t)$ and the current density $\boldsymbol{j}(x, y, z, t)$ everywhere, we can immediately write down the solutions of Eqs. (21.4) and (21.5). They are

$$\phi(1, t) = \int \frac{\rho(2, t - r_{12}/c)}{4\pi\epsilon_0 r_{12}}\,dV_2 \tag{21.15}$$

and

$$\boldsymbol{A}(1, t) = \int \frac{\boldsymbol{j}(2, t - r_{12}/c)}{4\pi\epsilon_0 c^2 r_{12}}\,dV_2. \tag{21.16}$$

The fields $\boldsymbol{E}$ and $\boldsymbol{B}$ can then be found by differentiating the potentials, using Eqs. (21.2) and (21.3). [Incidentally, it is possible to verify that the ϕ and $\boldsymbol{A}$ obtained from Eqs. (21.15) and (21.16) do satisfy the equality (21.6).]

We have solved Maxwell's equations. Given the currents and charges in any circumstance, we can find the potentials directly from these integrals and then differentiate and get the fields. So we have finished with the Maxwell theory. Also this permits us to close the ring back to our theory of light, because to connect with our earlier work on light, we need only calculate the electric field from a moving charge. All that remains is to take a moving charge, calculate the potentials from these integrals, and then differentiate to find $\boldsymbol{E}$ from $-\nabla\phi - \partial \boldsymbol{A}/\partial t$. We should get Eq. (21.1). It turns out to be lots of work, but that's the principle.

So here is the center of the universe of electromagnetism—the complete theory of electricity and magnetism, and of light; a complete description of the fields produced by any moving charges; and more. It is all here. Here is the structure built by Maxwell, complete in all its power and beauty. It is probably one of the greatest accomplishments of physics. To remind you of its importance, we will put it all together in a nice frame.

Maxwell's equations:

$$\nabla \cdot \boldsymbol{E} = \frac{\rho}{\epsilon_0} \qquad \nabla \cdot \boldsymbol{B} = 0$$

$$\nabla \times \boldsymbol{E} = -\frac{\partial \boldsymbol{B}}{\partial t} \qquad c^2 \nabla \times \boldsymbol{B} = \frac{\boldsymbol{j}}{\epsilon_0} + \frac{\partial \boldsymbol{E}}{\partial t}$$

Their solutions:

$$\boldsymbol{E} = -\nabla\phi - \frac{\partial \boldsymbol{A}}{\partial t}$$

$$\boldsymbol{B} = \nabla \times \boldsymbol{A}$$

$$\phi(1, t) = \int \frac{\rho(2, t - r_{12}/c)}{4\pi\epsilon_0 r_{12}}\, dV_2$$

$$\boldsymbol{A}(1, t) = \int \frac{\boldsymbol{j}(2, t - r_{12}/c)}{4\pi\epsilon_0 c^2 r_{12}}\, dV_2$$

21–4 The fields of an oscillating dipole

We have still not lived up to our promise to derive Eq. (21.1) for the electric field of a point charge in motion. Even with the results we already have, it is a relatively complicated thing to derive. We have not found Eq. (21.1) anywhere in the published literature except in Vol. I of these lectures.* So you can see that it is not easy to derive. (The fields of a moving charge have been written in many other forms that are equivalent, of course.) We will have to limit ourselves here just to showing that, in a few examples, Eqs. (21.15) and (21.16) give the same results as Eq. (21.1). First, we will show that Eq. (21.1) gives the correct fields with only the restriction that the motion of the charged particle is nonrelativistic. (Just this special case will take care of 90 percent, or more, of what we said about light.)

We consider a situation in which we have a blob of charge that is moving about in some way, in a small region, and we will find the fields far away. To put it another way, we are finding the field at any distance from a point charge that is shaking up and down in very small motion. Since light is usually emitted from neutral objects such as atoms, we will consider that our wiggling charge q is located near an equal and opposite charge at rest. If the separation between the centers of the charges is $\boldsymbol{d}$, the charges will have a dipole moment $\boldsymbol{p} = q\boldsymbol{d}$, which we take to be a function of time. Now we should expect that if we look at the fields close to the charges, we won't have to worry about the delay; the electric field will be exactly the same as the one we have calculated earlier for an electrostatic dipole

---

* The formula was worked out by R. P. Feynman, in about 1950, and given in some lectures as a good way of thinking about synchrotron radiation.

—using, of course, the instantaneous dipole moment $\boldsymbol{p}(t)$. But if we go very far out, we ought to find a term in the field that goes as $1/r$ and depends on the acceleration of the charge perpendicular to the line of sight. Let's see if we get such a result.

We begin by calculating the vector potential $\boldsymbol{A}$, using Eq. (21.16). Suppose that our moving charge is in a small blob whose charge density is given by $\rho(x, y, z)$, and the whole thing is moving at any instant with the velocity $\boldsymbol{v}$. Then the current density $\boldsymbol{j}(x, y, z)$ will be equal to $\boldsymbol{v}\rho(x, y, z)$. It will be convenient to take our coordinate system so that the z-axis is in the direction of $\boldsymbol{v}$; then the geometry of our problem is as shown in Fig. 21–2. We want the integral

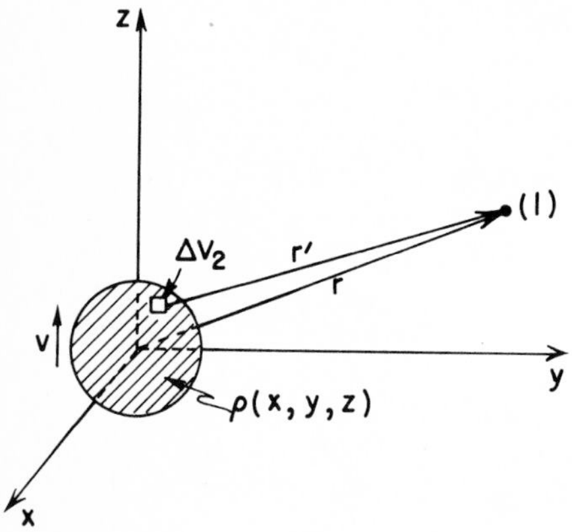

Fig. 21–2. The potentials at (1) are given by integrals over the charge density ρ.

$$\int \frac{\boldsymbol{j}(2, t - r_{12}/c)}{r_{12}}\, dV_2. \tag{21.17}$$

Now if the size of the charge-blob is really very small compared with r_{12}, we can set the r_{12} term in the denominator equal to r, the distance to the center of the blob, and take r outside the integral. Next, we are also going to set $r_{12} = r$ in the numerator, although that is not really quite right. It is not right because we should take $\boldsymbol{j}$ at, say, the top of the blob at a slightly different time than we used for $\boldsymbol{j}$ at the bottom of the blob. When we set $r_{12} = r$ in $\boldsymbol{j}(t - r_{12}/c)$, we are taking the current density for the whole blob at the same time $(t - r/c)$. That is an approximation that will be good only if the velocity v of the charge is much less than c. So we are making a nonrelativistic calculation. Replacing $\boldsymbol{j}$ by $\rho\boldsymbol{v}$, the integral (21.17) becomes

$$\frac{1}{r}\int \boldsymbol{v}\rho(2, t - r/c)\, dV_2.$$

Since all the charge has the same velocity, this integral is just $\boldsymbol{v}/r$ times the total charge q. But $q\boldsymbol{v}$ is just $\partial\boldsymbol{p}/\partial t$, the rate of change of the dipole moment—which is, of course, to be evaluated at the retarded time $(t - r/c)$. We will write it as $\dot{\boldsymbol{p}}(t - r/c)$. So we get for the vector potential

$$\boldsymbol{A}(1, t) = \frac{1}{4\pi\epsilon_0 c^2}\, \frac{\dot{\boldsymbol{p}}(t - r/c)}{r}. \tag{21.18}$$

Our result says that the current in a varying dipole produces a vector potential in the form of spherical waves whose source strength is $\dot{\boldsymbol{p}}/4\pi\epsilon_0 c^2$.

We can now get the magnetic field from $\boldsymbol{B} = \boldsymbol{\nabla} \times \boldsymbol{A}$. Since $\dot{\boldsymbol{p}}$ is totally in the z-direction, $\boldsymbol{A}$ has only a z-component; there are only two nonzero derivatives in the curl. So $B_x = \partial A_z/\partial y$ and $B_y = -\partial A_z/\partial x$. Let's first look at B_x:

$$B_x = \frac{\partial A_z}{\partial y} = \frac{1}{4\pi\epsilon_0 c^2}\, \frac{\partial}{\partial y}\, \frac{\dot{p}(t - r/c)}{r}. \tag{21.19}$$

To carry out the differentiation, we must remember that $r = \sqrt{x^2 + y^2 + z^2}$, so

$$B_x = \frac{1}{4\pi\epsilon_0 c^2}\, \dot{p}(t - r/c)\, \frac{\partial}{\partial y}\left(\frac{1}{r}\right) + \frac{1}{4\pi\epsilon_0 c^2}\, \frac{1}{r}\, \frac{\partial}{\partial y}\, \dot{p}(t - r/c). \tag{21.20}$$

Remembering that $\partial r/\partial y = y/r$, the first term gives

$$-\frac{1}{4\pi\epsilon_0 c^2}\, \frac{y\dot{p}(t - r/c)}{r^3}, \tag{21.21}$$

which drops off as $1/r^2$ like the fields of a static dipole (because y/r is constant for a given direction).

The second term in Eq. (21.20) gives us the new effects. Carrying out the differentiation, we get

$$-\frac{1}{4\pi\epsilon_0 c^2}\, \frac{y}{cr^2}\, \ddot{p}(t - r/c), \tag{21.22}$$

where $\ddot{p}$ means, of course, the second derivative of p with respect to t. This term, which comes from differentiating the numerator, is responsible for radiation. First, it describes a field which decreases with distance only as $1/r$. Second, it depends on the *acceleration* of the charge. You can begin to see how we are going to get a result like Eq. (21.1′), which describes the radiation of light.

Let's examine in a little more detail how this radiation term comes about—it is such an interesting and important result. We start with the expression (21.18), which has a $1/r$ dependence and is therefore like a Coulomb potential, except for the delay term in the numerator. Why is it then that when we differentiate with respect to space coordinates to get the fields, we don't just get a $1/r^2$ field—with, of course, the corresponding time delays?

We can see why in the following way: Suppose that we let our dipole oscillate up and down in a sinusoidal motion. Then we would have

$$p = p_z = p_0 \sin \omega t$$

and

$$A_z = \frac{1}{4\pi\epsilon_0 c^2} \frac{\omega p_0 \cos \omega(t - r/c)}{r}.$$

If we plot a graph of A_z as a function of r at a given instant, we get the curve shown in Fig. 21–3. The peak amplitude decreases as $1/r$, but there is, in addition, an oscillation in space, bounded by the $1/r$ envelope. When we take the spatial derivatives, they will be proportional to the *slope* of the curve. From the figure we see that there are slopes much steeper than the slope of the $1/r$ curve itself. It is, in fact, evident that for a given frequency the peak slopes are proportional to the amplitude of the wave, which varies as $1/r$. So that explains the drop-off rate of the radiation term.

It all comes about because the variations *with time* at the source are translated into variations *in space* as the waves are propagated outward, and the magnetic fields depend on the *spatial* derivatives of the potential.

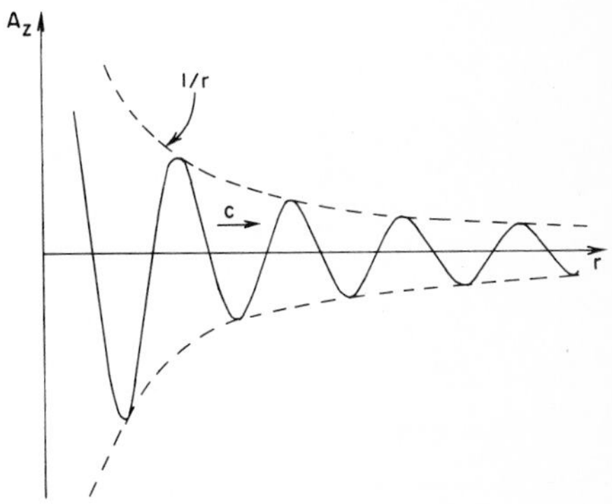

Fig. 21–3. The magnitdue of **A** as a function of r at the instant t for the spherical wave from an oscillating dipole.

Let's go back and finish our calculation of the magnetic field. We have for B_x the two terms (21.21) and (21.22), so

$$B_x = \frac{1}{4\pi\epsilon_0 c^2}\left[-\frac{y\dot{p}(t - r/c)}{r^3} - \frac{y\ddot{p}(t - r/c)}{cr^2}\right].$$

With the same kind of mathematics, we get

$$B_y = \frac{1}{4\pi\epsilon_0 c^2}\left[\frac{x\dot{p}(t - r/c)}{r^3} + \frac{x\ddot{p}(t - r/c)}{cr^2}\right].$$

Or we can put it all together in a nice vector formula:

$$\boldsymbol{B} = \frac{1}{4\pi\epsilon_0 c^2} \frac{[\dot{\boldsymbol{p}} + (r/c)\ddot{\boldsymbol{p}}]_{t-r/c} \times \boldsymbol{r}}{r^3}. \tag{21.23}$$

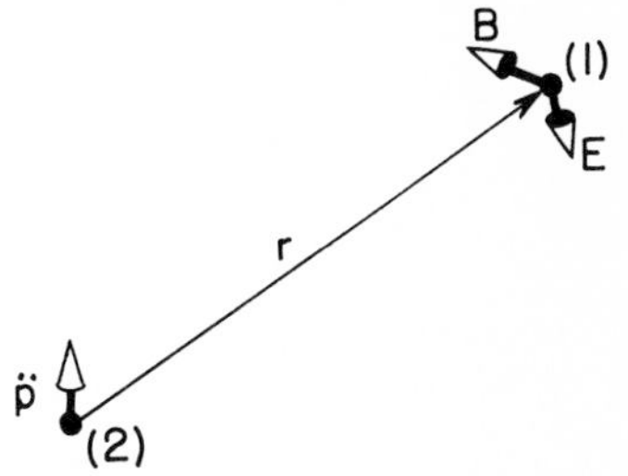

Fig. 21–4. The radiation fields **B** and **E** of an oscillating dipole.

Now let's look at this formula. First of all, if we go very far out in r, only the $\ddot{\boldsymbol{p}}$ term counts. The direction of $\boldsymbol{B}$ is given by $\boldsymbol{p} \times \boldsymbol{r}$, which is at right angles to the radius $\boldsymbol{r}$ and also at right angles to the acceleration, as in Fig. 21–4. Everything is coming out right; that is also the result we get from Eq. (21.1′).

Now let's look at what we are not used to—at what happens closer in. In Section 14–9 we worked out the law of Biot and Savart for the magnetic field of an element of current. We found that a current element $\boldsymbol{j}\,dV$ contributes to the magnetic field the amount

$$d\boldsymbol{B} = \frac{1}{4\pi\epsilon_0 c^2} \frac{\boldsymbol{j} \times \boldsymbol{r}}{r^3}\, dV. \tag{21.24}$$

You see that this formula looks very much like the first term of Eq. (21.23), if we remember that $\dot{\boldsymbol{p}}$ is the current. But there is one difference. In Eq. (21.23), the current is to be evaluated at the time $(t - r/c)$, which doesn't appear in Eq. (21.24). Actually, however, Eq. (21.24) is still very good for small r, because the *second* term of Eq. (21.23) tends to cancel out the effect of the retardation in the first term. The two *together* give a result very near to Eq. (21.24) when r is small.

We can see that this way: When r is small, $(t - r/c)$ is not very different from t, so we can expand the bracket in Eq. (21.23) in a Taylor series. For the first term,

$$\dot{p}(t - r/c) = \dot{p}(t) - \frac{r}{c}\,\ddot{p}(t) + \text{etc.},$$

and to the same order in r/c,

$$\ddot{p}(t - r/c) = \ddot{p}(t).$$

When we take the sum, the two terms in $\ddot{\boldsymbol{p}}$ cancel, and we are left with the *unretarded* current $\dot{\boldsymbol{p}}$: that is, $\dot{p}(t)$—plus terms of order $(r/c)^2$ or higher [e.g., $\frac{1}{2}(r/c)^2\dddot{p}$] which will be very small for r small enough that $\dot{\boldsymbol{p}}$ does not alter markedly in the time r/c.

So Eq. (21.23) gives fields very much like the instantaneous theory—much closer than the instantaneous theory with a delay; the first-order effects of the delay are taken out by the second term. The static formulas are very accurate, much more accurate than you might think. Of course, the compensation only works for points close in. For points far out the correction becomes very bad, because the time delays produce a very large effect, and we get the important $1/r$ term of the radiation.

We still have the problem of computing the electric field and demonstrating that it is the same as Eq. (21.1′). For large distances we can see that the answer is going to come out all right. We know that far from the sources, where we have a propagating wave, $\boldsymbol{E}$ is perpendicular to $\boldsymbol{B}$ (and also to $\boldsymbol{r}$), as in Fig. 21–4, and that $cB = E$. So $\boldsymbol{E}$ is proportional to the acceleration $\ddot{p}$, as expected from Eq. (21.1′).

To get the electric field completely for all distances, we need to solve for the electrostatic potential. When we computed the current integral for $\boldsymbol{A}$ to get Eq. (21.18), we made an approximation by disregarding the slight variation of r in the delay terms. This will not work for the electrostatic potential, because we would then get $1/r$ times the integral of the charge density, which is a constant. This approximation is too rough. We need to go to one higher order. Instead of getting involved in that higher-order computation directly, we can do something else—we can determine the scalar potential from Eq. (21.6), using the vector potential we have already found. The divergence of $\boldsymbol{A}$, in our case, is just $\partial A_z/\partial z$—since A_x and A_y are identically zero. Differentiating in the same way that we did above to find $\boldsymbol{B}$,

$$\begin{aligned}\nabla\cdot\boldsymbol{A} &= \frac{1}{4\pi\epsilon_0 c^2}\left[\dot{p}(t - r/c)\,\frac{\partial}{\partial z}\left(\frac{1}{r}\right) + \frac{1}{r}\,\frac{\partial}{\partial z}\,\dot{p}(t - r/c)\right]\\ &= \frac{1}{4\pi\epsilon_0 c^2}\left[-\frac{z\dot{p}(t - r/c)}{r^3} - \frac{z\ddot{p}(t - r/c)}{cr^2}\right].\end{aligned}$$

Or, in vector notation,

$$\nabla\cdot\boldsymbol{A} = -\frac{1}{4\pi\epsilon_0 c^2}\,\frac{[\dot{\boldsymbol{p}} + (r/c)\ddot{\boldsymbol{p}}]_{t-r/c}\cdot\boldsymbol{r}}{r^3}.$$

Using Eq. (21.6), we have an equation for ϕ:

$$\frac{\partial\phi}{\partial t} = \frac{1}{4\pi\epsilon_0}\,\frac{[\dot{\boldsymbol{p}} + (r/c)\ddot{\boldsymbol{p}}]_{t-r/c}\cdot\boldsymbol{r}}{r^3}.$$

Integrating with respect to t just removes one dot from each of the p's, so

$$\phi(r, t) = \frac{1}{4\pi\epsilon_0}\,\frac{[\boldsymbol{p} + (r/c)\dot{\boldsymbol{p}}]_{t-r/c}\cdot\boldsymbol{r}}{r^3}. \tag{21.25}$$

(The constant of integration would correspond to some superposed static field which could, of course, exist. For the oscillating dipole we have taken, there is no static field.)

We are now able to find the electric field $\boldsymbol{E}$ from

$$\boldsymbol{E} = -\nabla\phi - \frac{\partial \boldsymbol{A}}{\partial t}.$$

Since the steps are tedious but straightforward [providing you remember that $\boldsymbol{p}(t - r/c)$ and its time derivatives depend on x, y, and z through the retardation r/c], we will just give the result:

$$\boldsymbol{E}(r, t) = \frac{-1}{4\pi\epsilon_0 r^3}\left[-\boldsymbol{p}^* - 3\,\frac{(\boldsymbol{p}^*\cdot\boldsymbol{r})\boldsymbol{r}}{r^2} + \frac{1}{c^2}\{\ddot{\boldsymbol{p}}(t - r/c)\times\boldsymbol{r}\}\times\boldsymbol{r}\right] \tag{21.26}$$

with

$$\boldsymbol{p}^* = \boldsymbol{p}(t - r/c) + \frac{r}{c}\,\dot{\boldsymbol{p}}(t - r/c). \tag{21.27}$$

Although it looks rather complicated, the result is easily interpreted. The vector $\boldsymbol{p}^*$ is the dipole moment retarded and then "corrected" for the retardation, so the two terms with $\boldsymbol{p}^*$ give just the static dipole field when r is small. [See Chapter 6, Eq. (6.14).] When r is large, the term in $\ddot{\boldsymbol{p}}$ dominates, and the electric field is proportional to the acceleration of the charges, at right angles to $\boldsymbol{r}$, and, in fact, directed along the projection of $\ddot{\boldsymbol{p}}$ in a plane perpendicular to $\boldsymbol{r}$.

This result agrees with what we would have gotten using Eq. (21.1). Of course, Eq. (21.1) is more general; it works with any motion, while Eq. (21.26) is valid only for small motions for which we can take the retardation r/c as constant over the source. At any rate, we have now provided the underpinnings for our entire previous discussion of light (excepting some matters discussed in Chapter 36 of Vol. I), for it all hinged on the last term of Eq. (21.26). We will discuss next how the fields can be obtained for more rapidly moving charges (leading to the relativistic effects of Chapter 36 of Vol. I).

21–5 The potentials of a moving charge; the general solution of Liénard and Wiechert

In the last section we made a simplification in calculating our integral for $\boldsymbol{A}$ by considering only low velocities. But in doing so we missed an important point and also one where it is easy to go wrong. We will therefore take up now a calculation of the potentials for a point charge moving in any way whatever—even with a relativistic velocity. Once we have this result, we will have the complete electromagnetism of electric charges. Even Eq. (21.1) can then be derived by taking derivatives. The story will be complete. So bear with us.

Let's try to calculate the scalar potential $\phi(1)$ at the point (x_1, y_1, z_1) produced by a *point* charge, such as an electron, moving in any manner whatsoever. By a "point" charge we mean a very small ball of charge, shrunk down as small as you like, with a charge density $\rho(x, y, z)$. We can find ϕ from Eq. (21.15):

$$\phi(1, t) = \frac{1}{4\pi\epsilon_0}\int\frac{\rho(2, t - r_{12}/c)}{r_{12}}\,dV_2. \tag{21.28}$$

The answer would seem to be—and almost everyone would, at first, think—that the integral of ρ over such a "point" charge is just the total charge q, so that

$$\phi(1, t) = \frac{1}{4\pi\epsilon_0}\frac{q}{r'_{12}} \qquad \text{(wrong)}.$$

By $\boldsymbol{r}'_{12}$ we mean the radius vector from the charge at point (2) to point (1) at the retarded time $(t - r_{12}/c)$. It is wrong.

The correct answer is

$$\phi(1, t) = \frac{1}{4\pi\epsilon_0}\frac{q}{r'_{12}}\cdot\frac{1}{1 - v_{r'}/c}, \tag{21.29}$$

where $v_{r'}$ is the component of the velocity of the charge parallel to r'_{12}—namely, toward point (1). We will now show you why. To make the argument easier to

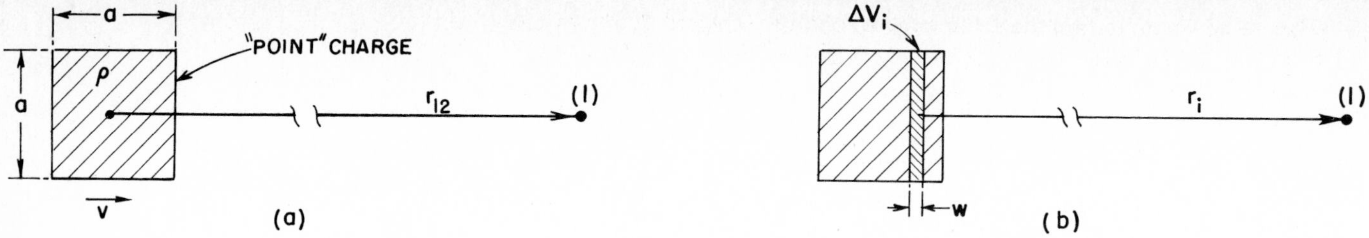

Fig. 21–5. (a) A "point" charge—considered as a small cubical distribution of charge—moving with the speed v toward point (1). (b) The volume element ΔV_i used for calculating the potentials.

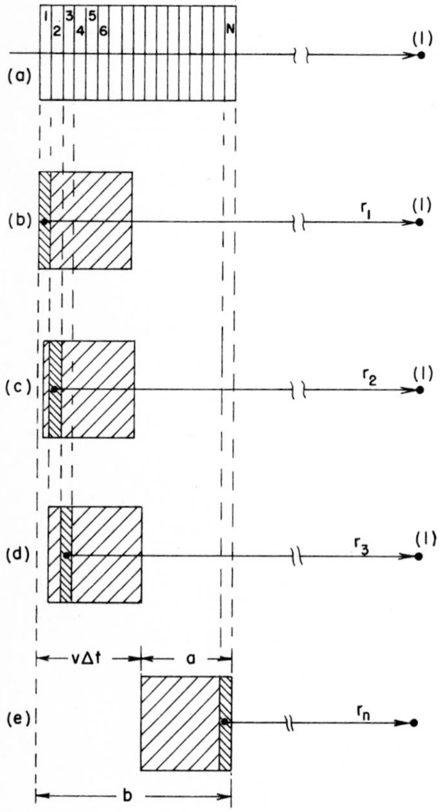

Fig. 21–6. Integrating $\rho(t - r'/c)\,dV$ for a moving charge.

follow, we will make the calculation first for a "point" charge which is in the form of a little cube of charge moving toward the point (1) with the speed v, as shown in Fig. 21–5(a). Let the length of a side of the cube be a, which we take to be much, much less than r_{12}, the distance from the center of the charge to the point (1).

Now to evaluate the integral of Eq. (21.28), we will return to basic principles; we will write it as the sum

$$\sum_i \frac{\rho_i\,\Delta V_i}{r_i}, \tag{21.30}$$

where r_i is the distance from point (1) to the ith volume element ΔV_i and ρ_i is the charge density at ΔV_i at the time $t_i = t - r_i/c$. Since $r_i \gg a$, always, it will be convenient to take our ΔV_i in the form of thin, rectangular slices perpendicular to $\boldsymbol{r}_{12}$, as shown in Fig. 21–5(b).

Suppose we start by taking the volume elements ΔV_i with some thickness w much less than a. The individual elements will appear as shown in Fig. 21–6(a), where we have put in more than enough to cover the charge. But we have *not* shown the charge, and for a good reason. Where should we draw it? For each volume element ΔV_i, we are to take ρ at the time $t_i = (t - r_i/c)$, but since the charge is *moving*, it is in a *different place for each volume element* ΔV_i!

Let's say that we begin with the volume element labeled "1" in Fig. 21–6(a), chosen so that at the time $t_1 = (t - r_1/c)$ the "back" edge of the charge occupies ΔV_1, as shown in Fig. 21–6(b). Then when we evalute $\rho_2\,\Delta V_2$, we must use the position of the charge at the slightly *later* time $t_2 = (t - r_2/c)$, when the charge will be in the position shown in Fig. 21–6(c). And so on, for ΔV_3, ΔV_4, etc. Now we can evaluate the sum.

Since the thickness of each ΔV_i is w, its volume is wa^2. Then each volume element that overlaps the charge distribution contains the amount of charge $wa^2\rho$, where ρ is the density of charge within the cube—which we take to be uniform. When the distance from the charge to point (1) is large, we will make a negligible error by setting all the r_i's in the denominators equal to some average value, say the retarded position r' of the center of the charge. Then the sum (21.30) is

$$\sum_{i=1}^{N} \frac{\rho w a^2}{r'},$$

where ΔV_N is the last ΔV_i that overlaps the charge distributions, as shown in Fig. 21–6(e). The sum is, clearly,

$$N\,\frac{\rho w a^2}{r'} = \frac{\rho a^3}{r'}\left(\frac{Nw}{a}\right).$$

Now ρa^3 is just the total charge q and Nw is the length b shown in part (e) of the figure. So we have

$$\phi = \frac{q}{4\pi\epsilon_0 r'}\left(\frac{b}{a}\right). \tag{21.31}$$

What is b? It is the length of the cube of charge *increased* by the distance moved by the charge between $t_1 = (t - r_1/c)$ and $t_N = (t - r_N/c)$—which is the distance the charge moves in the time

$$\Delta t = t_N - t_1 = (r_1 - r_N)/c = b/c.$$

Since the speed of the charge is v, the distance moved is $v\,\Delta t = vb/c$. But the length b is this distance added to a:

$$b = a + \frac{v}{c}\,b.$$

Solving for b, we get

$$b = \frac{a}{1 - (v/c)}.$$

Of course by v we mean the velocity at the retarded time $t' = (t - r'/c)$, which we can indicate by writing $[1 - v/c]_{\text{ret}}$, and Eq. (21.31) for the potential becomes

$$\phi(1, t) = \frac{q}{4\pi\epsilon_0 r'}\,\frac{1}{[1 - (v/c)]_{\text{ret}}}.$$

This result agrees with our assertion, Eq. (21.29). There is a correction term which comes about because the charge is moving as our integral "sweeps over the charge." When the charge is moving toward the point (1), its contribution to the integral is increased by the ratio b/a. Therefore the correct integral is q/r' multiplied by b/a, which is $1/[1 - v/c]_{\text{ret}}$.

If the velocity of the charge is not directed toward the observation point (1), you can see that what matters is the *component* of its velocity toward point (1). Calling this velocity component v_r, the correction factor is $1/[1 - v_r/c]_{\text{ret}}$. Also, the analysis we have made goes exactly the same way for a charge distribution of *any* shape—it doesn't have to be a cube. Finally, since the "size" of the charge q doesn't enter into the final result, the same result holds when we let the charge shrink to any size—even to a point. The general result is that the scalar potential for a point charge moving with any velocity is

$$\phi(t) = \frac{q}{4\pi\epsilon_0 r'[1 - (v_r/c)]_{\text{ret}}}. \tag{21.32}$$

This equation is often written in the equivalent form

$$\phi(1, t) = \frac{q}{4\pi\epsilon_0[r - (\boldsymbol{v}\cdot\boldsymbol{r}/c)]_{\text{ret}}}, \tag{21.33}$$

where $\boldsymbol{r}$ is the vector from the charge to the point (1), where ϕ is being evaluated, and all the quantities in the bracket are to have their values at the retarded time $t' = t - r'/c$.

The same thing happens when we compute $\boldsymbol{A}$ for a point charge, from Eq. (21.16). The current density is $\rho\boldsymbol{v}$ and the integral over ρ is the same as we found for ϕ. The vector potential is

$$\boldsymbol{A}(1, t) = \frac{q\boldsymbol{v}}{4\pi\epsilon_0 c^2[r - (\boldsymbol{v}\cdot\boldsymbol{r}/c)]_{\text{ret}}}. \tag{21.34}$$

The potentials for a point charge were first deduced in this form by Liénard and Wiechert and are called the *Liénard-Wiechert potentials.*

To close the ring back to Eq. (21.1) it is only necessary to compute $\boldsymbol{E}$ and $\boldsymbol{B}$ from these potentials (using $\boldsymbol{B} = \boldsymbol{\nabla}\times\boldsymbol{A}$ and $\boldsymbol{E} = -\boldsymbol{\nabla}\phi - \partial\boldsymbol{A}/\partial t$). It is now only arithmetic. The arithmetic, however, is fairly involved, so we will not write out the details. Perhaps you will take our word for it that Eq. (21.1) is equivalent to the Liénard-Wiechert potentials we have derived.*

---

* If you have a lot of paper and time you can try to work it through yourself. We would, then, make two suggestions: First, don't forget that the derivatives of r' are complicated, since it is a function of t'. Second, don't try to *derive* (21.1), but carry out all of the derivatives in it, and then compare what you get with the $\boldsymbol{E}$ obtained from the potentials (21.33) and (21.34).

21–6 The potentials for a charge moving with constant velocity; the Lorentz formula

We want next to use the Liénard-Wiechert potentials for a special case—to find the fields of a charge moving with uniform velocity in a straight line. We will do it again later, using the principle of relativity. We already know what the potentials are when we are standing in the rest frame of a charge. When the charge is moving, we can figure everything out by a relativistic transformation from one system to the other. But relativity had its origin in the theory of electricity and magnetism. The formulas of the Lorentz transformation (Chapter 15, Vol. I) were discoveries made by Lorentz when he was studying the equations of electricity and magnetism. So that you can appreciate where things have come from, we would like to show that the Maxwell equations do lead to the Lorentz transformation. We begin by calculating the potentials of a charge moving with uniform velocity, directly from the electrodynamics of Maxwell's equations. We have shown that Maxwell's equations lead to the potentials for a moving charge that we got in the last section. So when we use these potentials, we are using Maxwell's theory.

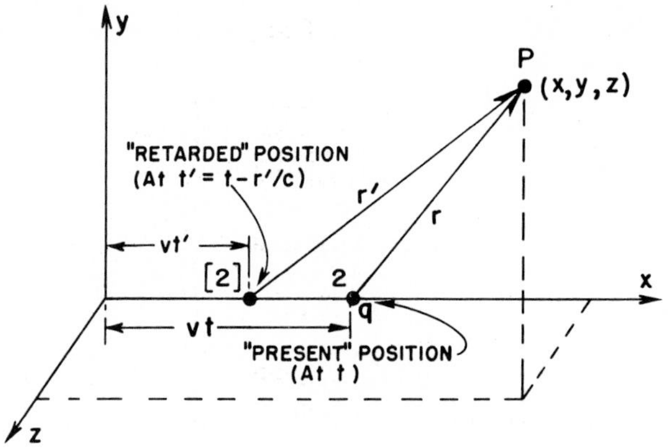

Fig. 21–7. Finding the potential at P of a charge moving with uniform velocity along the x-axis.

Suppose we have a charge moving along the x-axis with the speed v. We want the potentials at the point $P(x, y, z)$, as shown in Fig. 21–7. If $t = 0$ is the moment when the charge is at the origin, at the time t the charge is at $x = vt$, $y = z = 0$. What we need to know, however, is its position at the retarded time

$$t' = t - \frac{r'}{c}, \tag{21.35}$$

where r' is the distance to the point P from the charge *at the retarded time*. At the earlier time t', the charge was at $x = vt'$, so

$$r' = \sqrt{(x - vt')^2 + y^2 + z^2}. \tag{21.36}$$

To find r' or t' we have to combine this equation with Eq. (21.35). First, we eliminate r' by solving Eq. (21.35) for r' and substituting in Eq. (21.36). Then, squaring both sides, we get

$$c^2(t - t')^2 = (x - vt')^2 + y^2 + z^2,$$

which is a quadratic equation in t'. Expanding the squared binomials and collecting like terms in t', we get

$$(v^2 - c^2)t'^2 - 2(xv - c^2t)t' + x^2 + y^2 + z^2 - (ct)^2 = 0.$$

Solving for t',

$$\left(1 - \frac{v^2}{c^2}\right)t' = t - \frac{vx}{c^2} - \frac{1}{c}\sqrt{(x - vt)^2 + \left(1 - \frac{v^2}{c^2}\right)(y^2 + z^2)}\cdot \tag{21.37}$$

To get r' we have to substitute this expression for t' into

$$r' = c(t - t').$$

Now we are ready to find ϕ from Eq. (21.33), which, since $\boldsymbol{v}$ is constant, becomes

$$\phi(x, y, z, t) = \frac{q}{4\pi\epsilon_0} \frac{1}{r' - (\boldsymbol{v} \cdot \boldsymbol{r}'/c)}. \tag{21.38}$$

The component of $\boldsymbol{v}$ in the direction of $\boldsymbol{r}'$ is $v \times (x - vt)/r'$, so $\boldsymbol{v} \cdot \boldsymbol{r}'$ is just $v \times (x - vt')$, and the whole denominator is

$$c(t - t') - \frac{v}{c}(x - vt') = c\left[t - \frac{vx}{c^2} - \left(1 - \frac{v^2}{c^2}\right)t'\right].$$

Substituting for $(1 - v^2/c^2)t'$ from Eq. (21.37), we get for ϕ

$$\phi(x, y, z, t) = \frac{q}{4\pi\epsilon_0} \frac{1}{\sqrt{(x - vt)^2 + \left(1 - \frac{v^2}{c^2}\right)(y^2 + z^2)}}.$$

This equation is more understandable if we rewrite it as

$$\phi(x, y, z, t) = \frac{q}{4\pi\epsilon_0} \frac{1}{\sqrt{1 - \frac{v^2}{c^2}}} \frac{1}{\left[\left(\frac{x - vt}{\sqrt{1 - v^2/c^2}}\right)^2 + y^2 + z^2\right]^{1/2}}. \tag{21.39}$$

The vector potential $\boldsymbol{A}$ is the same expression with an additional factor of $\boldsymbol{v}/c^2$:

$$\boldsymbol{A} = \frac{\boldsymbol{v}}{c^2}\phi.$$

In Eq. (21.39) you can clearly see the beginning of the Lorentz transformation. If the charge were at the origin in its own rest frame, its potential would be

$$\phi(x, y, z) = \frac{q}{4\pi\epsilon_0} \frac{1}{[x^2 + y^2 + z^2]^{1/2}}.$$

We are seeing it in a moving coordinate system, and it appears that the coordinates should be transformed by

$$x \to \frac{x - vt}{\sqrt{1 - v^2/c^2}},$$
$$y \to y,$$
$$z \to z.$$

That is just the Lorentz transformation, and what we have done is essentially the way Lorentz discovered it.

But what about that extra factor $1/\sqrt{1 - v^2/c^2}$ that appears at the front of Eq. (21.39)? Also, how does the vector potential $\boldsymbol{A}$ appear, when it is everywhere zero in the rest frame of the particle? We will soon show that $\boldsymbol{A}$ and ϕ *together* constitute a four-vector, like the momentum $\boldsymbol{p}$ and the total energy U of a particle. The extra $1/\sqrt{1 - v^2/c^2}$ in Eq. (21.39) is the same factor that always comes in when one transforms the components of a four-vector—just as the charge density ρ transforms to $\rho/\sqrt{1 - v^2/c^2}$. In fact, it is almost apparent from Eqs. (21.4) and (21.5) that $\boldsymbol{A}$ and ϕ are components of a four-vector, because we have already shown in Chapter 13 that $\boldsymbol{j}$ and ρ are the components of a four-vector.

Later we will take up in more detail the relativity of electrodynamics; here we only wished to show how naturally the Maxwell equations lead to the Lorentz transformation. You will not, then, be surprised to find that the laws of electricity and magnetism are already correct for Einstein's relativity. We will not have to "fix them up," as we had to do for Newton's laws of mechanics.

22

AC Circuits

Review: Chapter 22, Vol. I, *Algebra*
Chapter 23, Vol. I, *Resonance*
Chapter 25, Vol. I, *Linear Systems and Review*

22–1 Impedances

Most of our work in this course has been aimed at reaching the complete equations of Maxwell. In the last two chapters we have been discussing the consequences of these equations. We have found that the equations contain all the static phenomena we had worked out earlier, as well as the phenomena of electromagnetic waves and light that we had gone over in some detail in Volume I. The Maxwell equations give both phenomena, depending upon whether one computes the fields close to the currents and charges, or very far from them. There is not much interesting to say about the intermediate region; no special phenomena appear there.

There still remain, however, several subjects in electromagnetism that we want to take up. We want to discuss the question of relativity and the Maxwell equations—what happens when one looks at the Maxwell equations with respect to moving coordinate systems. There is also the question of the conservation of energy in electromagnetic systems. Then there is the broad subject of the electromagnetic properties of materials; so far, except for the study of the properties of dielectrics, we have considered only the electromagnetic fields in free space. And although we covered the subject of light in some detail in Volume I, there are still a few things we would like to do again from the point of view of the field equations.

In particular, we want to take up again the subject of the index of refraction, particularly for dense materials. Finally, there are the phenomena associated with waves confined in a limited region of space. We touched on this kind of problem briefly when we were studying sound waves. Maxwell's equations lead also to solutions which represent confined waves of the electric and magnetic fields. We will take up this subject, which has important technical applications, in some of the following chapters. In order to lead up to that subject, we will begin by considering the properties of electrical circuits at low frequencies. We will then be able to make a comparison between those situations in which the almost static approximations of Maxwell's equations are applicable and those situations in which high-frequency effects are dominant.

So we descend from the great and esoteric heights of the last few chapters and turn to the relatively low-level subject of electrical circuits. We will see, however, that even such a mundane subject, when looked at in sufficient detail, can contain great complications.

We have already discussed some of the properties of electrical circuits in Chapters 23 and 25 of Vol. I. Now we will cover some of the same material again, but in greater detail. Again we are going to deal only with linear systems and with voltages and currents which all vary sinusoidally; we can then represent all voltages and currents by complex numbers, using the exponential notation described in Chapter 22 of Vol. I. Thus a time-varying voltage $V(t)$ will be written

$$V(t) = \hat{V}e^{i\omega t}, \tag{22.1}$$

where $\hat{V}$ represents a complex number that is independent of t. It is, of course, understood that the actual time-varying voltage $V(t)$ is given by the *real part* of the complex function on the right-hand side of the equation.

Similarly, all of our other time-varying quantities will be taken to vary sinusoidally at the same frequency ω. So we write

$$
\begin{aligned}
I &= \hat{I}\, e^{i\omega t} \quad \text{(current)},\\
\mathcal{E} &= \hat{\mathcal{E}}\, e^{i\omega t} \quad \text{(emf)},\\
\boldsymbol{E} &= \hat{\boldsymbol{E}}\, e^{i\omega t} \quad \text{(electric field)},
\end{aligned}
\tag{22.2}
$$

and so on.

Most of the time we will write our equations in terms of $V, I, \mathcal{E}, \ldots$ (instead of in terms of $\hat{V}, \hat{I}, \hat{\mathcal{E}}, \ldots$), remembering, though, that the time variations are as given in (22.2).

In our earlier discussion of circuits we assumed that such things as inductances, capacitances, and resistances were familiar to you. We want now to look in a little more detail at what is meant by these idealized circuit elements. We begin with the inductance.

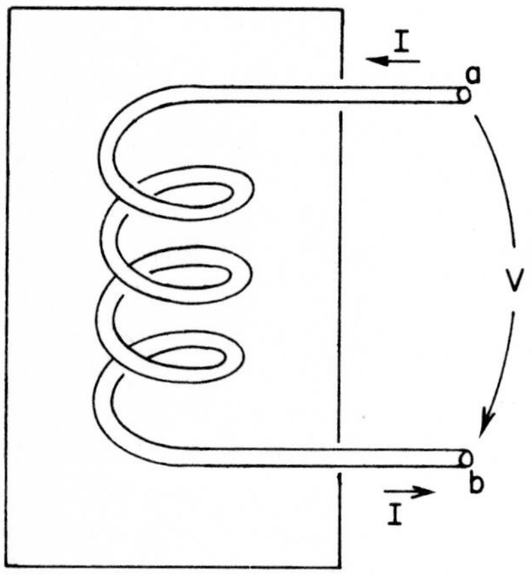

Fig. 22-1. An inductance.

An inductance is made by winding many turns of wire in the form of a coil and bringing the two ends out to terminals at some distance from the coil, as shown in Fig. 22–1. We want to assume that the magnetic field produced by currents in the coil does not spread out strongly all over space and interact with other parts of the circuit. This is usually arranged by winding the coil in a doughnut-shaped form, or by confining the magnetic field by winding the coil on a suitable iron core, or by placing the coil in some suitable metal box, as indicated schematically in Fig. 22–1. In any case, we assume that there is a negligible magnetic field in the external region near the terminals a and b. We are also going to assume that we can neglect any electrical resistance in the wire of the coil. Finally, we will assume that we can neglect the amount of electrical charge that appears on the surface of a wire in building up the electric fields.

With all these approximations we have what we call an "ideal" inductance. (We will come back later and discuss what happens in a real inductance.) For an ideal inductance we say that the voltage across the terminals is equal to $L(dI/dt)$. Let's see why that is so. When there is a current through the inductance, a magnetic field proportional to the current is built up inside the coil. If the current changes with time, the magnetic field also changes. In general, the curl of $\boldsymbol{E}$ is equal to $-d\boldsymbol{B}/dt$; or, put differently, the line integral of $\boldsymbol{E}$ all the way around any closed path is equal to the negative of the rate of change of the flux of $\boldsymbol{B}$ through the loop. Now suppose we consider the following path: Begin at terminal a and go along the coil (staying always inside the wire) to terminal b; then return from terminal b to terminal a through the air in the space outside the inductance. The line integral of $\boldsymbol{E}$ around this closed path can be written as the sum of two parts:

$$
\oint \boldsymbol{E}\cdot d\boldsymbol{s} = \int_{\substack{a\\ \text{via}\\ \text{coil}}}^{b} \boldsymbol{E}\cdot d\boldsymbol{s} + \int_{\substack{b\\ \text{outside}}}^{a} \boldsymbol{E}\cdot d\boldsymbol{s}. \tag{22.3}
$$

As we have seen before, there can be no electric fields inside a perfect conductor. (The smallest fields would produce infinite currents.) Therefore the integral from a to b via the coil is zero. The whole contribution to the line integral of $\boldsymbol{E}$ comes from the path outside the inductance from terminal b to terminal a. Since we have assumed that there are no magnetic fields in the space outside of the "box," this part of the integral is independent of the path chosen and we can define the potentials of the two terminals. The difference of these two potentials is what we call the voltage difference, or simply the voltage V, so we have

$$
V = -\int_b^a \boldsymbol{E}\cdot d\boldsymbol{s} = -\oint \boldsymbol{E}\cdot d\boldsymbol{s}.
$$

The complete line integral is what we have before called the electromotive force $\mathcal{E}$ and is, of course, equal to the rate of change of the magnetic flux in the coil. We have seen earlier that this emf is equal to the negative rate of change of the current, so we have

$$V = -\mathcal{E} = L\frac{dI}{dt},$$

where L is the inductance of the coil. Since $dI/dt = i\omega I$, we have

$$V = i\omega LI. \tag{22.4}$$

The way we have described the ideal inductance illustrates the general approach to other ideal circuit elements—usually called "lumped" elements. The properties of the element are described completely in terms of currents and voltages that appear at the terminals. By making suitable approximations, it is possible to ignore the great complexities of the fields that appear inside the object. A separation is made between what happens inside and what happens outside.

For all the circuit elements we will find a relation like the one in Eq. (22.4), in which the voltage is proportional to the current with a proportionality constant that is, in general, a complex number. This complex coefficient of proportionality is called the *impedance* and is usually written as z (not to be confused with the z-coordinate). It is, in general, a function of the frequency ω. So for any lumped element we write

$$\frac{V}{I} = \frac{\hat{V}}{\hat{I}} = z. \tag{22.5}$$

For an inductance, we have

$$z(\text{inductance}) = z_L = i\omega L. \tag{22.6}$$

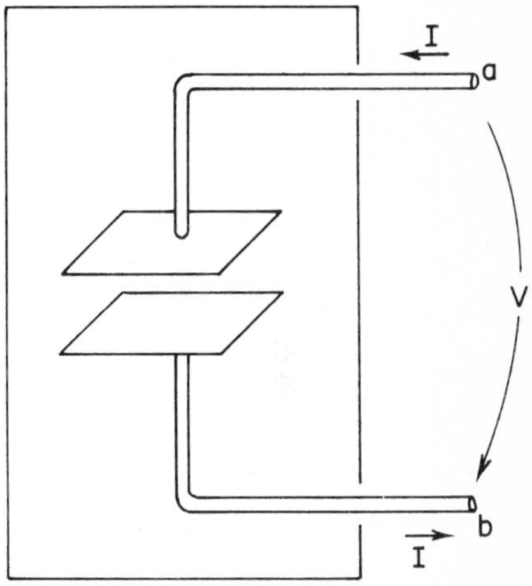

Fig. 22–2. A capacitor (or condenser).

Now let's look at a capacitor from the same point of view.* A capacitor consists of a pair of conducting plates from which two wires are brought out to suitable terminals. The plates may be of any shape whatsoever, and are often separated by some dielectric material. We illustrate such a situation schematically in Fig. 22–2. Again we make several simplifying assumptions. We assume that the plates and the wires are perfect conductors. We also assume that the insulation between the plates is perfect, so that no charges can flow across the insulation from one plate to the other. Next, we assume that the two conductors are close to each other but far from all others, so that all field lines which leave one plate end up on the other. Then there are always equal and opposite charges on the two plates and the charges on the plates are much larger than the charges on the surfaces of the lead-in wires. Finally, we assume that there are no magnetic fields close to the capacitor.

Suppose now we consider the line integral of $\boldsymbol{E}$ around a closed loop which starts at terminal a, goes along inside the wire to the top plate of the capacitor, jumps across the space between the plates, passes from the lower plate to terminal b through the wire, and returns to terminal a in the space outside the capacitor. Since there is no magnetic field, the line integral of $\boldsymbol{E}$ around this closed path is zero. The integral can be broken down into three parts:

$$\oint \boldsymbol{E}\cdot d\boldsymbol{s} = \int_{\substack{\text{along}\\ \text{wires}}} \boldsymbol{E}\cdot d\boldsymbol{s} + \int_{\substack{\text{between}\\ \text{plates}}} \boldsymbol{E}\cdot d\boldsymbol{s} + \int_{b\atop \text{outside}}^{a} \boldsymbol{E}\cdot d\boldsymbol{s}. \tag{22.7}$$

The integral along the wires is zero, because there are no electric fields inside perfect conductors. The integral from b to a outside the capacitor is equal to the negative of the potential difference between the terminals. Since we imagined that the two plates are in some way isolated from the rest of the world, the total charge on

* There are people who say we should call the *objects* by the names "inductor" and "capacitor" and call their *properties* "inductance" and "capacitance" (by analogy with "resistor" and "resistance"). We would rather use the words you will hear in the laboratory. Most people still say "inductance" for both the physical coil and its inductance L. The word "capacitor" seems to have caught on—although you will still hear "condenser" fairly often—and most people still prefer the sound of "capacity" to "capacitance."

the two plates must be zero; if there is a charge Q on the upper plate, there is an equal, opposite charge $-Q$ on the lower plate. We have seen earlier that if two conductors have equal and opposite charges, plus and minus Q, the potential difference between the plates is equal to Q/C, where C is called the capacity of the two conductors. From Eq. (22.7) the potential difference between the terminals a and b is equal to the potential difference between the plates. We have, therefore, that

$$V = \frac{Q}{C}.$$

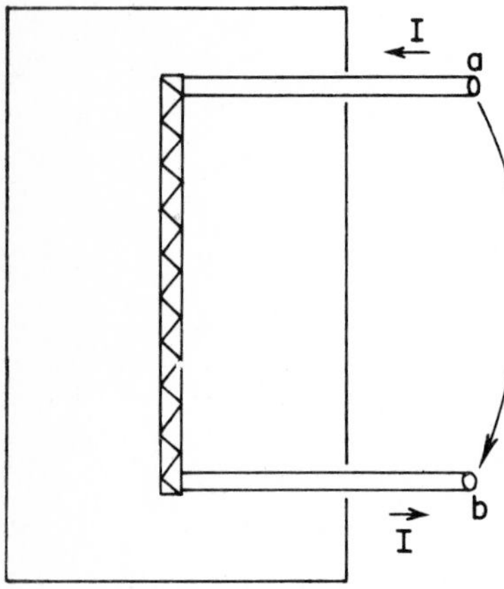

Fig. 22–3. A resistor.

The electric current I entering the capacitor through terminal a (and leaving through terminal b) is equal to dQ/dt, the rate of change of the electric charge on the plates. Writing dV/dt as $i\omega V$, we can put the voltage current relationship for a capacitor in the following way:

$$i\omega V = \frac{I}{C},$$

or

$$V = \frac{I}{i\omega C}. \tag{22.8}$$

The impedance z of a capacitor, is then

$$z\text{ (capacitor)} = z_C = \frac{1}{i\omega C}. \tag{22.9}$$

The third element we want to consider is a resistor. However, since we have not yet discussed the electrical properties of real materials, we are not yet ready to talk about what happens inside a real conductor. We will just have to accept as fact that electric fields can exist inside real materials, that these electric fields give rise to a flow of electric charge—that is, to a current—and that this current is proportional to the integral of the electric field from one end of the conductor to the other. We then imagine an ideal resistor constructed as in the diagram of Fig. 22–3. Two wires which we take to be perfect conductors go from the terminals a and b to the two ends of a bar of resistive material. Following our usual line of argument, the potential difference between the terminals a and b is equal to the line integral of the external electric field, which is also equal to the line integral of the electric field through the bar of resistive material. It then follows that the current I through the resistor is proportional to the terminal voltage V:

$$I = \frac{V}{R},$$

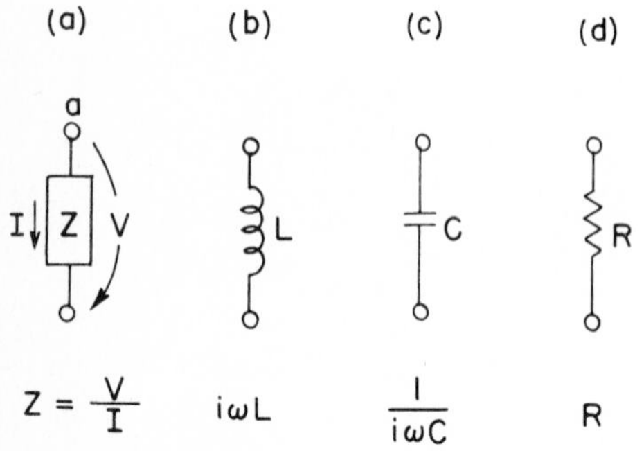

Fig. 22–4. The ideal lumped circuit elements (passive).

where R is called the resistance. We will see later that the relation between the current and the voltage for real conducting materials is only approximately linear. We will also see that this approximate proportionality is expected to be independent of the frequency of variation of the current and voltage only if the frequency is not too high. For alternating currents then, the voltage across a resistor is in phase with the current, which means that the impedance is a real number.

$$z\text{ (resistance)} = z_R = R. \tag{22.10}$$

Our results for the three lumped circuit elements—the inductor, the capacitor, and the resistor—are summarized in Fig. 22–4. In this figure, as well as in the preceding ones, we have indicated the voltage by an arrow that is directed from one terminal to another. If the voltage is "positive"—that is, if the terminal a is at a *higher* potential than the terminal b—the arrow indicates the direction of a positive "voltage drop."

Although we are talking about alternating currents, we can of course include the special case of circuits with steady currents by taking the limit as the frequency ω goes to zero. For zero frequency—that is, for DC—the impedance of an inductance goes to zero; it becomes a short circuit. For DC, the impedance of a condenser goes to infinity; it becomes an open circuit. Since the impedance of a resistor is independent of frequency, it is the only element left when we analyze a circuit for DC.

In the circuit elements we have described so far, the current and voltage are proportional to each other. If one is zero, so also is the other. We usually think in terms like these: An applied voltage is "responsible" for the current, or a current "gives rise to" a voltage across the terminals; so in a sense the elements "respond" to the "applied" external conditions. For this reason these elements are called *passive elements*. They can thus be contrasted with the active elements, such as the generators we will consider in the next section, which are the *sources* of the oscillating currents or voltages in a circuit.

22–2 Generators

Now we want to talk about an *active* circuit element—one that is a source of the currents and voltages in a circuit—namely, a *generator*.

Suppose that we have a coil like an inductance except that it has very few turns, so that we may neglect the magnetic field of its own current. This coil, however, sits in a changing magnetic field such as might be produced by a rotating magnet, as sketched in Fig. 22–5. (We have seen earlier that such a rotating magnetic field can also be produced by a suitable set of coils with alternating currents.) Again we must make several simplifying assumptions. The assumptions we will make are all the ones that we described for the case of the inductance. In particular, we assume that the varying magnetic field is restricted to a definite region in the vicinity of the coil and does not appear outside the generator in the space between the terminals.

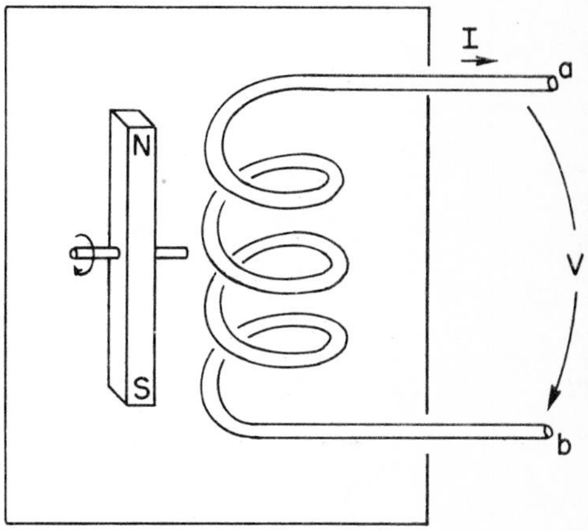

Fig. 22–5. A generator consisting of a fixed coil and a rotating magnetic field.

Following closely the analysis we made for the inductance, we consider the line integral of $\boldsymbol{E}$ around a complete loop that starts at terminal a, goes through the coil to terminal b and returns to its starting point in the space between the two terminals. Again we conclude that the potential difference between the terminals is equal to the total line integral of $\boldsymbol{E}$ around the loop:

$$V = -\oint \boldsymbol{E} \cdot d\boldsymbol{s}.$$

This line integral is equal to the emf in the circuit, so the potential difference V across the terminals of the generator is also equal to the rate of change of the magnetic flux linking the coil:

$$V = -\mathcal{E} = \frac{d}{dt}\,(\text{flux}). \tag{22.11}$$

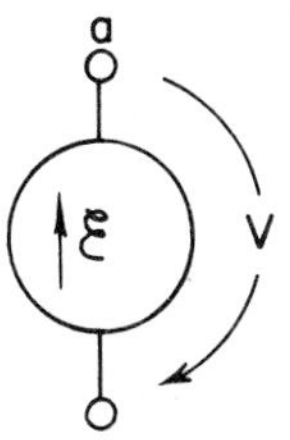

Fig. 22–6. Symbol for an ideal generator.

For an ideal generator we assume that the magnetic flux linking the coil is determined by external conditions—such as the angular velocity of a rotating magnetic field—and is not influenced in any way by the currents through the generator. Thus a generator—at least the *ideal* generator we are considering—is not an impedance. The potential difference across its terminals is determined by the arbitrarily assigned electromotive force $\mathcal{E}(t)$. Such an ideal generator is represented by the symbol shown in Fig. 22–6. The little arrow represents the direction of the emf when it is positive. A positive emf in the generator of Fig. 22–6 will produce a voltage $V = \mathcal{E}$, with the terminal a at a higher potential than the terminal b.

There is another way to make a generator which is quite different on the inside but which is indistinguishable from the one we have just described insofar as what happens beyond its terminals. Suppose we have a coil of wire which is rotated in a *fixed* magnetic field, as indicated in Fig. 22–7. We show a bar magnet to indicate the presence of a magnetic field; it could, of course, be replaced by any other source of a steady magnetic field, such as an additional coil carrying a steady current. As shown in the figure, connections from the rotating coil are made to the outside world by means of sliding contacts or "slip rings." Again, we are interested in the potential difference that appears across the two terminals

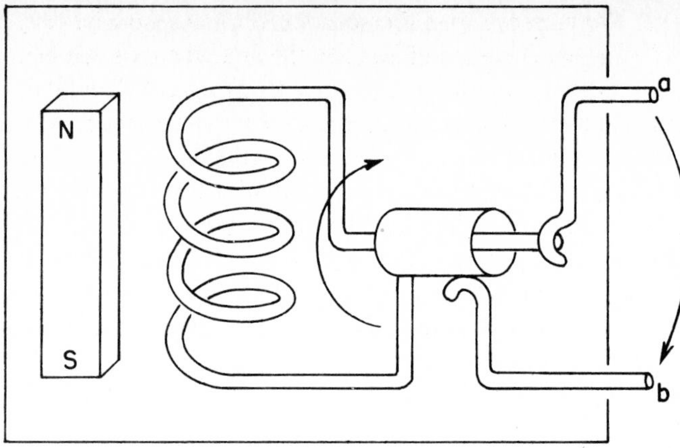

Fig. 22–7. A generator consisting of a coil rotating in a fixed magnetic field.

a and b, which is of course the integral of the electric field from terminal a to terminal b along a path outside the generator.

Now in the system of Fig. 22–7 there are no changing magnetic fields, so we might at first wonder how any voltage could appear at the generator terminals. In fact, there are no electric fields anywhere inside the generator. We are, as usual, assuming for our ideal elements that the wires inside are made of a perfectly conducting material, and as we have said many times, the electric field inside a perfect conductor is equal to zero. But that is not true. It is not true when a conductor is moving in a magnetic field. The true statement is that the total *force* on any charge inside a perfect conductor must be zero. Otherwise there would be an infinite flow of the free charges. So what is always true is that the sum of the electric field $\boldsymbol{E}$ and the cross product of the velocity of the conductor and the magnetic field $\boldsymbol{B}$—which is the total force on a unit charge—must have the value zero inside the conductor:

$$F/\text{unit charge} = \boldsymbol{E} + \boldsymbol{v} \times \boldsymbol{B} = 0 \quad \text{(in a perfect conductor)}, \tag{22.12}$$

where $\boldsymbol{v}$ represents the velocity of the conductor. Our earlier statement that there is no electric field inside a perfect conductor is all right if the velocity $\boldsymbol{v}$ of the conductor is zero; otherwise the correct statement is given by Eq. (22.12).

Returning to our generator of Fig. 22–7, we now see that the line integral of the electric field $\boldsymbol{E}$ from terminal a to terminal b through the conducting path of the generator must be equal to the line integral of $\boldsymbol{v} \times \boldsymbol{B}$ on the same path,

$$\underset{\substack{\text{inside}\\ \text{conductor}}}{\int_a^b} \boldsymbol{E} \cdot d\boldsymbol{s} = - \underset{\substack{\text{inside}\\ \text{conductor}}}{\int_a^b} (\boldsymbol{v} \times \boldsymbol{B}) \cdot d\boldsymbol{s}. \tag{22.13}$$

It is still true, however, that the line integral of $\boldsymbol{E}$ around a complete loop, including the return from b to a outside the generator, must be zero, because there are no changing magnetic fields. So the first integral in Eq. (22.13) is also equal to V, the voltage between the two terminals. It turns out that the right-hand integral of Eq. (22.13) is just the rate of change of the flux linkage through the coil and is therefore—by the flux rule—equal to the emf in the coil. So we have again that the potential difference across the terminals is equal to the electromotive force in the circuit, in agreement with Eq. (22.11). So whether we have a generator in which a magnetic field changes near a fixed coil, or one in which a coil moves in a fixed magnetic field, the external properties of the generators are the same. There is a voltage difference V across the terminals, which is independent of the current in the circuit but depends only on the arbitrarily assigned conditions inside the generator.

So long as we are trying to understand the operation of generators from the point of view of Maxwell's equations, we might also ask about the ordinary chemical cell, like a flashlight battery. It is also a generator, i.e., a voltage source, although it will of course only appear in DC circuits. The simplest kind of cell to understand is shown in Fig. 22–8. We imagine two metal plates immersed in some chemical solution. We suppose that the solution contains positive and negative ions. We suppose also that one kind of ion, say the negative, is much heavier than the one of opposite polarity, so that its motion through the solution by the process of diffusion is much slower. We suppose next that by some means or other it is arranged that the concentration of the solution is made to vary from one part of the liquid to the other, so that the number of ions of both polarities near, say, the lower plate is much larger than the concentration of ions near the upper plate. Because of their rapid mobility the positive ions will drift more readily into the region of lower concentration, so that there will be a slight excess of positive charge arriving at the upper plate. The upper plate will become positively charged and the lower plate will have a net negative charge.

As more and more charges diffuse to the upper plate, the potential of this plate will rise until the resulting electric field between the plates produces forces on the ions which just compensate for their excess mobility, so the two plates of the cell quickly reach a potential difference which is characteristic of the internal construction.

Arguing just as we did for the ideal capacitor, we see that the potential difference between the terminals a and b is just equal to the line integral of the electric field between the two plates when there is no longer any net diffusion of the ions. There is, of course, an essential difference between a capacitor and such a chemical cell. If we short-circuit the terminals of a condenser for a moment, the capacitor is discharged and there is no longer any potential difference across the terminals. In the case of the chemical cell a current can be drawn from the terminals continuously without any change in the emf—until, of course, the chemicals inside the cell have been used up. In a real cell it is found that the potential difference across the terminals decreases as the current drawn from the cell increases. In keeping with the abstractions we have been making, however, we may imagine an ideal cell in which the voltage across the terminals is independent of the current. A real cell can then be looked at as an ideal cell in series with a resistor.

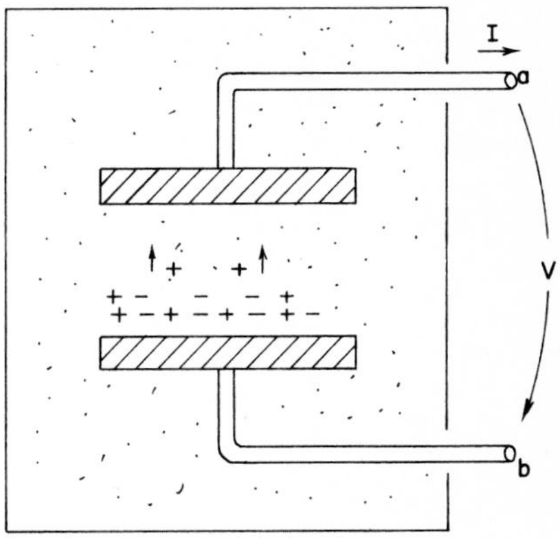

Fig. 22–8. A chemical cell.

22–3 Networks of ideal elements; Kirchhoff's rules

As we have seen in the last section, the description of an ideal circuit element in terms of what happens outside the element is quite simple. The current and the voltage are linearly related. But what is actually happening inside the element is quite complicated, and it is quite difficult to give a precise description in terms of Maxwell's equations. Imagine trying to give a precise description of the electric and magnetic fields of the inside of a radio which contains hundreds of resistors, capacitors, and inductors. It would be an impossible task to analyze such a thing by using Maxwell's equations. But by making the many approximations we have described in Section 22–2 and summarizing the essential features of the real circuit elements in terms of idealizations, it becomes possible to analyze an electrical circuit in a relatively straightforward way. We will now show how that is done.

Suppose we have a circuit consisting of a generator and several impedances connected together, as shown in Fig. 22–9. According to our approximations there is no magnetic field in the region outside the individual circuit elements. Therefore the line integral of $\boldsymbol{E}$ around any curve which does not pass through any of the elements is zero. Consider then the curve Γ shown by the broken line which goes all the way around the circuit in Fig. 22–9. The line integral of $\boldsymbol{E}$ around this curve is made up of several pieces. Each piece is the line integral from one terminal of a circuit element to the other. This line integral we have called the voltage drop across the circuit element. The complete line integral is then just the sum of the voltage drops across all of the elements in the circuit:

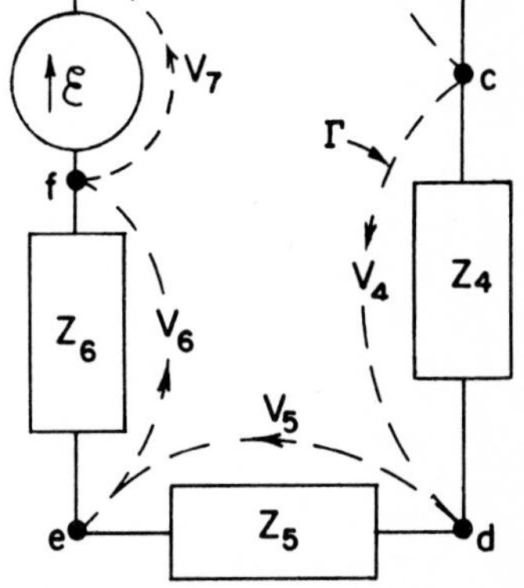

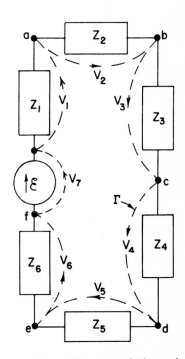

Fig. 22–9. The sum of the voltage drops around any closed path is zero.

$$\oint \boldsymbol{E}\cdot d\boldsymbol{s} = \sum V_n.$$

Since the line integral is zero, we have that the sum of the potential differences around a complete loop of a circuit is equal to zero:

$$\sum_{\substack{\text{around}\\ \text{any loop}}} V_n = 0. \tag{22.14}$$

This result follows from one of Maxwell's equations—that in a region where there are no magnetic fields the line integral of $\boldsymbol{E}$ around any complete loop is zero.

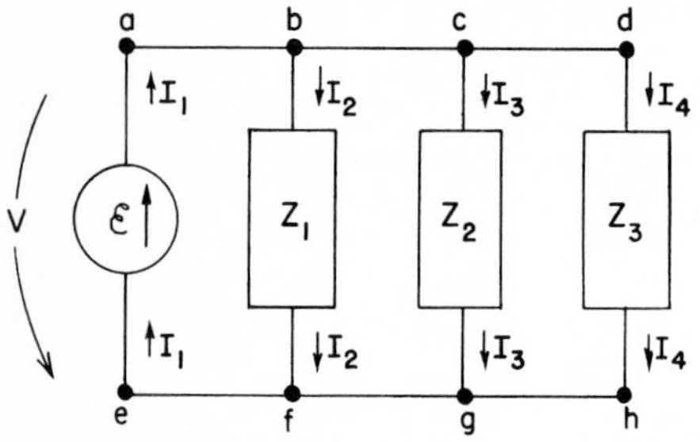

Fig. 22–10. The sum of the currents into any node is zero.

Suppose we consider now a circuit like that shown in Fig. 22–10. The horizontal line joining the terminals a, b, c, and d is intended to show that these terminals are all connected, or that they are joined by wires of negligible resistance. In any case, the drawing means that terminals a, b, c, and d are all at the same potential and, similarly, that the terminals e, f, g, and h are also at one common potential. Then the voltage drop V across each of the four elements is the same.

Now one of our idealizations has been that negligible electrical charges accumulate on the terminals of the impedances. We now assume further that any electrical charges on the wires joining terminals can also be neglected. Then the conservation of charge requires that any charge which leaves one circuit element immediately enters some other circuit element. Or, what is the same thing, we require that the algebraic sum of the currents which enter any given junction must be zero. By a junction, of course, we mean any set of terminals such as a, b, c, and d which are connected. Such a set of connected terminals is usually called a "node." The conservation of charge then requires that for the circuit of Fig. 22–10,

$$I_1 - I_2 - I_3 - I_4 = 0. \tag{22.15}$$

The sum of the currents entering the node which consists of the four terminals e, f, g, and h must also be zero:

$$-I_1 + I_2 + I_3 + I_4 = 0. \tag{22.16}$$

This is, of course, the same as Eq. (22.15). The two equations are not independent. The general rule is that *the sum of the currents into any node must be zero.*

$$\sum_{\substack{\text{into}\\ \text{a node}}} I_n = 0. \tag{22.17}$$

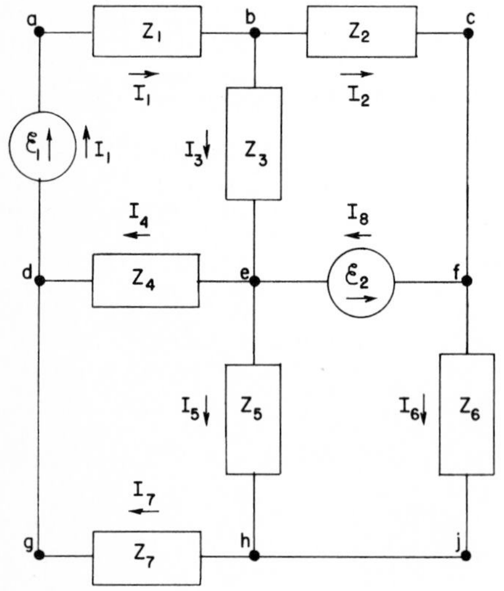

Fig. 22–11. Analyzing a circuit with Kirchhoff's rules.

Our earlier conclusion that the sum of the voltage drops around a closed loop is zero must apply to any loop in a complicated circuit. Also, our result that the sum of the currents into a node is zero must be true for any node. These two equations are known as *Kirchhoff's rules.* With these two rules it is possible to solve for the currents and voltages in any network whatever.

Suppose we consider the more complicated circuit of Fig. 22–11. How shall we find the currents and voltages in this circuit? We can find them in the following straightforward way. We consider separately each of the four subsidiary closed loops which appear in the circuit. (For instance, one loop goes from terminal a to terminal b to terminal e to terminal d and back to terminal a.) For each of the loops we write the equation for the first of Kirchhoff's rules—that the sum of the voltages around each loop is equal to zero. We must remember to count the voltage drop as positive if we are going *in* the direction of the current and negative if we are going across an element in the direction *opposite* to the current; and we must remember that the voltage drop across a generator is the *negative* of the emf in that direction. Thus if we consider the small loop that starts and ends at terminal a we have the equation

$$z_1 I_1 + z_3 I_3 + z_4 I_4 - \mathcal{E}_1 = 0.$$

Applying the same rule to the remaining loops, we would get three more equations of the same kind.

Next, we must write the current equation for each of the nodes in the circuit. For example, summing the currents into the node at terminal b gives the equation

$$I_1 - I_3 - I_2 = 0.$$

Similarly, for the node labeled e we would have the current equation

$$I_3 - I_4 + I_8 - I_5 = 0.$$

For the circuit shown there are five such current equations. It turns out, however, that any one of these equations can be derived from the other four; there are, therefore, only four independent current equations. We thus have a total of eight independent, linear equations: the four voltage equations and the four current equations. With these eight equations we can solve for the eight unknown currents. Once the currents are known the circuit is solved. The voltage drop across any element is given by the current through that element times its impedance (or, in the case of the voltage sources, it is already known).

We have seen that when we write the current equations, we get one equation which is not independent of the others. Generally it is also possible to write down too many voltage equations. For example, in the circuit of Fig. 22–11, although we have considered only the four small loops, there are a large number of other loops for which we could write the voltage equation. There is, for example, the loop along the path *abcfeda*. There is another loop which follows the path *abcfehgda*. You can see that there are many loops. In analyzing complicated circuits it is very easy to get too many equations. There are rules which tell us how to proceed so that only the minimum number of equations is written down, but usually with a little thought it is possible to see how to get the right number of equations in the simplest form. Besides, writing an extra equation or two doesn't do any harm. They will not lead to any wrong answers, only perhaps a little unnecessary algebra.

In Chapter 25 of Vol. I we showed that if the two impedances z_1 and z_2 are in *series*, they are equivalent to a single impedance z_s given by

$$z_s = z_1 + z_2. \tag{22.18}$$

We also showed that if the two impedances are connected in *parallel*, they are equivalent to the single impedance z_p given by

$$z_p = \frac{1}{(1/z_1) + (1/z_2)} = \frac{z_1 z_2}{z_1 + z_2}. \tag{22.19}$$

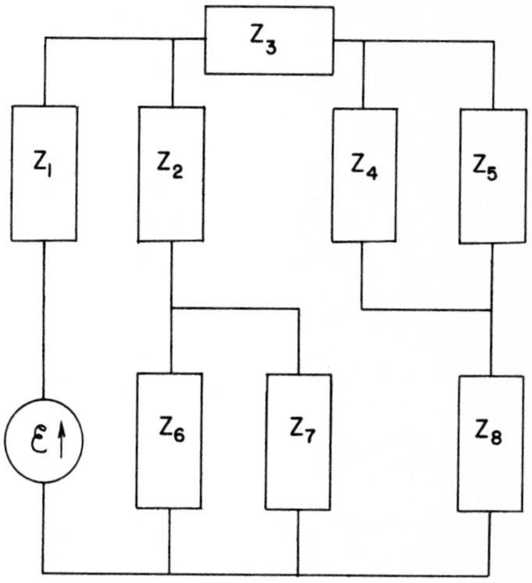

Fig. 22–12. A circuit which can be analyzed in terms of series and parallel combinations.

If you look back you will see that in deriving these results we were in effect making use of Kirchhoff's rules. It is often possible to analyze a complicated circuit by repeated application of the formulas for series and parallel impedances. For instance, the circuit of Fig. 22–12 can be analyzed that way. First, the impedances z_4 and z_5 can be replaced by their parallel equivalent, and so also can z_6 and z_7. Then the impedance z_2 can be combined with the parallel equivalent of z_6 and z_7 by the series rule. Proceeding in this way, the whole circuit can be reduced to a generator in series with a single impedance Z. The current through the generator is then just $\mathcal{E}/Z$. Then by working backward one can solve for the currents in each of the impedances.

There are, however, quite simple circuits which cannot be analyzed by this method, as for example the circuit of Fig. 22–13. To analyze this circuit we must

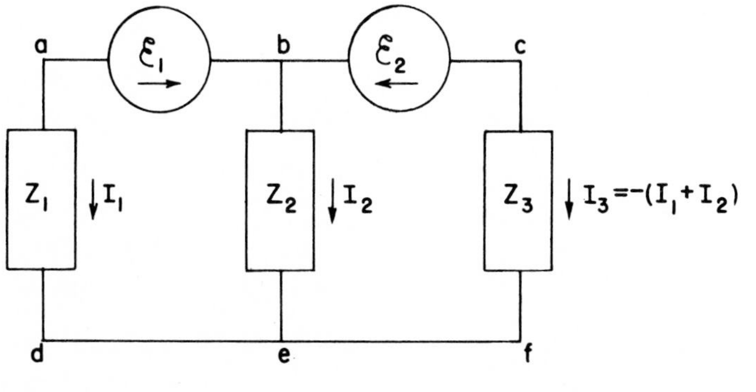

Fig. 22–13. A circuit that cannot be analyzed in terms of series and parallel combinations.

write down the current and voltage equations from Kirchhoff's rules. Let's do it. There is just one current equation:

$$I_1 + I_2 + I_3 = 0,$$

so we know immediately that

$$I_3 = -(I_1 + I_2).$$

We can save ourselves some algebra if we immediately make use of this result in writing the voltage equations. For this circuit there are two independent voltage equations; they are

$$-\mathcal{E}_1 + I_2 z_2 - I_1 z_1 = 0$$

and

$$\mathcal{E}_2 - (I_1 + I_2) z_3 - I_2 z_2 = 0.$$

There are two equations and two unknown currents. Solving these equations for I_1 and I_2, we get

$$I_1 = \frac{z_2\mathcal{E}_2 - (z_2 + z_3)\mathcal{E}_1}{z_1(z_2 + z_3) + z_2 z_3} \tag{22.20}$$

and

$$I_2 = \frac{z_1\mathcal{E}_2 + z_3\mathcal{E}_1}{z_1(z_2 + z_3) + z_2 z_3}. \tag{22.21}$$

The third current is obtained from the sum of these two.

Another example of a circuit that cannot be analyzed by using the rules for series and parallel impedance is shown in Fig. 22–14. Such a circuit is called a "bridge." It appears in many instruments used for measuring impedances. With such a circuit one is usually interested in the question: How must the various impedances be related if the current through the impedance z_3 is to be zero? We leave it for you to find the conditions for which this is so.

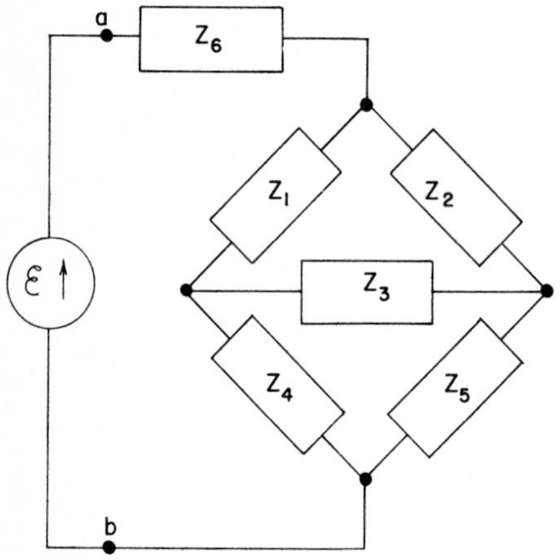

Fig. 22–14. A bridge circuit.

22–4 Equivalent circuits

Suppose we connect a generator $\mathcal{E}$ to a circuit containing some complicated interconnection of impedances, as indicated schematically in Fig. 22–15(a). All of the equations we get from Kirchhoff's rules are linear, so when we solve them for the current I through the generator, we will get that I is proportional to $\mathcal{E}$. We can write

$$I = \frac{\mathcal{E}}{z_{\text{eff}}},$$

where now z_{eff} is some complex number, an algebraic function of all the elements in the circuit. (If the circuit contains no generators other than the one shown, there is no additional term independent of $\mathcal{E}$.) But this equation is just what we would write for the circuit of Fig. 22–15(b). So long as we are interested only in what happens *to the left* of the two terminals a and b, the two circuits of Fig. 22–15 are *equivalent*. We can, therefore, make the general statement that *any* two-terminal network of passive elements can be replaced by a single impedance z_{eff} without changing the currents and voltages in the rest of the circuit. This statement is, of course, just a remark about what comes out of Kirchhoff's rules—and ultimately from the linearity of Maxwell's equations.

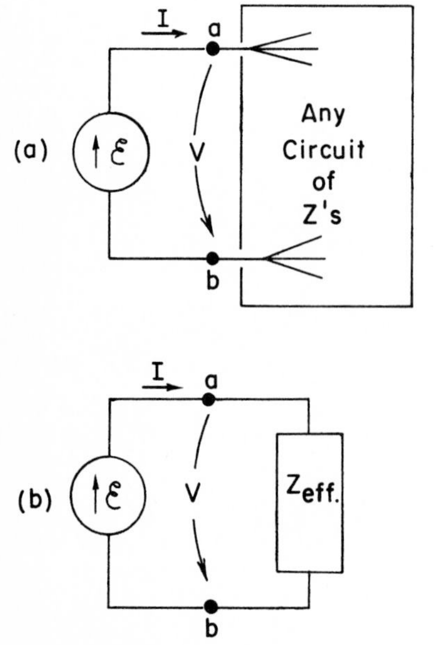

Fig. 22–15. Any two-terminal network of passive elements is equivalent to an effective impedance.

The idea can be generalized to a circuit that contains generators as well as impedances. Suppose we look at such a circuit "from the point of view" of one of the impedances, which we will call z_n, as in Fig. 22–16(a). If we were to solve the equation for the whole circuit, we would find that the voltage V_n between the two terminals a and b is a linear function of I, which we can write

$$V_n = A - BI_n, \tag{22.22}$$

where A and B depend on the generators and impedances in the circuit to the left of the terminals. For instance, for the circuit of Fig. 22–13, we find $V_1 = I_1 z_1$. This can be written (by rearranging Eq. (22.20)] as

$$V_1 = \left[\left(\frac{z_2}{z_2 + z_3}\right)\mathcal{E}_2 - \mathcal{E}_1\right] - \frac{z_2 z_3}{z_2 + z_3} I_1. \tag{22.23}$$

The complete solution is then obtained by combining this equation with the one for the impedance z_1, namely, $V_1 = I_1 z_1$, or in the general case, by combining Eq. (22.22) with

$$V_n = I_n z_n.$$

If now we consider that z_n is attached to a simple series circuit of a generator and a current, as in Fig. 22–15(b), the equation corresponding to Eq. (22.22) is

$$V_n = \mathcal{E}_{\text{eff}} - I_n z_{\text{eff}},$$

which is identical to Eq. (22.22) provided we set $\mathcal{E}_{\text{eff}} = A$ and $z_{\text{eff}} = B$. So if we are interested only in what happens *to the right* of the terminals a and b, the arbitrary circuit of Fig. 22–16 can always be replaced by an equivalent combination of a generator in series with an impedance.

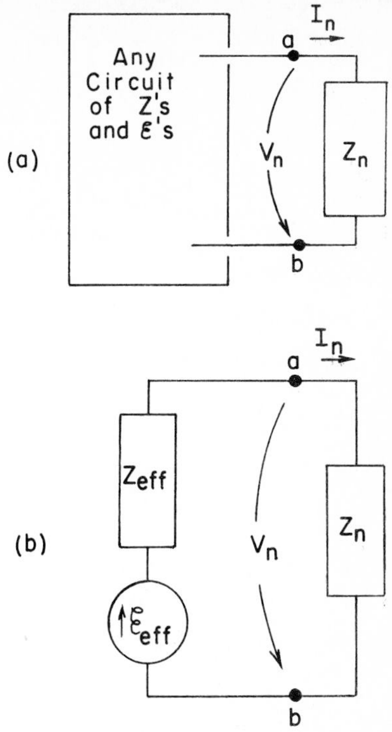

Fig. 22–16. Any two-terminal network can be replaced by a generator in series with an impedance.

22–5 Energy

We have seen that to build up the current I in an inductance, the energy $U = \frac{1}{2}LI^2$ must be provided by the external circuit. When the current falls back to zero, this energy is delivered back to the external circuit. There is no energy-loss mechanism in an ideal inductance. When there is an alternating current through an inductance, energy flows back and forth between it and the rest of the circuit, but the *average* rate at which energy is delivered to the circuit is zero. We say that an inductance is a *nondissipative* element; no electrical energy is dissipated—that is, "lost"—in it.

Similarly, the energy of a condenser, $U = \frac{1}{2}CV^2$, is returned to the external circuit when a condenser is discharged. When a condenser is in an AC circuit energy flows in and out of it, but the net energy flow in each cycle is zero. An ideal condenser is also a nondissipative element.

We know that an emf is a source of energy. When a current I flows in the direction of the emf, energy is delivered to the external circuit at the rate $dU/dt = \mathcal{E}I$. If current is driven *against* the emf—by other generators in the circuit—the emf will absorb energy at the rate $\mathcal{E}I$; since I is negative, dU/dt will also be negative.

If a generator is connected to a resistor R, the current through the resistor is $I = \mathcal{E}/R$. The energy being supplied by the generator at the rate $\mathcal{E}I$ is being absorbed by the resistor. This energy goes into heat in the resistor and is lost from the electrical energy of the circuit. We say that electrical energy is *dissipated* in a resistor. The rate at which energy is dissipated in a resistor is $dU/dt = RI^2$.

In an AC circuit the average rate of energy lost to a resistor is the average of RI^2 over one cycle. Since $I = \hat{I}e^{i\omega t}$—by which we really mean that I varies as $\cos\omega t$—the average of I^2 over one cycle is $|\hat{I}|^2/2$, since the peak current is $|\hat{I}|$ and the average of $\cos^2\omega t$ is 1/2.

What about the energy loss when a generator is connected to an arbitrary impedance z? (By "loss" we mean, of course, conversion of electrical energy into thermal energy.) Any impedance z can be written as the sum of its real and imaginary parts. That is,

$$z = R + iX, \tag{22.24}$$

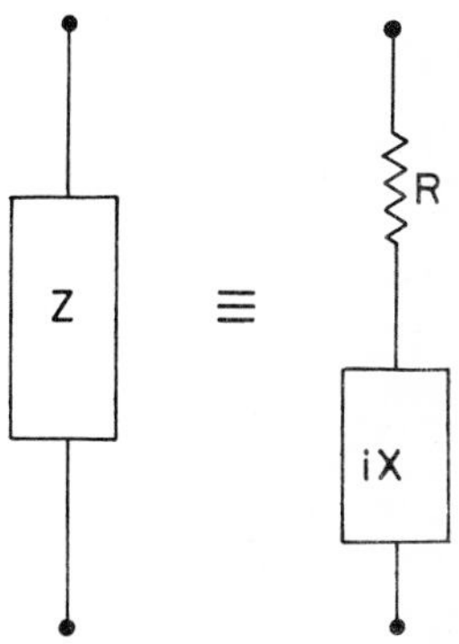

Fig. 22–17. Any impedance is equivalent to a series combination of a pure resistance and a pure reactance.

where R and X are real numbers. From the point of view of equivalent circuits we can say that any impedance is equivalent to a resistance in series with a pure imaginary impedance—called a *reactance*—as shown in Fig. 22–17.

We have seen earlier that any circuit that contains only L's and C's has an impedance that is a pure imaginary number. Since there is no energy loss into any of the L's and C's on the average, a pure reactance containing only L's and C's will have no energy loss. We can see that this must be true in general for a reactance.

If a generator with the emf $\mathcal{E}$ is connected to the impedance z of Fig. 22–17, the emf must be related to the current I from the generator by

$$\mathcal{E} = I(R + iX). \tag{22.25}$$

To find the average rate at which energy is delivered, we want the average of the product $\mathcal{E}I$. Now we must be careful. When dealing with such products, we must deal with the real quantities $\mathcal{E}(t)$ and $I(t)$. (The real parts of the complex functions will represent the actual physical quantities only when we have *linear* equations; now we are concerned with *products*, which are certainly not linear.)

Suppose we choose our origin of t so that the amplitude $\hat{I}$ is a real number, let's say I_0; then the actual time variation I is given by

$$I = I_0 \cos \omega t.$$

The emf of Eq. (22.25) is the real part of

$$I_0 e^{i\omega t}(R + iX)$$

or

$$\mathcal{E} = I_0 R \cos \omega t - I_0 X \sin \omega t. \tag{22.26}$$

The two terms in Eq. (22.26) represent the voltage drops across R and X in Fig. 22–17. We see that the voltage drop across the resistance is *in phase* with the current, while the voltage drop across the purely reactive part is *out of phase* with the current.

The *average rate* of energy loss, $\langle P \rangle_{\text{av}}$, from the generator is the integral of the product $\mathcal{E}I$ over one cycle divided by the period T; in other words,

$$\langle P \rangle_{\text{av}} = \frac{1}{T}\int_0^T \mathcal{E}I\,dt = \frac{1}{T}\int_0^T I_0^2 R \cos^2 \omega t\,dt - \frac{1}{T}\int_0^T I_0^2 X \cos \omega t \sin \omega t\,dt.$$

The first integral is $\frac{1}{2}I_0^2 R$, and the second integral is zero. So the average energy loss in an impedance $z = R + iX$ depends only on the real part of z, and is $I_0^2 R/2$, which is in agreement with our earlier result for the energy loss in a resistor. There is no energy loss in the reactive part.

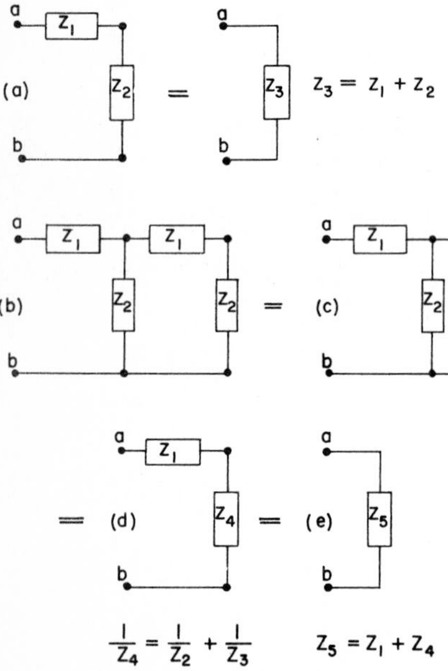

Fig. 22–18. The effective impedance of a ladder.

22–6 A ladder network

We would like now to consider an interesting circuit which can be analyzed in terms of series and parallel combinations. Suppose we start with the circuit of Fig. 22–18(a). We can see right away that the impedance from terminal a to terminal b is simply $z_1 + z_2$. Now let's take a little harder circuit, the one shown in Fig. 22–18(b). We could analyze this circuit using Kirchhoff's rules, but it is also easy to handle with series and parallel combinations. We can replace the two impedances on the right-hand end by a single impedance $z_3 = z_1 + z_2$, as in part (c) of the figure. Then the two impedances z_2 and z_3 can be replaced by their equivalent parallel impedance z_4, as shown in part (d) of the figure. Finally, z_1 and z_4 are equivalent to a single impedance z_5, as shown in part (e).

Now we may ask an amusing question: What would happen if in the network of Fig. 22–18(b) we kept on adding more sections *forever*—as we indicate by the dashed lines in Fig. 22–19(a)? Can we solve such an infinite network? Well, that's

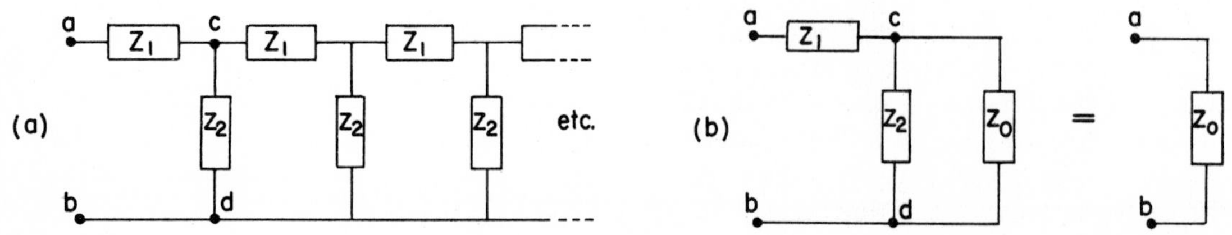

Fig. 22–19. The effective impedance of an infinite ladder.

not so hard. First, we notice that such an infinite network is unchanged if we add one more section at the "front" end. Surely, if we add one more section to an infinite network it is still the same infinite network. Suppose we call the impedance between the two terminals a and b of the infinite network z_0; then the impedance of all the stuff to the right of the two terminals c and d is also z_0. Therefore, so far as the front end is concerned, we can represent the network as shown in Fig. 22–19(b). Combining the parallel combinations $z_2 z_0$ and adding the result in series with z_1, we can immediately write down the impedance of this combination:

$$z = z_1 + \frac{1}{(1/z_2) + (1/z_0)} \quad \text{or} \quad z = z_1 + \frac{z_2 z_0}{z_2 + z_0}.$$

But this impedance is also equal to z_0, so we have the equation

$$z_0 = z_1 + \frac{z_2 z_0}{z_2 + z_0}.$$

We can solve for z_0 to get

$$z_0 = \frac{z_1}{2} + \sqrt{(z_1^2/4) + z_1 z_2}. \tag{22.27}$$

So we have found the solution for the impedance of an infinite ladder of repeated series and parallel impedances. The impedance z_0 is called the *characteristic impedance* of such an infinite network.

Let's now consider a specific example in which the series element is an inductance L and the shunt element is a capacitance C, as shown in Fig. 22–20(a). In this case we find the impedance of the infinite network by setting $z_1 = i\omega L$ and $z_2 = 1/i\omega C$. Notice that the first term, $z_1/2$, in Eq. (22.27) is just one-half the impedance of the first element. It would therefore seem more natural, or at least somewhat simpler, if we were to draw our infinite network as shown in Fig. 22–20(b). Looking at the infinite network from the terminal a' we would see the characteristic impedance

$$z_0 = \sqrt{(L/C) - (\omega^2 L^2/4)}. \tag{22.28}$$

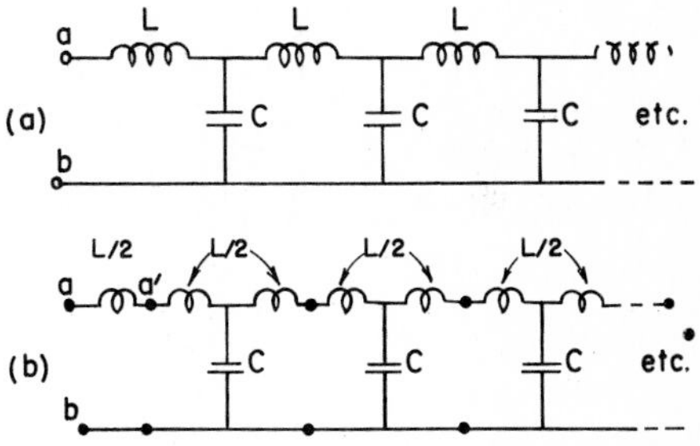

Fig. 22–20. An L-C ladder drawn in two equivalent ways.

Now there are two interesting cases, depending on the frequency ω. If ω^2 is less than $4/LC$, the second term in the radical will be smaller than the first, and the impedance z_0 will be a real number. On the other hand, if ω^2 is greater than $4/LC$ the impedance z_0 will be a pure imaginary number which we can write as

$$z_0 = i\sqrt{(\omega^2 L^2/4) - (L/C)}.$$

We have said earlier that a circuit which contains only imaginary impedances, such as inductances and capacitances, will have an impedance which is purely imaginary. How can it be then that for the circuit we are now studying—which has only L's and C's—the impedance is a pure resistance for frequencies below $\sqrt{4/LC}$? For higher frequencies the impedance is purely imaginary, in agreement with our earlier statement. For lower frequencies the impedance is a pure resistance and will therefore absorb energy. But how can the circuit continuously absorb energy, as a resistance does, if it is made only of inductances and capacitances? *Answer:* Because there is an infinite number of inductances and capacitances, so that when a source is connected to the circuit, it supplies energy to the first inductance and capacitance, then to the second, to the third, and so on. In a circuit of this kind, energy is continually absorbed from the generator at a constant rate and flows constantly out into the network, supplying energy which is stored in the inductances and capacitances down the line.

This idea suggests an interesting point about what is happening in the circuit. We would expect that if we connect a source to the front end, the effects of this source will be propagated through the network toward the infinite end. The propagation of the waves down the line is much like the radiation from an antenna which absorbs energy from its driving source; that is, we expect such a propagation to occur when the impedance is real, which occurs if ω is less than $\sqrt{4/LC}$. But when the impedance is purely imaginary, which happens for ω greater than $\sqrt{4/LC}$, we would not expect to see any such propagation.

22–7 Filters

We saw in the last section that the infinite ladder network of Fig. 22–20 absorbs energy continuously if it is driven at a frequency below a certain critical frequency $\sqrt{4/LC}$, which we will call the *cutoff frequency* ω_0. We suggested that this effect could be understood in terms of a continuous transport of energy down the line. On the other hand, at high frequencies, for $\omega > \omega_0$, there is no continuous absorption of energy; we should then expect that perhaps the currents don't "penetrate" very far down the line. Let's see whether these ideas are right.

Suppose we have the front end of the ladder connected to some AC generator and we ask what the voltage looks like at, say, the 754th section of the ladder. Since the network is infinite, whatever happens to the voltage from one section to the next is always the same; so let's just look at what happens when we go from some section, say the nth to the next. We will define the currents I_n and voltages V_n as shown in Fig. 22–21(a).

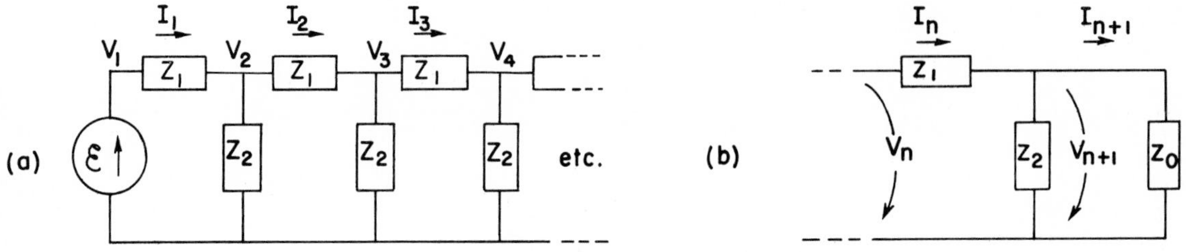

Fig. 22–21. Finding the propagation factor of a ladder.

We can get the voltage V_{n+1} from V_n by remembering that we can always replace the rest of the ladder after the nth section by its characteristic impedance z_0; then we need only analyze the circuit of Fig. 22–21(b). First, we notice that any V_n, since it is across z_0, must equal $I_n z_0$. Also, the difference between V_n and V_{n+1} is just $I_n z_1$:

$$V_n - V_{n+1} = I_n z_1 = V_n \frac{z_1}{z_0}.$$

So we get the ratio

$$\frac{V_{n+1}}{V_n} = 1 - \frac{z_1}{z_0} = \frac{z_0 - z_1}{z_0}.$$

We can call this ratio the *propagation factor* for one section of the ladder; we'll call it α. It is, of course, the same for all sections:

$$\alpha = \frac{z_0 - z_1}{z_0}. \tag{22.29}$$

The voltage after the nth section is then

$$V_n = \alpha^n \mathcal{E}. \tag{22.30}$$

You can now find the voltage after 754 sections; it is just α to the 754th power times $\mathcal{E}$.

Suppose we see what α is like for the L-C ladder of Fig. 22–20(a). Using z_0 from Eq. (22.27), and $z_1 = i\omega L$, we get

$$\alpha = \frac{\sqrt{(L/C) - (\omega^2L^2/4)} - i(\omega L/2)}{\sqrt{(L/C) - (\omega^2L^2/4)} + i(\omega L/2)}. \tag{22.31}$$

If the driving frequency is below the cutoff frequency $\omega_0 = \sqrt{4/LC}$, the radical is a real number, and the magnitudes of the complex numbers in the numerator and denominator are equal. Therefore, the magnitude of α is one; we can write

$$\alpha = e^{i\delta},$$

which means that the magnitude of the voltage is the same at every section; only its phase changes. The phase change δ is, in fact, a negative number and represents the "delay" of the voltage as it passes along the network.

For frequencies above the cutoff frequency ω_0 it is better to factor out an i from the numerator and denominator of Eq. (22.31) and rewrite it as

$$\alpha = \frac{\sqrt{(\omega^2L^2/4) - (L/C)} - (\omega L/2)}{\sqrt{(\omega^2L^2/4) - (L/C)} + (\omega L/2)}. \tag{22.32}$$

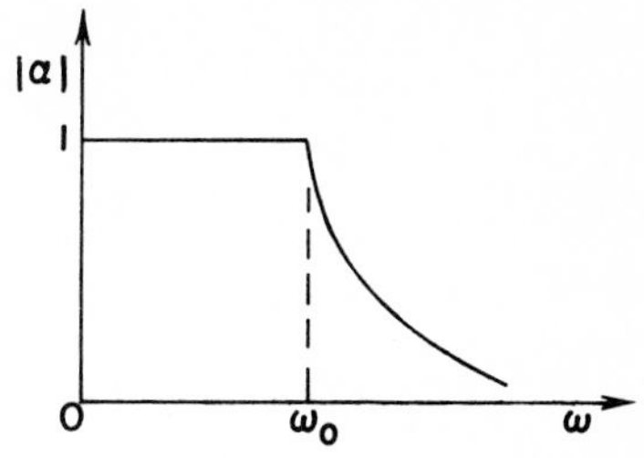

Fig. 22–22. The propagation factor of a section of an L-C ladder.

The propagation factor α is now a *real* number, and a number *less than one*. That means that the voltage at any section is always less than the voltage at the preceding section by the factor α. For any frequency above ω_0, the voltage dies away rapidly as we go along the network. A plot of the absolute value of α as a function of frequency looks like the graph in Fig. 22–22.

We see that the behavior of α, both above and below ω_0, agrees with our interpretation that the network propagates energy for $\omega < \omega_0$ and blocks it for $\omega > \omega_0$. We say that the network "passes" low frequencies and "rejects" or "filters out" the high frequencies. Any network designed to have its characteristics vary in a prescribed way with frequency is called a "filter." We have been analyzing a "low-pass filter."

You may be wondering why all this discussion of an infinite network which obviously cannot actually occur. The point is that the same characteristics are found in a finite network if we finish it off at the end with an impedence equal to the characteristic impedence z_0. Now in practice it is not possible to *exactly* reproduce the characteristic impedance with a few simple elements—like R's, L's, and C's. But it is often possible to do so with a fair approximation for a certain range of frequencies. In this way one can make a finite filter network whose properties are very nearly the same as those for the infinite case. For instance, the L-C ladder behaves much as we have described it if it is terminated in the pure resistance $R = \sqrt{L/C}$.

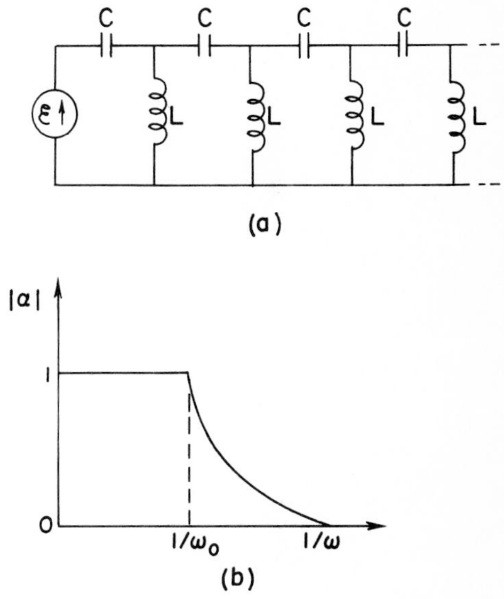

Fig. 22–23. (a) A high-pass filter; (b) its propagation factor as a function of $1/\omega$.

If in our L-C ladder we interchange the positions of the L's and C's, to make the ladder shown in Fig. 22–23(a), we can have a filter that propagates *high* frequencies and rejects *low* frequencies. It is easy to see what happens with this network by using the results we already have. You will notice that whenever we change an L to a C and *vice versa*, we also change every $i\omega$ to $1/i\omega$. So whatever happened at ω before will now happen at $1/\omega$. In particular, we can see how α will vary with frequency by using Fig. 22–22 and changing the label on the axis to $1/\omega$, as we have done in Fig. 22–23(b).

The low-pass and high-pass filters we have described have various technical applications. An L-C low-pass filter is often used as a "smoothing" filter in a DC power supply. If we want to manufacture DC power from an AC source, we begin with a rectifier which permits current to flow only in one direction. From the rectifier we get a series of pulses that look like the function $V(t)$ shown in Fig. 22–24, which is lousy DC, because it wobbles up and down. Suppose we would like a nice pure DC, such as a battery provides. We can come close to that by putting a low-pass filter between the rectifier and the load.

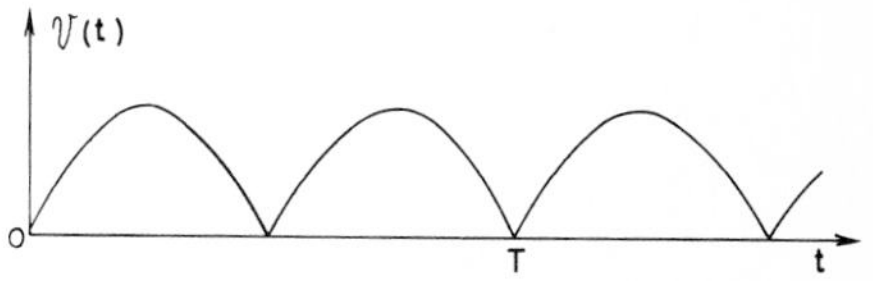

Fig. 22–24. The output voltage of a full-wave rectifier.

We know from Chapter 50 of Vol. I that the time function in Fig. 22–24 can be represented as a superposition of a constant voltage plus a sine wave, plus a higher-frequency sine wave, plus a still higher-frequency sine wave, etc.—by a Fourier series. If our filter is linear (if, as we have been assuming, the L's and C's don't vary with the currents or voltages) then what comes out of the filter is the superposition of the outputs for each component at the input. If we arrange that the cutoff frequency ω_0 of our filter is well below the lowest frequency in the function $V(t)$, the DC (for which $\omega = 0$) goes through fine, but the amplitude of the first harmonic will be cut down a lot. And amplitudes of the higher harmonics will be cut down even more. So we can get the output as smooth as we wish, depending only on how many filter sections we are willing to buy.

A high-pass filter is used if one wants to reject certain low frequencies. For instance, in a phonograph amplifier a high-pass filter may be used to let the music through, while keeping out the low-pitched rumbling from the motor of the turntable.

It is also possible to make "band-pass" filters that reject frequencies below some frequency ω_1 and above another frequency ω_2 (greater than ω_1), but pass the frequencies between ω_1 and ω_2. This can be done simply by putting together a high-pass and a low-pass filter, but it is more usually done by making a ladder in which the impedances z_1 and z_2 are more complicated—being each a combination of L's and C's. Such a band-pass filter might have a propagation constant like that shown in Fig. 22–25(a). It might be used, for example, in separating signals that occupy only an interval of frequencies, such as each of the many voice channels in a high-frequency telephone cable, or the modulated carrier of a radio transmission.

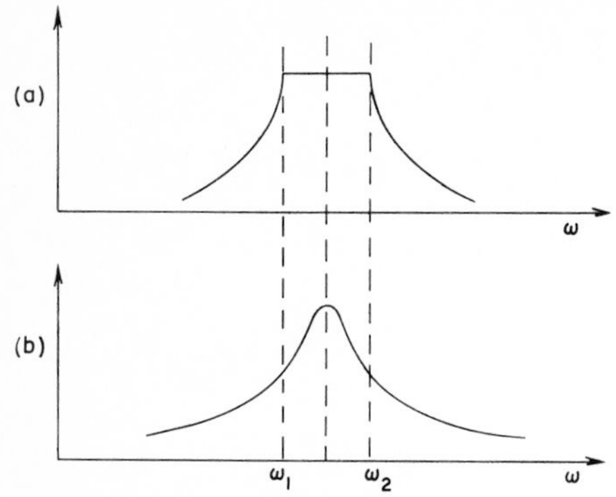

Fig. 22–25. (a) A band-pass filter. (b) A simple resonant filter.

We have seen in Chapter 25 of Vol. I that such filtering can also be done using the selectivity of an ordinary resonance curve, which we have drawn for comparison in Fig. 22–25(b). But the resonant filter is not as good for some purposes as the band-pass filter. You will remember (Chapter 48, Vol. I) that when a carrier of frequency ω_c is modulated with a "signal" frequency ω_s, the total signal contains not only the carrier frequency but also the two side-band frequencies $\omega_c + \omega_s$ and $\omega_c - \omega_s$. With a resonant filter, these side-bands are always attentuated somewhat, and the attenuation is more, the higher the signal frequency, as you can see from the figure. So there is a poor "frequency response." The higher musical tones don't get through. But if the filtering is done with a band-pass filter designed so that the width $\omega_2 - \omega_1$ is at least twice the highest signal frequency, the frequency response will be "flat" for the signals wanted.

We want to make one more point about the ladder filter: the L-C ladder of Fig. 22–20 is also an approximate representation of a transmission line. If we have a long conductor that runs parallel to another conductor—such as a wire in a coaxial cable, or a wire suspended above the earth—there will be some capacitance between the two conductors and also some inductance due to the magnetic field between them. If we imagine the line as broken up into small lengths $\Delta\ell$, each length will look like one section of the L-C ladder with a series inductance ΔL and a shunt capacitance ΔC. We can then use our results for the ladder filter. If we take the limit as $\Delta\ell$ goes to zero, we have a good description of the transmission line. Notice that as $\Delta\ell$ is made smaller and smaller, both ΔL and ΔC decrease, but in the same proportion, so that the ratio $\Delta L/\Delta C$ remains constant. So if we take the limit of Eq. (22.28) as ΔL and ΔC go to zero, we find that the characteristic impedance z_0 is a pure resistance whose magnitude is $\sqrt{\Delta L/\Delta C}$. We can also write the ratio $\Delta L/\Delta C$ as L_0/C_0, where L_0 and C_0 are the inductance and capacitance of a unit length of the line; then we have

$$z_0 = \sqrt{\frac{L_0}{C_0}}. \tag{22.33}$$

You will also notice that as ΔL and ΔC go to zero, the cutoff frequency $\omega_0 = \sqrt{4/LC}$ goes to infinity. There is no cutoff frequency for an ideal transmission line.

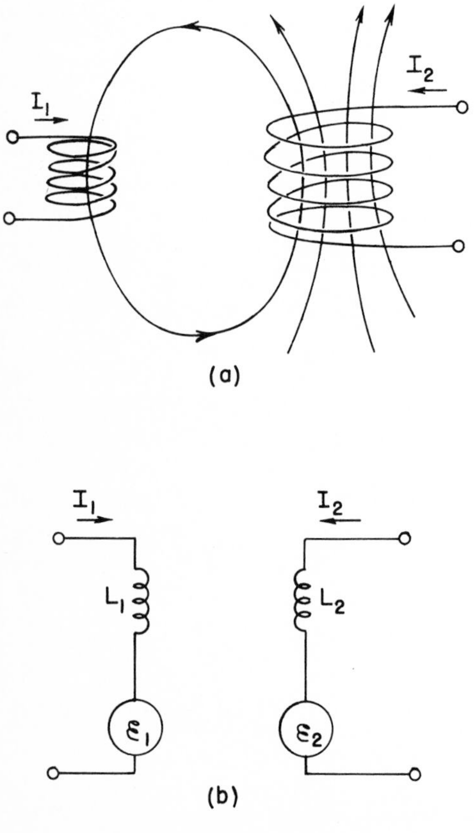

Fig. 22–26. Equivalent circuit of a mutual inductance.

22–8 Other circuit elements

We have so far defined only the ideal circuit impedances—the inductance, the capacitance, and the resistance—as well as the ideal voltage generator. We want now to show that other elements, such as mutual inductances or transistors or vacuum tubes, can be described by using only the same basic elements. Suppose that we have two coils and that on purpose, or otherwise, some flux from one of the coils links the other, as shown in Fig. 22–26(a). Then the two coils will have a mutual inductance M such that when the current varies in one of the coils, there will be a voltage generated in the other. Can we take into account such an effect in our equivalent circuits? We can in the following way. We have seen that the induced emf's in each of two interacting coils can be written as the sum of two parts:

$$\mathcal{E}_1 = -L_1 \frac{dI_1}{dt} \pm M \frac{dI_2}{dt},$$
$$\mathcal{E}_2 = -L_2 \frac{dI_2}{dt} \pm M \frac{dI_1}{dt}. \tag{22.34}$$

The first term comes from the self-inductance of the coil, and the second term comes from its mutual inductance with the other coil. The sign of the second term can be plus or minus, depending on the way the flux from one coil links the other. Making the same approximations we used in describing an ideal inductance, we would say that the potential difference across the terminals of each coil is equal to the electromotive force in the coil. Then the two equations of (22.34) are the same as the ones we would get from the circuit of Fig. 22–26(b), provided the electromotive force in each of the two circuits shown depends on the current in the opposite circuit according to the relations

$$\mathcal{E}_1 = \pm i\omega M I_2, \qquad \mathcal{E}_2 = \pm i\omega M I_1. \tag{22.35}$$

So what we can do is represent the effect of the self-inductance in a normal way but replace the effect of the mutual inductance by an auxiliary ideal voltage generator. We must in addition, of course, have the equation that relates this emf to the current in some other part of the circuit; but so long as this equation is linear, we have just added more linear equations to our circuit equations, and all of our earlier conclusions about equivalent circuits and so forth are still correct.

In addition to mutual inductances there may also be mutual capacitances. So far, when we have talked about condensers we have always imagined that there were only two electrodes, but in many situations, for example in a vacuum tube, there may be many electrodes close to each other. If we put an electric charge on any one of the electrodes, its electric field will induce charges on each of the other electrodes and affect its potential. As an example, consider the arrangement of four plates shown in Fig. 22–27(a). Suppose these four plates are connected to external circuits by means of the wires A, B, C, and D. So long as we are only worried about electrostatic effects, the equivalent circuit of such an arrangement of electrodes is as shown in part (b) of the figure. The electrostatic interaction of any electrode with each of the others is equivalent to a capacity between the two electrodes.

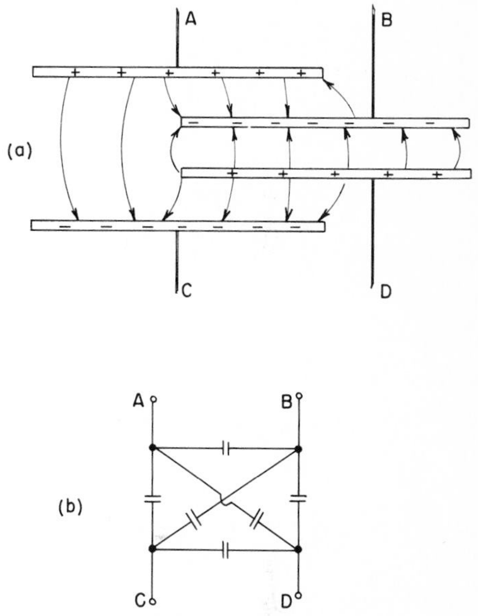

Fig. 22–27. Equivalent circuit of mutual capacitance.

Finally, let's consider how we should represent such complicated devices as transistors and radio tubes in an AC circuit. We should point out at the start that such devices are often operated in such a way that the relationship between the currents and voltages is not at all linear. In such cases, those statements we have made which depend on the linearity of equations are, of course, no longer correct. On the other hand, in many applications the operating characteristics are sufficiently linear that we may consider the transistors and tubes to be linear devices. By this we mean that the alternating currents in, say, the plate of a vacuum tube are linearly proportional to the voltages that appear on the other electrodes, say the grid voltage and the plate voltage. When we have such linear relationships, we can incorporate the device into our equivalent circuit representation.

As in the case of the mutual inductance, our representation will have to include auxiliary voltage generators which describe the influence of the voltages or currents in one part of the device on the currents or voltages in another part. For example, the plate circuit of a triode can usually be represented by a resistance in series with an ideal voltage generator whose source strength is proportional to the grid voltage. We get the equivalent circuit shown in Fig. 22–28.* Similarly, the collector circuit

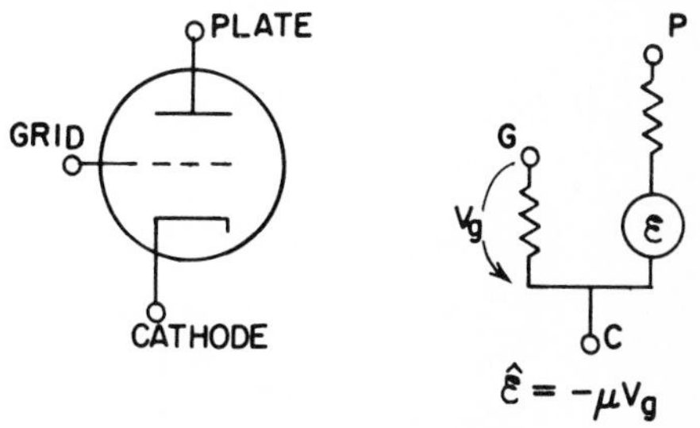

Fig. 22–28. A low-frequency equivalent circuit of a vacuum triode.

* The equivalent circuit shown is correct only for low frequencies. For high frequencies the equivalent circuit gets much more complicated and will include various so-called "parasitic" capacitances and inductances.

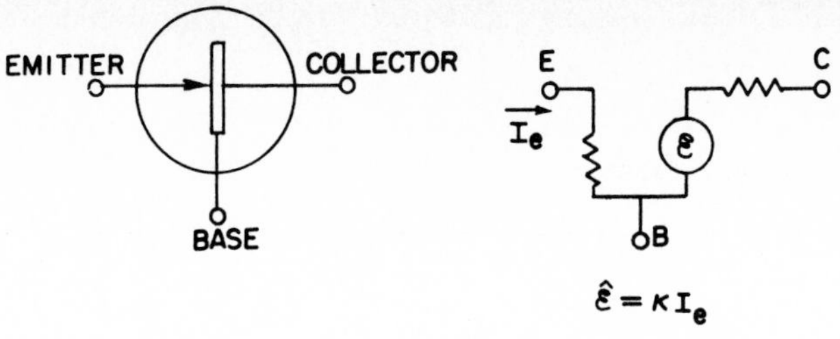

Fig. 22–29. A low-frequency equivalent circuit of a transistor.

of a transistor is conveniently represented as a resistor in series with an ideal voltage generator whose source strength is proportional to the current from the emitter to the base of the transistor. The equivalent circuit is then like that in Fig. 22–29. So long as the equations which describe the operation are linear, we can use such representations for tubes or transistors. Then, when they are incorporated in a complicated network, our general conclusions about the equivalent representation of any arbitrary connection of elements is still valid.

There is one remarkable thing about transistor and radio tube circuits which is different from circuits containing only impedances: the real part of the effective impedance z_{eff} can become negative. We have seen that the real part of z represents the loss of energy. But it is the important characteristic of transistors and tubes that they *supply* energy to the circuit. (Of course they don't just "make" energy; they take energy from the DC circuits of the power supplies and convert it into AC energy.) So it is possible to have a circuit with a negative resistance. Such a circuit has the property that if you connect it to an impedance with a positive real part, i.e., a positive resistance, and arrange matters so that the sum of the two real parts is exactly zero, then there is no dissipation in the combined circuit. If there is no loss of energy, any alternating voltage once started will remain forever. This is the basic idea behind the operation of an oscillator or signal generator which can be used as a source of alternating voltage at any desired frequency.

23

Cavity Resonators

Review: Chapter 23, Vol. I, *Resonance*
Chapter 49, Vol. I, *Modes*

23–1 Real circuit elements

When looked at from any one pair of terminals, any arbitrary circuit made up of ideal impedances and generators is, at any given frequency, equivalent to a generator $\mathcal{E}$ in series with an impedance z. That comes about because if we put a voltage V across the terminals and solve all the equations to find the current I, we must get a linear relation between the current and the voltage. Since all the equations are linear, the result for I must also depend only linearly on V. The most general linear form can be expressed as

$$I = \frac{1}{z}(V - \mathcal{E}). \tag{23.1}$$

In general, both z and $\mathcal{E}$ may depend in some complicated way on the frequency ω. Equation (23.1), however, is the relation we would get if behind the two terminals there was just the generator $\mathcal{E}(\omega)$ in series with the impedance $z(\omega)$.

There is also the opposite kind of question: If we have any electromagnetic device at all with two terminals and we *measure* the relation between I and V to determine $\mathcal{E}$ and z as functions of frequency, can we find a combination of our ideal elements that is equivalent to the internal impedance z? The answer is that for any reasonable—that is, physically meaningful—function $z(\omega)$, it *is* possible to *approximate* the situation to as high an accuracy as you wish with a circuit containing a finite set of ideal elements. We don't want to consider the general problem now, but only look at what might be expected from physical arguments for a few cases.

If we think of a real resistor, we know that the current through it will produce a magnetic field. So any real resistor should also have some inductance. Also, when a resistor has a potential difference across it, there must be charges on the ends of the resistor to produce the necessary electric fields. As the voltage changes, the charges will change in proportion, so the resistor will also have some capacitance. We expect that a *real* resistor might have the equivalent circuit shown in Fig. 23–1. In a well-designed resistor, the so-called "parasitic" elements L and C are small, so that at the frequencies for which it is intended, ωL is much less than R, and $1/\omega C$ is much greater than R. It may therefore be possible to neglect them. As the frequency is raised, however, they will eventually become important, and a resistor begins to look like a resonant circuit.

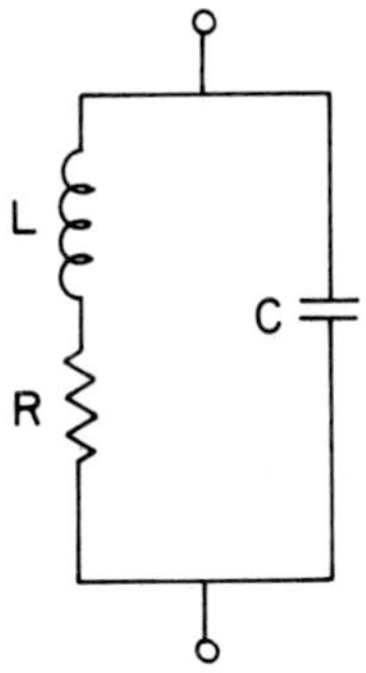

Fig. 23–1. Equivalent circuit of a real resistor.

A real inductance is also not equal to the idealized inductance, whose impedance is $i\omega L$. A real coil of wire will have some resistance, so at low frequencies the coil is really equivalent to an inductance in series with some resistance, as shown in Fig. 23–2(a). But, you are thinking, the resistance and inductance are *together* in a real coil—the resistance is spread all along the wire, so it is mixed in with the inductance. We should probably use a circuit more like the one in Fig. 23–2(b), which has several little R's and L's in series. But the total impedance of such a circuit is just $\sum R + \sum i\omega L$, which is equivalent to the simpler diagram of part (a).

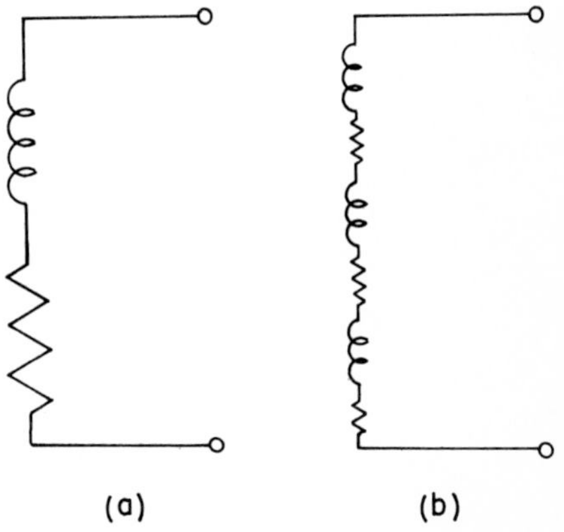

(a) (b)

Fig. 23–2. The equivalent circuit of a real inductance at low frequencies.

As we go up in frequency with a real coil, the approximation of an inductance plus a resistance is no longer very good. The charges that must build up on the wires to make the voltages will become important. It is as if there were little condensers across the turns of the coil, as sketched in Fig. 23–3(a). We might try to approximate the real coil by the circuit in Fig. 23–3(b). At low frequencies, this circuit can be imitated fairly well by the simpler one in part (c) of the figure (which is again the same resonant circuit we found for the high-frequency model of a resistor). For higher frequencies, however, the more complicated circuit of

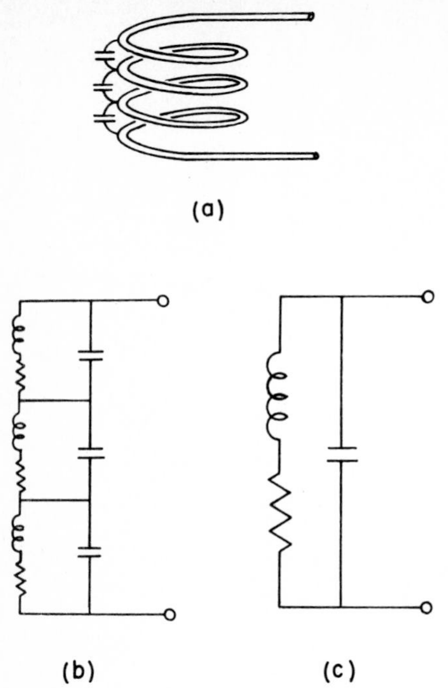

Fig. 23–3. The equivalent circuit of a real inductance at higher frequencies.

Fig. 23–3(b) is better. In fact, the more accurately you wish to represent the actual impedance of a real, physical inductance, the more ideal elements you will have to use in the artificial model of it.

Let's look a little more closely at what goes on in a real coil. The impedance of an inductance goes as ωL, so it becomes zero at low frequencies—it is a "short circuit": all we see is the resistance of the wire. As we go up in frequency, ωL soon becomes much larger than R, and the coil looks pretty much like an ideal inductance. As we go still higher, however, the capacities become important. Their impedance is proportional to $1/\omega C$, which is large for small ω. For small enough frequencies a condenser is an "open circuit," and when it is in parallel with something else, it draws no current. But at high frequencies, the current prefers to flow into the capacitance between the turns, rather than through the inductance. So the current in the coil jumps from one turn to the other and doesn't bother to go around and around where it has to buck the emf. So although we may have *intended* that the current should go around the loop, it will take the easier path—the path of least impedance.

If the subject had been one of popular interest, this effect would have been called "the high-frequency barrier," or some such name. The same kind of thing happens in all subjects. In aerodynamics, if you try to make things go faster than the speed of sound when they were designed for lower speeds, they don't work. It doesn't mean that there is a great "barrier" there; it just means that the object should be redesigned. So this coil which we designed as an "inductance" is not going to work as a good inductance, but as some other kind of thing at very high frequencies. For high frequencies, we have to find a new design.

23–2 A capacitor at high frequencies

Now we want to discuss in detail the behavior of a capacitor—a geometrically ideal capacitor—as the frequency gets larger and larger, so we can see the transition of its properties. (We prefer to use a capacitor instead of an inductance, because the geometry of a pair of plates is much less complicated than the geometry of a coil.) We consider the capacitor shown in Fig. 23–4(a), which consists of two parallel circular plates connected to an external generator by a pair of wires. If we charge the capacitor with DC, there will be a positive charge on one plate and a negative charge on the other; and there will be a uniform electric field between the plates.

Now suppose that instead of DC, we put an AC of low frequency on the plates. (We will find out later what is "low" and what is "high".) Say we connect the capacitor to a lower-frequency generator. As the voltage alternates, the positive charge on the top plate is taken off and negative charge is put on. While that is happening, the electric field disappears and then builds up in the opposite direction.

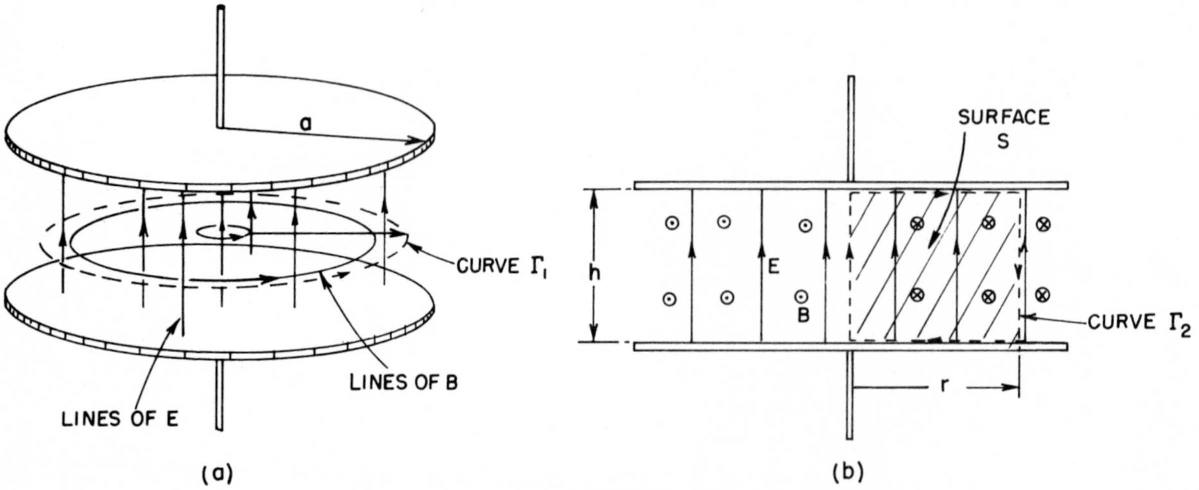

Fig. 23–4. The electric and magnetic fields between the plates of a capacitor.

As the charge sloshes back and forth slowly, the electric field follows. At each instant the electric field is uniform, as shown in Fig. 23–4(b), except for some edge effects which we are going to disregard. We can write the magnitude of the electric field as

$$E = E_0 e^{i\omega t}, \tag{23.2}$$

where E_0 is a constant.

Now will that continue to be right as the frequency goes up? No, because as the electric field is going up and down, there is a flux of electric field through any loop like Γ_1 in Fig. 23–4(a). And, as you know, a changing electric field acts to produce a magnetic field. One of Maxwell's equations says that when there is a varying electric field, as there is here, there has got to be a line integral of the magnetic field. The integral of the magnetic field around a closed ring, multiplied by c^2, is equal to the time rate-of-change of the electric flux through the area inside the ring (if there are no currents):

$$c^2 \oint_\Gamma \boldsymbol{B} \cdot d\boldsymbol{s} = \frac{\partial}{\partial t} \int_{\text{inside } \Gamma} \boldsymbol{E} \cdot \boldsymbol{n}\, da. \tag{23.3}$$

So how much magnetic field is there? That's not very hard. Suppose that we take the loop Γ_1, which is a circle of radius r. We can see from symmetry that the magnetic field goes around as shown in the figure. Then the line integral of $\boldsymbol{B}$ is $2\pi rB$. And, since the electric field is uniform, the flux of the electric field is simply E multiplied by πr^2, the area of the circle:

$$c^2 B \cdot 2\pi r = \frac{\partial}{\partial t} E \cdot \pi r^2. \tag{23.4}$$

The derivative of E with respect to time is, for our alternating field, simply $i\omega E_0 e^{i\omega t}$. So we find that our capacitor has the magnetic field

$$B = \frac{i\omega r}{2c^2} E_0 e^{i\omega t}. \tag{23.5}$$

In other words, the magnetic field also oscillates and has a strength proportional to r.

What is the effect of that? When there is a magnetic field that is varying, there will be induced electric fields and the capacitor will begin to act a little bit like an inductance. As the frequency goes up, the magnetic field gets stronger; it is proportional to the rate of change of E, and so to ω. The impedance of the capacitor will no longer be simply $1/i\omega C$.

Let's continue to raise the frequency and to analyze what happens more carefully. We have a magnetic field that goes sloshing back and forth. But then the electric field cannot be uniform, as we have assumed! When there is a varying magnetic field, there must be a line integral of the electric field—because of Faraday's law. So if there is an appreciable magnetic field, as begins to happen at high frequencies, the electric field cannot be the same at all distances from the center. The electric field must change with r so that the line integral of the electric field can equal the changing flux of the magnetic field.

Let's see if we can figure out the correct electric field. We can do that by computing a "correction" to the uniform field we originally assumed for low frequencies. Let's call the uniform field E_1, which will still be $E_0 e^{i\omega t}$, and write the correct field as

$$E = E_1 + E_2,$$

where E_2 is the correction due to the changing magnetic field. For any ω we will write the field at the center of the condenser as $E_0 e^{i\omega t}$ (thereby defining E_0), so that we have no correction at the center; $E_2 = 0$ at $r = 0$.

To find E_2 we can use the integral form of Faraday's law:

$$\oint_\Gamma \boldsymbol{E} \cdot d\boldsymbol{s} = -\frac{\partial}{\partial t} (\text{flux of } B).$$

The integrals are simple if we take them for the curve Γ_2, shown in Fig. 23–4(b), which goes up along the axis, out radially the distance r along the top plate, down vertically to the bottom plate, and back to the axis. The line integral of E_1 around this curve is, of course, zero; so only E_2 contributes, and its integral is just $-E_2(r) \cdot h$, where h is the spacing between the plates. (We call E positive if it points upward.) This is equal to the rate of change of the flux of B, which we have to get by an integral over the shaded area S inside Γ_2 in Fig. 23–4(b). The flux through a vertical strip of width dr is $B(r)h\,dr$, so the total flux is

$$h\int B(r)\,dr.$$

Setting $-\partial/\partial t$ of the flux equal to the line integral of E_2, we have

$$E_2(r) = \frac{\partial}{\partial t}\int B(r)\,dr. \tag{23.6}$$

Notice that the h cancels out; the fields don't depend on the separation of the plates.

Using Eq. (23.5) for $B(r)$, we have

$$E_2(r) = \frac{\partial}{\partial t}\frac{i\omega r^2}{4c^2}\,E_0e^{i\omega t}.$$

The time derivative just brings down another factor $i\omega$; we get

$$E_2(r) = -\frac{\omega^2 r^2}{4c^2}\,E_0e^{i\omega t}. \tag{23.7}$$

As we expect, the induced field tends to *reduce* the electric field farther out. The corrected field $E = E_1 + E_2$ is then

$$E = E_1 + E_2 = \left(1 - \frac{1}{4}\frac{\omega^2 r^2}{c^2}\right)E_0e^{i\omega t}. \tag{23.8}$$

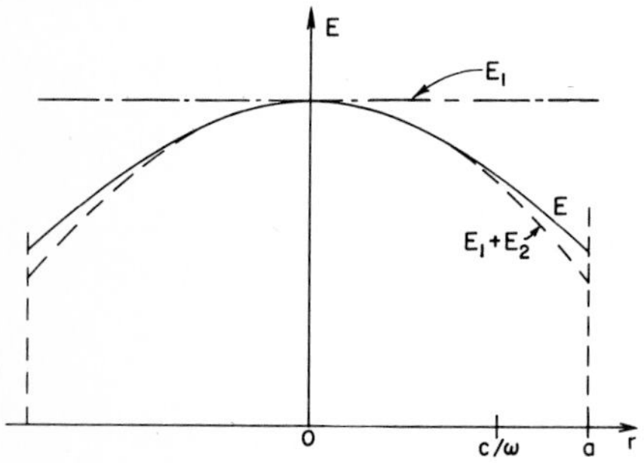

Fig. 23–5. The electric field between the capacitor plates at high frequency. (Edge effects are neglected.)

The electric field in the capacitor is no longer uniform; it has the parabolic shape shown by the broken line in Fig. 23–5. You see that our simple capacitor is getting slightly complicated.

We could now use our results to calculate the impedance of the capacitor at high frequencies. Knowing the electric field, we could compute the charges on the plates and find out how the current through the capacitor depends on the frequency ω, but we are not interested in that problem for the moment. We are more interested in seeing what happens as we continue to go up with the frequency —to see what happens at even higher frequencies. Aren't we already finished? No, because we have corrected the electric field, which means that the magnetic field we have calculated is no longer right. The magnetic field of Eq. (23.5) is approximately right, but it is only a first approximation. So let's call it B_1. We should then rewrite Eq. (23.5) as

$$B_1 = \frac{i\omega r}{2c^2}\,E_0e^{i\omega t}. \tag{23.9}$$

You will remember that this field was produced by the variation of E_1. Now the correct magnetic field will be that produced by the total electric field $E_1 + E_2$. If we write the magnetic field as $B = B_1 + B_2$, the second term is just the additional field produced by E_2. To find B_2 we can go through the same arguments we have used to find B_1; the line integral of B_2 around the curve Γ_1 is equal to the rate of change of the flux of E_2 through Γ_1. We will just have Eq. (23.4) again with B replaced by B_2 and E replaced by E_2:

$$c^2B_2 \cdot 2\pi r = \frac{\partial}{\partial t}\,(\text{flux of } E_2 \text{ through } \Gamma_1).$$

Since E_2 varies with radius, to obtain its flux we must integrate over the circular surface inside Γ_1. Using $2\pi r\,dr$ as the element of area, this integral is

$$\int_0^r E_2(r)\cdot 2\pi r\,dr.$$

So we get for $B_2(r)$

$$B_2(r) = \frac{1}{rc^2}\frac{\partial}{\partial t}\int E_2(r)r\,dr. \tag{23.10}$$

Using $E_2(r)$ from Eq. (23.7), we need the integral of $r^3\,dr$, which is, of course, $r^4/4$. Our correction to the magnetic field becomes

$$B_2(r) = -\frac{i\omega^3 r^3}{16c^4}E_0e^{i\omega t}. \tag{23.11}$$

But we are still not finished! If the magnetic field B is not the same as we first thought, then we have incorrectly computed E_2. We must make a further correction to E, which comes from the extra magnetic field B_2. Let's call this additional correction to the electric field E_3. It is related to the magnetic field B_2 in the same way that E_2 was related to B_1. We can use Eq. (23.6) all over again just by changing the subscripts:

$$E_3(r) = \frac{\partial}{\partial t}\int B_2(r)\,dr. \tag{23.12}$$

Using our result, Eq. (23.11), for B_2, the new correction to the electric field is

$$E_3(r) = +\frac{\omega^4 r^4}{64c^4}E_0e^{i\omega t}. \tag{23.13}$$

Writing our doubly corrected electric field as $E = E_1 + E_2 + E_3$, we get

$$E = E_0e^{i\omega t}\left[1 - \frac{1}{2^2}\left(\frac{\omega r}{c}\right)^2 + \frac{1}{2^2\cdot 4^2}\left(\frac{\omega r}{c}\right)^4\right]. \tag{23.14}$$

The variation of the electric field with radius is no longer the simple parabola we drew in Fig. 23–5, but at large radii lies slightly above the curve $(E_1 + E_2)$.

We are not quite through yet. The new electric field produces a new correction to the magnetic field, and the newly corrected magnetic field will produce a further correction to the electric field, and on and on. However, we already have all the formulas that we need. For B_3 we can use Eq. (23.10), changing the subscripts of B and E from 2 to 3.

The next correction to the electric field is

$$E_4 = -\frac{1}{2^2\cdot 4^2\cdot 6^2}\left(\frac{\omega r}{c}\right)^6 E_0e^{i\omega t}.$$

So to this order we have that the complete electric field is given by

$$E = E_0e^{i\omega t}\left[1 - \frac{1}{(1!)^2}\left(\frac{\omega r}{2c}\right)^2 + \frac{1}{(2!)^2}\left(\frac{\omega r}{2c}\right)^4 - \frac{1}{(3!)^2}\left(\frac{\omega r}{2c}\right)^6 + \cdots\right], \tag{23.15}$$

where we have written the numerical coefficients in such a way that it is obvious how the series is to be continued.

Our final result is that the electric field between the plates of the capacitor, for any frequency, is given by $E_0e^{i\omega t}$ times the infinite series which contains only the variable $\omega r/c$. If we wish, we can define a special function, which we will call $J_0(x)$, as the infinite series that appears in the brackets of Eq. (23.15):

$$J_0(x) = 1 - \frac{1}{(1!)^2}\left(\frac{x}{2}\right)^2 + \frac{1}{(2!)^2}\left(\frac{x}{2}\right)^4 - \frac{1}{(3!)^2}\left(\frac{x}{2}\right)^6 + \cdots \tag{23.16}$$

Then we can write our solution as $E_0e^{i\omega t}$ times this function, with $x = \omega r/c$:

$$E = E_0e^{i\omega t}J_0\left(\frac{\omega r}{c}\right). \tag{23.17}$$

The reason we have called our special function J_0 is that, naturally, this is not the first time anyone has ever worked out a problem with oscillations in a cylinder. The function has come up before and is usually called J_0. It always comes up whenever you solve a problem about waves with cylindrical symmetry. The function J_0 is to cylindrical waves what the cosine function is to waves on a straight line. So it is an important function, invented a long time ago. Then a man named Bessel got his name attached to it. The subscript zero means that Bessel invented a whole lot of different functions and this is just the first of them.

The other functions of Bessel—J_1, J_2, and so on—have to do with cylindrical waves which have a variation of their strength with the angle around the axis of the cylinder.

The completely corrected electric field between the plates of our circular capacitor, given by Eq. (23.17), is plotted as the solid line in Fig. 23–5. For frequencies that are not too high, our second approximation was already quite good. The third approximation was even better—so good, in fact, that if we had plotted it, you would not have been able to see the difference between it and the solid curve. You will see in the next section, however, that the complete series is needed to get an accurate description for large radii, or for high frequencies.

23–3 A resonant cavity

We want to look now at what our solution gives for the electric field between the plates of the capacitor as we continue to go to higher and higher frequencies. For large ω, the parameter $x = \omega r/c$ also gets large, and the first few terms in the series for J_0 of x will increase rapidly. That means that the parabola we have drawn in Fig. 23–5 curves downward more steeply at higher frequencies. In fact, it looks as though the field would fall all the way to zero at some high frequency, perhaps when c/ω is approximately one-half of a. Let's see whether J_0 does indeed go through zero and become negative. We begin by trying $x = 2$:

$$J_0(2) = 1 - 1 + \tfrac{1}{4} - \tfrac{1}{36} = 0.22.$$

The function is still not zero, so let's try a higher value of x, say, $x = 2.5$. Putting in numbers, we write

$$J_0(2.5) = 1 - 1.56 + 0.61 - 0.09 = -0.04.$$

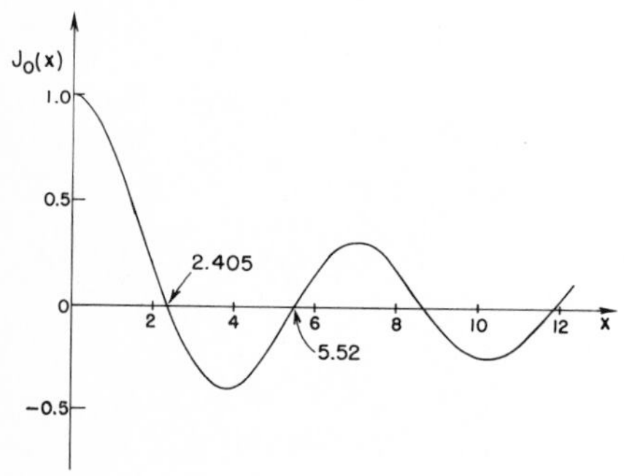

Fig. 23–6. The Bessel function $J_0(x)$.

The function J_0 has already gone through zero by the time we get to $x = 2.5$. Comparing the results for $x = 2$ and $x = 2.5$, it looks as though J_0 goes through zero at one-fifth of the way from 2.5 to 2. We would guess that the zero occurs for x approximately equal to 2.4. Let's see what that value of x gives:

$$J_0(2.4) = 1 - 1.44 + 0.52 - 0.08 = 0.00.$$

We get zero to the accuracy of our two decimal places. If we make the calculation more accurate (or since J_0 is a well-known function, if we look it up in a book), we find that it goes through zero at $x = 2.405$. We have worked it out by hand to show you that you too could have discovered these things rather than having to borrow them from a book.

As long as we are looking up J_0 in a book, it is interesting to notice how it goes for larger values of x; it looks like the graph in Fig. 23–6. As x increases, $J_0(x)$ oscillates between positive and negative values with a decreasing amplitude of oscillation.

We have gotten the following interesting result: If we go high enough in frequency, the electric field at the center of our condenser will be one way and the electric field near the edge will point in the opposite direction. For example, suppose that we take an ω high enough so that $x = \omega r/c$ at the outer edge of the capacitor is equal to 4; then the edge of the capacitor corresponds to the abscissa $x = 4$ in Fig. 23–6. This means that our capacitor is being operated at the frequency $\omega = 4c/a$. At the edge of the plates, the electric field will have a rather high magnitude opposite the direction we would expect. That is the terrible thing that can happen to a capacitor at high frequencies. If we go to very high frequencies, the direction of the electric field oscillates back and forth many times as we go out from the center of the capacitor. Also there are the magnetic fields associated with these electric fields. It is not surprising that our capacitor doesn't look like the ideal capacitance for high frequencies. We may even start to wonder whether it looks more like a capacitor or an inductance. We should emphasize that there are even more complicated effects that we have neglected which happen at the edges of the capacitor. For instance, there will be a radiation of waves out past the edges, so the fields are even more complicated than the ones we have computed, but we will not worry about those effects now.

We could try to figure out an equivalent circuit for the capacitor, but perhaps it is better if we just admit that the capacitor we have designed for low-frequency fields is just no longer satisfactory when the frequency is too high. If we want to treat the operation of such an object at high frequencies, we should abandon the approximations to Maxwell's equations that we have made for treating circuits and return to the complete set of equations which describe completely the fields in space. Instead of dealing with idealized circuit elements, we have to deal with the real conductors as they are, taking into account all the fields in the spaces in between. For instance, if we want a resonant circuit at high frequencies we will not try to design one using a coil and a parallel-plate capacitor.

We have already mentioned that the parallel-plate capacitor we have been analyzing has some of the aspects of both a capacitor and an inductance. With the electric field there are charges on the surfaces of the plates, and with the magnetic fields there are back emf's. Is it possible that we already have a resonant circuit? We do indeed. Suppose we pick a frequency for which the electric field pattern falls to zero at some radius inside the edge of the disc; that is, we choose $\omega a/c$ greater than 2.405. Everywhere on a circle coaxial with the plates the electric field will be zero. Now suppose we take a thin metal sheet and cut a strip just wide enough to fit between the plates of the capacitor. Then we bend it into a cylinder that will go around at the radius where the electric field is zero. Since there are no electric fields there, when we put this conducting cylinder in place, no currents will flow in it; and there will be no changes in the electric and magnetic fields. We have been able to put a direct short circuit across the capacitor without changing anything. And look what we have; we have a complete cylindrical can with electrical and magnetic fields inside and no connection at all to the outside world. The fields inside won't change even if we throw away the edges of the plates outside our can, and also the capacitor leads. All we have left is a closed can with electric and magnetic fields inside, as shown in Fig. 23–7(a). The electric fields are oscillating back and forth at the frequency ω—which, don't forget, determined the diameter of the can. The amplitude of the oscillating $\boldsymbol{E}$ field varies with the distance from the axis of the can, as shown in the graph of Fig. 23–7(b). This curve is just the first arch of the Bessel function of zero order. There is also a magnetic field which goes in circles around the axis and oscillates in time 90° out of phase with the electric field.

We can also write out a series for the magnetic field and plot it, as shown in the graph of Fig. 23–7(c).

How is it that we can have an electric and magnetic field inside a can with no external connections? It is because the electric and magnetic fields maintain themselves: the changing $\boldsymbol{E}$ makes a $\boldsymbol{B}$ and the changing $\boldsymbol{B}$ makes an $\boldsymbol{E}$—all according to the equations of Maxwell. The magnetic field has an inductive aspect, and the electric field a capacitive aspect; together they make something like a resonant circuit. Notice that the conditions we have described would only happen if the radius of the can is exactly 2.405 c/ω. For a can of a given radius, the oscillating electric and magnetic fields will maintain themselves—in the way we have described

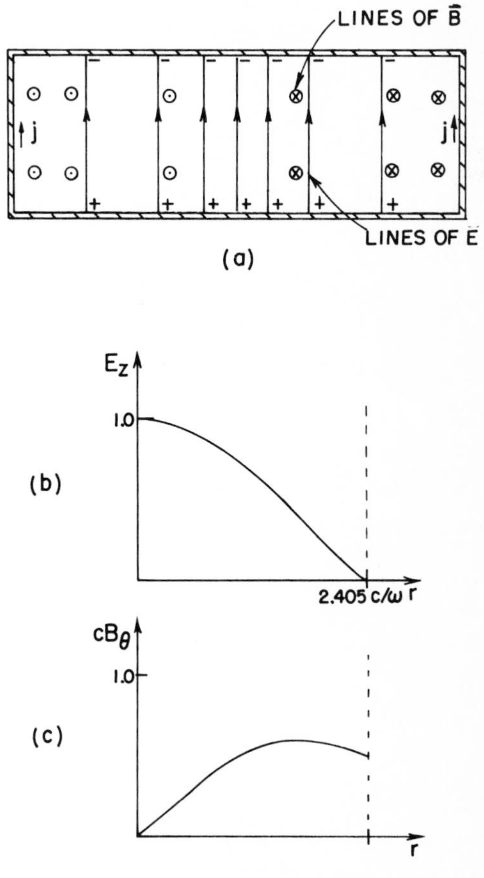

Fig. 23–7. The electric and magnetic fields in an enclosed cylindrical can.

—only at that particular frequency. So a cylindrical can of radius r is *resonant* at the frequency

$$\omega_0 = 2.405 \frac{c}{r}. \tag{23.18}$$

We have said that the fields continue to oscillate in the same way after the can is completely closed. That is not exactly right. It would be possible if the walls of the can were perfect conductors. For a real can, however, the oscillating currents which exist on the inside walls of the can lose energy because of the resistance of the material. The oscillations of the fields will gradually die away. We can see from Fig. 23–7 that there must be strong currents associated with electric and magnetic fields inside the cavity. Because the vertical electrical field stops suddenly at the top and bottom plates of the can, it has a large divergence there; so there must be positive and negative electric charges on the inner surfaces of the can, as shown in Fig. 23–7(a). When the electric field reverses, the charges must reverse also, so there must be an alternating current between the top and bottom plates of the can. These charges will flow in the sides of the can, as shown in the figure. We can also see that there must be currents in the sides of the can by considering what happens to the magnetic field. The graph of Fig. 23–7(c) tells us that the magnetic field suddenly drops to zero at the edge of the can. Such a sudden change in the magnetic field can happen only if there is a current in the wall. This current is what gives the alternating electric charges on the top and bottom plates of the can.

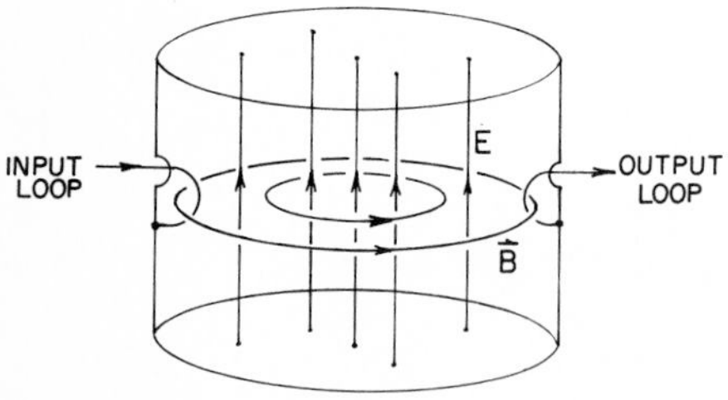

Fig. 23–8. Coupling into and out of a resonant cavity.

You may be wondering about our discovery of currents in the vertical sides of the can. What about our earlier statement that nothing would be changed when we introduced these vertical sides in a region where the electric field was zero? Remember, however, that when we first put in the sides of the can, the top and bottom plates extended out beyond them, so that there were also magnetic fields on the outside of our can. It was only when we threw away the parts of the capacitor plates beyond the edges of the can that net currents had to appear on the insides of the vertical walls.

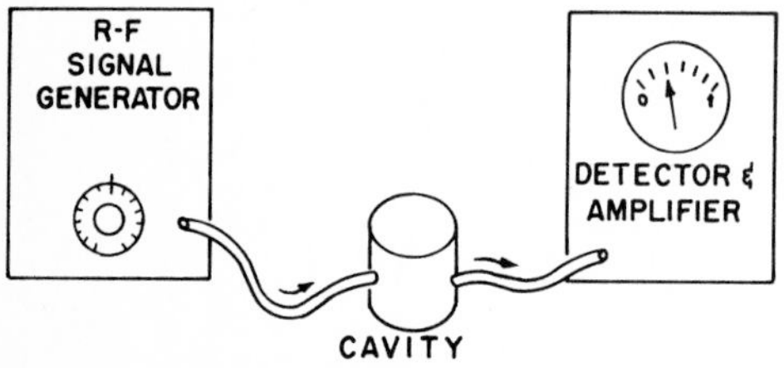

Fig. 23–9. A setup for observing the cavity resonance.

Although the electric and magnetic fields in the completely enclosed can will gradually die away because of the energy losses, we can stop this from happening if we make a little hole in the can and put in a little bit of electrical energy to make up the losses. We take a small wire, poke it through the hole in the side of the can, and fasten it to the inside wall so that it makes a small loop, as shown in Fig. 23–8. If we now connect this wire to a source of high-frequency alternating current, this current will couple energy into the electric and magnetic fields of the cavity and keep the oscillations going. This will happen, of course, only if the frequency of the driving source is at the resonant frequency of the can. If the source is at the wrong frequency, the electric and magnetic fields will not resonate, and the fields in the can will be very weak.

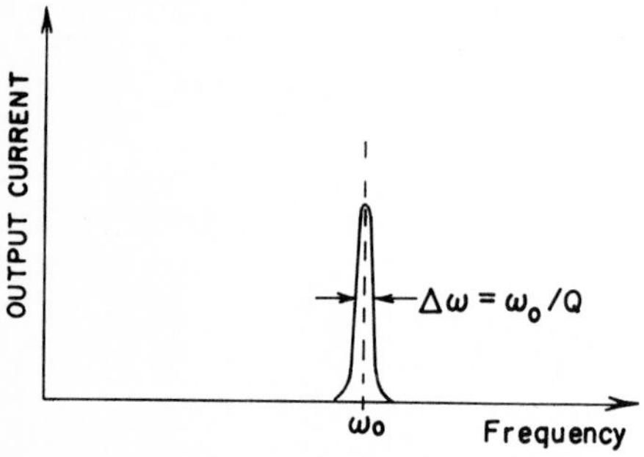

Fig. 23–10. The frequency response curve of a resonant cavity.

The resonant behavior can easily be seen by making another small hole in the can and hooking in another coupling loop, as we have also drawn in Fig. 23–8. The changing magnetic field through this loop will generate an induced electromotive force in the loop. If this loop is now connected to some external measuring circuit, the currents will be proportional to the strength of the fields in the cavity. Suppose we now connect the input loop of our cavity to an RF signal generator, as shown in Fig. 23–9. The signal generator contains a source of alternating current whose frequency can be varied by varying the knob on the front of the generator. Then we connect the output loop of the cavity to a "detector," which is an instrument that measures the current from the output loop. It gives a meter reading proportional to this current. If we now measure the output current as a function of the frequency of the signal generator, we find a curve like that shown in Fig. 23–10. The output current is small for all frequencies except those very near the frequency ω_0, which is the resonant frequency of the cavity. The resonance curve is very much like those we described in Chapter 23 of Vol. I. The width of the resonance is, however, much narrower than we usually find for resonant circuits made of inductances and capacitors; that is, the Q of the cavity is very high. It is not unusual to find Q's as high as 100,000 or more if the inside walls of the cavity are made of some material with a very good conductivity, such as silver.

23-4 Cavity modes

Suppose we now try to check our theory by making measurements with an actual can. We take a can which is a cylinder with a diameter of 3.0 inches and a height of about 2.5 inches. The can is fitted with an input and output loop, as shown in Fig. 23–8. If we calculate the resonant frequency expected for this can according to Eq. (23.18), we get that $f_0 = \omega_0/2\pi = 3010$ megacycles. When we set the frequency of our signal generator near 3000 megacycles and vary it slightly until we find the resonance, we observe that the maximum output current occurs for a frequency of 3050 megacycles, which is quite close to the predicted resonant frequency, but not exactly the same. There are several possible reasons for the discrepancy. Perhaps the resonant frequency is changed a little bit because of the holes we have cut to put in the coupling loops. A little thought, however, shows that the holes should lower the resonant frequency a little bit, so that cannot be the reason. Perhaps there is some slight error in the frequency calibration of the signal generator, or perhaps our measurement of the diameter of the cavity is not accurate enough. Anyway, the agreement is fairly close.

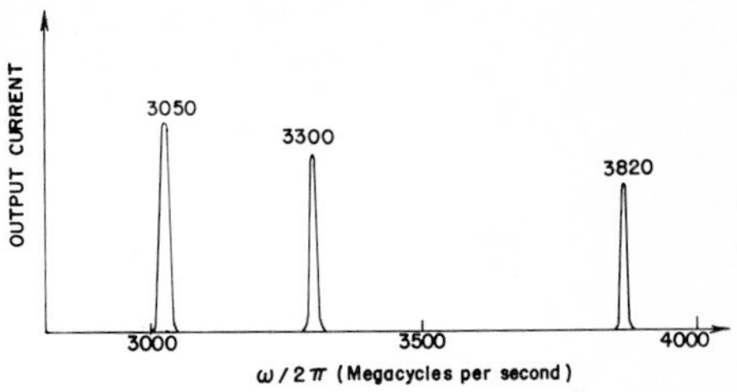

Fig. 23–11. Observed resonant frequencies of a cylindrical cavity.

Much more important is something that happens if we vary the frequency of our signal generator somewhat further from 3000 megacycles. When we do that we get the results shown in Fig. 23–11. We find that, in addition to the resonance we expected near 3000 megacycles, there is also a resonance near 3300 megacycles and one near 3820 megacycles. What do these extra resonances mean? We might get a clue from Fig. 23–6. Although we have been assuming that the first zero of the Bessel function occurs at the edge of the can, it could also be that the second zero of the Bessel function corresponds to the edge of the can, so that there is one complete oscillation of the electric field as we move from the center of the can out to the edge, as shown in Fig. 23–12. This is another possible mode for the oscillating fields. We should certainly expect the can to resonate in such a mode. But notice, the second zero of the Bessel function occurs at $x = 5.52$, which is over twice as large as the value at the first zero. The resonant frequency of this mode should therefore be higher than 6000 megacycles. We would, no doubt, find it there, but it doesn't explain the resonance we observe at 3300.

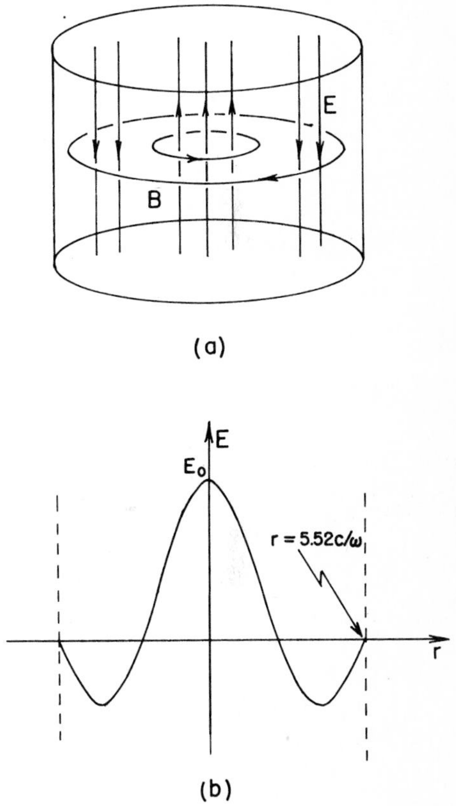

Fig. 23–12. A higher-frequency mode.

The trouble is that in our analysis of the behavior of a resonant cavity we have considered only one possible geometric arrangement of the electric and magnetic fields. We have assumed that the electric fields are vertical and that the magnetic fields lie in horizontal circles. But other fields are possible. The only requirements are that the fields should satisfy Maxwell's equations inside the can and that the electric field should meet the wall at right angles. We have considered the case in which the top and the bottom of the can are flat, but things would not be completely different if the top and bottom were curved. In fact, how is the can supposed to know which is its top and bottom, and which are its sides? It is, in fact, possible to show that there is a mode of oscillation of the fields inside the can in which the electric fields go more or less across the diameter of the can, as shown in Fig. 23–13.

It is not too hard to understand why the natural frequency of this mode should be not very different from the natural frequency of the first mode we have considered. Suppose that instead of our cylindrical cavity we had taken a cavity which was a cube 3 inches on a side. It is clear that this cavity would have three different modes, but all with the same frequency. A mode with the electric field going more or less up and down would certainly have the same frequency as the mode in which the electric field was directed right and left. If we now distort the cube into a cylinder, we will change these frequencies somewhat. We would still expect them not to be changed too much, provided we keep the dimensions of the cavity more or less the same. So the frequency of the mode of Fig. 23–13 should not be too different from the mode of Fig. 23–8. We could make a detailed calculation of the natural frequency of the mode shown in Fig. 23–13, but we will not do that now. When the calculations are carried through, it is found that, for the dimensions we have assumed, the resonant frequency comes out very close to the observed resonance at 3300 megacycles.

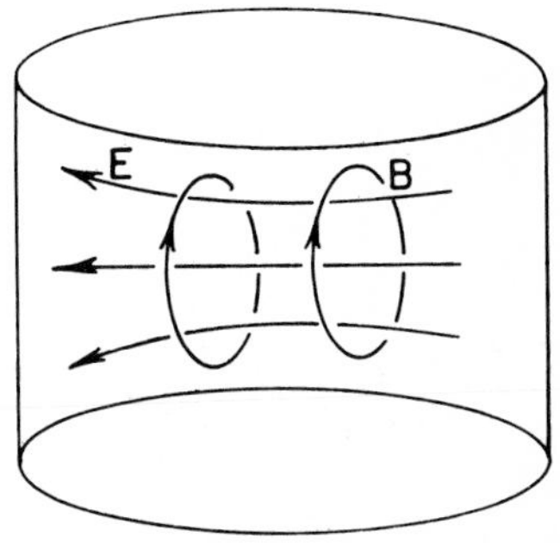

Fig. 23–13. A transverse mode of the cylindrical cavity.

By similar calculations it is possible to show that there should be still another mode at the other resonant frequency we found near 3800 megacycles. For this mode, the electric and magnetic fields are as shown in Fig. 23–14. The electric field does not bother to go all the way across the cavity. It goes from the sides to the ends, as shown.

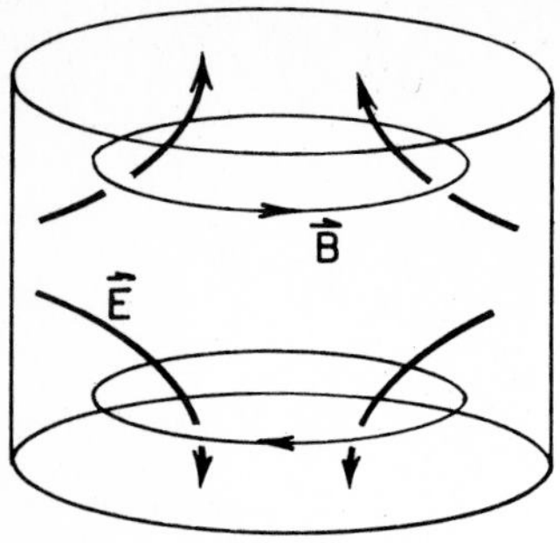

Fig. 23–14. Another mode of a cylindrical cavity.

As you will probably now believe, if we go higher and higher in frequency we should expect to find more and more resonances. There are many different modes, each of which will have a different resonant frequency corresponding to some particular complicated arrangement of the electric and magnetic fields. Each of these field arrangements is called a resonant *mode*. The resonance frequency of each mode can be calculated by solving Maxwell's equations for the electric and magnetic fields in the cavity.

When we have a resonance at some particular frequency, how can we know which mode is being excited? One way is to poke a little wire into the cavity through a small hole. If the electric field is along the wire, as in Fig. 23–15(a), there will be relatively large currents in the wire, sapping energy from the fields, and the resonance will be suppressed. If the electric field is as shown in Fig. 23–15(b), the wire will have a much smaller effect. We could find which way the field points in this mode by bending the end of the wire, as shown in Fig. 23–15(c). Then, as we rotate the wire, there will be a big effect when the end of the wire is parallel to $\boldsymbol{E}$ and a small effect when it is rotated so as to be at 90° to $\boldsymbol{E}$.

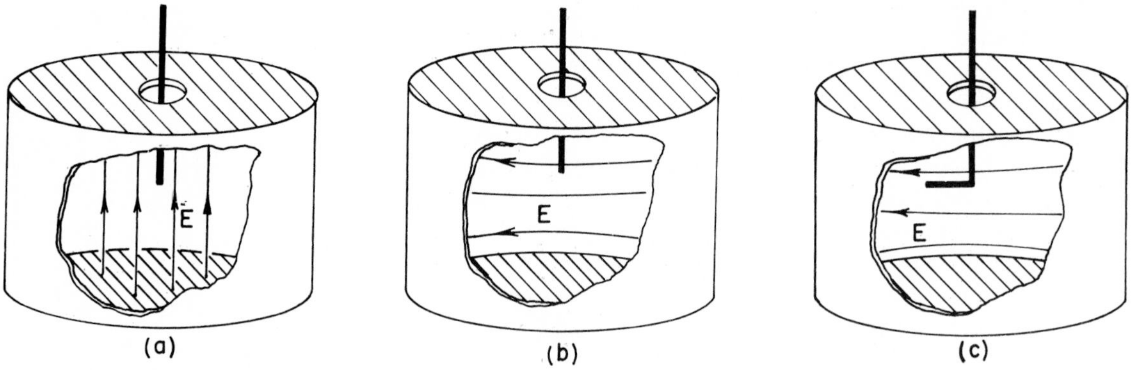

Fig. 23–15. A short metal wire inserted into a cavity will disturb the resonance much more when it is parallel to **E** than when it is at right angles.

23–5 Cavities and resonant circuits

Although the resonant cavity we have been describing seems to be quite different from the ordinary resonant circuit consisting of an inductance and a capacitor, the two resonant systems are, of course, closely related. They are both members of the same family; they are just two extreme cases of electromagnetic resonators—and there are many intermediate cases between these two extremes. Suppose we start by considering the resonant circuit of a capacitor in parallel with an inductance, as shown in Fig. 23–16(a). This circuit will resonate at the frequency $\omega_0 = 1/\sqrt{LC}$. If we want to raise the resonant frequency of this circuit, we can do so by lowering the inductance L. One way is to decrease the number of turns in the coil. We can, however, go only so far in this direction. Eventually we will get down to the last turn, and we will have just a piece of wire joining the top and bottom plates of the condenser. We could raise the resonant frequency still further by making the capacitance smaller; however, we can also continue to decrease the inductance by putting several inductances in parallel. Two one-turn inductances in parallel will have only half the inductance of each turn. So when our inductance has been reduced to a single turn, we can continue to raise the resonant frequency by adding other single loops from the top plate to the bottom plate of the condenser. For instance, Fig. 23–16(b) shows the condenser plates connected by six such "single-turn inductances." If we continue to add many such pieces of wire, we can make the transition to the completely enclosed resonant system shown in part (c) of the figure, which is a drawing of the cross section of a cylindrically symmetrical

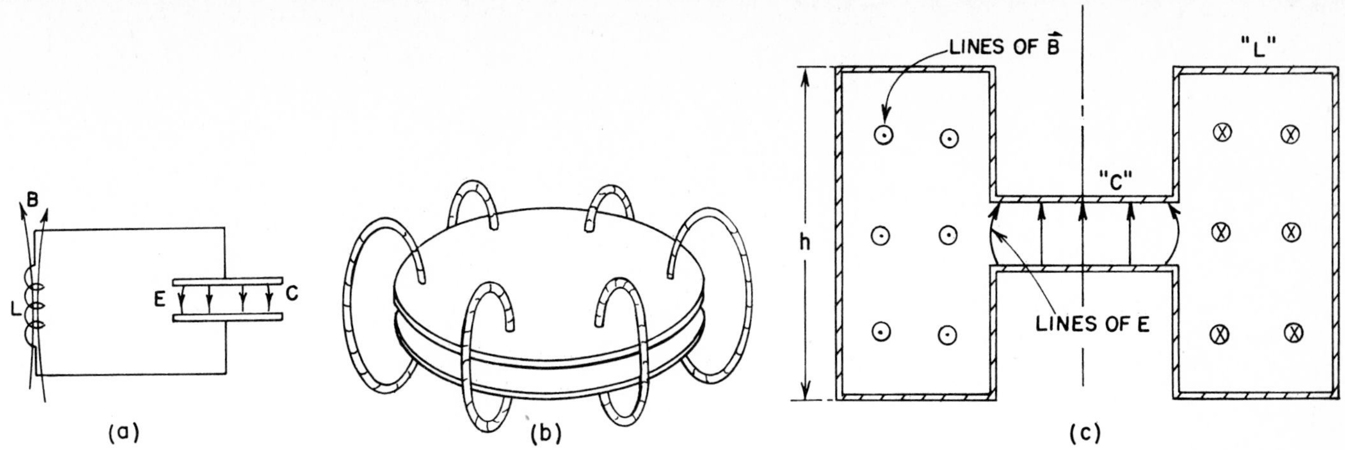

Fig. 23–16. Resonators of progressively higher resonant frequencies.

object. Our inductance is now a cylindrical hollow can attached to the edges of the condenser plates. The electric and magnetic fields will be as shown in the figure. Such an object is, of course, a resonant cavity. It is called a "loaded" cavity. But we can still think of it as an *L-C* circuit in which the capacity section is the region where we find most of the electric field and the inductance section is that region where we find most of the magnetic field.

If we want to make the frequency of the resonator in Fig. 23–16(c) still higher, we can do so by continuing to decrease the inductance L. To do that, we must decrease the geometric dimensions of the inductance section, for example by decreasing the dimension h in the drawing. As h is decreased, the resonant frequency will be increased. Eventually, of course, we will get to the situation in which the height h is just equal to the separation between the condenser plates. We then have just a cylindrical can; our resonant circuit has become the cavity resonator of Fig. 23–7.

You will notice that in the original *L-C* resonant circuit of Fig. 23–16 the electric and magnetic fields are quite separate. As we have gradually modified the resonant system to make higher and higher frequencies, the magnetic field has been brought closer and closer to the electric field until in the cavity resonator the two are quite intermixed.

Although the cavity resonators we have talked about in this chapter have been cylindrical cans, there is nothing magic about the cylindrical shape. A can of any shape will have resonant frequencies corresponding to various possible modes of oscillations of the electric and magnetic fields. For example, the "cavity" shown in Fig. 23–17 will have its own particular set of resonant frequencies—although they would be rather difficult to calculate.

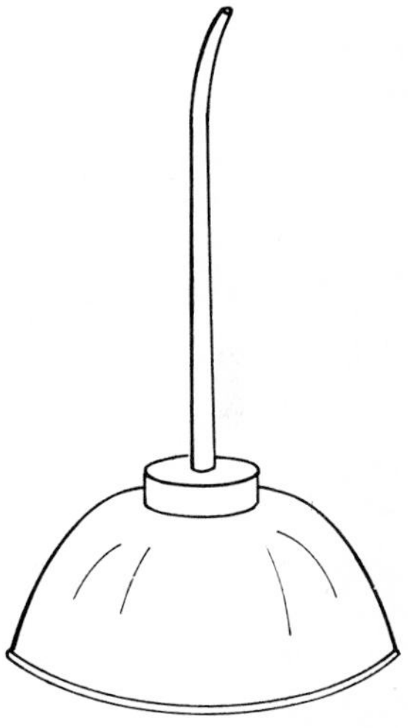

Fig. 23–17. Another resonant cavity.

24

Waveguides

24–1 The transmission line

In the last chapter we studied what happened to the lumped elements of circuits when they were operated at very high frequencies, and we were led to see that a resonant circuit could be replaced by a cavity with the fields resonating inside. Another interesting technical problem is the connection of one object to another, so that electromagnetic energy can be transmitted between them. In low-frequency circuits the connection is made with wires, but this method doesn't work very well at high frequencies because the circuits would radiate energy into all the space around them, and it is hard to control where the energy will go. The fields spread out around the wires; the currents and voltages are not "guided" very well by the wires. In this chapter we want to look into the ways that objects can be interconnected at high frequencies. At least, that's one way of presenting our subject.

Another way is to say that we have been discussing the behavior of waves in free space. Now it is time to see what happens when oscillating fields are confined in one or more dimensions. We will discover the interesting new phenomenon when the fields are confined in only two dimensions and allowed to go free in the third dimension, they propagate in waves. These are "guided waves"—the subject of this chapter.

We begin by working out the general theory of the *transmission line*. The ordinary power transmission line that runs from tower to tower over the country-side radiates away some of its power, but the power frequencies (50–60 cycles/sec) are so low that this loss is not serious. The radiation could be stopped by surrounding the line with a metal pipe, but this method would not be practical for power lines because the voltages and currents used would require a very large, expensive, and heavy pipe. So simple "open lines" are used.

For somewhat higher frequencies—say a few kilocycles—radiation can already be serious. However, it can be reduced by using "twisted-pair" transmission lines, as is done for short-run telephone connections. At higher frequencies, however, the radiation soon becomes intolerable, either because of power losses or because the energy appears in other circuits where it isn't wanted. For frequencies from a few kilocycles to some hundreds of megacycles, electromagnetic signals and power are usually transmitted via coaxial lines consisting of a wire inside a cylindrical "outer conductor" or "shield." Although the following treatment will apply to a transmission line of two parallel conductors of any shape, we will carry it out referring to a coaxial line.

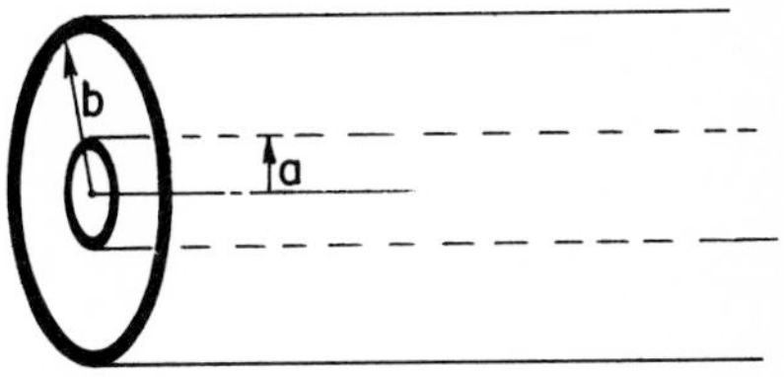

Fig. 24–1. A coaxial transmission line.

We take the simplest coaxial line that has a central conductor, which we suppose is a thin hollow cylinder, and an outer conductor which is another thin cylinder on the same axis as the inner conductor, as in Fig. 24–1. We begin by figuring out approximately how the line behaves at relatively low frequencies. We have already described some of the low-frequency behavior when we said earlier that two such conductors had a certain amount of inductance per unit length or a certain capacity per unit length. We can, in fact, describe the low-frequency behavior of any transmission line by giving its inductance per unit length, L_0 and its capacity per unit length, C_0. Then we can analyze the line as the limiting case of the L-C filter as discussed in Section 22–6. We can make a filter which imitates the line by taking small series elements $L_0\,\Delta x$ and small shunt capacities $C_0\,\Delta x$, where Δx is an element of length of the line. Using our results for the infinite filter, we see that there would be a propagation of electric signals along the line. Rather than following that approach, however, we would now rather look at the line from the point of view of a differential equation.

Suppose that we see what happens at two neighboring points along the transmission line, say at the distances x and $x + \Delta x$ from the beginning of the line. Let's call the voltage difference between the two conductors $V(x)$, and the current along the "hot" conductor $I(x)$ (see Fig. 24–2). If the current in the line is varying, the inductance will give us a voltage drop across the small section of line from x to $x + \Delta x$ in the amount

$$\Delta V = V(x + \Delta x) - V(x) = -L_0 \,\Delta x \,\frac{dI}{dt}.$$

Or, taking the limit as $\Delta x \to 0$, we get

$$\frac{\partial V}{\partial x} = -L_0 \frac{\partial I}{\partial t}. \tag{24.1}$$

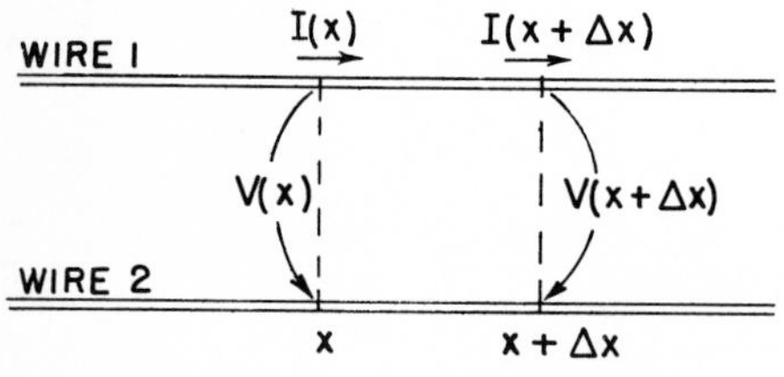

Fig. 24–2. The currents and voltages of a transmission line.

The changing current gives a gradient of the voltage.

Referring again to the figure, if the voltage at x is changing, there must be some charge supplied to the capacity in that region. If we take the small piece of line between x and $x + \Delta x$, the charge on it is $q = C_0 \,\Delta x V$. The time rate-of-change of this charge is $C_0 \,\Delta x \, dV/dt$, but the charge changes only if the current $I(x)$ into the element is different from the current $I(x + \Delta x)$ out. Calling the difference ΔI, we have

$$\Delta I = -C_0 \,\Delta x \,\frac{dV}{dt}.$$

Taking the limit as $\Delta x \to 0$, we get

$$\frac{\partial I}{\partial x} = -C_0 \frac{\partial V}{\partial t}. \tag{24.2}$$

So the conservation of charge implies that the gradient of the current is proportional to the time rate-of-change of the voltage.

Equations (24.1) and (24.2) are then the basic equations of a transmission line. If we wish, we could modify them to include the effects of resistance in the conductors or of leakage of charge through the insulation between the conductors, but for our present discussion we will just stay with the simple example.

The two transmission line equations can be combined by differentiating one with respect to t and the other with respect to x and eliminating either V or I. Then we have either

$$\frac{\partial^2 V}{\partial x^2} = C_0 L_0 \frac{\partial^2 V}{\partial t^2} \tag{24.3}$$

or

$$\frac{\partial^2 I}{\partial x^2} = C_0 L_0 \frac{\partial^2 I}{\partial t^2}. \tag{24.4}$$

Once more we recognize the wave equation in x. For a uniform transmission line, the voltage (and current) propagates along the line as a wave. The voltage along the line must be of the form $V(x, t) = f(x - vt)$ or $V(x, t) = g(x + vt)$, or a sum of both. Now what is the velocity v? We know that the coefficient of the $\partial^2/\partial t^2$ term is just $1/v^2$, so

$$v = \frac{1}{\sqrt{L_0 C_0}}. \tag{24.5}$$

We will leave it for you to show that the voltage *for each wave* in a line is proportional to the current of that wave and that the constant of proportionality is just the characteristic impedance z_0. Calling V_+ and I_+ the voltage and current for a wave going in the plus x-direction, you should get

$$V_+ = z_0 I_+. \tag{24.6}$$

Similary, for the wave going toward minus x the relation is

$$V_- = -z_0 I_-.$$

The characteristic impedance—as we found out from our filter equations—is given by

$$z_0 = \sqrt{\frac{L_0}{C_0}}, \tag{24.7}$$

and is, therefore, a pure resistance.

To find the propagation speed v and the characteristic impedance z_0 of a transmission line, we have to know the inductance and capacity per unit length. We can calculate them easily for a coaxial cable, so we will see how that goes. For the inductance we follow the ideas of Section 17–8, and set $\frac{1}{2}LI^2$ equal to the magnetic energy which we get by integrating $\epsilon_0 c^2 B^2/2$ over the volume. Suppose that the central conductor carries the current I; then we know that $B = I/2\pi\epsilon_0 c^2 r$, where r is the distance from the axis. Taking as a volume element a cylindrical shell of thickness dr and of length l, we have for the magnetic energy

$$U = \frac{\epsilon_0 c^2}{2} \int_a^b \left(\frac{I}{2\pi\epsilon_0 c^2 r}\right)^2 l\, 2\pi r\, dr,$$

where a and b are the radii of the inner and outer conductors, respectively. Carrying out the integral, we get

$$U = \frac{I^2 l}{4\pi\epsilon_0 c^2} \ln \frac{b}{a}. \tag{24.8}$$

Setting the energy equal to $\frac{1}{2}LI^2$, we find

$$L = \frac{l}{2\pi\epsilon_0 c^2} \ln \frac{b}{a}. \tag{24.9}$$

It is, as it should be, proportional to the length l of the line, so the inductance per unit length L_0 is

$$L_0 = \frac{\ln (b/a)}{2\pi\epsilon_0 c^2}. \tag{24.10}$$

We have worked out the charge on a cylindrical condenser (see Section 12–2). Now, dividing the charge by the potential difference, we get

$$C = \frac{2\pi\epsilon_0 l}{\ln (b/a)}.$$

The capacity per unit length C_0 is C/l. Combining this result with Eq. (24.10), we see that the product $L_0 C_0$ is just equal to $1/c^2$, so $v = 1/\sqrt{L_0 C_0}$ is equal to c. The wave travels down the line with the speed of light. We point out that this result depends on our assumptions: (a) that there are no dielectrics or magnetic materials in the space between the conductors, and (b) that the currents are all on the surfaces of the conductors (as they would be for perfect conductors). We will see later that for good conductors at high frequencies, all currents distribute themselves on the surfaces as they would for a perfect conductor, so this assumption is then valid.

Now it is interesting that so long as assumptions (a) and (b) are correct, the product $L_0 C_0$ is equal to $1/c^2$ for *any* parallel pair of conductors—even, say, for a hexagonal inner conductor anywhere inside an elliptical outer conductor. So long as the cross section is constant and the space between has no material, waves are propagated at the velocity of light.

No such general statement can be made about the characteristic impedance. For the coaxial line, it is

$$z_0 = \frac{\ln (b/a)}{2\pi\epsilon_0 c}. \tag{24.11}$$

The factor $1/\epsilon_0 c$ has the dimensions of a resistance and is equal to 120π ohms. The geometric factor $\ln{(b/a)}$ depends only logarithmically on the dimensions, so for the coaxial line—and most lines—the characteristic impedance has typical values of from 50 ohms or so to a few hundred ohms.

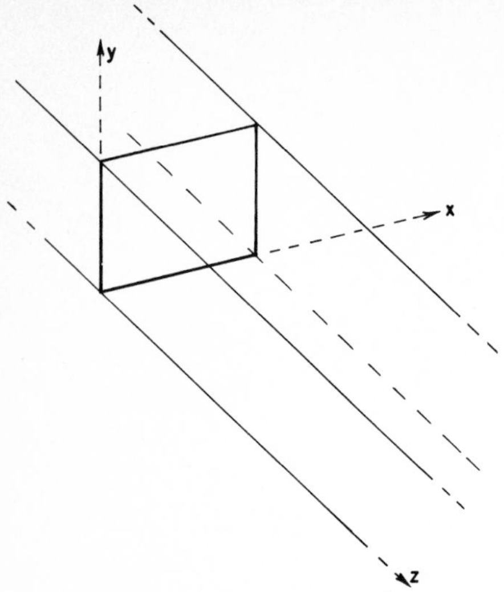

Fig. 24–3. Coordinates chosen for the rectangular waveguide.

24–2 The rectangular waveguide

The next thing we want to talk about seems, at first sight, to be a striking phenomenon: if the central conductor is removed from the coaxial line, it can still carry electromagnetic power. In other words, at high enough frequencies a hollow tube will work just as well as one with wires. It is related to the mysterious way in which a resonant circuit of a condenser and inductance gets replaced by nothing but a can at high frequencies.

Although it may seem to be a remarkable thing when one has been thinking in terms of a transmission line as a distributed inductance and capacity, we all know that electromagnetic waves can travel along inside a hollow metal pipe. If the pipe is straight, we can *see* through it! So certainly electromagnetic waves go through a pipe. But we also know that it is not possible to transmit low-frequency waves (power or telephone) through the inside of a single metal pipe. So it must be that electromagnetic waves will go through if their wavelength is short enough. Therefore we want to discuss the limiting case of the longest wavelength (or the lowest frequency) that can get through a pipe of a given size. Since the pipe is then being used to carry waves, it is called a *waveguide*.

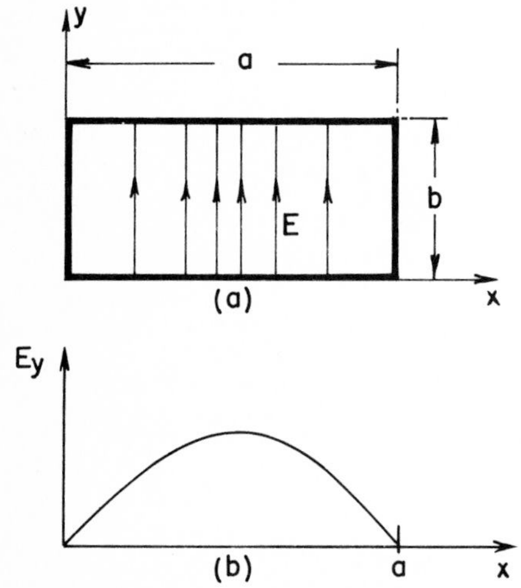

Fig. 24–4. The electric field in the waveguide at some value of z.

We will begin with a rectangular pipe, because it is the simplest case to analyze. We will first give a mathematical treatment and come back later to look at the problem in a much more elementary way. The more elementary approach, however, can be applied easily only to a rectangular guide. The basic phenomena are the same for a general guide of arbitrary shape, so the mathematical argument is fundamentally more sound.

Our problem, then, is to find what kind of waves can exist inside a rectangular pipe. Let's first choose some convenient coordinates; we take the z-axis along the length of the pipe, and the x- and y-axes parallel to the two sides, as shown in Fig. 24–3.

We know that when light waves go down the pipe, they have a transverse electric field; so suppose we look first for solutions in which $\boldsymbol{E}$ is perpendicular to z, say with only a y-component, E_y. This electric field will have some variation across the guide; in fact, it must go to zero at the sides parallel to the y-axis, because the currents and charges in a conductor always adjust themselves so that there is no tangential component of the electric field at the surface of a conductor. So E_y will vary with x in some arch, as shown in Fig. 24–4. Perhaps it is the Bessel function we found for a cavity? No, because the Bessel function has to do with cylindrical geometries. For a rectangular geometry, waves are usually simple harmonic functions, so we should try something like $\sin k_x x$.

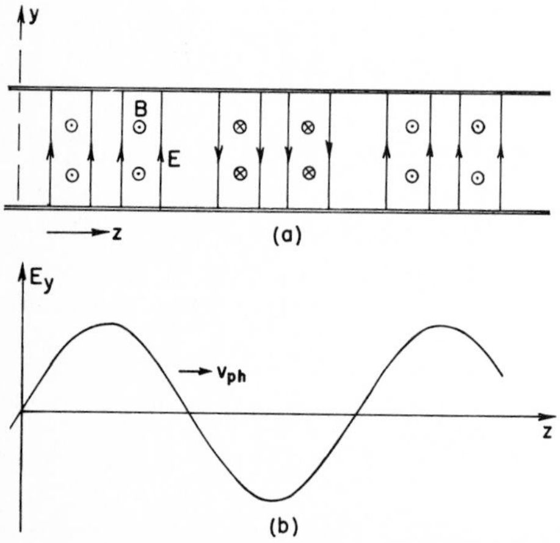

Fig. 24–5. The z-dependence of the field in the waveguide.

Since we want waves that propagate down the guide, we expect the field to alternate between positive and negative values as we go along in z, as in Fig. 24–5, and these oscillations will travel along the guide with some velocity v. If we have oscillations at some definite frequency ω, we would guess that the wave might vary with z like $\cos{(\omega t - k_z z)}$, or to use the more convenient mathematical form, like $e^{i(\omega t - k_z z)}$. This z-dependence represents a wave travelling with the speed $v = \omega/k_z$ (see Chapter 29, Vol. I).

So we might guess that the wave in the guide would have the following mathematical form:

$$E_y = E_0 \sin k_x x e^{i(\omega t - k_z z)}. \tag{24.12}$$

Let's see whether this guess satisfies the correct field equations. First, the electric field should have no tangential components at the conductors. Our field satisfies this requirement; it is perpendicular to the top and bottom faces and is zero at the two side faces. Well, it is if we choose k_x so that one-half a cycle of sin $k_x x$ just fits in the width of the guide—that is, if

$$k_x a = \pi. \tag{24.13}$$

There are other possibilities, like $k_x a = 2\pi, 3\pi, \ldots$, or, in general,

$$k_x a = n\pi, \tag{24.14}$$

where n is any integer. These represent various complicated arrangements of the field, but for now let's take only the simplest one, where $k_x = \pi/a$, where a is the width of the inside of the guide.

Next, the divergence of $\boldsymbol{E}$ must be zero in the free space inside the guide, since there are no charges there. Our $\boldsymbol{E}$ has only a y-component, and it doesn't change with y, so we do have that $\nabla \cdot \boldsymbol{E} = 0$.

Finally, our electric field must agree with the rest of Maxwell's equations in the free space inside the guide. That is the same thing as saying that it must satisfy the wave equation

$$\frac{\partial^2 E_y}{\partial x^2} + \frac{\partial^2 E_y}{\partial y^2} + \frac{\partial^2 E_y}{\partial z^2} - \frac{1}{c^2}\frac{\partial^2 E_y}{\partial t^2} = 0. \tag{24.15}$$

We have to see whether our guess, Eq. (24.12), will work. The second derivative of E_y with respect to x is just $-k_x^2 E_y$. The second derivative with respect to y is zero, since nothing depends on y. The second derivative with respect to z is $-k_z^2 E_y$, and the second derivative with respect to t is $-\omega^2 E_y$. Equation (24.15) then says that

$$k_x^2 E_y + k_z^2 E_y - \frac{\omega^2}{c^2} E_y = 0.$$

Unless E_y is zero everywhere (which is not very interesting), this equation is correct if

$$k_x^2 + k_z^2 - \frac{\omega^2}{c^2} = 0. \tag{24.16}$$

We have already fixed k_x, so this equation tells us that there can be waves of the type we have assumed if k_z is related to the frequency ω so that Eq. (24.16) is satisfied—in other words, if

$$k_z = \sqrt{(\omega^2/c^2) - (\pi^2/a^2)}. \tag{24.17}$$

The waves we have described are propagated in the z-direction with this value of k_z.

The wave number k_z we get from Eq. (24.17) tells us, for a given frequency ω, the speed with which the nodes of the wave propagate down the guide. The phase velocity is

$$v = \frac{\omega}{k_z}. \tag{24.18}$$

You will remember that the wavelength λ of a travelling wave is given by $\lambda = 2\pi v/\omega$, so k_z is also equal to $2\pi/\lambda_g$, where λ_g is the wavelength of the oscillations along the z-direction—the "guide wavelength." The wavelength in the guide is different ,of course, from the free-space wavelength of electromagnetic waves of the same frequency. If we call the free-space wavelength λ_0, which is equal to $2\pi c/\omega$, we can write Eq. (24.17) as

$$\lambda_g = \frac{\lambda_0}{\sqrt{1 - (\lambda_0/2a)^2}}. \tag{24.19}$$

Besides the electric fields there are magnetic fields that will travel with the wave, but we will not bother to work out an expression for them right now. Since $c^2\nabla \times \boldsymbol{B} = \partial\boldsymbol{E}/\partial t$, the lines of $\boldsymbol{B}$ will circulate around the regions in which $\partial\boldsymbol{E}/\partial t$ is largest, that is, halfway between the maximum and minimum of $\boldsymbol{E}$. The loops of $\boldsymbol{B}$ will lie parallel to the xz-plane and between the crests and troughs of $\boldsymbol{E}$, as shown in Fig. 24–6.

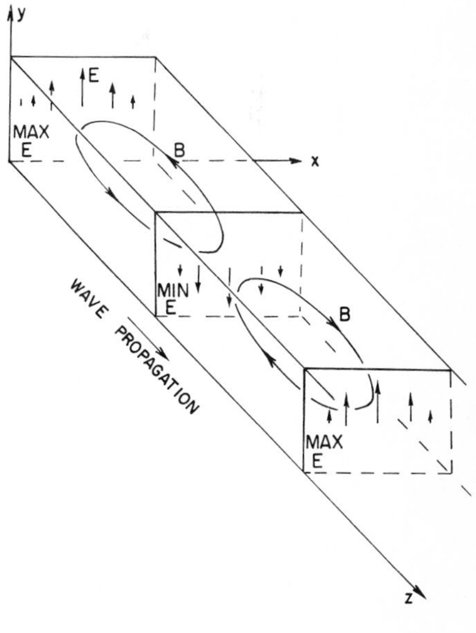

Fig. 24–6. The magnetic field in the waveguide.

24–3 The cutoff frequency

In solving Eq. (24.16) for k_z, there should really be two roots—one plus and one minus. We should write

$$k_z = \pm\sqrt{(\omega^2/c^2) - (\pi^2/a^2)}. \tag{24.20}$$

The two signs simply mean that there can be waves which propagate with a negative phase velocity (toward $-z$), as well as waves which propagate in the positive direction in the guide. Naturally, it should be possible for waves to go in either direction. Since both types of waves can be present at the same time, there will be the possibility of standing-wave solutions.

Our equation for k_z also tells us that higher frequencies give larger values of k_z, and therefore smaller wavelengths, until in the limit of large ω, k becomes equal to ω/c, which is the value we would expect for waves in free space. The light we "see" through a pipe still travels at the speed c. But now notice that if we go toward low frequencies, something strange happens. At first the wavelength gets longer and longer, but if ω gets too small the quantity inside the square root of Eq. (24.20) suddenly becomes negative. This will happen as soon as ω gets to be less than $\pi c/a$—or when λ_0 becomes greater than $2a$. In other words, when the frequency gets smaller than a certain critical frequency $\omega_c = \pi c/a$, the wave number k_z (and also λ_g) becomes imaginary and we haven't got a solution any more. Or do we? Who said that k_z has to be real? What if it does come out imaginary? Our field equations are still satisfied. Perhaps an imaginary k_z also represents a wave.

Suppose ω *is* less than ω_c; then we can write

$$k_z = \pm ik', \tag{24.21}$$

where k' is a positive real number:

$$k' = \sqrt{(\pi^2/a^2) - (\omega^2/c^2)}. \tag{24.22}$$

If we now go back to our expression, Eq. (24.12), for E_y, we have

$$E_y = E_0 \sin k_x x e^{i(\omega t \mp ik'z)}, \tag{24.23}$$

which we can write as

$$E_y = E_0 \sin k_x x e^{\pm k'z} e^{i\omega t}. \tag{24.24}$$

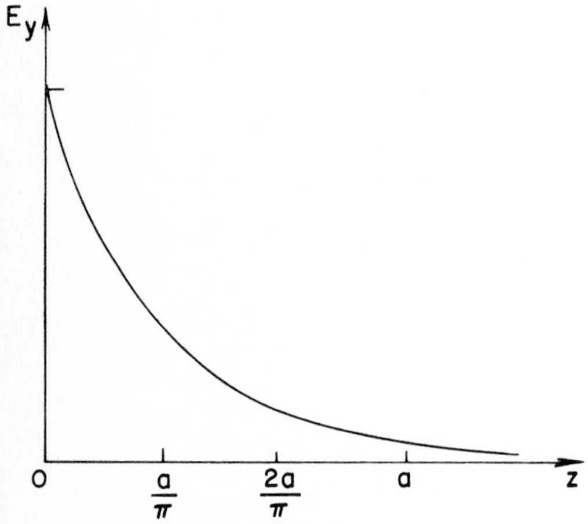

Fig. 24–7. The variation of E_y with z for $\omega \ll \omega_c$.

This expression gives an $\boldsymbol{E}$-field that oscillates with time as $e^{i\omega t}$ but which varies with z as $e^{\pm k'z}$. It decreases or increases with z smoothly as a real exponential. In our derivation we didn't worry about the sources that started the waves, but there must, of course, be a source someplace in the guide. The sign that goes with k' must be the one that makes the field decrease with increasing distance from the source of the waves.

So for frequencies below $\omega_c = \pi c/a$, waves do *not* propagate down the guide; the oscillating fields penetrate into the guide only a distance of the order of $1/k'$. For this reason, the frequency ω_c is called the "cutoff frequency" of the guide. Looking at Eq. (24.22), we see that for frequencies just a little below ω_c, the number k' is small and the fields can penetrate a long distance into the guide. But if ω is much less than ω_c, the exponential coefficient k' is equal to π/a and the field dies off extremely rapidly, as shown in Fig. 24–7. The field decreases by $1/e$ in the distance a/π, or in only about one-third of the guide width. The fields penetrate very little distance from the source.

We want to emphasize an interesting feature of our analysis of the guided waves—the appearance of the imaginary wave number k_z. Normally, if we solve an equation in physics and get an imaginary number, it doesn't mean anything physical. For *waves*, however, an imaginary wave number *does* mean something. The wave equation is still satisfied; it only means that the solution gives exponentially decreasing fields instead of propagating waves. So in any wave problem where k becomes imaginary for some frequency, it means that the form of the wave changes—the sine wave changes into an exponential.

24–4 The speed of the guided waves

The wave velocity we have used above is the phase velocity, which is the speed of a node of the wave; it is a function of frequency. If we combine Eqs. (24.17) and (24.18), we can write

$$v_{\text{phase}} = \frac{c}{\sqrt{1 - (\omega_c/\omega)^2}}. \tag{24.25}$$

For frequencies above cutoff—where travelling waves exist—ω_c/ω is less than one, and v_{phase} is real and *greater than* the speed of light. We have already seen in Chapter 48 of Vol. I that *phase* velocities greater than light are possible, because it is just the nodes of the wave which are moving and not energy or information. In order to know how fast *signals* will travel, we have to calculate the speed of pulses or modulations made by the interference of a wave of one frequency with one or more waves of slightly different frequencies (see Chapter 48, Vol. I). We have called the speed of the envelope of such a group of waves the group velocity; it is not ω/k but $d\omega/dk$:

$$v_{\text{group}} = \frac{d\omega}{dk}. \tag{24.26}$$

Taking the derivative of Eq. (24.17) with respect to ω and inverting to get $d\omega/dk$, we find that

$$v_{\text{group}} = c\sqrt{1 - (\omega_c/\omega)^2}, \tag{24.27}$$

which is less than the speed of light.

The geometric mean of v_{phase} and v_{group} is just c, the speed of light:

$$v_{\text{phase}}v_{\text{group}} = c^2. \tag{24.28}$$

This is curious, because we have seen a similar relation in quantum mechanics. For a particle with any velocity—even relativistic—the momentum p and energy U are related by

$$U^2 = p^2c^2 + m^2c^4. \tag{24.29}$$

But in quantum mechanics the energy is $\hbar\omega$, and the momentum is $\hbar/\lambda\!\!\!^-$, which is equal to $\hbar k$; so Eq. (24.29) can be written

$$\frac{\omega^2}{c^2} = k^2 + \frac{m^2c^2}{\hbar^2}, \tag{24.30}$$

or

$$k = \sqrt{(\omega^2/c^2) - (m^2c^2/\hbar^2)}, \tag{24.31}$$

which looks very much like Eq. (24.17) . . . Interesting!

The group velocity of the waves is also the speed at which energy is transported along the guide. If we want to find the energy flow down the guide, we can get it from the energy density times the group velocity. If the root mean square electric field is E_0, then the average density of electric energy is $\epsilon_0E_0^2/2$. There is also some energy associated with the magnetic field. We will not prove it here, but in any cavity or guide the magnetic and electric energies are equal, so the total electromagnetic energy density is $\epsilon_0E_0^2$. The power dU/dt transmitted by the guide is then

$$\frac{dU}{dt} = \epsilon_0E_0^2abv_{\text{group}}. \tag{24.32}$$

(We will see later another, more general way of getting the energy flow.)

24–5 Observing guided waves

Energy can be coupled into a waveguide by some kind of an "antenna." For example, a little vertical wire or "stub" will do. The presence of the guided waves can be observed by picking up some of the electromagnetic energy with a little receiving "antenna," which again can be a little stub of wire or a small loop.

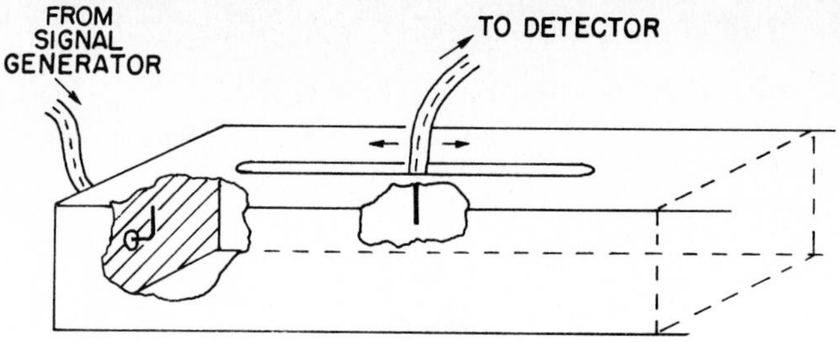

Fig. 24–8. A waveguide with a driving stub and a pickup probe.

In Fig. 24–8, we show a guide with some cutaways to show a driving stub and a pickup "probe". The driving stub can be connected to a signal generator via a coaxial cable, and the pickup probe can be connected by a similar cable to a detector. It is usually convenient to insert the pickup probe via a long thin slot in the guide, as shown in Fig. 24–8. Then the probe can be moved back and forth along the guide to sample the fields at various positions.

If the signal generator is set at some frequency ω greater than the cutoff frequency ω_c, there will be waves propagated down the guide from the driving stub. These will be the only waves present if the guide is infinitely long, which can effectively be arranged by terminating the guide with a carefully designed absorber in such a way that there are no reflections from the far end. Then, since the detector measures the time average of the fields near the probe, it will pick up a signal which is independent of the position along the guide; its output will be proportional to the power being transmitted.

If now the far end of the guide is finished off in some way that produces a reflected wave—as an extreme example, if we closed it off with a metal plate—there will be a reflected wave in addition to the original forward wave. These two waves will interfere and produce a standing wave in the guide similar to the standing waves on a string which we discussed in Chapter 49 of Vol. I. Then, as the pickup probe is moved along the line, the detector reading will rise and fall periodically, showing a maximum in the fields at each loop of the standing wave and a minimum at each node. The distance between two successive nodes (or loops) is just $\lambda_g/2$. This gives a convenient way of measuring the guide wavelength. If the frequency is now moved closer to ω_c, the distances between nodes increase, showing that the guide wavelength increases as predicted by Eq. (24.19).

Suppose now the signal generator is set at a frequency just a little below ω_c. Then the detector output will decrease gradually as the pickup probe is moved down the guide. If the frequency is set somewhat lower, the field strength will fall rapidly, following the curve of Fig. 24–7, and showing that waves are not propagated.

24–6 Waveguide plumbing

An important practical use of waveguides is for the transmission of high-frequency power, as, for example, in coupling the high-frequency oscillator or output amplifier of a radar set to an antenna. In fact, the antenna itself usually consists of a parabolic reflector fed at its focus by a waveguide flared out at the end to make a "horn" that radiates the waves coming along the guide. Although high frequencies can be transmitted along a coaxial cable, a waveguide is better for transmitting large amounts of power. First, the maximum power that can be transmitted along a line is limited by the breakdown of the insulation (solid or gas) between the conductors. For a given amount of power, the field strengths in a guide are usually less than they are in a coaxial cable, so higher powers can be transmitted before breakdown occurs. Second, the power losses in the coaxial cable are usually greater than in a waveguide. In a coaxial cable there must be insulating material to support the central conductor, and there is an energy loss in this material—particularly at high frequencies. Also, the current densities on the central conductor are quite high, and since the losses go as the *square* of the current density, the lower currents that appear on the walls of the guide result in lower

Fig. 24–9. Sections of waveguide connected with flanges.

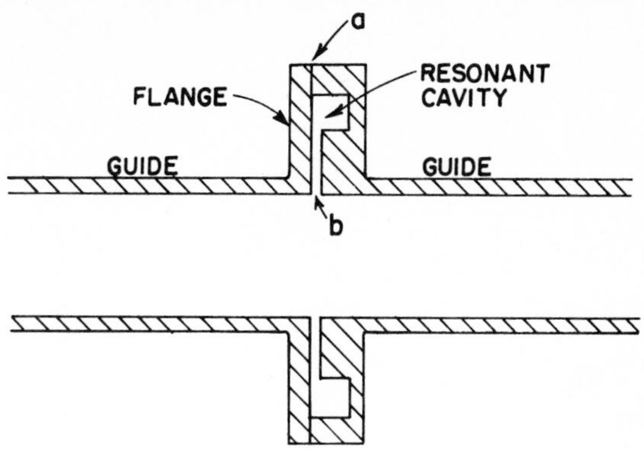

Fig. 24–10. A low-loss connection between two sections of waveguide.

energy losses. To keep these losses to a minimum, the inner surfaces of the guide are often plated with a material of high conductivity, such as silver.

The problem of connecting a "circuit" with waveguides is quite different from the corresponding circuit problem at low frequencies, and is usually called microwave "plumbing." Many special devices have been developed for the purpose. For instance, two sections of waveguide are usually connected together by means of flanges, as can be seen in Fig. 24–9. Such connections can, however, cause serious energy losses, because the surface currents must flow across the joint, which may have a relatively high resistance. One way to avoid such losses is to make the flanges as shown in the cross section drawn in Fig. 24–10. A small space is left between the adjacent sections of the guide, and a groove is cut in the face of one of the flanges to make a small cavity of the type shown in Fig. 23–16(c). The dimensions are chosen so that this cavity is resonant at the frequency being used. This resonant cavity presents a high "impedance" to the currents, so relatively little current flows across the metallic joints (at a in Fig. 24–10). The high guide currents simply charge and discharge the "capacity" of the gap (at b in the figure), where there is little dissipation of energy.

Suppose you want to stop a waveguide in a way that won't result in reflected waves. Then you must put something at the end that imitates an infinite length of guide. You need a "termination" which acts for the guide like the characteristic impedance does for a transmission line—something that absorbs the arriving waves without making reflections. Then the guide will act as though it went on forever. Such terminations are made by putting inside the guide some wedges of resistance material carefully designed to absorb the wave energy while generating almost no reflected waves.

If you want to connect *three* things together—for instance, one source to two different antennas—then you can use a "T" like the one shown in Fig. 24–11. Power fed in at the center section of the "T" will be split and go out the two side arms (and there may also be some reflected waves). You can see qualitatively from the sketches in Fig. 24–12 that the fields would spread out when they get to the end of the input section and make electric fields that will start waves going out the two arms. Depending on whether electric fields in the guide are parallel or perpendicular to the "top" of the "T," the fields at the junction would be roughly as shown in (a) or (b) of Fig. 24–12.

Finally, we would like to describe a device called an "unidirectional coupler," which is very useful for telling what is going on after you have connected a complicated arrangement of waveguides. Suppose you want to know which way the waves are going in a particular section of guide—you might be wondering, for instance, whether or not there is a strong reflected wave. The unidirectional coupler takes out a small fraction of the power of a guide if there is a wave going one way, but none if the wave is going the other way. By connecting the output of the coupler to a detector, you can measure the "one-way" power in the guide.

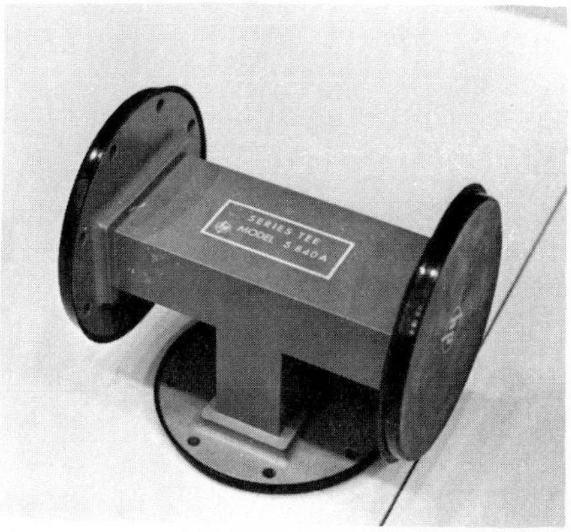

Fig. 24–11. A waveguide "T." (The flanges have plastic end caps to keep the inside clean while the "T" is not being used.)

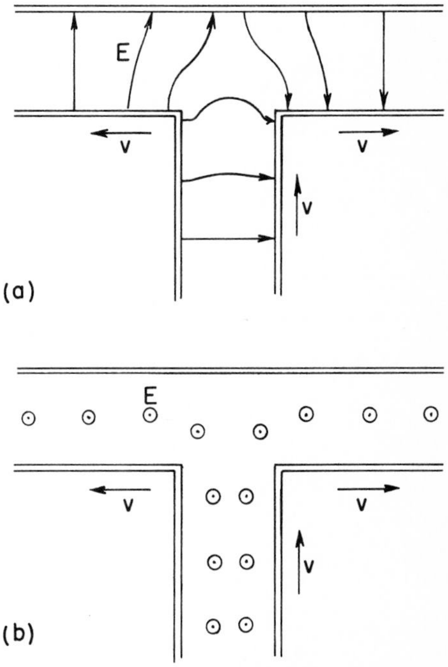

Fig. 24–12. The electric fields in a waveguide "T" for two possible field orientations.

Figure 24–13 is a drawing of a unidirectional coupler; a piece of waveguide *AB* has another piece of waveguide *CD* soldered to it along one face. The guide *CD* is curved away so that there is room for the connecting flanges. Before the guides are soldered together, two (or more) holes have been drilled in each guide (matching each other) so that some of the fields in the main guide *AB* can be coupled into the secondary guide *CD*. Each of the holes acts like a little antenna that produces a wave in the secondary guide. If there were only one hole, waves would be sent in both directions and would be the same no matter which way the wave was going in the primary guide. But when there are *two* holes with a separation space equal to one-quarter of the guide wavelength, they will make two sources 90° out of phase. Do you remember that we considered in Chapter 29 of Vol. I the interference of the waves from two antennas spaced $\lambda/4$ apart and excited 90° out of phase in time? We found that the waves subtract in one direction and add in the opposite direction. The same thing will happen here. The wave produced in the guide *CD* will be going in the same direction as the wave in *AB*.

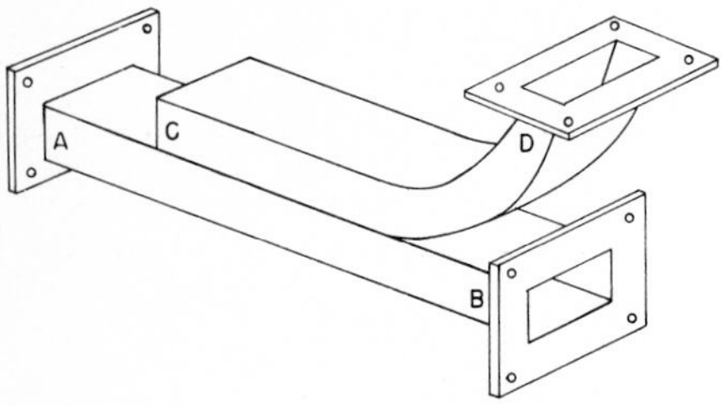

Fig. 24–13. A unidirectional coupler.

If the wave in the primary guide is travelling from *A* toward *B*, there will be a wave at the output *D* of the secondary guide. If the wave in the primary guide goes from *B* toward *A*, there will be a wave going toward the end *C* of the secondary guide. This end is equipped with a termination, so that this wave is absorbed and there is no wave at the output of the coupler.

24–7 Waveguide modes

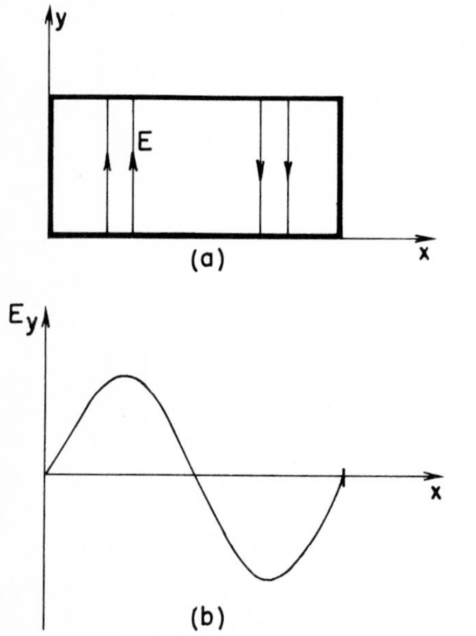

Fig. 24–14. Another possible variation of E_y with x.

The wave we have chosen to analyze is a special solution of the field equations. There are many more. Each solution is called a waveguide "mode." For example, our x-dependence of the field was just one-half a cycle of a sine wave. There is an equally good solution with a full cycle; then the variation of E_y with x is as shown in Fig. 24–14. The k_x for such a mode is twice as large, so the cutoff frequency is much higher. Also, in the wave we studied $\boldsymbol{E}$ has only a y-component, but there are other modes with more complicated electric fields. If the electric field has components only in x and y—so that the total electric field is always at right angles to the z-direction—the mode is called a "transverse electric" (or TE) mode. The magnetic field of such modes will always have a z-component. It turns out that if $\boldsymbol{E}$ has a component in the z-direction (along the direction of propagation), then the magnetic field will always have only transverse components. So such fields are called transverse magnetic (TM) modes. For a rectangular guide, all the other modes have a higher cutoff frequency than the simple TE mode we have described. It is, therefore, possible—and usual—to use a guide with a frequency just above the cutoff for this lowest mode but below the cutoff frequency for all the others, so that just the one mode is propagated. Otherwise, the behavior gets complicated and difficult to control.

24–8 Another way of looking at the guided waves

We want now to show you another way of understanding why a waveguide attenuates the fields rapidly for frequencies below the cutoff frequency ω_c. Then you will have a more "physical" idea of why the behavior changes so drastically between low and high frequencies. We can do this for the rectangular guide by analyzing the fields in terms of reflections—or images—in the walls of the guide. The approach only works for rectangular guides, however; that's why we started with the more mathematical analysis which works, in principle, for guides of any shape.

For the mode we have described, the vertical dimension (in y) had no effect, so we can ignore the top and bottom of the guide and imagine that the guide is extended indefinitely in the vertical direction. We imagine then that the guide just consists of two vertical plates with the separation a.

Let's say that the source of the fields is a vertical wire placed in the middle of the guide, with the wire carrying a current that oscillates at the frequency ω. In the absence of the guide walls such a wire would radiate cylindrical waves.

Now we consider that the guide walls are perfect conductors. Then, just as in electrostatics, the conditions at the surface will be correct if we add to the field of the wire the field of one or more suitable image wires. The image idea works just as well for electrodynamics as it does for electrostatics, provided, of course, that we also include the retardations. We know that is true because we have often seen a mirror producing an image of a light source. And a mirror is just a "perfect" conductor for electromagnetic waves with optical frequencies.

Now let's take a horizontal cross section, as shown in Fig. 24–15, where W_1 and W_2 are the two guide walls and S_0 is the source wire. We call the direction of the current in the wire positive. Now if there were only one wall, say W_1, we could remove it if we placed an image source (with opposite polarity) at the position marked S_1. But with both walls in place there will also be an image of S_0 in the wall W_2, which we show as the image S_2. This source, too, will have an image in W_1, which we call S_3. Now both S_1 and S_3 will have images in W_2 at the positions marked S_4 and S_6, and so on. For our two plane conductors with the source halfway between, the fields are the same as those produced by an infinite line of sources, all separated by the distance a. (It is, in fact just what you would *see* if you looked at a wire placed halfway between two parallel mirrors.) For the fields to be zero at the walls, the polarity of the currents in the images must alternate from one image to the next. In other words, they oscillate 180° out of phase. The waveguide field is, then, just the superposition of the fields of such an infinite set of line sources.

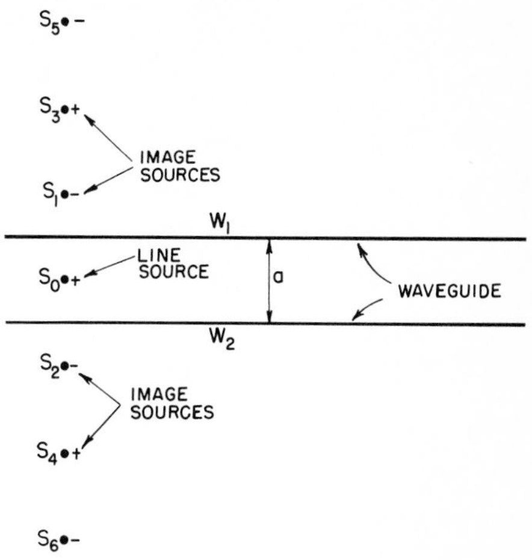

Fig. 24–15. The line source S_0 between the conducting plane walls W_1 and W_2. The walls can be replaced by the infinite sequence of image sources.

We know that if we are close to the sources, the field is very much like the static fields. We considered in Section 7–5 the static field of a grid of line sources and found that it is like the field of a charged plate except for terms that decrease exponentially with the distance from the grid. Here the average source strength is zero, because the sign alternates from one source to the next. Any fields which exist should fall off exponentially with distance. Close to the source, we see the field mainly of the nearest source; at large distances, many sources contribute and their average effect is zero. So now we see why the waveguide below cutoff frequency gives an exponentially decreasing field. At low frequencies, in particular, the static approximation is good, and it predicts a rapid attenuation of the fields with distance.

Now we are faced with the opposite question: Why are waves propagated at all? That is the mysterious part! The reason is that at high frequencies the retardation of the fields can introduce additional changes in phase which can cause the fields of the out-of-phase sources to add instead of cancelling. In fact, in Chapter 29 of Vol. I we have already studied, just for this problem, the fields generated by an array of antennas or by an optical grating. There we found that when several radio antennas are suitably arranged, they can give an interference pattern that has a strong signal in some direction but no signal in another.

Suppose we go back to Fig. 24–15 and look at the fields which arrive at a large distance from the array of image sources. The fields will be strong only in certain directions which depend on the frequency—only in those directions for which the fields from all the sources add in phase. At a reasonable distance from the sources the field propagates in these special directions as plane waves. We have sketched such a wave in Fig. 24–16, where the solid lines represent the wave crests and the dashed lines represent the troughs. The wave direction will be the one for which the difference in the retardation for two neighboring sources to the crest of a wave corresponds to one-half a period of oscillation. In other words, the difference between r_2 and r_0 in the figure is one-half of the free-space wavelength:

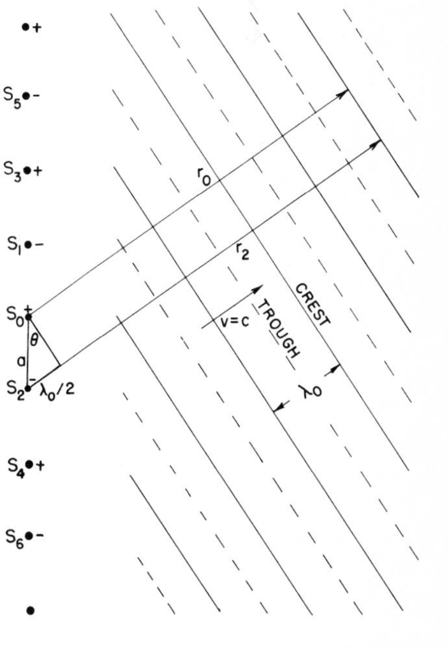

Fig. 24–16. One set of coherent waves from an array of line sources.

$$r_2 - r_0 = \frac{\lambda_0}{2}.$$

The angle θ is then given by

$$\sin\theta = \frac{\lambda_0}{2a}. \tag{24.33}$$

There is, of course, another set of waves travelling downward at the symmetric angle with respect to the array of sources. The complete waveguide field (not too close to the source) is the superposition of these two sets of waves, as shown in Fig. 24–17. The actual fields are really like this, of course, only between the two walls of the waveguide.

At points like A and C, the crests of the two wave patterns coincide, and the field will have a maximum; at points like B, both waves have their peak negative value, and the field has its minimum (largest negative) value. As time goes on the field in the guide appears to be travelling along the guide with a wavelength λ_g, which is the distance from A to C. That distance is related to θ by

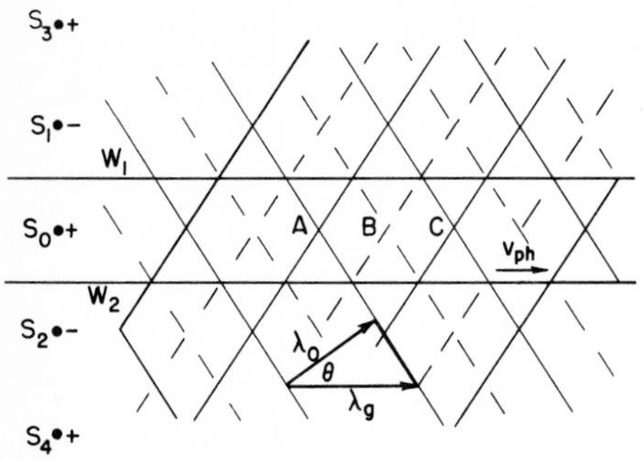

Fig. 24–17. The waveguide field can be viewed as the superposition of two trains of plane waves.

$$\cos\theta = \frac{\lambda_0}{\lambda_g}. \tag{24.34}$$

Using Eq. (24.33) for θ, we get that

$$\lambda_g = \frac{\lambda_0}{\cos\theta} = \frac{\lambda_0}{\sqrt{1 - (\lambda_0/2a)^2}}, \tag{24.35}$$

which is just what we found in Eq. (24.19).

Now we see why there is only wave propagation above the cutoff frequency ω_0. If the free-space wavelength is longer than $2a$, there is no angle where the waves shown in Fig. 24–16 can appear. The necessary constructive interference appears suddenly when λ_0 drops below $2a$, or when ω goes above $\omega_0 = \pi c/a$.

If the frequency is high enough, there can be two or more possible directions in which the waves will appear. For our case, this will happen if $\lambda_0 < \frac{2}{3}a$. In general, however, it could also happen when $\lambda_0 < a$. These additional waves correspond to the higher guide modes we have mentioned.

It has also been made evident by our analysis why the phase velocity of the guided waves is greater than c and why this velocity depends on ω. As ω is changed, the angle of the free waves of Fig. 24–16 changes, and therefore so does the velocity along the guide.

Although we have described the guided wave as the superposition of the fields of an infinite array of line sources, you can see that we would arrive at the same result if we imagined two sets of free-space waves being continually reflected back and forth between two perfect mirrors—remembering that a reflection means a reversal of phase. These sets of reflecting waves would all cancel each other unless they were going at just the angle θ given in Eq. (24.33). There are many ways of looking at the same thing.

Electrodynamics in Relativistic Notation

In this chapter: $c = 1$

Review: Chapter 15, Vol. I, *The Special Theory of Relativity*
Chapter 16, Vol. I, *Relativistic Energy and Momentum*
Chapter 17, Vol. I, *Space-Time*
Chapter 13, Vol. II, *Magnetostatics*

25–1 Four-vectors

We now discuss the application of the special theory of relativity to electrodynamics. Since we have already studied the special theory of relativity in Chapters 15 through 17 of Vol. I, we will just review quickly the basic ideas.

It is found experimentally that the laws of physics are unchanged if we move with uniform velocity. You can't tell if you are inside a spaceship moving with uniform velocity in a straight line, unless you look outside the spaceship, or at least make an observation having to do with the world outside. Any true law of physics we write down must be arranged so that this fact of nature is built in.

The relationship between the space and time of two systems of coordinates, one, S', in uniform motion in the x-direction with speed v relative to the other, S, is given by the *Lorentz transformation:*

$$t' = \frac{t - vx}{\sqrt{1 - v^2}}, \qquad y' = y,$$
$$x' = \frac{x - vt}{\sqrt{1 - v^2}}, \qquad z' = z. \tag{25.1}$$

The laws of physics must be such that after a Lorentz transformation, the new form of the laws looks just like the old form. This is just like the principle that the laws of physics don't depend on the *orientation* of our coordinate system. In Chapter 11 of Vol. I, we saw that the way to describe mathematically the invariance of physics with respect to rotations was to write our equations in terms of *vectors.*

For example, if we have two vectors

$$\boldsymbol{A} = (A_x, A_y, A_z) \qquad \text{and} \qquad \boldsymbol{B} = (B_x, B_y, B_z),$$

we found that the combination

$$\boldsymbol{A} \cdot \boldsymbol{B} = A_xB_x + A_yB_y + A_zB_z$$

was not changed if we transformed to a rotated coordinate system. So we know that if we have a scalar product like $\boldsymbol{A} \cdot \boldsymbol{B}$ on both sides of an equation, the equation will have exactly the same form in all rotated coordinate systems. We also discovered an operator (see Chapter 2),

$$\boldsymbol{\nabla} = \left(\frac{\partial}{\partial x}, \frac{\partial}{\partial y}, \frac{\partial}{\partial z}\right),$$

which, when applied to a scalar function, gave three quantities which transform just like a vector. With this operator we defined the gradient, and in combination with other vectors, the divergence and the Laplacian. Finally we discovered that by taking sums of certain products of pairs of the components of two vectors we could get three new quantities which behaved like a new vector. We called it the *cross product* of two vectors. Using the cross product with our operator $\boldsymbol{\nabla}$ we then defined the curl of a vector.

Since we will be referring back to what we have done in vector analysis, we have put in Table 25–1 a summary of all the important vector operations in three dimensions that we have used in the past. The point is that it must be possible to write the equations of physics so that both sides transform the same way under

Table 25–1

The important quantities and operations of vector analysis in three dimensions

Definition of a vector	$A = (A_x, A_y, A_z)$
Scalar product	$A \cdot B$
Differential vector operator	∇
Gradient	$\nabla\varphi$
Divergence	$\nabla \cdot A$
Laplacian	$\nabla \cdot \nabla = \nabla^2$
Cross product	$A \times B$
Curl	$\nabla \times A$

rotations. If one side is a vector, the other side must also be a vector, and both sides will change together in exactly the same way if we rotate our coordinate system. Similarly, if one side is a scalar, the other side must also be a scalar, so that neither side changes when we rotate coordinates, and so on.

Now in the case of special relativity, time and space are inextricably mixed, and we must do the analogous things for four dimensions. We want our equations to remain the same not only for rotations, but also for *any* inertial frame. That means that our equations should be invariant under the Lorentz transformation of equations (25.1). The purpose of this chapter is to show you how that can be done. Before we get started, however, we want to do something that makes our work a lot easier (and saves some confusion). And that is to choose our units of length and time so that the speed of light c is equal to 1. You can think of it as taking our unit of time to be *the time that it takes light to go one meter* (which is about 3×10^{-9} sec). We can even call this time unit "one meter." Using this unit, all of our equations will show more clearly the space-time symmetry. Also, all the c's will disappear from our relativistic equations. (If this bothers you, you can always put the c's back into any equation by replacing every t by ct, or, in general, by sticking in a c wherever it is needed to make the dimensions of the equations come out right.) With this groundwork we are ready to begin. Our program is to do in the four dimensions of space-time all of the things we did with vectors for three dimensions. It is really quite a simple game; we just work by analogy. The only real complications is the notation (we've already used up the vector symbol for three dimensions) and one slight twist of signs.

First, by analogy with vectors in three dimensions, we define a *four-vector* as a set of the four quantities a_t, a_x, a_y, and a_z, which transform like t, x, y, and z when we change to a moving coordinate system. There are several different notations people use for a four-vector; we will write a_μ, by which we mean the group of four numbers (a_t, a_x, a_y, a_z)—in other words, the subscript μ can take on the four "values" t, x, y, z. It will also be convenient, at times, to indicate the three space components by a three-vector, like this: $a_\mu = (a_t, \boldsymbol{a})$.

We have already encountered one four-vector, which consists of the energy and momentum of a particle (Chapter 17, Vol. I). In our new notation we write

$$p_\mu = (E, \boldsymbol{p}), \tag{25.2}$$

which means that the four-vector p_μ is made up of the energy E and the three components of the three-vector $\boldsymbol{p}$ of a particle.

It looks as though the game is really very simple—for each three-vector in physics all we have to do is find what the remaining component should be, and we have a four-vector. To see that this is not the case, consider the velocity vector with components

$$v_x = \frac{dx}{dt}, \qquad v_y = \frac{dy}{dt}, \qquad v_z = \frac{dz}{dt}.$$

The question is: What is the time component? Instinct should give the right answer. Since four-vectors are like t, x, y, z, we would guess that the time component is

$$v_t = \frac{dt}{dt} = 1.$$

This is wrong. The reason is that the t in each denominator is not an invariant when we make a Lorentz transformation. The numerators have the right behavior to make a four-vector, but the dt in the denominator spoils things; it is unsymmetric and is not the same in two different systems.

It turns out that the four "velocity" components which we have written down will become the components of a four-vector if we just divide by $\sqrt{1 - v^2}$. We can see that that is true because if we start with the momentum four-vector

$$p_\mu = (E, \boldsymbol{p}) = \left(\frac{m_0}{\sqrt{1 - v^2}}, \frac{m_0\boldsymbol{v}}{\sqrt{1 - v^2}}\right), \tag{25.3}$$

and divide it by the rest mass m_0, which is an invariant scalar in *four dimensions*, we have

$$\frac{p_\mu}{m_0} = \left(\frac{1}{\sqrt{1 - v^2}}, \frac{v}{\sqrt{1 - v^2}}\right), \tag{24.4}$$

which must still be a four-vector. (Dividing by an *invariant scalar* doesn't change the transformation properties.) So we can *define* the "*velocity four-vector*" u_μ by

$$\begin{aligned} u_t &= \frac{1}{\sqrt{1 - v^2}}, & u_y &= \frac{v_y}{\sqrt{1 - v^2}}, \\ u_x &= \frac{v_x}{\sqrt{1 - v^2}}, & u_z &= \frac{v_z}{\sqrt{1 - v^2}}, \end{aligned} \tag{25.5}$$

The four-velocity is a useful quantity; we can, for instance, write

$$p_\mu = m_0 u_\mu. \tag{25.6}$$

This is the typical sort of form an equation which is relativistically correct must have; each side is a four-vector. (The right-hand side is an invariant times a four-vector, which is still a four-vector.)

25–2 The scalar product

It is an accident of life, if you wish, that under coordinate rotations the distance of a point from the origin does not change. This means mathematically that $r^2 = x^2 + y^2 + z^2$ is an invariant. In other words, after a rotation $r'^2 = r^2$, or

$$x'^2 + y'^2 + z'^2 = x^2 + y^2 + z^2.$$

Now the question is: Is there a similar quantity which is invariant under the Lorentz transformation? There is. From Eq. (25.1) you can see that

$$t'^2 - x'^2 = t^2 - x^2.$$

That is pretty nice, except that it depends on a particular choice of the x-direction. We can fix that up by subtracting y^2 and z^2. Then any Lorentz transformation *plus* a rotation will leave the quantity unchanged. So the quantity which is analagous to r^2 for three dimensions, in four dimensions is

$$t^2 - x^2 - y^2 - z^2.$$

It is an invariant under what is called the "complete Lorentz group"—which means for transformation of both translations at constant velocity *and* rotations.

Now since this invariance is an algebraic matter depending only on the transformation rules of Eq. (25.1)—plus rotations—it is true for any four-vector (by definition they all transform the same). So for a four-vector a_μ we have that

$$a_t'^2 - a_x'^2 - a_y'^2 - a_z'^2 = a_t^2 - a_x^2 - a_y^2 - a_z^2.$$

We will call this quantity the square of "the length" of the four-vector a_μ. (Sometimes people change the sign of all the terms and call the length $a_x^2 + a_y^2 + a_z^2 - a_t^2$, so you'll have to watch out.)

Now if we have *two* vectors a_μ and b_μ, their corresponding components transform in the same way, so the combination

$$a_t b_t - a_x b_x - a_y b_y - a_z b_z$$

is also an invariant (scalar) quantity. (We have in fact already proved this in Chapter 17 of Vol. I.) Clearly this expression is quite analogous to the dot product for vectors. We will, in fact, call it the *dot product* or *scalar product* of two four-vectors. It would seem logical to write it as $a_\mu \cdot b_\mu$, so it would *look* like a dot product. But, unhappily, it's not done that way; it is usually written without the dot.

So we will follow the convention and write the dot product simply as $a_\mu b_\mu$. So, *by definition*,

$$a_\mu b_\mu = a_t b_t - a_x b_x - a_y b_y - a_z b_z. \tag{25.7}$$

Whenever you see two identical subscripts together (we will occasionally have to use ν or some other letter instead of μ) it means that you are to take the four products and sum, *remembering the minus sign* for the products of the space components. With this convention the invariance of the scalar product under a Lorentz transformation can be written as

$$a'_\mu b'_\mu = a_\mu b_\mu.$$

Since the last three terms in (25.7) are just the scalar dot product in three dimensions, it is often more convenient to write

$$a_\mu b_\mu = a_t b_t - \boldsymbol{a} \cdot \boldsymbol{b}.$$

It is also obvious that the four-dimensional length we described above can be written as $a_\mu a_\mu$:

$$a_\mu a_\mu = a_t^2 - a_x^2 - a_y^2 - a_z^2 = a_t^2 - \boldsymbol{a} \cdot \boldsymbol{a}. \tag{25.8}$$

It will also be convenient to sometimes write this quantity as a_μ^2:

$$a_\mu^2 \equiv a_\mu a_\mu.$$

We will now give you an illustration of the usefulness of four-vector dot products. Antiprotons ($\overline{\text{P}}$) are produced in large accelerators by the reaction

$$\text{P} + \text{P} \rightarrow \text{P} + \text{P} + \text{P} + \overline{\text{P}}.$$

That is, an energetic proton collides with a proton at rest (for example, in a hydrogen target placed in the beam), and if the incident proton has enough energy, a proton-antiproton pair may be produced, in addition to the two original protons.* The question is: How much energy must be given to the incident proton to make this reaction energetically possible?

The easiest way to get the answer is to consider what the reaction looks like in the center-of-mass (CM) system (see Fig. 25–1). We'll call the incident proton a and its four-momentum p_μ^a. Similarly, we'll call the target proton b and its four-

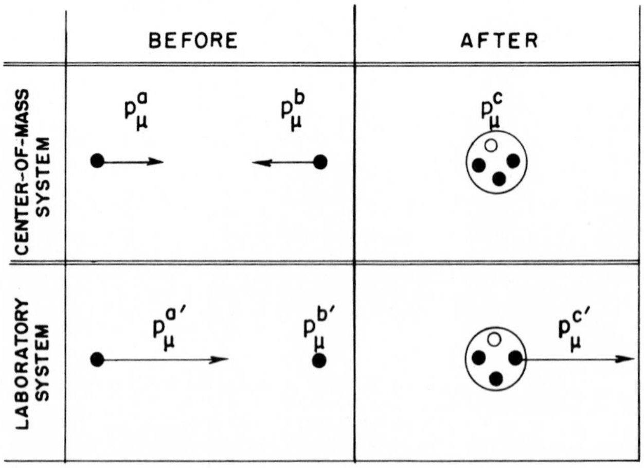

Fig. 25–1. The reaction $\text{P} + \text{P} \rightarrow 3\text{P} + \overline{\text{P}}$ viewed in the laboratory and CM systems. The incident proton is supposed to have just barely enough energy to make the reaction go. Protons are denoted by solid circles; antiprotons, by open circles.

---

* You may well ask: Why not consider the reactions

$$\text{P} + \text{P} \rightarrow \text{P} + \text{P} + \overline{\text{P}},$$

or even

$$\text{P} + \text{P} \rightarrow \text{P} + \overline{\text{P}}$$

which clearly require less energy? The answer is that a principle called *conservation of baryons* tells us the quantity "number of protons minus number of antiprotons" cannot change. This quantity is 2 on the left side of our reaction. Therefore, if we want an antiproton on the right side, we must have also *three* protons (or other baryons).

momentum p_μ^b. If the incident proton has *just barely* enough energy to make the reaction go, the final state—the situation after the collision—will consist of a glob containing three protons and an antiproton at rest in the CM system. If the incident energy were slightly higher, the final state particles would have some kinetic energy and be moving apart; if the incident energy were slightly lower, there would not be enough energy to make the four particles.

If we call p_μ^c the total four-momentum of the whole glob in the final state, conservation of energy and momentum tells us that

$$\boldsymbol{p}^a + \boldsymbol{p}^b = \boldsymbol{p}^c,$$

and

$$E^a + E^b = E^c.$$

Combining these two equations, we can write that

$$p_\mu^a + p_\mu^b = p_\mu^c. \tag{25.9}$$

Now the important thing is that this is an equation among four-vectors, and is, therefore, true in any inertial frame. We can use this fact to simplify our calculations. We start by taking the "length" of each side of Eq. (25.9); they are, of course, also equal. We get

$$(p_\mu^a + p_\mu^b)(p_\mu^a + p_\mu^b) = p_\mu^c p_\mu^c. \tag{25.10}$$

Since $p_\mu^c p_\mu^c$ is invariant, we can evaluate it in any coordinate system. In the CM system, the time component of p_μ^c is the rest energy of four protons, namely $4M$, and the space part $\boldsymbol{p}$ is zero; so $p_\mu^c = (4M, \boldsymbol{0})$. We have used the fact that the rest mass of an antiproton equals the rest mass of a proton, and we have called this common mass M.

Thus, Eq. (25.10) becomes

$$p_\mu^a p_\mu^a + 2p_\mu^a p_\mu^b + p_\mu^b p_\mu^b = 16M^2. \tag{25.11}$$

Now $p_\mu^a p_\mu^a$ and $p_\mu^b p_\mu^b$ are very easy, since the "length" of the momentum four-vector of any particle is just the mass of the particle squared:

$$p_\mu p_\mu = E^2 - \boldsymbol{p}^2 = M^2.$$

This can be shown by direct calculation or, more cleverly, by noting that for a particle *at rest* $p_\mu = (M, 0)$, so $p_\mu p_\mu = M^2$. But since it is an invariant, it is equal to M^2 in *any* frame. Using these results in Eq. (25.11), we have

$$2p_\mu^a p_\mu^b = 14M^2$$

or

$$p_\mu^a p_\mu^b = 7M^2. \tag{25.12}$$

Now we can also evaluate $p_\mu^a p_\mu^b$ in the laboratory system. The four-vector p_μ^a can be written $(E^a, \boldsymbol{p}^a)$, while $p_\mu^b = (M, \boldsymbol{0})$, since it describes a proton at rest. Thus, $p_\mu^a p_\mu^b$ must also be equal to ME^a; and since we know the scalar product is an invariant this must be numerically the same as what we found in (25.12). So we have that

$$E^a = 7M,$$

which is the result we were after. The *total* energy of the initial proton must be at least $7M$ (about 6.6 Gev since $M = 938$ Mev) or, subtracting the rest mass M, the *kinetic* energy must be at least $6M$ (about 5.6 Gev). The Bevatron accelerator at Berkeley was designed to give about 6.2 Gev of kinetic energy to the protons it accelerates, in order to be able to make antiprotons.

Since scalar products are invariant, they are always interesting to evaluate. What about the "length" of the four-velocity $u_\mu u_\mu$?

$$u_\mu u_\mu = u_t^2 - \boldsymbol{u}^2 = \frac{1}{1 - v^2} - \frac{v^2}{1 - v^2} = 1.$$

Thus, u_μ is the *unit four-vector.*

25–3 The four-dimensional gradient

The next thing that we have to discuss is the four-dimensional analog of the gradient. We recall (Chapter 14, Vol. I) that the three differential operators $\partial/\partial x$, $\partial/\partial y$, $\partial/\partial z$ transform like a three-vector and are called the gradient. The same scheme ought to work in four dimensions; that is, we might guess that the four-dimensional gradient should be $(\partial/\partial t, \partial/\partial x, \partial/\partial y, \partial/\partial z)$. *This is wrong.*

To see the error, consider a scalar function ϕ which depends only on x and t. The change in ϕ, if we make a small change Δt in t while holding x constant, is

$$\Delta\phi = \frac{\partial\phi}{\partial t}\,\Delta t. \tag{25.13}$$

On the other hand, according to a moving observer,

$$\Delta\phi = \frac{\partial\phi}{\partial x'}\,\Delta x' + \frac{\partial\phi}{\partial t'}\,\Delta t'.$$

We can express $\Delta x'$ and $\Delta t'$ in terms of Δt by using Eq. (25.1). Remembering that we are holding x constant, so that $\Delta x = 0$, we write

$$\Delta x' = -\frac{v}{\sqrt{1-v^2}}\,\Delta t; \qquad \Delta t' = \frac{\Delta t}{\sqrt{1-v^2}}.$$

Thus,

$$\begin{aligned}\Delta\phi &= \frac{\partial\phi}{\partial x'}\left(-\frac{v}{\sqrt{1-v^2}}\,\Delta t\right) + \frac{\partial\phi}{\partial t'}\left(\frac{\Delta t}{\sqrt{1-v^2}}\right)\\ &= \left(\frac{\partial\phi}{\partial t'} - v\,\frac{\partial\phi}{\partial x'}\right)\frac{\Delta t}{\sqrt{1-v^2}}.\end{aligned}$$

Comparing this result with Eq. (25.13), we learn that

$$\frac{\partial\phi}{\partial t} = \frac{1}{\sqrt{1-v^2}}\left(\frac{\partial\phi}{\partial t'} - v\,\frac{\partial\phi}{\partial x'}\right). \tag{25.14}$$

A similar calculation gives

$$\frac{\partial\phi}{\partial x} = \frac{1}{\sqrt{1-v^2}}\left(\frac{\partial\phi}{\partial x'} - v\,\frac{\partial\phi}{\partial t'}\right). \tag{25.15}$$

Now we can see that the gradient is rather strange. The formulas for x and t in terms of x' and t' [obtained by solving Eq. (25.1)] are:

$$t = \frac{t' + vx'}{\sqrt{1-v^2}}, \qquad x = \frac{x' + vt'}{\sqrt{1-v^2}}.$$

This is the way a four-vector *must* transform. But Eqs. (25.14) and (25.15) have a couple of signs wrong!

The answer is that instead of the *incorrect* $(\partial/\partial t, \boldsymbol{\nabla})$, we must *define* the *four-dimensional gradient* operator, which we will call ∇_μ, by

$$\nabla_\mu = \left(\frac{\partial}{\partial t}, -\boldsymbol{\nabla}\right) = \left(\frac{\partial}{\partial t}, -\frac{\partial}{\partial x}, -\frac{\partial}{\partial y}, -\frac{\partial}{\partial z}\right). \tag{25.16}$$

With this definition, the sign difficulties encountered above go away, and ∇_μ behaves as a four-vector should. (It's rather awkward to have those minus signs, but that's the way the world is.) Of course, what it means to say that ∇_μ "behaves like a four-vector" is simply that the four-gradient of a scalar is a four-vector. If ϕ is a true scalar invariant field (Lorentz invariant) then $\nabla_\mu\phi$ is a four-vector field.

All right, now that we have vectors, gradients, and dot products, the next thing is to look for an invariant which is analogous to the divergence of three-dimensional vector analysis. Clearly, the analog is to form the expression $\nabla_\mu b_\mu$, where b_μ is a four-vector field whose components are functions of space and time.

We *define* the *divergence* of the four-vector $b_\mu = (b_t, \boldsymbol{b})$ as the dot product of ∇_μ and b_μ:

$$\nabla_\mu b_\mu = \frac{\partial}{\partial t} b_t - \left(-\frac{\partial}{\partial x}\right) b_x - \left(-\frac{\partial}{\partial y}\right) b_y - \left(-\frac{\partial}{\partial z}\right) b_z$$
$$= \frac{\partial}{\partial t} b_t + \boldsymbol{\nabla} \cdot \boldsymbol{b}, \tag{25.17}$$

where $\boldsymbol{\nabla} \cdot \boldsymbol{b}$ is the ordinary three-divergence of the three-vector $\boldsymbol{b}$. Note that one has to be careful with the signs. Some of the minus signs come from the definition of the scalar product, Eq. (25.7); the others are required because the space components of ∇_μ are $-\partial/\partial x$, etc., as in Eq. (25.16). The divergence as defined by (25.17) is an invariant and gives the same answer in all coordinate systems which differ by a Lorentz transformation.

Let's look at a physical example in which the four-divergence shows up. We can use it to solve the problem of the fields around a moving wire. We have already seen (Section 13–7) that the electric charge density ρ and the current density $\boldsymbol{j}$ form a four-vector $j_\mu = (\rho, \boldsymbol{j})$. If an uncharged wire carries the current j_x, then in a frame moving past it with velocity v (along x), the wire will have the charge and current density [obtained from the Lorentz transformation Eqs. (25.1)] as follows:

$$\rho' = \frac{-vj_x}{\sqrt{1 - v^2}}, \qquad j'_x = \frac{j_x}{\sqrt{1 - v^2}}.$$

These are just what we found in Chapter 13. We can then use these sources in Maxwell's equation in *the moving system* to find the fields.

The charge conservation law, Section 13–2, also takes on a simple form in the four-vector notation. Consider the four divergence of j_μ:

$$\nabla_\mu j_\mu = \frac{\partial \rho}{\partial t} + \boldsymbol{\nabla} \cdot \boldsymbol{j}. \tag{25.18}$$

The law of the conservation of charge says that the outflow of current per unit volume must equal the negative rate of increase of charge density. In other words, that

$$\boldsymbol{\nabla} \cdot \boldsymbol{j} = -\frac{\partial \rho}{\partial t}.$$

Putting this into Eq. (25.18), the law of conservation of charge takes on the simple form

$$\nabla_\mu j_\mu = 0. \tag{25.19}$$

Since $\nabla_\mu j_\mu$ is an invariant scalar, if it is zero in one frame it is zero in all frames. We have the result that if charge is conserved in one coordinate system, it is conserved in all coordinate systems moving with uniform velocity.

As our last example we want to consider the scalar product of the gradient operator ∇_μ with itself. In three dimensions, such a product gives the Laplacian

$$\nabla^2 = \boldsymbol{\nabla} \cdot \boldsymbol{\nabla} = \frac{\partial^2}{\partial x^2} + \frac{\partial^2}{\partial y^2} + \frac{\partial^2}{\partial z^2}.$$

What do we get in four dimensions? That's easy. Following our rules for dot products and gradients, we get

$$\nabla_\mu \nabla_\mu = \frac{\partial}{\partial t}\frac{\partial}{\partial t} - \left(-\frac{\partial}{\partial x}\right)\left(-\frac{\partial}{\partial x}\right) - \left(-\frac{\partial}{\partial y}\right)\left(-\frac{\partial}{\partial y}\right) - \left(-\frac{\partial}{\partial z}\right)\left(-\frac{\partial}{\partial z}\right)$$
$$= \frac{\partial^2}{\partial t^2} - \nabla^2.$$

This operator, which is the analog of the three-dimensional Laplacian, is called the *D'Alembertian* and has a special notation:

$$\Box^2 = \nabla_\mu \nabla_\mu = \frac{\partial^2}{\partial t^2} - \nabla^2. \tag{25.20}$$

From its definition it is an invariant scalar operator; if it operates on a four-vector field, it produces a new four-vector field. (Some people define the D'Alembertian with the opposite sign to Eq. (25.20), so you will have to be careful when reading the literature.)

We have now found four-dimensional equivalents of most of the three-dimensional quantities we had listed in Table 25–1. (We do not yet have the equivalents of the cross product and the curl operation; we won't get to them until the next chapter.) It may help you remember how they go if we put all the important definitions and results together in one place, so we have made such a summary in Table 25–2.

Table 25–2

The important quantities of vector analysis in three and four dimensions.

	Three dimensions	Four dimensions
Vector	$\boldsymbol{A} = (A_x, A_y, A_z)$	$a_\mu = (a_t, a_x, a_y, a_z) = (a_t, \boldsymbol{a})$
Scalar product	$\boldsymbol{A} \cdot \boldsymbol{B} = A_x B_x + A_y B_y + A_z B_z$	$a_\mu b_\mu = a_t b_t - a_x b_x - a_y b_y - a_z b_z = a_t b_t - \boldsymbol{a} \cdot \boldsymbol{b}$
Vector operator	$\boldsymbol{\nabla} = (\partial/\partial x, \partial/\partial y, \partial/\partial z)$	$\nabla_\mu = (\partial/\partial t, -\partial/\partial x, -\partial/\partial y, -\partial/\partial z) = (\partial/\partial t, -\boldsymbol{\nabla})$
Gradient	$\boldsymbol{\nabla}\psi = \left(\frac{\partial\psi}{\partial x}, \frac{\partial\psi}{\partial y}, \frac{\partial\psi}{\partial z}\right)$	$\nabla_\mu \varphi = \left(\frac{\partial\varphi}{\partial t}, -\frac{\partial\varphi}{\partial x}, -\frac{\partial\varphi}{\partial y}, -\frac{\partial\varphi}{\partial z}\right) = \left(\frac{\partial\varphi}{\partial t}, -\boldsymbol{\nabla}\varphi\right)$
Divergence	$\boldsymbol{\nabla} \cdot \boldsymbol{A} = \frac{\partial A_x}{\partial x} + \frac{\partial A_y}{\partial y} + \frac{\partial A_z}{\partial z}$	$\nabla_\mu a_\mu = \frac{\partial a_t}{\partial t} + \frac{\partial a_x}{\partial x} + \frac{\partial a_y}{\partial y} + \frac{\partial a_z}{\partial z} = \frac{\partial a_t}{\partial t} + \boldsymbol{\nabla} \cdot \boldsymbol{a}$
Laplacian and D'Alembertian	$\boldsymbol{\nabla} \cdot \boldsymbol{\nabla} = \frac{\partial^2}{\partial x^2} + \frac{\partial^2}{\partial y^2} + \frac{\partial^2}{\partial z^2}$	$\nabla_\mu \nabla_\mu = \frac{\partial^2}{\partial t^2} - \frac{\partial^2}{\partial x^2} - \frac{\partial^2}{\partial y^2} - \frac{\partial^2}{\partial z^2} = \frac{\partial^2}{\partial t^2} - \nabla^2 = \Box^2$

25–4 Electrodynamics in four-dimensional notation

We have already encountered the D'Alembertian operator, without giving it that name, in Section 18–6; the differential equations we found there for the potentials can be written in the new notations as:

$$\Box^2 \phi = \frac{\rho}{\epsilon_0}, \qquad \Box^2 \boldsymbol{A} = \frac{\boldsymbol{j}}{\epsilon_0}. \tag{25.21}$$

The four quantities on the right-hand side of the two equations in (25.21) are ρ, j_x, j_y, j_z, divided by ϵ_0, which is a universal constant which will be the same in all coordinate systems if the same unit of charge is used in all frames. So the four quantities ρ/ϵ_0, j_x/ϵ_0, j_y/ϵ_0, j_z/ϵ_0 also transform as a four-vector. We can write them as j_μ/ϵ_0. The D'Alembertian doesn't change when the coordinate system is changed, so the quantities ϕ, A_x, A_y, A_z *must also transform* like a four-vector—which means that they *are* the components of a four-vector. In short,

$$A_\mu = (\phi, \boldsymbol{A})$$

is a four-vector. What we call the scalar and vector potentials are really different aspects of the same physical thing. They belong together. And if they are kept together the relativistic invariance of the world is obvious. We call A_μ the *four-potential.*

In the four-vector notation Eqs. (25.21) become simply

$$\Box^2 A_\mu = \frac{j_\mu}{\epsilon_0}, \tag{25.22}$$

The physics of this equation is just the same as Maxwell's equations. But there is some pleasure in being able to rewrite them in an elegant form. The pretty form is also meaningful; it shows directly the invariance of electrodynamics under the Lorentz transformation.

Remember that Eqs. (25.21) could be deduced from Maxwell's equations only if we imposed the gauge condition

$$\frac{\partial \phi}{\partial t} + \nabla \cdot A = 0, \tag{25.23}$$

which just says $\nabla_\mu A_\mu = 0$; the gauge condition says that the divergence of the four-vector A_μ is zero. This condition is called the *Lorentz condition.* It is very convenient because it is an invariant condition and therefore Maxwell's equations stay in the form of Eq. (25.22) for all frames.

25–5 The four-potential of a moving charge

Although it is implicit in what we have already said, let us write down the transformation laws which give ϕ and A in a moving system in terms of ϕ and A in a stationary system. Since $A_\mu = (\phi, A)$ is a four-vector, the equations must look just like Eqs. (25.1), except that t is replaced by ϕ, and x is replaced by A. Thus,

$$\begin{aligned} \phi' &= \frac{\phi - vA_x}{\sqrt{1 - v^2}}, & A'_y &= A_y, \\ A'_x &= \frac{A_x - v\phi}{\sqrt{1 - v^2}}, & A'_z &= A_z. \end{aligned} \tag{25.24}$$

This assumes that the primed coordinate system is moving with speed v in the positive x-direction, as measured in the unprimed coordinate system.

We will consider one example of the usefulness of the idea of the four-potential. What are the vector and scalar potentials of a charge q moving with speed v along the x-axis? The problem is easy in a coordinate system moving with the charge, since in this system the charge is standing still. Let's say that the charge is at the origin of the S'-frame, as shown in Fig. 25–2. The scalar potential in the moving system is then given by

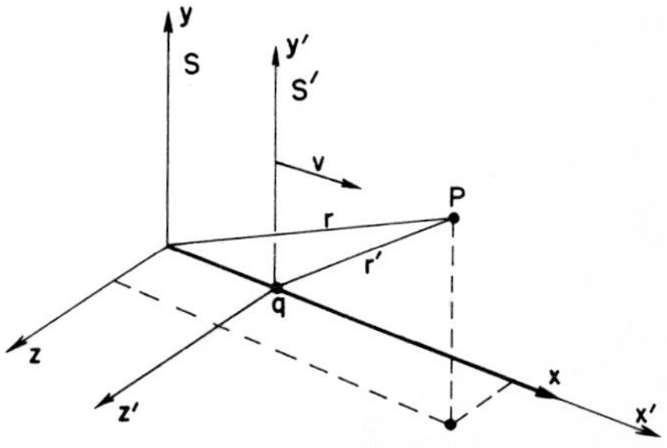

Fig. 25–2. The frame S' moves with velocity v (in the x-direction) with respect to S. A charge at rest at the origin of S' is at $x = vt$ in S. The potentials at P can be computed in either frame.

$$\phi' = \frac{q}{4\pi\epsilon_0 r'}, \tag{25.25}$$

r' being the distance from q to the field point, as measured in the moving system. The vector potential A' is, of course, zero.

Now it is straightforward to find ϕ and A, the potentials as measured in the stationary coordinates. The inverse relations to Eqs. (25.24) are

$$\begin{aligned} \phi &= \frac{\phi' + vA'_x}{\sqrt{1 - v^2}}, & A_y &= A'_y, \\ A_x &= \frac{A'_x + v\phi'}{\sqrt{1 - v^2}}, & A_z &= A'_z. \end{aligned} \tag{25.26}$$

Using the ϕ' given by Eq. (25.25), and $A' = 0$, we get

$$\begin{aligned} \phi &= \frac{q}{4\pi\epsilon_0} \frac{1}{r'\sqrt{1 - v^2}} \\ &= \frac{q}{4\pi\epsilon_0} \frac{1}{\sqrt{1 - v^2}\sqrt{x'^2 + y'^2 + z'^2}}. \end{aligned}$$

This gives us the scalar potential ϕ we would see in S, but, unfortunately, expressed in terms of the S' coordinates. We can get things in terms of t, x, y, z by substituting for t', x', y', and z', using (25.1). We get

$$\phi = \frac{q}{4\pi\epsilon_0} \frac{1}{\sqrt{1 - v^2}} \frac{1}{\sqrt{[(x - vt)/\sqrt{1 - v^2}]^2 + y^2 + z^2}} \cdot \tag{25.27}$$

Following the same procedure for the components of $\boldsymbol{A}$, you can show that

$$\boldsymbol{A} = \boldsymbol{v}\phi. \tag{25.28}$$

These are the same formulas we derived by a different method in Chapter 21.

25–6 The invariance of the equations of electrodynamics

We have found that the potentials ϕ and $\boldsymbol{A}$ taken together form a four-vector which we call A_μ, and that the wave equations—the full equations which determine the A_μ in terms of the j_μ—can be written as in Eq. (25.22). This equation, together with the conservation of charge, Eq. (25.19), gives us the fundamental law of the electromagnetic field:

$$\Box^2 A_\mu = \frac{1}{\epsilon_0} j_\mu, \qquad \nabla_\mu j_\mu = 0. \tag{25.29}$$

There, in one tiny space on the page, are all of the Maxwell equations—beautiful and simple. Did we learn anything from writing the equations this way, besides that they are beautiful and simple? In the first place, is it anything different from what we had before when we wrote everything out in all the various components? Can we from this equation deduce something that could not be deduced from the wave equations for the potentials in terms of the charges and currents? The answer is definitely no. The only thing we have been doing is changing the names of things —using a new notation. We have written a square symbol to represent the derivatives, but it still means nothing more nor less than the second derivative with respect to t, minus the second derivative with respect to x, minus the second derivative with respect to y, minus the second derivative with respect to z. And the μ means that we have four equations, one each for $\mu = t, x, y$, or z. What then is the significance of the fact that the equations can be written in this simple form? From the point of view of deducing anything directly, it doesn't mean anything. Perhaps, though, the simplicity of the equations means that nature also has a certain simplicity.

Let us show you something interesting that we have recently discovered: *All of the laws of physics can be contained in one equation.* That equation is

$$\mathsf{U} = 0. \tag{25.30}$$

What a simple equation! Of course, it is necessary to know what the symbol means. U is a physical quantity which we will call the "unworldliness" of the situation. And we have a formula for it. Here is how you calculate the unworldliness. You take all of the known physical laws and write them in a special form. For example, suppose you take the law of mechanics, $\boldsymbol{F} = m\boldsymbol{a}$, and rewrite it as $\boldsymbol{F} - m\boldsymbol{a} = 0$. Then you can call $(\boldsymbol{F} - m\boldsymbol{a})$—which should, of course, be zero—the "mismatch," of mechanics. Next, you take the *square* of this mismatch and call it U_1, which can be called the "unworldliness of mechanical effects." In other words, you take

$$\mathsf{U}_1 = (\boldsymbol{F} - m\boldsymbol{a})^2. \tag{25.31}$$

Now you write another physical law, say, $\boldsymbol{\nabla} \cdot \boldsymbol{E} = \rho/\epsilon_0$ and define

$$\mathsf{U}_2 = \left(\boldsymbol{\nabla} \cdot \boldsymbol{E} - \frac{\rho}{\epsilon_0}\right)^2,$$

which you might call "the gaussian unworldliness of electricity." You continue to write U_3, U_4, and so on—one for every physical law there is.

Finally you call the *total* unworldliness U of the world the sum of the various unworldlinesses U_i from all the subphenomena that are involved; that is, $\mathsf{U} = \sum \mathsf{U}_i$. Then the great "law of nature" is

$$\boxed{\mathsf{U} = 0.} \tag{25.32}$$

This "law" means, of course, that the sum of the squares of all the individual mismatches is zero, and the only way the sum of a lot of squares can be zero is for each one of the terms to be zero.

So the "beautifully simple" law in Eq. (25.32) is equivalent to the whole series of equations that you originally wrote down. It is therefore absolutely obvious that a simple notation that just hides the complexity in the definitions of symbols is not real simplicity. *It is just a trick.* The beauty that appears in Eq. (25.32)—just from the fact that several equations are hidden within it—is no more than a trick. When you unwrap the whole thing, you get back where you were before.

However, there *is* more to the simplicity of the laws of electromagnetism written in the form of Eq. (25.29). It means more, just as a theory of vector analysis means more. The fact that the electromagnetic equations can be written in a very particular notation *which was designed* for the four-dimensional geometry of the Lorentz transformations—in other words, as a vector equation in the four-space—means that it is invariant under the Lorentz transformations. It is because the Maxwell equations are invariant under those transformations that they can be written in a beautiful form.

It is no accident that the equations of electrodynamics can be written in the beautifully elegant form of Eq. (25.29). The theory of relativity was developed *because it was found experimentally* that the phenomena predicted by Maxwell's equations were the same in all inertial systems. And it was precisely by studying the transformation properties of Maxwell's equations that Lorentz discovered his transformation as the one which left the equations invariant.

There is, however, another reason for writing our equations this way. It has been discovered—after Einstein guessed that it might be so— that *all* of the laws of physics are invariant under the Lorentz transformation. That is the principle of relativity. Therefore, if we invent a notation which shows immediately when a law is written down whether it is invariant or not, we can be sure that in trying to make new theories we will write only equations which are consistent with the principle of relativity.

The fact that the Maxwell equations are simple in this particular notation is not a miracle, because the notation was invented with them in mind. But the interesting physical thing is that *every law* of physics—the propagation of meson waves or the behavior of neutrinos in beta decay, and so forth—must have this same invariance under the same transformation. Then when you are moving at a uniform velocity in a spaceship, all of the laws of nature transform together in such a way that no new phenomenon will show up. It is because the principle of relativity is a fact of nature that in the notation of four-dimensional vectors the equations of the world will look simple.

Lorentz Transformations of the Fields

In this chapter: $c = 1$

Review: Chapter 20, Vol. II, *Solution of Maxwell's Equations in Free Space*

26–1 The four-potential of a moving charge

We saw in the last chapter that the potential $A_\mu = (\phi, A)$ is a four-vector. The time component is the scalar potential ϕ, and the three space components are the vector potential A. We also worked out the potentials of a particle moving with uniform speed on a straight line by using the Lorentz transformation. (We had already found them by another method in Chapter 21.) For a point charge whose position at the time t is $(vt, 0, 0)$, the potentials at the point (x, y, z) are

$$\phi = \frac{q}{4\pi\epsilon_0\sqrt{1-v^2}\left[\dfrac{(x-vt)^2}{1-v^2}+y^2+z^2\right]^{1/2}}$$

$$A_x = \frac{qv}{4\pi\epsilon_0\sqrt{1-v^2}\left[\dfrac{(x-vt)^2}{1-v^2}+y^2+z^2\right]^{1/2}} \tag{26.1}$$

$$A_y = A_z = 0$$

Equations (26.1) give the potentials at x, y, and z at the time t, for a charge whose "present" position (by which we mean the position *at the time* t) is at $x = vt$. Notice that the equations are in terms of $(x - vt)$, y, and z, which are the coordinates measured *from the current position* P of the moving charge (see Fig. 26–1). The actual influence we know really travels at the speed c, so it is the behavior of the charge back at the retarded position P' that really counts.† The point P' is at $x = vt'$ (where, $t' = t - r'/c$ is the retarded time). But we said that the charge was moving with uniform velocity in a straight line, so naturally the behavior at P' and the current position are directly related. In fact, if we make the added assumption that the potentials depend only upon the position and the velocity at the retarded moment, we have in equations (26.1) a *complete* formula for the potentials for a charge moving *any* way. It works this way. Suppose that you have a charge moving in some arbitrary fashion, say with the trajectory in Fig. 26–2, and you are trying to find the potentials at the point (x, y, z). First, you find the retarded position P' and the velocity v' at that point. Then you imgaine that the charge would keep on moving with this velocity during the delay time $(t' - t)$, so that it would then appear at an imaginary position P_{proj}, which we can call the "projected position," and would arrive there with the velocity v'. (Of course, it doesn't do that; its real position at t is at P.) Then the potentials at (x, y, z) are just what equations (26.1) would give for the imaginary charge at the projected position P_{proj}. What we are saying is that since the potentials depend only on what the charge is doing at the *retarded* time, the potentials will be the same whether the charge continued moving at a constant velocity or whether it changed its velocity after t'—that is, after the potentials that were going to appear at (x, y, z) at the time t were already determined.

You know, of course, that the moment that we have the formula for the potentials from a charge moving in any manner whatsoever, we have the complete electrodynamics; we can get the potentials of any charge distribution by super-

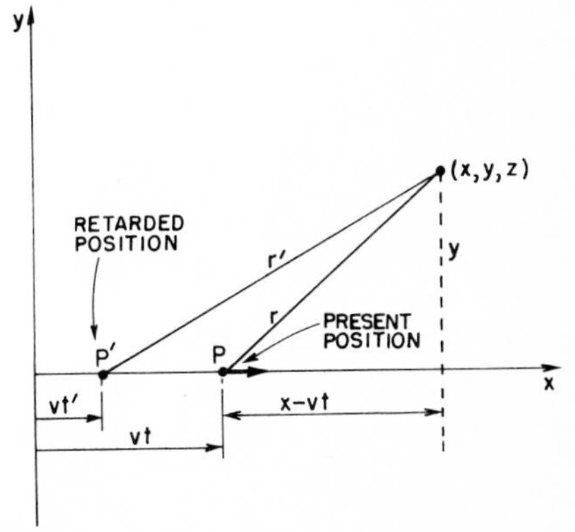

Fig. 26–1. Finding the fields at P due to a charge q moving along the x-axis with the constant speed v. The field "now" at the point (x, y, z) can be expressed in terms of the "present" position P, as well as in terms of P', the "retarded" position (at $t' = t - r'/c$).

† The primes used here to indicate the *retarded* positions and times should not be confused with the primes referring to a Lorentz-transformed frame in the preceding chapter.

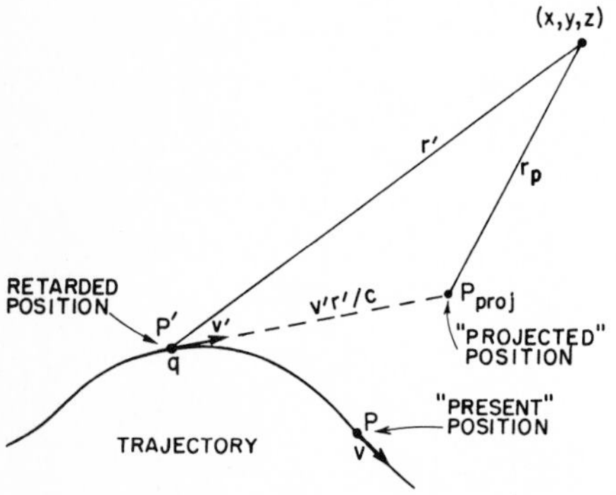

Fig. 26–2. A charge moves on an arbitrary trajectory. The potentials at (x, y, z) at the time t are determined by the position P' and velocity v' at the retarded time $t' - r'/c$. They are conveniently expressed in terms of the coordinates from the "projected" position P_{proj}. (The actual position at t is P.)

position. Therefore we can summarize all the phenomena of electrodynamics either by writing Maxwell's equations or by the following series of remarks. (Remember them in case you are ever on a desert island. From them, all can be reconstructed. You will, of course, know the Lorentz transformation; you will never forget *that* on a desert island or anywhere else.)

First, A_μ is a four-vector. *Second*, the Coulomb potential for a stationary charge is $q/4\pi\epsilon_0 r$. *Third*, the potentials produced by a charge moving in any way depend only upon the velocity and position at the retarded time. With those three facts we have everything. From the fact that A_μ is a four-vector, we transform the Coulomb potential, which we know, and get the potentials for a constant velocity. Then, by the last statement that potentials depend only upon the past velocity at the retarded time, we can use the projected position game to find them. It is not a particularly useful way of doing things, but it is interesting to show that the laws of physics can be put in so many different ways.

It is sometimes said, by people who are careless, that all of electrodynamics can be deduced solely from the Lorentz transformation and Coulomb's law. Of course, that is completely false. First, we have to suppose that there is a scalar potential and a vector potential that together make a four-vector. That tells us how the potentials transform. Then why is it that the effects at the retarded time are the only things that count? Better yet, why is it that the potentials depend only on the position and the velocity and not, for instance, on the acceleration? The *fields* $\boldsymbol{E}$ and $\boldsymbol{B}$ *do* depend on the acceleration. If you try to make the same kind of an argument with respect to them, you would say that they depend only upon the position and velocity at the retarded time. But then the fields from an accelerating charge would be the same as the fields from a charge at the projected position—which is false. The *fields* depend not only on the position and the velocity along the path but also on the acceleration. So there are several additional tacit assumptions in this great statement that everything can be deduced from the Lorentz transformation. (Whenever you see a sweeping statement that a tremendous amount can come from a very small number of assumptions, you always find that it is false. There are usually a large number of implied assumptions that are far from obvious if you think about them sufficiently carefully.)

26–2 The fields of a point charge with a constant velocity

Now that we have the potentials from a point charge moving at constant velocity, we ought to find the fields—for practical reasons. There are many cases where we have uniformly moving particles—for instance, cosmic rays going through a cloud chamber, or even slow-moving electrons in a wire. So let's at least see what the fields actually do look like for any speed—even for speeds nearly that of light—assuming only that there is no acceleration. It is an interesting question.

We get the fields from the potentials by the usual rules:

$$\boldsymbol{E} = -\boldsymbol{\nabla}\phi - \frac{\partial \boldsymbol{A}}{\partial t}, \qquad \boldsymbol{B} = \boldsymbol{\nabla} \times \boldsymbol{A}.$$

First, for E_z

$$E_z = -\frac{\partial\phi}{\partial z} - \frac{\partial A_z}{\partial t}.$$

But A_z is zero; so differentiating ϕ in equations (26.1), we get

$$E_z = \frac{q}{4\pi\epsilon_0\sqrt{1 - v^2}} \frac{z}{\left[\dfrac{(x - vt)^2}{1 - v^2} + y^2 + z^2\right]^{3/2}}. \tag{26.2}$$

Similarly, for E_y,

$$E_y = \frac{q}{4\pi\epsilon_0\sqrt{1 - v^2}} \frac{y}{\left[\dfrac{(x - vt)^2}{1 - v^2} + y^2 + z^2\right]^{3/2}}. \tag{26.3}$$

The x-component is a little more work. The derivative of ϕ is more complicated and A_x is not zero. First,

$$-\frac{\partial\phi}{\partial x} = \frac{q}{4\pi\epsilon_0\sqrt{1-v^2}} \frac{(x-vt)/(1-v^2)}{\left[\frac{(x-vt)^2}{1-v^2}+y^2+z^2\right]^{3/2}}. \tag{26.4}$$

Then, differentiating A_x with respect to t, we find

$$-\frac{\partial A_x}{\partial t} = \frac{q}{4\pi\epsilon_0\sqrt{1-v^2}} \frac{-v^2(x-vt)/(1-v^2)}{\left[\frac{(x-vt)^2}{1-v^2}+y^2+z^2\right]^{3/2}}. \tag{26.5}$$

And finally, taking the sum,

$$E_x = \frac{q}{4\pi\epsilon_0\sqrt{1-v^2}} \frac{x-vt}{\left[\frac{(x-vt)^2}{1-v^2}+y^2+z^2\right]^{3/2}}. \tag{26.6}$$

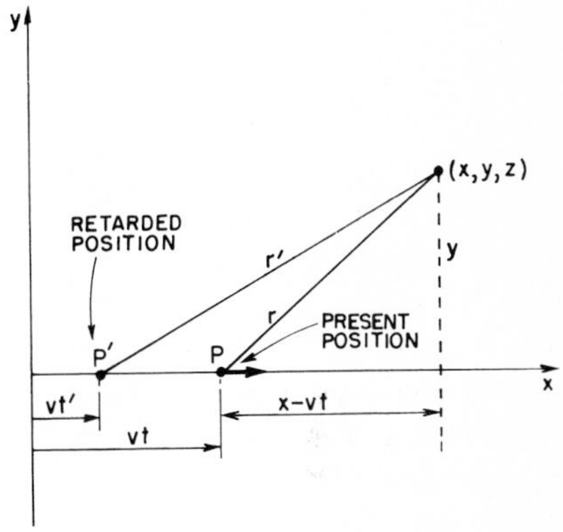

Fig. 26–3. For a charge moving with constant speed, the electric field points radially from the "present" position of the charge.

We'll look at the physics of $\boldsymbol{E}$ in a minute; let's first find $\boldsymbol{B}$. For the z-component,

$$B_z = \frac{\partial A_y}{\partial x} - \frac{\partial A_x}{\partial y}.$$

Since A_y is zero, we have just one derivative to get. Notice, however, that A_x is just $v\phi$, and $\partial/\partial y$ of $v\phi$ is just $-vE_y$. So

$$B_z = vE_y. \tag{26.7}$$

Similarly,

$$B_y = \frac{\partial A_x}{\partial z} - \frac{\partial A_z}{\partial x} = +v\frac{\partial\phi}{\partial z},$$

and

$$B_y = -vE_z. \tag{26.8}$$

Finally, B_x is zero, since A_y and A_z are both zero. We can write the magnetic field simply as

$$\boldsymbol{B} = \boldsymbol{v} \times \boldsymbol{E}. \tag{26.9}$$

Now let's see what the fields look like. We will try to draw a picture of the field at various positions around the present position of the charge. It is true that the influence of the charge comes, in a certain sense, from the retarded position; but because the motion is exactly specified, the retarded position is uniquely given in terms of the present position. For uniform velocities, it's nicer to relate the fields to the current position, because the field components at (x, y, z) depend only on $(x - vt)$, y, and z—which are the components of the displacements $\boldsymbol{r}_P$ from the present position to (x, y, z) (see Fig. 26–3).

Consider first a point with $z = 0$. Then $\boldsymbol{E}$ has only x- and y-components. From Eqs. (26.3) and (26.6), the ratio of these components is just equal to the ratio of the x- and y-components of the displacement. That means that $\boldsymbol{E}$ is in *the same direction* as $\boldsymbol{r}_P$, as shown in Fig. 26–3. Since E_z is also proportional to z, it is clear that this result holds in three dimensions. In short, the electric field is radial from the charge, and the field lines radiate directly out of the charge, just as they do for a stationary charge. Of course, the field isn't exactly the same as for the stationary charge, because of all the extra factors of $(1 - v^2)$. But we can show something rather interesting. The difference is just what you would get if you were to draw the Coulomb field with a peculiar set of coordinates in which the scale of x was squashed up by the factor $\sqrt{1-v^2}$. If you do that, the field lines will be spread out ahead and behind the charge and will be squeezed together around the sides, as shown in Fig. 26–4.

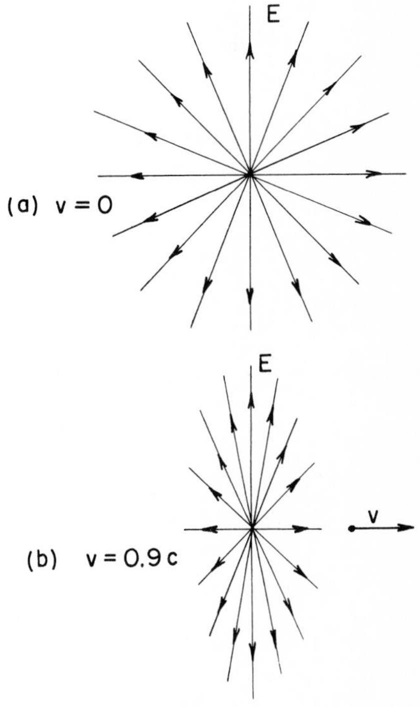

Fig. 26–4. The electric field of a charge moving with the constant speed $v = 0.9c$, part (b), compared with the field of a charge at rest, part (a).

If we relate the strength of $\boldsymbol{E}$ to the density of the field lines in the conventional way, we see a stronger field at the sides and a weaker field ahead and behind, which is just what the equations say. First, if we look at the strength of the field at right angles to the line of motion, that is, for $(x - vt) = 0$, the distance from the charge is $(y^2 + z^2)$. Here the total field strength is $\sqrt{E_y^2 + E_z^2}$, which is

$$E = \frac{q}{4\pi\epsilon_0\sqrt{1 - v^2}} \frac{1}{y^2 + z^2}. \tag{26.10}$$

The field is proportional to the inverse square of the distance—just like the Coulomb field except increased by the constant, extra factor $1/\sqrt{1 - v^2}$, which is always greater than one. So at the *sides* of a moving charge, the electric field is stronger than you get from the Coulomb law. In fact, the field in the sidewise direction is bigger than the Coulomb potential by the ratio of the energy of the particle to its rest mass.

Ahead of the charge (and behind), y and z are zero and

$$E = E_x = \frac{q(1 - v^2)}{4\pi\epsilon_0(x - vt)^2}. \tag{26.11}$$

The field again varies as the inverse square of the distance from the charge but is now *reduced* by the factor $(1 - v^2)$, in agreement with the picture of the field lines. If v/c is small, v^2/c^2 is still smaller, and the effect of the $(1 - v^2)$ terms is very small; we get back to Coulomb's law. But if a particle is moving very close to the speed of light, the field in the forward direction is enormously reduced, and the field in the sidewise direction is enormously increased.

Our results for the electric field of a charge can be put this way: Suppose you were to draw on a piece of paper the field lines for a charge at rest, and then set the picture to travelling with the speed v. Then, of course, the whole picture would be compressed by the Lorentz contraction; that is, the carbon granules on the paper would appear in different places. The miracle of it is that the picture you would see as the page flies by would still represent the field lines of the point charge. The contraction moves them closer together at the sides and spreads them out ahead and behind, just in the right way to give the correct line densities. We have emphasized before that field lines are not real but are only one way of representing the field. However, here they almost seem to be real. In this particular case, if you make the mistake of thinking that the field lines are somehow really there in space, and transform them, you get the correct field. That doesn't, however, make the field lines any more real. All you need do to remind yourself that they aren't real is to think about the electric fields produced by a charge together with a magnet; when the magnet moves, new electric fields are produced, and destroy the beautiful picture. So the neat idea of the contracting picture doesn't work in general. It is, however, a handy way to remember what the fields from a fast-moving charge are like.

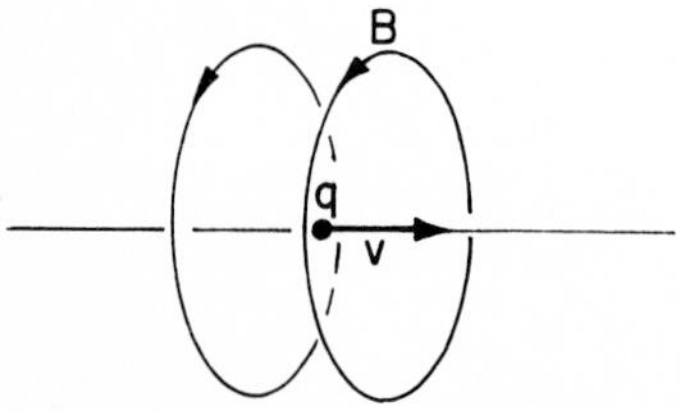

Fig. 26–5. The magnetic field near a moving charge is $\boldsymbol{v} \times \boldsymbol{E}$. (Compare with Fig. 26–4.)

The magnetic field is $\boldsymbol{v} \times \boldsymbol{E}$ [from Eq. (26.9)]. If you take the velocity crossed into a radial $\boldsymbol{E}$-field, you get a $\boldsymbol{B}$ which circles around the line of motion, as shown in Fig. 26–5. If we put back the c's, you will see that it's the same result we had for low-velocity charges. A good way to see where the c's must go is to refer back to the force law,

$$\boldsymbol{F} = q(\boldsymbol{E} + \boldsymbol{v} \times \boldsymbol{B}).$$

You see that a velocity times the magnetic field has the same dimensions as an electric field. So the right-hand side of Eq. (26.9) must have a factor $1/c^2$:

$$\boldsymbol{B} = \frac{\boldsymbol{v} \times \boldsymbol{E}}{c^2}. \tag{26.12}$$

For a slow-moving charge ($v \ll c$), we can take for $\boldsymbol{E}$ the Coulomb field; then

$$\boldsymbol{B} = \frac{q}{4\pi\epsilon_0 c^2} \frac{\boldsymbol{v} \times \boldsymbol{r}}{r^3}. \tag{26.13}$$

This formula corresponds exactly to equations for the magnetic field of a current that we found in Section 14–7.

We would like to point out, in passing, something interesting for you to think about. (We will come back to discuss it again later.) Imagine two electrons with velocities at right angles, so that one will cross over the path of the other, but in front of it, so they don't collide. At some instant, their relative positions will be as in Fig. 26–6(a). We look at the force on q_1 due to q_2 and vice versa. On q_2 there is only the electric force from q_1, since q_1 makes no magnetic field along its line of motion. On q_1, however, there is again the electric force but, in addition, a magnetic force, since it is moving in a $\boldsymbol{B}$-field made by q_2. The forces are as drawn in Fig. 26–6(b). The electric forces on q_1 and q_2 are equal and opposite. However, there is a sidewise (magnetic) force on q_1 *and no sidewise force on* q_2. Does action not equal reaction? We leave it for you to worry about.

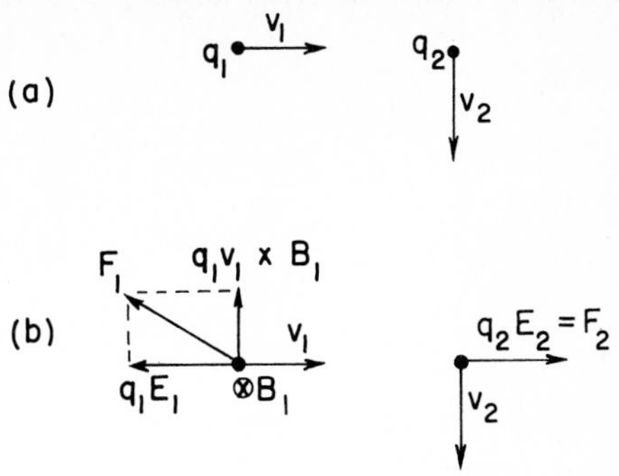

Fig. 26–6. The forces between two moving charges are not always equal and opposite. It appears that "action" is not equal to "reaction."

26–3 Relativistic transformation of the fields

In the last section we calculated the electric and magnetic fields from the transformed potentials. The fields are important, of course, in spite of the arguments given earlier that there is physical meaning and reality to the potentials. The fields, too, are real. It would be convenient for many purposes to have a way to compute the fields in a moving system if you already know the fields in some "rest" system. We have the transformation laws for ϕ and $\boldsymbol{A}$, because A_μ is a four-vector. Now we would like to know the transformation laws of $\boldsymbol{E}$ and $\boldsymbol{B}$. Given $\boldsymbol{E}$ and $\boldsymbol{B}$ in one frame, how do they look in another frame moving past? It is a convenient transformation to have. We could always work back through the potentials, but it is useful sometimes to be able to transform the fields directly. We will now see how that goes.

How can we find the transformation laws of the fields? We know the transformation laws of the ϕ and $\boldsymbol{A}$, and we know how the fields are given in terms of ϕ and $\boldsymbol{A}$—it should be easy to find the transformation for the $\boldsymbol{B}$ and $\boldsymbol{E}$. (You might think that with every vector there should be something to make it a four-vector, so with $\boldsymbol{E}$ there's got to be something else we can use for the fourth component. And also for $\boldsymbol{B}$. But it's not so. It's quite different from what you would expect.) To begin with, let's take just a magnetic field $\boldsymbol{B}$, which is, of course $\nabla \times \boldsymbol{A}$. Now we know that the vector potential with its x-, y-, and z-components is only a piece of something; there is also a t-component. Also we know that for derivatives like ∇, besides the x, y, z parts, there is also a derivative with respect to t. So let's try to figure out what happens if we replace a "y" by a "t", or a "z" by a "t," or something like that.

First, notice the form of the terms in $\nabla \times \boldsymbol{A}$ when we write out the components:

$$B_x = \frac{\partial A_z}{\partial y} - \frac{\partial A_y}{\partial z}, \quad B_y = \frac{\partial A_x}{\partial z} - \frac{\partial A_z}{\partial x}, \quad B_z = \frac{\partial A_y}{\partial x} - \frac{\partial A_x}{\partial y}. \tag{26.14}$$

The x-component is equal to a couple of terms that involve only y- and z-components. Suppose we call this combination of derivatives and components a "zy-thing," and give it a shorthand name, F_{zy}. We simply mean that

$$F_{zy} \equiv \frac{\partial A_z}{\partial y} - \frac{\partial A_y}{\partial z}. \tag{26.15}$$

Similarly, B_y is equal to the same kind of "thing," but this time it is an "xz-thing." And B_z is, of course, the corresponding "yx-thing." We have

$$B_x = F_{zy}, \quad B_y = F_{xz}, \quad B_z = F_{yx}. \tag{26.16}$$

Now what happens if we simply try to concoct also some "t"-type things like F_{xt} and F_{tz} (since nature should be nice and symmetric in x, y, z, and t)? For instance, what is F_{tz}? It is, of course,

$$\frac{\partial A_t}{\partial z} - \frac{\partial A_z}{\partial t}.$$

But remember that $A_t = \phi$, so it is also

$$\frac{\partial\phi}{\partial z} - \frac{\partial A_z}{\partial t}.$$

You've seen that before. It is the z-component of $\boldsymbol{E}$. Well, almost—there is a sign wrong. But we forgot that in the four-dimensional gradient the t-derivative comes with the opposite sign from x, y, and z. So we should really have taken the more consistent extension of F_{tz} as

$$F_{tz} = \frac{\partial A_t}{\partial z} + \frac{\partial A_z}{\partial t}. \tag{26.17}$$

Then it is exactly equal to $-E_z$. Trying also F_{tx} and F_{ty}, we find that the three possibilities give

$$F_{tx} = -E_x, \qquad F_{ty} = -E_y, \qquad F_{tz} = -E_z. \tag{26.18}$$

What happens if both subscripts are t? Or, for that matter, if both are x? We get things like

$$F_{tt} = \frac{\partial A_t}{\partial t} - \frac{\partial A_t}{\partial t},$$

and

$$F_{xx} = \frac{\partial A_x}{\partial x} - \frac{\partial A_x}{\partial x},$$

which give nothing but zero.

We have then six of these F-things. There are six more which you get by reversing the subscripts, but they give nothing really new, since

$$F_{xy} = -F_{yx},$$

and so on. So, out of sixteen possible combinations of the four subscripts taken in pairs, we get only six different physical objects; *and they are the components of* $\boldsymbol{B}$ *and* $\boldsymbol{E}$.

To represent the general term of F, we will use the general subscripts μ and ν, where each can stand for 0, 1, 2, or 3—meaning in our usual four-vector notation t, x, y, and z. Also, everything will be consistent with our four-vector notation if we define $F_{\mu\nu}$ by

$$F_{\mu\nu} = \nabla_\mu A_\nu - \nabla_\nu A_\mu, \tag{26.19}$$

remembering that $\nabla_\mu = (\partial/\partial t, -\partial/\partial x, -\partial/\partial y, -\partial/\partial z)$ and that $A_\mu = (\phi, A_x, A_y, A_z)$.

Table 26–1

The components of $F_{\mu\nu}$

$F_{\mu\nu} = -F_{\nu\mu}$	
$F_{\mu\mu} = 0$	
$F_{xy} = -B_z$	$F_{xt} = E_x$
$F_{yz} = -B_x$	$F_{yt} = E_y$
$F_{zx} = -B_y$	$F_{zt} = E_z$

What we have found is that there are six quantities that belong together in nature—that are different aspects of the same thing. The electric and magnetic fields which we have considered as separate vectors in our slow-moving world (where we don't worry about the speed of light) are not vectors in four-space. They are parts of a new "thing." Our physical "field" is really the six-component object $F_{\mu\nu}$. That is the way we must look at it for relativity. We summarize our results on $F_{\mu\nu}$ in Table 26–1.

You see that what we have done here is to generalize the cross product. We began with the curl operation, and the fact that the transformation properties of the curl are the same as the transformation properties of *two* vectors—the ordinary three-dimensional vector $\boldsymbol{A}$ and the gradient operator which we know also behaves like a vector. Let's look for a moment at an ordinary cross product in three dimensions, for example, the angular momentum of a particle. When an object is moving in a plane, the quantity $(xv_y - yv_x)$ is important. For motion in three dimensions, there are three such important quantities, which we call the angular momentum:

$$L_{xy} = m(xv_y - yv_x), \qquad L_{yz} = m(yv_z - zv_y), \qquad L_{zx} = m(zv_x - xv_z).$$

Then (although you may have forgotten by now) we discovered in Chapter 20 of Vol. I the miracle that these three quantities could be identified with the components of a vector. In order to do so, we had to make an artificial rule with a right-hand convention. It was just luck. It was luck because L_{ij} (with i and j equal to x, y, or z) was an antisymmetric object:

$$L_{ij} = -L_{ji}, \qquad L_{ii} = 0.$$

Of the nine possible quantities, there are only three independent numbers. And it just happens that when you change coordinate systems these three objects transform in exactly the same way as the components of a vector.

The same thing lets us represent an element of surface as a vector. A surface element has two parts—say dx and dy—which we can represent by the vector $\boldsymbol{da}$ normal to the surface. But we can't do that in four dimensions. What is the "normal" to $dx\,dy$? Is it along z or along t?

In short, for three dimensions it happens by luck that after you've taken a combination of two vectors like L_{ij}, you can represent it again by another vector because there are just three terms that happen to transform like the components of a vector. But in four dimensions that is evidently impossible, because there are six independent terms, and you can't represent six things by four things.

Even in three dimensions it is possible to have combinations of vectors that can't be represented by vectors. Suppose we take any two vectors $\boldsymbol{a} = (a_x, a_y, a_z)$ and $\boldsymbol{b} = (b_x, b_y, b_z)$, and make the various possible combinations of components, like a_xb_x, a_xb_y, etc. There would be nine possible quantities:

$$\begin{matrix} a_xb_x, & a_xb_y, & a_xb_z, \\ a_yb_x, & a_yb_y, & a_yb_z, \\ a_zb_x, & a_zb_y, & a_zb_z. \end{matrix}$$

We might call these quantities T_{ij}.

If we now go to a rotated coordinate system (say rotated about the z-axis), the components of $\boldsymbol{a}$ and $\boldsymbol{b}$ are changed. In the new system, a_x, for example, gets replaced by

$$a'_x = a_x \cos\theta + a_y \sin\theta,$$

and b_y gets replaced by

$$b'_y = b_y \cos\theta - b_x \sin\theta.$$

And similarly for other components. The nine components of the product quantity T_{ij} we have invented are all changed too, of course. For instance, $T_{xy} = a_xb_y$ gets changed to

$$T'_{xy} = a_xb_y(\cos^2\theta) - a_xb_x(\cos\theta\sin\theta) + a_yb_y(\sin\theta\cos\theta) - a_yb_x(\sin^2\theta),$$

or

$$T'_{xy} = T_{xy}\cos^2\theta - T_{xx}\cos\theta\sin\theta + T_{yy}\sin\theta\cos\theta - T_{yx}\sin^2\theta.$$

Each component of T'_{ij} is a linear combination of the components of T_{ij}.

So we discover that it is not only possible to have a "vector product" like $\boldsymbol{a} \times \boldsymbol{b}$ which has three components that transform like a vector, but we can—artificially—also make another kind of "product" of two vectors T_{ij} with *nine* components that transform under a rotation by a complicated set of rules that we could figure out. Such an object which has two indices to describe it, instead of one, is called a *tensor*. It is a tensor of the "second rank," because you can play this game with three vectors too and get a tensor of the third rank,—or with four, to get a tensor of the fourth rank, and so on. A tensor of the first rank is a vector.

The point of all this is that our electromagnetic quantity $F_{\mu\nu}$ is also a tensor of the second rank, because it has two indices in it. It is, however, a tensor in four dimensions. It transforms in a special way which we will work out in a moment—it is just the way a product of vectors transforms. For $F_{\mu\nu}$ it happens that if you change the indices around, $F_{\mu\nu}$ changes sign. That's a special case—it is an *antisymmetric tensor.* So we say: the electric and magnetic fields are both part of an antisymmetric tensor of the second rank in four dimensions.

You've come a long way. Remember way back when we defined what a velocity meant? Now we are talking about "an antisymmetric tensor of the second rank in four dimensions."

Now we have to find the law of the transformation of $F_{\mu\nu}$. It isn't at all difficult to do; it's just laborious—the brains involved are nil, but the work is not. What we want is the Lorentz transformation of $\nabla_\mu A_\nu - \nabla_\nu A_\mu$. Since ∇_μ is just a special case of a vector, we will work with the general antisymmetric vector combination, which we can call $G_{\mu\nu}$:

$$G_{\mu\nu} = a_\mu b_\nu - a_\nu b_\mu. \tag{26.20}$$

(For our purposes, a_μ will eventually be replaced by ∇_μ and b_μ will be replaced by the potential A_μ.) The components of a_μ and b_μ transform by the Lorentz formulas, which are

$$\begin{aligned}
a'_t &= \frac{a_t - va_x}{\sqrt{1-v^2}}, & b'_t &= \frac{b_t - vb_x}{\sqrt{1-v^2}},\\
a'_x &= \frac{a_x - va_t}{\sqrt{1-v^2}}, & b'_x &= \frac{b_x - vb_t}{\sqrt{1-v^2}},\\
a'_y &= a_y, & b'_y &= b_y,\\
a'_z &= a_z. & b'_z &= b_z.
\end{aligned} \tag{26.21}$$

Now let's transform the components of $G_{\mu\nu}$. We start with G_{tx}:

$$\begin{aligned}
G'_{tx} &= a'_t b'_x - a'_x b'_t\\
&= \left(\frac{a_t - va_x}{\sqrt{1-v^2}}\right)\left(\frac{b_x - vb_t}{\sqrt{1-v^2}}\right) - \left(\frac{a_x - va_t}{\sqrt{1-v^2}}\right)\left(\frac{b_t - vb_x}{\sqrt{1-v^2}}\right)\\
&= a_t b_x - a_x b_t.
\end{aligned}$$

But that is just G_{tx}; so we have the simple result

$$G'_{tx} = G_{tx}.$$

We will do one more.

$$G'_{ty} = \frac{a_t - va_x}{\sqrt{1-v^2}}\, b_y - a_y \frac{b_t - vb_x}{\sqrt{1-v^2}} = \frac{(a_t b_y - a_y b_t) - v(a_x b_y - a_y b_x)}{\sqrt{1-v^2}}.$$

So we get that

$$G'_{ty} = \frac{G_{ty} - vG_{xy}}{\sqrt{1-v^2}}.$$

And, of course, in the same way,

$$G'_{tz} = \frac{G_{tz} - vG_{xz}}{\sqrt{1-v^2}}.$$

It is clear how the rest will go. Let's make a table of all six terms; only now we may as well write them for $F_{\mu\nu}$:

$$\begin{aligned}
F'_{tx} &= F_{tx}, & F'_{xy} &= \frac{F_{xy} - vF_{ty}}{\sqrt{1-v^2}},\\
F'_{ty} &= \frac{F_{ty} - vF_{xy}}{\sqrt{1-v^2}}, & F'_{yz} &= F_{yz},\\
F'_{tz} &= \frac{F_{tz} - vF_{xz}}{\sqrt{1-v^2}}, & F'_{zx} &= \frac{F_{zx} - vF_{zt}}{\sqrt{1-v^2}}.
\end{aligned} \tag{26.22}$$

Of course, we still have $F'_{\mu\nu} = -F'_{\nu\mu}$ and $F'_{\mu\mu} = 0$.

So we have the transformation of the electric and magnetic fields. All we have to do is look at Table 26–1 to find out what our grand notation in terms of $F_{\mu\nu}$ means in terms of $\boldsymbol{E}$ and $\boldsymbol{B}$. It's just a matter of substitution. So that we can see how it looks in the ordinary symbols, we'll rewrite our transformation of the field components in Table 26–2.

Table 26–2

The Lorentz transformation of the electric and magnetic fields (Note: $c = 1$)

$$E'_x = E_x \qquad B'_x = B_x$$

$$E'_y = \frac{E_y - vB_z}{\sqrt{1 - v^2}} \qquad B'_y = \frac{B_y + vE_z}{\sqrt{1 - v^2}}$$

$$E'_z = \frac{E_z + vB_y}{\sqrt{1 - v^2}} \qquad B'_z = \frac{B_z - vE_y}{\sqrt{1 - v^2}}.$$

The equations in Table 26–2 tell us how $\boldsymbol{E}$ and $\boldsymbol{B}$ change if we go from one inertial frame to another. If we know $\boldsymbol{E}$ and $\boldsymbol{B}$ in one system, we can find what they are in another that moves by with the speed v.

We can write these equations in a form that is easier to remember if we notice that since v is in the x-direction, all the terms with v are components of the cross products $\boldsymbol{v} \times \boldsymbol{E}$ and $\boldsymbol{v} \times \boldsymbol{B}$. So we can rewrite the transformations as shown in Table 26–3.

Table 26–3

An alternative form for the field transformations (Note: $c = 1$)

$$E'_x = E_x \qquad B'_x = B_x$$

$$E'_y = \frac{(\boldsymbol{E} + \boldsymbol{v} \times \boldsymbol{B})_y}{\sqrt{1 - v^2}} \qquad B'_y = \frac{(\boldsymbol{B} - \boldsymbol{v} \times \boldsymbol{E})_y}{\sqrt{1 - v^2}}$$

$$E'_z = \frac{(\boldsymbol{E} + \boldsymbol{v} \times \boldsymbol{B})_z}{\sqrt{1 - v^2}} \qquad B'_z = \frac{(\boldsymbol{B} - \boldsymbol{v} \times \boldsymbol{E})_z}{\sqrt{1 - v^2}}$$

It is now easier to remember which components go where. In fact, the transformation can be written even more simply if we define the field components along x as the "parallel" components $E_{||}$ and $B_{||}$ (because they are parallel to the relative velocity of S and S'), and the total transverse components—the vector sums of the y- and z-components—as the "perpendicular" components $E_{\perp}$ and $B_{\perp}$. Then we get the equations in Table 26–4. (We have also put back the c's, so it will be more convenient when we want to refer back later.)

Table 26–4

Still another form for the Lorentz transformation of $\boldsymbol{E}$ and $\boldsymbol{B}$

$$E'_{||} = E \qquad B'_{||} = B$$

$$E'_{\perp} = \frac{(\boldsymbol{E} + \boldsymbol{v} \times \boldsymbol{B})_{\perp}}{\sqrt{1 - v^2/c^2}} \qquad B'_{\perp} = \frac{\left(\boldsymbol{B} - \dfrac{\boldsymbol{v} \times \boldsymbol{E}}{c^2}\right)_{\perp}}{\sqrt{1 - v^2/c^2}}$$

The field transformations give us another way of solving some problems we have done before—for instance, for finding the fields of a moving point charge. We have worked out the fields before by differentiating the potentials. But we could now do it by transforming the Coulomb field. If we have a point charge at rest in the S-frame, then there is only the simple radial $\boldsymbol{E}$-field. In the S'-frame we will see a point charge moving with the velocity u, if the S'-frame moves by the

S-frame with the speed $v = -u$. We will let you show that the transformations of Tables 26–3 and 26–4 give the same electric and magnetic fields we got in Section 26–2.

The transformation of Table 26–2 gives us an interesting and simple answer for what we see if we move past *any* system of fixed charges. For example, suppose we want to know the fields in *our* frame S' if we are moving along between the plates of a condenser, as shown in Fig. 26–7. (It is, of course, the same thing if we say that a charged condenser is moving past *us*.) What do we see? The transformation is easy in this case because the $\boldsymbol{B}$-field in the original system is zero. Suppose, first, that our motion is perpendicular to $\boldsymbol{E}$; then we will see an $\boldsymbol{E}' = \boldsymbol{E}/\sqrt{1 - v^2/c^2}$ which is still completely transverse. We will see, in addition, a magnetic field $\boldsymbol{B}' = -\boldsymbol{v} \times \boldsymbol{E}'/c^2$. (The $\sqrt{1 - v^2}$ doesn't appear in our formula for $\boldsymbol{B}'$ because we wrote it in terms of $\boldsymbol{E}'$ rather than $\boldsymbol{E}$; but it's the same thing.) So when we move along perpendicular to a static electric field, we see a reduced $\boldsymbol{E}$ and an added transverse $\boldsymbol{B}$. If our motion is not perpendicular to $\boldsymbol{E}$, we break $\boldsymbol{E}$ into $\boldsymbol{E}_{||}$ and $\boldsymbol{E}_{\perp}$. The parallel part is unchanged, $E'_{||} = E_{||}$, and the perpendicular component does as just described.

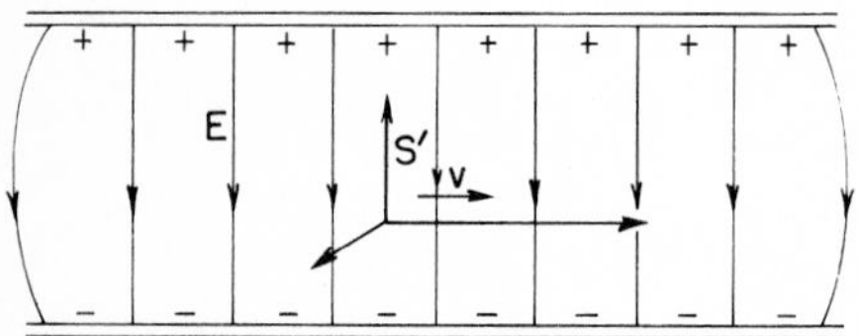

Fig. 26–7. The coordinate frame S' moving through a static electric field.

Let's take the opposite case, and imagine we are moving through a pure static *magnetic* field. This time we would see an *electric* field $\boldsymbol{E}'$ equal to $\boldsymbol{v} \times \boldsymbol{B}'$, and the magnetic field changed by the factor $1/\sqrt{1 - v^2/c^2}$ (assuming it is transverse). So long as v is much less than c, we can neglect the change in the magnetic field, and the main effect is that an electric field appears. As one example of this effect, consider this once famous problem of determining the speed of an airplane. It's no longer famous, since radar can now be used to determine the air speed from ground reflections, but for many years it was very hard to find the speed of an airplane in bad weather. You could not see the ground and you didn't know which way was up, and so on. Yet it was important to know how fast you were moving relative to the earth. How can this be done without seeing the earth? Many who knew the transformation formulas thought of the idea of using the fact that the airplane moves in the magnetic field of the earth. Suppose that an airplane is flying where there is a magnetic field more or less known. Let's just take the simple case where the magnetic field is vertical. If we were flying through it with a horizontal velocity v, then, according to our formula, we should see an electric field which is $\boldsymbol{v} \times \boldsymbol{B}$, i.e., perpendicular to the line of motion. If we hang an insulated wire across the airplane, this electric field will induce charges on the ends of the wire. That is nothing new. From the point of view of someone on the ground, we are moving a wire through a field, and the $\boldsymbol{v} \times \boldsymbol{B}$ force causes charges to move to the ends of the wire. The transformation equations just say the same thing in a different way. (The fact that we can say the thing more than one way doesn't mean that one way is better than another. We are getting so many different methods and tools that we can usually get the same result in 65 different ways!)

So to measure v, all we have to do is measure the voltage between the ends of the wire. We can't do it with a voltmeter because the same fields will act on the wires in the voltmeter, but there are ways of measuring such fields. We talked about some of them when we discussed atmospheric electricity in Chapter 9. So it should be possible to measure the speed of the airplane.

This important problem was, however, never solved this way. The reason is that the electric field that is developed is of the order of millivolts per meter. It is possible to measure such fields, but the trouble is that these fields are, unfortunately, not any different from any other electric fields. The field that is produced by motion through the magnetic field can't be distinguished from some electric field that was already in the air from another cause, say from electrostatic charges in the air, or on the clouds. We described in Chapter 9 that there are, typically, electric fields above the surface of the earth with strengths of about 100 volts per meter. But they are quite irregular. So as the airplane flies through the air, it sees fluctuations of atmospheric electric fields which are enormous in comparison to the tiny fields produced by the $\boldsymbol{v} \times \boldsymbol{B}$ term, and it turns out for practical reasons to be impossible to measure speeds of an airplane by its motion through the earth's magnetic field.

26–4 The equations of motion in relativistic notation*

It doesn't do much good to find electric and magnetic fields from Maxwell's equations unless we know what the fields do when we have them. You may remember that the fields are required to find the forces on charges, and that those forces determine the motion of the charge. So, of course, part of the theory of electrodynamics is the relation between the motion of charges and the forces.

For a single charge in the fields $\boldsymbol{E}$ and $\boldsymbol{B}$, the force is

$$\boldsymbol{F} = q(\boldsymbol{E} + \boldsymbol{v} \times \boldsymbol{B}). \tag{26.23}$$

This force is equal to the mass times the acceleration for low velocities, but the correct law for any velocity is that the force is equal to $d\boldsymbol{p}/dt$. Writing $\boldsymbol{p} = m_0\boldsymbol{v}/\sqrt{1 - v^2/c^2}$, we find that the relativistically correct equation of motion is

$$\frac{d}{dt}\left(\frac{m_0\boldsymbol{v}}{\sqrt{1 - v^2/c^2}}\right) = \boldsymbol{F} = q(\boldsymbol{E} + \boldsymbol{v} \times \boldsymbol{B}). \tag{26.24}$$

We would like now to discuss this equation from the point of view of relativity. Since we have put our Maxwell equations in relativistic form, it would be interesting to see what the equations of motion would look like in relativistic form. Let's see whether we can rewrite the equation in a four-vector notation.

We know that the momentum is part of a four-vector p_μ whose time component is the energy $m_0/\sqrt{1 - v^2/c^2}$. So we might think to replace the left-hand side of Eq. (26.24) by dp_μ/dt. Then we need only find a fourth component to go with F. This fourth component must equal the rate-of-change of the energy, or the rate of doing work, which is $\boldsymbol{F} \cdot \boldsymbol{v}$. We would then like to write the right-hand side of Eq. (26.24) as a four-vector like $(\boldsymbol{F} \cdot \boldsymbol{v}, F_x, F_y, F_z)$. But this does not make a four-vector.

The *time* derivative of a four-vector is no longer a four-vector, because the d/dt requires the choice of some special frame for measuring t. We got into that trouble before when we tried to make $\boldsymbol{v}$ into a four-vector. Our first guess was that the time component would be $c\,dt/dt = c$. But the quantities

$$\left(c, \frac{dx}{dt}, \frac{dy}{dt}, \frac{dz}{dt}\right) = (c, \boldsymbol{v}) \tag{26.25}$$

are *not* the components of a four-vector. We found that they could be made into one by multiplying each component by $1/\sqrt{1 - v^2/c^2}$. The "four-velocity" u_μ is the four-vector

$$u_\mu = \left(\frac{c}{\sqrt{1 - v^2/c^2}}, \frac{\boldsymbol{v}}{\sqrt{1 - v^2/c^2}}\right). \tag{26.26}$$

So it appears that the trick is to multiply d/dt by $1/\sqrt{1 - v^2/c^2}$, if we want the derivatives to make a four-vector.

Our second guess then is that

$$\frac{1}{\sqrt{1 - v^2/c^2}}\frac{d}{dt}(p_\mu) \tag{26.27}$$

should be a four-vector. But what is $\boldsymbol{v}$? It is the velocity of the particle—not of a coordinate frame! Then the quantity f_μ defined by

$$f_\mu = \left(\frac{\boldsymbol{F} \cdot \boldsymbol{v}}{\sqrt{1 - v^2/c^2}}, \frac{\boldsymbol{F}}{\sqrt{1 - v^2/c^2}}\right) \tag{26.28}$$

is the extension into four dimensions of a force—we can call it the "four-force." It is indeed a four-vector, and its space components are not the components of $\boldsymbol{F}$ but of $\boldsymbol{F}/\sqrt{1 - v^2/c^2}$.

---

* In this section we will put back all of the c's.

The question is—why is f_μ a four-vector? It would be nice to get a little understanding of that $1/\sqrt{1 - v^2/c^2}$ factor. Since it has come up twice now, it is time to see why the d/dt can always be fixed by the same factor. The answer is in the following: When we take the time derivative of some function x, we compute the increment Δx in a small interval Δt in the variable t. But in another frame, the interval Δt might correspond to a change in both t' and x', so if we vary only t', the change in x will be different. We have to find a variable for our differentiation that is a measure of an "interval" in *space-time*, which will then be the same in all coordinate systems. When we take Δx for that interval, it will be the same for all coordinate frames. When a particle "moves" in four-space, there are the changes Δt, Δx, Δy, Δz. Can we make an invariant interval out of them? Well, they are the components of the four-vector $x_\mu = (ct, x, y, z)$ so if we define a quantity Δs by

$$(\Delta s)^2 = \frac{1}{c^2}\,\Delta x_\mu\,\Delta x_\mu = \frac{1}{c^2}\,(c^2\,\Delta t^2 - \Delta x^2 - \Delta y^2 - \Delta z^2) \tag{26.29}$$

—which is a four-dimensional dot product—we then have a good four-scalar to use as a measure of a four-dimensional interval. From Δs—or its limit ds—we can define a parameter $s = \int ds$. And a derivative with respect to s, d/ds, is a nice four-dimensional operation, because it is invariant with respect to a Lorentz transformation.

It is easy to relate ds to dt for a moving particle. For a moving point particle,

$$dx = v_x\,dt, \qquad dy = v_y\,dt, \qquad dz = v_z\,dt, \tag{26.30}$$

and

$$ds = \sqrt{(dt^2/c^2)(c^2 - v_x^2 - v_y^2 - v_z^2)} = dt\sqrt{1 - v^2/c^2}. \tag{26.31}$$

So the operator

$$\frac{1}{\sqrt{1 - v^2/c^2}}\,\frac{d}{dt}$$

is an *invariant operator*. If we operate on any four-vector with it, we get another four-vector. For instance, if we operate on (ct, x, y, z), we get the four-velocity u_μ:

$$\frac{dx_\mu}{ds} = u_\mu.$$

We see now why the factor $\sqrt{1 - v^2/c^2}$ fixes things up.

The invariant variable s is a useful physical quantity. It is called the "proper time" along the path of a particle, because ds is always an interval of time in a frame that is moving with the particle at any particluar instant. (Then, $\Delta x = \Delta y = \Delta z = 0$, and $\Delta s = \Delta t$.) If you can imagine some "clock" whose rate doesn't depend on the acceleration, such a clock carried along with the particle would show the time s.

We can now go back and write Newton's law (as corrected by Einstein) in the neat form

$$\frac{dp_\mu}{ds} = f_\mu, \tag{26.32}$$

where f_μ is given in Eq. (26.28). Also, the momentum p_μ can be written as

$$p_\mu = m_0 u_\mu = m_0\,\frac{dx_\mu}{ds}, \tag{26.33}$$

where the coordinates $x_\mu = (ct, x, y, z)$ now describe the trajectory of the particle. Finally, the four-dimensional notation gives us this very simple form of the equations of motion:

$$f_\mu = m_0\,\frac{d^2x_\mu}{ds^2}, \tag{26.34}$$

which is reminiscent of $F = ma$. It is important to notice that Eq. (26.34) is *not* the same as $F = ma$, because the four-vector formula Eq. (26.34) has in it the relativistic mechanics which are different from Newton's law for high velocities. It is unlike the case of Maxwell's equations, where we were able to rewrite the equations in the relativistic form *without any change in the meaning at all*—but with just a change of notation.

Now let's return to Eq. (26.24) and see how we can write the right-hand side in four-vector notation. The three components—when divided by $\sqrt{1 - v^2/c^2}$—are the components of f_μ, so

$$f_x = \frac{q(\boldsymbol{E} + \boldsymbol{v} \times \boldsymbol{B})_x}{\sqrt{1 - v^2/c^2}} = q\left[\frac{E_x}{\sqrt{1 - v^2/c^2}} + \frac{v_y B_z}{\sqrt{1 - v^2/c^2}} - \frac{v_z B_y}{\sqrt{1 - v^2/c^2}}\right]. \tag{26.35}$$

Now we must put all quantities in their relativistic notation. First, $c/\sqrt{1 - v^2/c^2}$ and $v_y/\sqrt{1 - v^2/c^2}$ and $v_z/\sqrt{1 - v^2/c^2}$ are the t-, y-, and z-components of the four-velocity u_μ. And the components of $\boldsymbol{E}$ and $\boldsymbol{B}$ are components of the second-rank tensor of the fields $F_{\mu\nu}$. Looking back in Table 26–1 for the components of $F_{\mu\nu}$ that correspond to E_x, B_z, and B_y, we get

$$f_x = q(u_t F_{xt} - u_y F_{xy} - u_z F_{xz}),$$

which begins to look interesting. Every term has the subscript x, which is reasonable, since we're finding an x-component. Then all the others appear in pairs: tt, yy, zz—except that the xx-term is missing. So we just stick it in, and write

$$f_x = q(u_t F_{xt} - u_x F_{xx} - u_y F_{xy} - u_z F_{xz}). \tag{26.36}$$

We haven't changed anything because $F_{\mu\nu}$ is antisymmetric, and F_{xx} is zero. The reason for wanting to put in the xx-term is so that we can write Eq. (26.36) in the short-hand form

$$f_\mu = q u_\nu F_{\mu\nu}. \tag{26.37}$$

This equation is the same as Eq. (26.36) if we make the *rule* that whenever any subscript occurs *twice* (as ν does here), you automatically sum over terms in the same way as for the scalar product, *using the same convention for the signs.*

You can easily believe that (26.37) works equally well for $\mu = y$ or $\mu = z$, but what about $\mu = t$? Let's see, for fun, what it says:

$$f_t = q(u_t F_{tt} - u_x F_{tx} - u_y F_{ty} - u_z F_{tz}).$$

Now we have to translate back to E's and B's. We get

$$f_t = q\left(0 + \frac{v_x}{\sqrt{1 - v^2/c^2}}\,E_x + \frac{v_y}{\sqrt{1 - v^2/c^2}}\,E_y + \frac{v_z}{\sqrt{1 - v^2/c^2}}\,E_z\right),$$

or

$$f_t = \frac{q\boldsymbol{v} \cdot \boldsymbol{E}}{\sqrt{1 - v^2/c^2}}. \tag{26.38}$$

But from Eq. (26.28), f_t is supposed to be

$$\frac{\boldsymbol{F} \cdot \boldsymbol{v}}{\sqrt{1 - v^2/c^2}} = \frac{q(\boldsymbol{E} + \boldsymbol{v} \times \boldsymbol{B}) \cdot \boldsymbol{v}}{\sqrt{1 - v^2/c^2}}.$$

This is the same thing as Eq. (26.38), since $(\boldsymbol{v} \times \boldsymbol{B}) \cdot \boldsymbol{v}$ is zero. So everything comes out all right.

Summarizing, our equation of motion can be written in the elegant form

$$m_0 \frac{d^2 x_\mu}{ds^2} = f_\mu = q u_\nu F_{\mu\nu}. \tag{26.39}$$

Although it is nice to see that the equations can be written that way, this form is not particularly useful. It's usually more convenient to solve for particle motions by using the original equations (26.24), and that's what we will usually do.

27

Field Energy and Field Momentum

27–1 Local conservation

It is clear that the energy of matter is not conserved. When an object radiates light it loses energy. However, the energy lost is possibly describable in some other form, say in the light. Therefore the theory of the conservation of energy is incomplete without a consideration of the energy which is associated with the light or, in general, with the electromagnetic field. We take up now the law of conservation of energy and, also, of momentum for the fields. Certainly, we cannot treat one without the other, because in the relativity theory they are different aspects of the same four-vector.

Very early in Volume I, we discussed the conservation of energy; we said then merely that the total energy in the world is constant. Now we want to extend the idea of the energy conservation law in an important way—in a way that says something in *detail* about *how* energy is conserved. The new law will say that if energy goes away from a region, it is because it *flows* away through the boundaries of that region. It is a somewhat stronger law than the conservation of energy without such a restriction.

To see what the statement means, let's look at how the law of the conservation of charge works. We described the conservation of charge by saying that there is a current density $\boldsymbol{j}$ and a charge density ρ, and that when the charge decreases at some place there must be a flow of charge away from that place. We call that the conservation of charge. The mathematical form of the conservation law is

$$\nabla \cdot \boldsymbol{j} = -\frac{\partial \rho}{\partial t}. \tag{27.1}$$

This law has the consequence that the total charge in the world is always constant—there is never any net gain or loss of charge. However, the total charge in the world could be constant in another way. Suppose that there is some charge Q_1 near some point (1) while there is no charge near some point (2) some distance away (Fig. 27–1). Now suppose that, as time goes on, the charge Q_1 were to gradually fade away and that *simultaneously* with the decrease of Q_1 some charge Q_2 would appear near point (2), and in such a way that at every instant the sum of Q_1 and Q_2 was a constant. In other words, at any intermediate state the amount of charge lost by Q_1 would be added to Q_2. Then the total amount of charge in the world would be conserved. That's a "world-wide" conservation, but not what we will call a "local" conservation, because in order for the charge to get from (1) to (2), it didn't have to appear anywhere in the space between point (1) and point (2). Locally, the charge was just "lost."

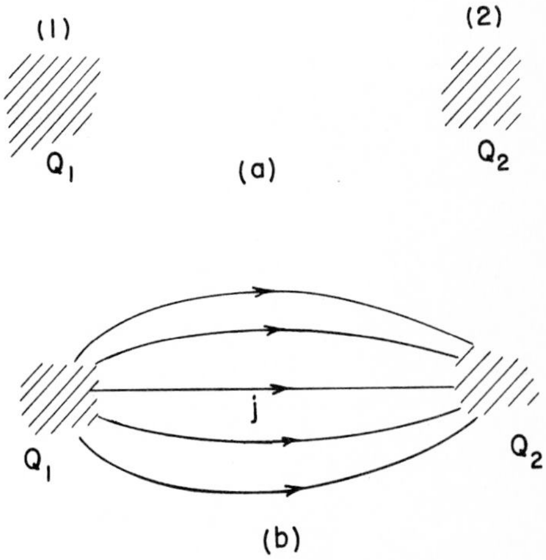

Fig. 27–1. Two ways to conserve charge: (a) $Q_1 + Q_2$ is constant; (b) $dQ_1/dt = \int \boldsymbol{j} \cdot \boldsymbol{n}\, da = -dQ_2/dt$.

There is a difficulty with such a "world-wide" conservation law in the theory of relativity. The concept of "simultaneous moments" at distant points is one which is not equivalent in different systems. Two events that are simultaneous in one system are not simultaneous for another system moving past. For "world-wide" conservation of the kind described, it is necessary that the charge lost from Q_1 should appear *simultaneously* in Q_2. Otherwise there would be some moments when the charge was not conserved. There seems to be no way to make the law of charge conservation relativistically invariant without making it a "local" conservation law. As a matter of fact, the requirement of the Lorentz relativistic invariance seems to restrict the possible laws of nature in surprising ways. In modern quantum field theory, for example, people have often wanted to alter the theory by allowing what we call a "nonlocal" interaction—where something *here* has a direct effect on something *there*—but we get in trouble with the relativity principle.

"Local" conservation involves another idea. It says that a charge can get from one place to another only if there is something happening in the space between. To describe the law we need not only the density of charge, ρ, but also another kind of quantity, namely $\boldsymbol{j}$, a vector giving the rate of flow of charge across a surface. Then the flow is related to the rate of change of the density by Eq. (27.1). This is the more extreme kind of a conservation law. It says that charge is conserved in a special way—conserved "locally."

It turns out that energy conservation is also a *local* process. There is not only an energy density in a given region of space but also a vector to represent the rate of flow of the energy through a surface. For example, when a light source radiates, we can find the light energy moving out from the source. If we imagine some mathematical surface surrounding the light source, the energy lost from inside the surface is equal to the energy that flows out through the surface.

27–2 Energy conservation and electromagnetism

We want now to write quantitatively the conservation of energy for electromagnetism. To do that, we have to describe how much energy there is in any volume element of space, and also the rate of energy flow. Suppose we think first only of the electromagnetic field energy. We will let u represent the *energy density* in the field (that is, the amount of energy per unit volume in space) and let the vector $\boldsymbol{S}$ represent the *energy flux* of the field (that is, the flow of energy per unit time across a unit area perpendicular to the flow). Then, in perfect analogy with the conservation of charge, Eq. (27.1), we can write the "local" law of energy conservation in the field as

$$\frac{\partial u}{\partial t} = -\boldsymbol{\nabla}\cdot\boldsymbol{S}. \tag{27.2}$$

Of course, this law is not true in general; it is not true that the field energy is conserved. Suppose you are in a dark room and then turn on the light switch. All of a sudden the room is full of light, so there is energy in the field, although there wasn't any energy there before. Equation (27.2) is not the complete conservation law, because the *field* energy *alone* is not conserved, only the total energy in the world—there is also the energy of matter. The field energy will change if there is some work being done by matter on the field or by the field on matter.

However, if there is matter inside the volume of interest, we know how much energy it has: Each particle has the energy $m_0c^2/\sqrt{1-v^2/c^2}$. The total energy of the matter is just the sum of all the particle energies, and the flow of this energy through a surface is just the sum of the energy carried by each particle that crosses the surface. We want now to talk only about the energy of the electromagnetic field. So we must write an equation which says that the total *field* energy in a given volume decreases *either* because field energy flows out of the volume *or* because the field loses energy to matter (or gains energy, which is just a negative loss). The field energy inside a volume V is

$$\int_V u\,dV,$$

and its rate of decrease is minus the time derivative of this integral. The flow of field energy out of the volume V is the integral of the normal component of $\boldsymbol{S}$ over the surface Σ that encloses V,

$$\int_\Sigma \boldsymbol{S}\cdot\boldsymbol{n}\,da.$$

So

$$-\frac{\partial}{\partial t}\int_V u\,dV = \int_\Sigma \boldsymbol{S}\cdot\boldsymbol{n}\,da + (\text{work done on matter inside } V). \tag{27.3}$$

We have seen before that the field does work on each unit volume of matter at the rate $\boldsymbol{E}\cdot\boldsymbol{j}$. [The force on a particle is $\boldsymbol{F} = q(\boldsymbol{E} + \boldsymbol{v}\times\boldsymbol{B})$, and the rate of doing work is $\boldsymbol{F}\cdot\boldsymbol{v} = q\boldsymbol{E}\cdot\boldsymbol{v}$. If there are N particles per unit volume, the rate of doing work per unit volume is $Nq\boldsymbol{E}\cdot\boldsymbol{v}$, but $Nq\boldsymbol{v} = \boldsymbol{j}$.] So the quantity $\boldsymbol{E}\cdot\boldsymbol{j}$ must be equal to the loss of energy per unit time and per unit volume *by* the field. Equation (27.3) then becomes

$$-\frac{\partial}{\partial t}\int_V u\,dV = \int_\Sigma \boldsymbol{S}\cdot\boldsymbol{n}\,da + \int_V \boldsymbol{E}\cdot\boldsymbol{j}\,dV. \tag{27.4}$$

This is our conservation law for energy in the field. We can convert it into a differential equation like Eq. (27.2) if we can change the second term to a volume integral. That is easy to do with Gauss' theorem. The surface integral of the normal component of $\boldsymbol{S}$ is the integral of its divergence over the volume inside. So Eq. (27.3) is equivalent to

$$-\int_V \frac{du}{dt}\,dV = \int_V \boldsymbol{\nabla}\cdot\boldsymbol{S}\,dV + \int_V \boldsymbol{E}\cdot\boldsymbol{j}\,dV,$$

where we have put the time derivative of the first term inside the integral. Since this equation is true for any volume, we can take away the integrals and we have the energy equation for the electromagnetic fields:

$$-\frac{\partial u}{\partial t} = \boldsymbol{\nabla}\cdot\boldsymbol{S} + \boldsymbol{E}\cdot\boldsymbol{j}. \tag{27.5}$$

Now this equation doesn't do us a bit of good unless we know what u and $\boldsymbol{S}$ are. Perhaps we should just tell you what they are in terms of $\boldsymbol{E}$ and $\boldsymbol{B}$, because all we really want is the result. However, we would rather show you the kind of argument that was used by Poynting in 1884 to obtain formulas for $\boldsymbol{S}$ and u, so you can see where they come from. (You won't, however, need to learn this derivation for our later work.)

27–3 Energy density and energy flow in the electromagnetic field

The idea is to suppose that there is a field energy density u and a flux $\boldsymbol{S}$ that depend only upon the fields $\boldsymbol{E}$ and $\boldsymbol{B}$. (For example, we know that in electrostatics, at least, the energy density can be written $\frac{1}{2}\epsilon_0\boldsymbol{E}\cdot\boldsymbol{E}$.) Of course, the u and $\boldsymbol{S}$ might depend on the potentials or something else, but let's see what we can work out. We can try to rewrite the quantity $\boldsymbol{E}\cdot\boldsymbol{j}$ in such a way that it becomes the sum of two terms: one that is the time derivative of one quantity and another that is the divergence of a second quantity. The first quantity would then be u and the second would be $\boldsymbol{S}$ (with suitable signs). Both quantities must be written in terms of the fields only; that is, we want to write our equality as

$$\boldsymbol{E}\cdot\boldsymbol{j} = -\frac{\partial u}{\partial t} - \boldsymbol{\nabla}\cdot\boldsymbol{S}. \tag{27.6}$$

The left-hand side must first be expressed in terms of the fields only. How can we do that? By using Maxwell's equations, of course. From Maxwell's equation for the curl of $\boldsymbol{B}$,

$$\boldsymbol{j} = \epsilon_0 c^2\boldsymbol{\nabla}\times\boldsymbol{B} - \epsilon_0\frac{\partial\boldsymbol{E}}{\partial t}\cdot$$

Substituting this in (27.6) we will have only $\boldsymbol{E}$'s and $\boldsymbol{B}$'s:

$$\boldsymbol{E}\cdot\boldsymbol{j} = \epsilon_0 c^2\boldsymbol{E}\cdot(\boldsymbol{\nabla}\times\boldsymbol{B}) - \epsilon_0\boldsymbol{E}\cdot\frac{\partial\boldsymbol{E}}{\partial t}\cdot \tag{27.7}$$

We are already partly finished. The last term is a time derivative—it is $(\partial/\partial t)(\frac{1}{2}\epsilon_0\boldsymbol{E}\cdot\boldsymbol{E})$. So $\frac{1}{2}\epsilon_0\boldsymbol{E}\cdot\boldsymbol{E}$ is at least one part of u. It's the same thing we found in electrostatics. Now, all we have to do is to make the other term into the divergence of something.

Notice that the first term on the right-hand side of (27.7) is the same as

$$(\nabla \times \boldsymbol{B}) \cdot \boldsymbol{E}. \tag{27.8}$$

And, as you know from vector algebra, $(\boldsymbol{a} \times \boldsymbol{b}) \cdot \boldsymbol{c}$ is the same as $\boldsymbol{a} \cdot (\boldsymbol{b} \times \boldsymbol{c})$; so our term is also the same as

$$\nabla \cdot (\boldsymbol{B} \times \boldsymbol{E}) \tag{27.9}$$

and we have the divergence of "something," just as we wanted. Only that's wrong! We warned you before that ∇ is "like" a vector, but not "exactly" the same. The reason it is not is because there is an additional *convention* from calculus: when a derivative operator is in front of a product, it works on everything to the right. In Eq. (27.7), the ∇ operates only on $\boldsymbol{B}$, not on $\boldsymbol{E}$. But in the form (27.9), the normal convention would say that ∇ operates on both $\boldsymbol{B}$ and $\boldsymbol{E}$. So it's *not* the same thing. In fact, if we work out the components of $\nabla \cdot (\boldsymbol{B} \times \boldsymbol{E})$ we can see that it is equal to $\boldsymbol{E} \cdot (\nabla \times \boldsymbol{B})$ *plus* some other terms. It's like what happens when we take a derivative of a product in algebra. For instance,

$$\frac{d}{dx}(fg) = \frac{df}{dx}g + f\frac{dg}{dx}.$$

Rather than working out all the components of $\nabla \cdot (\boldsymbol{B} \times \boldsymbol{E})$, we would like to show you a trick that is very useful for this kind of problem. It is a trick that allows you to use all the rules of vector algebra on expressions with the ∇ operator, without getting into trouble. The trick is to throw out—for a while at least—the rule of the calculus notation about what the derivative operator works on. You see, ordinarily, the order of terms is used for *two* separate purposes. One is for calculus: $f(d/dx)g$ is not the same as $g(d/dx)f$; and the other is for vectors: $\boldsymbol{a} \times \boldsymbol{b}$ is different from $\boldsymbol{b} \times \boldsymbol{a}$. We can, if we want, choose to abandon momentarily the calculus rule. Instead of saying that a derivative operates on everything to the right, we make a *new* rule that doesn't depend on the order in which terms are written down. Then we can juggle terms around without worrying.

Here is our new convention: we show, by a subscript, what a differential operator works on; the *order* has no meaning. Suppose we let the operator D stand for $\partial/\partial x$. Then D_f means that only the derivative of the variable quantity f is taken. Then

$$D_f f = \frac{\partial f}{\partial x}.$$

But if we have $D_f fg$, it means

$$D_f fg = \left(\frac{\partial f}{\partial x}\right) g.$$

But notice now that according to our new rule, $f\,D_f g$ means the same thing. We can write the same thing any which way:

$$D_f fg = gD_f f = f\,D_f g = fg\,D_f.$$

You see, the D_f can even come *after* everything. (It's surprising that such a handy notation is never taught in books on mathematics or physics.)

You may wonder: What if I *want* to write the derivative of fg? I *want* the derivative of *both* terms. That's easy, you just say so; you write $D_f(fg) + D_g(fg)$. That is just $g(\partial f/\partial x) + f(\partial g/\partial x)$, which is what you mean in the old notation by $\partial(fg)/\partial x$.

You will see that it is now going to be very easy to work out a new expression for $\nabla \cdot (\boldsymbol{B} \times \boldsymbol{E})$. We start by changing to the new notation; we write

$$\nabla \cdot (\boldsymbol{B} \times \boldsymbol{E}) = \nabla_B \cdot (\boldsymbol{B} \times \boldsymbol{E}) + \nabla_E \cdot (\boldsymbol{B} \times \boldsymbol{E}). \tag{27.10}$$

The moment we do that we don't have to keep the order straight any more. We always know that ∇_E operates on $\boldsymbol{E}$ only, and ∇_B operates on $\boldsymbol{B}$ only. In these circumstances, we can use ∇ as though it were an ordinary vector. (Of course, when we are finished, we will want to return to the "standard" notation that everybody usually uses.) So now we can do the various things like interchanging dots and crosses and making other kinds of rearrangements of the terms. For instance, the middle term of Eq. (27.10) can be rewritten as $\boldsymbol{E}\cdot\boldsymbol{\nabla}_B\times\boldsymbol{B}$. (You remember that $\boldsymbol{a}\cdot\boldsymbol{b}\times\boldsymbol{c}=\boldsymbol{b}\cdot\boldsymbol{c}\times\boldsymbol{a}$.) And the last term is the same as $\boldsymbol{B}\cdot\boldsymbol{E}\times\boldsymbol{\nabla}_E$. It looks freakish, but it is all right. Now if we try to go back to the ordinary convention, we have to arrange that the $\boldsymbol{\nabla}$ operates only on its "own" variable. The first one is already that way, so we can just leave off the subscript. The second one needs some rearranging to put the $\boldsymbol{\nabla}$ in front of the $\boldsymbol{E}$, which we can do by reversing the cross product and changing sign:

$$\boldsymbol{B}\cdot(\boldsymbol{E}\times\boldsymbol{\nabla}_E)=-\boldsymbol{B}\cdot(\boldsymbol{\nabla}_E\times\boldsymbol{E}).$$

Now it is in a conventional order, so we can return to the usual notation. Equation (27.10) is equivalent to

$$\boldsymbol{\nabla}\cdot(\boldsymbol{B}\times\boldsymbol{E})=\boldsymbol{E}\cdot(\boldsymbol{\nabla}\times\boldsymbol{B})-\boldsymbol{B}\cdot(\boldsymbol{\nabla}\times\boldsymbol{E}).\tag{27.11}$$

(A quicker way would have been to use components in this special case, but it was worth taking the time to show you the mathematical trick. You probably won't see it anywhere else, and it is very good for unlocking vector algebra from the rules about the order of terms with derivatives.)

We now return to our energy conservation discussion and use our new result, Eq. (27.11), to transform the $\boldsymbol{\nabla}\times\boldsymbol{B}$ term of Eq. (27.7). That energy equation becomes

$$\boldsymbol{E}\cdot\boldsymbol{j}=\epsilon_0c^2\boldsymbol{\nabla}\cdot(\boldsymbol{B}\times\boldsymbol{E})+\epsilon_0c^2\boldsymbol{B}\cdot(\boldsymbol{\nabla}\times\boldsymbol{E})-\frac{\partial}{\partial t}\left(\tfrac{1}{2}\epsilon_0\boldsymbol{E}\cdot\boldsymbol{E}\right)\tag{27.12}$$

Now you see we're almost finished. We have one term which is a nice derivative with respect to t to use for u and another that is a beautiful divergence to represent $\boldsymbol{S}$. Unfortunately, there is the center term left over, which is neither a divergence nor a derivative with respect to t. So we almost made it, but not quite. After some thought, we look back at the differential equations of Maxwell and discover that $\boldsymbol{\nabla}\times\boldsymbol{E}$ is, fortunately, equal to $-\partial\boldsymbol{B}/\partial t$, which means that we can turn the extra term into something that is a pure time derivative:

$$\boldsymbol{B}\cdot(\boldsymbol{\nabla}\times\boldsymbol{E})=\boldsymbol{B}\cdot\left(-\frac{\partial\boldsymbol{B}}{\partial t}\right)=-\frac{\partial}{\partial t}\left(\frac{\boldsymbol{B}\cdot\boldsymbol{B}}{2}\right).$$

Now we have exactly what we want. Our energy equation reads

$$\boldsymbol{E}\cdot\boldsymbol{j}=\boldsymbol{\nabla}\cdot(\epsilon_0c^2\boldsymbol{B}\times\boldsymbol{E})-\frac{\partial}{\partial t}\left(\frac{\epsilon_0c^2}{2}\boldsymbol{B}\cdot\boldsymbol{B}+\frac{\epsilon_0}{2}\boldsymbol{E}\cdot\boldsymbol{E}\right),\tag{27.13}$$

which is exactly like Eq. (27.6), if we make the *definitions*

$$u=\frac{\epsilon_0}{2}\boldsymbol{E}\cdot\boldsymbol{E}+\frac{\epsilon_0c^2}{2}\boldsymbol{B}\cdot\boldsymbol{B}\tag{27.14}$$

and

$$\boldsymbol{S}=\epsilon_0c^2\boldsymbol{E}\times\boldsymbol{B}.\tag{27.15}$$

(Reversing the cross product makes the signs come out right.)

Our program was successful. We have an expression for the energy density that is the sum of an "electric" energy density and a "magnetic" energy density, whose forms are just like the ones we found in statics *when we worked out the energy in terms of the fields*. Also, we have found a formula for the energy flow vector of the electromagnetic field. This new vector, $\boldsymbol{S}=\epsilon_0c^2\boldsymbol{E}\times\boldsymbol{B}$, is called "Poynting's vector," after its discoverer. It tells us the rate at which the field energy moves around in space. The energy which flows through a small area da per second is $\boldsymbol{S}\cdot\boldsymbol{n}\,da$, where $\boldsymbol{n}$ is the unit vector perpendicular to da. (Now that we have our formulas for u and $\boldsymbol{S}$, you can forget the derivations if you want.)

27–4 The ambiguity of the field energy

Before we take up some applications of the Poynting formulas [Eqs. (27.14) and (27.15)], we would like to say that we have not really "proved" them. All we did was to find a *possible* "u" and a *possible* "$\boldsymbol{S}$." How do we know that by juggling the terms around some more we couldn't find another formula for "u" and another formula for "$\boldsymbol{S}$"? The new $\boldsymbol{S}$ and the new u would be different, but they would still satisfy Eq. (27.6). It's possible. It can be done, but the forms that have been found always involve various *derivatives* of the field (and always with second-order terms like a second derivative or the square of a first derivative). There are, in fact, an infinite number of different possibilities for u and $\boldsymbol{S}$, and so far no one has thought of an experimental way to tell which one is right! People have guessed that the simplest one is probably the correct one, but we must say that we do not know for certain what is the actual location in space of the electromagnetic field energy. So we too will take the easy way out and say that the field energy is given by Eq. (27.14). Then the flow vector $\boldsymbol{S}$ must be given by Eq. (27.15).

It is interesting that there seems to be no unique way to resolve the indefiniteness in the location of the field energy. It is sometimes claimed that this problem can be resolved by using the theory of gravitation in the following argument. In the theory of gravity, all energy is the source of gravitational attraction. Therefore the energy density of electricity must be located properly if we are to know in which direction the gravity force acts. As yet, however, no one has done such a delicate experiment that the precise location of the gravitational influence of electromagnetic fields could be determined. That electromagnetic fields alone can be the source of gravitational force is an idea it is hard to do without. It has, in fact, been observed that light is deflected as it passes near the sun—we could say that the sun pulls the light down toward it. Do you not want to allow that the light pulls equally on the sun? Anyway, everyone always accepts the simple expressions we have found for the location of electromagnetic energy and its flow. And although sometimes the results obtained from using them seem strange, noboby has ever found anything wrong with them—that is, no disagreement with experiment. So we will follow the rest of the world—besides, we believe that it is probably perfectly right.

We should make one further remark about the energy formula. In the first place, the energy per unit volume in the field is very simple: It is the electrostatic energy plus the magnetic energy, *if* we write the electrostatic energy in terms of E^2 and the magnetic energy as B^2. We found two such expressions as *possible* expressions for the energy when we were doing static problems. We also found a number of other formulas for the energy in the electrostatic field, such as $\rho\phi$, which is *equal* to the integral of $\boldsymbol{E}\cdot\boldsymbol{E}$ in the electrostatic case. However, in an electrodynamic field the equality failed, and there was no obvious choice as to which was the right one. Now we know which is the right one. Similarly, we have found the formula for the magnetic energy that is correct in general. The right formula for the energy density of *dynamic* fields is Eq. (27.14).

27–5 Examples of energy flow

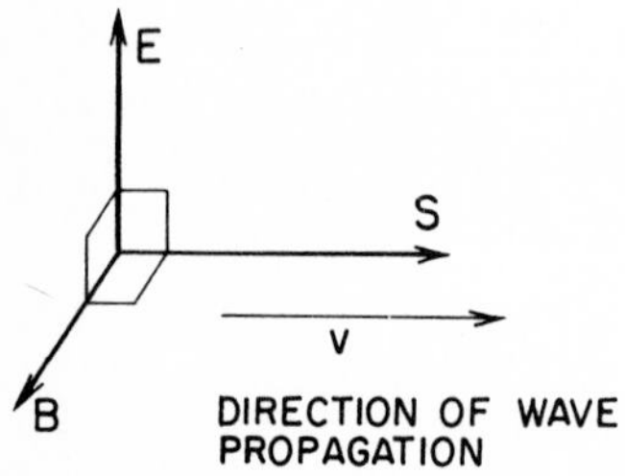

Fig. 27–2. The vectors ***E***, ***B***, and ***S*** for a light wave.

Our formula for the energy flow vector $\boldsymbol{S}$ is something quite new. We want now to see how it works in some special cases and also to see whether it checks out with anything that we knew before. The first example we will take is light. In a light wave we have an $\boldsymbol{E}$ vector and a $\boldsymbol{B}$ vector at right angles to each other and to the direction of the wave propagation. (See Fig. 27–2.) In an electromagnetic wave, the magnitude of $\boldsymbol{B}$ is equal to $1/c$ times the magnitude of $\boldsymbol{E}$, and since they are at right angles,

$$|\boldsymbol{E}\times\boldsymbol{B}| = \frac{E^2}{c}.$$

Therefore, for light, the flow of energy per unit area per second is

$$S = \epsilon_0 c E^2. \tag{27.16}$$

For a light wave in which $E = E_0 \cos \omega(t - x/c)$, the average rate of energy flow per unit area, $\langle S\rangle_{\text{av}}$—which is called the "intensity" of the light—is the mean value of the square of the electric field times $\epsilon_0 c$:

$$\text{Intensity} = \langle S\rangle_{\text{av}} = \epsilon_0 c\langle E^2\rangle_{\text{av}}. \tag{27.17}$$

Believe it or not, we have already derived this result in Section 31–3 of Vol. I, when we were studying light. We can believe that it is right because it also checks against something else. When we have a light beam, there is an energy density in space given by Eq. (27.14). Using $cB = E$ for a light wave, we get that

$$u = \frac{\epsilon_0}{2}E^2 + \frac{\epsilon_0 c^2}{2}\left(\frac{E^2}{c_2}\right) = \epsilon_0 E^2.$$

But $\boldsymbol{E}$ varies in space, so the average energy density is

$$\langle u\rangle_{\text{av}} = \epsilon_0\langle E^2\rangle_{\text{av}}. \tag{27.18}$$

Now the wave travels at the speed c, so we should think that the energy that goes through a square meter in a second is c times the amount of energy in one cubic meter. So we would say that

$$\langle S\rangle_{\text{av}} = \epsilon_0 c\langle E^2\rangle_{\text{av}}.$$

And it's right; it is the same as Eq. (27.17).

Now we take another example. Here is a rather curious one. We look at the energy flow in a capacitor that we are charging slowly. (We don't want frequencies so high that the capacitor is beginning to look like a resonant cavity, but we don't want DC either.) Suppose we use a circular parallel plate capacitor of our usual kind, as shown in Fig. 27–3. There is a nearly uniform electric field inside which is changing with time. At any instant the total electromagnetic energy inside is u times the volume. If the plates have a radius a and a separation h, the total energy between the plates is

$$U = \left(\frac{\epsilon_0}{2}E^2\right)(\pi a^2 h). \tag{27.19}$$

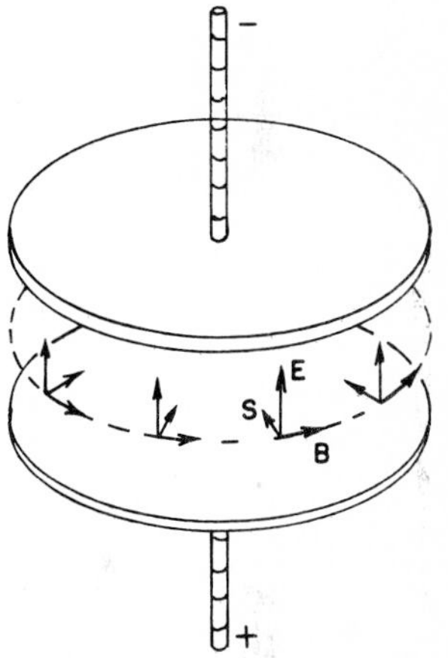

Fig. 27–3. Near a charging capacitor, the Poynting vector $\boldsymbol{S}$ points inward toward the axis.

This energy changes when E changes. When the capacitor is being charged, the volume between the plates is receiving energy at the rate

$$\frac{dU}{dt} = \epsilon_0 \pi a^2 h E\dot{E}. \tag{27.20}$$

So there must be a flow of energy into that volume from somewhere. Of course you know that it must come in on the charging wires—not at all! It can't enter the space between the plates from that direction, because $\boldsymbol{E}$ is perpendicular to the plates; $\boldsymbol{E} \times \boldsymbol{B}$ must be *parallel* to the plates.

You remember, of course, that there is a magnetic field that circles around the axis when the capacitor is charging. We discussed that in Chapter 23. Using the last of Maxwell's equations, we found that the magnetic field at the edge of the capacitor is given by

$$2\pi a c^2 B = \dot{E}\cdot \pi a^2,$$

or

$$B = \frac{a}{2c^2}\dot{E}.$$

Its direction is shown in Fig. 27–3. So there is an energy flow proportional to $\boldsymbol{E} \times \boldsymbol{B}$ that comes in all around the edges, as shown in the figure. The energy isn't actually coming down the wires, but from the space surrounding the capacitor.

Let's check whether or not the total amount of flow through the whole surface between the edges of the plates checks with the rate of change of the energy inside—it had better; we went through all that work proving Eq. (27.15) to make sure, but let's see. The area of the surface is $2\pi ah$, and $\boldsymbol{S} = \epsilon_0 c^2 \boldsymbol{E} \times \boldsymbol{B}$ is in magnitude

$$\epsilon_0 c^2 E\left(\frac{a}{2c^2}\,\dot{E}\right),$$

so the total flux of energy is

$$\pi a^2 h\epsilon_0 E\dot{E}.$$

It does check with Eq. (27.20). But it tells us a peculiar thing: that when we are charging a capacitor, the energy is not coming down the wires; it is coming in through the edges of the gap. That's what this theory says!

How can that be? That's *not* an easy question, but here is one way of thinking about it. Suppose that we had some charges above and below the capacitor and far away. When the charges are far away, there is a weak but enormously spread-out field that surrounds the capacitor. (See Fig. 27–4.) Then, as the charges come together, the field gets stronger nearer to the capacitor. So the field energy which is way out moves toward the capacitor and eventually ends up between the plates.

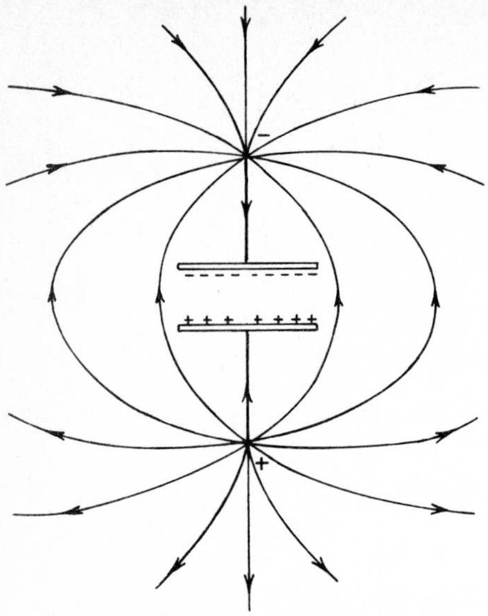

Fig. 27–4. The fields outside a capacitor when it is being charged by bringing two charges from a large distance.

As another example, we ask what happens in a piece of resistance wire when it is carrying a current. Since the wire has resistance, there is an electric field along it, driving the current. Because there is a potential drop along the wire, there is also an electric field just outside the wire, parallel to the surface. (See Fig. 27–5.) There is, in addition, a magnetic field which goes around the wire because of the current. The $\boldsymbol{E}$ and $\boldsymbol{B}$ are at right angles; therefore there is a Poynting vector directed radially inward, as shown in the figure. There is a flow of energy into the wire all around. It is, of course, equal to the energy being lost in the wire in the form of heat. So our "crazy" theory says that the electrons are getting their energy to generate heat because of the energy flowing into the wire from the field outside. Intuition would seem to tell us that the electrons get their energy from being pushed along the wire, so the energy should be flowing down (or up) along the wire. But the theory says that the electrons are really being pushed by an electric field, which has come from some charges very far away, and that the electrons get their energy for generating heat from these fields. The energy somehow flows from the distant charges into a wide area of space and then inward to the wire.

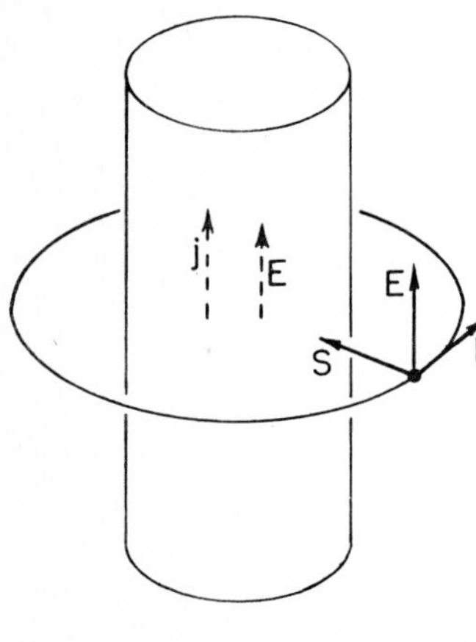

Fig. 27–5 The Poynting vector **S** near a wire carrying a current.

Finally, in order to really convince you that this theory is obviously nuts, we will take one more example—an example in which an electric charge and a magnet are *at rest* near each other—both sitting quite still. Suppose we take the example of a point charge sitting near the center of a bar magnet, as shown in Fig. 27–6. Everything is at rest, so the energy is not changing with time. Also, $\boldsymbol{E}$ and $\boldsymbol{B}$ are quite static. But the Poynting vector says that there is a flow of energy, because there is an $\boldsymbol{E} \times \boldsymbol{B}$ that is not zero. If you look at the energy flow, you find that it just circulates around and around. There isn't any change in the energy anywhere—everything which flows into one volume flows out again. It is like incompressible water flowing around. So there is a circulation of energy in this so-called static condition. How absurd it gets!

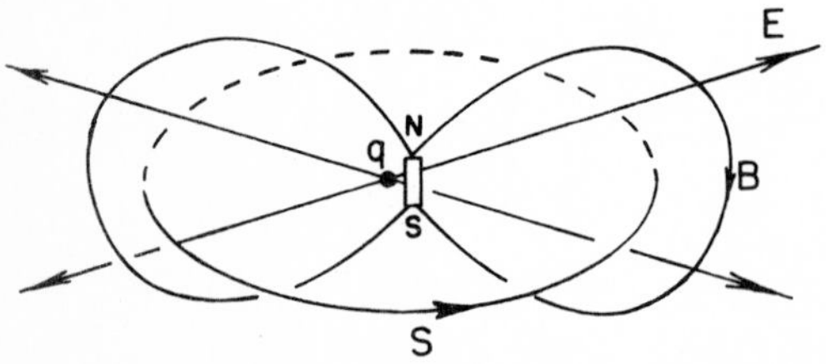

Fig. 27–6. A charge and a magnet produce a Poynting vector that circulates in closed loops.

Perhaps it isn't so terribly puzzling, though, when you remember that what we called a "static" magnet is really a circulating permanent current. In a permanent magnet the electrons are spinning permanently inside. So maybe a circulation of the energy outside isn't so queer after all.

You no doubt begin to get the impression that the Poynting theory at least partially violates your intuition as to where energy is located in an electromagnetic field. You might believe that you must revamp all your intuitions, and, therefore have a lot of things to study here. But it seems really not necessary. You don't need to feel that you will be in great trouble if you forget once in a while that the energy in a wire is flowing into the wire from the outside, rather than along the wire. It seems to be only rarely of value, when using the idea of energy conservation, to notice in detail what path the energy is taking. The circulation of energy around a magnet and a charge seems, in most circumstances, to be quite unimportant. It is not a vital detail, but it is clear that our ordinary intuitions are quite wrong.

27–6 Field momentum

Next we would like to talk about the *momentum* in the electromagnetic field. Just as the field has energy, it will have a certain momentum per unit volume. Let us call that momentum density $\boldsymbol{g}$. Of course, momentum has various possible directions, so that $\boldsymbol{g}$ must be a vector. Let's talk about one component at a time; first, we take the x-component. Since each component of momentum is conserved we should be able to write down a law that looks something like this:

$$-\frac{\partial}{\partial t}\begin{pmatrix}\text{momentum}\\ \text{of matter}\end{pmatrix}_x = \frac{\partial g_x}{\partial t} + \begin{pmatrix}\text{momentum}\\ \text{outflow}\end{pmatrix}_x.$$

The left side is easy. The rate-of-change of the momentum of matter is just the force on it. For a particle, it is $\boldsymbol{F} = q(\boldsymbol{E} + \boldsymbol{v} \times \boldsymbol{B})$; for a distribution of charges, the force per unit volume is $(\rho\boldsymbol{E} + \boldsymbol{j} \times \boldsymbol{B})$. The "momentum outflow" term, however, is strange. It cannot be the divergence of a vector because it is not a scalar; it is, rather, an x-component of some vector. Anyway, it should probably look something like

$$\frac{\partial a}{\partial x} + \frac{\partial b}{\partial y} + \frac{\partial c}{\partial z},$$

because the x-momentum could be flowing in any one of the three directions. In any case, whatever a, b, and c are, the combination is supposed to equal the outflow of the x-momentum.

Now the game would be to write $\rho\boldsymbol{E} + \boldsymbol{j} \times \boldsymbol{B}$ in terms only of $\boldsymbol{E}$ and $\boldsymbol{B}$—eliminating ρ an $\boldsymbol{j}$ by using Maxwell's equations—and then to juggle terms and make substitutions to get it into a form that looks like

$$\frac{\partial g_x}{\partial t} + \frac{\partial a}{\partial x} + \frac{\partial b}{\partial y} + \frac{\partial c}{\partial z}.$$

Then, by identifying terms, we would have expressions for g_x, a, b, and c. It's a lot of work, and we are not going to do it. Instead, we are only going to find an expression for $\boldsymbol{g}$, the momentum density—and by a different route.

There is an important theorem in mechanics which is this: whenever there is a flow of energy in any circumstance at all (field energy or any other kind of energy), the energy flowing through a unit area per unit time, when multiplied by $1/c^2$, is equal to the momentum per unit volume in the space. In the special case of electrodynamics, this theorem gives the result that $\boldsymbol{g}$ is $1/c^2$ times the Poynting vector:

$$\boldsymbol{g} = \frac{1}{c^2}\boldsymbol{S}. \tag{27.21}$$

So the Poynting vector gives not only energy flow but, if you divide by c^2, also the momentum density. The same result would come out of the other analysis we suggested, but it is more interesting to notice this more general result. We will now give a number of interesting examples and arguments to convince you that the general theorem is true.

First example: Suppose that we have a lot of particles in a box—let's say N per cubic meter—and that they are moving along with some velocity $\boldsymbol{v}$. Now let's consider an imaginary plane surface perpendicular to $\boldsymbol{v}$. The energy flow through a unit area of this surface per second is equal to Nv, the number which flow through the surface per second, times the energy carried by each one. The energy in each particle is $m_0c^2/\sqrt{1 - v^2/c^2}$. So the energy flow per second is

$$Nv\,\frac{m_0c^2}{\sqrt{1 - v^2/c^2}}.$$

But the momentum of each particle is $m_0v/\sqrt{1 - v^2/c^2}$, so the *density* of momentum is

$$N\,\frac{m_0v}{\sqrt{1 - v^2/c^2}},$$

which is just $1/c^2$ times the energy flow—as the theorem says. So the theorem is true for a bunch of particles.

It is also true for light. When we studied light in Volume I, we saw that when the energy is absorbed from a light beam, a certain amount of momentum is delivered to the absorber. We have, in fact, shown in Chapter 36 of Vol. I that the momentum is $1/c$ times the energy absorbed [Eq. (36.24) of Vol. I]. If we let U_0 be the energy arriving at a unit area per second, then the momentum arriving at a unit area per second is U_0/c. But the momentum is travelling at the speed c, so its *density* in front of the absorber must be U_0/c^2. So again the theorem is right.

Finally we will give an argument due to Einstein which demonstrates the same thing once more. Suppose that we have a railroad car on wheels (assumed frictionless) with a certain big mass M. At one end there is a device which will shoot out some particles or light (or anything, it doesn't make any difference what it is), which are then stopped at the opposite end of the car. There was some energy originally at one end—say the energy U indicated in Fig. 27–7(a)—and then later it is at the opposite end, as shown in Fig. 27–7(c). The energy U has been displaced the distance L, the length of the car. Now the energy U has the mass U/c^2, so if the car stayed still, the center of gravity of the car would be moved. Einstein didn't like the idea that the center of gravity of an object could be moved by fooling around only on the inside, so he assumed that it is impossible to move the center of gravity by doing anything inside. But if that is the case, when we moved the energy U from one end to the other, the whole car must have recoiled some distance x, as shown in part (c) of the figure. You can see, in fact, that the total mass of the car, times x, must equal the mass of the energy moved, U/c^2 times L (assuming that U/c^2 is much less than M):

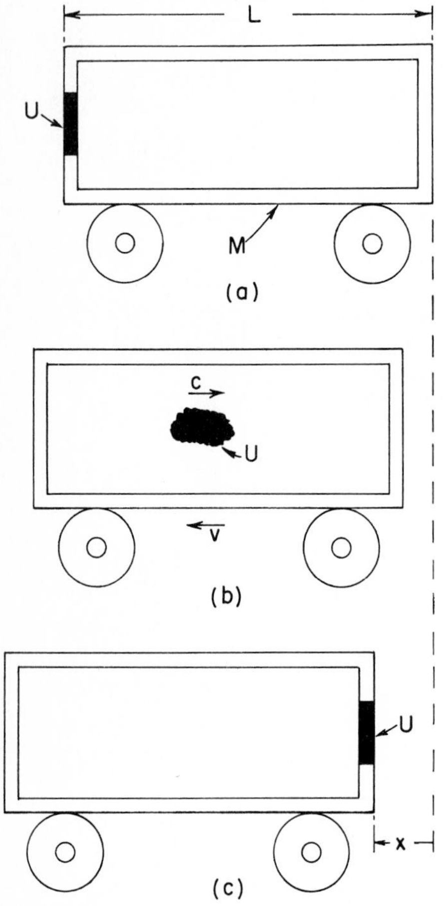

Fig. 27–7. The energy U in motion at the speed c carries the momentum U/c.

$$Mx = \frac{U}{c^2}\,L. \tag{27.22}$$

Let's now look at the special case of the energy being carried by a light flash. (The argument would work as well for particles, but we will follow Einstein, who was interested in the problem of light.) What causes the car to be moved? Einstein argued as follows: When the light is emitted there must be a recoil, some unknown recoil with momentum p. It is this recoil which makes the car roll backward. The recoil velocity v of the car will be this momentum divided by the mass of the car:

$$v = \frac{p}{M}\cdot$$

The car moves with this velocity until the light energy U gets to the opposite end. Then, when it hits, it gives back its momentum and stops the car. If x is small, then the time the car moves is nearly equal to L/c; so we have that

$$x = vt = v\,\frac{L}{c} = \frac{p}{M}\,\frac{L}{c}\cdot$$

Putting this x in Eq. (27.22), we get that

$$p = \frac{U}{c}\cdot$$

Again we have the relation of energy and momentum for light. Dividing by c to get the momentum density $g = p/c$, we get once more that

$$g = \frac{U}{c^2}\cdot \tag{27.23}$$

You may well wonder: What is so important about the center-of-gravity theorem? Maybe *it* is wrong. Perhaps, but then we would also lose the conservation of angular momentum. Suppose that our boxcar is moving along a track at some speed v and that we shoot some light energy from the *top* to the *bottom* of the car—say, from A to B in Fig. 27–8. Now we look at the angular momentum of the system about the point P. Before the energy U leaves A, it has the mass

$m = U^2/c$ and the velocity v, so it has the angular momentum mvr_a. When it arrives at B, it has the same mass and, if the *linear* momentum of the whole boxcar is not to change, it must still have the velocity v. It's angular momentum about P is then mvr_B. The angular momentum will be changed *unless* the right recoil momentum was given to the car when the light was emitted—that is, unless the light carries the momentum U/c. It turns out that the angular momentum conservation and the theorem of center-of-gravity are closely related in the relativity theory. So the conservation of angular momentum would also be destroyed if our theorem were not true. At any rate, it does turn out to be a true general law, and in the case of electrodynamics we can use it to get the momentum in the field.

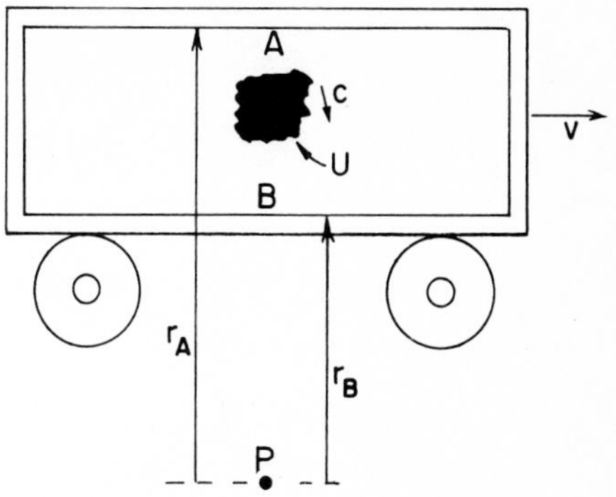

Fig. 27–8. The energy U must carry the momentum U/c if the angular momentum about P is to be conserved.

We will mention two further examples of momentum in the electromagnetic field. We pointed out in Section 26–2 the failure of the law of action and reaction when two charged particles were moving on orthogonal trajectories. The forces on the two particles don't balance out, so the action and reaction are not equal; therefore the net momentum of the matter must be changing. It is not conserved. But the momentum in the field is also changing in such a situation. If you work out the amount of momentum given by the Poynting vector, it is not constant. However, the change of the particle momenta is just made up by the field momentum, so the total momentum of particles plus field is conserved.

Finally, another example is the situation with the magnet and the charge, shown in Fig. 27–6. We were unhappy to find that energy was flowing around in circles, but now, since we know that energy flow and momentum are proportional, we know also that there is momentum circulating in the space. But a *circulating* momentum means that there is *angular* momentum. So there is *angular* momentum in the field. Do you remember the paradox we described in Section 17–4 about a solenoid and some charges mounted on a disc? It seemed that when the current turned off, the whole disc should start to turn. The puzzle was: Where did the angular momentum come from? The answer is that if you have a magnetic field and some charges, there will be some angular momentum in the field. It must have been put there when the field was built up. When the field is turned off, the angular momentum is given back. So the disc in the paradox *would* start rotating. This mystic circulating flow of energy, which at first seemed so ridiculous, is absolutely necessary. There is really a momentum flow. It is needed to maintain the conservation of angular momentum in the whole world.

28

Electromagnetic Mass

28–1 The field energy of a point charge

In bringing together relativity and Maxwell's equations, we have finished our main work on the theory of electromagnetism. There are, of course, some details we have skipped over and one large area that we will be concerned with in the future—the interaction of electromagnetic fields with matter. But we want to stop for a moment to show you that this tremendous edifice, which is such a beautiful success in explaining so many phenomena, ultimately falls on its face. When you follow any of our physics too far, you find that it always gets into some kind of trouble. Now we want to discuss a serious trouble—the failure of the classical electromagnetic theory. You can appreciate that there is a failure of all classical physics because of the quantum-mechanical effects. Classical mechanics is a mathematically consistent theory; it just doesn't agree with experience. It is interesting, though, that the classical theory of electromagnetism is an unsatisfactory theory all by itself. There are difficulties associated with the *ideas* of Maxwell's theory which are not solved by and not directly associated with quantum mechanics. You may say, "Perhaps there's no use worrying about these difficulties. Since the quantum mechanics is going to change the laws of electrodynamics, we should wait to see what difficulties there are after the modification." However, when electromagnetism is joined to quantum mechanics, the difficulties remain. So it will not be a waste of our time now to look at what these difficulties are. Also, they are of great historical importance. Furthermore, you may get some feeling of accomplishment from being able to go far enough with the theory to see everything—including all of its troubles.

The difficulty we speak of is associated with the concepts of electromagnetic momentum and energy, when applied to the electron or any charged particle. The concepts of simple charged particles and the electromagnetic field are in some way inconsistent. To describe the difficulty, we begin by doing some exercises with our energy and momentum concepts.

First, we compute the energy of a charged particle. Suppose we take a simple model of an electron in which all of its charge q is uniformly distributed on the surface of a sphere of radius a, which we may take to be zero for the special case of a point charge. Now let's calculate the energy in the electromagnetic field. If the charge is standing still, there is no magnetic field, and the energy per unit volume is proportional to the square of the electric field. The magnitude of the electric field is $q/4\pi\epsilon_0 r^2$, and the energy density is

$$u = \frac{\epsilon_0}{2}\,E^2 = \frac{q^2}{32\pi^2\epsilon_0 r^4}\,.$$

To get the total energy, we must integrate this density over all space. Using the volume element $4\pi r^2\,dr$, the total energy, which we will call U_{elec}, is

$$U_{\text{elec}} = \int \frac{q^2}{8\pi\epsilon_0 r^2}\,dr.$$

This is readily integrated. The lower limit is a, and the upper limit is ∞, so

$$U_{\text{elec}} = \frac{1}{2}\,\frac{q^2}{4\pi\epsilon_0}\,\frac{1}{a}\,. \tag{28.1}$$

If we use the electronic charge q_e for q and the symbol e^2 for $q_e^2/4\pi\epsilon_0$, then

$$U_{\text{elec}} = \frac{1}{2}\frac{e^2}{a}. \tag{28.2}$$

It is all fine until we set a equal to zero for a point charge—there's the great difficulty. Because the energy of the field varies inversely as the fourth power of the distance from the center, its volume integral is infinite. There is an infinite amount of energy in the field surrounding a point charge.

What's wrong with an infinite energy? If the energy can't get out, but must stay there forever, is there any real difficulty with an infinite energy? Of course, a quantity that comes out infinite may be annoying, but what really matters is only whether there are any *observable* physical effects. To answer that question, we must turn to something else besides the energy. Suppose we ask how the energy *changes* when we *move* the charge. Then, if the *changes* are infinite, we will be in trouble.

28–2 The field momentum of a moving charge

Suppose an electron is moving at a uniform velocity through space, assuming for a moment that the velocity is low compared with the speed of light. Associated with this moving electron there is a momentum—even if the electron had no mass before it was charged—because of the momentum in the electromagnetic field. We can show that the field momentum is in the direction of the velocity $\boldsymbol{v}$ of the charge and is, for small velocities, proportional to v. For a point P at the distance r from the center of the charge and at the angle θ with respect to the line of motion (see Fig. 28–1) the electric field is radial and, as we have seen, the magnetic field is $\boldsymbol{v}\times\boldsymbol{E}/c^2$. The momentum density, Eq. (27.21), is

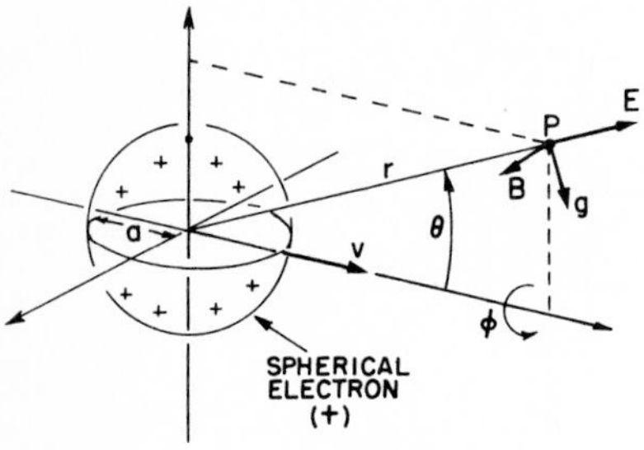

Fig. 28–1. The fields ***E*** and ***B*** and the momentum density ***g*** for a positive electron. For a negative electron, ***E*** and ***B*** are reversed but ***g*** is not.

$$\boldsymbol{g} = \epsilon_0\boldsymbol{E}\times\boldsymbol{B}.$$

It is directed obliquely toward the line of motion, as shown in the figure, and has the magnitude

$$g = \frac{\epsilon_0 v}{c^2}E^2\sin\theta.$$

The fields are symmetric about the line of motion, so when we integrate over space, the transverse components will sum to zero, giving a resultant momentum parallel to $\boldsymbol{v}$. The component of $\boldsymbol{g}$ in this direction is $g\sin\theta$, which we must integrate over all space. We take as our volume element a ring with its plane perpendicular to $\boldsymbol{v}$, as shown in Fig. 28–2. Its volume is $2\pi r^2\sin\theta\,d\theta\,dr$. The total momentum is then

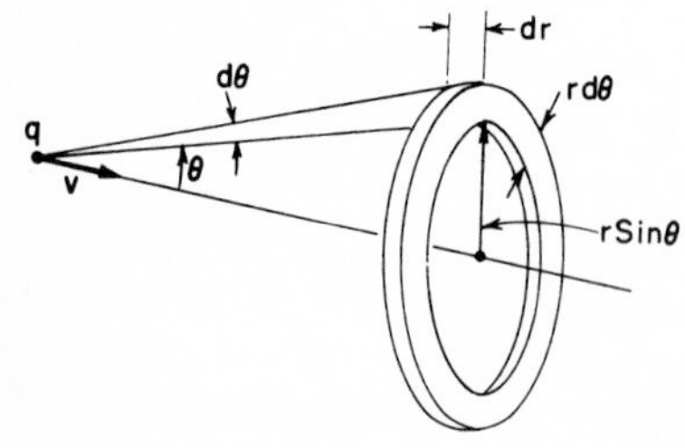

Fig. 28–2. The volume element $2\pi r^2\sin\theta\,d\theta\,dr$ used for calculating the field momentum.

$$p = \int\frac{\epsilon_0 v}{c^2}E^2\sin^2\theta\,2\pi r^2\sin\theta\,d\theta\,dr.$$

Since E is independent of θ (for $v \ll c$), we can immediately integrate over θ; the integral is

$$\int\sin^3\theta\,d\theta = -\int(1-\cos^2\theta)\,d(\cos\theta) = -\cos\theta + \frac{\cos^3\theta}{3}.$$

The limits of θ are 0 and π, so the θ-integral gives merely a factor of 4/3, and

$$p = \frac{8\pi}{3}\frac{\epsilon_0 v}{c^2}\int E^2 r^2\,dr.$$

The integral (for $v \ll c$) is the one we have just evaluated to find the energy; it is $q^2/16\pi^2\epsilon_0^2 a$, and

$$p = \frac{2}{3}\frac{q^2}{4\pi\epsilon_0}\frac{v}{ac^2},$$

or

$$p = \frac{2}{3}\frac{e^2}{ac^2}v. \tag{28.3}$$

The momentum in the field—the electromagnetic momentum—is proportional to $\boldsymbol{v}$. It is just what we should have for a particle with the mass equal to the coefficient of $\boldsymbol{v}$. We can, therefore, call this coefficient the *electromagnetic mass*, m_{elec}, and write it as

$$m_{\text{elec}} = \frac{2}{3}\frac{e^2}{ac^2}. \tag{28.4}$$

28–3 Electromagnetic mass

Where does the mass come from? In our laws of mechanics we have supposed that every object "carries" a thing we call the mass—which also means that it "carries" a momentum proportional to its velocity. Now we discover that it is understandable that a charged particle carries a momentum proportional to its velocity. It might, in fact, be that the mass is just the effect of electrodynamics. The origin of mass has until now been unexplained. We have at last in the theory of electrodynamics a grand opportunity to understand something that we never understood before. It comes out of the blue—or rather, from Maxwell and Poynting—that any charged particle will have a momentum proportional to its velocity just from electromagnetic influences.

Let's be conservative and say, for a moment, that there are two kinds of mass—that the total momentum of an object could be the sum of a mechanical momentum and the electromagnetic momentum. The mechanical momentum is the "mechanical" mass, m_{mech}, times $\boldsymbol{v}$. In experiments where we measure the mass of a particle by seeing how much momentum it has, or how it swings around in an orbit, we are measuring the total mass. We say generally that the momentum is the total mass ($m_{\text{mech}} + m_{\text{elec}}$) times the velocity. So the observed mass can consist of two pieces (or possibly more if we include other fields): a mechanical piece plus an electromagnetic piece. We know that there is definitely an electromagnetic piece, and we have a formula for it. And there is the thrilling possibility that the mechanical piece is not there at all—that the mass is all electromagnetic.

Let's see what size the electron must have if there is to be no mechanical mass. We can find out by setting the electromagnetic mass of Eq. (28.4) equal to the observed mass m_e of an electron. We find

$$a = \frac{2}{3}\frac{e^2}{m_ec^2}. \tag{28.5}$$

The quantity

$$r_0 = \frac{e^2}{m_ec^2} \tag{28.6}$$

is called the "classical electron radius"; it has the numerical value 2.82×10^{-13} cm, about one one-hundred-thousandth of the diameter of an atom.

Why is r_0 called the electron radius, rather than our a? Because we could equally well do the same calculation with other assumed distributions of charges—the charge might be spread uniformly through the volume of a sphere or it might be smeared out like a fuzzy ball. For any particular assumption the factor 2/3 would change to some other fraction. For instance, for a charge uniformly distributed throughout the volume of a sphere, the 2/3 gets replaced by 4/5. Rather than to argue over which distribution is correct, it was decided to define r_0 as the "nominal" radius. Then different theories could supply their pet coefficients.

Let's pursue our electromagnetic theory of mass. Our calculation was for $v \ll c$; what happens if we go to high velocities? Early attempts led to a certain amount of confusion, but Lorentz realized that the charged sphere would contract into a ellipsoid at high velocities and that the fields would change in accordance with the formulas (26.6) and (26.7) we derived for the relativistic case in Chapter 26. If you carry through the integrals for $\boldsymbol{p}$ in that case, you find that for an arbitrary velocity $\boldsymbol{v}$, the momentum is altered by the factor $1/\sqrt{1 - v^2/c^2}$:

$$\boldsymbol{p} = \frac{2}{3}\frac{e^2}{ac^2}\frac{\boldsymbol{v}}{\sqrt{1 - v^2/c^2}}. \tag{28.7}$$

In other words, the electromagnetic mass rises with velocity inversely as $\sqrt{1 - v^2/c^2}$—a discovery that was made before the theory of relativity.

Early experiments were proposed to measure the changes with velocity in the observed mass of a particle in order to determine how much of the mass was mechanical and how much was electrical. It was believed at the time that the electrical part *would* vary with velocity, whereas the mechanical part would *not*. But while the experiments were being done, the theorists were also at work. Soon the theory of relativity was developed, which proposed that no matter what the origin of the mass, it *all* should vary as $m_0/\sqrt{1 - v^2/c^2}$. Equation (28.7) was the beginning of the theory that mass depended on velocity.

Let's now go back to our calculation of the energy in the field, which led to Eq. (28.2). According to the theory of relativity, the energy U will have the mass U/c^2; Eq. (28.2) then says that the field of the electron should have the mass

$$m'_{\text{elec}} = \frac{U_{\text{elec}}}{c^2} = \frac{1}{2}\frac{e^2}{ac^2}, \tag{28.8}$$

which is not the same as the electromagnetic mass, m_{elec}, of Eq. (28.4). In fact, if we just combine Eqs. (28.2) and (28.4), we would write

$$U_{\text{elec}} = \frac{3}{4} m_{\text{elec}} c^2.$$

This formula was discovered before relativity, and when Einstein and others began to realize that it must always be that $U = mc^2$, there was great confusion.

28–4 The force of an electron on itself

The discrepancy between the two formulas for the electromagnetic mass is especially annoying, because we have carefully proved that the theory of electrodynamics is consistent with the principle of relativity. Yet the theory of relativity implies without question that the momentum must be the same as the energy times v/c^2. So we are in some kind of trouble; we must have made a mistake. We did not make an algebraic mistake in our calculations, but we have left something out.

In deriving our equations for energy and momentum, we assumed the conservation laws. We assumed that *all* forces were taken into account and that any work done and any momentum carried by other "nonelectrical" machinery was included. Now if we have a sphere of charge, the electrical forces are all repulsive and an electron would tend to fly apart. Because the system has unbalanced forces, we can get all kinds of errors in the laws relating energy and momentum. To get a *consistent* picture, we must imagine that something holds the electron together. The charges must be *held* to the sphere by some kind of rubber bands—something that keeps the charges from flying off. It was first pointed out by Poincaré that the rubber bands—or whatever it is that holds the electron together—must be included in the energy and momentum calculations. For this reason the extra nonelectrical forces are also known by the more elegant name "the Poincaré stresses." If the extra forces are included in the calculations, the masses obtained in two ways are changed (in a way that depends on the detailed assumptions). And the results are consistent with relativity; i.e., the mass that comes out from the momentum calculation is the same as the one that comes from the energy calculation. However, both of them contain *two* contributions: an electromagnetic mass and contribution from the Poincaré stresses. Only when the two are added together do we get a consistent theory.

It is therefore impossible to get all the mass to be electromagnetic in the way we hoped. It is not a legal theory if we have nothing but electrodynamics. Something else has to be added. Whatever you call them—"rubber bands," or "Poincaré stresses," or something else—there have to be other forces in nature to make a consistent theory of this kind.

Clearly, as soon as we have to put forces on the inside of the electron, the beauty of the whole idea begins to disappear. Things get very complicated. You would want to ask: How strong are the stresses? How does the electron shake? Does it oscillate? What are all its internal properties? And so on. It might be possible that an electron does have some complicated internal properties. If we made a theory of the electron along these lines, it would predict odd properties, like modes of oscillation, which haven't apparently been observed. We say "apparently" because we observe a lot of things in nature that still do not make sense. We may someday find out that one of the things we don't understand today (for example, the muon) can, in fact, be explained as an oscillation of the Poincaré stresses. It doesn't seem likely, but no one can say for sure. There are so many things about fundamental particles that we still don't understand. Anyway, the complex structure implied by this theory is undesirable, and the attempt to explain all mass in terms of electromagnetism—at least in the way we have described—has led to a blind alley.

We would like to think a little more about why we say we have a mass when the momentum in the field is proportional to the velocity. Easy! The mass is the coefficient between momentum and velocity. But we can look at the mass in another way: a particle has mass if you have to exert a force in order to accelerate it. So it may help our understanding if we look a little more closely at where the forces come from. How do we know that there has to be a force? Because we have proved the law of the conservation of momentum for the fields. If we have a charged particle and push on it for awhile, there will be some momentum in the electromagnetic field. Momentum must have been poured into the field somehow. Therefore there must have been a force pushing on the electron in order to get it going—a force in addition to that required by its mechanical inertia, a force due to its electromagnetic interaction. And there must be a corresponding force back on the "pusher." But where does that force come from?

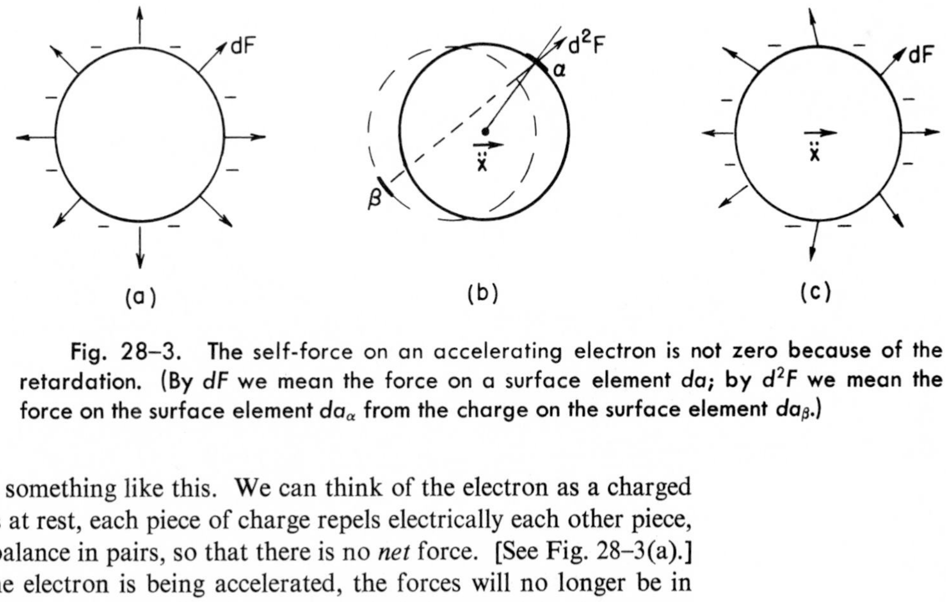

Fig. 28–3. The self-force on an accelerating electron is not zero because of the retardation. (By dF we mean the force on a surface element da; by d^2F we mean the force on the surface element da_α from the charge on the surface element da_β.)

The picture is something like this. We can think of the electron as a charged sphere. When it is at rest, each piece of charge repels electrically each other piece, but the forces all balance in pairs, so that there is no *net* force. [See Fig. 28–3(a).] However, when the electron is being accelerated, the forces will no longer be in balance because of the fact that the electromagnetic influences take time to go from one piece to another. For instance, the force on the piece α in Fig. 28–3(b) from a piece β on the opposite side depends on the position of β at an earlier time, as shown. Both the magnitude and direction of the force depend on the motion of the charge. If the charge is accelerating, the forces on various parts of the electron might be as shown in Fig. 28–3(c). When all these forces are added up, they don't cancel out. They would cancel for a uniform velocity, even though it looks at first glance as though the retardation would give an unbalanced force even for a uniform velocity. But it turns out that there is no net force unless the electron is being accelerated. With acceleration, if we look at the forces between the various parts of the electron, action and reaction are not exactly equal, and the electron exerts a force *on itself* that tries to hold back the acceleration. It holds itself back by its own bootstraps.

It is possible, but difficult, to calculate this self-reaction force; however, we don't want to go into such an elaborate calculation here. We will tell you what the result is for the special case of relatively uncomplicated motion in one dimension, say x. Then, the self-force can be written in a series. The first term in the series depends on the acceleration $\ddot{x}$, the next term is proportional to $\dddot{x}$, and so on.* The result is

$$F = \alpha \frac{e^2}{ac^2}\ddot{x} - \frac{2}{3}\frac{e^2}{c^3}\dddot{x} + \gamma \frac{e^2 a}{c^4}\ddddot{x} + \cdots, \tag{28.9}$$

where α and γ are numerical coefficients of the order of 1. The coefficient α of the $\ddot{x}$ term depends on what charge distribution is assumed; if the charge is distributed uniformly on a sphere, then $\alpha = 2/3$. So there is a term, proportional to the acceleration, which varies inversely as the radius a of the electron and agrees exactly with the value we got in Eq. (28.4) for m_{elec}. If the charge distribution is chosen to be different, so that α is changed, the fraction 2/3 in Eq. (28.4) would be changed in the same way. The term in $\dddot{x}$ is *independent* of the assumed radius a, and also of the assumed distribution of the charge; its coefficient is *always* 2/3. The next term is proportional to the radius a, and its coefficient γ depends on the charge distribution. You will notice that if we let the electron radius a go to zero, the last term (and all higher terms) will go to zero; the second term remains constant, but the first term—the electromagnetic mass—goes to infinity. And we can see that the infinity arises because of the force of one part of the electron on another—because we have allowed what is perhaps a silly thing, the possibility of the "point" electron acting on itself.

28–5 Attempts to modify the Maxwell theory

We would like now to discuss how it might be possible to modify Maxwell's theory of electrodynamics so that the idea of an electron as a simple point charge could be maintained. Many attempts have been made, and some of the theories were even able to arrange things so that all the electron mass was electromagnetic. But all of these theories have died. It is still interesting to discuss some of the possibilities that have been suggested—to see the struggles of the human mind.

We started out our theory of electricity by talking about the interaction of one charge with another. Then we made up a theory of these interacting charges and ended up with a field theory. We believe it so much that we allow it to tell us about the force of one part of an electron on another. Perhaps the entire difficulty is that electrons do not act on themselves; perhaps we are making too great an extrapolation from the interaction of separate electrons to the idea that an electron interacts with itself. Therefore some theories have been proposed in which the possibility that an electron acts on itself is ruled out. Then there is no longer the infinity due to the self-action. Also, there is no longer any electromagnetic mass associated with the particle; all the mass is back to being mechanical, but there are new difficulties in the theory.

We must say immediately that such theories require a modification of the idea of the electromagnetic field. You remember we said at the start that the force on a particle at any point was determined by just two quantities—$\boldsymbol{E}$ and $\boldsymbol{B}$. If we abandon the "self-force" this can no longer be true, because if there is an electron in a certain place, the force isn't given by the total $\boldsymbol{E}$ and $\boldsymbol{B}$, but by only those parts due to *other* charges. So we have to keep track always of how much of $\boldsymbol{E}$ and $\boldsymbol{B}$ is due to the charge on which you are calculating the force and how much is due to the other charges. This makes the theory much more elaborate, but it gets rid of the difficulty of the infinity.

---

* We are using the notation: $\dot{x} = dx/dt$, $\ddot{x} = d^2x/dt^2$, $\dddot{x} = d^3x/dt^3$, etc.

So we can, *if we want to*, say that there is no such thing as the electron acting upon itself, and throw away the whole set of forces in Eq. (28.9). However, we have then thrown away the baby with the bath! Because the second term in Eq. (28.9), the term in $\dddot{x}$, is needed. That force does something very definite. If you throw it away, you're in trouble again. When we accelerate a charge, it radiates electromagnetic waves, so it loses energy. Therefore, to accelerate a charge, we must require more force than is required to accelerate a neutral object of the same mass; otherwise energy wouldn't be conserved. The rate at which we do work on an accelerating charge must be equal to the rate of loss of energy per second by radiation. We have talked about this effect before—it is called the radiation resistance. We still have to answer the question: Where does the extra force, against which we must do this work, come from? When a big antenna is radiating, the forces come from the influence of one part of the antenna current on another. For a single accelerating electron radiating into otherwise empty space, there would seem to be only one place the force could come from—the action of one part of the electron on another part.

We found back in Chapter 32 of Vol. I that an oscillating charge radiates energy at the rate

$$\frac{dW}{dt} = \frac{2}{3}\frac{e^2(\ddot{x})^2}{c^3}. \tag{28.10}$$

Let's see what we get for the rate of doing work *on* an electron against the bootstrap force of Eq. (28.9). The rate of work is the force times the velocity, or $F\dot{x}$:

$$\frac{dW}{dt} = \alpha\frac{e^2}{ac^2}\ddot{x}\,\dot{x} - \frac{2}{3}\frac{e^2}{c^3}\dddot{x}\,\dot{x} + \cdots \tag{28.11}$$

The first term is proportional to $d\dot{x}^2/dt$, and therefore just corresponds to the rate of change of the kinetic energy $\frac{1}{2}mv^2$ associated with the electromagnetic mass. The second term should correspond to the radiated power in Eq. (28.10). But it is different. The discrepancy comes from the fact that the term in Eq. (28.11) is generally true, whereas Eq. (28.10) is right only for an *oscillating* charge. We can show that the two are equivalent if the motion of the charge is periodic. To do that, we rewrite the second term of Eq. (28.11) as

$$-\frac{2}{3}\frac{e^2}{c^3}\frac{d}{dt}(\dot{x}\,\ddot{x}) + \frac{2}{3}\frac{e^2}{c^3}(\ddot{x})^2,$$

which is just an algebraic transformation. If the motion of the electron is periodic, the quantity $\dot{x}\,\ddot{x}$ returns periodically to the same value, so that if we take the *average* of its time derivative, we get zero. The second term, however, is always positive (it's a square), so its average is also positive. This term gives the net work done and is just equal to Eq. (28.10).

The term in $\dddot{x}$ of the bootstrap force is required in order to have energy conservation in radiating systems, and we can't throw it away. It was, in fact, one of the triumphs of Lorentz to show that there is such a force and that it comes from the action of the electron on itself. We must believe in the idea of the action of the electron on itself, and we *need* the term in $\dddot{x}$. The problem is how we can get that term without getting the first term in Eq. (28.9), which gives all the trouble. We don't know how. You see that the classical electron theory has pushed itself into a tight corner.

There have been several other attempts to modify the laws in order to straighten the thing out. One way, proposed by Born and Infeld, is to change the Maxwell equations in a complicated way so that they are no longer linear. Then the electromagnetic energy and momentum can be made to come out finite. But the laws they suggest predict phenomena which have never been observed. Their theory also suffers from another difficulty we will come to later, which is common to all the attempts to avoid the troubles we have described.

The following peculiar possibility was suggested by Dirac. He said: Let's admit that an electron acts on itself through the *second* term in Eq. (28.9) but not through the first. He then had an ingenious idea for getting rid of one but not the other. Look, he said, we made a special assumption when we took only the *retarded* wave solutions of Maxwell's equations; if we were to take the *advanced* waves instead, we would get something different. The formula for the self-force would be

$$F = \alpha \frac{e^2}{ac^2} \ddot{x} + \frac{2}{3} \frac{e^2}{c^3} \dddot{x} + \gamma \frac{e^2 a}{c^4} \ddddot{x}. \tag{28.12}$$

This equation is just like Eq. (28.9) except for the sign of the second term—and some higher terms—of the series. [Changing from retarded to advanced waves is just changing the *sign* of the delay which, it is not hard to see, is equivalent to changing the sign of t everywhere. The only effect on Eq. (28.9) is to change the sign of all the odd time derivatives.] So, Dirac said, let's make the new rule that an electron acts on itself by one-half the *difference* of the retarded and advanced fields which it produces. The difference of Eqs. (28.9) and (28.12), divided by two, is then

$$F = -\frac{2}{3} \frac{e^2}{c^3} \dddot{x} + \text{higher terms.}$$

In all the higher terms, the radius a appears to some positive power in the numerator. Therefore, when we go to the limit of a point charge, we get only the one term—just what is needed. In this way, Dirac got the radiation resistance force and none of the inertial forces. There is no electromagnetic mass, and the classical theory is saved—but at the expense of an arbitrary assumption about the self-force.

The arbitrariness of the extra assumption of Dirac was removed, to some extent at least, by Wheeler and Feynman, who proposed a still stranger theory. They suggest that point charges interact *only* with other charges, but that the interaction is half through the advanced and half through the retarded waves. It turns out, most surprisingly, that in most situations you won't see any effects of the advanced waves, but they do have the effect of producing just the radiation reaction force. The radiation resistance is *not* due to the electron acting on itself, but from the following peculiar effect. When an electron is accelerated at the time t, it shakes all the other charges in the world at a *later* time $t' = t + r/c$ (where r is the distance to the other charge), because of the *retarded* waves. But then these other charges react back on the original electron through their *advanced* waves, which will arrive at the time t'', equal to t' *minus* r/c, which is, of course, just t. (They also react back with their retarded waves too, but that just corresponds to the normal "reflected" waves.) The combination of the advanced and retarded waves means that at the instant it is accelerated an oscillating charge feels a force from all the charges that are "going to" absorb its radiated waves. You see what tight knots people have gotten into in trying to get a theory of the electron!

We'll describe now still another kind of theory, to show the kind of things that people think of when they are stuck. This is another modification of the laws of electrodynamics, proposed by Bopp. You realize that once you decide to change the equations of electromagnetism you can start anywhere you want. You can change the force law for an electron, or you can change the Maxwell equations (as we saw in the examples we have described), or you can make a change somewhere else. One possibility is to change the formulas that give the potentials in terms of the charges and currents. One of our formulas has been that the potentials at some point are given by the current density (or charge) at each other point at an earlier time. Using our four-vector notation for the potentials, we write

$$A_\mu(1, t) = \frac{1}{4\pi\epsilon_0 c^2} \int \frac{j_\mu(2, t - r_{12}/c)}{r_{12}} \, dV_2. \tag{28.13}$$

Bopp's beautifully simple idea is that: Maybe the trouble is in the $1/r$ factor in the integral. Suppose we were to start out by assuming only that the potential at one point depends on the charge density at any other point as *some* function of the distance between the points, say as $f(r_{12})$. The total potential at point (1) will then be given by the integral of j_μ times this function over all space:

$$A_\mu(1) = \int j_\mu(2) f(r_{12})\, dV_2.$$

That's all. No differential equation, nothing else. Well, one more thing. We also ask that the result should be relativistically invariant. So by "distance" we should take the invariant "distance" between two points in space-time. This distance squared (within a sign which doesn't matter) is

$$\begin{aligned} s_{12}^2 &= c^2(t_1 - t_2)^2 - r_{12}^2 \\ &= c^2(t_1 - t_2)^2 - (x_1 - x_2)^2 - (y_1 - y_2)^2 - (z_1 - z_2)^2. \end{aligned} \tag{28.14}$$

So, for a relativistically invariant theory, we should take some function of the magnitude of s_{12}, or what is the same thing, some function of s_{12}^2. So Bopp's theory is that

$$A_\mu(1, t_1) = \int j_\mu(2, t_2) F(s_{12}^2)\, dV_2\, dt_2. \tag{28.15}$$

(The integral must, of course, be over the four-dimensional volume $dt_2\, dx_2\, dy_2\, dz_2$.)

All that remains is to choose a suitable function for F. We assume only one thing about F—that it is very small except when its argument is near zero—so that a graph of F would be a curve like the one in Fig. 28–4. It is a narrow spike with a finite area centered at $s^2 = 0$, and with a width which we can say is roughly a^2. We can say, crudely, that when we calculate the potential at point (1), only those points (2) produce any appreciable effect if $s_{12}^2 = c^2(t_2 - t_1)^2 - r_{12}^2$ is within $\pm a^2$ of zero. We can indicate this by saying that F is important only for

$$s_{12}^2 = c^2(t_1 - t_2)^2 - r_{12}^2 \approx \pm a^2. \tag{28.16}$$

You can make it more mathematical if you want to, but that's the idea.

Now suppose that a is very small in comparison with the size of ordinary objects like motors, generators, and the like so that for normal problems $r_{12} \gg a$. Then Eq. (28.16) says that charges contribute to the integral of Eq. (28.15) only when $t_1 - t_2$ is in the small range

$$c(t_1 - t_2) \approx \sqrt{r_{12}^2 \pm a^2} \approx r_{12}\sqrt{1 \pm \frac{a^2}{r_{12}^2}}.$$

Since $a^2/r_{12}^2 \ll 1$, the square root can be approximated by $1 \pm a^2/2r_{12}^2$, so

$$t_1 - t_2 = \frac{r_{12}}{c}\left(1 \pm \frac{a^2}{2r_{12}^2}\right) = \frac{r_{12}}{c} \pm \frac{a^2}{2r_{12}c}.$$

What is the significance? This result says that the only *times* t_2 that are important in the integral of A_μ are those which differ from the time t_1, at which we want the potential, by the delay r_{12}/c—with a negligible correction so long as $r_{12} \gg a$. In other words, this theory of Bopp approaches the Maxwell theory—so long as we are far away from any particular charge—in the sense that it gives the retarded wave effects.

We can, in fact, see approximately what the integral of Eq. (28.15) is going to give. If we integrate first over t_2 from $-\infty$ to $+\infty$—keeping r_{12} fixed—then s_{12}^2 is also going to go from $-\infty$ to $+\infty$. The integral will all come from t_2's in a small interval of width $\Delta t_2 = 2 \times a^2/2r_{12}c$, centered at $t_1 - r_{12}/c$. Say that the function $F(s^2)$ has the value K at $s^2 = 0$; then the integral over t_2 gives approximately $Kj_\mu\Delta t_2$, or

$$\frac{Ka^2}{c}\frac{j_\mu}{r_{12}}.$$

We should, of course, take the value of j_μ at $t_2 = t_1 - r_{12}/c$, so that Eq. (28.15) becomes

$$A_\mu(1, t_1) = \frac{Ka^2}{c}\int \frac{j_\mu(2, t_1 - r_{12}/c)}{r_{12}}\, dV_2.$$

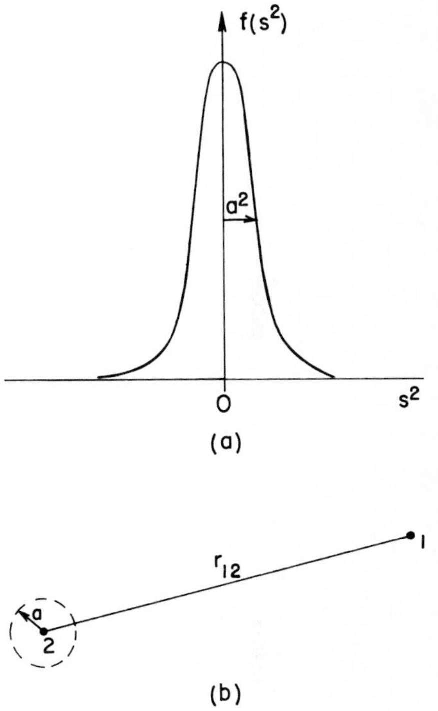

Fig. 28–4. The function $F(s^2)$ used in the nonlocal theory of Bopp.

If we pick $K = q^2c/4\pi\epsilon_0 a^2$, we are right back to the retarded potential solution of Maxwell's equations—including automatically the $1/r$ dependence! And it all came out of the simple proposition that the potential at one point in space-time depends on the current density at all other points in space-time, but with a weighting factor that is some narrow function of the four-dimensional distance between the two points. This theory again predicts a finite electromagnetic mass for the electron, and the energy and mass have the right relation for the relativity theory. They must, because the theory is relativistically invariant from the start, and everything seems to be all right.

There is, however, one fundamental objection to this theory and to all the other theories we have described. All particles we know obey the laws of quantum mechanics, so a quantum-mechanical modification of electrodynamics has to be made. Light behaves like photons. It isn't 100 percent like the Maxwell theory. So the electrodynamic theory has to be changed. We have already mentioned that it might be a waste of time to work so hard to straighten out the classical theory, because it could turn out that in quantum electrodynamics the difficulties will disappear or may be resolved in some other fashion. But the difficulties do not disappear in quantum electrodynamics. That is one of the reasons that people have spent so much effort trying to straighten out the classical difficulties, hoping that if they *could* straighten out the classical difficulty and *then* make the quantum modifications, everything would be straightened out. The Maxwell theory still has the difficulties after the quantum mechanics modifications are made.

The quantum effects do make some changes—the formula for the mass is modified, and Planck's constant $\hbar$ appears—but the answer still comes out infinite unless you cut off an integration somehow—just as we had to stop the classical integrals at $r = a$. And the answers depend on how you stop the integrals. We cannot, unfortunately, demonstrate for you here that the difficulties are really basically the same, because we have developed so little of the theory of quantum mechanics and even less of quantum electrodynamics. So you must just take our word that the quantized theory of Maxwell's electrodynamics gives an infinite mass for a point electron.

It turns out, however, that nobody has ever succeeded in making a *self-consistent* quantum theory out of *any* of the modified theories. Born and Infeld's ideas have never been satisfactorily made into a quantum theory. The theories with the advanced and retarded waves of Dirac, or of Wheeler and Feynman, have never been made into a satisfactory quantum theory. The theory of Bopp has never been made into a satisfactory quantum theory. So today, there is no known solution to this problem. We do not know how to make a consistent theory—including the quantum mechanics—which does not produce an infinity for the self-energy of an electron, or any point charge. And at the same time, there is no satisfactory theory that describes a non-point charge. It's an unsolved problem.

In case you are deciding to rush off to make a theory in which the action of an electron on itself is completely removed, so that electromagnetic mass is no longer meaningful, and then to make a quantum theory of it, you should be warned that you are certain to be in trouble. There is definite experimental evidence of the existence of electromagnetic inertia—there is evidence that some of the mass of charged particles is electromagnetic in origin.

It used to be said in the older books that since Nature will obviously not present us with two particles—one neutral and the other charged, but otherwise the same—we will never be able to tell how much of the mass is electromagnetic and how much is mechanical. But it turns out that Nature *has* been kind enough to present us with just such objects, so that by comparing the observed mass of the charged one with the observed mass of the neutral one, we can tell whether there is any electromagnetic mass. For example, there are the neutrons and protons. They interact with tremendous forces—the nuclear forces—whose origin is unknown. However, as we have already described, the nuclear forces have one remarkable property. So far as they are concerned, the neutron and proton are exactly the same. The *nuclear* forces between neutron and neutron, neutron and proton, and proton and proton are all identical as far as we can tell. Only the little electromagnetic forces are different; electrically the proton and neutron are as different as night and day. This is just what we wanted. There are two particles, identical from the point of view of the strong interactions, but different electrically. And they have a small difference in mass. The mass difference between the proton and the neutron—expressed as the difference in the rest-energy mc^2 in units of Mev—is about 1.3 Mev, which is about 2.6 times the electron mass. The classical theory would then predict a radius of about $\frac{1}{3}$ to $\frac{1}{2}$ the classical electron radius, or about 10^{-13} cm. Of course, one should really use the quantum theory, but by some strange accident, all the constants—2π's and $\hbar$'s, etc.—come out so that the quantum theory gives roughly the same radius as the classical theory. The only trouble is that the *sign* is wrong! The neutron is *heavier* than the proton.

Table 28–1

Particle Masses

Particle	Charge (electronic)	Mass (Mev)	Δm* (Mev)
n (neutron)	0	939.5	
p (proton)	+1	938.2	−1.3
π (π-meson)	0	135.0	
	±1	139.6	+4.6
K (K-meson)	0	497.8	
	±1	493.9	−3.9
Σ (sigma)	0	1191.5	
	+1	1189.4	−2.1
	−1	1196.0	+4.5

* Δm = (mass of charged) − (mass of neutral).

Nature has also given us several other pairs—or triplets—of particles which appear to be exactly the same except for their electrical charge. They interact with protons and neutrons, through the so-called "strong" interactions of the nuclear forces. In such interactions, the particles of a given kind—say the π-mesons—behave in every way like one object *except* for their electrical charge. In Table 28–1 we give a list of such particles, together with their measured masses. The charged π-mesons—positive or negative—have a mass of 139.6 Mev, but the neutral π-meson is 4.6 Mev lighter. We believe that this mass difference is electromagnetic; it would correspond to a particle radius of 3 to 4×10^{-14} cm. You will see from the table that the mass differences of the other particles are usually of the same general size.

Now the size of these particles can be determined by other methods, for instance by the diameters they appear to have in high-energy collisions. So the electromagnetic mass seems to be in general agreement with electromagnetic theory, if we stop our integrals of the field energy at the same radius obtained by these other methods. That's why we believe that the differences do represent electromagnetic mass.

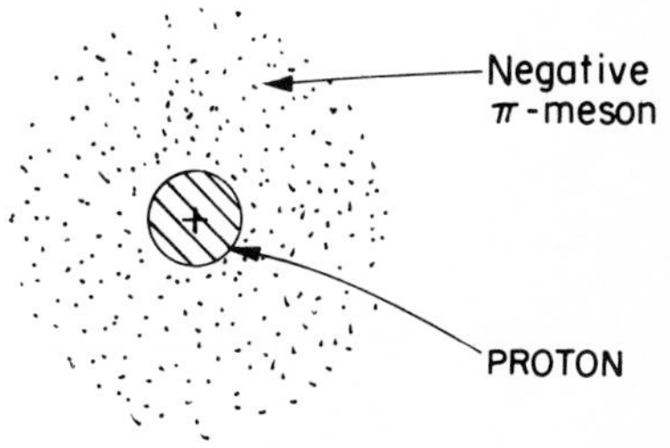

Fig. 28–5. A neutron may exist, at times, as a proton surrounded by a negative π-meson.

You are no doubt worried about the different signs of the mass differences in the table. It is easy to see why the charged ones should be heavier than the neutral ones. But what about those pairs like the proton and the neutron, where the measured mass comes out the other way? Well, it turns out that these particles are complicated, and the computation of the electromagnetic mass must be more elaborate for them. For instance, although the neutron has no *net* charge, it *does* have a charge distribution inside it—it is only the *net* charge that is zero. In fact, we believe that the neutron looks—at least sometimes—like a proton with a negative π-meson in a "cloud" around it, as shown in Fig. 28–5. Although the neutron is "neutral," because its total charge is zero, there are still electromagnetic energies (for example, it has a magnetic moment), so it's not easy to tell the sign of the electromagnetic mass difference without a detailed theory of the internal structure.

We only wish to emphasize here the following points: (1) the electromagnetic theory predicts the existence of an electromagnetic mass, but it also falls on its face in doing so, because it does not produce a consistent theory—and the same is true with the quantum modifications; (2) there is experimental evidence for the existence of electromagnetic mass; and (3) all these masses are roughly the same as the mass of an electron. So we come back again to the original idea of Lorentz—maybe all the mass of an electron is purely electromagnetic, maybe the whole 0.511 Mev is due to electrodynamics. Is it or isn't it? We haven't got a theory, so we cannot say.

We must mention one more piece of information, which is the most annoying. There is another particle in the world called a *muon*—or μ-meson—which, so far as we can tell, differs in no way whatsoever from an electron except for its mass. It acts in every way like an electron: it interacts with neutrinos and with the electromagnetic field, and it has no nuclear forces. It does nothing different from what an electron does—at least, nothing which cannot be understood as merely a consequence of its higher mass (206.77 times the electron mass). Therefore, whenever someone finally gets the explanation of the mass of an electron, he will then have the puzzle of where a muon gets its mass. Why? Because whatever the electron does, the muon does the same—so the mass ought to come out the same. There are those who believe faithfully in the idea that the muon and the electron are the same particle and that, in the final theory of the mass, the formula for the mass will be a quadratic equation with two roots—one for each particle. There are also those who propose it will be a transcendental equation with an infinite number of roots, and who are engaged in guessing what the masses of the other particles in the series must be, and why these particles haven't been discovered yet.

28–6 The nuclear force field

We would like to make some further remarks about the part of the mass of nuclear particles that is not electromagnetic. Where does this other large fraction come from? There are other forces besides electrodynamics—like nuclear forces—that have their own field theories, although no one knows whether the current theories are right. These theories also predict a field energy which gives the nuclear particles a mass term analogous to electromagnetic mass; we could call it the "π-mesic-field-mass." It is presumably very large, because the forces are great, and it is the possible origin of the mass of the heavy particles. But the meson field theories are still in a most rudimentary state. Even with the well-developed theory of electromagnetism, we found it impossible to get beyond first base in explaining the electron mass. With the theory of the mesons, we strike out.

We may take a moment to outline the theory of the mesons, because of its interesting connection with electrodynamics. In electrodynamics, the field can be described in terms of a four-potential that satisfies the equation

$$\Box^2 A_\mu = \text{sources}.$$

Now we have seen that pieces of the field can be radiated away so that they exist separated from the sources. These are the photons of light, and they are described by a differential equation without sources:

$$\Box^2 A_\mu = 0.$$

People have argued that the field of nuclear forces ought also to have its own "photons"—they would presumably be the π-mesons—and that they should be described by an analogous differential equation. (Because of the weakness of the human brain, we can't think of something really new; so we argue by analogy with what we know.) So the meson equation might be

$$\Box^2 \phi = 0,$$

where ϕ could be a different four-vector or perhaps a scalar. It turns out that the pion has no polarization, so ϕ should be a scalar. With the simple equation $\Box^2\phi = 0$, the meson field would vary with distance from a source as $1/r^2$, just as the electric field does. But we know that nuclear forces have much shorter distances of action, so the simple equation won't work. There is one way we can change things without disrupting the relativistic invariance: we can add or subtract from the D'Alembertian a constant, times ϕ. So Yukawa suggested that the free quanta of the nuclear force field might obey the equation

$$\Box^2\phi - \mu^2\phi = 0, \tag{28.17}$$

where μ^2 is a constant—that is, an invariant scalar. (Since $\Box^2$ is a scalar differential operator in four dimensions, its invariance is unchanged if we add another scalar to it.)

Let's see what Eq. (28.17) gives for the nuclear force when things are not changing with time. We want a spherically symmetric solution of

$$\nabla^2\phi - \mu^2\phi = 0$$

around some point source at, say, the origin. If ϕ depends only on r, we know that

$$\nabla^2\phi = \frac{1}{r}\frac{\partial^2}{\partial r^2}(r\phi).$$

So we have the equation

$$\frac{1}{r}\frac{\partial^2}{\partial r^2}(r\phi) - \mu^2\phi = 0$$

or

$$\frac{\partial^2}{\partial r^2}(r\phi) = \mu^2(r\phi).$$

Thinking of $(r\phi)$ as our dependent variable, this is an equation we have seen many times. It's solution is

$$r\phi = Ke^{\pm\mu r}.$$

Clearly, ϕ cannot become infinite for large r, so the + sign in the exponent is ruled out. The solution is

$$\phi = K\frac{e^{-\mu r}}{r}. \tag{28.18}$$

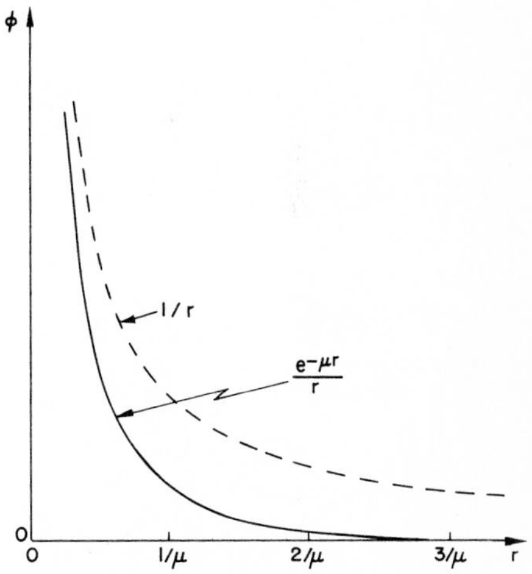

Fig. 28–6. The Yukawa potential $e^{-\mu r}/r$, compared with the Coulomb potential $1/r$.

This function is called the *Yukawa potential.* For an attractive force, K is a negative number whose magnitude must be adjusted to fit the experimentally observed strength of the forces.

The Yukawa potential of the nuclear forces dies off more rapidly than $1/r$ by the exponential factor. The potential—and therefore the force—falls to zero much more rapidly than $1/r$ for distances beyond $1/\mu$, as shown in Fig. 28–6. The "range" of nuclear forces is much less than the "range" of electrostatic forces. It is found experimentally that the nuclear forces do not extend beyond about 10^{-13} cm, so $\mu \approx 10^{15}\ \text{m}^{-1}$.

Finally, let's look at the free-wave solution of Eq. (28.17). If we substitute

$$\phi = \phi_0 e^{i(\omega t - kz)}$$

into Eq. (28.17), we get that

$$\frac{\omega^2}{c^2} - k^2 - \mu^2 = 0.$$

Relating frequency to energy and wave number to momentum, as we did at the end of Chapter 36 of Vol. I, we get that

$$\frac{E^2}{c^2} - p^2 = \mu^2\hbar^2,$$

which says that the Yukawa "photon" has a mass equal to $\mu\hbar/c$. If we use for μ

the estimate $10^{15}\ \mathrm{m}^{-1}$, which gives the observed range of the nuclear forces, the mass comes out to 3×10^{-25} gm, or 170 Mev, which is roughly the observed mass of the π-meson. So, by an analogy with electrodynamics, we would say that the π-meson is the "photon" of the nuclear force field. But now we have pushed the ideas of electrodynamics into regions where they may not really be valid—we have gone beyond electrodynamics to the problem of the nuclear forces.

29

The Motion of Charges in Electric and Magnetic Fields

Review: Chapter 30, Vol. I, *Diffraction*

29–1 Motion in a uniform electric or magnetic field

We want now to describe—mainly in a qualitative way—the motions of charges in various circumstances. Most of the interesting phenomena in which charges are moving in fields occur in very complicated situations, with many, many charges all interacting with each other. For instance, when an electromagnetic wave goes through a block of material or a plasma, billions and billions of charges are interacting with the wave and with each other. We will come to such problems later, but now we just want to discuss the much simpler problem of the motions of a single charge in a *given* field. We can then disregard all other charges —except, of course, those charges and currents which exist somewhere to produce the fields we will assume.

We should probably ask first about the motion of a particle in a uniform electric field. At low velocities, the motion is not particularly interesting—it is just a uniform acceleration in the direction of the field. However, if the particle picks up enough energy to become relativistic, then the motion gets more complicated. But we will leave the solution for that case for you to play with.

Next, we consider the motion in a uniform magnetic field with zero electric field. We have already solved this problem—one solution is that the particle goes in a circle. The magnetic force $q\boldsymbol{v} \times \boldsymbol{B}$ is always at right angles to the motion, so $d\boldsymbol{p}/dt$ is perpendicular to $\boldsymbol{p}$ and has the magnitude vp/R, where R is the radius of the circle:

$$F = qvB = \frac{vp}{R}.$$

The radius of the circular orbit is then

$$R = \frac{p}{qB}. \tag{29.1}$$

That is only one possibility. If the particle has a component of its motion along the field direction, that motion is constant, since there can be no component of the magnetic force in the direction of the field. The general motion of a particle in a uniform magnetic field is a constant velocity parallel to $\boldsymbol{B}$ and a circular motion at right angles to $\boldsymbol{B}$—the trajectory is a cylindrical helix (Fig. 29–1). The radius of the helix is given by Eq. (29.1) if we replace p by $p_\perp$, the component of momentum at right angles to the field.

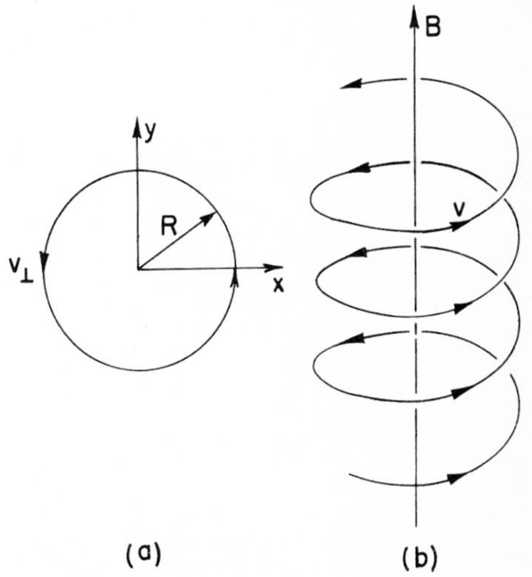

Fig. 29–1. Motion of a particle in a uniform magnetic field.

29–2 Momentum analysis

A uniform magnetic field is often used in making a "momentum analyzer," or "momentum spectrometer," for high-energy charged particles. Suppose that charged particles are shot into a uniform magnetic field at the point A in Fig. 29–2(a), the magnetic field being perpendicular to the plane of the drawing. Each particle will go into an orbit which is a circle whose radius is proportional to its momentum. If all the particles enter perpendicular to the edge of the field, they will leave the field at a distance x (from A) which is proportional to their momentum p. A counter placed at some point such as C will detect only those particles whose momentum is in an interval Δp near the momentum $p = qBx/2$.

It is, of course, not necessary that the particles go through 180° before they are counted, but the so-called "180° spectrometer" has a special property. It is not

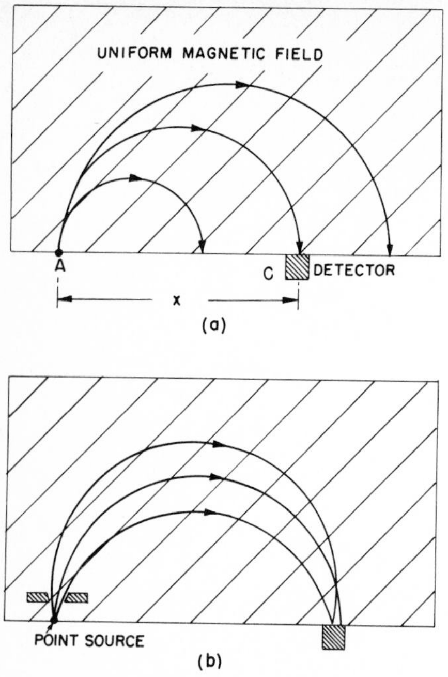

Fig. 29–2. A uniform-field, momentum spectrometer with 180° focusing: (a) different momenta; (b) different angles. (The magnetic field is directed perpendicular to the plane of the figure.)

necessary that all the particles enter at right angles to the field edge. Figure 29–2(b) shows the trajectories of three particles, all with the *same* momentum but entering the field at different angles. You see that they take different trajectories, but all leave the field very close to the point C. We say that there is a "focus." Such a focusing property has the advantage that larger angles can be accepted at A—although some limit is usually imposed, as shown in the figure. A larger angular acceptance usually means that more particles are counted in a given time, decreasing the time required for a given measurement.

By varying the magnetic field, or moving the counter along in x, or by using many counters to cover a range of x, the "spectrum" of momenta in the incoming beam can be measured. [By the "momentum spectrum" $f(p)$, we mean that the number of particles with momenta between p and $(p + dp)$ is $f(p)\,dp$.] Such measurements have been made, for example, to determine the distribution of energies in the β-decay of various nuclei.

There are many other forms of momentum spectrometers, but we will describe just one more, which has an especially large *solid* angle of acceptance. It is based on the helical orbits in a uniform field, like the one shown in Fig. 29–1. Let's think of a cylindrical coordinate system—ρ, θ, z—set up with the z-axis along the direction of the field. If a particle is emitted from the origin at some angle α with respect to the z-axis, it will move along a spiral whose equation is

$$\rho = a \sin kz, \qquad \theta = bz,$$

where a, b, and k are parameters you can easily work out in terms of p, α, and the magnetic field B. If we plot the distance ρ from the axis as a function of z for a given momentum, but for several starting angles, we will get curves like the solid ones drawn in Fig. 29–3. (Remember that this is just a kind of projection of a helical trajectory.) When the angle between the axis and the starting direction is larger, the peak value of ρ is large but the longitudinal velocity is less, so the trajectories for different angles tend to come to a kind of "focus" near the point A in the figure. If we put a narrow aperture of A, particles with a range of initial angles can still get through and pass on to the axis, where they can be counted by the long detector D.

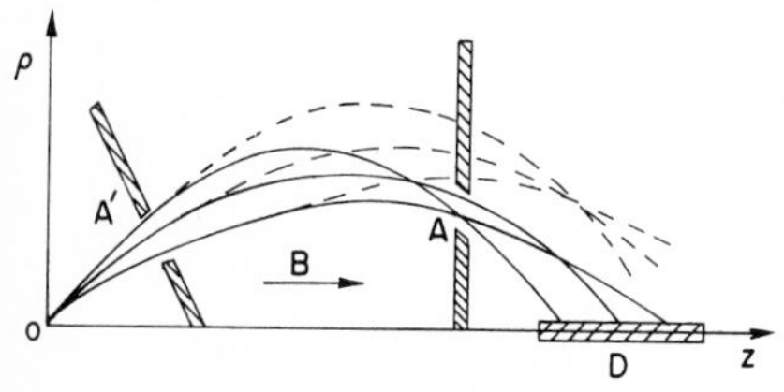

Fig. 29–3. An axial-field spectrometer.

Particles which leave the source at the origin with a higher momentum but at the same angles, follow the paths shown by the broken lines and do not get through the aperture at A. So the apparatus selects a small interval of momenta. The advantage over the first spectrometer described is that the aperture A—and the aperture A'—can be an annulus, so that particles which leave the source in a rather large solid angle are accepted. A large fraction of the particles from the source are used—an important advantage for weak sources or for very precise measurements.

One pays a price for this advantage, however, because a large volume of uniform magnetic field is required, and this is usually only practical for low-energy particles. One way of making a uniform field, you remember, is to wind a coil on a sphere, with a surface current density proportional to the sine of the angle. You can also show that the same thing is true for an ellipsoid of rotation. So such spectrometers are often made by winding an elliptical coil on a wooden (or aluminum) frame. All that is required is that the current in each interval of axial distance Δx be the same, as shown in Fig. 29–4.

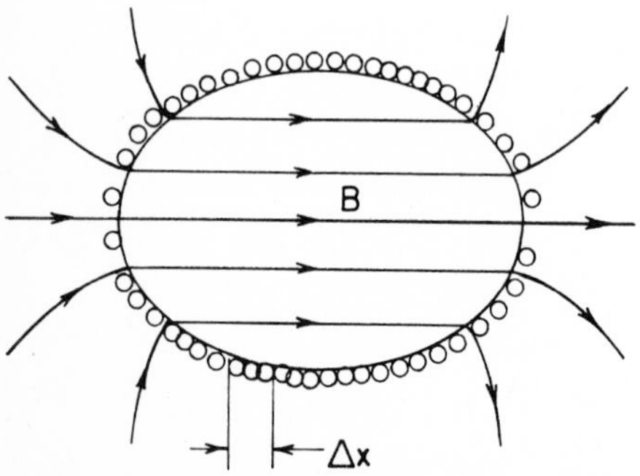

Fig. 29–4. An ellipsoidal coil with equal currents in each axial interval Δx produces a uniform magnetic field inside.

29–3 An electrostatic lens

Particle focusing has many applications. For instance, the electrons that leave the cathode in a TV picture tube are brought to a focus at the screen—to make a fine spot. In this case, one wants to take electrons all of the same energy but with different initial angles and bring them together in a small spot. The problem is like focusing light with a lens, and devices which do the corresponding job for particles are also called lenses.

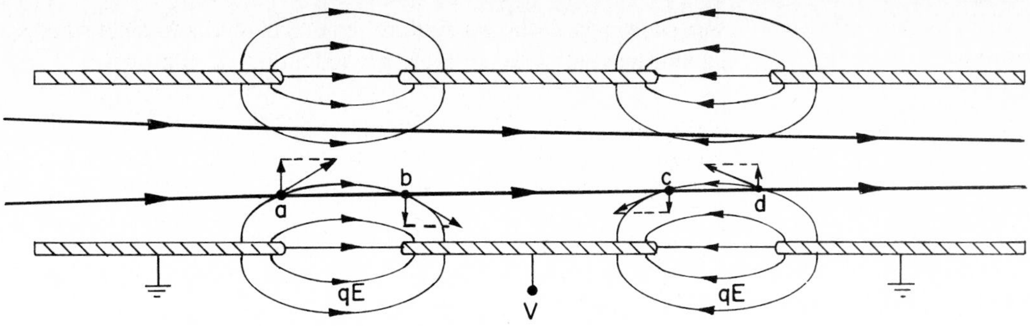

Fig. 29–5. An electrostatic lens. The field lines shown are "lines of force," that is, of $q\boldsymbol{E}$.

One example of an electron lens is sketched in Fig. 29–5. It is an "electrostatic" lens whose operation depends on the electric field between two adjacent electrodes. Its operation can be understood by considering what happens to a parallel beam that enters from the left. When the electrons arrive at the region a, they feel a force with a sidewise component and get a certain impulse that bends them toward the axis. You might think that they would get an equal and opposite impulse in the region b, but that is not so. By the time the electrons reach b they have gained energy and so *spend less time* in the region b. The forces are the same, but the time is shorter, so the impulse is less. In going through the regions a and b, there is a net axial impulse, and the electrons are bent toward a common point. In leaving the high-voltage region, the particles get another kick toward the axis. The force is outward in region c and inward in region d, but the particles stay longer in the latter region, so there is again a net impulse. For distances not too far from the axis, the total impulse through the lens is proportional to the distance from the axis (Can you see why?), and this is just the condition necessary for lens-type focusing.

You can use the same arguments to show that there is focusing if the potential of the middle electrode is either positive or negative with respect to the other two. Electrostatic lenses of this type are commonly used in cathode-ray tubes and in some electron microscopes.

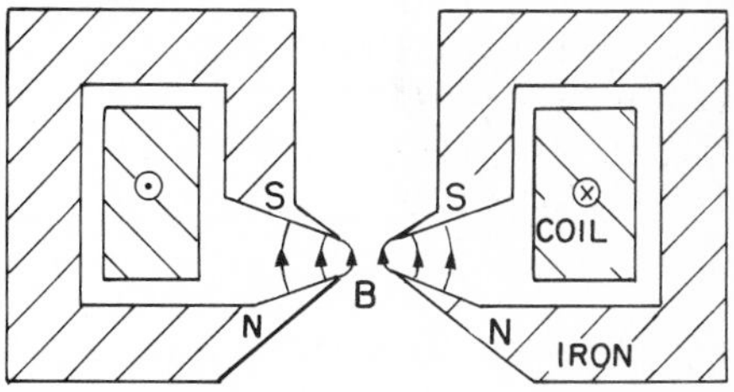

Fig. 29–6. A magnetic lens.

29–4 A magnetic lens

Another kind of lens—often found in electron microscopes—is the magnetic lens sketched schematically in Fig. 29–6. A cylindrically symmetric electromagnet has very sharp circular pole tips which produce a strong, nonuniform field in a small region. Electrons which travel vertically through this region are focused. You can understand the mechanism by looking at the magnified view of the pole-tip region drawn in Fig. 29–7. Consider two electrons a and b that leave the source S at some angle with respect to the axis. As electron a reaches the beginning of the field, it is deflected *away from you* by the horizontal component of the field. But then it will have a lateral velocity, so that when it passes through the strong vertical field, it will get an impulse toward the axis. Its lateral motion is taken out by the magnetic force as it leaves the field, so the net effect is an impulse toward the axis, plus a "rotation" about the axis. All the forces on particle b are opposite, so it also is deflected toward the axis. In the figure, the divergent electrons are brought into parallel paths. The action is like a lens with an object at the focal point. Another similar lens upstream can be used to focus the electrons back to a single point, making an image of the source S.

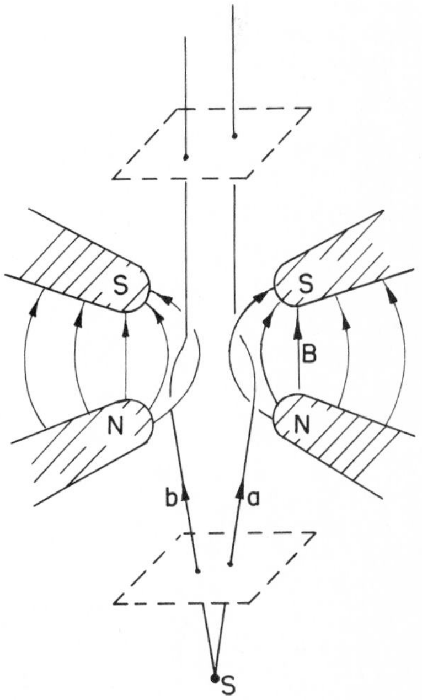

Fig. 29–7. Electron motion in the magnetic lens.

29–5 The electron microscope

You know that electron microscopes can "see" objects too small to be seen by optical microscopes. We discussed in Chapter 30 of Vol. I the basic limitations of any optical system due to diffraction of the lens opening. If a lens opening sub-

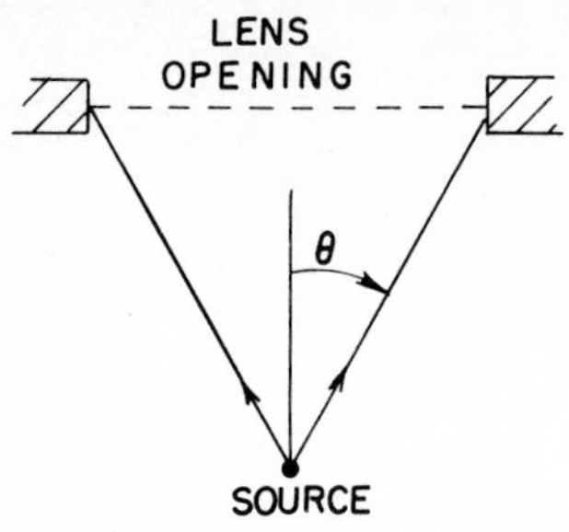

Fig. 29–8. The resolution of a microscope is limited by the angle subtended from the source.

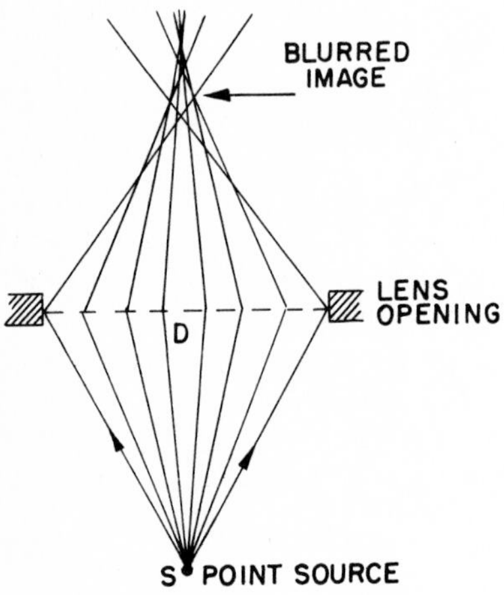

Fig. 29–9. Spherical aberration of a lens.

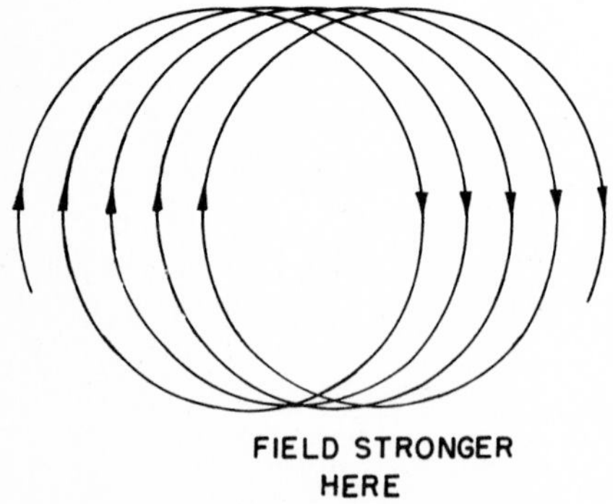

Fig. 29–10. Particle motion in a slightly nonuniform field.

tends the angle 2θ from a source (see Fig. 29–8), two neighboring spots at the source cannot be seen as separate if they are closer than about

$$\delta \approx \frac{\lambda}{\sin \theta},$$

where λ is the wavelength of the light. With the best optical microscope, θ approaches the theoretical limit of 90°, so δ is about equal to λ, or approximately 5000 angstroms.

The same limitation would also apply to an electron microscope, but there the wavelength is—for 50-kilovolt electrons—about 0.05 angstrom. If one could use a lens opening of near 30°, it would be possible to see objects only $\frac{1}{5}$ of an angstrom apart. Since the atoms in molecules are typically 1 or 2 angstroms apart, we could get photographs of molecules. Biology would be easy; we would have a photograph of the DNA structure. What a tremendous thing that would be! Most of present-day research in molecular biology is an attempt to figure out the shapes of complex organic molecules. If we could only see them!

Unfortunately, the best resolving power that has been achieved in an electron microscope is more like 20 angstroms. The reason is that no one has yet designed a lens with a large opening. All lenses have "spherical aberration," which means that rays at large angles from the axis have a different point of focus than the rays nearer the axis, as shown in Fig. 29–9. By special techniques, optical microscope lenses can be made with a negligible spherical aberration, but no one has yet been able to make an electron lens which avoids spherical aberration.

In fact, one can show that any electrostatic or magnetic lens of the types we have described must have an irreducible amount of spherical aberration. This aberration—together with diffraction—limits the resolving power of electron microscopes to their present value.

The limitation we have mentioned does not apply to electric and magnetic fields which are not axially symmetric or which are not constant in time. Perhaps some day someone will think of a new kind of electron lens that will overcome the inherent aberration of the simple electron lens. Then we will be able to photograph atoms directly. Perhaps one day chemical compounds will be analyzed by looking at the positions of the atoms rather than by looking at the color of some precipitate!

29–6 Accelerator guide fields

Magnetic fields are also used to produce special particle trajectories in high-energy particle accelerators. Machines like the cyclotron and synchrotron bring particles to high energies by passing the particles repeatedly through a strong electric field. The particles are held in their cyclic orbits by a magnetic field.

We have seen that a particle in a uniform magnetic field will go in a circular orbit. This, however, is true only for a perfectly uniform field. Imagine a field B which is nearly uniform over a large area but which is slightly stronger in one region than in another. If we put a particle of momentum p in this field, it will go in a nearly circular orbit with the radius $R = p/qB$. The radius of curvature will, however, be slightly smaller in the region where the field is stronger. The orbit is not a closed circle but will "walk" through the field, as shown in Fig. 29–10. We can, if we wish, consider that the slight "error" in the field produces an extra angular kick which sends the particle off on a new track. If the particles are to make millions of revolutions in an accelerator, some kind of "radial focusing" is needed which will tend to keep the trajectories close to some design orbit.

Another difficulty with a uniform field is that the particles do not remain in a plane. If they start out with the slightest angle—or are given a slight angle by any small error in the field—they will go in a helical path that will eventually take them into the magnet pole or the ceiling or floor of the vacuum tank. Some arrangement must be made to inhibit such vertical drifts; the field must provide "vertical focusing" as well as radial focusing.

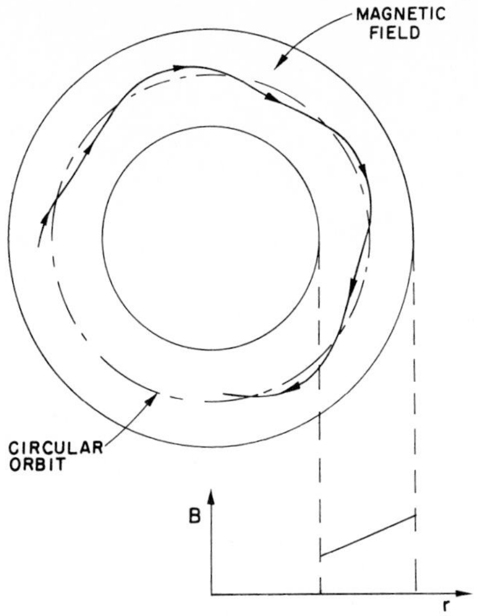

Fig. 29–11. Radial motion of a particle in a magnetic field with a large positive slope.

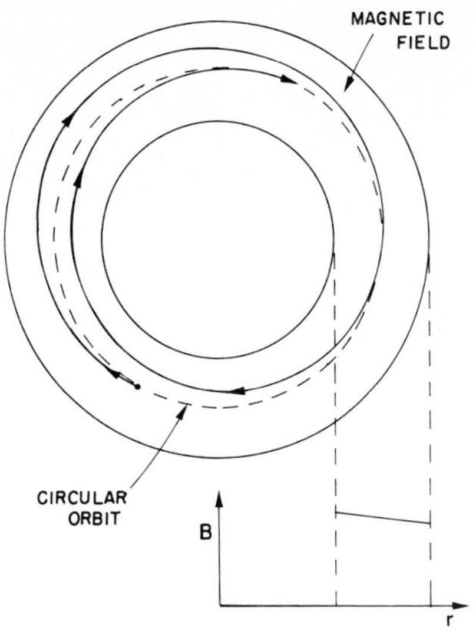

Fig. 29–12. Radial motion of a particle in a magnetic field with a small negative slope.

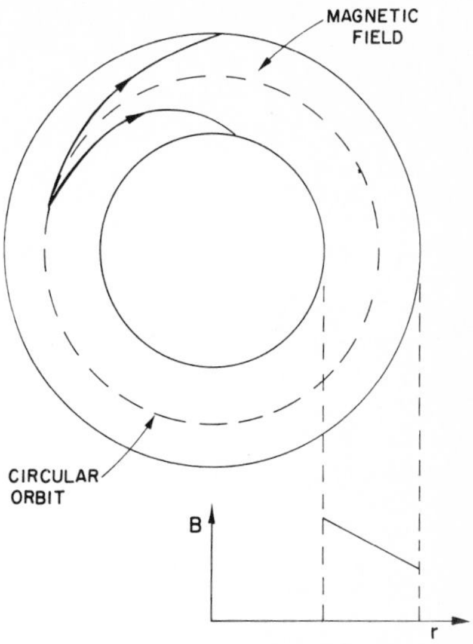

Fig. 29–13. Radial motion of a particle in a magnetic field with a large negative slope.

One would, at first, guess that radial focusing could be provided by making a magnetic field which increases with increasing distance from the center of the design path. Then if a particle goes out to a large radius, it will be in a stronger field which will bend it back toward the correct radius. If it goes to too small a radius, the bending will be less, and it will be returned toward the design radius. If a particle is once started at some angle with respect to the ideal circle, it will oscillate about the ideal circular orbit, as shown in Fig. 29–11. The radial focusing would keep the particles near the circular path.

Actually there is still some radial focusing even with the *opposite* field slope. This can happen if the radius of curvature of the trajectory does not increase more rapidly than the increase in the distance of the particle from the center of the field. The particle orbits will be as drawn in Fig. 29–12. If the gradient of the field is too large, however, the orbits will not return to the design radius but will spiral inward or outward, as shown in Fig. 29–13.

We usually describe the slope of the field in terms of the "relative gradient" or *field index*, n:

$$n = \frac{dB/B}{dr/r}. \tag{29.2}$$

A guide field gives radial focusing if this relative gradient is greater than -1.

A radial field gradient will also produce *vertical* forces on the particles. Suppose we have a field that is stronger nearer to the center of the orbit and weaker at the outside. A vertical cross section of the magnet at right angles to the orbit might be as shown in Fig. 29–14. (For protons the orbits would be coming out of the page.) If the field is to be stronger to the left and weaker to the right, the lines of the magnetic field must be curved as shown . We can see that this must be so by using the law that the circulation of $\boldsymbol{B}$ is zero in free space. If we take coordinates as shown in the figure, then

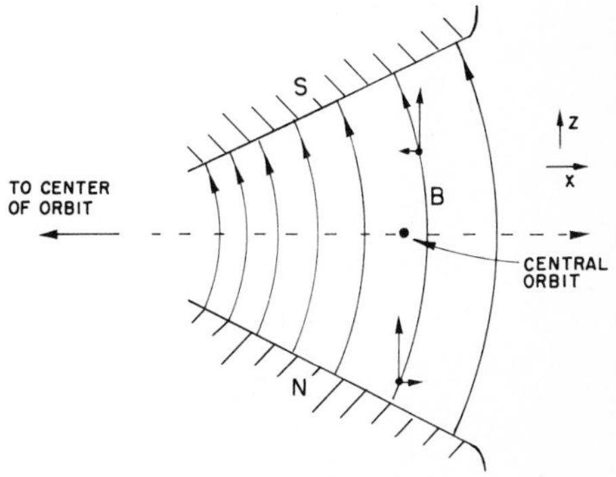

Fig. 29–14. A vertical guide field as seen in a cross section perpendicular to the orbits.

$$(\boldsymbol{\nabla} \times \boldsymbol{B})_y = \frac{\partial B_x}{\partial z} - \frac{\partial B_z}{\partial x} = 0,$$

or

$$\frac{\partial B_x}{\partial z} = \frac{\partial B_z}{\partial x}. \tag{29.3}$$

Since we assume that $\partial B_z/\partial x$ is negative, there must be an equal negative $\partial B_x/\partial z$.

If the "nominal" plane of the orbit is a plane of symmetry where $B_x = 0$, then the radial component B_x will be negative above the plane and positive below. The lines must be curved as shown.

Such a field will have vertical focusing properties. Imagine a proton that is travelling more or less parallel to the central orbit but above it. The horizontal component of $\boldsymbol{B}$ will exert a downward force on it. If the proton is below the central orbit, the force is reversed. So there is an effective "restoring force" toward the central orbit. From our arguments there will be vertical focusing, provided that the *vertical* field decreases with increasing radius; but if the field gradient is positive, there will be "vertical defocusing." So for vertical focusing, the field index n must be less than zero. We found above that for radial focusing n had to be greater than -1. The two conditions together give the condition that

$$-1 < n < 0$$

if the particles are to be kept in stable orbits. In cyclotrons, values very near zero are used; in betatrons and synchrotrons, the value $n = -0.6$ is typically used.

29–7 Alternating-gradient focusing

Such small values of n give rather "weak" focusing. It is clear that much more effective radial focusing would be given by a large positive gradient ($n \gg 1$), but then the vertical forces would be strongly defocusing. Similarly, large negative slopes ($n \ll -1$) would give stronger vertical forces but would cause radial defocusing. It was realized about 10 years ago, however, that a force that alternates between strong focusing and strong defocusing can still have a *net* focusing force.

To explain how *alternating-gradient focusing* works, we will first describe the operation of a quadrupole lens, which is based on the same principle. Imagine that a uniform negative magnetic field is added to the field of Fig. 29–14, with the strength adjusted to make zero field at the orbit. The resulting field—for small displacements from the neutral point—would be like the field shown in Fig. 29–15. Such a four-pole magnet is called a "quadrupole lens." A positive particle that enters (from the reader) to the right or left of the center is pushed back toward the center. If the particle enters above or below, it is pushed *away* from the center. This is a horizontal focusing lens. If the horizontal gradient is reversed—as can be done by reversing all the polarities—the signs of all the forces are reversed and we have a vertical focusing lens, as in Fig. 29–16. For such lenses, the field strength—and therefore the focusing forces—increase linearly with the distance of the lens from the axis.

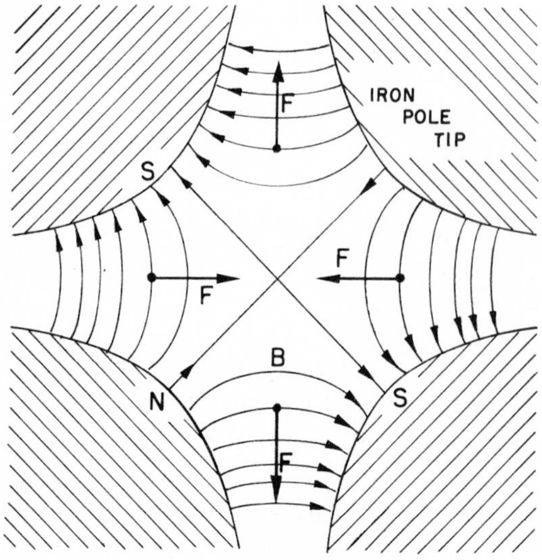

Fig. 29–15. A horizontal focusing quadrupole lens.

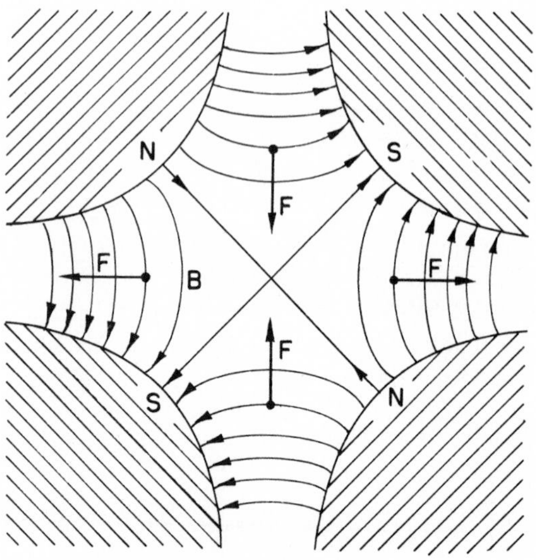

Fig. 29–16. A vertical focusing quadrupole lens.

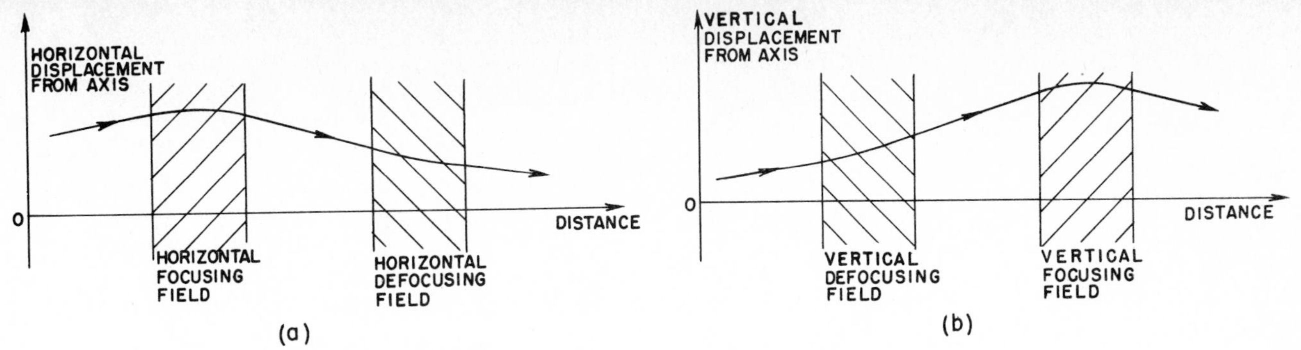

Fig. 29–17. Horizontal and vertical focusing with a pair of quadrupole lenses.

Now imagine that two such lenses are placed in series. If a particle enters with some horizontal displacement from the axis, as shown in Fig. 29–17(a), it will be deflected toward the axis in the first lens. When it arrives at the second lens it is closer to the axis, so the force outward is less and the outward deflection is less. There is a net bending toward the axis; the *average* effect is horizontally focusing. On the other hand, if we look at a particle which enters off the axis in the vertical direction, the path will be as shown in Fig. 29–17(b). The particle is first deflected *away* from the axis, but then it arrives at the second lens with a larger displacement, feels a stronger force, and so is bent toward the axis. Again the net effect is focusing. Thus a pair of quadrupole lenses acts independently for horizontal and vertical motion—very much like an optical lens. Quadrupole lenses are used to form and control beams of particles in much the same way that optical lenses are used for light beams.

We should point out that an alternating-gradient system does not *always* produce focusing. If the gradients are too large (in relation to the particle momentum or to the spacing between the lenses), the net effect can be a defocusing one. You can see how that could happen if you imagine that the spacing between the two lenses of Fig. 29–17 were increased, say, by a factor of three or four.

Let's return now to the synchrotron guide magnet. We can consider that it consists of an alternating sequence of "positive" and "negative" lenses with a superimposed uniform field. The uniform field serves to bend the particles, on the average, in a horizontal circle (with no effect on the vertical motion), and the alternating lenses act on any particles that might tend to go astray—pushing them always toward the central orbit (on the average).

There is a nice mechanical analog which demonstrates that a force which alternates between a "focusing" force and a "defocusing" force can have a net "focusing" effect. Imagine a mechanical "pendulum" which consists of a *solid* rod with a weight on the end, suspended from a pivot which is arranged to be moved rapidly up and down by a motor driven crank. Such a pendulum has *two* equilibrium positions. Besides the normal, downward-hanging position, the pendulum is also in equilibrium "hanging upward"—with its "bob" *above* the pivot! Such a pendulum is drawn in Fig. 29–18.

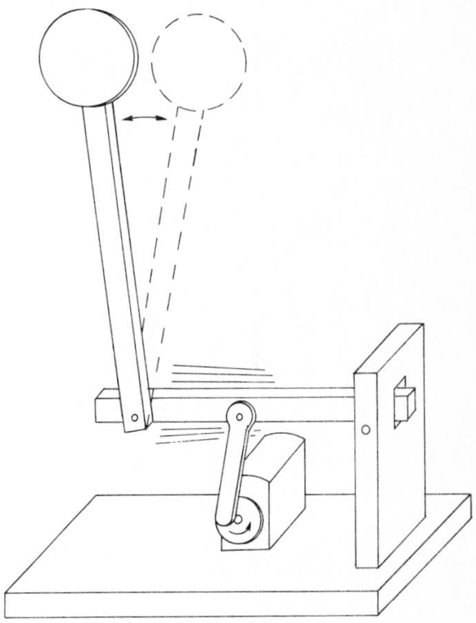

Fig. 29–18. A pendulum with an oscillating pivot can have a stable position with the bob above the pivot.

By the following argument you can see that the vertical pivot motion is equivalent to an alternating focusing force. When the pivot is accelerated downward, the "bob" tends to move inward, as indicated in Fig. 29–19. When the pivot is accelerated upward, the effect is reversed. The force restoring the "bob" toward the axis alternates, but the average effect is a force toward the axis. So the pendulum will swing back and forth about a neutral position which is just opposite the normal one.

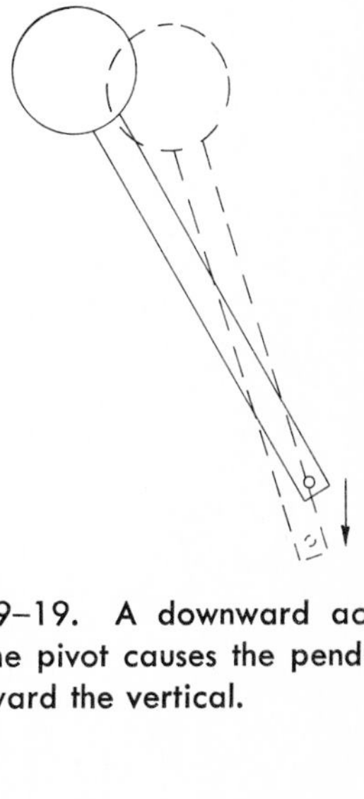

Fig. 29–19. A downward acceleration of the pivot causes the pendulum to move toward the vertical.

There is, of course, a much easier way of keeping a pendulum upside down, and that is by *balancing* it on your finger! But try to balance *two independent* sticks on the *same finger*! Or one stick with your eyes closed! Balancing involves making a correction for what is going wrong. And this is not possible, in general, if there are several things going wrong at once. In a synchrotron there are billions of particles going around together, each one of which may start out with a different "error." The kind of focusing we have been describing works on them all.

29–8 Motion in crossed electric and magnetic fields

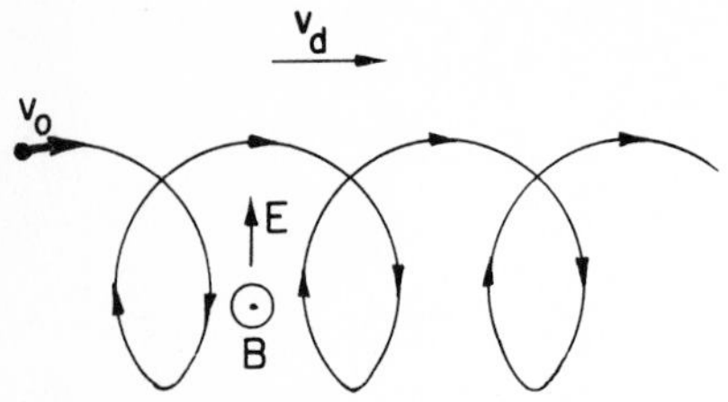

Fig. 29–20. Path of a particle in crossed electric and magnetic fields.

So far we have talked about particles in electric fields only or in magnetic fields only. There are some interesting effects when there are both kinds of fields at the same time. Suppose we have a uniform magnetic field $\boldsymbol{B}$ and an electric field $\boldsymbol{E}$ at right angles. Particles that start out perpendicular to $\boldsymbol{B}$ will move in a curve like the one in Fig. 29–20. (The figure is a *plane* curve, *not* a helix!) We can understand this motion qualitatively. When the particle (assumed positive) moves in the direction of $\boldsymbol{E}$, it picks up speed, and so it is bent less by the magnetic field. When it is going against the $\boldsymbol{E}$-field, it loses speed and is continually bent more by the magnetic field. The net effect is that it has an average "drift" in the direction of $\boldsymbol{E} \times \boldsymbol{B}$.

We can, in fact, show that the motion is a uniform circular motion superimposed on a uniform sidewise motion at the speed $v_d = E/B$—the trajectory in Fig. 29–20 is a cycloid. Imagine an observer who is moving to the right at a constant speed. In his frame our magnetic field gets transformed to a new magnetic field *plus* an electric field in the *downward* direction. If he has just the right speed, his total electric field will be zero, and he will see the electron going in a circle. So the motion *we* see is a circular motion, plus a translation at the drift speed $v_d = E/B$. The motion of electrons in crossed electric and magnetic fields is the basis of the *magnetron* tubes, i.e., oscillators used for generating microwave energy.

There are many other interesting examples of particle motions in electric and magnetic fields—such as the orbits of the electrons and protons trapped in the Van Allen belts—but we do not, unfortunately, have the time to deal with them here.

The Internal Geometry of Crystals

Reference: C. Kittel, *Introduction to Solid State Physics*, John Wiley and Sons, Inc., New York, 2nd ed., 1956.

30–1 The internal geometry of crystals

We have finished the study of the basic laws of electricity and magnetism, and we are now going to study the electromagnetic properties of matter. We begin by describing solids—that is, crystals. When the atoms of matter are not moving around very much, they get stuck together and arrange themselves in a configuration with as low an energy as possible. If the atoms in a certain place have found a pattern which seems to be of low energy, then the atoms somewhere else will probably make the same arrangement. For these reasons, we have in a solid material a repetitive pattern of atoms.

In other words, the conditions in a crystal are this way: The environment of a particular atom in a crystal has a certain arrangement, and if you look at the same kind of an atom at another place farther along, you will find one whose surroundings are exactly the same. If you pick an atom farther along by the same distance, you will find the conditions exactly the same once more. The pattern is repeated over and over again—and, of course, in three dimensions.

Imagine the problem of designing a wallpaper—or a cloth, or some geometric design for a plane area—in which you are supposed to have a design element which repeats and repeats and repeats, so that you can make the area as large as you want. This is the two-dimensional analog of a problem which a crystal solves in three dimensions. For example, Fig. 30–1(a) shows a common kind of wallpaper design. There is a single element repeated in a pattern that can go on forever. The geometric characteristics of this wallpaper design, considering only its repetition properties and not worrying about the geometry of the flower itself or its artistic merit, are contained in Fig. 30–1(b). If you start at any point, you can find the *corresponding* point by moving the distance a along the direction of arrow 1. You can also get to a corresponding point if you move the distance b in the direction of the other arrow. There are, of course, many other directions. You can go, for example, from point α to point β and reach a corresponding position, but such a step can be considered as a combination of a step along direction 1, followed by a step along direction 2. One of the basic properties of the pattern can be described by the two shortest steps to nearby equal positions. By "equal" positions we mean that if you were to stand in any one of them and look around you, you would see exactly the same thing as if you were to stand in another one. That's the fundamental property of a crystal. The only difference is that a crystal is a three-dimensional arrangement instead of a two-dimensional arrangement; and naturally, instead of flowers, each element of the lattice is some kind of an arrangement of atoms—perhaps six hydrogen atoms and two carbon atoms—in some kind of pattern. The pattern of atoms in a crystal can be found out experimentally by x-ray diffraction. We have mentioned this method briefly before, and won't say any more now except that the precise arrangement of the atoms in space has been worked out for most simple crystals and also for some fairly complex ones.

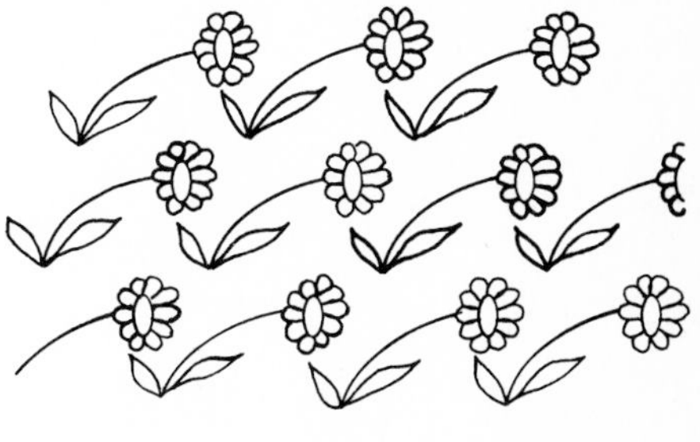

(a)

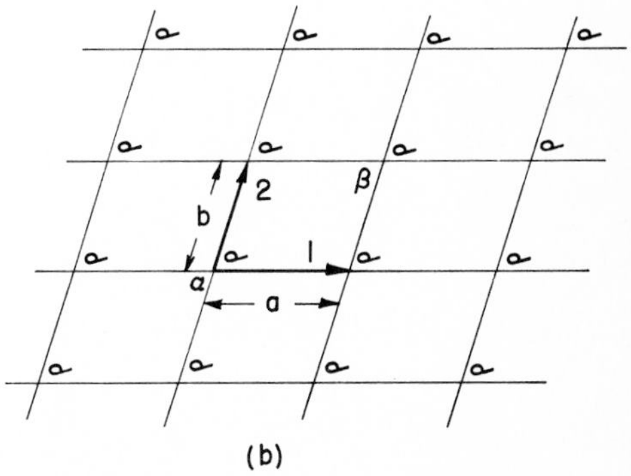

(b)

Fig. 30–1. A repeating pattern in two dimensions.

The internal pattern of a crystal shows up in several ways. First, the binding strength of the atoms in certain directions is usually stronger than in other directions. This means that there are certain planes through the crystal where it is more easily broken than others. They are called the *cleavage* planes. If you crack a crystal with a knife blade it will often split apart along such a plane. Second, the internal structure often appears at the surface because of the way the crystal was formed. Imagine a crystal being deposited out of a solution. There are the atoms floating around in the solution and finally settling down when they find a position of lowest energy. (It's as if the wallpaper got made by flowers drifting around until one drifted accidentally into place and got stuck, and then the next, and the next so that the pattern gradually grows.) You can appreciate that there will be certain directions in which it will grow at a different speed than in other directions, thereby growing into some kind of geometrical shape. Because of such effects, the outside surfaces of many crystals show some of the character of the internal arrangement of the atoms.

For example, Fig. 30–2(a) shows the shape of a typical quartz crystal whose internal pattern is hexagonal. If you look closely at such a crystal, you will notice that the outside does not make a very good hexagon because the sides are not all of equal length—they are, in fact, often very unequal. But in one respect it is a very good hexagon: the *angles* between the faces are exactly 120°. Clearly, the size of any particular face is an accident of the growth, but the *angles* are a representation of the internal geometry. So every crystal of quartz has a different shape, even though the angles between corresponding faces are always the same.

(a)

(b)

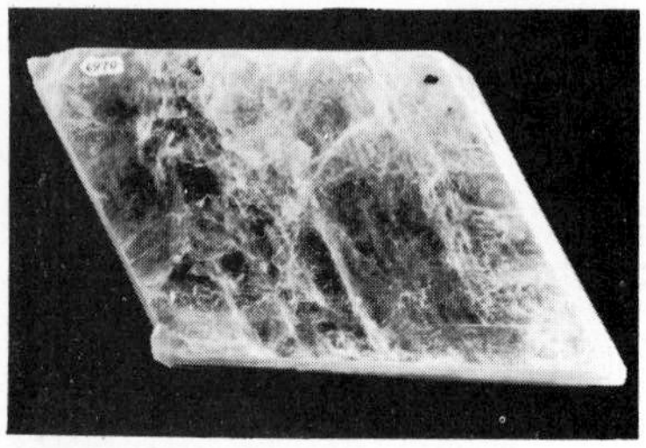

(c)

Fig. 30–2. Natural crystals: (a) quartz, (b) sodium chloride, (c) mica.

The internal geometry of a crystal of sodium chloride is also evident from its external shape. Figure 30–2(b) shows the shape of a typical grain of salt. Again the crystal is not a perfect cube, but the faces *are* exactly at right angles to one another.

A more complicated crystal is mica, which has the shape shown in Fig. 30–2(c). It is a highly anisotropic crystal, as is easily seen from the fact that it is very tough if you try to pull it apart in one direction (horizontally in the figure), but very easy to split by pulling apart in the other direction (vertically). It has commonly been used to obtain very tough, thin sheets. Mica and quartz are two examples of natural minerals containing silica. A third example of a mineral with silica is asbestos, which has the interesting property that it is easily pulled apart in two directions but not in the third. It appears to be made of very strong, *linear* fibers.

30–2 Chemical bonds in crystals

The mechanical properties of crystals clearly depend on the kind of chemical bindings between the atoms. The strikingly different strength of mica along different directions depends on the kinds of interatomic binding in the different directions. You have already learned in chemistry, no doubt, about the different kinds of chemical bonds. First, there are ionic bonds, as we have already discussed for sodium chloride. Roughly speaking, the sodium atoms have lost an electron and become positive ions; the chlorine atoms have gained an electron and become negative ions. The positive and negative ions are arranged in a three-dimensional checkerboard and are held together by electrical forces.

The covalent bond—in which electrons are shared between two atoms—is more common and is usually very strong. In a diamond, for example, the carbon atoms have covalent bonds in all four directions to the nearest neighbors, so the crystal is very hard indeed. There is also covalent bonding between silicon and oxygen in a quartz crystal, but there the bond is really only partially covalent. Because there is not complete sharing of the electrons, the atoms are partly charged, and the crystal is somewhat ionic. Nature is not as simple as we try to make it; there are really all possible gradations between covalent and ionic bonding.

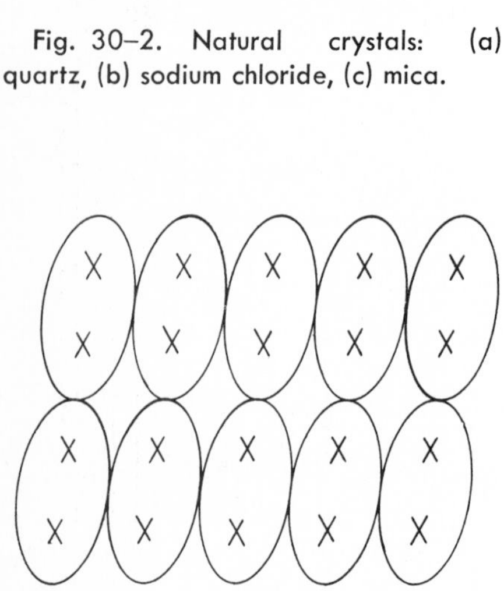

Fig. 30–3. The lattice of a molecular crystal.

A sugar crystal has still another kind of binding. In it there are large molecules in which the atoms are held strongly together by covalent bonds, so that the molecule is a tough structure. But since the strong bonds are completely satisfied, there are only relatively weak attractions between the separate, individual molecules. In such *molecular* crystals the molecules keep their individual identity, so to speak, and the internal arrangement might be as shown in Fig. 30–3. Since the molecules are not held strongly to each other, the crystals are easy to break. They are quite different from something like diamond, which is really one giant molecule that cannot be broken anywhere without disrupting strong covalent bonds. Pariffin is another example of a molecular crystal.

An extreme example of a molecular crystal occurs in a substance like solid argon. There is very little attraction between the atoms—each atom is a completely saturated monatomic molecule. But at very low temperatures, the thermal motion is very small, so the slight interatomic forces can cause the atoms to settle down into a regular array like a pile of closely packed spheres.

The metals form a completely different class of substances. The bonding is of an entirely different kind. In a metal the bonding is not between adjacent atoms but is a property of the whole crystal. The valence electrons are not attached to one atom or to a pair of atoms but are shared throughout the crystal. Each atom contributes an electron to a universal pool of electrons, and the atomic positive ions reside in the sea of negative electrons. The electron sea holds the ions together like some kind of glue.

In the metals, since there are no special bonds in any particular direction, there is no strong directionality in the binding. They are still crystalline, however, because the total energy is lowest when the atomic ions are arranged in some definite array—although the energy of the preferred arrangement is not usually much lower than other possible ones. To a first approximation, the atoms of many metals are like small spheres packed in as tightly as possible.

30–3 The growth of crystals

Try to imagine the natural formation of crystals in the earth. In the earth's surface there is a big mixture of all kinds of atoms. They are being continually churned about by volcanic action, by wind, and by water—continually being moved about and mixed. Yet, by some trick, silicon atoms gradually begin to find each other, and to find oxygen atoms, to make silica. One atom at a time is added to the others to build up a crystal—the mixture gets unmixed. And somewhere nearby, sodium and chlorine atoms are finding each other and building up a crystal of salt.

How does it happen that once a crystal is started, it permits only a particular kind of atom to join on? It happens because the whole system is working toward the lowest possible energy. A growing crystal will accept a new atom if it is going to make the energy as low as possible. But how does it *know* that a silicon—or an oxygen—atom at some particular spot is going to result in the lowest possible energy? It does it by trial and error. In the liquid, all of the atoms are in perpetual motion. Each atom bounces against its neighbors about 10^{13} times every second. If it hits against the right spot of growing crystal, it has a somewhat smaller chance of jumping off again if the energy is low. By continually testing over periods of millions of years at a rate of 10^{13} tests per second, the atoms gradually build up at the places where they find their lowest energy. Eventually they grow into big crystals.

(a)

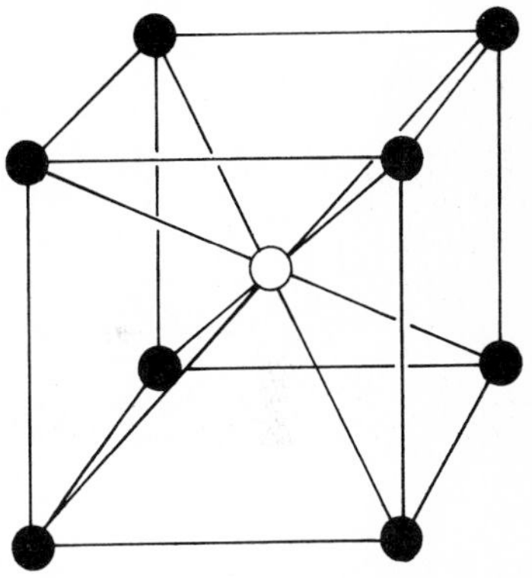

(b)

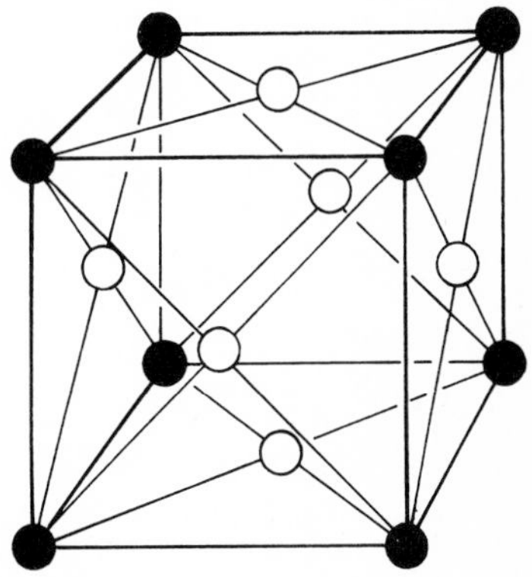

Fig. 30–4. The unit cell of cubic crystals: (a) body-centered, (b) face-centered.

30–4 Crystal lattices

The arrangement of the atoms in a crystal—the crystal *lattice*—can take on many geometric forms. We would like to describe first the simplest lattices, which are characteristic of most of the metals and of the solid form of the inert gases. They are the cubic lattices which can occur in two forms: the body-centered cubic, shown in Fig. 30–4(a), and the face-centered cubic shown in Fig. 30–4(b). The drawings show, of course, only one cube of the lattice; you are to imagine that the pattern is repeated indefinitely in three dimensions. Also, to make the drawing clearer, only the "centers" of the atoms are shown. In an actual crystal, the atoms are more like spheres in contact with each other. The dark and light spheres in the drawings may, in general, stand for different kinds of atoms or may be the same kind. For instance, iron has a body-centered cubic lattice at low temperatures, but a face-centered cubic lattice at higher temperatures. The physical properties are quite different in the two crystalline forms.

How do such forms come about? Imagine that you have the problem of packing spherical atoms together as tightly as possible. One way would be to start by making a layer in a "hexagonal close-packed array," as shown in Fig. 30–5(a). Then you could build up a second layer like the first, but displaced horizontally,

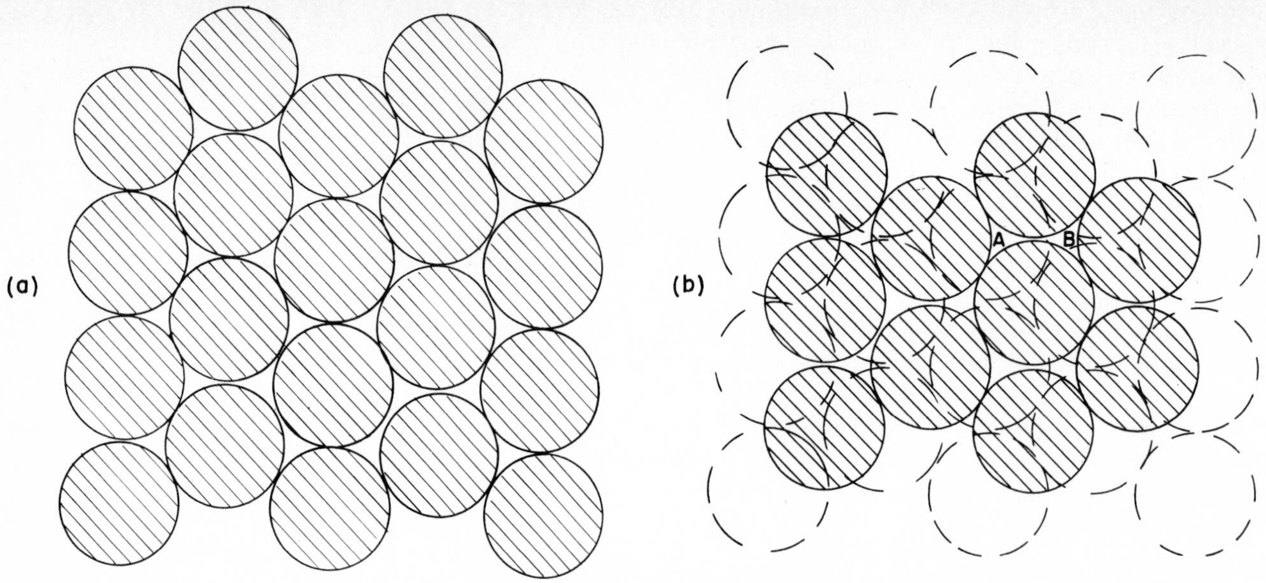

Fig. 30–5. Building up a hexagonal close-packed lattice.

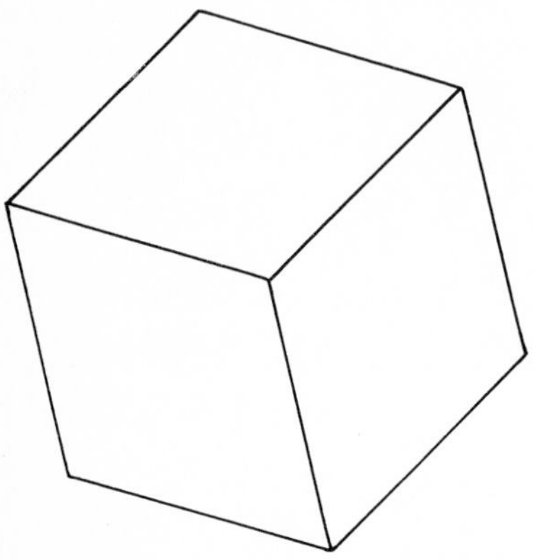

Fig. 30–6. Is this a hexagon or a cube seen from one corner?

as shown in Fig. 30–5(b). Next, you can put on the third layer. But notice! There are *two* distinct ways of placing the *third* layer. If you start the third layer by placing an atom at A in Fig. 30–5(b), each atom in the third layer is directly above an atom of the bottom layer. On the other hand, if you start the third layer by putting an atom at the position B, the atoms of the third layer will be centered at points exactly in the middle of a triangle formed by three atoms of the bottom layer. Any other starting place is equivalent to A or B, so there are only two ways of placing the third layer.

If the third layer has an atom at point B, the crystal lattice is a face-centered cubic—but seen at an angle. It seems funny that starting with hexagons you can end up with cubes. But notice that a cube looked at from a corner has a hexagonal outline. For instance, Fig. 30–6 could represent a plane hexagon or a cube seen in perspective!

If a third layer is added to Fig. 30–5(b) by starting with an atom at A, there is no cubical structure, and the lattice has instead only a hexagonal symmetry. It is clear that both possibilities we have described are equally close-packed.

Some metals—for example, copper and silver—choose the first alternative, the face-centered cubic. Others—for example, beryllium and magnesium—choose the other alternatives; they form hexagonal crystals. Clearly, which crystal lattice appears cannot depend only on the packing of little spheres, but must also be determined in part by other factors. In particular, it depends on the slight remaining angular dependence of the interatomic forces (or, in the case of the metals, on the energy of the electron pool). You will, no doubt, learn all about such things in your chemistry courses.

30–5 Symmetries in two dimensions

We would now like to discuss some of the properties of crystals from the point of view of their internal symmetries. The main feature of a crystal is that if you start at one atom and move to a corresponding atom one lattice unit away, you are again in the same kind of an environment. That's the fundamental proposition. But if you were an atom, there would be another kind of change that could take you again to the same environment—that is, another possible "symmetry." Figure 30–7(a) shows another possible "wallpaper-type" design (though one you have probably never seen). Suppose we compare the environments for points A and B. You might, at first, think that they are the same—but not quite. Points C and D are equivalent to A, but the environment of B is like that of A only if the surroundings are reversed, as in a mirror reflection.

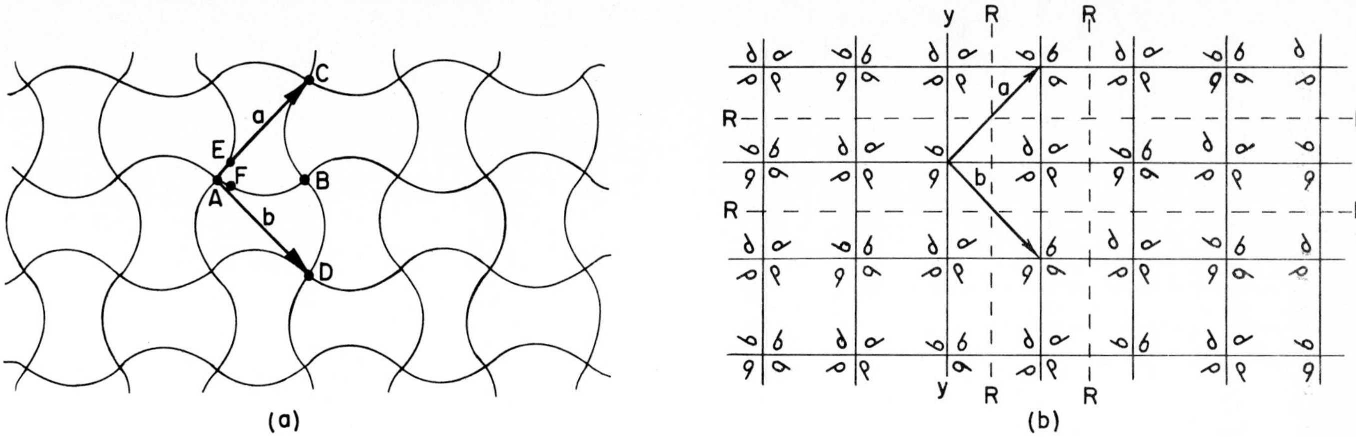

Fig. 30–7. A pattern of high symmetry.

There are other kinds of "equivalent" points in the pattern. For instance, the points E and F have the "same" environments except that one is rotated 90° with respect to the other. The pattern is quite special. A rotation of 90°—or any multiple of it—about a vertex such as A gives the same pattern all over again. A crystal with such a structure would have square corners on the outside, but inside it is more complicated than a simple cube.

Now that we have described some special examples, let's try to figure out all the possible symmetries a crystal can have. First, we consider what happens in a plane. A *plane* lattice can be defined by the two so-called *primitive* vectors that go from one point of the lattice to the two *nearest* equivalent points. The two vectors **1** and **2** are the primitive vectors of the lattice of Fig. 30–1. The two vectors $\boldsymbol{a}$ and $\boldsymbol{b}$ of Fig. 30–7(a) are the primitive vectors of the pattern there. We could, of course, equally well replace $\boldsymbol{a}$ by $-\boldsymbol{a}$, or $\boldsymbol{b}$ by $-\boldsymbol{b}$. Since $\boldsymbol{a}$ and $\boldsymbol{b}$ are equal in magnitude and at right angles, a rotation of 90° turns $\boldsymbol{a}$ into $\boldsymbol{b}$, and $\boldsymbol{b}$ into $-\boldsymbol{a}$, giving the same lattice once again.

We see that there are lattices which have a "four-sided" symmetry. And we have described earlier a close-packed array based on a hexagon which could have a six-sided symmetry. A rotation of the array of circles in Fig. 30–5(a) by an angle of 60° about the center of any circle brings the pattern back to itself.

What other kinds of rotational symmetry are there? Can we have, for example, a fivefold or an eightfold rotational symmetry? It is easy to see that they are impossible. *The only symmetry with more sides than four is a six-sided symmetry.* First, let's show that more than sixfold symmetry is impossible. Suppose we try to imagine a lattice with two equal primitive vectors with an enclosed angle less than 60°, as in Fig. 30–8(a). We are to suppose that points B and C are equivalent to A, and that $\boldsymbol{a}$ and $\boldsymbol{b}$ are the two *shortest* vectors from A to its equivalent neighbors. But that is clearly wrong, because the distance between B and C is shorter than from either one to A. There must be a neighbor at D equivalent to A which is closer than B or C. We should have chosen $\boldsymbol{b}'$ as one of our primitive vectors. So the angle between the two primitive vectors must be 60° or larger. Octagonal symmetry is not possible.

What about fivefold symmetry? If we assume that the primitive vectors $\boldsymbol{a}$ and $\boldsymbol{b}$ have equal lengths and make an angle of $2\pi/5 = 72°$, as in Fig. 30–8(b), then there should also be an equivalent lattice point at D, at 72° from C. But the vector $\boldsymbol{b}'$ from E to D is then less than $\boldsymbol{b}$, so $\boldsymbol{b}$ is not a primitive vector. There can be no fivefold symmetry. The only possibilities that do not get us into this kind of difficulty are $\theta = 60°$, 90°, or 120°. Zero or 180° are also clearly possible. One way of stating our result is that the pattern can be left unchanged by a rotation of one full turn (no change at all), one-half of a turn, one-third, one-fourth, or one-sixth of a turn. And those are all the possible rotational symmetries in a plane—a total of five. If $\theta = 2\pi/n$, we speak of an "n-fold" symmetry. We say

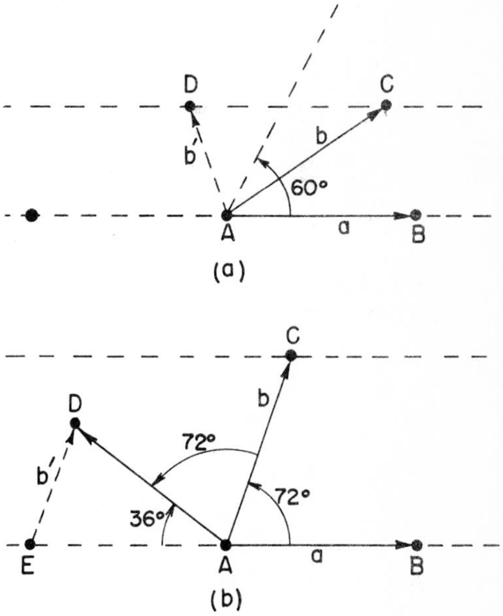

Fig. 30–8. (a) Rotational symmetries greater than sixfold are not possible. (b) Fivefold rotational symmetry is not possible.

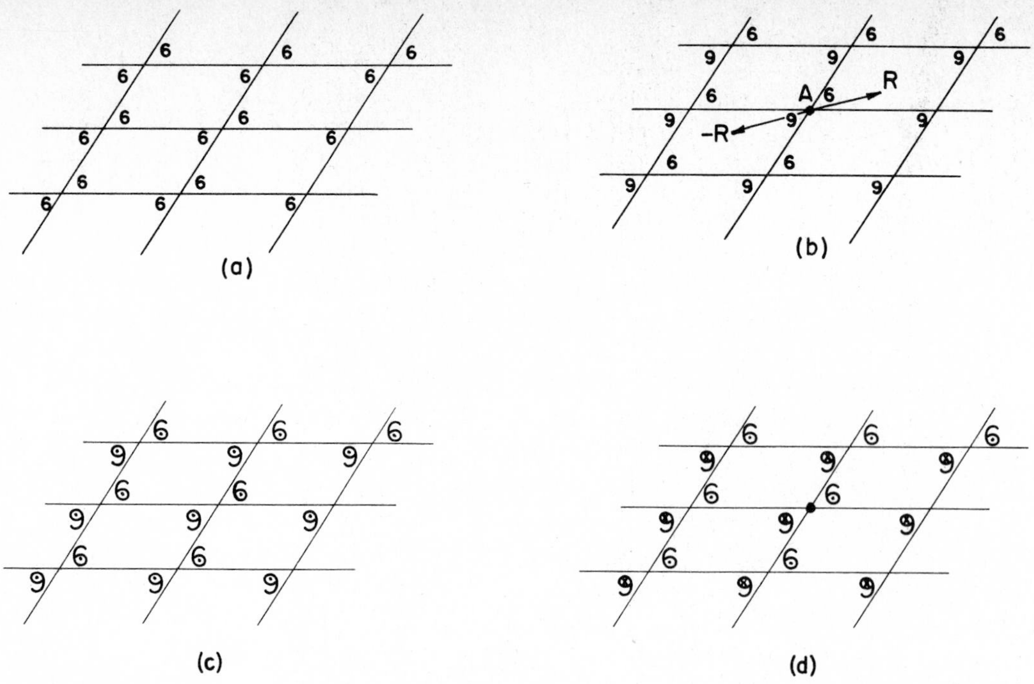

Fig. 30–9. Symmetry under inversion. Pattern (b) is unchanged if $\boldsymbol{R} \to -\boldsymbol{R}$, but pattern (a) is changed. In three dimensions pattern (d) *is* symmetric under an inversion but (c) is not.

that a pattern with n equal to 4 or to 6 has a "higher symmetry" than one with n equal to 1 or to 2.

Returning to Fig. 30–7(a), we see that the pattern has a fourfold rotational symmetry. We have drawn in Fig. 30–7(b) another design which has the same symmetry properties as part (a). The little comma-like figures are asymmetric objects which serve to define the symmetry of the design inside of each square. Notice that the commas are reversed in alternate squares, so that the unit cell is larger than one of the small squares. If there were no commas, the pattern would still have fourfold symmetry, but the unit cell would be smaller. The patterns of Fig. 30–7 also have other symmetry properties. For instance, a reflection about any of the broken lines R–R reproduces the same pattern.

The patterns of Fig. 30–7 have still another kind of symmetry. If the pattern is reflected about the line Y–Y *and* shifted one square to the right (or left), we get back the original pattern. The line Y–Y is called a "glide" line.

These are all the possible symmetries in two dimensions. There is one more spatial symmetry operation which is equivalent *in two dimensions* to a 180° rotation, but which is a quite distinct operation in three dimensions. It is *inversion*. By an inversion we mean that any point at the vector displacement $\boldsymbol{R}$ from some origin [for instance, the point A in Fig. 30–9(b)] is moved to the point at $-\boldsymbol{R}$.

An inversion of pattern (a) of Fig. 30–9 produces a new pattern, but an inversion of pattern (b) reproduces the same pattern. For a two-dimensional pattern (as you can see from the figure), an inversion of the pattern (b) through the point A is equivalent to a rotation of 180° about the same point. Suppose, however, we make the pattern in Fig. 30–9(b) three dimensional by imagining that the little 6's and 9's each have an "arrow" *pointing out of the page*. After an inversion in three dimensions all the arrows will be reversed, so the pattern is *not* reproduced. If we indicate the heads and tails of the arrows by dots and crosses, respectively, we can make a *three-dimensional* pattern, as in Fig. 30–9(c), which is *not* symmetric under an inversion, or we can make a pattern like the one shown in (d), which *does* have such a symmetry. Notice that it is *not* possible to imitate a three-dimensional inversion by any combination of rotations.

If we characterize the "symmetry" of a pattern—or lattice—by the kinds of symmetry operations we have been describing, it turns out that for two dimensions 17 distinct patterns are possible. We have drawn one pattern of the lowest possible symmetry in Fig. 30–1, and one of high symmetry in Fig. 30–7. We will leave you with the game of trying to figure out all of the 17 possible patterns.

It is peculiar how few of the 17 possible patterns are used in making wallpaper and fabrics. One always sees the same three or four basic patterns. Is this because of a lack of imagination of designers, or because many of the possible patterns are not pleasing to the eye?

30–6 Symmetries in three dimensions

So far we have talked only about patterns in two dimensions. What we are really interested in, however, are patterns of atoms in three dimensions. First, it is clear that a three-dimensional crystal will have *three* primitive vectors. If we then ask about the possible symmetry operations in three dimensions, we find that there are 230 different possible symmetries! For some purposes, these 230 types can be grouped into seven classes, which are drawn in Fig. 30–10. The lattice with the least symmetry is called the *triclinic*. Its unit cell is a parallelepiped. The primitive vectors are of different lengths, and no two of the angles between them are equal. There is no possibility of any rotational or reflection symmetry. There are, however, still two possible symmetries—the unit cell is, or is not, changed by an inversion through the vertex. (By an inversion in three dimensions, we again mean that spatial displacements $\boldsymbol{R}$ are replaced by $-\boldsymbol{R}$—in other words, that (x, y, z) goes into $(-x, -y, -z)$. So the triclinic lattice has only two possible symmetries, unless there is some special relation among the primitive vectors. For example, if all the vectors are equal and are separated by equal angles, one has the *trigonal* lattice shown in the figure. This figure can have an additional symmetry; it may be unchanged by a rotation about the long, body diagonal.

If one of the primitive vectors, say $\boldsymbol{c}$, is at right angles to the other two, we get a *monoclinic* unit cell. A new symmetry is possible—a rotation by 180° about $\boldsymbol{c}$. The *hexagonal* cell is a special case in which the vectors $\boldsymbol{a}$ and $\boldsymbol{b}$ are equal and the angle between them is 60°, so that a rotation of 60°, or 120°, or 180° about the vector $\boldsymbol{c}$ repeats the same lattice (for certain internal symmetries).

If all three primitive vectors are at right angles, but of different lengths, we get the *orthorhombic* cell. The figure is symmetric for rotations of 180° about the three axes. Higher-order symmetries are possible with the *tetragonal* cell, which has all right angles and two equal primitive vectors. Finally, there is the *cubic* cell, which is the most symmetric of all.

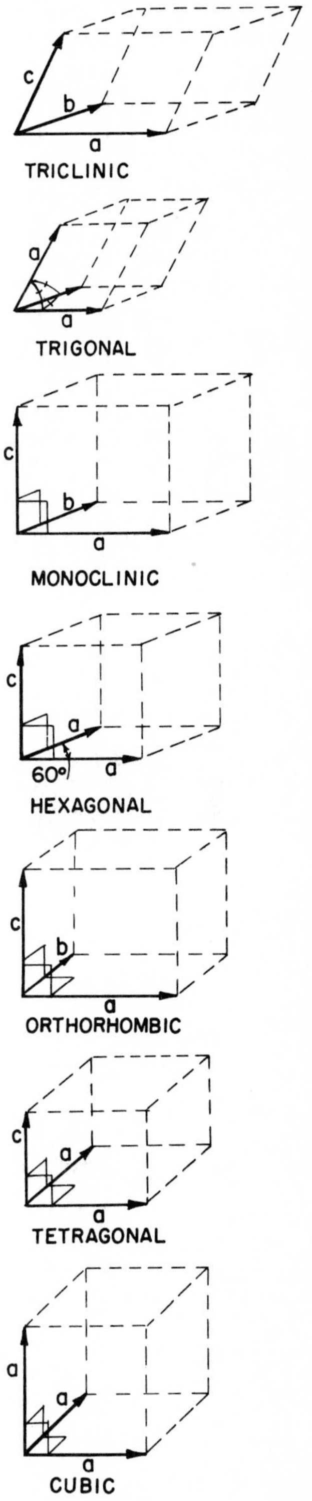

Fig. 30–10. The seven classes of crystal lattices.

The point of all this discussion about symmetries is that the internal symmetries of the crystals show up—sometimes in subtle ways—in the macroscopic physical properties of the crystal. For instance, a crystal will, in general, have a tensor electric polarizability. If we describe the tensor in terms of the ellipsoid of polarization, we should expect that some of the crystal symmetries should show up also in the ellipsoid. For example, a cubic crystal is symmetric with respect to a rotation of 90° about any one of three orthogonal directions. Clearly, the only ellipsoid with this property is a sphere. *A cubic crystal must be an isotropic dielectric.*

On the other hand, a tetragonal crystal has a fourfold rotational symmetry. Its ellipsoid must have two of its principal axes equal, and the third must be parallel to the axis of the crystal. Similarly, since the orthorhombic crystal has twofold rotational symmetry about three orthogonal axes, its axes must coincide with the axes of the polarization ellipsoid. In a like manner, *one* of the axes of a monoclinic crystal must be parallel to *one* of the principal axes of the ellipsoid, though we can't say anything about the other axes. Since a triclinic crystal has no rotational symmetry, the ellipsoid can have any orientation at all.

As you can see, we can make a big game of figuring out the possible symmetries and relating them to the possible physical tensors. We have considered only the polarization tensor, but things get more complicated for others—for instance, for the tensor of elasticity. There is a branch of mathematics called "group theory" that deals with such subjects, but usually you can figure out what you want with common sense.

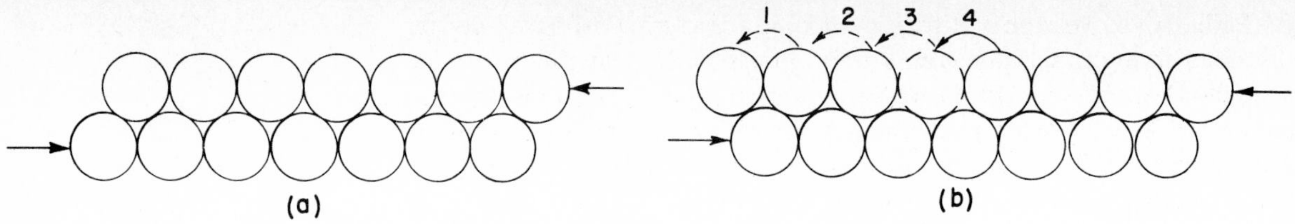

Fig. 30–11. Slippage of crystal planes.

30–7 The strength of metals

We have said that metals usually have a simple cubic crystal structure; we want now to discuss their mechanical properties—which depend on this structure. Metals are, generally speaking, very "soft," because it is easy to slide one layer of the crystal over the next. You may think: "That's ridiculous; metals are strong." Not so, a *single crystal* of a metal can be distorted very easily.

Suppose we look at two layers of a crystal subjected to a shear force, as shown in the diagram of Fig. 30–11(a). You might at first think the whole layer would resist motion until the force was big enough to push the whole layer "over the hump," so that it shifted one notch to the left. Although slipping does occur along a plane, it doesn't happen that way. (If it did, you would calculate that the metal is much stronger than it really is.) What happens is more like one atom going at a time; first the atom on the left makes its jump, then the next, and so on, as indicated in Fig. 30–11(b). In effect it is the vacant space between two atoms that quickly travels to the right, with the net result that the whole second layer has moved over one atomic spacing. The slipping goes this way because it takes much less energy to lift one atom at a time over the hump than to lift a whole row. Once the force is enough to start the process, it goes the rest of the way very fast.

Fig. 30–12. A photograph of a small crystal of copper after stretching. [Courtesy of S. S. Brenner, Senior Scientist, United States Steel Research Center, Monroeville, Pa.]

It turns out that in a real crystal, slipping will occur repeatedly at one plane, then will stop there and start at some other plane. The details of why it starts and stops are quite mysterious. It is, in fact, quite strange that successive regions of slip are often fairly evenly spaced. Figure 30–12 shows a photograph of a tiny, thin copper crystal that has been stretched. You can see the various planes where slipping has occurred.

The sudden slipping of individual crystal planes is quite apparent if you take a piece of tin wire that has large crystals in it and stretch it while holding it next to your ear. You can hear a rush of "ticks" as the planes snap to their new positions, one after the other.

The problem of having a "missing" atom in one row is somewhat more difficult than it might appear from Fig. 30–11. When there are more layers, the situation must be something like that shown in Fig. 30–13. Such an imperfection in a crystal is called a *dislocation*. It is presumed that such dislocations are either present when the crystal was formed or are generated at some notch or crack at the surface. Once they are produced, they can move relatively freely through the crystal. The gross distortions result from the motions of many of such dislocations.

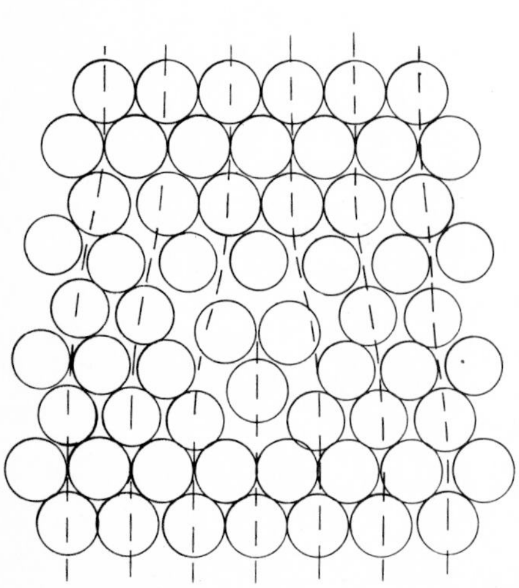

Fig. 30–13. A dislocation in a crystal.

Dislocations can move freely—that is, they require little extra energy—so long as the rest of the crystal has a perfect lattice. But they may get "stuck" if they encounter some other kind of imperfection in the crystal. If it takes a lot of energy for them to pass the imperfection, they will be stopped. This is precisely the mechanism that gives strength to *imperfect* metal crystals. Pure iron crystals are quite soft, but a small concentration of impurity atoms may cause enough imperfections to effectively immobilize the dislocations. As you know, steel, which is primarily iron, is very hard. To make steel, a small amount of carbon is dissolved in the iron melt; if the melt is cooled rapidly, the carbon precipitates out in little grains, making many microscopic distortions in the lattice. The dislocations can no longer move about, and the metal is hard.

Pure copper is very soft, but can be "work-hardened." This is done by hammering on it or bending it back and forth. In this case, many new dislocations of various kinds are made which interfere with one another, cutting down their mobility. Perhaps you've seen the trick of taking a bar of "dead soft" copper and gently bending it around someone's wrist as a bracelet. In the process, it becomes work-hardened and cannot easily be unbent again! A work-hardened metal like copper can be made soft again by annealing at a high temperature. The thermal motion of the atoms "irons out" the dislocations and makes large single crystals again. We have, so far, described only the so-called *slip* dislocation. There are many other kinds, one of which is the *screw* dislocation shown in Fig. 30–14. Such dislocations often play an important part in crystal growth.

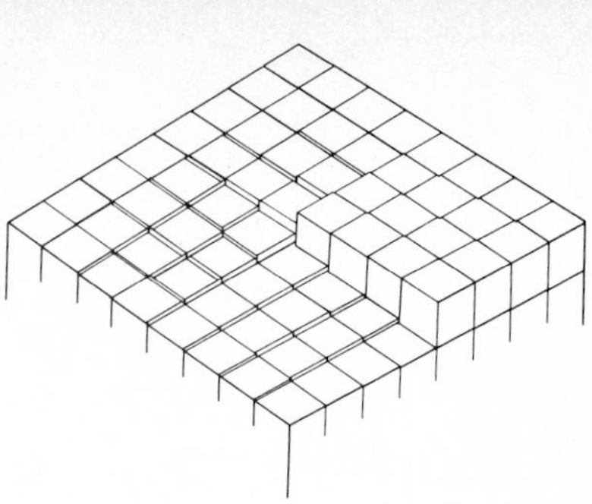

Fig. 30–14. A screw dislocation. [From Charles Kittel, *Introduction to Solid State Physics*, John Wiley and Sons, Inc., New York, 2nd ed., 1956.]

30–8 Dislocations and crystal growth

One of the great puzzles for a long time was how crystals can possibly grow. We have described how it is that each atom might, by repeated testing, determine whether it was better to be in the crystal or not. But that means that each atom must find a place of low energy. However, an atom put on a new surface is only bound by one or two bonds from below, and doesn't have the same energy it would have if it were placed in a corner, where it would have atoms on three sides. Suppose we imagine a growing crystal as a stack of blocks, as shown in Fig. 30–15. If we try a new block at, say, position *A*, it will have only one of the six neighbors it should ultimately get. With so many bonds lacking, its energy is not very low. It would be better off at position *B*, where it already has one-half of its quota of bonds. Crystals do indeed grow by attaching new atoms at places like *B*.

What happens, though, when that line is finished? To start a new line, an atom must come to rest with only two sides attached, and that is again not very likely. Even if it did, what would happen when the layer was finished? How could a new layer get started? One answer is that the crystal prefers to grow at a dislocation, for instance around a screw dislocation like the one shown in Fig. 30–14. As blocks are added to this crystal, there is always some place where there are three available bonds. The crystal prefers, therefore, to grow with a dislocation built in. Such a spiral pattern of growth is shown in Fig. 30–16, which is a photograph of a single crystal of paraffin.

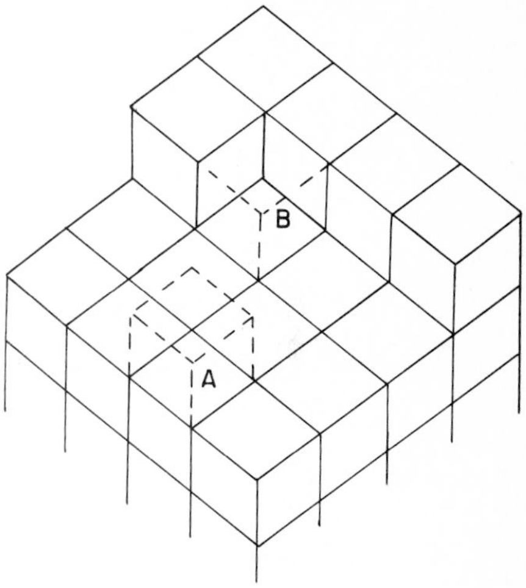

Fig. 30–15. Crystal growth.

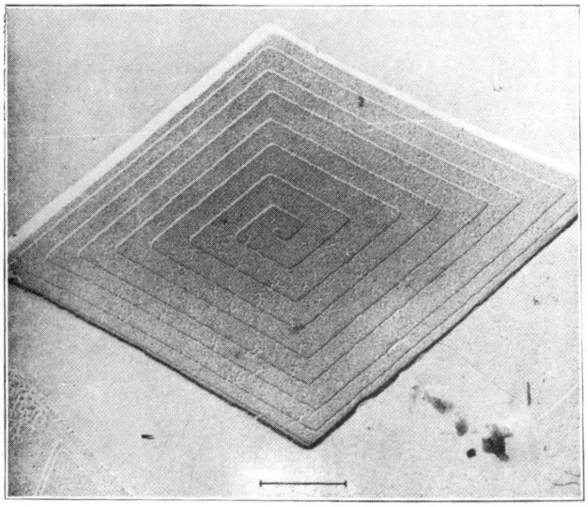

Fig. 30–16. A paraffin crystal which has grown around a screw dislocation. [From Charles Kittel, *Introduction to Solid State Physics*, John Wiley and Sons, Inc., New York, 2nd ed., 1956.]

30–9 The Bragg-Nye crystal model

We cannot, of course, see what goes on with the individual atoms in a crystal. Also, as you realize by now, there are many complicated phenomena that are not easy to treat quantitatively. Sir Lawrence Bragg and J. F. Nye have devised a scheme for making a model of a metallic crystal which shows in a striking way many of the phenomena that are believed to occur in a real metal. In the following pages we have reproduced their original article, which describes their method and shows some of the results they obtained with it. (The article is reprinted from the *Proceedings of the Royal Society of London*, Vol. 190, September 1947, pp. 474–481 —with the permission of the authors and of the Royal Society.)

A dynamical model of a crystal structure

By Sir Lawrence Bragg, F.R.S. and J. F. Nye
Cavendish Laboratory, University of Cambridge

(*Received* 9 *January* 1947—*Read* 19 *June* 1947)

[Plates 8 to 21]

The crystal structure of a metal is represented by an assemblage of bubbles, a millimetre or less in diameter, floating on the surface of a soap solution. The bubbles are blown from a fine pipette beneath the surface with a constant air pressure, and are remarkably uniform in size. They are held together by surface tension, either in a single layer on the surface or in a three-dimensional mass. An assemblage may contain hundreds of thousands of bubbles and persists for an hour or more. The assemblages show structures which have been supposed to exist in metals, and simulate effects which have been observed, such as grain boundaries, dislocations and other types of fault, slip, recrystallization, annealing, and strains due to 'foreign' atoms.

1. The bubble model

Models of crystal structure have been described from time to time in which the atoms are represented by small floating or suspended magnets, or by circular disks floating on a water surface and held together by the forces of capillary attraction. These models have certain disadvantages; for instance, in the case of floating objects in contact, frictional forces impede their free relative movement. A more serious disadvantage is that the number of components is limited, for a large number of components is required in order to approach the state of affairs in a real crystal. The present paper describes the behaviour of a model in which the atoms are represented by small bubbles from 2·0 to 0·1 mm. in diameter floating on the surface of a soap solution. These small bubbles are sufficiently persistent for experiments lasting an hour or more, they slide past each other without friction, and they can be produced in large numbers. Some of the illustrations in this paper were taken from assemblages of bubbles numbering 100,000 or more. The model most nearly represents the behaviour of a metal structure, because the bubbles are of one type only and are held together by a general capillary attraction, which represents the binding force of the free electrons in the metal. A brief description of the model has been given in the *Journal of Scientific Instruments* (Bragg 1942*b*).

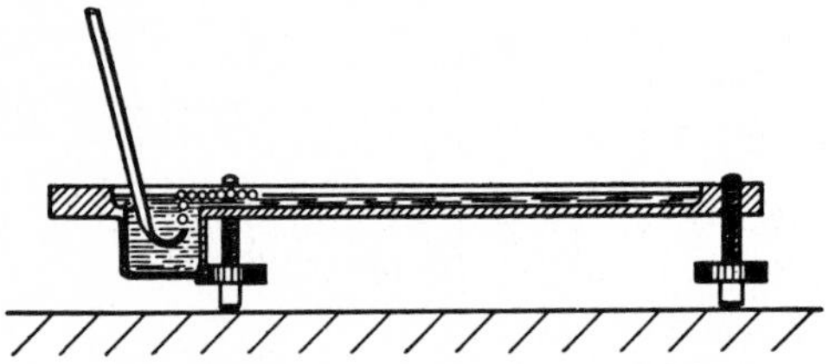

Figure 1. Apparatus for producing rafts of bubbles.

2. Method of formation

The bubbles are blown from a fine orifice, beneath the surface of a soap solution. We have had the best results with a solution the formula of which was given to us by Mr Green of the Royal Institution. 15·2 c.c. of oleic acid (pure redistilled) is well shaken in 50 c.c. of distilled water. This is mixed thoroughly with 73 c.c. of 10 % solution of tri-ethanolamine and the mixture made up to 200 c.c. To this is added 164 c.c. of pure glycerine. It is left to stand and the clear liquid is drawn off from below. In some experiments this was diluted in three times its volume of water to reduce viscosity. The orifice of the jet is about 5 mm. below the surface. A constant air pressure of 50 to 200 cm. of water is supplied by means of two Winchester flasks. Normally the bubbles are remarkably uniform in size. Occasionally they issue in an irregular manner, but this can be corrected by a change of jet or of pressure. Unwanted bubbles can easily be destroyed by playing a small flame over the surface. Figure 1 shows the apparatus. We have found it of advantage to blacken the bottom of the vessel, because details of structure, such as grain boundaries and dislocations, then show up more clearly.

Figure 2, plate 8, shows a portion of a raft' or two-dimensional crystal of bubbles. Its regularity can be judged by looking at the figure in a glancing direction. The size of the bubbles varies with the aperture, but does not appear to vary to any marked degree with the pressure or the depth of the orifice beneath the surface. The main effect of increasing the pressure is to increase the rate of issue of the bubbles. As an example, a thick-walled jet of 49μ bore with a pressure of 100 cm. produced bubbles of 1·2 mm. in diameter. A thin-walled jet of 27μ diameter and a pressure of 180 cm. produced bubbles of 0·6 mm. diameter. It is convenient to refer to bubbles of 2·0 to 1·0 mm. diameter as 'large' bubbles, those from 0·8 to 0·6 mm. diameter as 'medium' bubbles, and those from 0·3 to 0·1 mm. diameter as 'small' bubbles, since their behaviour varies with their size.

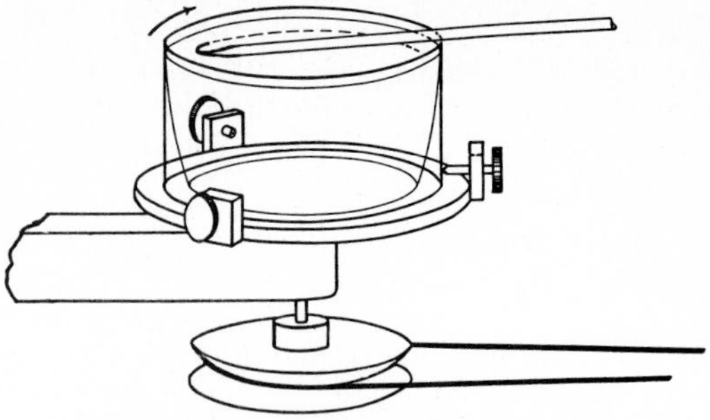

Figure 3. Apparatus for producing bubbles of small size.

With this apparatus we have not found it possible to reduce the size of the jet and so produce bubbles of smaller diameter than 0·6 mm. As it was desired to experiment with very small bubbles, we had recourse to placing the soap solution in a rotating vessel and introducing a fine jet as nearly as possible parallel to a stream line. The bubbles are swept away as they form, and under steady conditions are reasonably uniform. They issue at a rate of one thousand or more per second, giving a high-pitched note. The soap solution mounts up in a steep wall around the perimeter of the vessel while it is rotating, but carries back most of the bubbles with it when rotation ceases. With this device, illustrated in figure 3, bubbles down to 0·12 mm. in diameter can be obtained. As an example, an orifice 38μ across in a thin-walled jet, with a pressure of 190 cm. of water, and a speed of the fluid of 180 cm./sec. past the orifice, produced bubbles of 0·14 mm. diameter. In this case a dish of diameter 9·5 cm. and speed of 6 rev./sec. was used. Figure 4, plate 8, is an enlarged picture of these 'small' bubbles and shows their degree of regularity; the pattern is not as perfect with a rotating as with a stationary vessel, the rows being seen to be slightly irregular when viewed in a glancing direction.

These two-dimensional crystals show structures which have been supposed to exist in metals, and simulate effects which have been observed, such as grain boundaries, dislocations and other types of fault, slip, recrystallization, annealing, and strains due to 'foreign' atoms.

3. Grain boundaries

Figures 5*a*, 5*b* and 5*c*, plates 9 and 10, show typical grain boundaries for bubbles of 1·87, 0·76 and 0·30 mm. diameter respectively. The width of the disturbed area at the boundary, where the bubbles have an irregular distribution, is in general greater the smaller the bubbles. In figure 5*a*, which shows portions of several adjacent grains, bubbles at a boundary between two grains adhere definitely to one crystalline arrangement or the other. In figure 5*c* there is a marked 'Beilby layer' between the two grains. The small bubbles, as will be seen, have a greater rigidity than the large ones, and this appears to give rise to more irregularity at the interface.

Separate grains show up distinctly when photographs of polycrystalline rafts such as figures 5*a* to 5*c*, plates 9 and 10, and figures 12*a* to 12*e*, plates 14 to 16, are viewed obliquely. With suitable lighting, the floating raft of bubbles itself when viewed obliquely resembles a polished and etched metal in a remarkable way.

It often happens that some 'impurity atoms', or bubbles which are markedly larger or smaller than the average, are found in a polycrystalline raft, and when this is so a large proportion of them are situated at the grain boundaries. It would be incorrect to say that the irregular bubbles make their way to the boundaries; it is a defect of the model that no diffusion of bubbles through the structure can take place, mutual adjustments of neighbours alone being possible. It appears that the boundaries tend to readjust themselves by the growth of one crystal at the expense of another till they pass through the irregular atoms.

4. Dislocations

When a single crystal or polycrystalline raft is compressed, extended, or otherwise deformed it exhibits a behaviour very similar to that which has been pictured for metals subjected to strain. Up to a certain limit the model is within its elastic range. Beyond that point it yields by slip along one of the three equally inclined directions of closely packed rows. Slip takes place by the bubbles in one row moving forward over those in the next row by an amount equal to the distance between neighbours. It is very interesting to watch this process taking place. The movement is not simultaneous along the whole row but begins at one end with the appearance of a 'dislocation', where there is locally one more bubble in the rows on one side of the slip line as compared with those on the other. This dislocation then runs along the slip line from one side of the crystal to the other, the final result being a slip by one 'inter-atomic' distance. Such a process has been invoked by Orowan, by Polanyi and by Taylor to explain the small forces required to produce plastic gliding in metal structures. The theory put forward by Taylor (1934) to explain the mechanism of plastic deformation of crystals considers the mutual action and equilibrium of such dislocations. The bubbles afford a very striking picture of what has been supposed to take place in the metal. Sometimes the dislocations run along quite slowly, taking a matter of seconds to cross a crystal; stationary dislocations also are to be seen in crystals which are not homogeneously strained. They appear as short black lines, and can be seen in the series of photographs, figures 12*a* to 12*e*, plates 14 to 16. When a polycrystalline raft is compressed, these dark lines are seen to be dashing about in all directions across the crystals.

Figures 6*a*, 6*b* and 6*c*, plates 10 and 11, show examples of dislocations. In figure 6*a*, where the diameter of the bubbles is 1·9 mm., the dislocation is very local, extending over about six bubbles. In figure 6*b* (diameter 0·76 mm.) it extends over twelve bubbles, and in figure 6*c* (diameter 0·30 mm.) its influence can be traced for a length of about fifty bubbles. The greater rigidity of the small bubbles leads to longer dislocations. The study of any mass of bubbles shows, however, that there is not a standard length of dislocation for each size. The length depends upon the nature of the strain in the crystal. A boundary between two crystals with corresponding axes at approximately 30° (the maximum angle which can occur) may be regarded as a series of dislocations in alternate rows, and in this case the dislocations are very short. As the angle between the neighbouring crystals decreases, the dislocations occur at wider intervals and at the same time become longer, till one finally has single dislocations in a large body of perfect structure as shown in figures 6*a*, 6*b* and 6*c*.

Figure 7, plate 11, shows three parallel dislocations. If we call them positive and negative (following Taylor) they are positive, negative, positive, reading from left to right. The strip between the last two has three bubbles in excess, as can be seen by looking along the rows in a horizontal direction. Figure 8, plate 12, shows a dislocation projecting from a grain boundary, an effect often observed.

Figure 9, plate 12, shows a place where two bubbles take the place of one. This may be regarded as a limiting case of positive and negative dislocations on neighbouring rows, with the compressive sides of the dislocations facing each other. The contrary case would lead to a hole in the structure, one bubble being missing at the point where the dislocations met.

5. Other types of fault

Figure 10, plate 12, shows a narrow strip between two crystals of parallel orientation, the strip being crossed by a number of fault lines where the bubbles are not in close packing. It is in such places as these that recrystallization may be expected. The boundaries approach and the strip is absorbed into a wider area of perfect crystal.

Figures 11*a* to 11*g*, plates 13 and 14 are examples of arrangements which frequently appear in places where there is a local deficiency of bubbles. While a dislocation is seen as a dark stripe in a general view, these structures show up in the shape of the letter V or as triangles. A typical V structure is seen in figure 11*a*. When the model is being distorted, a V structure is formed by two dislocations meeting at an inclination of 60°; it is destroyed by the dislocations continuing along their paths. Figure 11*b* shows a small triangle, which also embodies a dislocation, for it will be noticed that the rows below the fault have one more bubble than those below. If a mild amount of 'thermal movement' is imposed by gentle agitation of one side of the crystal, such faulty places disappear and a perfect structure is formed.

Here and there in the crystals there is a blank space where a bubble is missing, showing as a black dot in a general view. Examples occur in figure 11*g*. Such a gap cannot be closed by a local readjustment, since filling the hole causes another to appear. Such holes both appear and disappear when the crystal is 'cold-worked'.

These structures in the model suggest that similar local faults may exist in an actual metal. They may play a part in processes such as diffusion or the order-disorder change by reducing energy barriers in their neighbourhood, and act as nuclei for crystallization in an allotropic change.

6. Recrystallization and annealing

Figures 12*a* to 12*e*, plates 14 to 16, show the same raft of bubbles at successive times. A raft covering the surface of the solution was given a vigorous stirring with a glass rake, and then left to adjust itself. Figure 12*a* shows its aspect about 1 sec. after stirring has ceased. The raft is broken into a number of small 'crystallites'; these are in a high state of non-homogeneous strain as is shown by the numerous dislocations and other faults. The following photograph (figure 12*b*) shows the same raft 32 sec. later. The small grains have coalesced to form larger grains, and much of the strain has disappeared in the process. Recrystallization takes place right through the series, the last three photographs of which show the appearance of the raft 2, 14 and 25 min. after the initial stirring. It is not possible to follow the rearrangement for much longer times, because the bubbles shrink after long standing, apparently due to the diffusion of air through their walls, and they also become thin and tend to burst. No agitation was given to the model during this process. An ever slower process of rearrangement goes on, the movement of the bubbles in one part of the raft setting up strains which activate a rearrangement in a neighbouring part, and that in its turn still another.

A number of interesting points are to be seen in this series. Note the three small grains at the points indicated by the co-ordinates *AA*, *BB*, *CC*. *A* persists, though changed in form, throughout the whole series. *B* is still present after 14 min., but has disappeared in 25 min., leaving behind it four dislocations marking internal strain in the grain. Grain *C* shrinks and finally disappears in figure 12*d*, leaving a hole and a V which has disappeared in figure 12*e*. At the same time the ill-defined boundary in figure 12*d* at *DD* has become a definite one in figure 12*e*. Note also the straightening out of the grain boundary in the neighbourhood of *EE* in figures 12*b* to 12*e*. Dislocations of various lengths can be seen, marking all stages between a slight warping of the structure and a definite boundary. Holes where bubbles are missing show up as black dots. Some of these holes are formed or filled up by movements of dislocations, but others represent places where a bubble has burst. Many examples of V's and some of triangles can be seen. Other interesting points will be apparent from a study of this series of photographs.

Figures 13*a*, 13*b* and 13*c*, plate 17, show a portion of a raft 1 sec., 4 sec. and 4 min. after the stirring process, and is interesting as showing two successive stages in the relaxation towards a more perfect arrangement. The changes show up well when one looks in a glancing direction across the page. The arrangement is very broken in figure 13*a*. In figure 13*b* the bubbles have grouped themselves in rows, but the curvature of these rows indicates a high degree of internal strain. In figure 13*c* this strain has been relieved by the formation of a new boundary at *A–A*, the rows on either side now being straight. It would appear that the energy of this strained crystal is greater than that of the intercrystalline boundary. We are indebted to Messrs Kodak for the photographs of figure 13, which were taken when the cinematograph film referred to below was produced.

7. Effect of impurity atom

Figure 14, plate 18, shows the widespread effect of a bubble which is of the wrong size. If this figure is compared with the perfect rafts shown in figures 2 and 4, plate 8, it will be seen that three bubbles, one larger and two smaller than normal, disturb the regularity of the rows over the whole of the figure. As has been mentioned above, bubbles of the wrong size are generally found in the grain boundaries, where holes of irregular size occur which can accommodate them.

8. Mechanical properties of the two-dimensional model

The mechanical properties of a two-dimensional perfect raft have been described in the paper referred to above (Bragg 1942*b*). The raft lies between two parallel springs dipping horizontally in the surface of the soap solution. The pitch of the springs is adjusted to fit the spacing of the rows of bubbles, which then adhere firmly to them. One spring can be translated parallel to itself by a micrometer screw, and the other is supported by two thin vertical glass fibres. The shearing stress can be measured by noting the deflexion of the glass fibres. When subjected to a shearing strain, the raft obeys Hooke's law of elasticity up to the point where the elastic limit is reached. It then slips along some intermediate row by an amount equal to the width of one bubble. The elastic shear and slip can be repeated several times. The elastic limit is approximately reached when one side of the raft has been sheared by an amount equal to a bubble width past the other side. This feature supports the basic assumption made by one of us in the calculation of the elastic limit of a metal (Bragg 1942*a*), in which it is supposed that each crystallite in a cold-worked metal only yields when the strain in it has reached such a value that energy is released by the slip.

A calculation has been made by M. M. Nicolson of the forces between the bubbles, and will be published shortly. It shows two interesting points. The curve for the variation of potential energy with distance between centres is very similar to those which have been plotted for atoms. It has a minimum for a distance between centres slightly less than a free bubble diameter, and rises sharply for smaller distances. Further, the rise is extremely sharp for bubbles of 0·1 mm. diameter but much less so for bubbles of 1 mm. diameter, thus confirming the impression given by the model that the small bubbles behave as if they were much more rigid than the large ones.

9. Three-dimensional assemblages

If the bubbles are allowed to accumulate in multiple layers on the surface, they form a mass of three-dimensional 'crystals' with one of the arrangements of closest packing. Figure 15, plate 18, shows an oblique view of such a mass; its resemblance to a polished and etched metal surface is noticeable. In figure 16, plate 20, a similar mass is seen viewed normally. Parts of the structure are definitely in cubic closest packing, the outer surface being the (111) face or (100) face. Figure 17*a*, plate 19, shows a (111) face. The outlines of the three bubbles on which each upper bubble rests can be clearly seen, and the next layer of these bubbles is faintly visible in a position not beneath the uppermost layer, showing that the packing of the (111) planes has the well-known cubic succession. Figure 17*b*, plate 19, shows a (100) face with each bubble resting on four others. The cubic axes are of course inclined at 45° to the close-packed rows of the surface layer. Figure 17*c*, plate 19, shows a twin in the cubic structure across the face (111). The uppermost faces are (111) and (100), and they make a small angle with each other, though this is not apparent in the figure; it shows up in an oblique view. Figure 17*d*, plate 19, appears to show both the cubic and hexagonal succession of closely packed planes, but it is difficult to verify whether the left-hand side follows the true hexagonal close-packed structure because it is not certain that the assemblage had a depth of more than two layers at this point. Many instances of twins, and of intercrystalline boundaries, can be seen in figure 16, plate 20.

Figure 18, plate 21, shows several dislocations in a three-dimensional structure subjected to a bending strain.

10. Demonstration of the model

With the co-operation of Messrs Kodak, a 16 mm. cinematograph film has been made of the movements of the dislocations and grain boundaries when single crystal and polycrystalline rafts are sheared, compressed, or extended. Moreover, if the soap solution is placed in a glass vessel with a flat bottom, the model lends itself to projection on a large scale by transmitted light. Since a certain depth is required for producing the bubbles, and the solution is rather opaque, it is desirable to make the projection through a glass block resting on the bottom of the vessel and just submerged beneath the surface.

In conclusion, we wish to express our thanks to Mr C. E. Harrold, of King's College, Cambridge, who made for us some of the pipettes which were used to produce the bubbles.

References

Bragg, W. L. 1942*a* *Nature*, **149**, 511.
Bragg, W. L. 1942*b* *J. Sci. Instrum.* **19**, 148.
Taylor, G. I. 1934 *Proc. Roy. Soc.* A, **145**, 362.

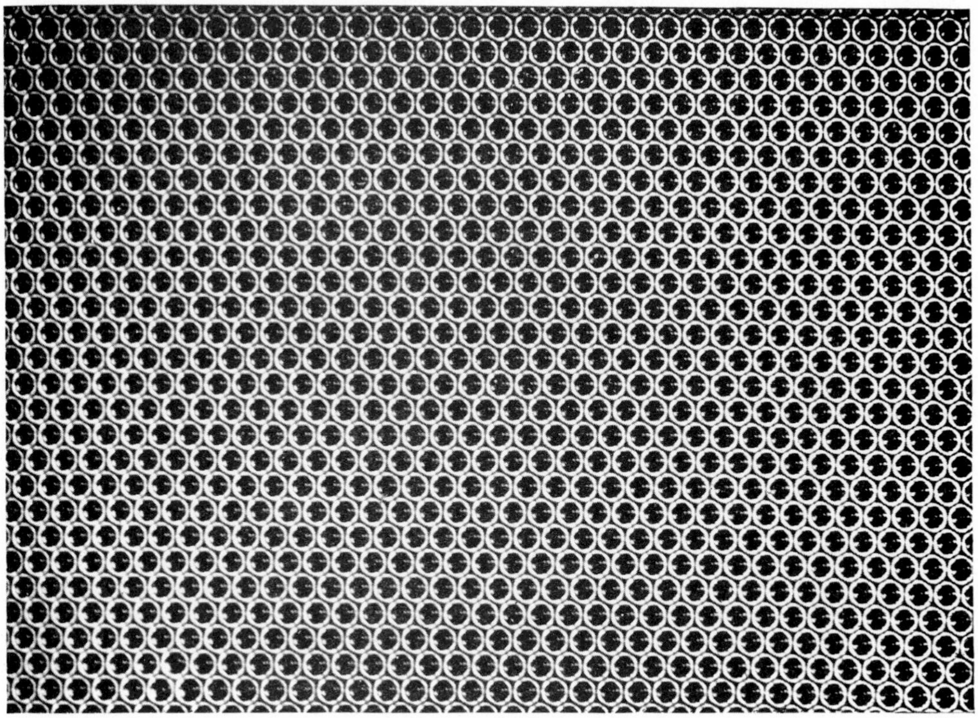

FIGURE 2. Perfect crystalline raft of bubbles. Diameter 1·41 mm.

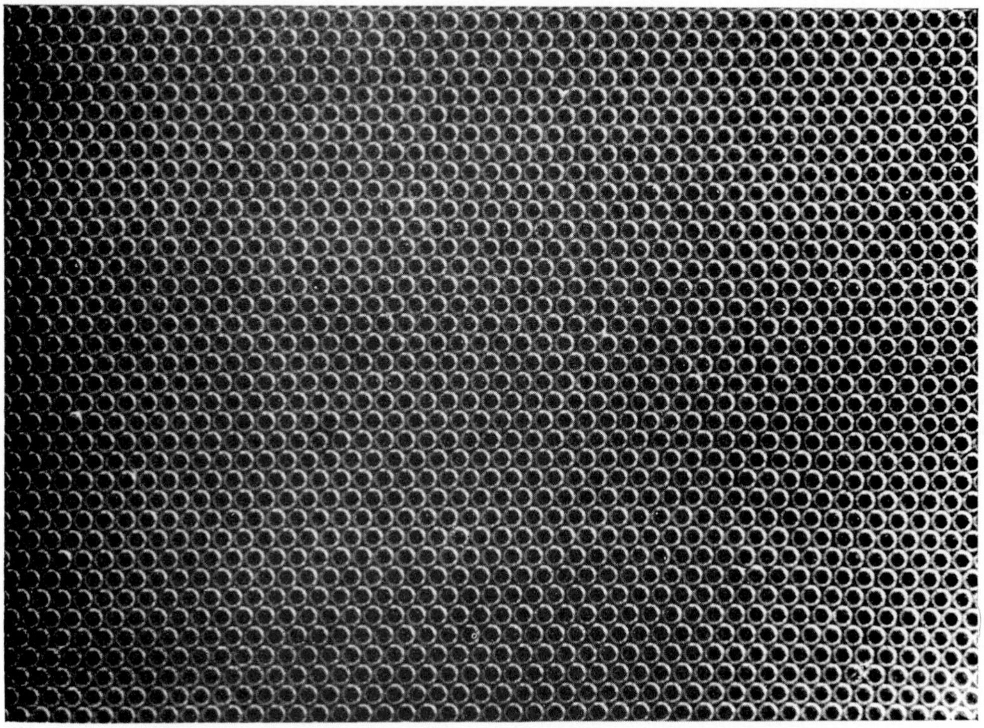

FIGURE 4. Perfect crystalline raft of bubbles. Diameter 0·30 mm.

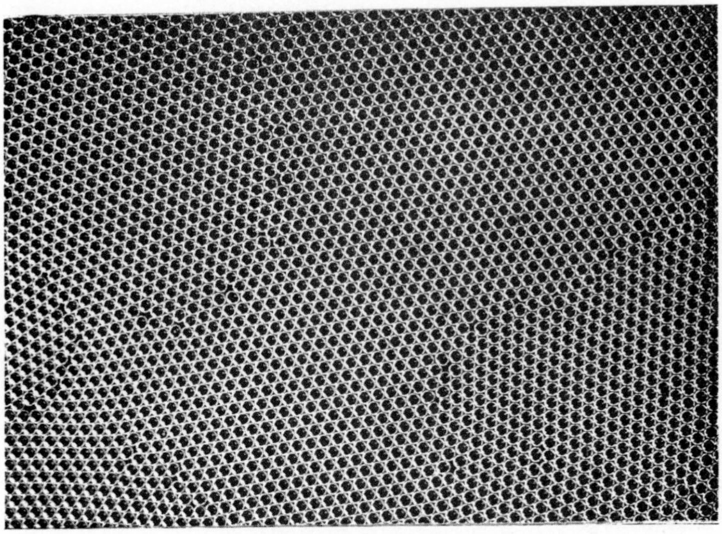

FIGURE 5*a*. Diameter 1·87 mm.

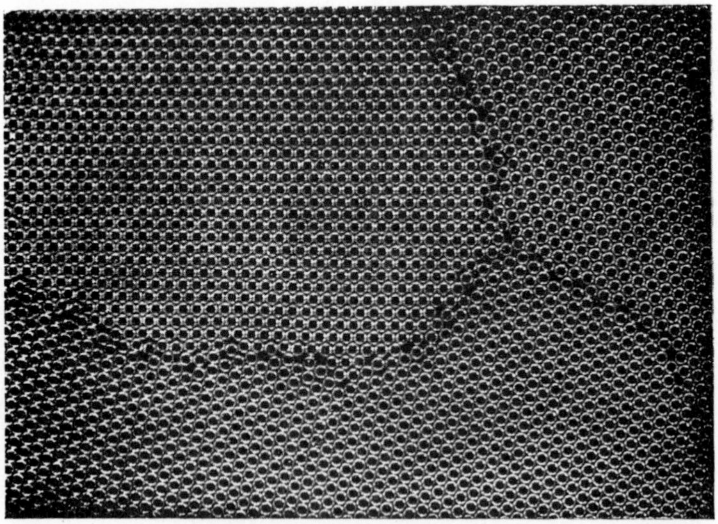

FIGURE 5*b*. Diameter 0·76 mm.

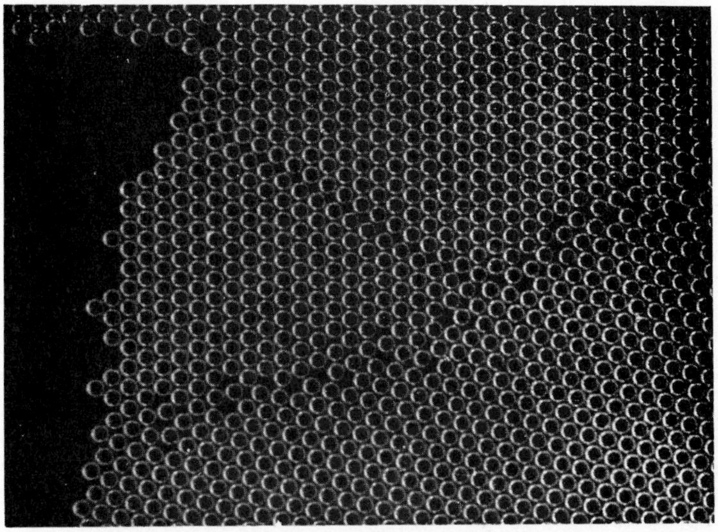

FIGURE 5c. A grain boundary. Diameter 0·30 mm.

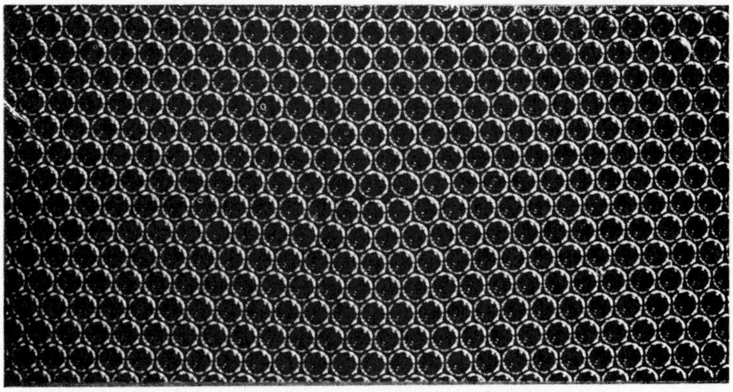

FIGURE 6a. A dislocation. Diameter 1·9 mm.

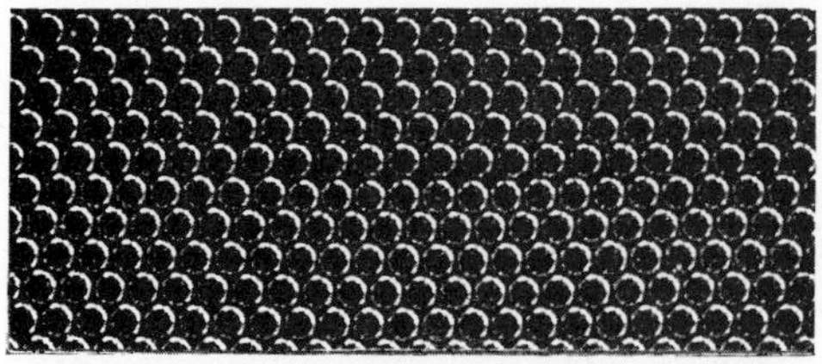

Figure 6b. Diameter 0·76 mm.

Figure 6c. Diameter 0·30 mm.

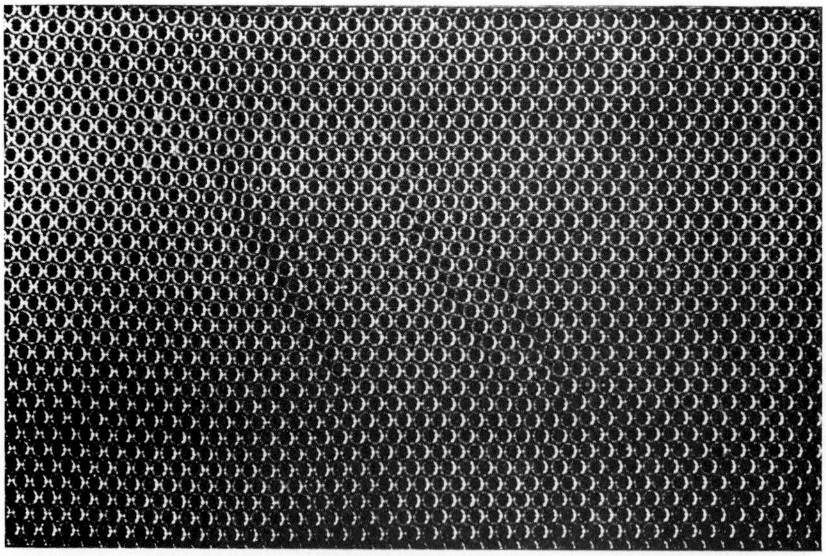

Figure 7. Parallel dislocations. Diameter 0·76 mm.

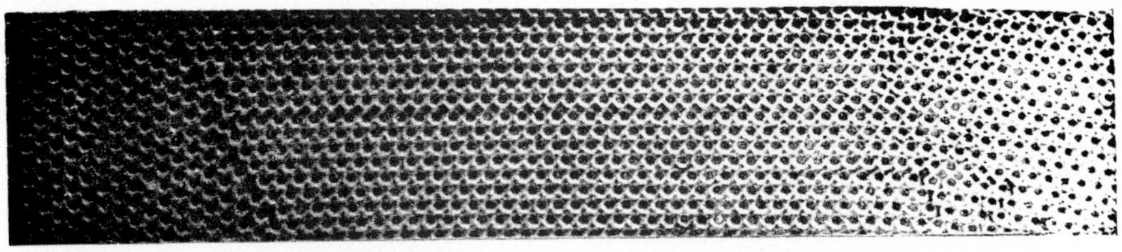

FIGURE 8. Dislocation projecting from a grain boundary. Diameter 0·30 mm.

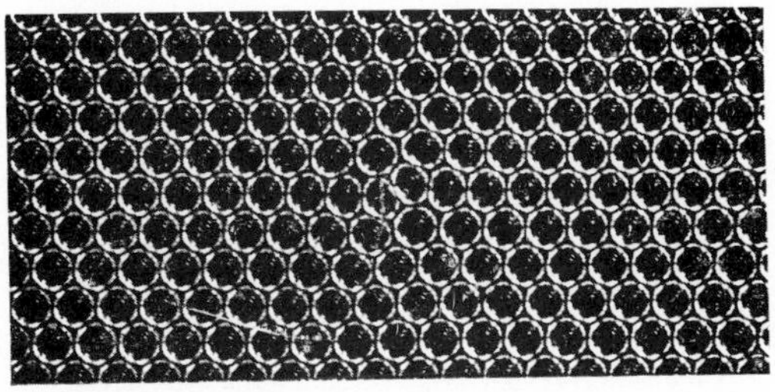

FIGURE 9. Dislocations in adjacent rows. Diameter 1·9 mm.

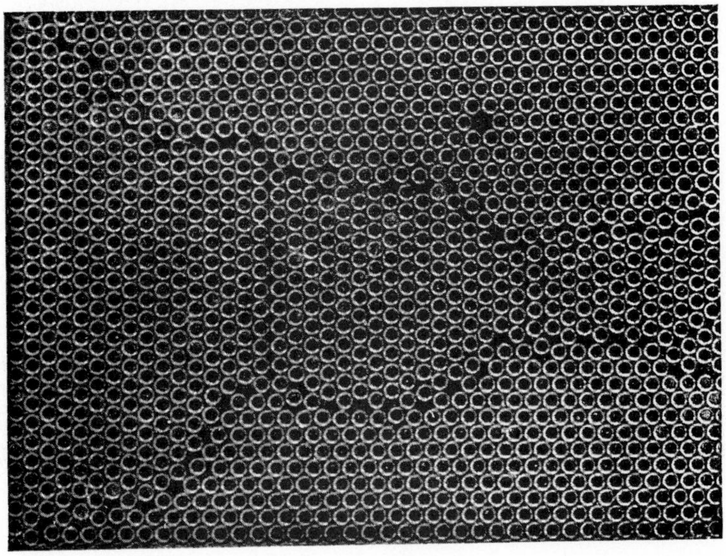

FIGURE 10. Series of fault lines between two areas of parallel orientation. Diameter 0·30 mm.

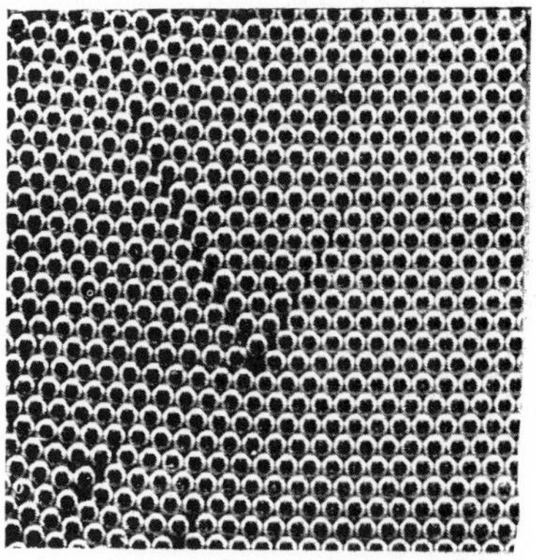

Diameter 0·68 mm.

a

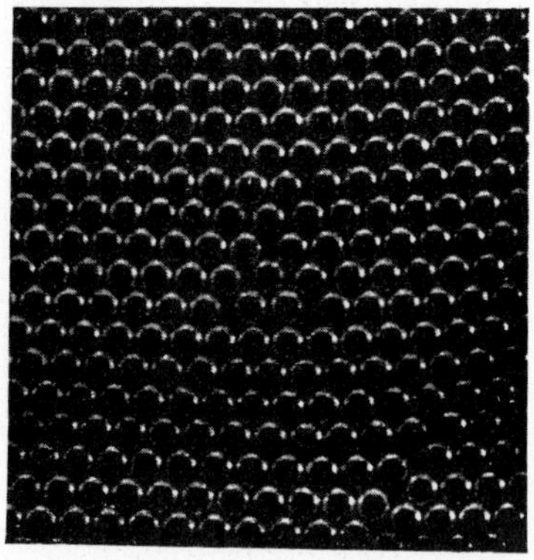

Diameter 0·68 mm.

b

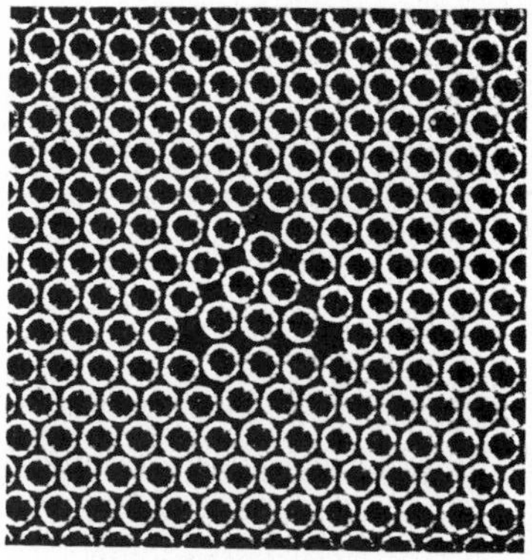

Diameter 0·6 mm.

c

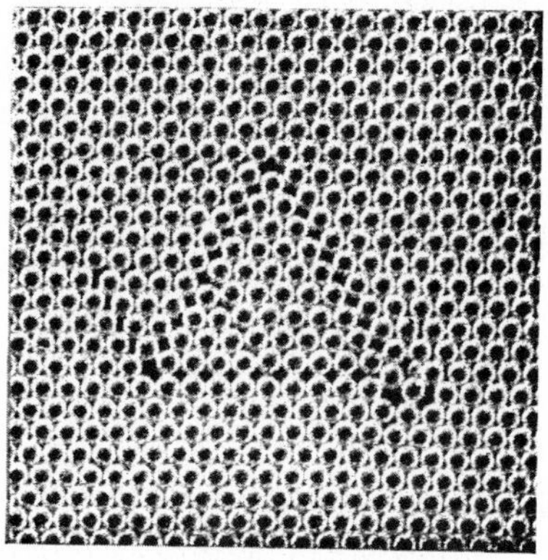

Diameter 0·30 mm.

d

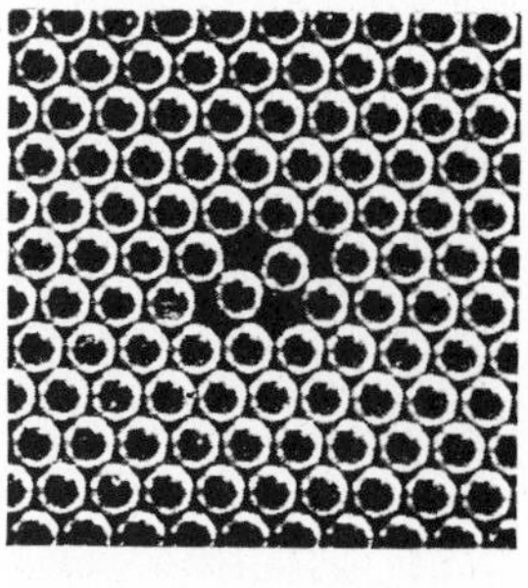

Diameter 0·6 mm.

e

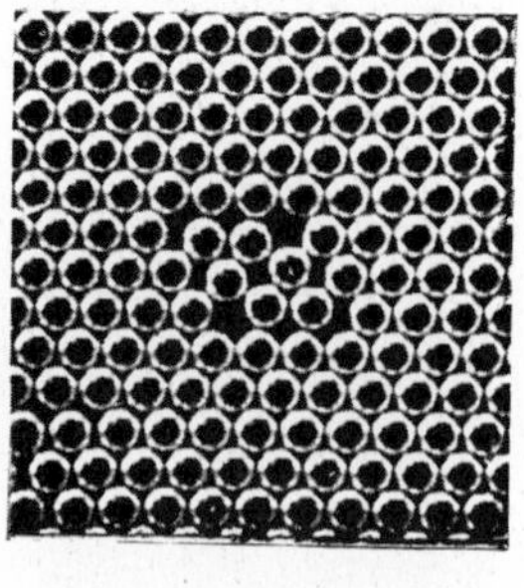

Diameter 0·6 mm.

f

FIGURE 11. Types of fault.

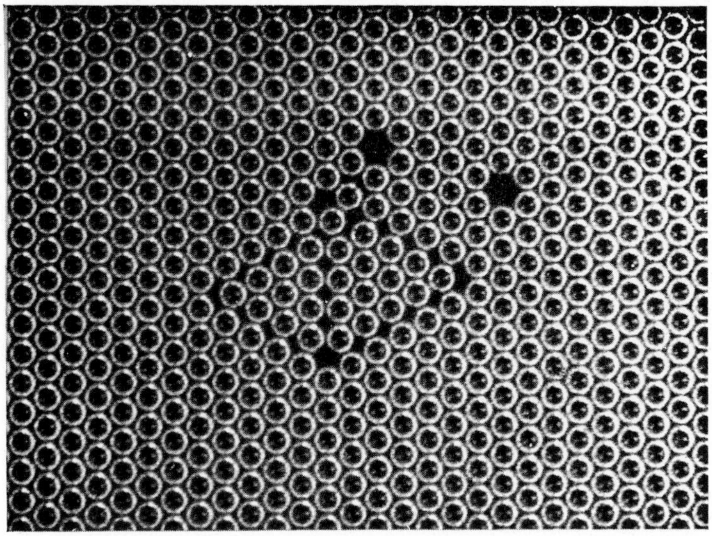

g

FIGURE 11. Types of fault. Diameter 0·68 mm.

a. Immediately after stirring.

FIGURE 12. Recrystallization. Diameter 0·60 mm.

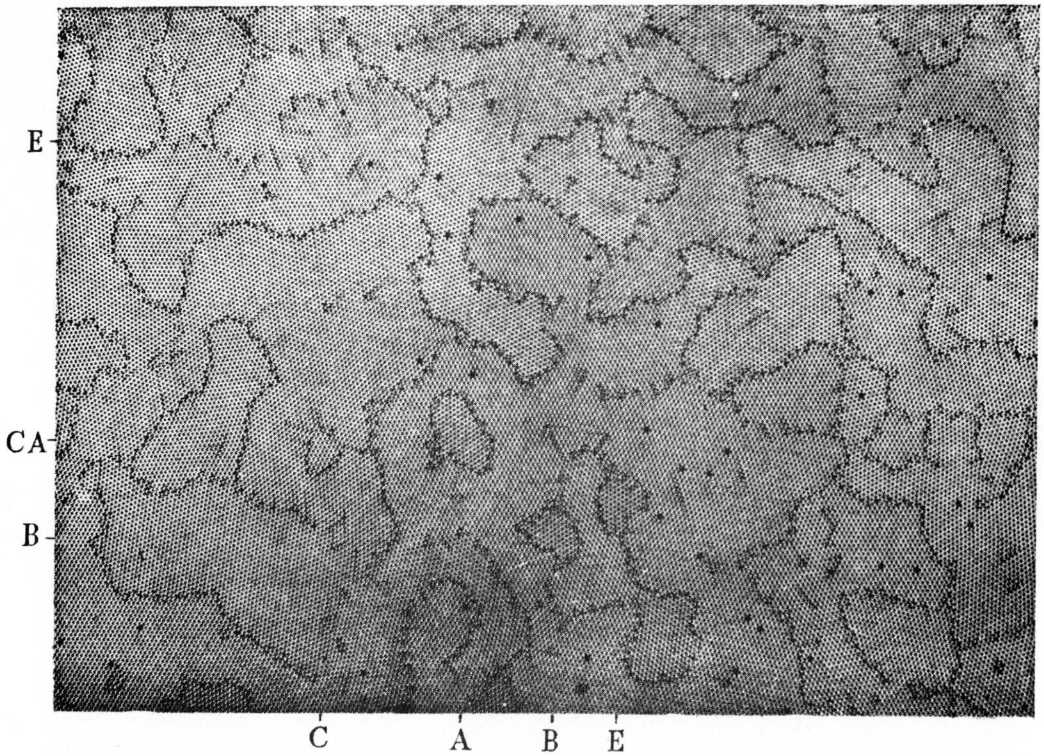

b. After 33 sec.

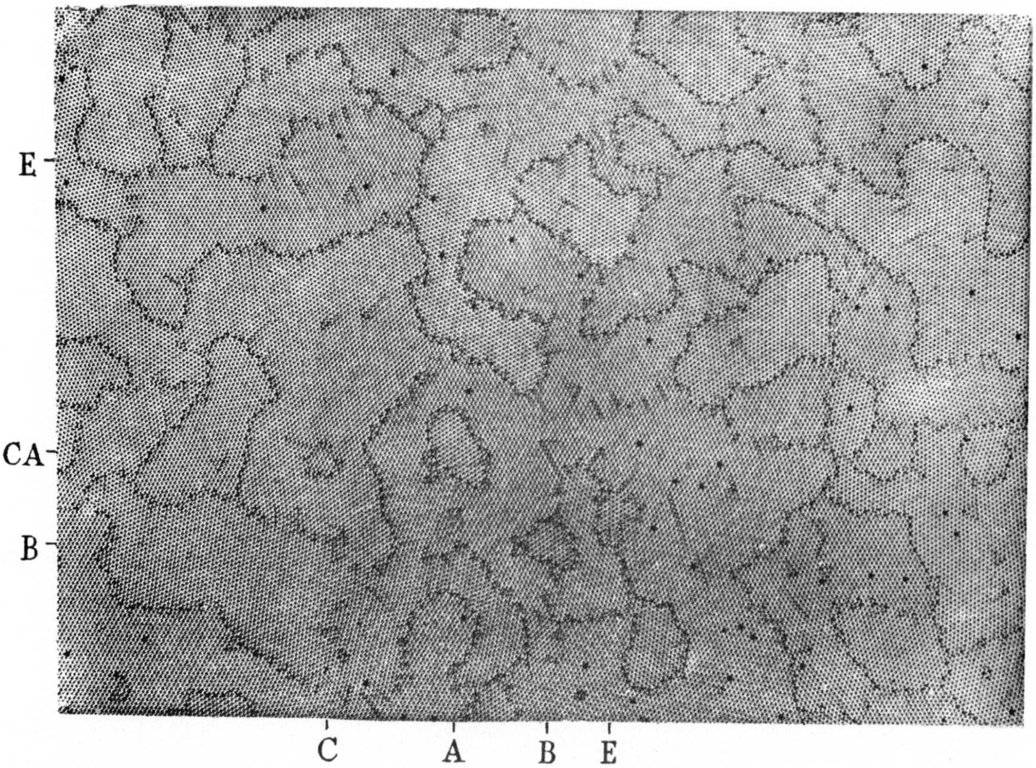

c. After 2 min.

FIGURE 12

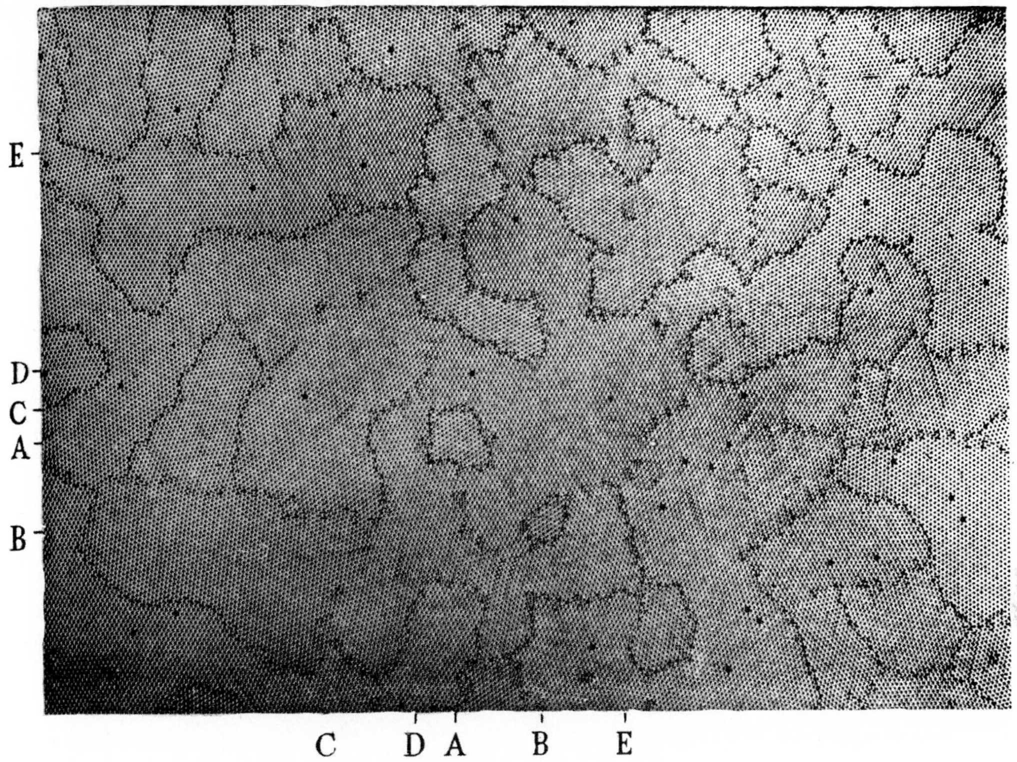

d. After 14 min.

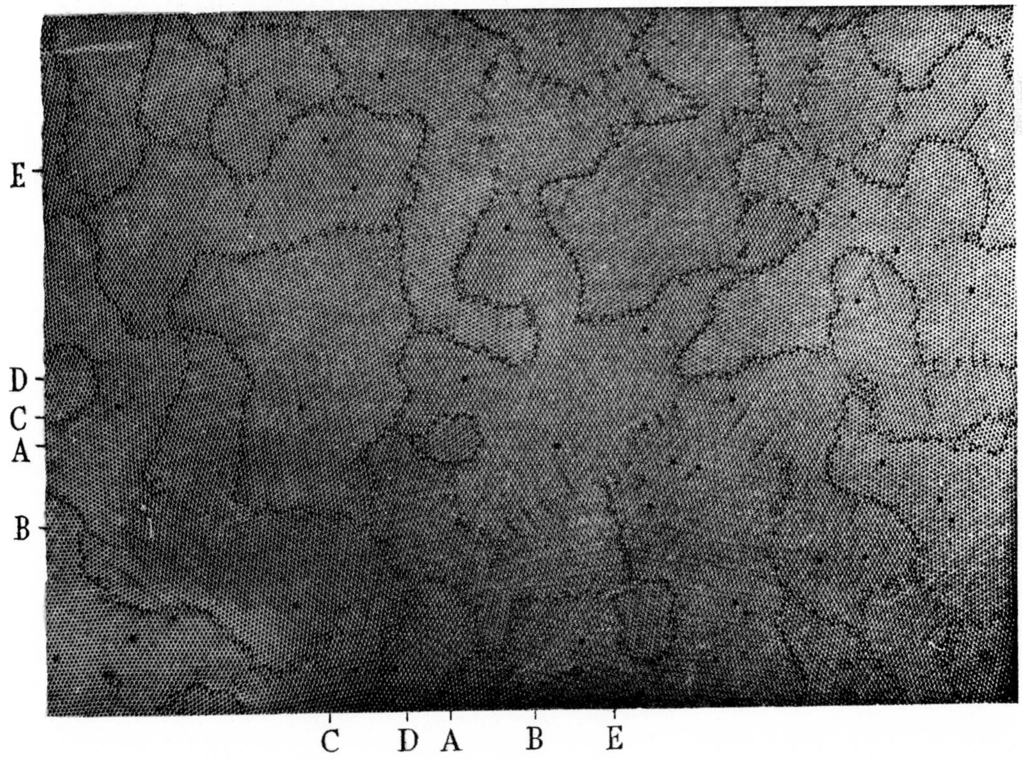

e. After 25 min.

FIGURE 12

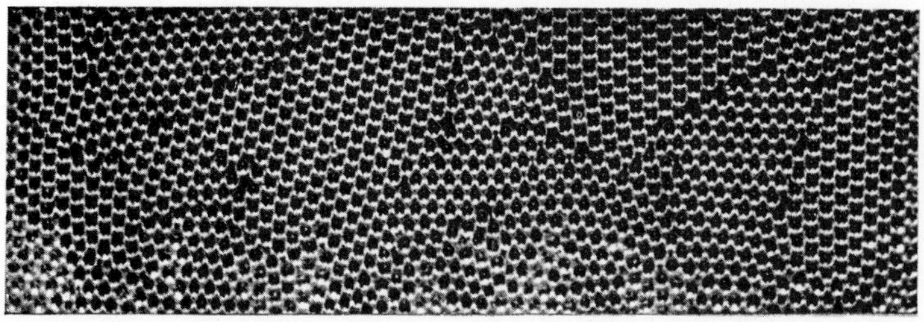

a After 1 sec.

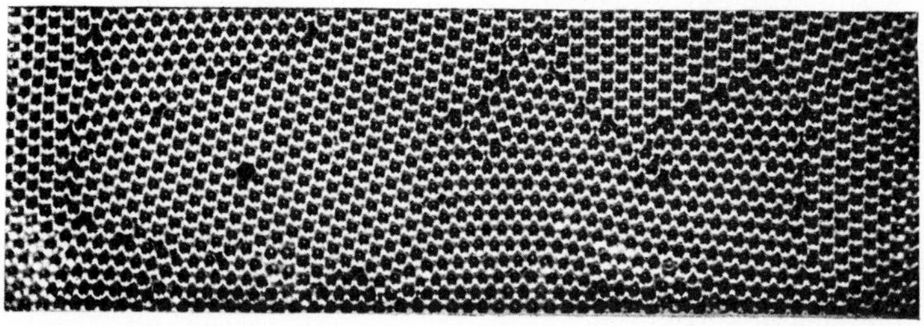

b After 4 sec.

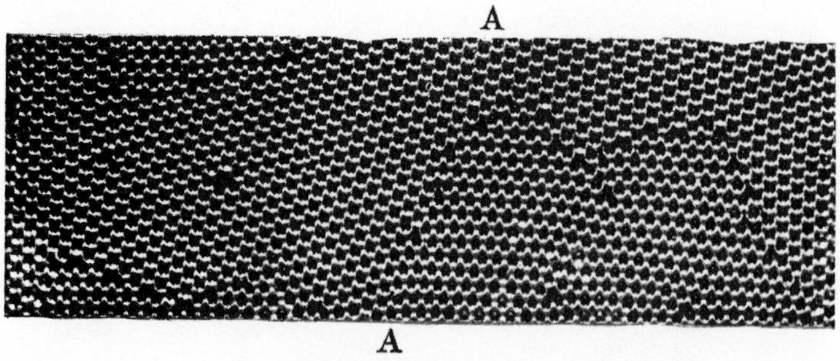

c After 4 min.

FIGURE 13. Two stages of recrystallization. Diameter 1·64 mm.

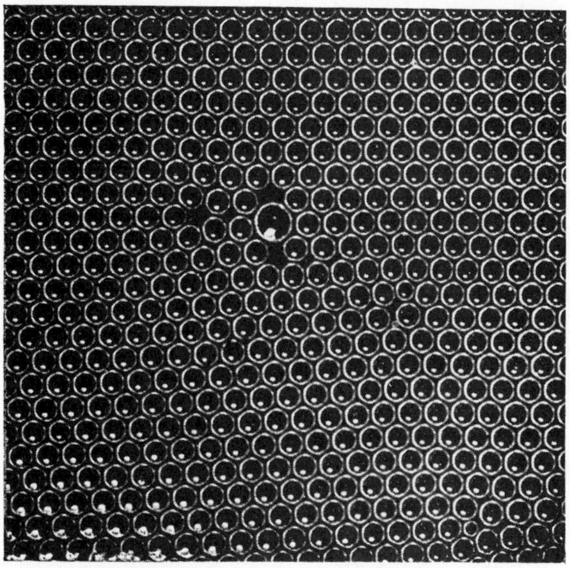

FIGURE 14. Effect of atoms of impurity. Diameter of uniform bubbles about 1·3 mm.

FIGURE 15. Oblique view of three-dimensional raft.

a. (111) face.

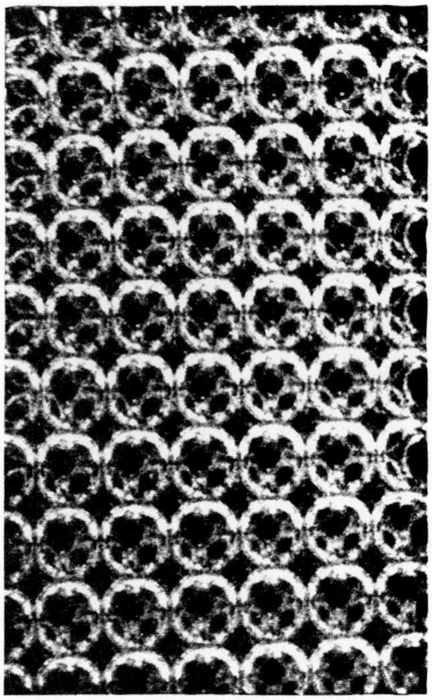

b. (100) face.

Face-centred cubic structure.

c. Twin across (111), cubic structure.

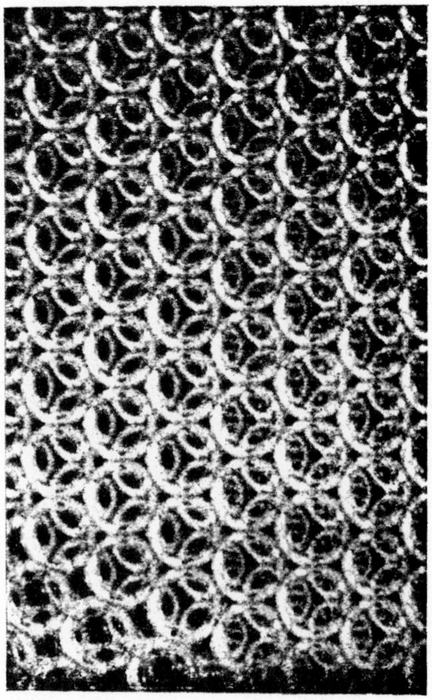

d. Possible example of hexagonal close-packing.

Diameter 0·70 mm.

FIGURE 17

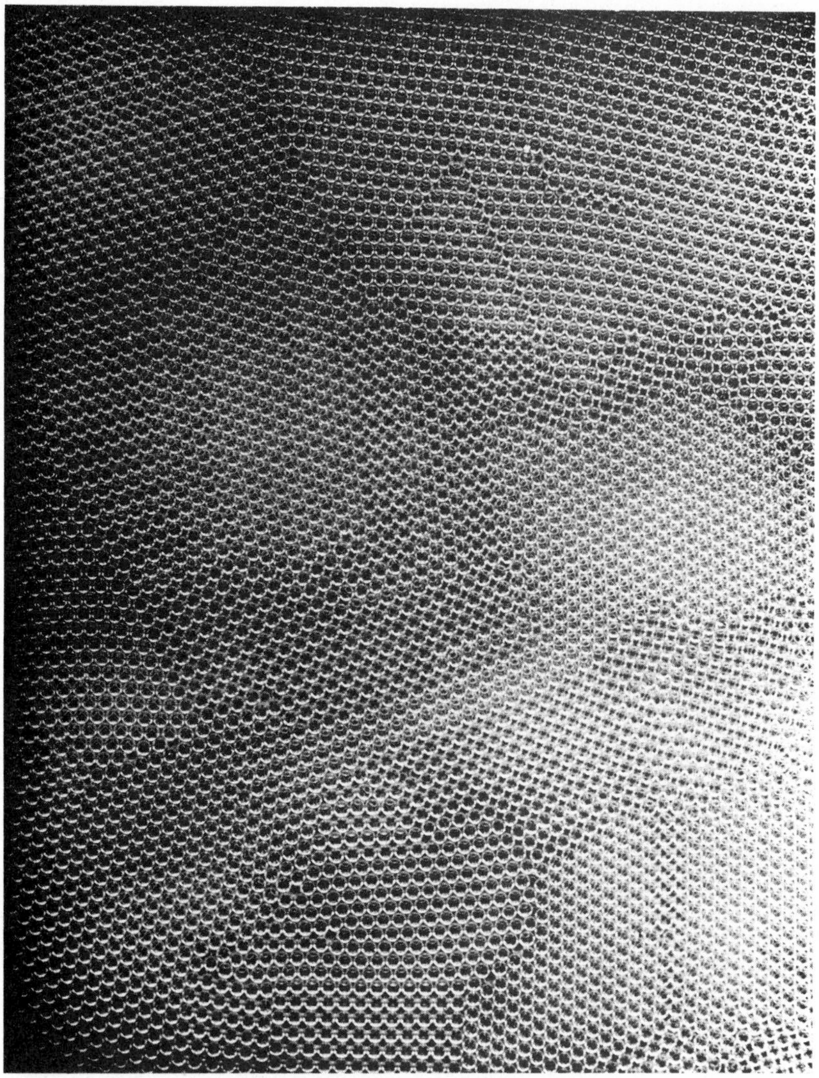

FIGURE 16. A three-dimensional raft viewed normally. Diameter 0·70 mm.

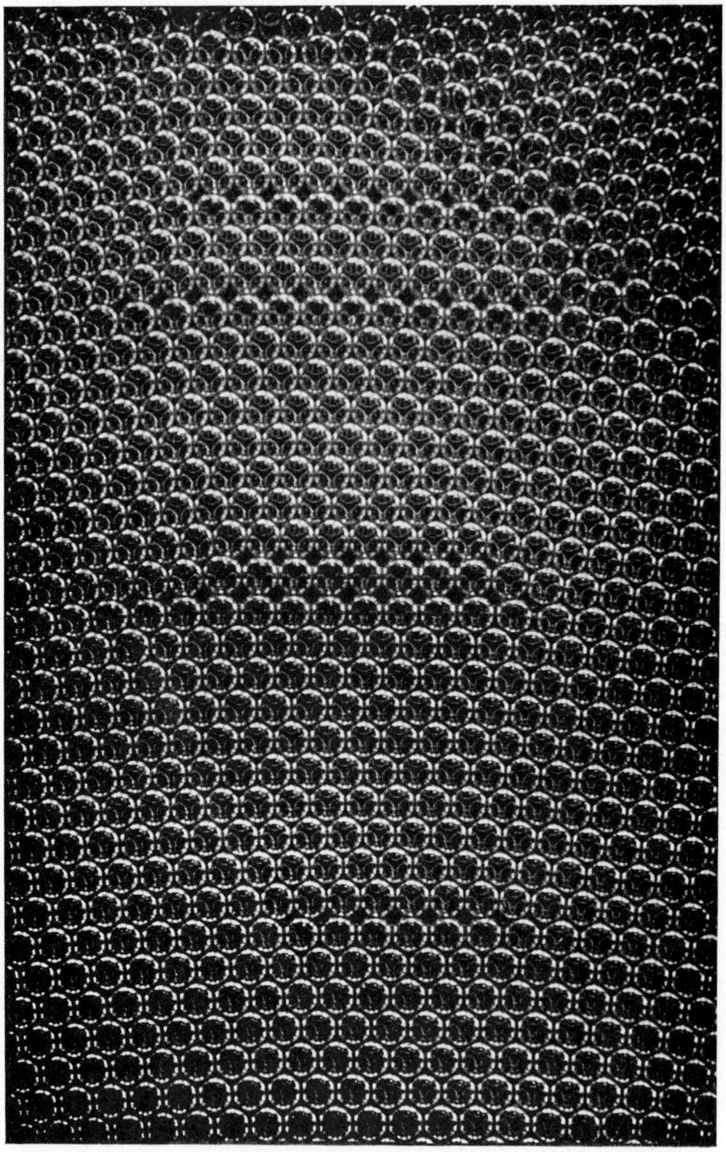

FIGURE 18. Dislocations in three-dimensional structure. Diameter 0·70 mm.

31

Tensors

Review: Chapter 11, Vol. I, *Vectors*
Chapter 20, Vol. I, *Rotation in Space*

31–1 The tensor of polarizability

Physicists always have a habit of taking the simplest example of any phenomenon and calling it "physics," leaving the more complicated examples to become the concern of other fields—say of applied mathematics, electrical engineering, chemistry, or crystallography. Even solid-state physics is almost only half physics because it worries too much about special substances. So in these lectures we will be leaving out many interesting things. For instance, one of the important properties of crystals—or of most substances—is that their electric polarizability is different in different directions. If you apply a field in any direction, the atomic charges shift a little and produce a dipole moment, but the magnitude of the moment depends very much on the direction of the field. That is, of course, quite a complication. But in physics we usually start out by talking about the special case in which the polarizability is the same in all directions, to make life easier. We leave the other cases to some other field. Therefore, for our later work, we will not need at all what we are going to talk about in this chapter.

The mathematics of tensors is particularly useful for describing properties of substances which vary in direction—although that's only one example of their use. Since most of you are not going to become physicists, but are going to go into the *real* world, where things depend severely upon direction, sooner or later you will need to use tensors. In order not to leave anything out, we are going to describe tensors, although not in great detail. We want the feeling that our treatment of physics is complete. For example, our electrodynamics is complete—as complete as any electricity and magnetism course, even a graduate course. Our mechanics is not complete, because we studied mechanics when you didn't have a high level of mathematical sophistication, and we were not able to discuss subjects like the principle of least action, or Lagrangians, or Hamiltonians, and so on, which are *more elegant ways* of describing mechanics. Except for general relativity, however, we do have the complete *laws* of mechanics. Our electricity and magnetism is complete, and a lot of other things are quite complete. The quantum mechanics, naturally, will not be—we have to leave something for the future. But you should at least know what a tensor is.

We emphasized in Chapter 30 that the properties of crystalline substances are different in different directions—we say they are *anisotropic.* The variation of the induced dipole moment with the direction of the applied electric field is only one example, the one we will use for our example of a tensor. Let's say that for a given direction of the electric field the induced dipole moment per unit volume $\boldsymbol{P}$ is proportional to the strength of the applied field $\boldsymbol{E}$. (This is a good approximation for many substances if E is not too large.) We will call the proportionality constant α.* We want now to consider substances in which α depends on the direction of the applied field, as, for example, in crystals like calcite, which make double images when you look through them.

Suppose, in a particular crystal, we find that an electric field $\boldsymbol{E}_1$ in the x-direction produces the polarization $\boldsymbol{P}_1$ in the x-direction. Then we find that an electric field $\boldsymbol{E}_2$ in the y-direction, with the same *strength,* as $\boldsymbol{E}_1$ produces a different polar-

* In Chapter 10 we followed the usual convention and wrote $P = \epsilon_0\chi E$ and called χ ("khi") the "susceptibility." Here, it will be more convenient to use a single letter, so we write α for $\epsilon_0\chi$. For isotropic dielectrics, $\alpha = (\kappa-1)\epsilon_0$, where κ is the dielectric constant (see Section 10–4).

ization $\boldsymbol{P}_2$ in the y-direction. What would happen if we put an electric field at 45°? Well, that's a superposition of two fields along x and y, so the polarization $\boldsymbol{P}$ will be the vector sum of $\boldsymbol{P}_1$ and $\boldsymbol{P}_2$, as shown in Fig. 31–1(a). The polarization is no longer in the same direction as the electric field. You can see how that might come about. There may be charges which can move easily up and down, but which are rather stiff for sidewise motions. When a force is applied at 45°, the charges move farther up than they do toward the side. The displacements are not in the direction of the external force, because there are asymmetric internal elastic forces.

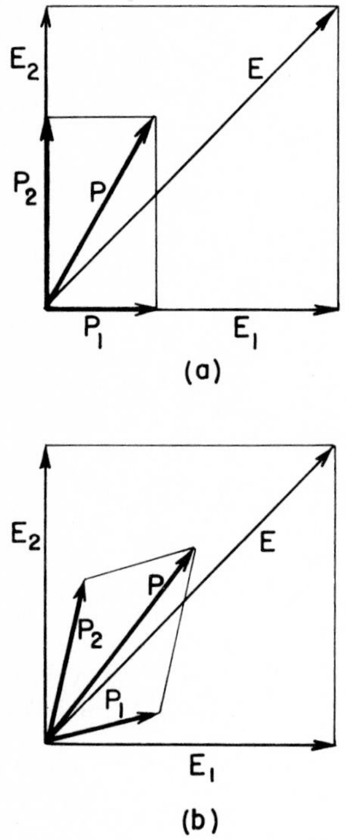

Fig. 31–1. The vector addition of polarizations in an anisotropic crystal.

There is, of course, nothing special about 45°. It is *generally* true that the induced polarization of a crystal is *not* in the direction of the electric field. In our example above, we happened to make a "lucky" choice of our x- and y-axes, for which $\boldsymbol{P}$ was along $\boldsymbol{E}$ for both the x- and y-directions. If the crystal were rotated with respect to the coordinate axes, the electric field $\boldsymbol{E}_2$ in the y-direction would have produced a polarization $\boldsymbol{P}$ with both an x- and a y-component. Similarly, the polarization due to an electric field in the x-direction would have produced a polarization with an x-component and a y-component. Then the polarizations would be as shown in Fig. 31–1(b), instead of as in part (a). Things get more complicated—but for any field $\boldsymbol{E}$, the *magnitude* of $\boldsymbol{P}$ is still proportional to the magnitude of $\boldsymbol{E}$.

We want now to treat the general case of an arbitrary orientation of a crystal with respect to the coordinate axes. An electric field in the x-direction will produce a polarization $\boldsymbol{P}$ with x-, y-, and z-components; we can write

$$P_x = \alpha_{xx}E_x, \qquad P_y = \alpha_{yx}E_x, \qquad P_z = \alpha_{zx}E_x. \tag{31.1}$$

All we are saying here is that if the electric field is in the x-direction, the polarization does not have to be in that same direction, but rather has an x-, a y-, and a z-component—each proportional to E_x. We are calling the constants of proportionality α_{xx}, α_{yx}, and α_{zx}, respectively (the first letter to tell us which component of P is involved, the last to refer to the direction of the electric field).

Similarly, for a field in the y-direction, we can write

$$P_x = \alpha_{xy}E_y, \qquad P_y = \alpha_{yy}E_y, \qquad P_z = \alpha_{zy}E_y; \tag{31.2}$$

and for a field in the z-direction,

$$P_x = \alpha_{xz}E_z, \qquad P_y = \alpha_{yz}E_z, \qquad P_z = \alpha_{zz}E_z. \tag{31.3}$$

Now we have said that polarization depends linearly on the fields, so if there is an electric field $\boldsymbol{E}$ that has both an x- and a y-component, the resulting x-component of $\boldsymbol{P}$ will be the sum of the two P_x's of Eqs. (31.1) and (31.2). If $\boldsymbol{E}$ has components along x, y, and z, the resulting components of $\boldsymbol{P}$ will be the sum of the three contributions in Eqs. (31.1), (31.2), and (31.3). In other words, $\boldsymbol{P}$ will be given by

$$\begin{aligned} P_x &= \alpha_{xx}E_x + \alpha_{xy}E_y + \alpha_{xz}E_z,\\ P_y &= \alpha_{yx}E_x + \alpha_{yy}E_y + \alpha_{yz}E_z,\\ P_z &= \alpha_{zx}E_x + \alpha_{zy}E_y + \alpha_{zz}E_z. \end{aligned} \tag{31.4}$$

The dielectric behavior of the crystal is then completely described by the nine quantities (α_{xx}, α_{xy}, α_{xz}, α_{yz}, . . .), which we can represent by the symbol α_{ij}. (The subscripts i and j each stand for any one of the three possible letters x, y, and z.) Any arbitrary electric field $\boldsymbol{E}$ can be resolved with the components E_x, E_y, and E_z; from these we can use the α_{ij} to find P_x, P_y, and P_z, which together give the total polarization $\boldsymbol{P}$. The set of nine coefficients α_{ij} is called a *tensor*—in this instance, the *tensor of polarizability*. Just as we say that the three numbers (E_x, E_y, E_z) "form the vector $\boldsymbol{E}$," we say that the nine numbers (α_{xx}, α_{xy}, . . .) "form the tensor α_{ij}."

31–2 Transforming the tensor components

You know that when we change to a different coordinate system x', y', and z', the components $E_{x'}$, $E_{y'}$, and $E_{z'}$ of the vector will be quite different—as will also *the components* of $\boldsymbol{P}$. So all the coefficients α_{ij} will be different for a different set of coordinates. You can, in fact, see how the α's must be changed by changing the components of $\boldsymbol{E}$ and $\boldsymbol{P}$ in the proper way, because if we describe the *same physical* electric field in the new coordinate system we should get the same polarization. For any new set of coordinates, $P_{x'}$ is a linear combination of P_x, P_y, and P_z:

$$P_{x'} = aP_x + bP_y + cP_z,$$

and similarly for the other components. If you substitute for P_x, P_y, and P_z in terms of the E's, using Eq. (31.4), you get

$$\begin{aligned} P_{x'} &= a(\alpha_{xx}E_x + \alpha_{xy}E_y + \alpha_{xz}E_z) \\ &\quad + b(\alpha_{yx}E_x + \alpha_{yy}E_y + \cdots) \\ &\quad + c(\alpha_{zx}E_x + \quad\cdots\quad + \cdots). \end{aligned}$$

Then you write E_x, E_y, and E_z in terms of $E_{x'}$, $E_{y'}$, and $E_{z'}$; for instance,

$$E_x = a'E_{x'} + b'E_{y'} + c'E_{z'},$$

where a', b', c' are related to, but not equal to, a, b, c. So you have $P_{x'}$, expressed in terms of the components $E_{x'}$, $E_{y'}$, and $E_{z'}$; that is, you have the new α_{ij}. It is fairly messy, but quite straightforward.

When we talk about changing the axes we are assuming that the crystal stays put *in space*. If the crystal were rotated *with* the axes, the α's would not change. Conversely, if the orientation of the crystal were changed with respect to the axes, we would have a new set of α's. But if they are known for *any* one orientation of the crystal, they can be found for any other orientation by the transformation we have just described. In other words, the dielectric property of a crystal is described *completely* by giving the components of the polarization tensor α_{ij} with respect to any arbitrarily chosen set of axes. Just as we can associate a vector velocity $\boldsymbol{v} = (v_x, v_y, v_z)$ with a particle, knowing that the three components will change in a certain definite way if we change our coordinate axes, so with a crystal we associate its polarization tensor α_{ij}, whose nine components will transform in a certain definite way if the coordinate system is changed.

The relation between $\boldsymbol{P}$ and $\boldsymbol{E}$ written in Eq. (31.4) can be put in the more compact notation:

$$P_i = \sum_j \alpha_{ij}E_j, \tag{31.5}$$

where it is understood that i represents either x, y, or z and that the sum is taken on $j = x$, y, and z. Many special notations have been invented for dealing with tensors, but each of them is convenient only for a limited class of problems. One common convention is to omit the sum sign ($\sum$) in Eq. (31.5), leaving it *understood* that whenever the same subscript occurs twice (here j), a sum is to be taken over that index. Since we will be using tensors so little, we will not bother to adopt any such special notations or conventions.

31–3 The energy ellipsoid

We want now to get some experience with tensors. Suppose we ask the interesting question: What energy is required to polarize the crystal (in addition to the energy in the electric field which we know is $\epsilon_0 E^2/2$ per unit volume)? Consider for a moment the atomic charges that are being displaced. The work done in displacing the charge the distance dx is $qE_x\,dx$, and if there are N charges per unit volume, the work done is $qE_xN\,dx$. But $qN\,dx$ is the change dP_x in the dipole moment per unit volume. So the energy required *per unit volume* is

$$E_x\,dP_x.$$

Combining the work for the three components of the field, the work per unit volume is found to be

$$\boldsymbol{E}\cdot d\boldsymbol{P}.$$

Since the magnitude of $\boldsymbol{P}$ is proportional to $\boldsymbol{E}$, the work done per unit volume in bringing the polarization from 0 to $\boldsymbol{P}$ is the integral of $\boldsymbol{E}\cdot d\boldsymbol{P}$. Calling this work u_P,* we write

$$u_P = \tfrac{1}{2}\boldsymbol{E}\cdot\boldsymbol{P} = \tfrac{1}{2}\sum_i E_iP_i. \tag{31.6}$$

Now we can express $\boldsymbol{P}$ in terms of $\boldsymbol{E}$ by Eq. (31.5), and we have that

$$u_P = \tfrac{1}{2}\sum_i\sum_j \alpha_{ij}E_iE_j. \tag{31.7}$$

The energy density u_P is a number independent of the choice of axes, so it is a scalar. A tensor has then the property that when it is summed over one index (with a vector), it gives a new vector; and when it is summed over *both* indexes (with *two* vectors), it gives a scalar.

The tensor α_{ij} should really be called a "tensor of second rank," because it has two indexes. A vector—with *one* index—is a tensor of the first rank, and a scalar—with no index—is a tensor of zero rank. So we say that the electric field $\boldsymbol{E}$ is a tensor of the first rank and that the energy density u_P is a tensor of zero rank. It is possible to extend the ideas of a tensor to three or more indexes, and so to make tensors of ranks higher than two.

The subscripts of the polarization tensor range over three possible values—they are tensors in three dimensions. The mathematicians consider tensors in four, five, or more dimensions. We have already used a four-dimensional tensor $F_{\mu\nu}$ in our relativistic description of the electromagnetic field (Chapter 26).

The polarization tensor α_{ij} has the interesting property that it is *symmetric*, that is, that $\alpha_{xy} = \alpha_{yx}$, and so on for any pair of indexes. (This is a *physical* property of a real crystal and not necessary for all tensors.) You can prove for yourself that this must be true by computing the change in energy of a crystal through the following cycle: (1) Turn on a field in the x-direction; (2) turn on a field in the y-direction; (3) turn *off* the x-field; (4) turn off the y-field. The crystal is now back where it started, and the net work done on the polarization must be back to zero. You can show, however, that for this to be true, α_{xy} must be equal to α_{yx}. The same kind of argument can, of course, be given for α_{xz}, etc. So the polarization tensor is symmetric.

This also means that the polarization tensor can be measured by just measuring the energy required to polarize the crystal in various directions. Suppose we apply an $\boldsymbol{E}$-field with only an x- and a y-component; then according to Eq. (31.7),

$$u_P = \tfrac{1}{2}\,[\alpha_{xx}E_x^2 + (\alpha_{xy}+\alpha_{yx})E_xE_y + \alpha_{yy}E_y^2]. \tag{31.8}$$

With an E_x alone, we can determine α_{xx}; with an E_y alone, we can determine α_{yy}; with both E_x and E_y, we get an extra energy due to the term with $(\alpha_{xy}+\alpha_{yx})$. Since the α_{xy} and α_{yx} are equal, this term is $2\alpha_{xy}$ and can be related to the energy.

The energy expression, Eq. (31.8), has a nice geometric interpretation. Suppose we ask what fields E_x and E_y correspond to some *given* energy density—say u_0. That is just the mathematical problem of solving the equation

$$\alpha_{xx}E_x^2 + 2\alpha_{xy}E_xE_y + \alpha_{yy}E_y^2 = 2u_0.$$

This is a quadratic equation, so if we plot E_x and E_y, the solutions of this equation

* This work done in *producing* the polarization by an electric field is not to be confused with the potential energy $-\boldsymbol{p}_0\cdot\boldsymbol{E}$ of a permanent dipole moment p_0.

are all the points on an ellipse (Fig. 31–2). (It must be an ellipse, rather than a parabola or a hyperbola, because the energy for any field is always positive and finite.) The vector $\boldsymbol{E}$ with components E_x and E_y can be drawn from the origin to the ellipse. So such an "energy ellipse" is a nice way of "visualizing" the polarization tensor.

If we now generalize to include all three components, the electric vector $\boldsymbol{E}$ in *any* direction required to give a unit energy density gives a point which will be on the surface of an ellipsoid, as shown in Fig. 31–3. The shape of this ellipsoid of constant energy uniquely characterizes the tensor polarizability.

Now an ellipsoid has the nice property that it can always be described simply by giving the directions of three "principal axes" and the diameters of the ellipse along these axes. The "principal axes" are the directions of the longest and shortest diameters and the direction at right angles to both. They are indicated by the axes a, b, and c in Fig. 31–3. With respect to these axes, the ellipsoid has the particularly simple equation

$$\alpha_{aa}E_a^2 + \alpha_{bb}E_b^2 + \alpha_{cc}E_c^2 = 2u_0.$$

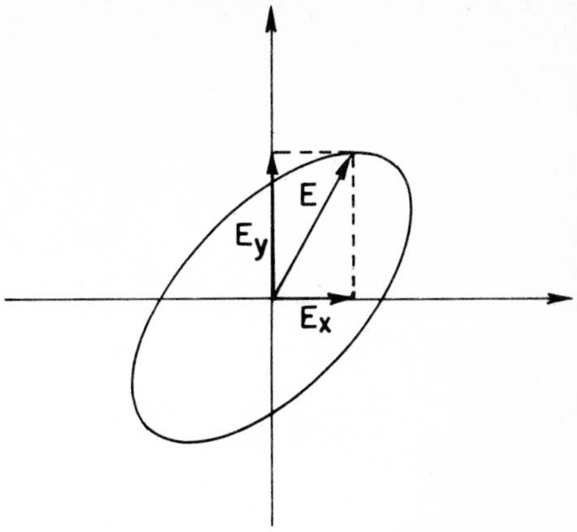

Fig. 31–2. Locus of the vector $\boldsymbol{E} = (E_x, E_y)$ that gives a constant energy of polarization.

So with respect to these axes, the dielectric tensor has only three components that are not zero: α_{aa}, α_{bb}, and α_{cc}. That is to say, no matter how complicated a crystal is, it is always possible to choose a set of axes (not necessarily the crystal axes) for which the polarization tensor has only three components. With such a set of axes, Eq. (31.4) becomes simply

$$P_a = \alpha_{aa}E_a, \qquad P_b = \alpha_{bb}E_b, \qquad P_c = \alpha_{cc}E_c. \tag{31.9}$$

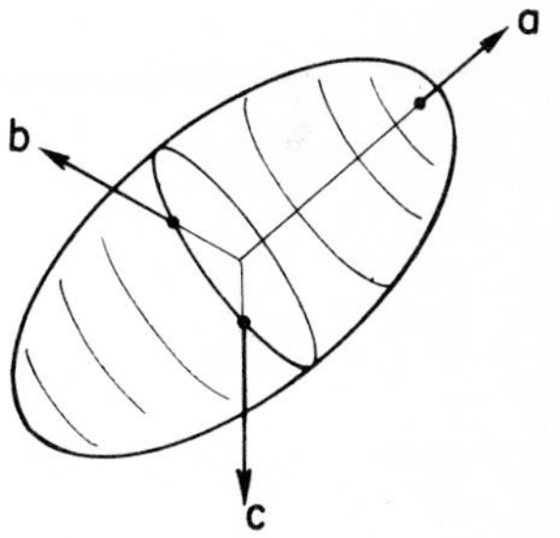

Fig. 31–3. The energy ellipsoid of the polarization tensor.

An electric field along any one of the principal axes produces a polarization along the same axis, but the coefficients for the three axes may, of course, be different.

Often, a tensor is described by listing the nine coefficients in a table inside of a pair of brackets:

$$\begin{bmatrix} \alpha_{xx} & \alpha_{xy} & \alpha_{xz} \\ \alpha_{yx} & \alpha_{yy} & \alpha_{yz} \\ \alpha_{zx} & \alpha_{zy} & \alpha_{zz} \end{bmatrix}. \tag{31.10}$$

For the principal axes a, b, and c, only the diagonal terms are not zero; we say then that "the tensor is diagonal." The complete tensor is

$$\begin{bmatrix} \alpha_{aa} & 0 & 0 \\ 0 & \alpha_{bb} & 0 \\ 0 & 0 & \alpha_{cc} \end{bmatrix}. \tag{31.11}$$

The important point is that any polarization tensor (in fact, *any symmetric* tensor of rank two in any number of dimensions) can be put in this form by choosing a suitable set of coordinate axes.

If the three elements of the polarization tensor in diagonal form are all equal, that is, if

$$\alpha_{aa} = \alpha_{bb} = \alpha_{cc} = \alpha, \tag{31.12}$$

the energy ellipsoid becomes a sphere, and the polarizability is the same in all directions. The material is isotropic. In the tensor notation,

$$\alpha_{ij} = \alpha\delta_{ij}, \tag{31.13}$$

where δ_{ij} is the *unit tensor*

$$\delta_{ij} = \begin{bmatrix} 1 & 0 & 0 \\ 0 & 1 & 0 \\ 0 & 0 & 1 \end{bmatrix}. \tag{31.14}$$

That means, of course,

$$\begin{aligned} \delta_{ij} &= 1, \quad \text{if} \quad i = j; \\ \delta_{ij} &= 0, \quad \text{if} \quad i \neq j. \end{aligned} \tag{31.15}$$

The tensor δ_{ij} is often called the "Kronecker delta." You may amuse yourself by proving that the tensor (31.14) has exactly the same form if you change the coordinate system to any other rectangular one. The polarization tensor of Eq. (31.13) gives

$$P_i = \alpha \sum_j \delta_{ij} E_j = \alpha E_i,$$

which means the same as our old result for isotropic dielectrics:

$$\boldsymbol{P} = \alpha \boldsymbol{E}.$$

The shape and orientation of the polarization ellipsoid can sometimes be related to the symmetry properties of the crystal. We have said in Chapter 30 that there are 230 different possible internal symmetries of a three-dimensional lattice and that they can, for many purposes, be conveniently grouped into seven classes, according to the shape of the unit cell. Now the ellipsoid of polarizability must share the internal geometric symmetries of the crystal. For example, a triclinic crystal has low symmetry—the ellipsoid of polarizability will have unequal axes, and its orientation will not, in general, be aligned with the crystal axes. On the other hand, a monoclinic crystal has the property that its properties are unchanged if the crystal is rotated 180° about one axis. So the polarization tensor must be the same after such a rotation. It follows that the ellipsoid of the polarizability must return to itself after a 180° rotation. That can happen only if one of the axes of the ellipsoid is in the same direction as the symmetry axis of the crystal. Otherwise, the orientation and dimensions of the ellipsoid are unrestricted.

For an orthorhombic crystal, however, the axes of the ellipsoid must correspond to the crystal axes, because a 180° rotation about any one of the three axes repeats the same lattice. If we go to a tetragonal crystal, the ellipse must have the same symmetry, so it must have two equal diameters. Finally, for a cubic crystal, all three diameters of the ellipsoid must be equal; it becomes a sphere, and the polarizability of the crystal is the same in all directions.

There is a big game of figuring out the possible kinds of tensors for all the possible symmetries of a crystal. It is called a "group-theoretical" analysis. But for the simple case of the polarizability tensor, it is relatively easy to see what the relations must be.

31–4 Other tensors; the tensor of inertia

There are many other examples of tensors appearing in physics. For example, in a metal, or in any conductor, one often finds that the current density $\boldsymbol{j}$ is approximately proportional to the electric field $\boldsymbol{E}$; the proportionality constant is called the conductivity σ:

$$\boldsymbol{j} = \sigma \boldsymbol{E}.$$

For crystals, however, the relation between $\boldsymbol{j}$ and $\boldsymbol{E}$ is more complicated; the conductivity is not the same in all directions. The conductivity is a tensor, and we write

$$j_i = \sum \sigma_{ij} E_j.$$

Another example of a physical tensor is the moment of inertia. In Chapter 18 of Volume I we saw that a solid object rotating about a fixed axis has an angular momentum L proportional to the angular velocity ω, and we called the proportionality factor I, the moment of inertia:

$$L = I\omega.$$

For an arbitrarily shaped object, the moment of inertia depends on its orientation with respect to the axis of rotation. For instance, a rectangular block will have different moments about each of its three orthogonal axes. Now angular velocity $\boldsymbol{\omega}$ and angular momentum $\boldsymbol{L}$ are both vectors. For rotations about one of the axes of symmetry, they are parallel. But if the moment of inertia is different for the three principal axes, then $\boldsymbol{\omega}$ and $\boldsymbol{L}$ are, in general, not in the same direction (see Fig. 31–4). They are related in a way analogous to the relation between $\boldsymbol{E}$ and $\boldsymbol{P}$. In general, we must write

$$\begin{aligned} L_x &= I_{xx}\omega_x + I_{xy}\omega_y + I_{xz}\omega_z, \\ L_y &= I_{xy}\omega_x + I_{yy}\omega_y + I_{yz}\omega_z, \\ L_z &= I_{zx}\omega_x + I_{zy}\omega_y + I_{zz}\omega_z. \end{aligned} \tag{31.16}$$

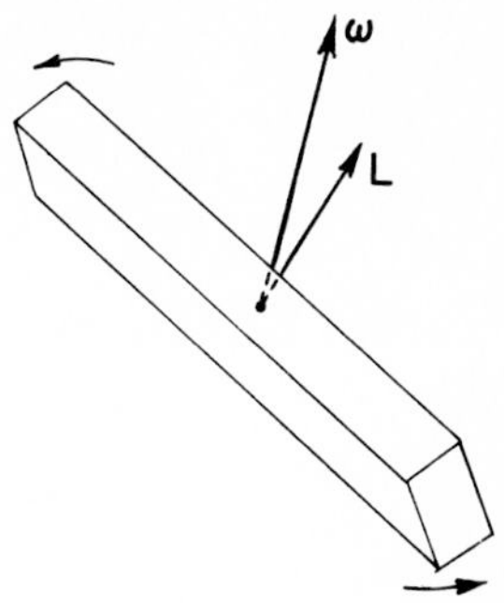

Fig. 31–4. The angular momentum $\boldsymbol{L}$ of a solid object is not, in general, parallel to its angular velocity $\boldsymbol{\omega}$.

The nine coefficients I_{ij} are called the tensor of inertia. Following the analogy with the polarization, the kinetic energy for any angular momentum must be some quadratic form in the components ω_x, ω_y, and ω_z:

$$\text{KE} = \tfrac{1}{2}\sum_{ij} I_{ij}\omega_i\omega_j. \tag{31.17}$$

We can use the energy to define the ellipsoid of inertia. Also, energy arguments can be used to show that the tensor is symmetric—that $I_{ij} = I_{ji}$.

The tensor of inertia for a rigid body can be worked out if the shape of the object is known. We need only to write down the total kinetic energy of all the particles in the body. A particle of mass m and velocity $\boldsymbol{v}$ has the kinetic energy $\frac{1}{2}mv^2$, and the total kinetic energy is just the sum

$$\sum \tfrac{1}{2}mv^2$$

over all of the particles of the body. The velocity $\boldsymbol{v}$ of each particle is related to the angular velocity $\boldsymbol{\omega}$ of the solid body. Let's assume that the body is rotating about its center of mass, which we take to be at rest. Then if $\boldsymbol{r}$ is the displacement of a particle from the center of mass, its velocity $\boldsymbol{v}$ is given by $\boldsymbol{\omega} \times \boldsymbol{r}$. So the total kinetic energy is

$$\text{KE} = \sum \tfrac{1}{2}m(\boldsymbol{\omega} \times \boldsymbol{r})^2. \tag{31.18}$$

Now all we have to do is write $\boldsymbol{\omega} \times \boldsymbol{r}$ out in terms of the components ω_x, ω_y, ω_z, and x, y, z, and compare the result with Eq. (31.17); we find I_{ij} by identifying terms. Carrying out the algebra, we write

$$\begin{aligned} (\boldsymbol{\omega} \times \boldsymbol{r})^2 &= (\boldsymbol{\omega} \times \boldsymbol{r})_x^2 + (\boldsymbol{\omega} \times \boldsymbol{r})_y^2 + (\boldsymbol{\omega} \times \boldsymbol{r})_z^2 \\ &= (\omega_y z - \omega_z y)^2 + (\omega_z x - \omega_x z)^2 + (\omega_x y - \omega_y x)^2 \\ &= +\,\omega_y^2 z^2 - 2\omega_y\omega_z zy + \omega_z^2 y^2 \\ &\quad +\,\omega_z^2 x^2 - 2\omega_z\omega_x xz + \omega_x^2 z^2 \\ &\quad +\,\omega_x^2 y^2 - 2\omega_x\omega_y yx + \omega_y^2 x^2. \end{aligned}$$

Multiplying this equation by $m/2$, summing over all particles, and comparing with Eq. (31.17), we see that I_{xx}, for instance, is given by

$$I_{xx} = \sum m(y^2 + z^2).$$

This is the formula we have had before (Chapter 19, Vol. I) for the moment of inertia of a body about the x-axis. Since $r^2 = x^2 + y^2 + z^2$, we can also write this term as

$$I_{xx} = \sum m(r^2 - x^2).$$

Working out all of the other terms, the tensor of inertia can be written as

$$I_{ij} = \begin{bmatrix} \sum m(r^2 - x^2) & -\sum mxy & -\sum mxz \\ -\sum myx & \sum m(r^2 - y^2) & -\sum myz \\ -\sum mzx & -\sum mzy & \sum m(r^2 - z^2) \end{bmatrix}. \tag{31.19}$$

If you wish, this may be written in "tensor notation" as

$$I_{ij} = \sum m(r^2\,\delta_{ij} - r_i r_j), \tag{31.20}$$

where the r_i are the components (x, y, z) of the position vector of a particle and the $\sum$ means to sum over all the particles. The moment of inertia, then, is a tensor of the second rank whose terms are a property of the body and relate $\boldsymbol{L}$ to $\boldsymbol{\omega}$ by

$$L_i = \sum_j I_{ij}\omega_j. \tag{31.21}$$

For a body of any shape whatever, we can find the ellipsoid of inertia and, therefore, the three principal axes. Referred to these axes, the tensor will be diagonal, so for any object there are always three orthogonal axes for which the angular velocity and angular momentum are parallel. They are called the principal axes of inertia.

31–5 The cross product

We should point out that we have been using tensors of the second rank since Chapter 20 of Volume I. There, we defined a "torque in a plane," such as τ_{xy}, by

$$\tau_{xy} = xF_y - yF_x.$$

Generalized to three dimensions, we could write

$$\tau_{ij} = r_iF_j - r_jF_i. \tag{31.22}$$

The quantity τ_{ij} is a tensor of the second rank. One way to see that this is so is by combining τ_{ij} with some vector, say the unit vector $\boldsymbol{e}$, according to

$$\sum_j \tau_{ij}e_j.$$

If this quantity is a *vector*, then τ_{ij} must transform as a tensor—this is our definition of a tensor. Substituting for τ_{ij}, we have

$$\begin{aligned}\sum_j \tau_{ij}e_j &= \sum_j r_iF_je_j - \sum_j r_je_jF_i\\ &= r_i(\boldsymbol{F}\cdot\boldsymbol{e}) - (\boldsymbol{r}\cdot\boldsymbol{e})F_i.\end{aligned}$$

Since the dot products are scalars, the two terms on the right-hand side are vectors, and likewise their difference. So τ_{ij} is a tensor.

But τ_{ij} is a special kind of tensor; it is *antisymmetric*, that is,

$$\tau_{ij} = -\tau_{ji},$$

so it has only three nonzero terms—τ_{xy}, τ_{yz}, and τ_{zx}. We were able to show in Chapter 20 of Volume I that these three terms, almost "by accident," transform like the three components of a vector, so that we could *define*

$$\boldsymbol{\tau} = (\tau_x, \tau_y, \tau_z) = (\tau_{yz}, \tau_{zx}, \tau_{xy}).$$

We say "by accident," because it happens only in three dimensions. In four dimensions, for instance, an antisymmetric tensor of the second rank has *six* nonezero terms and certainly cannot be replaced by a vector with *four* components.

Just as the axial vector $\boldsymbol{\tau} = \boldsymbol{r} \times \boldsymbol{F}$ is a tensor, so also is every cross product of two polar vectors—all the same arguments apply. By luck, however, they are also representable by vectors (really pseudovectors), so our mathematics has been made easier for us.

Mathematically, if $\boldsymbol{a}$ and $\boldsymbol{b}$ are any two vectors, the nine quantities a_ib_j form a tensor (although it may have no useful physical purpose). Thus, for the position vector r_i, r_ir_j is a tensor, and since δ_{ij} is also, we see that the right side of Eq. (31.20) is indeed a tensor. Likewise Eq. (31.22) is a tensor, since the two terms on the right-hand side are tensors.

31–6 The tensor of stress

The symmetric tensors we have described so far arose as coefficients in relating one vector to another. We would like to look now at a tensor which has a different physical significance—the tensor of *stress*. Suppose we have a solid object with various forces on it. We say that there are various "stresses" inside, by which we mean that there are internal forces between neighboring parts of the material. We have talked a little about such stresses in a two-dimensional case when we considered the surface tension in a stretched diaphragm in Section 12–3. We will now see that the internal forces in the material of a three-dimensional body can be described in terms of a tensor.

Consider a body of some elastic material—say a block of jello. If we make a cut through the block, the material on each side of the cut will, in general, get displaced by the internal forces. Before the cut was made, there must have been forces between the two parts of the block that kept the material in place; we can define the stresses in terms of these forces. Suppose we look at an imaginary plane perpendicular to the x-axis—like the plane σ in Fig. 31–5—and ask about the force across a small area $\Delta y\,\Delta z$ in this plane. The material on the left of the area exerts the force $\Delta\boldsymbol{F}_1$ on the material to the right, as shown in part (b) of the figure. There is, of course, the opposite reaction force $-\Delta\boldsymbol{F}_1$ exerted on the material to the left of the surface. If the area is small enough, we expect that $\Delta\boldsymbol{F}_1$ is proportional to the area $\Delta y\,\Delta z$.

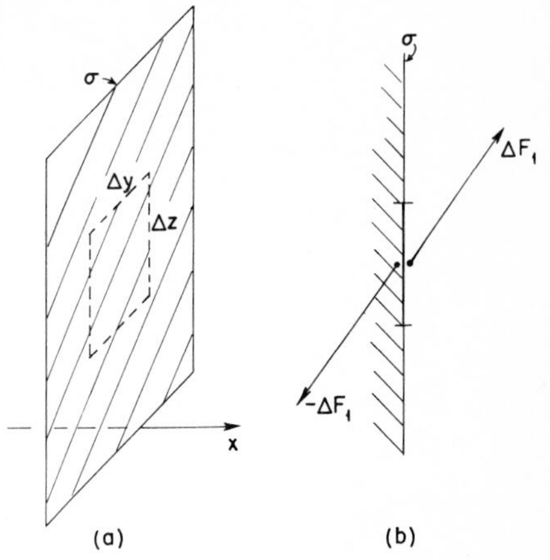

Fig. 31–5. The material to the left of the plane σ exerts across the area $\Delta y\,\Delta z$ the force $\Delta\boldsymbol{F}_1$ on the material to the right of the plane.

You are already familiar with one kind of stress—the pressure in a static liquid. There the force is equal to the pressure times the area and is at right angles to the surface element. For solids—also for viscous liquids in motion—the force need not be normal to the surface; there are *shear* forces in addition to pressures (positive or negative). (By a "shear" force we mean the *tangential* components of the force across a surface.) All three components of the force must be taken into account. Notice also that if we make our cut on a plane with some other orientation, the forces will be different. A complete description of the internal stress requires a tensor.

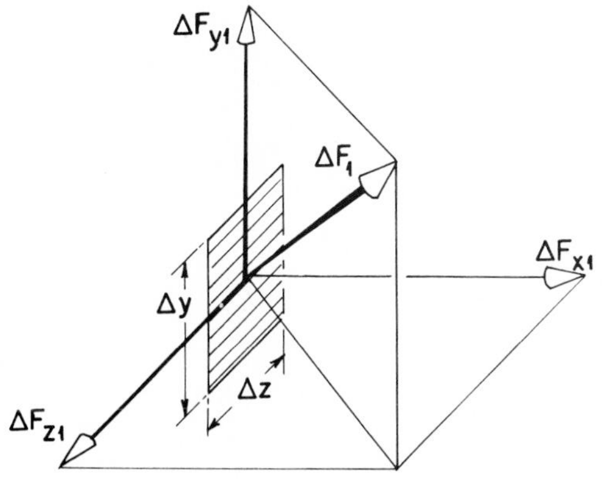

Fig. 31–6. The force $\Delta\boldsymbol{F}_1$ across an element of area $\Delta y\,\Delta z$ perpendicular to the x-axis is resolved into the three components ΔF_{x1}, ΔF_{y1}, and ΔF_{z1}.

We define the stress tensor in the following way: First, we imagine a cut perpendicular to the x-axis and resolve the force $\Delta\boldsymbol{F}_1$ across the cut into its components ΔF_{x1}, ΔF_{y1}, ΔF_{z1}, as in Fig. 31–6. The ratio of these forces to the area $\Delta y\,\Delta z$, we call S_{xx}, S_{yx}, and S_{zx}. For example,

$$S_{yx} = \frac{\Delta F_{y1}}{\Delta y\,\Delta z}.$$

The first index y refers to the direction force component; the second index x is normal to the area. If you wish, you can write the area $\Delta y\,\Delta z$ as Δa_x, meaning an element of area perpendicular to x. Then

$$S_{yx} = \frac{\Delta F_{y1}}{\Delta a_x}.$$

Next, we think of an imaginary cut perpendicular to the y-axis. Across a small

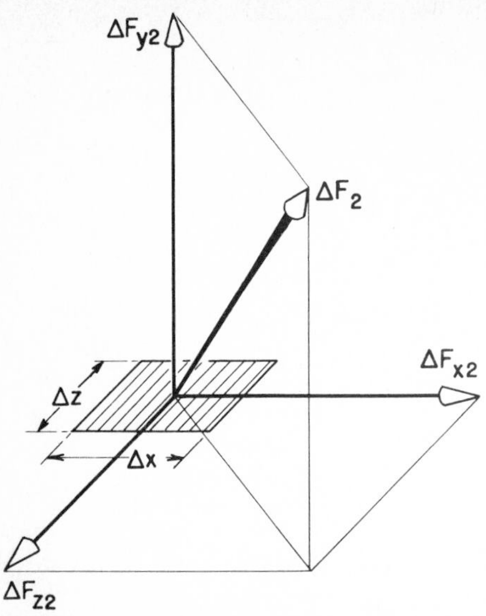

Fig. 31–7. The force across an element of area perpendicular to y is resolved into three rectangular components.

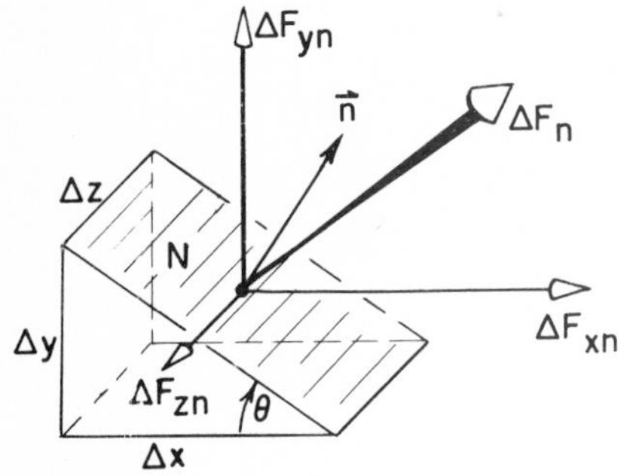

Fig. 31–8. The force F_n across the face N (whose unit normal is $\boldsymbol{n}$) is resolved into components.

area $\Delta x\,\Delta z$ there will be a force $\Delta \boldsymbol{F}_2$. Again we resolve this force into three components, as shown in Fig. 31–7, and define the three components of the stress, S_{xy}, S_{yy}, S_{zy}, as the force per unit area in the three directions. Finally, we make an imaginary cut perpendicular to z and define the three components S_{xz}, S_{yz}, and S_{zz}. So we have the nine numbers

$$S_{ij} = \begin{bmatrix} S_{xx} & S_{xy} & S_{xz} \\ S_{yx} & S_{yy} & S_{yz} \\ S_{zx} & S_{zy} & S_{zz} \end{bmatrix}. \tag{31.23}$$

We want to show now that these nine numbers are sufficient to describe completely the internal state of stress, and that S_{ij} is indeed a tensor. Suppose we want to know the force across a surface oriented at some arbitrary angle. Can we find it from S_{ij}? Yes, in the following way: We imagine a little solid figure which has one face N in the new surface, and the other faces parallel to the coordinate axes. If the face N happened to be parallel to the z-axis, we would have the triangular piece shown in Fig. 31–8. (This is a somewhat special case, but will illustrate well enough the general method.) Now the stress forces on the little solid triangle in Fig. 31–8 are in equilibrium (at least in the limit of infinitesimal dimensions), so the total force on it must be zero. We know the forces on the faces parallel to the coordinate axes directly from S_{ij}. Their vector sum must equal the force on the face N, so we can express this force in terms of S_{ij}.

Our assumption that the *surface* forces on the small triangular volume are in equilibrium neglects any other *body* forces that might be present, such as gravity or pseudo forces if our coordinate system is not an inertial frame. Notice, however, that such body forces will be proportional to the *volume* of the little triangle and, therefore, to Δx, Δy, Δz, whereas all the surface forces are proportional to the areas such as $\Delta x\,\Delta y$, $\Delta y\,\Delta z$, etc. So if we take the scale of the little wedge small enough, the body forces can always be neglected in comparison with the surface forces.

Let's now add up the forces on the little wedge. We take first the x-component, which is the sum of five parts—one from each face. However, if Δz is small enough, the forces on the triangular faces (perpendicular to the z-axis) will be equal and opposite, so we can forget them. The x-component of the force on the bottom rectangle is

$$\Delta F_{x2} = S_{xy}\,\Delta x\,\Delta z.$$

The x-component of the force on the vertical rectangle is

$$\Delta F_{x1} = S_{xx}\,\Delta y\,\Delta z.$$

These two must be equal to the x-component of the force *outward* across the face N. Let's call $\boldsymbol{n}$ the unit vector normal to the face N, and the force on it $\boldsymbol{F}_n$; then we have

$$\Delta F_{xn} = S_{xx}\,\Delta y\,\Delta z + S_{xy}\,\Delta x\,\Delta z.$$

The x-component S_{xn} of the stress across this plane is equal to ΔF_{xn} divided by the area, which is $\Delta z\sqrt{\Delta x^2 + \Delta y^2}$, or

$$S_{xn} = S_{xx}\frac{\Delta y}{\sqrt{\Delta x^2 + \Delta y^2}} + S_{xy}\frac{\Delta x}{\sqrt{\Delta x^2 + \Delta y^2}}.$$

Now $\Delta x/\sqrt{\Delta x^2 + \Delta y^2}$ is the cosine of the angle θ between $\boldsymbol{n}$ and the y-axis, as shown in Fig. 31–8, so it can also be written as n_y, the y-component of $\boldsymbol{n}$. Similarly, $\Delta y/\sqrt{\Delta x^2 + \Delta y^2}$ is $\sin\theta = n_x$. We can write

$$S_{xn} = S_{xx}n_x + S_{xy}n_y.$$

If we now generalize to an arbitrary surface element, we would get that

$$S_{xn} = S_{xx}n_x + S_{xy}n_y + S_{xz}n_z$$

or, in general,

$$S_{in} = \sum_j S_{ij} n_j. \tag{31.24}$$

We *can* find the force across any surface element in terms of the S_{ij}, so it does describe completely the state of internal stress of the material.

Equation (31.24) says that the tensor S_{ij} relates the force $\boldsymbol{S}_n$ to the unit vector $\boldsymbol{n}$, just as α_{ij} relates $\boldsymbol{P}$ to $\boldsymbol{E}$. Since $\boldsymbol{n}$ and $\boldsymbol{S}_n$ are vectors, the components of S_{ij} must transform as a tensor with changes in coordinate axes. So S_{ij} is indeed a tensor.

We can also show that S_{ij} is a *symmetric* tensor by looking at the forces on a little cube of material. Suppose we take a little cube, oriented with its faces parallel to our coordinate axes, and look at it in cross section, as shown in Fig. 31–9. If we let the edge of the cube be one unit, the x- and y-components of the forces on the faces normal to the x- and y-axes might be as shown in the figure. If the cube is small, the stresses do not change appreciably from one side of the cube to the opposite side, so the force components are equal and opposite as shown. Now there must be no torque on the cube, or it would start spinning. The total torque about the center is $(S_{yx} - S_{xy})$ (times the unit edge of the cube), and since the total is zero, S_{yx} is equal to S_{xy}, and the stress tensor is symmetric.

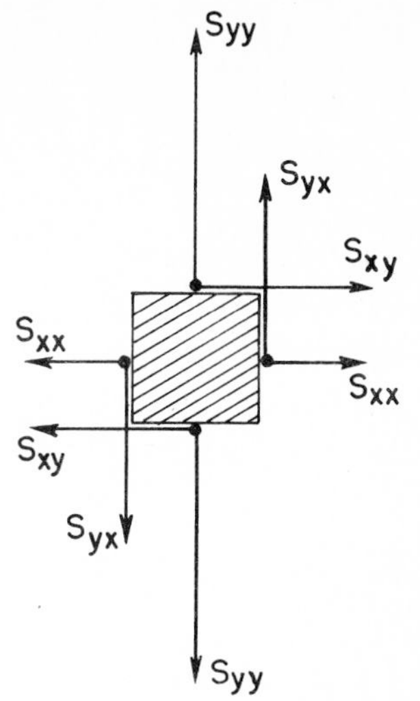

Fig. 31–9. The x- and y-forces on four faces of a small unit cube.

Since S_{ij} is a symmetric tensor, it can be described by an ellipsoid which will have three principal axes. For surfaces normal to these axes, the stresses are particularly simple—they correspond to pushes or pulls perpendicular to the surfaces. There are no shear forces along these faces. For *any* stress, we can always choose our axes so that the shear components are zero. If the ellipsoid is a sphere, there are only normal forces in *any* direction. This corresponds to a hydrostatic pressure (positive or negative). So for a hydrostatic pressure, the tensor is diagonal and all three components are equal; they are, in fact, just equal to the pressure p. We can write

$$S_{ij} = p\delta_{ij}. \tag{31.25}$$

The stress tensor—and also its ellipsoid—will, in general, vary from point to point in a block of material; to describe the whole block we need to give the value of each component of S_{ij} as a function of position. So the stress tensor is a *field.* We have had *scalar fields*, like the temperature $T(x, y, z)$, which give one number for each point in space, and *vector fields* like $\boldsymbol{E}(x, y, z)$, which give three numbers for each point. Now we have a *tensor field* which gives nine numbers for each point in space—or really six for the symmetric tensor S_{ij}. A complete description of the internal forces in an arbitrarily distorted solid requires six functions of x, y, and z.

31–7 Tensors of higher rank

The stress tensor S_{ij} describes the internal *forces* of matter. If the material is elastic, it is convenient to describe the internal *distortion* in terms of another tensor T_{ij}—called the *strain* tensor. For a simple object like a bar of metal, you know that the change in length, ΔL, is approximately proportional to the force, so we say it obeys Hooke's law:

$$\Delta L = \gamma F.$$

For a solid elastic body with arbitrary distortions, the strain T_{ij} is related to the stress S_{ij} by a set of linear equations:

$$T_{ij} = \sum_{k,l} \gamma_{ijkl} S_{kl}. \tag{31.26}$$

Also, you know that the potential energy of a spring (or bar) is

$$\tfrac{1}{2}F\,\Delta L = \tfrac{1}{2}\gamma F^2.$$

The generalization for the elastic energy *density* in a solid body is

$$U_{\text{elastic}} = \sum_{ijkl} \tfrac{1}{2}\gamma_{ijkl} S_{ij} S_{kl}. \tag{31.27}$$

The complete description of the elastic properties of a crystal must be given in terms of the coefficients γ_{ijkl}. This introduces us to a new beast. It is a tensor of the *fourth* rank. Since each index can take on any one of three values, x, y, or z, there are $3^4 = 81$ coefficients. But there are really only 21 *different* numbers. First, since S_{ij} is symmetric, it has only six different values, and only 36 *different* coefficients are needed in Eq. (31.27). But also, S_{ij} can be interchanged with S_{kl} without changing the energy, so γ_{ijkl} must be symmetric if we interchange ij and kl. This reduces the number of different coefficients to 21. So to describe the elastic properties of a crystal of the lowest possible symmetry requires 21 elastic constants! This number is, of course, reduced for crystals of higher symmetry. For example, a cubic crystal has only three elastic constants, and an isotropic substance has only two.

That the latter is true can be seen as follows. How can the components of γ_{ijkl} be independent of the direction of the axes, as they must be if the material is isotropic? *Answer:* They can be independent *only* if they are expressible in terms of the tensor δ_{ij}. There are two possible expressions, $\delta_{ij}\delta_{kl}$ and $\delta_{ik}\delta_{jl} + \delta_{il}\delta_{jk}$, which have the required symmetry, so γ_{ijkl} must be a linear combination of them. Therefore, for isotropic materials,

$$\gamma_{ijkl} = a(\delta_{ij}\delta_{kl}) + b(\delta_{ik}\delta_{jl} + \delta_{il}\delta_{jk}),$$

and the material requires two constants, a and b, to describe its elastic properties. We will leave it for you to show that a cubic crystal needs only three.

As a final example, this time of a third-rank tensor, we have the piezoelectric effect. Under stress, a crystal generates an electric field proportional to the stress; hence, in general, the law is

$$E_i = \sum_{j,k} P_{ijk}S_{jk},$$

where E_i is the electric field, and the P_{ijk} are the piezoelectric coefficients—or the piezoelectric tensor. Can you show that if the crystal has a center of inversion (invariant under $x, y, z \to -x, -y, -z$) the piezoelectric coefficients are all zero?

31-8 The four-tensor of electromagnetic momentum

All the tensors we have looked at so far in this chapter relate to the three dimensions of space; they are defined to have a certain transformation property under spatial rotations. In Chapter 26 we had occasion to use a tensor in the four dimensions of relativistic space-time—the electromagnetic field tensor $F_{\mu\nu}$. The components of such a four-tensor transform under a Lorentz transformation of the coordinates in a special way that we worked out. (Although we did not do it that way, we could have considered the Lorentz transformation as a "rotation" in a four-dimensional "space" called Minkowski space; then the analogy with what we are doing here would have been clearer.)

As our last example, we want to consider another tensor in the four dimensions (t, x, y, z) of relativity theory. When we wrote the stress tensor, we defined S_{ij} as a component of a force across a unit area. But a force is equal to the time rate of change of a momentum. Therefore, instead of saying "S_{xy} is the x-component of the force across a unit area perpendicular to y," we could equally well say, "S_{xy} is the rate of flow of the x-component of momentum through a unit area perpendicular to y." In other words, each term of S_{ij} also represents the flow of the i-component of momentum through a unit area perpendicular to the j-direction. These are pure space components, but they are parts of a "larger" tensor $S_{\mu\nu}$ in four dimensions (μ and $\nu = t, x, y, z$) containing additional components like S_{tx}, S_{yt}, S_{tt}, etc. We will now try to find the physical meaning of these extra components.

We know that the space components represent flow of momentum. We can get a clue on how to extend this to the time dimension by studying another kind of "flow"—the flow of electric charge. For the *scalar* quantity, charge, the rate of flow (per unit area perpendicular to the flow) is a space *vector*—the current density vector $\boldsymbol{j}$. We have seen that the time component of this flow vector is the density of the stuff that is flowing. For instance, $\boldsymbol{j}$ can be combined with a time component, $j_t = \rho$, the charge density, to make the four-vector $j_\mu = (\rho, \boldsymbol{j})$; that is, the μ in j_μ takes on the values t, x, y, z to mean "density, rate of flow in the x-direction, rate of flow in y, rate of flow in z" of the scalar charge.

Now by analogy with our statement about the time component of the flow of a scalar quantity, we might expect that with S_{xx}, S_{xy}, and S_{xz}, describing the flow of the x-component of momentum, there should be a time component S_{xt} which would be the density of whatever is flowing; that is, S_{xt} should be the density of x-momentum. So we can extend our tensor horizontally to include a t-component. We have

$$\begin{aligned} S_{xt} &= \text{density of } x\text{-momentum}, \\ S_{xx} &= x\text{-flow of } x\text{-momentum}, \\ S_{xy} &= y\text{-flow of } x\text{-momentum}, \\ S_{xz} &= z\text{-flow of } x\text{-momentum}. \end{aligned}$$

Similarly, for the y-component of momentum we have the three components of flow—S_{yx}, S_{yy}, S_{yz}—to which we should add a fourth term:

$$S_{yt} = \text{density of } y\text{-momentum}.$$

And, of course, to S_{zx}, S_{zy}, S_{zz} we would add

$$S_{zt} = \text{density of } z\text{-momentum}.$$

In four dimensions there is also a t-component of momentum, which is, we know, energy. So the tensor S_{ij} should be extended vertically with S_{tx}, S_{ty}, and S_{tz}, where

$$\begin{aligned} S_{tx} &= x\text{-flow of energy}, \\ S_{ty} &= y\text{-flow of energy}, \\ S_{tz} &= z\text{-flow of energy}; \end{aligned} \tag{31.28}$$

that is, S_{tx} is the flow of energy per unit area and per unit time across a surface perpendicular to the x-axis, and so on. Finally, to complete our tensor we need S_{tt}, which would be the *density* of *energy*. We have extended our stress tensor S_{ij} of three dimensions to the four-dimensional *stress-energy tensor* $S_{\mu\nu}$. The index μ can take on the four values t, x, y, and z, meaning, respectively, "density," "flow per unit area in the x-direction," "flow per unit area in the y-direction," and "flow per unit area in the z-direction." In the same way, ν takes on the four values t, x, y, z to tell us *what* flows, namely, "energy," "momentum in the x-direction," "momentum in the y-direction," and "momentum in the z-direction."

As an example, we will discuss this tensor not in matter, but in a region of free space in which there is an electromagnetic field. We know that the flow of energy is the Poynting vector $\boldsymbol{S} = \epsilon_0 c^2 \boldsymbol{E} \times \boldsymbol{B}$. So the x-, y-, and z-components of $\boldsymbol{S}$ are, from the relativistic point of view, the components S_{tx}, S_{ty}, and S_{tz} of our four-dimensional stress-energy tensor. The symmetry of the tensor S_{ij} carries over into the time components as well, so the four-dimensional tensor $S_{\mu\nu}$ is symmetric:

$$S_{\mu\nu} = S_{\nu\mu}. \tag{31.29}$$

In other words, the components S_{xt}, S_{yt}, S_{zt}, which are the *densities* of x, y, and z *momentum*, are also equal to the x-, y-, and z-components of the Poynting vector $\boldsymbol{S}$, the *energy flow*—as we have already shown in an earlier chapter by a different kind of argument.

The remaining components of the electromagnetic stress tensor $S_{\mu\nu}$ can also be expressed in terms of the electric and magnetic fields $\boldsymbol{E}$ and $\boldsymbol{B}$. That is to say, we must admit stress or, to put it less mysteriously, flow of momentum in the electromagnetic field. We discussed this in Chapter 27 in connection with Eq. (27.21), but did not work out the details.

Those who want to exercise their prowess in tensors in four dimensions might like to see the formula for $S_{\mu\nu}$ in terms of the fields:

$$S_{\mu\nu} = \frac{\epsilon_0}{2}\left(\sum_{\alpha} F_{\mu\alpha}F_{\nu\alpha} - \tfrac{1}{4}\,\delta_{\mu\nu}\sum_{\alpha,\beta} F_{\beta\alpha}F_{\beta\alpha}\right),$$

where sums on α, β are on t, x, y, z but (as usual in relativity) we adopt a special meaning for the sum sign $\sum$ and for the symbol δ. In the sums the x, y, z terms are to be *subtracted* and $\delta_{tt} = +1$, while $\delta_{xx} = \delta_{yy} = \delta_{zz} = -1$ and $\delta_{\mu\nu} = 0$ for $\mu \neq \nu$ ($c = 1$). Can you verify that it gives the energy density $S_{tt} = (\epsilon_0/2)\,(E^2 + B^2)$ and the Poynting vector $\epsilon_0 \boldsymbol{E} \times \boldsymbol{B}$? Can you show that in an electrostatic field with $\boldsymbol{B} = 0$ the principal axes of stress are in the direction of the electric field, that there is a *tension* $(\epsilon_0/2)E^2$ along the direction of the field, and that there is an equal *pressure* in directions perpendicular to the field direction?

32

Refractive Index of Dense Materials

Review: See Table 32–1.

32–1 Polarization of matter

We want now to discuss the phenomenon of the refraction of light—and also, therefore, the absorption of light—by dense materials. In Chapter 31 of Volume I we discussed the theory of the index of refraction, but because of our limited mathematical abilities at that time, we had to restrict ourselves to finding the index only for materials of low density, like gases. The physical principles that produced the index were, however, made clear. The electric field of the light wave polarizes the molecules of the gas, producing oscillating dipole moments. The acceleration of the oscillating charges radiates new waves of the field. This new field, interfering with the old field, produces a changed field which is equivalent to a phase shift of the original wave. Because this phase shift is proportional to the thickness of the material, the effect is equivalent to having a different phase velocity in the material. When we looked at the subject before, we neglected the complications that arise from such effects as the new wave changing the fields at the oscillating dipoles. We assumed that the forces on the charges in the atoms came just from the *incoming* wave, whereas, in fact, their oscillations are driven not only by the incoming wave but also by the radiated waves of all the other atoms. It would have been difficult for us at that time to include this effect, so we studied only the rarefied gas, where such effects are not important.

Now, however, we will find that it is very easy to treat the problem by the use of differential equations. This method obscures the physical origin of the index (as coming from the re-radiated waves interfering with the original waves), but it makes the theory for dense materials much simpler. This chapter will bring together a large number of pieces from our earlier work. We've taken up practically everything we will need, so there are relatively few really new ideas to be introduced. Since you may need to refresh your memory about what we are going to need, we give in Table 32–1 a list of the equations we are going to use, together with a reference to the place where each can be found. In most instances, we will not take the time to give the physical arguments again, but will just use the equations.

Table 32–1

Our work in this chapter will be based on the following material, already covered in earlier chapters

Subject	Reference	Equation
Damped oscillations	Vol. I, Chap. 23	$m(\ddot{x} + \gamma\dot{x} + \omega_0^2 x) = F$
Index of gases	Vol. I, Chap. 31	$n = 1 + \dfrac{1}{2}\dfrac{Nq_e^2}{\epsilon_0(\omega_0^2 - \omega^2)}$
		$n = n' - in''$
Mobility	Vol. I, Chap. 41	$m\ddot{x} + \mu\dot{x} = F$
Electrical conductivity	Vol. I, Chap. 43	$\mu = \dfrac{\tau}{m}\,;\ \sigma = \dfrac{Nq_e^2\tau}{m}$
Polarizability	Vol. II, Chap. 10	$\rho_{\text{pol}} = -\boldsymbol{\nabla}\cdot\boldsymbol{P}$
Inside dielectrics	Vol. II, Chap. 11	$\boldsymbol{E}_{\text{local}} = \boldsymbol{E} + \dfrac{1}{3\epsilon_0}\boldsymbol{P}$

We begin by recalling the machinery of the index of refraction for a gas. We suppose that there are N particles per unit volume and that each particle behaves as a harmonic oscillator. We use a model of an atom or molecule in which the electron is bound with a force proportional to its displacement (as though the electron were held in place by a spring). We emphasized that this was not a legitimate *classical* model of an atom, but we will show later that the correct quantum mechanical theory gives results equivalent to this model (in simple cases). In our earlier treatment, we did not include the possibility of a damping force in the atomic oscillators, but we will do so now. Such a force corresponds to a resistance to the motion, that is, to a force proportional to the velocity of the electron. Then the equation of motion is

$$F = q_eE = m(\ddot{x} + \gamma\dot{x} + \omega_0^2x), \tag{32.1}$$

where x is the displacement parallel to the direction of $\boldsymbol{E}$. (We are assuming an *isotropic* oscillator whose restoring force is the same in all directions. Also, we are taking, for the moment, a linearly polarized wave, so that $\boldsymbol{E}$ doesn't change direction.) If the electric field acting on the atom varies sinusoidally with time, we write

$$E = E_0e^{i\omega t}. \tag{32.2}$$

The displacement will then oscillate with the same frequency, and we can let

$$x = x_0e^{i\omega t}.$$

Substituting $\dot{x} = i\omega x$ and $\ddot{x} = -\omega^2x$, we can solve for x in terms of E:

$$x = \frac{q_e/m}{-\omega^2 + i\gamma\omega + \omega_0^2}\,E. \tag{32.3}$$

Knowing the displacement, we can calculate the acceleration $\ddot{x}$ and find the radiated wave responsible for the index. This was the way we computed the index in Chapter 31 of Volume I.

Now, however, we want to take a different approach. The induced dipole moment p of an atom is q_ex or, using Eq. (32.3),

$$\boldsymbol{p} = \frac{q_e^2/m}{-\omega^2 + i\gamma\omega + \omega_0^2}\,\boldsymbol{E}. \tag{32.4}$$

Since $\boldsymbol{p}$ is proportional to $\boldsymbol{E}$, we write

$$\boldsymbol{p} = \epsilon_0\alpha(\omega)\boldsymbol{E}, \tag{32.5}$$

where α is called the *atomic polarizability*.* With this definition, we have

$$\alpha = \frac{q_e^2/m\epsilon_0}{-\omega^2 + i\gamma\omega + \omega_0^2}\cdot \tag{32.6}$$

The quantum mechanical solution for the motions of electrons in atoms gives a similar answer except with the following modifications. The atoms have several natural frequencies, each frequency with its own dissipation constant γ. Also the effective "strength" of each mode is different, which we can represent by multiplying the polarizability for each frequency by a strength factor f, which is a number we expect to be of the order of 1. Representing the three parameters ω, γ, and f by ω_k, γ_k, and f_k for each mode of oscillation, and summing over the

---

* Throughout this chapter we follow the notation of Chapter 31 of Volume I, and let α represent the *atomic* polarizability as defined here. In the last chapter, we used α to represent the *volume* polarizability—the ratio of P to E. In the notation of *this* chapter $P = N\alpha\epsilon_0E$ (see Eq. 32.8).

various modes, we modify Eq. (32.6) to read

$$\alpha(\omega) = \frac{q_e^2}{\epsilon_0 m} \sum_k \frac{f_k}{-\omega^2 + i\gamma_k\omega + \omega_{0k}^2}. \tag{32.7}$$

If N is the number of atoms per unit volume in the material, the polarization P is just $Np = \epsilon_0 N\alpha E$, and is proportional to E:

$$\boldsymbol{P} = \epsilon_0 N\alpha(\omega)\boldsymbol{E}. \tag{32.8}$$

In other words, when there is a sinusoidal electric field acting in a material, there is an induced dipole moment per unit volume which is proportional to the electric field—with a proportionality constant α that, we emphasize, depends upon the frequency. At very high frequencies, α is small; there is not much response. However, at low frequencies there can be a strong response. Also, the proportionality constant is a complex number, which means that the polarization does not exactly follow the electric field, but may be shifted in phase to some extent. At any rate, there is a polarization per unit volume whose magnitude is proportional to the strength of the electric field.

32–2 Maxwell's equations in a dielectric

The existence of polarization in matter means that there are polarization charges and currents inside of the material, and these must be put into the complete Maxwell equations in order to find the fields. We are going to solve Maxwell's equations this time in a situation in which the charges and currents are not zero, as in a vacuum, but are given implicitly by the polarization vector. Our first step is to find explicitly the charge density ρ and current density $\boldsymbol{j}$, averaged over a small volume of the same size we had in mind when we defined $\boldsymbol{P}$. Then the ρ and $\boldsymbol{j}$ we need can be obtained from the polarization.

We have seen in Chapter 10 that when the polarization $\boldsymbol{P}$ varies from place to place, there is a charge density given by

$$\rho_{\text{pol}} = -\boldsymbol{\nabla}\cdot\boldsymbol{P}. \tag{32.9}$$

At that time, we were dealing with static fields, but the same formula is valid also for time-varying fields. However, when $\boldsymbol{P}$ varies with time, there are charges in motion, so there is also a polarization *current*. Each of the oscillating charges contributes a current equal to its charge q_e, times its velocity v. With N such charges per unit volume, the current density $\boldsymbol{j}$ is

$$\boldsymbol{j} = Nq_e\boldsymbol{v}.$$

Since we know that $v = dx/dt$, then $j = Nq_e(dx/dt)$, which is just dP/dt. Therefore the current density from the varying polarization is

$$\boldsymbol{j}_{\text{pol}} = \frac{d\boldsymbol{P}}{dt}. \tag{32.10}$$

Our problem is now direct and simple. We write Maxwell's equations with the charge density and current density expressed in terms of $\boldsymbol{P}$, using Eqs. (32.9) and (32.10). (We assume that there are no other currents and charges in the material.) We then relate $\boldsymbol{P}$ to $\boldsymbol{E}$ with Eq. (32.5), and we solve the equation for $\boldsymbol{E}$ and $\boldsymbol{B}$—looking for the wave solutions.

Before we do this, we would like to make an historical note. Maxwell originally wrote his equations in a form which was different from the one we have been using. Because the equations were written in this different form for many years—and are still written that way by many people—we will explain the difference. In the early days, the mechanism of the dielectric constant was not fully and clearly appreciated. The nature of atoms was not understood, nor that there was a polarization of the material. So people did not appreciate that there was a contribution to the charge density ρ from $\nabla \cdot \boldsymbol{P}$. They thought only in terms of charges that were not bound to atoms (such as the charges that flow in wires or are rubbed off surfaces).

Today, we prefer to let ρ represent the *total* charge density, including the part from the bound atomic charges. If we call that part ρ_{pol}, we can write

$$\rho = \rho_{\text{pol}} + \rho_{\text{other}},$$

where ρ_{other} is the charge density considered by Maxwell and refers to the charges not bound to individual atoms. We would then write

$$\nabla \cdot \boldsymbol{E} = \frac{\rho_{\text{pol}} + \rho_{\text{other}}}{\epsilon_0}.$$

Substituting ρ_{pol} from Eq. (32.9),

$$\nabla \cdot \boldsymbol{E} = \frac{\rho_{\text{other}}}{\epsilon_0} - \frac{1}{\epsilon_0} \nabla \cdot \boldsymbol{P}$$

or

$$\nabla \cdot (\epsilon_0 \boldsymbol{E} + \boldsymbol{P}) = \rho_{\text{other}}. \tag{32.11}$$

The current density in the Maxwell equations for $\nabla \times \boldsymbol{B}$ also has, in general, contributions from bound atomic currents. We can therefore write

$$\boldsymbol{j} = \boldsymbol{j}_{\text{pol}} + \boldsymbol{j}_{\text{other}},$$

and the Maxwell equation becomes

$$c^2 \nabla \times \boldsymbol{B} = \frac{\boldsymbol{j}_{\text{other}}}{\epsilon_0} + \frac{\boldsymbol{j}_{\text{pol}}}{\epsilon_0} + \frac{\partial \boldsymbol{E}}{\partial t}. \tag{32.12}$$

Using Eq. (32.10), we get

$$\epsilon_0 c^2 \nabla \times \boldsymbol{B} = \boldsymbol{j}_{\text{other}} + \frac{\partial}{\partial t} (\epsilon_0 \boldsymbol{E} + \boldsymbol{P}). \tag{32.13}$$

Now you can see that if we were to *define* a new vector $\boldsymbol{D}$ by

$$\boldsymbol{D} = \epsilon_0 \boldsymbol{E} + \boldsymbol{P}, \tag{32.14}$$

the two field equations would become

$$\nabla \cdot \boldsymbol{D} = \rho_{\text{other}} \tag{32.15}$$

and

$$\epsilon_0 c^2 \nabla \times \boldsymbol{B} = \boldsymbol{j}_{\text{other}} + \frac{\partial \boldsymbol{D}}{\partial t}. \tag{32.16}$$

These are actually the forms that Maxwell used for dielectrics. His two remaining equations were

$$\nabla \times \boldsymbol{E} = -\frac{\partial \boldsymbol{B}}{\partial t},$$

and

$$\nabla \cdot \boldsymbol{B} = 0,$$

which are the same as we have been using.

Maxwell and the other early workers also had a problem with magnetic materials (which we will take up soon). Because they did not know about the circulating currents responsible for atomic magnetism, they used a current density that was missing still another part. Instead of Eq. (32.16), they actually wrote

$$\nabla \times \boldsymbol{H} = \boldsymbol{j}' + \frac{\partial \boldsymbol{D}}{\partial t}, \tag{32.17}$$

where $\boldsymbol{H}$ differs from $\epsilon_0 c^2 \boldsymbol{B}$ because it includes the effects of atomic currents. (Then $\boldsymbol{j}'$ represents what is left of the currents.) So Maxwell had *four* field vectors—$\boldsymbol{E}$, $\boldsymbol{D}$, $\boldsymbol{B}$, and $\boldsymbol{H}$—the $\boldsymbol{D}$ and $\boldsymbol{H}$ were hidden ways of not paying attention to what was going on inside the material. You will find the equations written this way in many places.

To solve the equations, it is necessary to relate $\boldsymbol{D}$ and $\boldsymbol{H}$ to the other fields, and people used to write

$$\boldsymbol{D} = \epsilon \boldsymbol{E} \quad \text{and} \quad \boldsymbol{B} = \mu \boldsymbol{H}. \tag{32.18}$$

However, these relations are only approximately true for some materials and even then only if the fields are not changing rapidly with time. (For sinusoidally varying fields one often *can* write the equations this way by making ϵ and μ complex functions of the frequency, but not for an arbitrary time variation of the fields.) So there used to be all kinds of cheating in solving the equations. We think the right way is to keep the equations in terms of the fundamental quantities as we now understand them—and that's how we have done it.

32–3 Waves in a dielectric

We want now to find out what kind of electromagnetic waves can exist in a dielectric material in which there are no extra charges other than those bound in atoms. So we take $\rho = -\boldsymbol{\nabla} \cdot \boldsymbol{P}$ and $\boldsymbol{j} = \partial \boldsymbol{P}/\partial t$. Maxwell's equations then become

$$\begin{aligned} &\text{(a)} \quad \boldsymbol{\nabla} \cdot \boldsymbol{E} = -\frac{\boldsymbol{\nabla} \cdot \boldsymbol{P}}{\epsilon_0} \qquad &&\text{(b)} \quad c^2 \boldsymbol{\nabla} \times \boldsymbol{B} = \frac{\partial}{\partial t}\left(\frac{\boldsymbol{P}}{\epsilon_0} + \boldsymbol{E}\right) \\ &\text{(c)} \quad \boldsymbol{\nabla} \times \boldsymbol{E} = -\frac{\partial \boldsymbol{B}}{\partial t} \qquad &&\text{(d)} \quad \boldsymbol{\nabla} \cdot \boldsymbol{B} = 0 \end{aligned} \tag{32.19}$$

We can solve these equations as we have done before. We start by taking the curl of Eq. (32.19c):

$$\boldsymbol{\nabla} \times (\boldsymbol{\nabla} \times \boldsymbol{E}) = -\frac{\partial}{\partial t} \boldsymbol{\nabla} \times \boldsymbol{B}.$$

Next, we make use of the vector identity

$$\boldsymbol{\nabla} \times (\boldsymbol{\nabla} \times \boldsymbol{E}) = \boldsymbol{\nabla}(\boldsymbol{\nabla} \cdot \boldsymbol{E}) - \nabla^2 \boldsymbol{E},$$

and also substitute for $\boldsymbol{\nabla} \times \boldsymbol{B}$, using Eq. (32.19b); we get

$$\boldsymbol{\nabla}(\boldsymbol{\nabla} \cdot \boldsymbol{E}) - \nabla^2 \boldsymbol{E} = -\frac{1}{\epsilon_0 c^2} \frac{\partial^2 \boldsymbol{P}}{\partial t^2} - \frac{1}{c^2} \frac{\partial^2 \boldsymbol{E}}{\partial t^2}.$$

Using Eq. (32.19a) for $\boldsymbol{\nabla} \cdot \boldsymbol{E}$, we get

$$\nabla^2 \boldsymbol{E} - \frac{1}{c^2} \frac{\partial^2 \boldsymbol{E}}{\partial t^2} = -\frac{1}{\epsilon_0} \boldsymbol{\nabla}(\boldsymbol{\nabla} \cdot \boldsymbol{P}) + \frac{1}{\epsilon_0 c^2} \frac{\partial^2 \boldsymbol{P}}{\partial t^2}. \tag{32.20}$$

So instead of the wave equation, we now get that the D'Alembertian of $\boldsymbol{E}$ is equal to two terms involving the polarization $\boldsymbol{P}$.

Since $\boldsymbol{P}$ depends on $\boldsymbol{E}$, however, Eq. (32.20) can still have wave solutions. We will now limit ourselves to *isotropic* dielectrics, so that $\boldsymbol{P}$ is always in the same direction as $\boldsymbol{E}$. Let's try to find a solution for a wave going in the z-direction. Then, the electric field might vary as $e^{i(\omega t - kz)}$. We will also suppose that the wave is polarized in the x-direction—that the electric field has only an x-component. We write

$$E_x = E_0 e^{i(\omega t - kz)}. \tag{32.21}$$

You know that any function of $(z - vt)$ represents a wave that travels with the speed v. The exponent of Eq. (32.21) can be written as

$$-ik\left(z - \frac{\omega}{k} t\right),$$

so, Eq. (32.21) represents a wave with the phase velocity

$$v_{\text{ph}} = \omega/k.$$

The index of refraction n is defined (see Chapter 31, Vol. I) by letting

$$v_{\text{ph}} = \frac{c}{n}\cdot$$

Thus Eq. (32.21) becomes

$$E_x = E_0 e^{i\omega(t-nz/c)}.$$

So we can find n by finding what value of k is required if Eq. (32.21) is to satisfy the proper field equations, and then using

$$n = \frac{kc}{\omega}\cdot \tag{32.22}$$

In an isotropic material, there will be only an x-component of the polarization; then $\boldsymbol{P}$ has no variation with the x-coordinate, so $\boldsymbol{\nabla}\cdot\boldsymbol{P} = 0$, and we get rid of the first term on the right-hand side of Eq. (32.20). Also, since we are assuming a linear dielectric, P_x will vary as $e^{i\omega t}$, and $\partial^2 P_x/\partial t^2 = -\omega^2 P_x$. The Laplacian in Eq. (32.20) becomes simply $\partial^2 E_x/\partial z^2 = -k^2 E_x$, so we get

$$-k^2 E_x + \frac{\omega^2}{c^2} E_x = -\frac{\omega^2}{\epsilon_0 c^2} P_x. \tag{32.23}$$

Now let us assume for the moment that since $\boldsymbol{E}$ is varying sinusoidally, we can set $\boldsymbol{P}$ proportional to $\boldsymbol{E}$, as in Eq. (32.5). (We'll come back to discuss this assumption later.) We write

$$P_x = \epsilon_0 N\alpha E_x.$$

Then E_x drops out of Eq. (32.23), and we find

$$k^2 = \frac{\omega^2}{c^2}(1 + N\alpha). \tag{32.24}$$

We have found that a wave like Eq. (32.21), with the wave number k given by Eq. (32.24), will satisfy the field equations. Using Eq. (32.22), the index n is given by

$$n^2 = 1 + N\alpha. \tag{32.25}$$

Let's compare this formula with what we obtained in our theory of the index of a gas (Chapter 31, Vol. I). There, we got Eq. (31.29), which is

$$n = 1 + \frac{1}{2}\frac{Nq_e^2}{m\epsilon_0}\frac{1}{-\omega^2 + \omega_0^2}\cdot \tag{32.26}$$

Taking α from Eq. (32.6), Eq. (32.25) would give us

$$n^2 = 1 + \frac{Nq_e^2}{m\epsilon_0}\frac{1}{-\omega^2 + i\gamma\omega + \omega_0^2}\cdot \tag{32.27}$$

First, we have the new term in $i\gamma\omega$, because we are including the dissipation of the oscillators. Second, the left-hand side is n instead of n^2, and there is an extra factor of 1/2. But notice that if N is small enough so that n is close to one (as it is for a gas), then Eq. (32.27) says that n^2 is one plus a small number: $n^2 = 1 + \epsilon$. We can then write $n = \sqrt{1+\epsilon} \approx 1 + \epsilon/2$, and the two expressions are equivalent. Thus our new method gives for a gas the same result we found earlier.

Now you might think that Eq. (32.27) should give the index of refraction for dense materials also. It needs to be modified, however, for several reasons. First, the derivation of this equation assumes that the polarizing field on each atom is the field E_x. That assumption is *not* right, however, because in dense materials there is also the field produced by other atoms in the vicinity, which may be comparable to E_x. We considered a similar problem when we studied the static fields in dielectrics. (See Chapter 11.) You will remember that we estimated the field at a single atom by imagining that it sat in a spherical hole in the surrounding dielectric. The field in such a hole—which we called the *local* field—is increased over the average field $\boldsymbol{E}$ by the amount $P/3\epsilon_0$. (Remember, however, that this result is only strictly true in isotropic materials—including the special case of a cubic crystal.)

The same arguments will hold for the electric field in a wave, so long as the wavelength of the wave is much longer than the spacing between atoms. Limiting ourselves to such cases, we write

$$E_{\text{local}} = E + \frac{P}{3\epsilon_0}. \tag{32.28}$$

This local field is the one that should be used for E in Eq. (32.3); that is, Eq.(32.8) should be rewritten:

$$P = \epsilon_0 N\alpha E_{\text{local}}. \tag{32.29}$$

Using E_{local} from Eq. (32.28), we find

$$P = \epsilon_0 N\alpha \left(E + \frac{P}{3\epsilon_0}\right)$$

or

$$P = \frac{N\alpha}{1 - (N\alpha/3)}\, \epsilon_0 E. \tag{32.30}$$

In other words, for dense materials P is still proportional to E (for sinusoidal fields). However, the constant of proportionality is not $\epsilon_0 N\alpha$, as we wrote below Eq. (32.23), but should be $\epsilon_0 N\alpha/[1 - (N\alpha/3)]$. So we should correct Eq. (32.25) to read

$$n^2 = 1 + \frac{N\alpha}{1 - (N\alpha/3)}. \tag{32.31}$$

It will be more convenient if we rewrite this equation as

$$3\,\frac{n^2 - 1}{n^2 + 2} = N\alpha, \tag{32.32}$$

which is algebraically equivalent. This is known as the Clausius-Mosotti equation.

There is another complication in dense materials. Because neighboring atoms are so close, there are strong interactions between them. The internal modes of oscillation are, therefore, modified. The natural frequencies of the atomic oscillations are spread out by the interactions, and they are usually quite heavily damped —the resistance coefficient becomes quite large. So the ω_0's and γ's of the solid will be quite different from those of the free atoms. With these reservations, we can still represent α, at least approximately, by Eq. (32.7). We have then that

$$3\,\frac{n^2 - 1}{n^2 + 2} = \frac{Nq_e^2}{m\epsilon_0} \sum_k \frac{f_k}{-\omega^2 + i\gamma_k\omega + \omega_{0k}^2}. \tag{32.33}$$

One final complication. If the dense material is a mixture of several components, each will contribute to the polarization. The total α will be the sum of the contributions from each component of the mixture [except for the inaccuracy of the local field approximation, Eq. (32.28), in ordered crystals—effects we discussed when analyzing ferroelectrics]. Writing N_j as the number of atoms of each component per unit volume, we should replace Eq. (32.32) by

$$3\left(\frac{n^2 - 1}{n^2 + 2}\right) = \sum_j N_j\alpha_j, \tag{32.34}$$

where each α_j will be given by an expression like Eq. (32.7). Equation (32.34) completes our theory of the index of refraction. The quantity $3(n^2 - 1)/(n^2 + 2)$ is given by some complex function of frequency, which is the mean atomic polarizability $\alpha(\omega)$. The precise evaluation of $\alpha(\omega)$ (that is, finding f_k, γ_k and ω_{0k}) in dense substances is a difficult problem of quantum mechanics. It has been done from first principles only for a few especially simple substances.

32–4 The complex index of refraction

We want to look now at the consequences of our result, Eq. (32.33). First, we notice that α is complex, so the index n is going to be a complex number. What does that mean? Let's say that we write n as the sum of a real and an imaginary part:

$$n = n_R - in_I, \tag{32.35}$$

where n_R and n_I are real functions of ω. We write in_I with a minus sign, so that n_I will be a positive quantity in all ordinary optical materials. (In ordinary inactive materials—that are not, like lasers, light sources themselves—γ is a positive number, and that makes the imaginary part of n negative.) Our plane wave of Eq. (32.21) is written in terms of n as

$$E_x = E_0 e^{-i\omega(t-nz/c)}.$$

Writing n as in Eq. (32.35), we would have

$$E_x = E_0 e^{-\omega n_I z/c} e^{i\omega(t-n_R z/c)}. \tag{32.36}$$

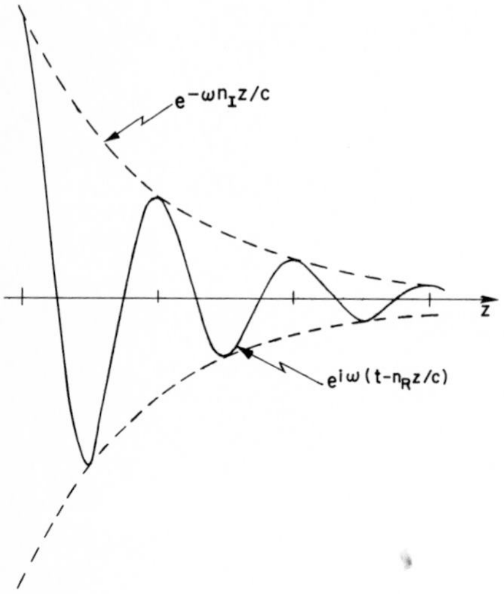

Fig. 32–1. A graph of E_x for some instant t, if $n_I \approx n_R/2\pi$.

The term $e^{i\omega(t-n_R z/c)}$ represents a wave travelling with the speed c/n_R, so n_R represents what we normally think of as the index of refraction. But the *amplitude* of this wave is

$$E_0 e^{-\omega n_I z/c},$$

which decreases exponentially with z. A graph of the strength of the electric field at some instant as a function of z is shown in Fig. 32–1, for $n_I \approx n_R/2\pi$. The imaginary part of the index represents the attenuation of the wave due to the energy losses in the atomic oscillators. The *intensity* of the wave is proportional to the square of the amplitude, so

$$\text{Intensity} \propto e^{-2\omega n_I z/c}.$$

This is often written as

$$\text{Intensity} \propto e^{-\beta z},$$

where $\beta = 2\omega n_I/c$ is called the *absorption coefficient*. Thus we have in Eq. (32.33) not only the theory of the index of refraction of materials, but the theory of their absorption of light as well.

In what we usually consider to be transparent material, the quantity $c/\omega n_I$—which has the dimensions of a length—is quite large in comparison with the thickness of the material.

32–5 The index of a mixture

There is another prediction of our theory of the index of refraction that we can check against experiment. Suppose we consider a mixture of two materials. The index of the mixture is not the average of the two indexes, but should be given in terms of the sum of the two polarizabilities, as in Eq. (32.34). If we ask about the index of, say, a sugar solution, the total polarizability is the sum of the polarizability of the water and that of the sugar. Each must, of course, be calculated using for N the number per unit volume of the molecules of the particular kind. In other words, if a given solution has N_1 molecules of water, whose polarizability is α_1, and N_2 molecules of sucrose ($C_{12}H_{22}O_{11}$), whose polarizability is α_2, we should have that

$$3\left(\frac{n^2-1}{n^2+2}\right) = N_1\alpha_1 + N_2\alpha_2. \tag{32.37}$$

We can use this formula to test our theory against experiment by measuring the index for various concentrations of sucrose in water. We are making several assumptions here, however. Our formula assumes that there is no chemical action when the sucrose is dissolved and that the disturbances to the individual atomic

Table 32–2

Refractive index of sucrose solutions, and comparison with predictions of Eq. (32.37).

Data from Handbook

A	B	C	D	E	F	G	H	J
Fraction of sucrose by weight	density (gm/cm^3)	n at 20°C	Moles of sucrose[d] per liter, N_2/N_0	Moles of water[e] per liter, N_1/N_0	$3\left(\frac{n^2-1}{n^2+2}\right)$	$N_1\alpha_1$	$N_2\alpha_2$	$N_0\alpha_2$ (gm/liter)
0[a]	0.9982	1.333	0	55.5	0.617	0.617	0	—
0.30	1.1270	1.3811	0.970	43.8	0.698	0.487	0.211	0.213
0.50	1.2296	1.4200	1.798	34.15	0.759	0.379	0.380	0.211
0.85	1.4454	1.5033	3.59	12.02	0.886	0.1335	0.752	0.210
1.00[b]	1.588	1.5577[c]	4.64	0	0.960	0	0.960	0.207

[a] pure water
[b] sugar crystals
[c] average (see text)
[d] molecular weight of sucrose = 342
[e] molecular weight of water = 18

oscillators are not too different for various concentrations. So our result is certainly only approximate. Anyway, let's see how good it is.

We have picked the example of a sugar solution because there is a good table of measurements of the index of refraction in the *Handbook of Chemistry and Physics* and also because sugar is a molecular crystal that goes into solution without ionizing or otherwise changing its chemical state.

We give in the first three columns of Table 32–2 the data from the handbook. Column A is the percent of sucrose by weight, column B is the measured density (gm/cm^3), and column C is the measured index of refraction for light whose wavelength is 589.3 millimicrons. For pure sugar we have taken the measured index of sugar crystals. The crystals are not isotropic, so the measured index is different along different directions. The handbook gives three values:

$$n_1 = 1.5376, \qquad n_2 = 1.5651, \qquad n_3 = 1.5705.$$

We have taken the average.

Now we could try to compute n for each concentration, but we don't know what value to take for α_1 or α_2. Let's test the theory this way: We will assume that the polarizability of water (α_1) is the same at all concentrations and compute the polarizability of sucrose by using the experiment of values for n and solving Eq. (38.27) for α_2. If the theory is correct, we should get the same α_2 for all concentrations.

First, we need to know N_1 and N_2: let's express them in terms of Avogadro's number, N_0. Let's take one liter (1000 cm^3) for our unit of volume. Then N_i/N_0 is the weight per liter divided by the gram-molecular weight. And the weight per liter is the density (multiplied by 1000 to get grams per liter) times the fractional weight of either the sucrose or the water. In this way, we get N_2/N_0 and N_1/N_0 as in columns D and E of the table.

In column F we have computed $3(n^2-1)/(n^2+2)$ from the experimental values of n in column C. For pure water, $3(n^2-1)/(n^2+2)$ is 0.617, which is equal to just $N_1\alpha_1$. We can then fill in the rest of Column G, since for each row G/E may be in the same ratio—namely, 0.617:55.5. Subtracting column G from column F, we get the contribution $N_2\alpha_2$ of the sucrose, shown in column H. Dividing these entries by the values of N_2/N_0 in column D, we get the value of $N_0\alpha_2$ shown in column J.

From our theory we would expect all the values of $N_0\alpha_2$ to be the same. They are not exactly equal, but pretty close. We can conclude that our ideas are fairly correct. Even more, we find that the polarizability of the sugar molecule doesn't seem to depend much on its surroundings—its polarizability is nearly the same in a dilute solution as it is in the crystal.

32–6 Waves in metals

The theory we have worked out in this chapter for solid materials can also be applied to good conductors, like metals, with very little modification. In metals some of the electrons have no binding force holding them to any particular atom; it is these "free" electrons which are responsible for the conductivity. There are other electrons which are bound, and the theory above is directly applicable to them. Their influence, however, is usually swamped by the effects of the conduction electrons. We will consider now only the effects of the free electrons.

If there is no restoring force on an electron—but still some resistance to its motion—its equation of motion differs from Eq. (32.1) only because the term in $\omega_0^2 x$ is lacking. So all we have to do is set $\omega_0^2 = 0$ in the rest of our derivations—except that there is one more difference. The reason that we had to distinguish between the average field and the local field in a dielectric is that in an insulator each of the dipoles is fixed in position, so that it has a definite relationship to the position of the others. But because the conduction electrons in a metal move around all over the place, the field on them *on the average* is just the average field $\boldsymbol{E}$. So the correction we made to Eq. (32.5) by using Eq. (32.28) should *not* be made for conduction electrons. Therefore the formula for the index of refraction for metals should look like Eq. (32.27), except with ω_0 set equal to zero, namely,

$$n^2 = 1 + \frac{Nq_e^2}{m\epsilon_0}\frac{1}{-\omega^2 + i\gamma\omega}. \tag{32.38}$$

This is only the contribution from the conduction electrons, which we will assume is the major term for metals.

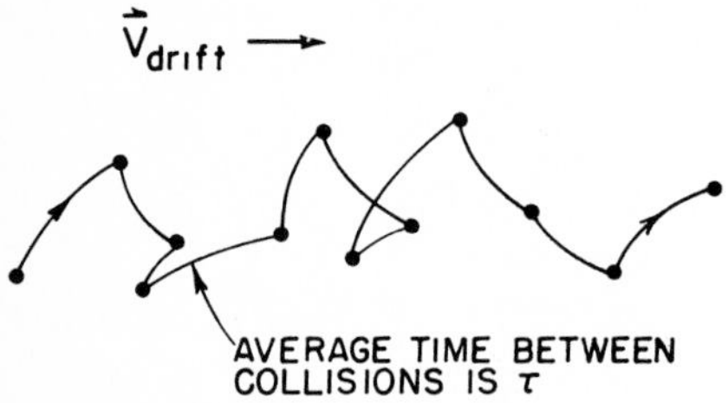

Fig. 32–2. The motion of a free electron.

Now we even know how to find what value to use for γ, because it is related to the conductivity of the metal. In Chapter 43 of Volume I we discussed how the conductivity of a metal comes from the diffusion of the free electrons through the crystal. The electrons go on a jagged path from one scattering to the next, and between scatterings they move freely except for an acceleration due to any average electric field (as shown in Fig. 32–2). We found in Chapter 43 of Volume I that the average drift velocity is just the acceleration times the average time τ between collisions. The acceleration is q_eE/m, so

$$v_{\text{drift}} = \frac{q_eE}{m}\tau. \tag{32.39}$$

This formula assumed that E was constant, so that v_{drift} was a steady velocity. Since there is no average acceleration, the drag force is equal to the applied force. We have defined γ by saying that γmv is the drag force [see Eq. (32.1)], which is q_eE; therefore we have that

$$\gamma = \frac{1}{\tau}. \tag{32.40}$$

Although we cannot easily measure τ directly, we can determine it by measuring the conductivity of the metal. It is found experimentally that an electric field $\boldsymbol{E}$ in a metal produces a current with the density $\boldsymbol{j}$ proportional to $\boldsymbol{E}$ (for isotropic materials):

$$\boldsymbol{j} = \sigma\boldsymbol{E}.$$

The proportionality constant σ is called the *conductivity*. This is just what we expect from Eq. (32.39) if we set

$$j = Nq_ev_{\text{drift}}.$$

Then

$$\sigma = \frac{Nq_e^2}{m}\tau. \tag{32.41}$$

So τ—and therefore γ—can be related to the observed electrical conductivity. Using Eqs. (32.40) and (32.41), we can rewrite our formula for the index, Eq.

(32.38), in the following form:

$$n^2 = 1 + \frac{\sigma/\epsilon_0}{i\omega(1 + i\omega\tau)}, \tag{32.42}$$

where

$$\tau = \frac{1}{\gamma} = \frac{m\sigma}{Nq_e^2}\cdot \tag{32.43}$$

This is a convenient formula for the index of refraction of metals.

32–7 Low-frequency and high-frequency approximations; the skin depth and the plasma frequency

Our result, Eq. (32.42), for the index of refraction for metals predicts quite different characteristics for wave propagation at different frequencies. Let's first see what happens at very *low* frequencies. If ω is small enough, we can approximate Eq. (32.42) by

$$n^2 = -i\,\frac{\sigma}{\epsilon_0\omega}\cdot \tag{32.44}$$

Now, as you can check by taking the square,*

$$\sqrt{-i} = \frac{1 - i}{\sqrt{2}};$$

so for low frequencies,

$$n = \sqrt{\sigma/2\epsilon_0\omega}\,(1 - i). \tag{32.45}$$

The real and imaginary parts of n have the same magnitude. With such a large imaginary part to n, the wave is rapidly attenuated in the metal. Referring to Eq. (32.36), the amplitude of a wave going in the z-direction decreases as

$$\exp\,[-\sqrt{\sigma\omega/2\epsilon_0c^2}\,z]. \tag{32.46}$$

Let's write this as

$$e^{-z/\delta}, \tag{32.47}$$

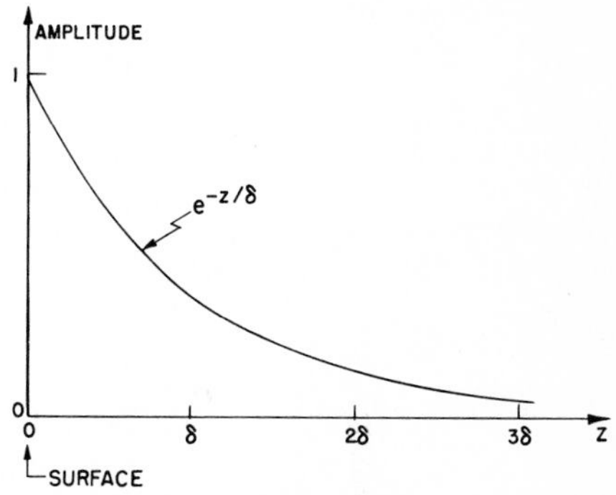

Fig. 32–3. The amplitude of a transverse electromagnetic wave as a function of distance into a metal.

where δ is then the distance in which the wave amplitude decreases by the factor $e^{-1} = 1/2.72$—or roughly one-third. The amplitude of such a wave as a function of z is shown in Fig. 32–3. Since electromagnetic waves will penetrate into a metal only this distance, δ is called the *skin depth*. It is given by

$$\delta = \sqrt{2\epsilon_0c^2/\sigma\omega}. \tag{32.48}$$

Now what do we mean by "low" frequencies? Looking at Eq. (32.42), we see that it can be approximated by Eq. (32.44) only if $\omega\tau$ is much less than one *and* if $\omega\epsilon_0/\sigma$ is also much less than one—that is, our low-frequency approximation applies when

$$\omega \ll \frac{1}{\tau}$$

and

$$\omega \ll \frac{\sigma}{\epsilon_0}\cdot \tag{32.49}$$

Let's see what frequencies these correspond to for a typical metal like copper. We compute τ by using Eq. (32.43), and σ/ϵ_0, by using the measured conductivity. We take the following data from a handbook:

$$\sigma = 5.76 \times 10^7\ (\text{ohm-meter})^{-1},$$
$$\text{atomic weight} = 63.5\ \text{grams},$$
$$\text{density} = 8.9\ \text{grams} - \text{cm}^{-3},$$
$$\text{Avogadro's number} = 6.02 \times 10^{23}\ (\text{gram atomic weight})^{-1}.$$

* Or writing $-i = e^{-i\pi/2}$; $\sqrt{-i} = e^{-i\pi/4} = \cos\pi/4 - i\sin\pi/4$, which gives the same result.

If we assume that there is one free electron per atom, then the number of electrons per cubic meter is

$$N = 8.5 \times 10^{28}\ \text{meter}^{-3}.$$

Using

$$q_e = 1.6 \times 10^{-19}\ \text{coulomb},$$
$$\epsilon_0 = 8.85 \times 10^{-12}\ \text{farad-meter}^{-1},$$
$$m = 9.11 \times 10^{-31}\ \text{kgm},$$

we get

$$\tau = 2.4 \times 10^{-14}\ \text{sec},$$
$$\frac{1}{\tau} = 4.1 \times 10^{13}\ \text{sec}^{-1},$$
$$\frac{\sigma}{\epsilon_0} = 6.5 \times 10^{18}\ \text{sec}^{-1}.$$

So for frequencies less than about 10^{12} cycles per second, copper will have the "low-frequency" behavior we describe (that means for waves whose free-space wavelength is longer than 0.3 millimeters—*very* short radio waves!).

For these waves, the skin depth in copper is

$$\delta = \sqrt{\frac{0.028\ \text{m}^2\text{-sec}^{-1}}{\omega}}\,.$$

For microwaves of 10,000 megacycles per second (3-cm waves)

$$\delta = 6.7 \times 10^{-4}\ \text{cm}.$$

The wave penetrates a very small distance.

We can see from this why in studying cavities (or waveguides) we needed to worry only about the fields inside the cavity, and not in the metal or outside the cavity. Also, we see why the losses in a cavity are reduced by a thin plating of silver or gold. The losses come from the current, which are appreciable only in a thin layer equal to the skin depth.

Suppose we look now at the index of a metal like copper at high frequencies. For very high frequencies $\omega\tau$ is much greater than one, and Eq. (32.42) is well approximated by

$$n^2 = 1 - \frac{\sigma}{\epsilon_0\omega^2\tau}\,. \tag{32.50}$$

For waves of high frequencies the index of a metal becomes real—and less than one! This is also evident from Eq. (32.38) if the dissipation term with γ is neglected, as can be done for very large ω. Equation (32.38) gives

$$n^2 = 1 - \frac{Nq_e^2}{m\epsilon_0\omega^2}\,, \tag{32.51}$$

which is, of course, the same as Eq. (32.50). We have seen before the quantity $Nq_e^2/m\epsilon_0$, which we called the square of the plasma frequency (Section 7–3):

$$\omega_p^2 = \frac{Nq_e^2}{\epsilon_0 m}\,,$$

so we can write Eq. (32.50) or Eq. (32.51) as

$$n^2 = 1 - \left(\frac{\omega_p}{\omega}\right)^2.$$

The plasma frequency is a kind of "critical" frequency.

For $\omega < \omega_p$ the index of a metal has an imaginary part, and waves are attenuated; but for $\omega \gg \omega_p$ the index is real, and the metal becomes transparent. You know, of course, that metals are reasonably transparent to x-rays. But some metals are even transparent in the ultraviolet. In Table 32–3 we give for several metals the experimental observed wavelength at which they begin to become transparent. In the second column we give the calculated critical wavelength $\lambda_p = 2\pi c/\omega_p$. Considering that the experimental wavelength is not too well defined, the fit of the theory is fairly good.

You may wonder why the plasma frequency ω_p should have anything to do with the propagation of electromagnetic waves in metals. The plasma frequency came up in Chapter 7 as the natural frequency of *density* oscillations of the free electrons. (A clump of electrons is repelled by electric forces, and the inertia of the electrons leads to an oscillation of density.) So *longitudinal* plasma waves are resonant at ω_p. But we are now talking about *transverse* electromagnetic waves, and we have found that transverse waves are absorbed for frequencies below ω_p. (It's an interesting and *not* accidental coincidence.)

Table 32–3*

Wavelengths below which the metal becomes transparent

Metal	λ (experimental)	$\lambda_p = 2\pi c/\omega_p$
Li	1550 A	1550 A
Na	2100	2090
K	3150	2870
Rb	3400	3220

* From: C. Kittel, *Introduction to Solid State Physics*, John Wiley and Sons, Inc., New York, 2nd ed., 1956, p. 266.

Although we have been talking about wave propagation in metals, you appreciate by this time the universality of the phenomena of physics—that it doesn't make any difference whether the free electrons are in a metal or whether they are in the plasma of the ionosphere of the earth, or in the atmosphere of a star. To understand radio propagation in the ionosphere, we can use the same expressions—using, of course, the proper values for N and τ. We can see now why long radio waves are absorbed or reflected by the ionosphere, whereas short waves go right through. (Short waves must be used for communication with satellites.)

We have talked about the high- and low-frequency extremes for wave propagation in metals. For the in-between frequencies the full-blown formula of Eq. (32.42) must be used. In general, the index will have real and imaginary parts; the wave is attenuated as it propagates into the metal. For very thin layers, metals are somewhat transparent even at optical frequencies. As an example, special goggles for people who work around high-temperature furnaces are made by evaporating a thin layer of gold on glass. The visible light is transmitted fairly well—with a strong green tinge—but the infrared is strongly absorbed.

Finally, it cannot have escaped the reader that many of these formulas resemble in some ways those for the dielectric constant κ discussed in Chapter 10. The dielectric constant κ measures the response of the material to a constant field, that is, for $\omega = 0$. If you look carefully at the definition of n and κ you see that κ is simply the limit of n^2 as $\omega \to 0$. Indeed, placing $\omega = 0$ and $n^2 = \kappa$ in equations of this chapter will reproduce the equations of the theory of the dielectric constant of Chapter 11.

33

Reflection from Surfaces

Review: Chapter 33, Vol. I, *Polarization*

33–1 Reflection and refraction of light

The subject of this chapter is the reflection and refraction of light—or electromagnetic waves in general—at surfaces. We have already discussed the laws of reflection and refraction in Chapter 35 of Volume I. Here's what we found out there:

1. The angle of reflection is equal to the angle of incidence. With the angles defined as shown in Fig. 33–1,
$$\theta_r = \theta_i. \tag{33.1}$$

2. The product $n \sin \theta$ is the same for the incident and transmitted beams (Snell's law):
$$n_1 \sin \theta_i = n_2 \sin \theta_t. \tag{33.2}$$

3. The intensity of the reflected light depends on the angle of incidence and also on the direction of polarization. For $\boldsymbol{E}$ perpendicular to the plane of incidence, the reflection coefficient $R_\perp$ is
$$R_\perp = \frac{I_r}{I_i} = \frac{\sin^2(\theta_i - \theta_t)}{\sin^2(\theta_i + \theta_t)}. \tag{33.3}$$
For $\boldsymbol{E}$ parallel to the plane of incidence, the reflection coefficient $R_{||}$ is
$$R_{||} = \frac{I_r}{I_i} = \frac{\tan^2(\theta_i - \theta_t)}{\tan^2(\theta_i + \theta_t)}. \tag{33.4}$$

4. For normal incidence (any polarization, of course!),
$$\frac{I_r}{I_i} = \left(\frac{n_2 - n_1}{n_2 + n_1}\right)^2 \tag{33.5}$$

(Earlier, we used i for the incident angle and r for the refracted angle. Since we can't use r for both "refracted" *and* "reflected" angles, we are now using $\theta_i =$ incident angle, $\theta_r =$ reflected angle, and $\theta_t =$ transmitted angle.)

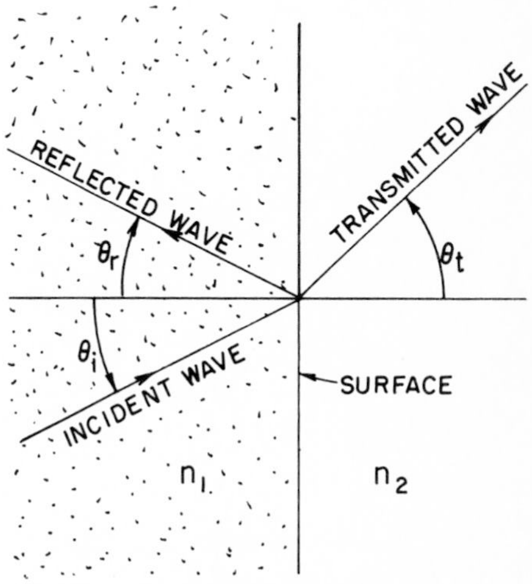

Fig. 33–1. Reflection and refraction of light waves at a surface. (The wave directions are normal to the wave crests.)

Our earlier discussion is really about as far as anyone would normally need to go with the subject, but we are going to do it all over again a different way. Why? One reason is that we assumed before that the indexes were real (no absorption in the materials). But another reason is that you should know how to deal with what happens to waves at surfaces from the point of view of Maxwell's equations. We'll get the same answers as before, but now from a straightforward solution of the wave problem, rather than by some clever arguments.

We want to emphasize that the amplitude of a surface reflection is not a property of the *material*, as is the index of refraction. It is a "surface property," one that depends precisely on how the surface is made. A thin layer of extraneous junk on the surface between two materials of indices n_1 and n_2 will usually change the reflection. (There are all kinds of possibilities of interference here—like the colors of oil films. Suitable thickness can even reduce the reflected amplitude to zero for a given frequency; that's how coated lenses are made.) The formulas we will derive are correct only if the change of index is sudden—within a distance very small compared with one wavelength. For light, the wavelength is about 5000 A, so by a "smooth" surface we mean one in which the conditions change in going a distance of only a few atoms (or a few angstroms). Our equations will work for light for highly polished surfaces. In general, if the index changes gradually over a distance of several wavelengths, there is very little reflection at all.

33–2 Waves in dense materials

First, we remind you about the convenient way of describing a sinusoidal plane wave we used in Chapter 36 of Volume I. Any field *component* in the wave (we use E as an example) can be written in the form

$$E = E_0 e^{i(\omega t - \boldsymbol{k}\cdot\boldsymbol{r})}, \tag{33.6}$$

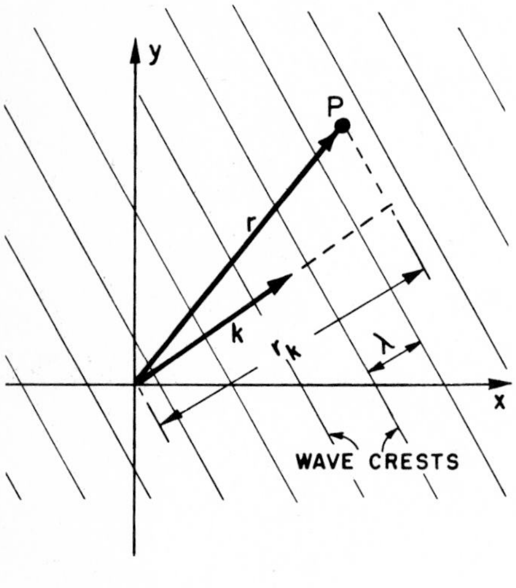

Fig. 33–2. For a wave moving in the direction $\boldsymbol{k}$, the phase at any point P is $(\omega t - \boldsymbol{k}\cdot\boldsymbol{r})$.

where E represents the amplitude at the point $\boldsymbol{r}$ (from the origin) at the time t. The vector $\boldsymbol{k}$ points in the direction the wave is travelling, and its magnitude $|\boldsymbol{k}| = k = 2\pi/\lambda$ is the wave number. The phase velocity of the wave is $v_{\text{ph}} = \omega/k$; for a light wave in a material of index n, $v_{\text{ph}} = c/n$, so

$$k = \frac{\omega n}{c}\cdot \tag{33.7}$$

Suppose $\boldsymbol{k}$ is in the z-direction; then $\boldsymbol{k}\cdot\boldsymbol{r}$ is just kz, as we have often used it. For $\boldsymbol{k}$ in any other direction, we should replace z by r_k, the distance from the origin in the $\boldsymbol{k}$-direction; that is, we should replace kz by kr_k, which is just $\boldsymbol{k}\cdot\boldsymbol{r}$. (See Fig. 33–2.) So Eq. (33.6) is a convenient representation of a wave in any direction. We must remember, of course, that

$$\boldsymbol{k}\cdot\boldsymbol{r} = k_x x + k_y y + k_z z,$$

where k_x, k_y, and k_z are the components of $\boldsymbol{k}$ along the three axes. In fact, we pointed out once that (ω, k_x, k_y, k_z) is a four-vector, and that its scalar product with (t, x, y, z) is an invariant. So the *phase* of a wave is an invariant, and Eq. (33.6) could be written

$$E = E_0 e^{ik_\mu x_\mu}.$$

But we don't need to be that fancy now.

For a sinusoidal E, as in Eq. (33.6), $\partial E/\partial t$ is the same as $i\omega E$, and $\partial E/\partial x$ is $-ik_x E$, and so on for the other components. You can see why it is very convenient to use the form in Eq. (33.6) when working with differential equations—differentiations are replaced by multiplications. One further useful point: The operation $\nabla = (\partial/\partial x, \partial/\partial y, \partial/\partial z)$ gets replaced by the three multiplications $(-ik_x, -ik_y, -ik_z)$. But these three factors transform as the components of the vector $\boldsymbol{k}$, so the operator ∇ gets replaced by multiplication with $-i\boldsymbol{k}$:

$$\frac{\partial}{\partial t} \to i\omega,$$

$$\nabla \to -i\boldsymbol{k}. \tag{33.8}$$

This remains true for any ∇ operation—whether it is the gradient, or the divergence, or the curl. For instance, the z-component of $\nabla \times \boldsymbol{E}$ is

$$\frac{\partial E_y}{\partial x} - \frac{\partial E_x}{\partial y}\cdot$$

If both E_y and E_x vary as $e^{-i\boldsymbol{k}\cdot\boldsymbol{r}}$, then we get

$$-ik_x E_y + ik_y E_x,$$

which is, you see, the z-component of $-i\boldsymbol{k} \times \boldsymbol{E}$.

So we have the very useful general fact that whenever you have to take the gradient of a vector that varies as a wave in three dimensions (they are an important part of physics), you can always take the derivations quickly and almost without thinking by remembering that the operation ∇ is equivalent to multiplication by $-i\boldsymbol{k}$.

For instance, the Faraday equation

$$\nabla \times \boldsymbol{E} = -\frac{\partial \boldsymbol{B}}{\partial t}$$

becomes for a wave

$$-i\boldsymbol{k} \times \boldsymbol{E} = -i\omega \boldsymbol{B}.$$

This tells us that

$$\boldsymbol{B} = \frac{\boldsymbol{k} \times \boldsymbol{E}}{\omega}, \tag{33.9}$$

which corresponds to the result we found earlier for waves in free space—that $\boldsymbol{B}$, in a wave, is at right angles to $\boldsymbol{E}$ and to the wave direction. (In free space, $\omega/k = c$.) You can remember the sign in Eq. (33.9) from the fact that $\boldsymbol{k}$ is in the direction of Poynting's vector $\boldsymbol{S} = \epsilon_0 c^2 \boldsymbol{E} \times \boldsymbol{B}$.

If you use the same rule with the other Maxwell equations, you get again the results of the last chapter and, in particular, that

$$\boldsymbol{k} \cdot \boldsymbol{k} = k^2 = \frac{\omega^2 n^2}{c^2}. \tag{33.10}$$

But since we know that, we won't do it again.

If you want to entertain yourself, you can try the following terrifying problem that was the ultimate test for graduate students back in 1890: solve Maxwell's equations for plane waves in an *anisotropic* crystal, that is, when the polarization $\boldsymbol{P}$ is related to the electric field $\boldsymbol{E}$ by a tensor of polarizability. You should, of course, choose your axes along the principal axes of the tensor, so that the relations are simplest (then $P_x = \alpha_a E_x$, $P_y = \alpha_b E_y$, and $P_z = \alpha_c E_z$), but let the waves have an arbitrary direction and polarization. You should be able to find the relations between $\boldsymbol{E}$ and $\boldsymbol{B}$, and how $\boldsymbol{k}$ varies with direction and wave polarization. Then you will understand the optics of an anisotropic crystal. It would be best to start with the simpler case of a birefringent crystal—like calcite—for which two of the polarizabilities are equal (say, $\alpha_b = \alpha_c$), and see if you can understand why you see double when you look through such a crystal. If you can do that, then try the hardest case, in which all three α's are different. Then you will know whether you are up to the level of a graduate student of 1890. In this chapter, however, we will consider only isotropic substances.

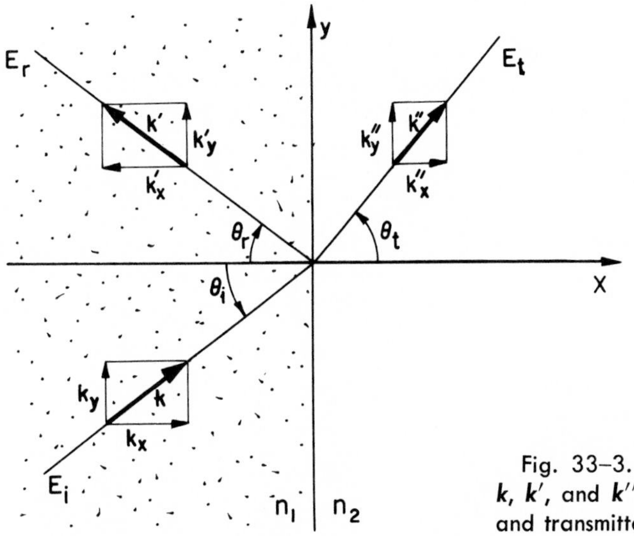

Fig. 33–3. The propagation vectors $\boldsymbol{k}$, $\boldsymbol{k}'$, and $\boldsymbol{k}''$ for the incident, reflected, and transmitted waves.

We know from experience that when a plane wave arrives at the boundary between two different materials—say, air and glass, or water and oil—there is a wave reflected and a wave transmitted. Suppose we assume no more than that and see what we can work out. We choose our axes with the *yz*-plane in the surface and the *xy*-plane perpendicular to the incident wave surfaces, as shown in Fig. 33–3.

The electric vector of the incident wave can then be written as

$$E_i = E_0 e^{i(\omega t - k \cdot r)}. \tag{33.11}$$

Since $\boldsymbol{k}$ is perpendicular to the z-axis,

$$k \cdot r = k_x x + k_y y. \tag{33.12}$$

We write the reflected wave as

$$E_r = E_0' e^{i(\omega' t - k' \cdot r)}, \tag{33.13}$$

so that its frequency is ω', its wave number is $\boldsymbol{k}'$, and its amplitude is E_0'. (We know, of course, that the frequency is the same and the magnitude of $\boldsymbol{k}$ is the same as for the incident wave, but we are not going to assume even that. We will let it come out of the mathematical machinery.) Finally, we write for the transmitted wave,

$$E_t = E_0'' e^{i(\omega'' t - k'' \cdot r)}. \tag{33.14}$$

We know that one of Maxwell's equations gives Eq. (33.9), so for each of the waves we have

$$B_i = \frac{k \times E_i}{\omega}, \qquad B_r = \frac{k' \times E_r}{\omega'}, \qquad B_t = \frac{k'' \times E_t}{\omega''}. \tag{33.15}$$

Also, if we call the indexes of the two media n_1 and n_2, we have from Eq. (33.10)

$$k^2 = k_x^2 + k_y^2 = \frac{\omega^2 n_1^2}{c^2}. \tag{33.16}$$

Since the reflected wave is in the same material, then

$$k'^2 = \frac{\omega'^2 n_1^2}{c^2}, \tag{33.17}$$

whereas for the transmitted wave,

$$k''^2 = \frac{\omega''^2 n_2^2}{c^2}. \tag{33.18}$$

33–3 The boundary conditions

All we have done so far is to describe the three waves; our problem now is to work out the parameters of the reflected and transmitted waves in terms of those of the incident wave. How can we do that? The three waves we have described satisfy Maxwell's equations in the uniform material, but Maxwell's equations must also be satisfied *at* the boundary between the two different materials. So we must now look at what happens right *at* the boundary. We will find that Maxwell's equations demand that the three waves fit together in a certain way.

As an example of what we mean, the y-component of the electric field $\boldsymbol{E}$ must be the *same* on both sides of the boundary. This is required by Faraday's law,

$$\nabla \times E = -\frac{\partial B}{\partial t}, \tag{33.19}$$

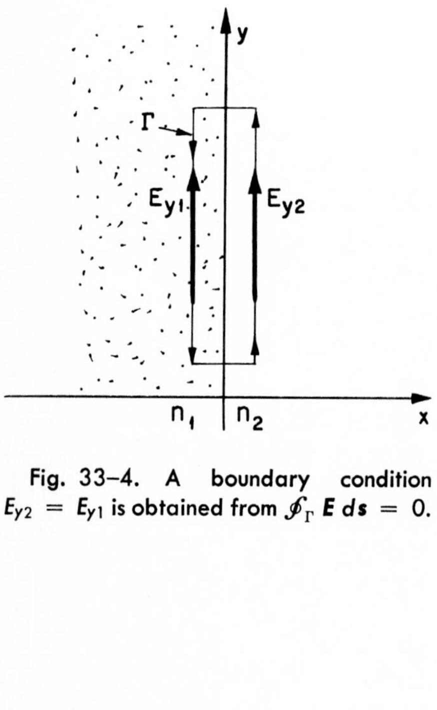

Fig. 33–4. A boundary condition $E_{y2} = E_{y1}$ is obtained from $\oint_\Gamma E \cdot ds = 0$.

as we can see in the following way. Consider a little rectangular loop Γ which straddles the boundary, as shown in Fig. 33–4. Equation (33.19) says that the line integral of $\boldsymbol{E}$ around Γ is equal to the rate of change of the flux of $\boldsymbol{B}$ through the loop:

$$\oint_\Gamma E \cdot ds = -\frac{\partial}{\partial t} \int B \cdot n \, da.$$

Now imagine that the rectangle is very narrow, so that the loop encloses an infinitesimal area. If $\boldsymbol{B}$ remains finite (and there's no reason it should be infinite at the boundary!) the flux through the area is zero. So the line integral of $\boldsymbol{E}$ must be zero. If E_{y1} and E_{y2} are the components of the field on the two sides of the boundary and if the length of the rectangle is l, we have

$$E_{y1}l - E_{y2}l = 0$$

or

$$E_{y1} = E_{y2}, \tag{33.20}$$

as we have said. This gives us one relation among the fields of the three waves.

The procedure of working out the consequences of Maxwell's equations at the boundary is called "determining the boundary conditions." Ordinarily, it is done by finding as many equations like Eq. (33.20) as one can, by making arguments about little rectangles like Γ in Fig. 33–4, or by using little gaussian surfaces that straddle the boundary. Although that is a perfectly good way of proceeding, it gives the impression that the problem of dealing with a boundary is different for every different physical problem.

For example, in a problem of heat flow across a boundary, how are the temperatures on the two sides related? Well, you could argue, for one thing, that the heat flow *to* the boundary from one side would have to equal the flow *away* from the other side. It is usually possible, and generally quite useful, to work out the boundary conditions by making such physical arguments. There may be times, however, when in working on some problem you have only some equations, and you may not see right away what physical arguments to use. So although we are at the moment interested only in an electromagnetic problem, where we *can* make the physical arguments, we want to show you a method that can be used for any problem—a *general* way of finding what happens at a boundary directly from the differential equations.

We begin by writing all the Maxwell equations for a dielectric—and this time we are very specific and write out explicitly all the components:

$$\nabla \cdot \boldsymbol{E} = -\frac{\nabla \cdot \boldsymbol{P}}{\epsilon_0}$$

$$\epsilon_0\left(\frac{\partial E_x}{\partial x} + \frac{\partial E_y}{\partial y} + \frac{\partial E_z}{\partial z}\right) = -\left(\frac{\partial P_x}{\partial x} + \frac{\partial P_y}{\partial y} + \frac{\partial P_z}{\partial z}\right) \tag{33.21}$$

$$\nabla \times \boldsymbol{E} = -\frac{\partial \boldsymbol{B}}{\partial t}$$

$$\frac{\partial E_z}{\partial y} - \frac{\partial E_y}{\partial z} = -\frac{\partial B_x}{\partial t} \tag{33.22a}$$

$$\frac{\partial E_x}{\partial z} - \frac{\partial E_z}{\partial x} = -\frac{\partial B_y}{\partial t} \tag{33.22b}$$

$$\frac{\partial E_y}{\partial x} - \frac{\partial E_x}{\partial y} = -\frac{\partial B_z}{\partial t} \tag{33.22c}$$

$$\nabla \cdot \boldsymbol{B} = 0$$

$$\frac{\partial B_x}{\partial x} + \frac{\partial B_y}{\partial y} + \frac{\partial B_z}{\partial z} = 0 \tag{33.23}$$

$$c^2\nabla \times \boldsymbol{B} = \frac{1}{\epsilon_0}\frac{\partial \boldsymbol{P}}{\partial t} + \frac{\partial \boldsymbol{E}}{\partial t}$$

$$c^2\left(\frac{\partial B_z}{\partial y} - \frac{\partial B_y}{\partial z}\right) = \frac{1}{\epsilon_0}\frac{\partial P_x}{\partial t} + \frac{\partial E_x}{\partial t} \tag{33.24a}$$

$$c^2\left(\frac{\partial B_x}{\partial z} - \frac{\partial B_z}{\partial x}\right) = \frac{1}{\epsilon_0}\frac{\partial P_y}{\partial t} + \frac{\partial E_y}{\partial t} \tag{33.24b}$$

$$c^2\left(\frac{\partial B_y}{\partial x} - \frac{\partial B_x}{\partial y}\right) = \frac{1}{\epsilon_0}\frac{\partial P_z}{\partial t} + \frac{\partial E_z}{\partial t} \tag{33.24c}$$

Now these equations must all hold in region 1 (to the left of the boundary) and in region 2 (to the right of the boundary). We have already written the solutions in regions 1 and 2. Finally, they must also be satisfied *in* the boundary, which we can call region 3. Although we usually think of the boundary as being sharply discontinuous, in reality it is not. The physical properties change very rapidly but not infinitely fast. In any case, we can imagine that there is a very rapid, but *continuous*, transition of the index between region 1 and 2, in a short distance we can call region 3. Also, any field quantity like P_x, or E_y, etc., will make a similar kind of transition in region 3. In this region, the differential equations must still be satisfied, and it is by following the differential equations in this region that we can arrive at the needed "boundary conditions."

For instance, suppose that we have a boundary between vacuum (region 1) and glass (region 2). There is nothing to polarize in the vacuum, so $\boldsymbol{P}_1 = 0$. Let's say there is some polarization $\boldsymbol{P}_2$ in the glass. Between the vacuum and the glass there is a smooth, but rapid, transition. If we look at any component of $\boldsymbol{P}$, say P_x, it might vary as drawn in Fig. 33–5(a). Suppose now we take the first of our equations, Eq. (33.21). It involves derivatives of the components of $\boldsymbol{P}$ with respect to x, y, and z. The y- and z-derivatives are not interesting; nothing spectacular is happening in those directions. But the x-derivative of P_x will have some very large values in region 3, because of the tremendous slope of P_x. The derivative $\partial P_x/\partial x$ will have a sharp spike at the boundary, as shown in Fig. 33–5(b). If we imagine squashing the boundary to an even thinner layer, the spike would get much higher. If the boundary is really sharp for the waves we are interested in, the magnitude of $\partial P_x/\partial x$ in region 3 will be much, much greater than any contributions we might have from the variation of P in the wave away from the boundary—so we ignore any variations other than those due to the boundary.

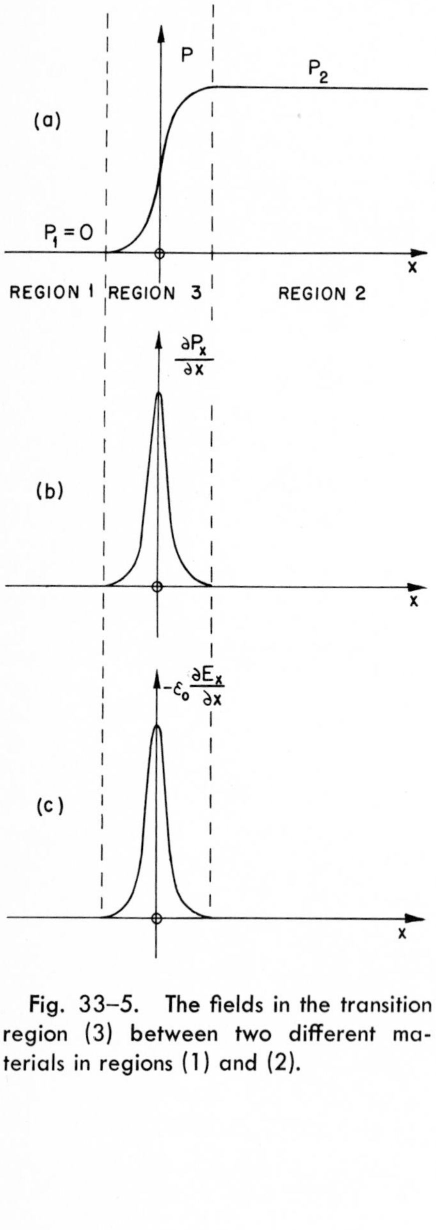

Fig. 33–5. The fields in the transition region (3) between two different materials in regions (1) and (2).

Now how can Eq. (33.21) be satisfied if there is a whopping big spike on the right-hand side? Only if there is an equally whopping big spike on the other side. Something on the left-hand side must also be big. The only candidate is $\partial E_x/\partial x$, because the variations with y and z are only those small effects in the wave we just mentioned. So $-\epsilon_0(\partial E/\partial x)$ must be as drawn in Fig. 33–5(c)—just a copy of $\partial P_x/\partial x$. We have that

$$\epsilon_0 \frac{\partial E_x}{\partial x} = -\frac{\partial P_x}{\partial x}.$$

If we integrate this equation with respect to x across region 3, we conclude that

$$\epsilon_0(E_{x2} - E_{x1}) = -(P_{x2} - P_{x1}). \tag{33.25}$$

In other words, the jump in $\epsilon_0 E_x$ in going from region 1 to region 2 must be equal to the jump in $-P_x$.

We can rewrite Eq. (33.25) as

$$\epsilon_0 E_{x2} + P_{x2} = \epsilon_0 E_{x1} + P_{x1}, \tag{33.26}$$

which says that the quantity $(\epsilon_0 E_x + P_x)$ has equal values in region 2 and region 1. People say: the quantity $(\epsilon_0 E_x + P_x)$ is *continuous* across the boundary. We have, in this way, one of our boundary conditions.

Although we took as an illustration the case in which $\boldsymbol{P}_1$ was zero because region 1 was a vacuum, it is clear that the same argument applies for any two materials in the two regions, so Eq. (33.26) is true in general.

Let's now go through the rest of Maxwell's equations and see what each of them tells us. We take next Eq. (33.22a). There are no x-derivatives, so it doesn't tell us anything. (Remember that the fields *themselves* do not get especially large at the boundary; only the derivatives with respect to x can become so huge that they dominate the equation.) Next, we look at Eq. (33.22b). Ah! There is an x-derivative! We have $\partial E_z/\partial x$ on the left-hand side. Suppose it has a huge derivative. But wait a moment! There is nothing on the right-hand side to match it with; therefore E_z *cannot* have any jump in going from region 1 to region 2. [If it did, there would be a spike on the left of Eq. (33.22a) but none on the right, and the equation would be false.] So we have a new condition:

$$E_{z2} = E_{z1}. \tag{33.27}$$

By the same argument, Eq. (33.22c) gives

$$E_{y2} = E_{y1}. \tag{33.28}$$

This last result is just what we got in Eq. (33.20) by a line integral argument.

We go on to Eq. (33.23). The only term that could have a spike is $\partial B_x/\partial x$. But there's nothing on the right to match it, so we conclude that

$$B_{x2} = B_{x1}. \tag{33.29}$$

On to the last of Maxwell's equations! Equation (33.24a) gives nothing, because there are no x-derivatives. Equation (33.23b) has one, $-c^2\,\partial B_z/\partial x$, but again, there is nothing to match it with. We get

$$B_{z2} = B_{z1}. \tag{33.30}$$

The last equation is quite similar, and gives

$$B_{y2} = B_{y1}. \tag{33.31}$$

The last three equations gives us that $\boldsymbol{B}_2 = \boldsymbol{B}_1$. We want to emphasize, however, that we get this result only when the materials on both sides of the boundary are nonmagnetic—or rather, when we can neglect any magnetic effects of the materials. This can usually be done for most materials, except ferromagnetic ones. (We will treat the magnetic properties of materials in some later chapters.)

Our program has netted us the six relations between the fields in region 1 and those in region 2. We have put them all together in Table 33–1. We can now use them to match the waves in the two regions. We want to emphasize, however, that the idea we have just used will work in *any* physical situation in which you have differential equations and you want a solution that crosses a sharp boundary between two regions where some property changes. For our present purposes, we could have easily derived the same equations by using arguments about the fluxes and circulations at the boundary. (You might see whether you can get the same result that way.) But now you have seen a method that will work in case you ever get stuck and don't see any easy argument about the physics of what is happening at the boundary—you can just work with the equations.

Table 33–1

Boundary conditions at the surface of a dielectric

$$(\epsilon_0\boldsymbol{E}_1 + \boldsymbol{P}_1)_x = (\epsilon_0\boldsymbol{E}_2 + \boldsymbol{P}_2)_x$$
$$(\boldsymbol{E}_1)_y = (\boldsymbol{E}_2)_y$$
$$(\boldsymbol{E}_1)_z = (\boldsymbol{E}_2)_z$$
$$\boldsymbol{B}_1 = \boldsymbol{B}_2$$

(The surface is in the yz-plane)

33–4 The reflected and transmitted waves

Now we are ready to apply our boundary conditions to the waves we wrote down in Section 33–2. We had:

$$\boldsymbol{E}_i = \boldsymbol{E}_0 e^{i(\omega t - k_x x - k_y y)}, \tag{33.32}$$

$$\boldsymbol{E}_r = \boldsymbol{E}_0' e^{i(\omega' t - k_x' x - k_y' y)}, \tag{33.33}$$

$$\boldsymbol{E}_t = \boldsymbol{E}_0'' e^{i(\omega'' t - k_x'' x - k_y'' y)}, \tag{33.34}$$

$$\boldsymbol{B}_i = \frac{\boldsymbol{k} \times \boldsymbol{E}_i}{\omega}, \tag{33.35}$$

$$\boldsymbol{B}_r = \frac{\boldsymbol{k}' \times \boldsymbol{E}_r}{\omega'}, \tag{33.36}$$

$$\boldsymbol{B}_t = \frac{\boldsymbol{k}'' \times \boldsymbol{E}_t}{\omega''}. \tag{33.37}$$

We have one further bit of knowledge: $\boldsymbol{E}$ is perpendicular to its propagation vector $\boldsymbol{k}$ for each wave.

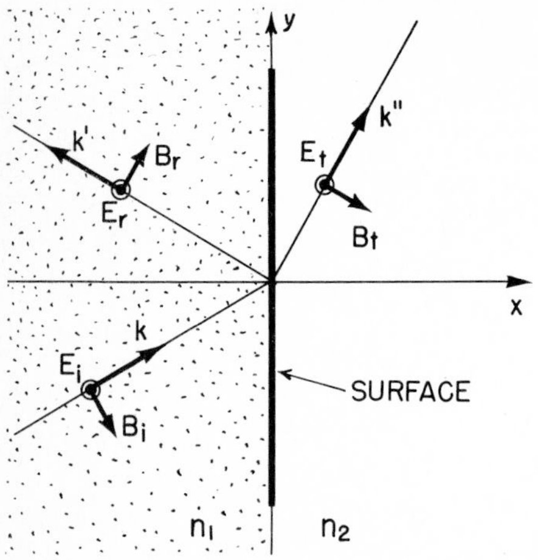

Fig. 33–6. Polarization of the reflected and transmitted waves when the **E**-field of the incident wave is perpendicular to the plane of incidence.

The results will depend on the direction of the **E**-vector (the "polarization") of the incoming wave. The analysis is much simplified if we treat separately the case of an incident wave with its **E**-vector *parallel* to the "plane of incidence" (that is, the xy-plane) and the case of an incident wave with the **E**-vector *perpendicular* to the plane of incidence. A wave of any other polarization is just a linear combination of two such waves. In other words, the reflected and transmitted intensities are different for different polarizations, and it is easiest to pick the two simplest cases and treat them separately.

We will carry through the analysis for an incoming wave polarized perpendicular to the plane of incidence and then just give you the result for the other. We are cheating a little by taking the simplest case, but the principle is the same for both. So we take that $\boldsymbol{E}_i$ has only a z-component, and since all the **E**-vectors are in the same direction we can leave off the vector signs.

So long as both materials are isotropic, the induced oscillations of charges in the material will also be in the z-direction, and the **E**-field of the transmitted and radiated waves will have only z-components. So for all the waves, E_x and E_y and P_x and P_y are zero. The waves will have their **E**- and **B**-vectors as drawn in Fig. 33–6. (We are cutting a corner here on our original plan of getting everything from the equations. This result would also come out of the boundary conditions, but we can save a lot of algebra by using the physical argument. When you have some spare time, see if you can get the same result from the equations. It is clear that what we have said agrees with the equations; it is just that we have not shown that there are no *other* possibilities.)

Now our boundary conditions, Eqs. (33.26) through (33.31), give relations between the components of **E** and **B** in regions 1 and 2. For region 2 we have only the transmitted wave, but in region 1 we have *two* waves. Which one do we use? The fields in region 1 are, of course, the superposition of the fields of the incident and reflected waves. (Since each satisfies Maxwell's equations, so does the sum.) So when we use the boundary conditions, we must use that

$$\boldsymbol{E}_1 = \boldsymbol{E}_i + \boldsymbol{E}_r, \qquad \boldsymbol{E}_2 = \boldsymbol{E}_t,$$

and similarly for the **B**'s.

For the polarization we are considering, Eqs. (33.26) and (33.28) give us no new information; only Eq. (33.27) is useful. It says that

$$E_i + E_r = E_t$$

at the boundary, that is, for $x = 0$. So we have that

$$E_0e^{i(\omega t-k_yy)} + E_0'e^{i(\omega' t-k_y' y)} = E_0''e^{i(\omega'' t-k_y'' y)}, \tag{33.38}$$

which must be true for *all* t and for *all* y. Suppose we look first at $y = 0$. Then we have

$$E_0e^{i\omega t} + E_0'e^{i\omega' t} = E_0''e^{i\omega'' t}.$$

This equation says that two oscillating terms are equal to a third oscillation. That can happen only if all the oscillations have the same frequency. (It is impossible for three—or any number—of such terms with different frequencies to add to zero for all times.) So

$$\omega'' = \omega' = \omega. \tag{33.39}$$

As we knew all along, the frequencies of the reflected and transmitted waves are the same as that of the incident wave.

We should really have saved ourselves some trouble by putting that in at the beginning, but we wanted to show you that it can also be got out of the equations. When you are doing a real problem, it is usually the best thing to put everything you know into the works right at the start and save yourself a lot of trouble.

By definition, the *magnitude* of k is given by $k^2 = n^2\omega^2/c^2$, so we have also that

$$\frac{k''^2}{n_2^2} = \frac{k'^2}{n_1^2} = \frac{k^2}{n_1^2}. \tag{33.40}$$

Now look at Eq. (33.38) for $t = 0$. Using again the same kind of argument we have just made, but this time based on the fact that the equation must hold for all values of y, we get that

$$k_y'' = k_y' = k_y. \tag{33.41}$$

From Eq. (33.40), $k'^2 = k^2$, so

$$k_x'^2 + k_y'^2 = k_x^2 + k_y^2.$$

Combining this with Eq. (33.41), we have that

$$k_x'^2 = k_x^2,$$

or that $k_x' = \pm k_x$. The positive sign makes no sense; that would not give a *reflected* wave, but another *incident* wave, and we said at the start that we were solving the problem of only one incident wave. So we have

$$k_x' = -k_x. \tag{33.42}$$

The two equations (33.41) and (33.42) give us that the angle of reflection is equal to the angle of incidence, as we expected. (See Fig. 33–3.) The reflected wave is

$$E_r = E_0' e^{i(\omega t - k_x x + k_y y)}. \tag{33.43}$$

For the transmitted wave we already have that

$$k_y'' = k_y,$$

and

$$\frac{k''^2}{n_2^2} = \frac{k^2}{n_1^2}; \tag{33.44}$$

so we can solve these to find k_x''. We get

$$k_x''^2 = k''^2 - k_y''^2 = \frac{n_2^2}{n_1^2} k^2 - k_y^2. \tag{33.45}$$

Suppose for a moment that n_1 and n_2 are real numbers (that the imaginary parts of the indexes are very small). Then all the k's are also real numbers, and from Fig. 33–3 we find that

$$\frac{k_y}{k} = \sin\theta_i, \qquad \frac{k_y''}{k''} = \sin\theta_t. \tag{33.46}$$

From (33.44) we get that

$$n_2 \sin\theta_t = n_1 \sin\theta_i, \tag{33.47}$$

which is Snell's law of refraction—again, something we already knew. If the indexes are not real, the wave numbers are complex, and we have to use Eq. (33.45). [We could still *define* the angles θ_i and θ_t by Eq. (33.46), and Snell's law, Eq. (33.47), would be true in general. But then the "angles" also are complex numbers, thereby losing their simple geometrical interpretation as angles. It is best then to describe the behavior of the waves by their complex k_x or k_x'' values.]

So far, we haven't found anything new. We have just had the simple-minded delight of getting some obvious answers from a complicated mathematical machinery. Now we are ready to find the amplitudes of the waves which we have not yet known. Using our results for the ω's and k's, the exponential factors in Eq. (33.38) can be cancelled, and we get

$$E_0 + E_0' = E_0''. \tag{33.48}$$

Since both E_0' and E_0'' are unknown, we need one more relationship. We must use another of the boundary conditions. The equations for E_x and E_y are no help, because all the $\boldsymbol{E}$'s have only a z-component. So we must use the conditions on $\boldsymbol{B}$. Let's try Eq. (33.29):

$$B_{x2} = B_{x1}.$$

From Eqs. (33.35) through (33.37),

$$B_{xi} = \frac{k_y E_i}{\omega}, \qquad B_{xr} = \frac{k'_y E_r}{\omega'}, \qquad B_{xt} = \frac{k''_y E_t}{\omega''}.$$

Recalling that $\omega'' = \omega' = \omega$ and $k''_y = k'_y = k_y$, we get that

$$E_0 + E'_0 = E''_0.$$

But this is just Eq. (33.48) all over again! We've just wasted time getting something we already knew.

We could try Eq. (33.30), $B_{z2} = B_{z1}$, but there are no z-components of $\boldsymbol{B}$! So there's only one equation left: Eq. (33.31), $B_{y2} = B_{y1}$. For the three waves:

$$B_{yi} = -\frac{k_x E_i}{\omega}, \qquad B_{yr} = -\frac{k'_x E_r}{\omega'}, \qquad B_{yt} = -\frac{k''_x E_t}{\omega''}. \tag{33.49}$$

Putting for E_i, E_r, and E_t the wave expression for $x = 0$ (to be at the boundary), the boundary condition is

$$\frac{k_x}{\omega} E_0 e^{i(\omega t - k_y y)} + \frac{k'_x}{\omega'} E'_0 e^{i(\omega' t - k'_y y)} = \frac{k''_x}{\omega''} E''_0 e^{i(\omega'' t - k''_y y)}.$$

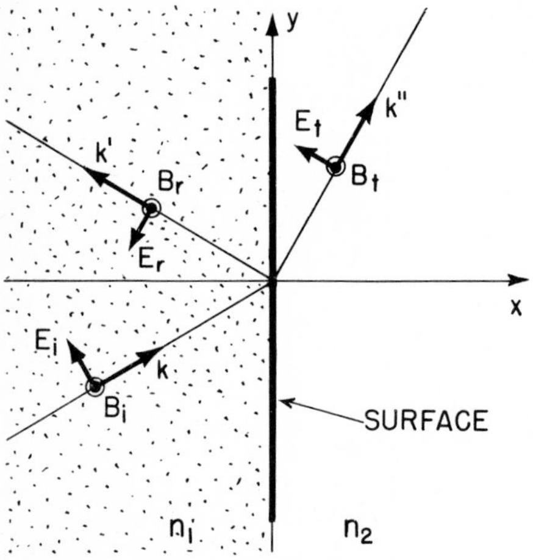

Fig. 33–7. Polarization of the waves when the **E**-field of the incident wave is parallel to the plane of incidence.

Again all ω's and k_y's are equal, so this reduces to

$$k_x E_0 + k'_x E'_0 = k''_x E''_0. \tag{33.50}$$

This gives us an equation for the E's that is different from Eq. (33.48). With the two, we can solve for E'_0 and E''_0. Remembering that $k'_x = -k_x$, we get

$$E'_0 = \frac{k_x - k''_x}{k_x + k''_x} E_0, \tag{33.51}$$

$$E''_0 = \frac{2k_x}{k_x + k''_x} E_0. \tag{33.52}$$

These, together with Eq. (33.45) or Eq. (33.46) for k''_x, give us what we wanted to know. We will discuss the consequences of this result in the next section.

If we begin with a wave polarized with its $\boldsymbol{E}$-vector *parallel* to the plane of incidence, $\boldsymbol{E}$ will have both x- and y-components, as shown in Fig. 33–7. The algebra is straightforward but more complicated. (The work can be somewhat reduced by expressing things in this case in terms of the *magnetic* fields, which are all in the z-direction.) One finds that

$$|E'_0| = \frac{n_2^2 k_x - n_1^2 k''_x}{n_2^2 k_x + n_1^2 k''_x} |E_0| \tag{33.53}$$

and

$$|E''_0| = \frac{2 n_1 n_2 k_x}{n_2^2 k_x + n_1^2 k''_x} |E_0|. \tag{33.54}$$

Let's see whether our results agree with those we got earlier. Equation (33.3) is the result we worked out in Chapter 35 of Volume I for the ratio of the intensity of the reflected wave to the intensity of the incident wave. Then, however, we were considering only *real* indexes. For real indexes (and k's), we can write

$$k_x = k \cos\theta_i = \frac{\omega n_1}{c} \cos\theta_i,$$

$$k''_x = k'' \cos\theta_t = \frac{\omega n_2}{c} \cos\theta_t.$$

Substituting in Eq. (33.51), we have

$$\frac{E'_0}{E_0} = \frac{n_1 \cos\theta_i - n_2 \cos\theta_t}{n_1 \cos\theta_i + n_2 \cos\theta_t}, \tag{33.55}$$

which does not look the same as Eq. (33.3). It will, however, if we use Snell's law to get rid of the n's. Setting $n_2 = n_1 \sin\theta_i/\sin\theta_t$, and multiplying the numerator and denominator by $\sin\theta_t$, we get

$$\frac{E_0'}{E_0} = \frac{\cos\theta_i\sin\theta_t - \sin\theta_i\cos\theta_t}{\cos\theta_i\sin\theta_t + \sin\theta_i\cos\theta_t}.$$

The numerator and denominator are just the sines of $(\theta_i - \theta_t)$ and $(\theta_i + \theta_t)$; we get

$$\frac{E_0'}{E_0} = \frac{\sin(\theta_i - \theta_t)}{\sin(\theta_i + \theta_t)}. \tag{33.56}$$

Since E_0' and E_0 are in the same material, the intensities are proportional to the squares of the electric fields, and we get the same result as before. Similarly, Eq. (33.53) is the same as Eq. (33.4).

For waves which arrive at normal incidence, $\theta_i = 0$ and $\theta_t = 0$. Equation (33.56) gives 0/0, which is not very useful. We can, however, go back to Eq. (33.55), which gives

$$\frac{I_r}{I_i} = \left(\frac{E_0'}{E_0}\right)^2 = \left(\frac{n_1 - n_2}{n_1 + n_2}\right)^2. \tag{33.57}$$

This result, naturally, applies for "either" polarization, since for normal incidence there is no special "plane of incidence."

33–5 Reflection from metals

We can now use our results to understand the interesting phenomenon of reflection from metals. Why is it that metals are shiny? We saw in the last chapter that metals have an index of refraction which, for some frequencies, has a large imaginary part. Let's see what we would get for the reflected intensity when light shines from air (with $n = 1$) onto a material with $n = -in_I$. Then Eq. (33.55) gives (for normal incidence)

$$\frac{E_0'}{E_0} = \frac{1 + in_I}{1 - in_I}.$$

For the *intensity* of the reflected wave, we want the square of the absolute values of E_0' and E_0:

$$\frac{I_r}{I_i} = \frac{|E_0'|^2}{|E_0|^2} = \frac{|1 + in_I|^2}{|1 - in_I|^2},$$

or

$$\frac{I_r}{I_i} = \frac{1 + n_I^2}{1 + n_I^2} = 1. \tag{33.58}$$

For a material with an index which is a pure imaginary number, there is 100 percent reflection!

Metals do not reflect 100 percent, but many do reflect visible light very well. In other words, the imaginary part of their indexes is very large. But we have seen that a large imaginary part of the index means a strong absorption. So there is a general rule that if *any* material gets to be a *very* good absorber at any frequency, the waves are strongly reflected at the surface and very little gets inside to be absorbed. You can see this effect with strong dyes. Pure crystals of the strongest dyes have a "metallic" shine. Probably you have noticed that at the edge of a bottle of purple ink the dried dye will give a golden metallic reflection, or that dried red ink will sometimes give a greenish metallic reflection. Red ink absorbs out the greens of *transmitted* light, so if the ink is very concentrated, it will exhibit a strong surface *reflection* for the frequencies of green light.

You can easily show this effect by coating a glass plate with red ink and letting it dry. If you direct a beam of white light at the back of the plate, as shown in Fig. 33–8, there will be a transmitted beam of réd light and a reflected beam of green light.

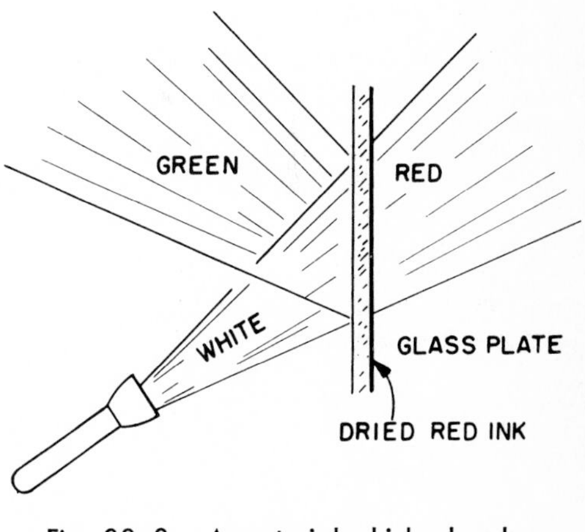

Fig. 33–8. A material which absorbs light strongly at the frequency ω also reflects light of that frequency.

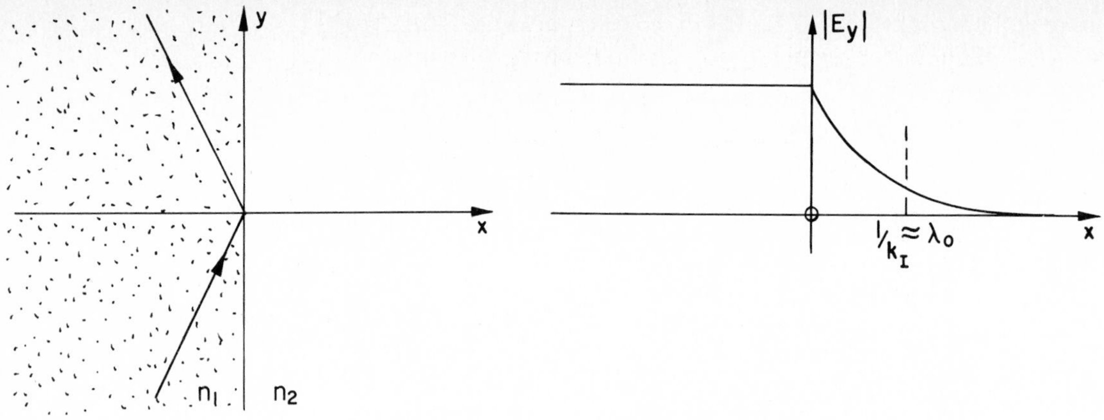

Fig. 33–9. Total internal reflection.

33–6 Total internal reflection

If light goes from a material like glass, with a real index n greater than 1, toward, say, air, with an index n_2 equal to 1, Snell's law says that

$$\sin \theta_t = n \sin \theta_i.$$

The angle θ_t of the transmitted wave becomes 90° when the incident angle θ_i is equal to the "critical angle" θ_c given by

$$n \sin \theta_c = 1. \tag{33.59}$$

What happens for θ_i greater than the critical angle? You know that there is total internal reflection. But how does that come about?

Let's go back to Eq. (33.45) which gives the wave number k_x'' for the transmitted wave. We would have

$$k_x''^2 = \frac{k^2}{n^2} - k_y^2.$$

Now $k_y = k \sin \theta_i$ and $k = \omega n/c$, so

$$k_x''^2 = \frac{\omega^2}{c^2}(1 - n^2 \sin^2 \theta_i).$$

If $n \sin \theta_i$ is greater than one, $k_x''^2$ is *negative* and k_x'' is a pure imaginary, say $\pm ik_I$. You know by now what that means! The "transmitted" wave (Eq. 33.34) will have the form

$$\boldsymbol{E}_t = \boldsymbol{E}_0'' e^{\pm k_I x} e^{i(\omega t - k_y y)}.$$

The wave amplitude either grows or drops off exponentially with increasing x. Clearly, what we want here is the negative sign. Then the *amplitude* of the wave to the right of the boundary will go as shown in Fig. 33–9. Notice that k_I is of the order ω/c—which is λ_0, the free-space wavelength of the light. When light is totally reflected from the inside of a glass-air surface, there are fields in the air, but they extend beyond the surface only a distance of the order of the wavelength of the light.

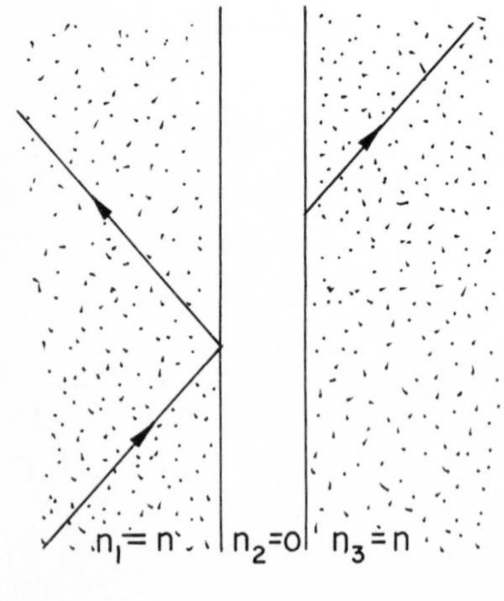

Fig. 33–10. If there is a small gap, internal reflection is not "total"; a transmitted wave appears beyond the gap.

We can now see how to answer the following question: If a light wave in glass arrives at the surface at a large enough angle, it is reflected; if another piece of glass is brought up to the surface (so that the "surface" in effect disappears) the light is transmitted. Exactly when does this happen? Surely there must be continuous change from total reflection to no reflection! The answer, of course, is that if the air gap is so small that the exponential tail of the wave in the air has an appreciable strength at the second piece of glass, it will shake the electrons there and generate a new wave, as shown in Fig. 33–10. Some light will be transmitted. (Clearly, our solution is incomplete; we should solve all the equations again for a thin layer of air between two regions of glass.)

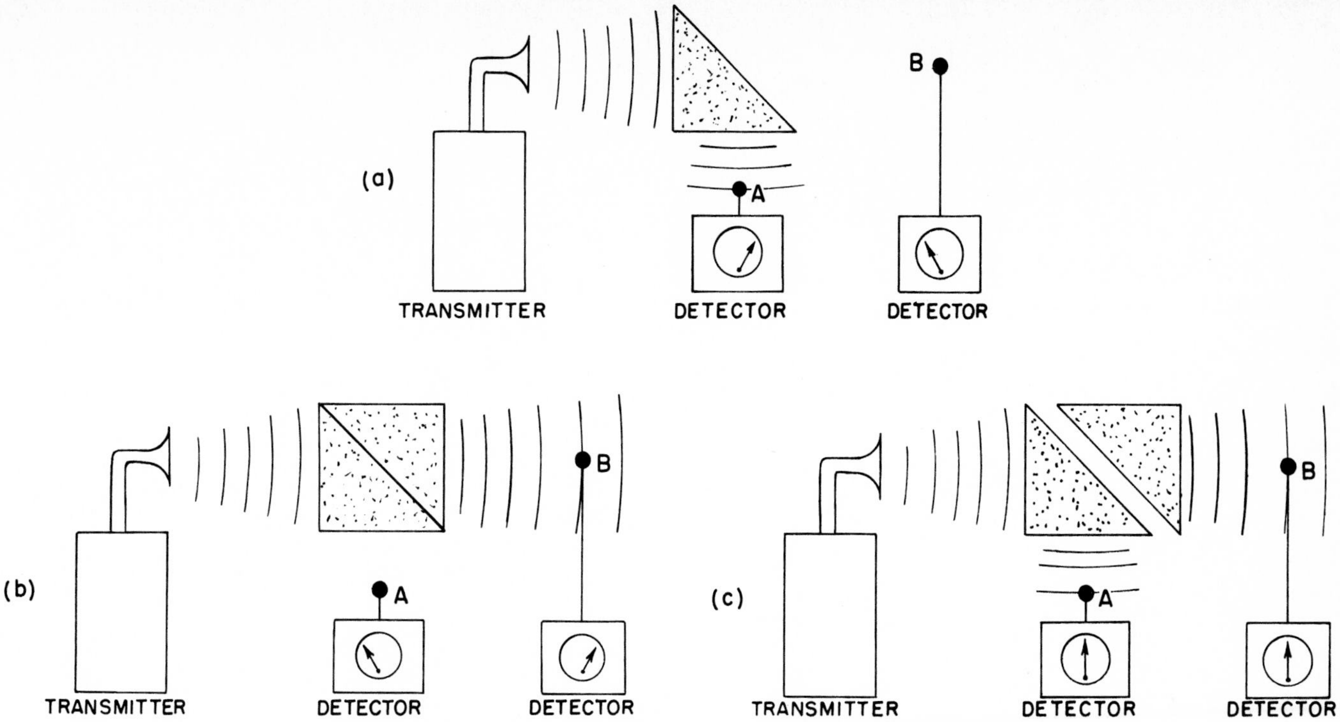

Fig. 33–11. A demonstration of the penetration of internally reflected waves.

This transmission effect can be observed with ordinary light only if the air gap is very small (of the order of the wavelength of light, like 10^{-5} cm), but it is easily demonstrated with three-centimeter waves. Then the exponentially decreasing field extends several centimeters. A microwave apparatus that shows the effect is drawn in Fig. 33–11. Waves from a small three-centimeter transmitter are directed at a 45° prism of paraffin. The index of refraction of paraffin for these frequencies is 1.50, and therefore the critical angle is 41.5°. So the wave is totally reflected from the 45° face and is picked up by detector A, as indicated in Fig. 33–11(a). If a second paraffin prism is placed in contact with the first, as shown in part (b) of the figure, the wave passes straight through and is picked up at detector B. If a gap of a few centimeters is left between the two prisms, as in part (c), there are both transmitted and reflected waves. The electric field outside the 45° face of the prism in Fig. 33–11(a) can also be shown by bringing detector B to within a few centimeters of the surface.

The Magnetism of Matter

34–1 Diamagnetism and paramagnetism

In this chapter we are going to talk about the magnetic properties of materials. The material which has the most striking magnetic properties is, of course, iron. Similar magnetic properties are shared also by the elements nickel, cobalt, and—at sufficiently low temperatures (below 16°C)—by gadolinium, as well as by a number of peculiar alloys. That kind of magnetism, called *ferromagnetism,* is sufficiently striking and complicated that we will discuss it in a special chapter. However, all ordinary substances do show some magnetic effects, although very small ones—a thousand to a million times less than the effects in ferromagnetic materials. Here we are going to describe ordinary magnetism, that is to say, the magnetism of substances other than the ferromagnetic ones.

This small magnetism is of two kinds. Some materials are *attracted* toward magnetic fields; others are *repelled.* Unlike the electrical effect in matter, which always causes dielectrics to be attracted, there are two signs to the magnetic effect. These two signs can be easily shown with the help of a strong electromagnet which has one sharply pointed pole piece and one flat pole piece, as drawn in Fig. 34–1. The magnetic field is much stronger near the pointed pole than near the flat pole. If a small piece of material is fastened to a long string and suspended between the poles, there will, in general, be a small force on it. This small force can be seen by the slight displacement of the hanging material when the magnet is turned on. The few ferromagnetic materials are attracted very strongly toward the pointed pole; all other materials feel only a very weak force. Some are weakly attracted to the pointed pole; and some are weakly repelled.

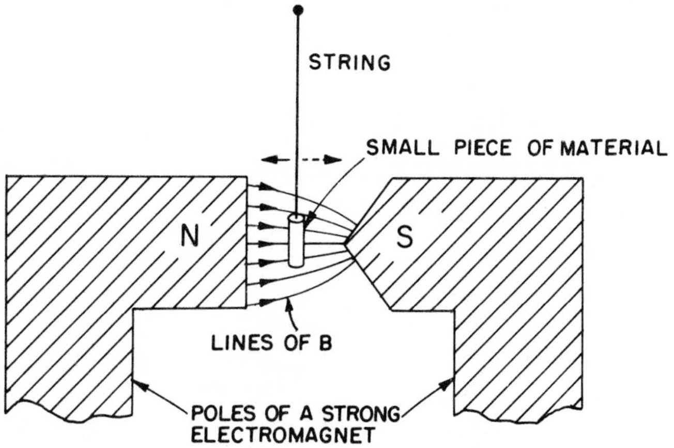

Fig. 34–1. A small cylinder of bismuth is weakly repelled by the sharp pole; a piece of aluminum is attracted.

The effect is most easily seen with a small cylinder of bismuth, which is *repelled* from the high-field region. Substances which are repelled in this way are called *diamagnetic.* Bismuth is one of the strongest diamagnetic materials, but even with it, the effect is still quite weak. Diamagnetism is always very weak. If a small piece of aluminum is suspended between the poles, there is also a weak force, but *toward* the pointed pole. Substances like aluminum are called *paramagnetic.* (In such an experiment, eddy-current forces arise when the magnet is turned on and off, and these can give off strong impulses. You must be careful to look for the net displacement after the hanging object settles down.)

We want now to describe briefly the mechanisms of these two effects. First, in many substances the atoms have no permanent magnetic moments, or rather, all the magnets within each atom balance out so that the *net* moment of the atom is zero. The electron spins and orbital motions all exactly balance out, so that any particular atom has no average magnetic moment. In these circumstances, when you turn on a magnetic field little extra currents are generated inside the atom by induction. According to Lenz's law, these currents are in such a direction as to oppose the increasing field. So the induced magnetic moments of the atoms are directed *opposite* to the magnetic field. This is the mechanism of diamagnetism.

Then there are some substances for which the atoms do have a permanent magnetic moment—in which the electron spins and orbits have a net circulating current that is not zero. So besides the diamagnetic effect (which is always present), there is also the possibility of lining up the individual atomic magnetic moments. In this case, the moments try to line up *with* the magnetic field (in the way the permanent dipoles of a dielectric are lined up by the electric field), and the induced magnetism tends to enhance the magnetic field. These are the paramagnetic substances. Paramagnetism is generally fairly weak because the lining-up forces are relatively small compared with the forces from the thermal motions which try to derange the order. It also follows that paramagnetism is usually sensitive to the temperature. (The paramagnetism arising from the spins of the electrons responsible for conduction in a metal constitutes an exception. We will not be discussing this phenomenon here.) For ordinary paramagnetism, the lower the temperature, the stronger the effect. There is more lining-up at low temperatures when the deranging effects of the collisions are less. Diamagnetism, on the other hand, is more or less independent of the temperature. In any substance with built-in magnetic moments there is a diamagnetic as well as a paramagnetic effect, but the paramagnetic effect usually dominates.

In Chapter 11 we described a *ferroelectric* material, in which all the electric dipoles get lined up by their own mutual electric fields. It is also possible to imagine the magnetic analog of ferroelectricity, in which all the atomic moments would line up and lock together. If you make calculations of how this should happen, you will find that because the magnetic forces are so much smaller than the electric forces, thermal motions should knock out this alignment even at temperatures as low as a few tenths of a degree Kelvin. So it would be impossible at room temperature to have any permanent lining up of the magnets.

On the other hand, this is exactly what does happen in iron—it does get lined up. There is an effective force between the magnetic moments of the different atoms of iron which is much, much greater than the *direct magnetic* interaction. It is an indirect effect which can be explained only by quantum mechanics. It is about ten thousand times stronger than the direct magnetic interaction, and is what lines up the moments in ferromagnetic materials. We discuss this special interaction in a later chapter.

Now that we have tried to give you a qualitative explanation of diamagnetism and paramagnetism, we must correct ourselves and say that *it is not possible* to understand the magnetic effects of materials in any honest way from the point of view of classical physics. Such magnetic effects are a *completely quantum-mechanical phenomenon.* It is, however, possible to make some phoney classical arguments and to get some idea of what is going on. We might put it this way. You can make some classical arguments and get guesses as to the behavior of the material, but these arguments are not "legal" in any sense because it is absolutely essential that quantum mechanics be involved in every one of these magnetic phenomena. On the other hand, there are situations, such as in a plasma or a region of space with many free electrons, where the electrons do obey the laws of classical mechanics. And in those circumstances, some of the theorems from classical magnetism are worth while. Also, the classical arguments are of some value for historical reasons. The first few times that people were able to guess at the meaning and behavior of magnetic materials, they used classical arguments. Finally, as we have already illustrated, classical mechanics can give us some useful guesses as to what might happen—even though the really honest way to study this subject would be to learn quantum mechanics first and then to understand the magnetism in terms of quantum mechanics.

On the other hand, we don't want to wait until we learn quantum mechanics inside out to understand a simple thing like diamagnetism. We will have to lean on the classical mechanics as kind of half showing what happens, realizing, however, that the arguments are really not correct. We therefore make a series of theorems about classical magnetism that will confuse you because they will prove different things. Except for the last theorem, every one of them will be wrong. Furthermore, they will all be wrong as a description of the physical world, because quantum mechanics is left out.

34–2 Magnetic moments and angular momentum

The first theorem we want to prove from classical mechanics is the following: If an electron is moving in a circular orbit (for example, revolving around a nucleus under the influence of a central force), there is a definite ratio between the magnetic moment and the angular momentum. Let's call $\boldsymbol{J}$ the angular momentum and $\boldsymbol{\mu}$ the magnetic moment of the electron in the orbit. The magnitude of the angular momentum is the mass of the electron times the velocity times the radius. (See Fig. 34–2.) It is directed perpendicular to the plane of the orbit.

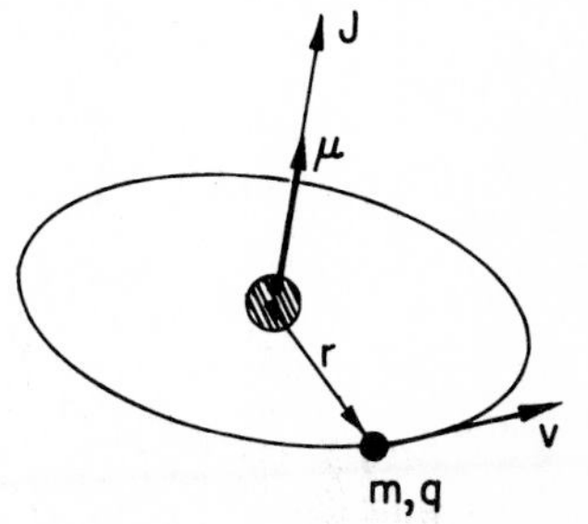

Fig. 34–2. For any circular orbit the magnetic moment $\boldsymbol{\mu}$ is $q/2m$ times the angular momentum $\boldsymbol{J}$.

$$J = mvr. \tag{34.1}$$

(This is, of course, a nonrelativistic formula, but it is a good approximation for atoms, because for the electrons involved v/c is generally of the order of $e^2/\hbar c = 1/137$, or about 1 percent.)

The magnetic moment of the same orbit is the current times the area. (See Section 14–5.) The current is the charge per unit time which passes any point on the orbit, namely, the charge q times the frequency of rotation. The frequency is the velocity divided by the circumference of the orbit; so

$$I = q\,\frac{v}{2\pi r}.$$

The area is πr^2, so the magnetic moment is

$$\mu = \frac{qvr}{2}. \tag{34.2}$$

It is also directed perpendicular to the plane of the orbit. So $\boldsymbol{J}$ and $\boldsymbol{\mu}$ are in the same direction:

$$\boldsymbol{\mu} = \frac{q}{2m}\,\boldsymbol{J}\ \text{(orbit)}. \tag{34.3}$$

Their ratio depends neither on the velocity nor on the radius. For any particle moving in a circular orbit the magnetic moment is equal to $q/2m$ times the angular momentum. For an electron, the charge is negative—we can call it $-q_e$; so for an electron

$$\boldsymbol{\mu} = -\frac{q_e}{2m}\,\boldsymbol{J}\ \text{(electron orbit)}. \tag{34.4}$$

That's what we would expect classically and, miraculously enough, it is also true quantum-mechanically. It's one of those things. However, if you keep going with the classical physics, you find other places where it gives the wrong answers, and it is a great game to try to remember which things are right and which things are wrong. We might as well give you immediately what is true *in general* in quantum mechanics. First, Eq. (34.4) is true for *orbital motion*, but that's not the only magnetism that exists. The electron also has a spin rotation about its own axis (something like the earth rotating on its axis), and as a result of that spin it has both an angular momentum and a magnetic moment. But for reasons that are purely quantum-mechanical—there is no classical explanation—the ratio of $\boldsymbol{\mu}$

to $\boldsymbol{J}$ for the electron spin is twice as large as it is for orbital motion of the spinning electron:

$$\boldsymbol{\mu} = -\frac{q_e}{m}\boldsymbol{J}\text{ (electron spin).} \tag{34.5}$$

In any atom there are, generally speaking, several electrons and some combination of spin and orbit rotations which builds up a total angular momentum and a total magnetic moment. Although there is no classical reason why it should be so, it is *always true* in quantum mechanics that (for an isolated atom) the direction of the magnetic moment is exactly opposite to the direction of the angular momentum. The ratio of the two is not necessarily either $-q_e/m$ or $-q_e/2m$, but somewhere in between, because there is a mixture of the contributions from the orbits and the spins. We can write

$$\boldsymbol{\mu} = -g\left(\frac{q_e}{2m}\right)\boldsymbol{J}, \tag{34.6}$$

where g is a factor which is characteristic of the state of the atom. It would be 1 for a pure orbital moment, or 2 for a pure spin moment, or some other number in between for a complicated system like an atom. This formula does not, of course, tell us very much. It says that the magnetic moment is *parallel to* the angular momentum, but can have any magnitude. The form of Eq. (34.6) is convenient, however, because g—called the "Landé g-factor"—is a dimensionless constant whose magnitude is of the order of one. It is one of the jobs of quantum mechanics to predict the g-factor for any particular atomic state.

You might also be interested in what happens in nuclei. In nuclei there are protons and neutrons which may move around in some kind of orbit and at the same time, like an electron, have an intrinsic spin. Again the magnetic moment is parallel to the angular momentum. Only now the order of magnitude of the ratio of the two is what you would expect for a *proton* going around in a circle, with m in Eq. (34.3) equal to the *proton* mass. Therefore it is usual to write for nuclei

$$\boldsymbol{\mu} = g\left(\frac{q_e}{2m_p}\right)\boldsymbol{J}, \tag{34.7}$$

where m_p is the mass of the proton, and g—called the *nuclear* g-factor—is a number near one, to be determined for each nucleus.

Another important difference for a nucleus is that the *spin* magnetic moment of the proton does *not* have a g-factor of 2, as the electron does. For a proton, $g = 2(2.79)$. Surprisingly enough, the *neutron* also has a spin magnetic moment, and its magnetic moment relative to its angular momentum is $2(-1.93)$. The neutron, in other words, is not exactly "neutral" in the magnetic sense. It is like a little magnet, and it has the kind of magnetic moment that a rotating *negative* charge would have.

34–3 The precession of atomic magnets

One of the consequences of having the magnetic moment proportional to the angular momentum is that an atomic magnet placed in a magnetic field will *precess*. First we will argue classically. Suppose that we have the magnetic moment $\boldsymbol{\mu}$ suspended freely in a uniform magnetic field. It will feel a torque $\boldsymbol{\tau}$, equal to $\boldsymbol{\mu}\times\boldsymbol{B}$, which tries to bring it in line with the field direction. But the atomic magnet is a gyroscope—it has the angular momentum $\boldsymbol{J}$. Therefore the torque due to the magnetic field will not cause the magnet to line up. Instead, the magnet will *precess*, as we saw when we analyzed a gyroscope in Chapter 20 of Volume I. The angular momentum—and with it the magnetic moment—precesses about an axis parallel to the magnetic field. We can find the rate of precession by the same method we used in Chapter 20 of the first volume.

Suppose that in a small time Δt the angular momentum changes from $\boldsymbol{J}$ to $\boldsymbol{J}'$, as drawn in Fig. 34–3, staying always at the same angle θ with respect to the direction of the magnetic field $\boldsymbol{B}$. Let's call ω_p the angular velocity of the precession, so that in the time Δt the angle *of precession* is $\omega_p\,\Delta t$. From the geometry of the figure, we see that the change of angular momentum in the time Δt is

$$\Delta J = (J\sin\theta)(\omega_p\,\Delta t).$$

So the rate of change of the angular momentum is

$$\frac{dJ}{dt} = \omega_p J\sin\theta, \tag{34.8}$$

which must be equal to the torque:

$$\tau = \mu B\sin\theta. \tag{34.9}$$

The angular velocity of precession is then

$$\omega_p = \frac{\mu}{J}\,B. \tag{34.10}$$

Substituting μ/J from Eq. (34.6), we see that for an atomic system

$$\omega_p = g\,\frac{q_eB}{2m}; \tag{34.11}$$

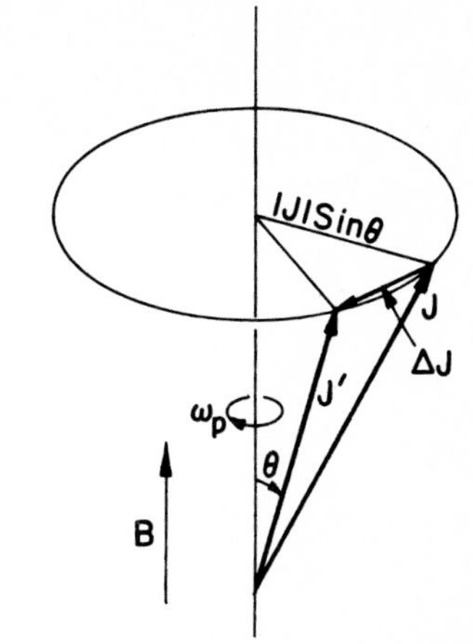

Fig. 34–3. An object with angular momentum $\boldsymbol{J}$ and a parallel magnetic moment $\boldsymbol{\mu}$ placed in a magnetic field $\boldsymbol{B}$ precesses with the angular velocity ω_p.

the precession frequency is proportional to B. It is handy to remember that for an atom (or electron)

$$f_p = \frac{\omega_p}{2\pi} = (1.4\text{ megacycles/gauss})gB, \tag{34.12}$$

and that for a nucleus

$$f_p = \frac{\omega_p}{2\pi} = (0.76\text{ kilocycles/gauss})gB. \tag{34.13}$$

(The formulas for atoms and nuclei are different only because of the different conventions for g for the two cases.)

According to the *classical* theory, then, the electron orbits—and spins—in an atom should precess in a magnetic field. Is it also true quantum-mechanically? It is essentially true, but the meaning of the "precession" is different. In quantum mechanics one cannot talk about the *direction* of the angular momentum in the same sense as one does classically; nevertheless, there is a very close analogy—so close that we continue to call it "precession." We will discuss it later when we talk about the quantum-mechanical point of view.

34–4 Diamagnetism

Next we want to look at *dia*magnetism from the classical point of view. It can be worked out in several ways, but one of the nice ways is the following. Suppose that we slowly turn on a magnetic field in the vicinity of an atom. As the magnetic field changes an *electric* field is generated by magnetic induction. From Faraday's law, the line integral of $\boldsymbol{E}$ around any closed path is the rate of change of the magnetic flux through the path. Suppose we pick a path Γ which is a circle of radius r concentric with the center of the atom, as shown in Fig. 34–4. The average tangential electric field E around this path is given by

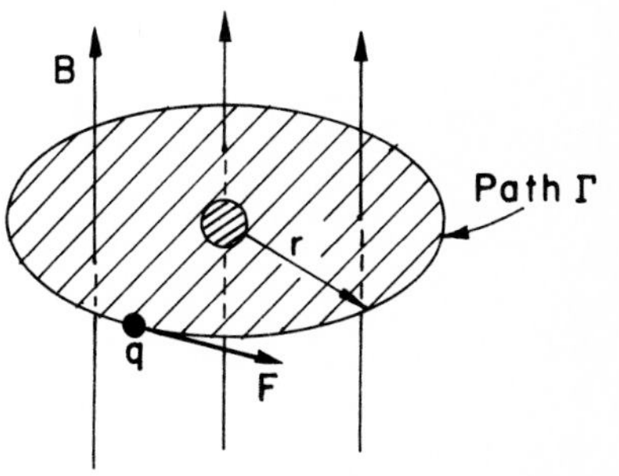

Fig. 34–4. The induced electric forces on the electrons in an atom.

$$E2\pi r = -\frac{d}{dt}(B\pi r^2),$$

and there is a circulating electric field whose strength is

$$E = -\frac{r}{2}\frac{dB}{dt}.$$

The induced electric field acting on an electron in the atom produces a torque equal to $-q_eEr$, which must equal the rate of change of the angular momentum dJ/dt:

$$\frac{dJ}{dt} = \frac{q_er^2}{2}\frac{dB}{dt}. \tag{34.14}$$

Integrating with respect to time from zero field, we find that the change in angular momentum due to turning on the field is

$$\Delta J = \frac{q_e r^2}{2} B. \tag{34.15}$$

This is the extra angular momentum from the twist given to the electrons as the field is turned on.

This added angular momentum makes an extra magnetic moment which, because it is an *orbital* motion, is just $-q_e/2m$ times the angular momentum. The induced diamagnetic moment is

$$\Delta\mu = -\frac{q_e}{2m}\,\Delta J = -\frac{q_e^2 r^2}{4m}\,B. \tag{34.16}$$

The minus sign (as you can see is right by using Lenz's law) means that the added moment is opposite to the magnetic field.

We would like to write Eq. (34.16) a little differently. The r^2 which appears is the radius from an axis through the atom parallel to B, so if B is along the z-direction, it is $x^2 + y^2$. If we consider spherically symmetric atoms (or average over atoms with their natural axes in all directions) the average of $x^2 + y^2$ is 2/3 of the average of the square of the true radial distance from the center *point* of the atom. It is therefore usually more convenient to write Eq. (34.16) as

$$\Delta\mu = -\frac{q_e^2}{6m}\,\langle r^2\rangle_{\text{av}} B. \tag{34.17}$$

In any case, we have found an induced atomic moment proportional to the magnetic field B and opposing it. This is diamagnetism of matter. It is this magnetic effect that is responsible for the small force on a piece of bismuth in a nonuniform magnetic field. (You could compute the force by working out the energy of the induced moments in the field and seeing how the energy changes as the material is moved into or out of the high-field region.)

We are still left with the problem: What is the mean square radius, $\langle r^2\rangle_{\text{av}}$? Classical mechanics cannot supply an answer. We must go back and start over with quantum mechanics. In an atom we cannot really say where an electron is, but only know the probability that it will be at some place. If we interpret $\langle r^2\rangle_{\text{av}}$ to mean the average of the square of the distance from the center for the probability distribution, the diamagnetic moment given by quantum mechanics is just the same as formula (34.17). This equation, of course, is the moment for one electron. The total moment is given by the sum over all the electrons in the atom. The surprising thing is that the classical argument and quantum mechanics give the same answer, although, as we shall see, the classical argument that gives Eq. (34.17) is not really valid in classical mechanics.

The same diamagnetic effect occurs even when an atom already has a permanent moment. Then the system will precess in the magnetic field. As the whole atom precesses, it takes up an additional small angular velocity, and that slow turning gives a small current which represents a correction to the magnetic moment. This is just the diamagnetic effect represented in another way. But we don't really have to worry about that when we talk about paramagnetism. If the diamagnetic effect is first computed, as we have done here, we don't have to worry about the fact that there is an extra little current from the precession. That has already been included in the diamagnetic term.

34–5 Larmor's theorem

We can already conclude something from our results so far. First of all, in the classical theory the moment $\boldsymbol{\mu}$ was always proportional to $\boldsymbol{J}$, with a given constant of proportionality for a particular atom. There wasn't any spin of the electrons, and the constant of proportionality was always $-q_e/2m$; that is to say, in Eq. (34.6) we should set $g = 1$. The ratio of $\boldsymbol{\mu}$ to $\boldsymbol{J}$ was independent of the internal motion of the electrons. Thus, according to the classical theory, all systems of electrons would precess with *the same* angular velocity. (This is *not* true in quantum mechanics.) This result is related to a theorem in classical mechanics that we would now like to prove. Suppose we have a group of electrons which are all held together by attraction toward a central point—as the electrons are attracted by a nucleus. The electrons will also be interacting with each other, and can, in general, have complicated motions. Suppose you have solved for the motions with *no* magnetic field and then want to know what the motions would be *with* a weak magnetic field. The theorem says that the motion with a weak magnetic field is always one of the no-field solutions with an added rotation, about the axis of the field, with the angular velocity $\omega_L = q_eB/2m$. (This is the same as ω_p, if $g = 1$.) There are, of course, many possible motions. The point is that for every motion without the magnetic field there is a corresponding motion in the field, which is the original motion plus a uniform rotation. This is called Larmor's theorem, and ω_L is called the *Larmor frequency*.

We would like to show how the theorem can be proved, but we will let you work out the details. Take, first, one electron in a central force field. The force on it is just $\boldsymbol{F}(r)$, directed toward the center. If we now turn on a uniform magnetic field, there is an additional force, $q\boldsymbol{v} \times \boldsymbol{B}$; so the total force is

$$\boldsymbol{F}(r) + q\boldsymbol{v} \times \boldsymbol{B}. \tag{34.18}$$

Now let's look at the same system from a coordinate system rotating with angular velocity ω about an axis through the center of force and parallel to $\boldsymbol{B}$. This is no longer an inertial system, so we have to put in the proper pseudoforces—the centrifugal and Coriolis forces we talked about in Chapter 19 of Volume I. We found there that in a frame rotating with angular velocity ω, there is an apparent *tangential* force proportional to v_r, the radial component of velocity:

$$F_t = -2m\omega v_r. \tag{34.19}$$

And there is an apparent radial force which is given by

$$F_r = m\omega^2 r + 2m\omega v_t, \tag{34.20}$$

where v_t is the tangential component of the velocity, measured *in* the rotating frame. (The radial component v_r for rotating and inertial frames is the same.)

Now for small enough angular velocities (that is, if $\omega r \ll v_t$), we can neglect the first term (centrifugal) in Eq. (34.20) in comparison with the second (Coriolis). Then Eqs. (34.19) and (34.20) can be written together as

$$\boldsymbol{F} = -2m\boldsymbol{\omega} \times \boldsymbol{v}. \tag{34.21}$$

If we now *combine* a rotation and a magnetic field, we must add the force in Eq. (34.21) to that in Eq. (34.18). The total force is

$$\boldsymbol{F}(r) + q\boldsymbol{v} \times \boldsymbol{B} + 2m\boldsymbol{v} \times \boldsymbol{\omega} \tag{34.22}$$

[we reverse the cross product and the sign of Eq. (34.21) to get the last term]. Looking at our result, we see that if

$$2m\boldsymbol{\omega} = -q\boldsymbol{B}$$

the two terms on the right cancel, and in the moving frame the only force is $\boldsymbol{F}(r)$. The motion of the electron is just the same as with no magnetic field—and, of course, no rotation. We have proved Larmor's theorem for one electron. Since the proof assumes a small ω, it also means that the theorem is true only for weak magnetic fields. The only thing we could ask you to improve on is to take the case of many electrons mutually interacting with each other, but all in the same central field, and prove the same theorem. So no matter how complex an atom is, if it has a central field the theorem is true. But that's the end of the classical mechanics, because it isn't true in fact that the motions precess in that way. The precession frequency ω_p of Eq. (34.11) is only equal to ω_L if g happens to be equal to 1.

34–6 Classical physics gives neither diamagnetism nor paramagnetism

Now we would like to demonstrate that according to classical mechanics there can be no diamagnetism and no paramagnetism at all. It sounds crazy—first, we have proved that there are paramagnetism, diamagnetism, precessing orbits, and so on, and now we are going to prove that it is all wrong. Yes!—We are going to prove that *if* you follow the *classical* mechanics far enough, there are no such magnetic effects—*they all cancel out.* If you start a classical argument in a certain place and don't go far enough, you can get any answer you want. But the only legitimate and correct proof shows that there is no magnetic effect whatever.

It is a consequence of classical mechanics that if you have any kind of system—a gas with electrons, protons, and whatever—kept in a box so that the whole thing can't turn, there will be no magnetic effect. It is possible to have a magnetic effect if you have an isolated system, like a star held together by itself, which can start rotating when you put on the magnetic field. But if you have a piece of material that is held in place so that it can't start spinning, then there will be no magnetic effects. What we mean by holding down the spin is summarized this way: At a given temperature we suppose that there is *only one state* of thermal equilibrium. The theorem then says that if you turn on a magnetic field and wait for the system to get into thermal equilibrium, there will be no paramagnetism or diamagnetism—there will be no induced magnetic moment. Proof: According to statistical mechanics, the probability that a system will have any given state of motion is proportional to $e^{-U/kt}$, where U is the energy of that motion. Now what is the energy of motion? For a particle moving in a constant magnetic field, the energy is the ordinary potential energy plus $mv^2/2$, with nothing additional for the magnetic field. [You know that the forces from electromagnetic fields are $q(\boldsymbol{E} + \boldsymbol{v} \times \boldsymbol{B})$, and that the rate of work $\boldsymbol{F} \cdot \boldsymbol{v}$ is just $q\boldsymbol{E} \cdot \boldsymbol{v}$, which is not affected by the magnetic field.] So the energy of a system, whether it is in a magnetic field or not, is always given by the kinetic energy plus the potential energy. Since the probability of any motion depends only on the energy—that is, on the velocity and position—it is the same whether or not there is a magnetic field. For *thermal* equilibrium, therefore, the magnetic field has no effect. If we have one system in a box, and then have another system in a second box, this time with a magnetic field, the probability of any particular velocity at any point in the first box is the same as in the second. If the first box has no average circulating current (which it will not have if it is in equilibrium with the stationary walls), there is no average magnetic moment. Since in the second box all the motions are the same, there is no average magnetic moment there either. Hence, if the temperature is kept constant and thermal equilibrium is re-established after the field is turned on, there can be no magnetic moment induced by the field—according to classical mechanics. We can only get a satisfactory understanding of magnetic phenomena from quantum mechanics.

Unfortunately, we cannot assume that you have a thorough understanding of quantum mechanics, so this is hardly the place to discuss the matter. On the other hand, we don't always have to learn something first by learning the exact rules and then by learning how they are applied in different cases. Almost every subject that we have taken up in this course has been treated in a different way. In the case of electricity, we wrote the Maxwell equations on "Page One" and then deduced all the consequences. That's one way. But we will *not* now try to begin a new "Page One," writing the equations of quantum mechanics and deducing everything from them. We will just have to tell you some of the consequences of quantum mechanics, before you learn where they come from. So here we go.

34–7 Angular momentum in quantum mechanics

We have already given you a relation between the magnetic moment and the angular momentum. That's pleasant. But what do the magnetic moment and the angular momentum *mean* in quantum mechanics? In quantum mechanics it turns out to be best to define things like magnetic moments in terms of the other concepts such as energy, in order to make sure that one knows what it means. Now, it is easy to define a magnetic moment in terms of energy, because the energy of a moment in a magnetic field is, in the classical theory, $\boldsymbol{\mu} \cdot \boldsymbol{B}$. Therefore, the following definition has been taken in quantum mechanics: If we calculate the energy of a system in a magnetic field and we find that it is proportional to the field strength (for small field), the coefficient is called the component of magnetic moment in the direction of the field. (We don't have to get so elegant for our work now; we can still think of the magnetic moment in the ordinary, to some extent classical, sense.)

Now we would like to discuss the idea of angular momentum in quantum mechanics—or rather, the characteristics of what, in quantum mechanics, is called angular momentum. You see, when you go to new kinds of laws, you can't just assume that each word is going to mean exactly the same thing. You may think, say, "Oh, I know what angular momentum is. It's that thing that is changed by a torque." But what's a torque? In quantum mechanics we have to have new definitions of old quantities. It would, therefore, be legally best to call it by some other name such as "quantangular momentum," or something like that, because it is the angular momentum as defined in quantum mechanics. But if we can find a quantity in quantum mechanics which is identical to our old idea of angular momentum when the system becomes large enough, there is no use in inventing an extra word. We might as well just call it angular momentum. With that understanding, this odd thing that we are about to describe *is* angular momentum. It is the thing which in a large system we recognize as angular momentum in classical mechanics.

First, we take a system in which angular momentum is conserved, such as an atom all by itself in empty space. Now such a thing (like the earth spinning on its axis) could, in the ordinary sense, be spinning around any axis one wished to choose. And for a given spin, there could be many different "states," all of the same energy, each "state" corresponding to a particular direction of the axis of the angular momentum. So in the classical theory, with a given angular momentum, there is an infinite number of possible states, all of the same energy.

It turns out in quantum mechanics, however, that several strange things happen. First, the number of states in which such a system *can exist* is limited—there is only a finite number. If the system is small, the finite number is very small, and if the system is large, the finite number gets very, very large. Second, we *cannot* describe a "state" by giving the *direction* of its angular momentum, but only by giving the *component* of the angular momentum along some direction—say in the z-direction. Classically, an object with a given total angular momentum J could have, for its z-component, any value from $+J$ to $-J$. But quantum-mechanically, the z-component of angular momentum can have only certain discrete values. Any given system—a particular atom, or a nucleus, or anything—with a given energy, has a characteristic number j, and its z-component of angular momentum can only be one of the following set of values:

$$\begin{array}{r} j\hbar \\ (j-1)\hbar \\ (j-2)\hbar \\ \vdots \\ -(j-2)\hbar \\ -(j-1)\hbar \\ -j\hbar \end{array} \tag{34.23}$$

The largest z-component is j times $\hbar$; the next smaller is one unit of $\hbar$ less, and so on down to $-j\hbar$. The number j is called "the spin of the system." (Some people call it the "total angular momentum quantum number"; but we'll call it the "spin.")

You may be worried that what we are saying can only be true for some "special" z-axis. But that is not so. For a system whose spin is j, the component of angular momentum along *any* axis can have only one of the values in (34.23). Although it is quite mysterious, we ask you just to accept it for the moment. We will come back and discuss the point later. You may at least be pleased to hear that the z-component goes from some number to minus the *same* number, so that we at least don't have to decide which is the plus direction of the z-axis. (Certainly, if we said that it went from $+j$ to minus a different amount, that would be infinitely mysterious, because we wouldn't have been able to define the z-axis, pointing the other way.)

Now if the z-component of angular momentum must go down by integers from $+j$ to $-j$, then j must be an integer. No! Not quite; twice j must be an integer. It is only the *difference* between $+j$ and $-j$ that must be an integer. So, in general, the spin j is either an integer or a half-integer, depending on whether $2j$ is even or odd. Take, for instance, a nucleus like lithium, which has a spin of three-halves, $j = 3/2$. Then the angular momentum around the z-axis, in units of $\hbar$, is one of the following:

$$\begin{gathered} +3/2 \\ +1/2 \\ -1/2 \\ -3/2. \end{gathered}$$

There are four possible states, each of the same energy, if the nucleus is in empty space with no external fields. If we have a system whose spin is two, then the z-component of angular momentum has only the values, in units of $\hbar$,

$$\begin{gathered} 2 \\ 1 \\ 0 \\ -1 \\ -2. \end{gathered}$$

If you count how many states there are for a given j, there are $(2j + 1)$ possibilities. In other words, if you tell me the energy and also the spin j, it turns out that there are exactly $(2j + 1)$ states with that energy, each state corresponding to one of the different possible values of the z-component of the angular momentum.

We would like to add one other fact. If you pick out any atom of known j at random and measure the z-component of the angular momentum, then you may get any one of the possible values, and each of the values is *equally* likely. All of the states are in fact single states, and each is just as good as any other. Each one has the same "weight" in the world. (We are assuming that nothing has been done to sort out a special sample.) This fact has, incidentally, a simple classical analog. If you ask the same question classically: What is the likelihood of a particular z-component of angular momentum if you take a random sample of systems, all with the same total angular momentum?—the answer is that all values from the maximum to the minimum are equally likely. (You can easily work that out.) The classical result corresponds to the equal probability of the $(2j + 1)$ possibilities in quantum mechanics.

From what we have so far, we can get another interesting and somewhat surprising conclusion. In certain classical calculations the quantity that appears in the final result is the *square* of the magnitude of the angular momentum $\boldsymbol{J}$—in other words, $\boldsymbol{J} \cdot \boldsymbol{J}$. It turns out that it is often possible to *guess* at the correct quantum-mechanical formula by using the classical calculation and the following simple rule: Replace $J^2 = \boldsymbol{J} \cdot \boldsymbol{J}$ by $j(j + 1)\hbar^2$. This rule is commonly used, and usually gives the correct result, but *not* always. We can give the following argument to show why you might expect this rule to work.

The scalar product $\boldsymbol{J} \cdot \boldsymbol{J}$ can be written as

$$\boldsymbol{J} \cdot \boldsymbol{J} = J_x^2 + J_y^2 + J_z^2.$$

Since it is a scalar, it should be the same for any orientation of the spin. Suppose we pick samples of any given atomic system at random and make measurements of J_x^2, or J_y^2, or J_z^2, the *average value* should be the same for each. (There is no special distinction for any one of the directions.) Therefore, the average of $\boldsymbol{J} \cdot \boldsymbol{J}$ is just equal to three times the average of any component squared, say of J_z^2;

$$\langle \boldsymbol{J}\cdot\boldsymbol{J}\rangle_{\text{av}} = 3\langle J_z^2\rangle.$$

But since $\boldsymbol{J}\cdot\boldsymbol{J}$ is the same for all orientations, its average is, of course, just its constant value; we have

$$\boldsymbol{J}\cdot\boldsymbol{J} = 3\langle J_z^2\rangle_{\text{av}}. \tag{34.24}$$

If we now say that we will use the same equation for quantum mechanics, we can easily find $\langle J_z^2\rangle_{\text{av}}$. We just have to take the sum of the $(2j+1)$ possible values of J_z^2, and divide by the total number;

$$\langle J_z^2\rangle_{\text{av}} = \frac{j^2+(j-1)^2+\cdots+(-j+1)^2+(-j)^2}{2j+1}\,\hbar^2. \tag{34.25}$$

For a system with a spin of 3/2, it goes like this:

$$\langle J_z^2\rangle_{\text{av}} = \frac{(3/2)^2+(1/2)^2+(-1/2)^2+(-3/2)^2}{4}\,\hbar^2 = \frac{5}{4}\,\hbar^2.$$

We conclude that

$$\boldsymbol{J}\cdot\boldsymbol{J} = 3\langle J_z^2\rangle_{\text{av}} = 3\tfrac{5}{4}\hbar^2 = \tfrac{3}{2}(\tfrac{3}{2}+1)\hbar^2.$$

We will leave it for you to show that Eq. (34.25), together with Eq. (34.24), gives the general result

$$\boldsymbol{J}\cdot\boldsymbol{J} = j(j+1)\hbar^2. \tag{34.26}$$

Although we would think classically that the largest possible value of the z-component of $\boldsymbol{J}$ is just the magnitude of $\boldsymbol{J}$—namely, $\sqrt{\boldsymbol{J}\cdot\boldsymbol{J}}$—quantum mechanically the maximum of J_z is always a little less than that, because $j\hbar$ is always less than $\sqrt{j(j+1)}\,\hbar$. The angular momentum is never "completely along the z-direction."

34–8 The magnetic energy of atoms

Now we want to talk again about the magnetic moment. We have said that in quantum mechanics the magnetic moment of a particular atomic system can be written in terms of the angular momentum by Eq. (34.6);

$$\boldsymbol{\mu} = -g\left(\frac{q_e}{2m}\right)\boldsymbol{J}, \tag{34.27}$$

where $-q_e$ and m are the charge and mass of the electron.

An atomic magnet placed in an external magnetic field will have an extra magnetic energy which depends on the component of its magnetic moment along the field direction. We know that

$$U_{\text{mag}} = -\boldsymbol{\mu}\cdot\boldsymbol{B}. \tag{34.28}$$

Choosing our z-axis along the direction of $\boldsymbol{B}$,

$$U_{\text{mag}} = -\mu_z B. \tag{34.29}$$

Using Eq. (34.27), we have that

$$U_{\text{mag}} = g\left(\frac{q_e}{2m}\right)J_z B.$$

Quantum mechanics says that J_z can have only certain values: $j\hbar$, $(j-1)\hbar, \ldots, -j\hbar$. Therefore, the magnetic energy of an atomic system is not arbitrary; it can have only certain values. Its maximum value, for instance, is

$$g\left(\frac{q_e}{2m}\right)\hbar jB.$$

The quantity $q_e\hbar/2m$ is usually given the name "the Bohr magneton" and written μ_B:

$$\mu_B = \frac{q_e\hbar}{2m}.$$

The possible values of the magnetic energy are

$$U_{\text{mag}} = g\mu_B B\frac{J_z}{\hbar},$$

where $J_z/\hbar$ takes on the possible values j, $(j - 1)$, $(j - 2)$, ..., $(-j + 1)$, $-j$.

In other words, the energy of an atomic system is changed when it is put in a magnetic field by an amount that is proportional to the field, and proportional to J_z. We say that the energy of an atomic system is "split into $2j + 1$ levels" by a magnetic field. For instance, an atom whose energy is U_0 outside a magnetic field and whose j is 3/2, will have four possible energies when placed in a field. We can show these energies by an energy-level diagram like that drawn in Fig. 34-5. Any particular atom can have only one of the four possible energies in any given field B. That is what quantum mechanics says about the behavior of an atomic system in a magnetic field.

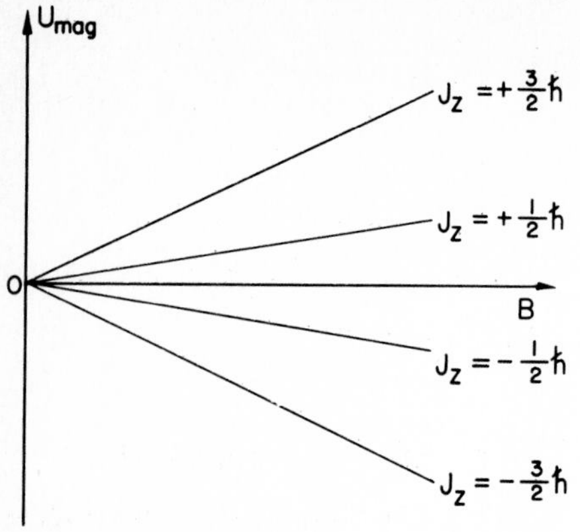

Fig. 34-5. The possible magnetic energies of an atomic system with a spin of 3/2 in a magnetic field *B*.

The simplest "atomic" system is a single electron. The spin of an electron is 1/2, so there are two possible states: $J_z = \hbar/2$ and $J_z = -\hbar/2$. For an electron at rest (no orbital motion), the spin magnetic moment has a g-value of 2, so the magnetic energy can be either $\pm\mu_B B$. The possible energies in a magnetic field are shown in Fig. 34-6. Speaking loosely we say that the electron either has its spin "up" (along the field) or "down" (opposite the field).

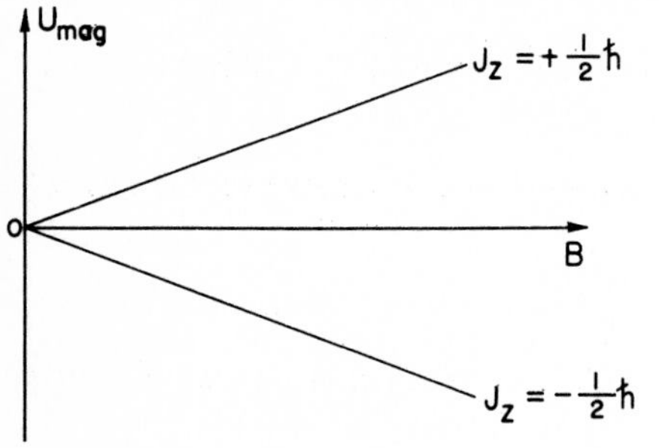

Fig. 34-6. The two possible energy states of an electron in a magnetic field *B*.

For systems with higher spins, there are more states. We can think that the spin is "up" or "down" or cocked at some "angle" in between, depending on the value of J_z.

We will use these quantum mechanical results to discuss the magnetic properties of materials in the next chapter.

35

Paramagnetism and Magnetic Resonance

Review: Chapter 11, *Inside Dielectrics*

35–1 Quantized magnetic states

In the last chapter we described how in quantum mechanics the angular momentum of a thing does not have an arbitrary direction, but its component along a given axis can take on only certain equally spaced, discrete values. It is a shocking and peculiar thing. You may think that perhaps we should not go into such things until your minds are more advanced and ready to accept this kind of an idea. Actually, your minds will never become more advanced—in the sense of being able to accept such a thing easily. There isn't any descriptive way of making it intelligible that isn't so subtle and advanced in its own form that it is more complicated than the thing you were trying to explain. The behavior of matter on a small scale—as we have remarked many times—is different from anything that you are used to and is very strange indeed. As we proceed with classical physics, it is a good idea to try to get a growing acquaintance with the behavior of things on a small scale, at first as a kind of experience without any deep understanding. Understanding of these matters comes very slowly, if at all. Of course, one does get better able to know what is going to happen in a quantum-mechanical situation—if that is what understanding means—but one never gets a comfortable feeling that these quantum-mechanical rules are "natural." Of course they *are*, but they are not natural to our own experience at an ordinary level. We should explain that the attitude that we are going to take with regard to this rule about angular momentum is quite different from many of the other things we have talked about. We are not going to try to "explain" it, but we must at least *tell* you what happens; it would be dishonest to describe the magnetic properties of materials without mentioning the fact that the classical description of magnetism—of angular momentum and magnetic moments—is incorrect.

One of the most shocking and disturbing features about quantum mechanics is that if you take the angular momentum along any particular axis you find that it is always an integer or half-integer times $\hbar$. This is so no matter which axis you take. The subtleties involved in that curious fact—that you can take any other axis and find that the component for it is also locked to the same set of values—we will leave to a later chapter, when you will experience the delight of seeing how this apparent paradox is ultimately resolved.

We will now just accept the fact that for every atomic system there is a number j, called the *spin* of the system—which must be an integer or a half-integer—and that the component of the angular momentum along any particular axis will always have one of the following values between $+j\hbar$ and $-j\hbar$:

$$J_z = \text{one of}\left\{\begin{matrix} j \\ j-1 \\ j-2 \\ \vdots \\ -j+2 \\ -j+1 \\ -j \end{matrix}\right\}\cdot\hbar. \tag{35.1}$$

We have also mentioned that every simple atomic system has a magnetic moment which has the same direction as the angular momentum. This is true not only for atoms and nuclei but also for the fundamental particles. Each fundamental particle has its own characteristic value of j and its magnetic moment.

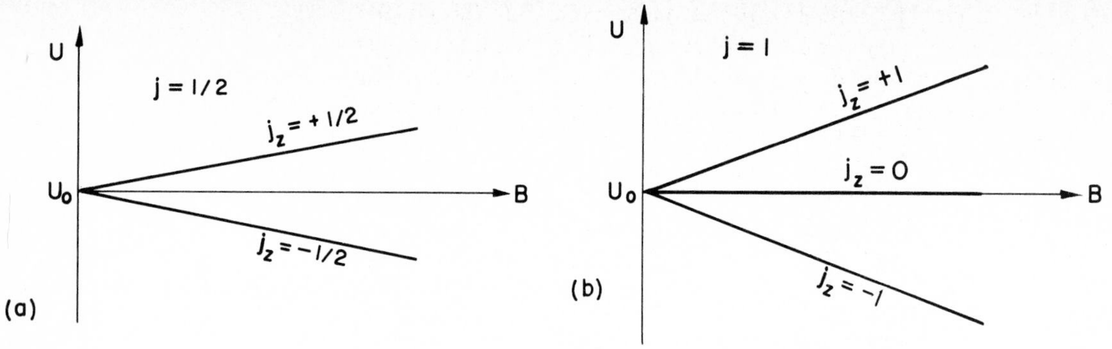

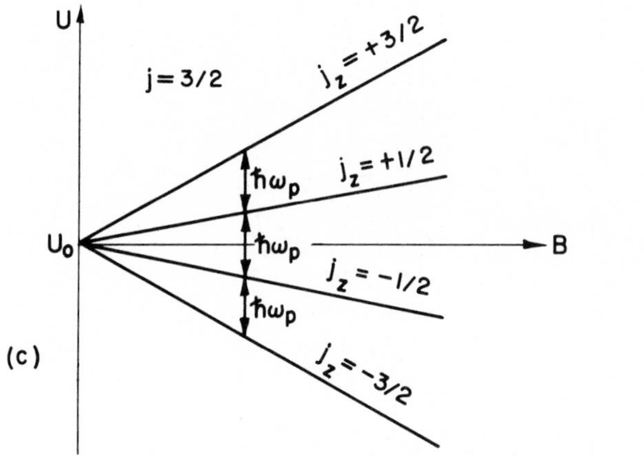

Fig. 35–1. An atomic system with spin j has $(2j + 1)$ possible energy values in a magnetic field $\boldsymbol{B}$. The energy splitting is proportional to B for small fields.

(For some particles, both are zero.) What we mean by "the magnetic moment" in this statement is that the energy of the system in a magnetic field, say in the z-direction, can be written as $-\mu_z B$ for small magnetic fields. We must have the condition that the field should not be too great, otherwise it could disturb the internal motions of the system and the energy would not be a measure of the magnetic moment that was there before the field was turned on. But if the field is sufficiently weak, the field changes the energy by the amount

$$\Delta U = -\mu_z B, \tag{35.2}$$

with the understanding that in this equation we are to replace μ_z by

$$\mu_z = g\left(\frac{q}{2m}\right) J_z, \tag{35.3}$$

where J_z has one of the values in Eq. (35.1).

Suppose we take a system with a spin $j = 3/2$. Without a magnetic field, the system has four different possible states corresponding to the different values of J_z, all of which have exactly the same energy. But the moment we turn on the magnetic field, there is an additional energy of interaction which separates these states into four slightly different energy levels. The energies of these levels are given by a certain energy proportional to B, multiplied by $\hbar$ times $3/2$, $1/2$, $-1/2$, and $-3/2$—the values of J_z. The splitting of the energy levels for atomic systems with spins of 1/2, 1, and 3/2 are shown in the diagrams of Fig. 35–1. (Remember that for any arrangement of electrons the magnetic moment is always directed opposite to the angular momentum.)

You will notice from the diagrams that the "center of gravity" of the energy levels is the same with and without a magnetic field. Also notice that the spacings from one level to the next are always equal for a given particle in a given magnetic field. We are going to write the energy spacing, for a given magnetic field B, as $\hbar\omega_p$—which is just a definition of ω_p. Using Eqs. (35.2) and (35.3), we have

$$\hbar\omega_p = g\,\frac{q}{2m}\,\hbar B$$

or

$$\omega_p = g\,\frac{q}{2m}\,B. \tag{35.4}$$

The quantity $g(q/2m)$ is just the ratio of the magnetic moment to the angular momentum—it is a property of the particle. Equation (35.4) is the same formula that we got in Chapter 34 for the angular velocity of precession in a magnetic field, for a gyroscope whose angular momentum is $\boldsymbol{J}$ and whose magnetic moment is $\boldsymbol{\mu}$.

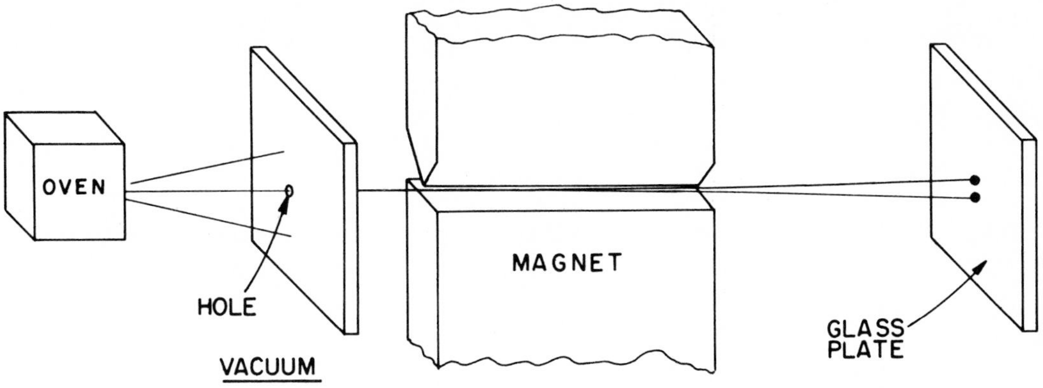

Fig. 35–2. The experiment of Stern and Gerlach.

35–2 The Stern-Gerlach experiment

The fact that the angular momentum is quantized is such a surprising thing that we will talk a little bit about it historically. It was a shock from the moment it was discovered (although it was expected theoretically). It was first observed in an experiment done in 1922 by Stern and Gerlach. If you wish, you can consider the experiment of Stern-Gerlach as a direct justification for a belief in the quantization of angular momentum. Stern and Gerlach devised an experiment for measuring the magnetic moment of individual silver atoms. They produced a beam of silver atoms by evaporating silver in a hot oven and letting some of them come out through a series of small holes. This beam was directed between the pole tips of a special magnet, as shown in Fig. 35–2. Their idea was the following. If the silver atom has a magnetic moment $\boldsymbol{\mu}$, then in a magnetic field $\boldsymbol{B}$ it has an energy $-\mu_z B$, where z is the direction of the magnetic field. In the classical theory, μ_z would be equal to the magnetic moment times the cosine of the angle between the moment and the magnetic field, so the extra energy in the field would be

$$\Delta U = -\mu B \cos\theta. \tag{35.5}$$

Of course, as the atoms come out of the oven, their magnetic moments would point in every possible direction, so there would be all values of θ. Now if the magnetic field varies very rapidly with z—if there is a strong field gradient—then the magnetic energy will also vary with position, and there will be a force on the magnetic moments whose direction will depend on whether cosine θ is positive or negative. The atoms will be pulled up or down by a force proportional to the derivative of the magnetic energy; from the principle of virtual work,

$$F_z = -\frac{\partial U}{\partial z} = \mu\cos\theta\,\frac{\partial B}{\partial z}\cdot \tag{35.6}$$

Stern and Gerlach made their magnet with a very sharp edge on one of the pole tips in order to produce a very rapid variation of the magnetic field. The beam of silver atoms was directed right along this sharp edge, so that the atoms would feel a vertical force in the inhomogeneous field. A silver atom with its magnetic moment directed horizontally would have no force on it and would go straight past the magnet. An atom whose magnetic moment was exactly vertical would have a force pulling it up toward the sharp edge of the magnet. An atom whose magnetic moment was pointed downward would feel a downward push. Thus, as they left the magnet, the atoms would be spread out according to their vertical components of magnetic moment. In the classical theory all angles are possible, so that when the silver atoms are collected by deposition on a glass plate, one should expect a smear of silver along a vertical line. The height of the line would be proportional to the magnitude of the magnetic moment. The abject failure of classical ideas was completely revealed when Stern and Gerlach saw what actually happened. They found on the glass plate two distinct spots. The silver atoms had formed two beams.

That a beam of atoms whose spins would apparently be randomly oriented gets split up into two separate beams is most miraculous. How does the magnetic moment *know* that it is only allowed to take on certain components in the direction of the magnetic field? Well, that was really the beginning of the discovery of the quantization of angular momentum, and instead of trying to give you a theoretical explanation, we will just say that you are stuck with the result of this experiment just as the physicists of that day had to accept the result when the experiment was done. It is an *experimental fact* that the energy of an atom in a magnetic field takes on a series of individual values. For each of these values the energy is proportional to the field strength. So in a region where the field varies, the principle of virtual work tells us that the possible magnetic force on the atoms will have a set of separate values; the force is different for each state, so the beam of atoms is split into a small number of separate beams. From a measurement of the deflection of the beams, one can find the strength of the magnetic moment.

35–3 The Rabi molecular-beam method

We would now like to describe an improved apparatus for the measurement of magnetic moments which was developed by I. I. Rabi and his collaborators. In the Stern-Gerlach experiment the deflection of atoms is very small, and the measurement of the magnetic moment is not very precise. Rabi's technique permits a fantastic precision in the measurement of the magnetic moments. The method is based on the fact that the original energy of the atoms in a magnetic field is split up into a finite number of energy levels. That the energy of an atom in the magnetic field can have only certain discrete energies is really not more surprising than the fact that atoms *in general* have only certain discrete energy levels—something we mentioned often in Volume I. Why should the same thing *not* hold for atoms in a magnetic field? It does. But it is the attempt to correlate this with the idea of an *oriented magnetic moment* that brings out some of the strange implications of quantum mechanics.

When an atom has two levels which differ in energy by the amount ΔU, it can make a transition from the upper level to the lower level by emitting a light quantum of frequency ω, where

$$\hbar\omega = \Delta U. \tag{35.7}$$

The same thing can happen with atoms in a magnetic field. Only then, the energy differences are so small that the frequency does not correspond to light, but to microwaves or to radiofrequencies. The transitions from the lower energy level to an upper energy level of an atom can also take place with the absorption of light or, in the case of atoms in a magnetic field, by the absorption of microwave energy. Thus if we have an atom in a magnetic field, we can cause transitions from one state to another by applying an additional electromagnetic field of the proper frequency. In other words, if we have an atom in a strong magnetic field and we "tickle" the atom with a weak varying electromagnetic field, there will be a certain probability of knocking it to another level if the frequency is near to the ω in Eq. (35.7). For an atom in a magnetic field, this frequency is just what we have earlier called ω_p and it is given in terms of the magnetic field by Eq. (35.4). If the atom is tickled with the wrong frequency, the chance of causing a transition is very small. Thus there is a sharp *resonance* at ω_p in the probability of causing a transition. By measuring the frequency of this resonance in a known magnetic field B, we can measure the quantity $g(q/2m)$—and hence the g-factor—with great precision.

It is interesting that one comes to the same conclusion from a classical point of view. According to the classical picture, when we place a small gyroscope with a magnetic moment μ and an angular momentum J in an external magnetic field, the gyroscope will precess about an axis parallel to the magnetic field. (See Fig. 35–3.) Suppose we ask: How can we change the angle of the classical gyroscope with respect to the field—namely, with respect to the z-axis? The magnetic field produces a torque around a *horizontal* axis. Such a torque you would think is *trying* to line up the magnet with the field, but it only causes the precession. If we want to change the angle of the gyroscope with respect to the z-axis, we must exert a torque on it *about the z-axis*. If we apply a torque which goes in the same direction as the precession, the angle of the gyroscope will change to give a smaller component of $\boldsymbol{J}$ in the z-direction. In Fig. 35–3, the angle between $\boldsymbol{J}$ and the z-axis would increase. If we try to hinder the precession, $\boldsymbol{J}$ moves toward the vertical.

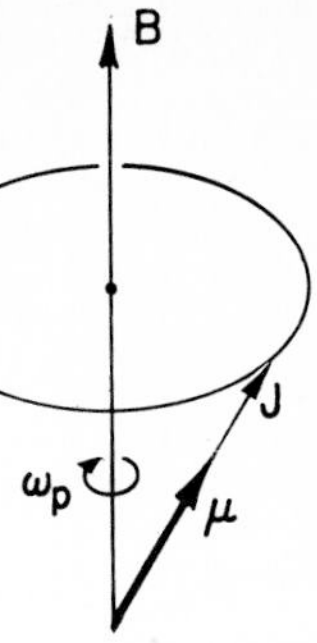

Fig. 35–3. The classical precession of an atom with the magnetic moment $\boldsymbol{\mu}$ and the angular momentum $\boldsymbol{J}$.

For our precessing atom in a uniform magnetic field, how can we apply the kind of torque we want? The answer is: with a weak magnetic field from the side. You might at first think that the direction of this magnetic field would have to rotate with the precession of the magnetic moment, so that it was always at right angles to the moment, as indicated by the field B' in Fig. 35–4(a). Such a field works very well, but an *alternating* horizontal field is almost as good. If we have a small horizontal field B', which is always in the x-direction (plus or minus) and which oscillates with the frequency ω_p, then on each one-half cycle the torque on the magnetic moment reverses, so that it has a cumulative effect which is almost as effective as a rotating magnetic field. Classically, then, we would expect the component of the magnetic moment along the z-direction to change if we have a very weak oscillating magnetic field at a frequency which is exactly ω_p. Classically, of course, μ_z would change continuously, but in quantum mechanics the z-component of the magnetic moment cannot adjust continuously. It must jump suddenly from one value to another. We have made the comparison between the consequences of classical mechanics and quantum mechanics to give you some clue as to what might happen classically and how it is related to what actually happens in quantum mechanics. You will notice, incidentally, that the expected resonant frequency is the same in both cases.

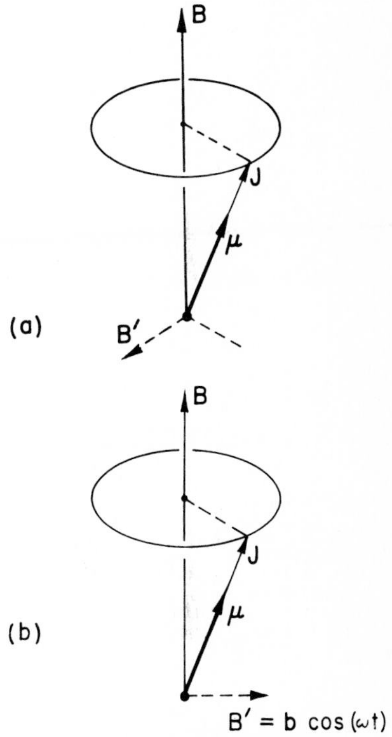

Fig. 35–4. The angle of precession of an atomic magnet can be changed by a horizontal magnetic field always at right angles to $\boldsymbol{\mu}$, as in (a), or by an oscillating field, as in (b).

One additional remark: From what we have said about quantum mechanics, there is no apparent reason why there couldn't also be transitions at the frequency $2\omega_p$. It happens that there isn't any analog of this in the classical case, and also it doesn't happen in the quantum theory either—at least not for the particular method of inducing the transitions that we have described. With an oscillating horizontal magnetic field, the probability that a frequency $2\omega_p$ would cause a jump of two steps at once is zero. It is only at the frequency ω_p that transitions, either upward or downward, are likely to occur.

Now we are ready to describe Rabi's method for measuring magnetic moments. We will consider here only the operation for atoms with a spin of 1/2. A diagram of the apparatus is shown in Fig. 35–5. There is an oven which gives out a stream of neutral atoms which passes down a line of three magnets. Magnet 1

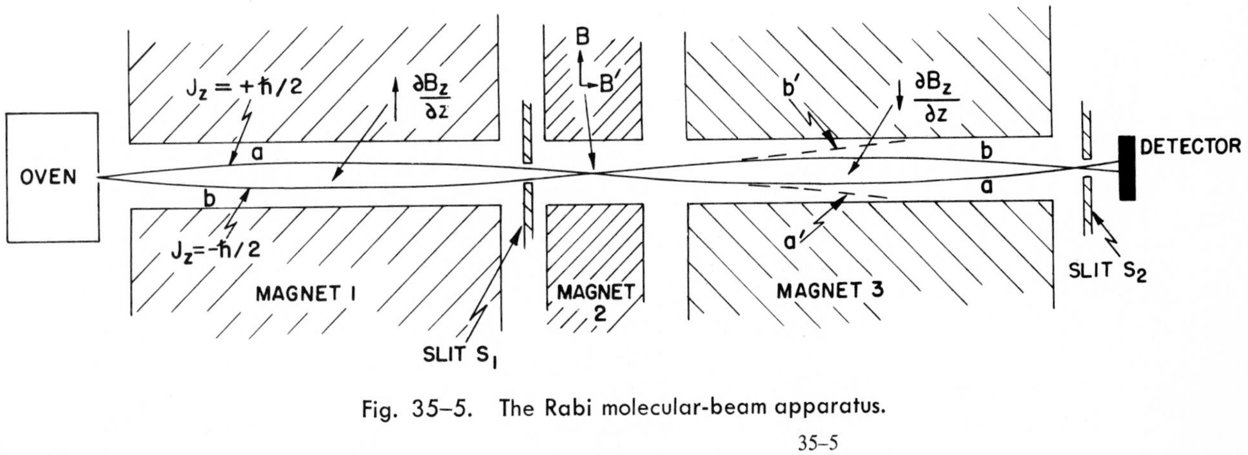

Fig. 35–5. The Rabi molecular-beam apparatus.

is just like the one in Fig. 35–2, and has a field with a strong field gradient—say, with $\partial B_z/\partial z$ positive. If the atoms have a magnetic moment, they will be deflected downward if $J_z = +\hbar/2$, or upward if $J_z = -\hbar/2$ (since for electrons $\boldsymbol{\mu}$ is directed opposite to $\boldsymbol{J}$). If we consider only those atoms which can get through the slit S_1, there are two possible trajectories, as shown. Atoms with $J_z = +\hbar/2$ must go along curve a to get through the slit, and those with $J_z = -\hbar/2$ must go along curve b. Atoms which start out from the oven along other paths will not get through the slit.

Magnet 2 has a uniform field. There are no forces on the atoms in this region, so they go straight through and enter magnet 3. Magnet 3 is just like magnet 1 but with the field *inverted*, so that $\partial B_z/\partial z$ has the opposite sign. The atoms with $J_z = +\hbar/2$ (we say "with spin up"), that felt a downward push in magnet 1, get an *upward* push in magnet 3; they continue on the path a and go through slit S_2 to a detector. The atoms with $J_z = -\hbar/2$ ("with spin down") also have opposite forces in magnets 1 and 3 and go along the path b, which also takes them through slit S_2 to the detector.

The detector may be made in various ways, depending on the atom being measured. For example, for atoms of an alkali metal like sodium, the detector can be a thin, hot tungsten wire connected to a sensitive current meter. When sodium atoms land on the wire, they are evaporated off as Na^+ ions, leaving an electron behind. There is a current from the wire proportional to the number of sodium atoms arriving per second.

In the gap of magnet 2 there is a set of coils that produces a small horizontal magnetic field $\boldsymbol{B}'$. The coils are driven with a current which oscillates at a variable frequency ω. So between the poles of magnet 2 there is a strong, constant, vertical field $\boldsymbol{B}_0$ and a weak, oscillating, horizontal field $\boldsymbol{B}'$.

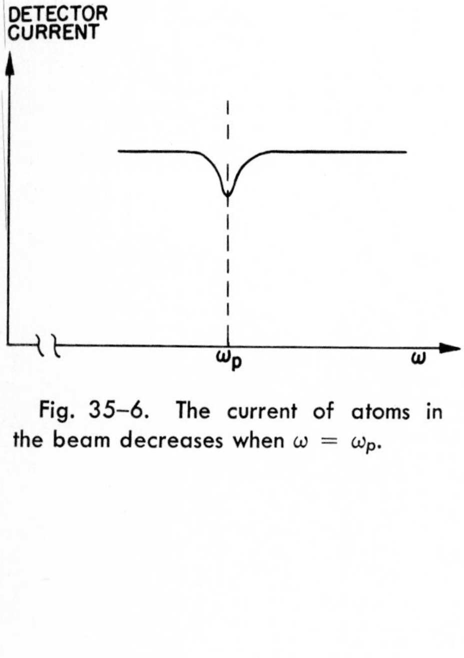

Fig. 35–6. The current of atoms in the beam decreases when $\omega = \omega_p$.

Suppose now that the frequency ω of the oscillating field is set at ω_p—the "precession" frequency of the atoms in the field $\boldsymbol{B}$. The alternating field will cause some of the atoms passing by to make transitions from one J_z to the other. An atom whose spin was initially "up" ($J_z = +\hbar/2$) may be flipped "down" ($J_z = -\hbar/2$). Now this atom has the direction of its magnetic moment reversed, so it will feel a *downward* force in magnet 3 and will move along the path a', shown in Fig. 35–5. It will no longer get through the slit S_2 to the detector. Similarly, some of the atoms whose spins were initially down ($J_z = -\hbar/2$) will have their spins flipped up ($J_z = +\hbar/2$) as they pass through magnet 2. They will then go along the path b' and will not get to the detector.

If the oscillating field $\boldsymbol{B}'$ has a frequency appreciably different from ω_p, it will not cause any spin flips, and the atoms will follow their undisturbed paths to the detector. So you can see that the "precession" frequency ω_p of the atoms in the field $\boldsymbol{B}_0$ can be found by varying the frequency ω of the field $\boldsymbol{B}'$ until a decrease is observed in the current of atoms arriving at the detector. A decrease in the current will occur when ω is "in resonance" with ω_p. A plot of the detector current as a function of ω might look like the one shown in Fig. 35–6. Knowing ω_p, we can obtain the g-value of the atom.

Such atomic-beam or, as they are usually called, "molecular" beam resonance experiments are a beautiful and delicate way of measuring the magnetic properties of atomic objects. The resonance frequency ω_p can be determined with great precision—in fact, with a greater precision than we can measure the magnetic field $\boldsymbol{B}_0$, which we must know to find g.

35–4 The paramagnetism of bulk materials

We would like now to describe the phenomenon of the paramagnetism of bulk materials. Suppose we have a substance whose atoms have permanent magnetic moments, for example a crystal like copper sulfate. In the crystal there are copper ions whose inner electron shells have a net angular momentum and a net magnetic moment. So the copper ion is an object which has a permanent magnetic moment. Let's say just a word about which atoms have magnetic moments and which ones don't. Any atom, like sodium for instance, which has an *odd* number of electrons, will have a magnetic moment. Sodium has one electron in its unfilled shell. This electron gives the atom a spin and a magnetic moment. Ordinarily, however, when compounds are formed the extra electrons in the outside shell are coupled together with other electrons whose spin directions are exactly opposite, so that all the angular momenta and magnetic moments of the valence electrons usually cancel out. That's why, in general, molecules do not have a magnetic moment. Of course if you have a gas of sodium atoms, there is no such cancellation.* Also, if you have what is called in chemistry a "free radical"—an object with an odd number of valence electrons—then the bonds are not completely satisfied, and there is a net angular momentum.

In most bulk materials there is a net magnetic moment only if there are atoms present whose *inner* electron shell is not filled. Then there can be a net angular momentum and a magnetic moment. Such atoms are found in the "transition element" part of the periodic table—for instance, chromium, manganese, iron, nickel, cobalt, palladium, and platinum are elements of this kind. Also, all of the rare earth elements have unfilled inner shells and permanent magnetic moments. There are a couple of other strange things that also happen to have magnetic moments, such as liquid oxygen, but we will leave it to the chemistry department to explain the reason.

Now suppose that we have a box full of atoms or molecules with permanent moments—say a gas, or a liquid, or a crystal. We would like to know what happens if we apply an external magnetic field. With *no* magnetic field, the atoms are kicked around by the thermal motions, and the moments wind up pointing in all directions. But when there is a magnetic field, it acts to line up the little magnets; then there are more moments lying toward the field than away from it. The material is "magnetized."

We define the *magnetization* $\boldsymbol{M}$ of a material as the net magnetic moment per unit volume, by which we mean the vector sum of all the atomic magnetic moments in a unit volume. If there are N atoms per unit volume and their *average* moment is $\langle\boldsymbol{\mu}\rangle_{\text{av}}$ then $\boldsymbol{M}$ can be written as N times the average atomic moment:

$$\boldsymbol{M} = N\langle\boldsymbol{\mu}\rangle_{\text{av}}. \tag{35.8}$$

The definition of $\boldsymbol{M}$ corresponds to the definition of the electric polarization $\boldsymbol{P}$ of Chapter 10.

The classical theory of paramagnetism is just like the theory of the dielectric constant we showed you in Chapter 11. One assumes that each of the atoms has a magnetic moment $\boldsymbol{\mu}$, which always has the same magnitude but which can point in any direction. In a field $\boldsymbol{B}$, the magnetic energy is $-\boldsymbol{\mu}\cdot\boldsymbol{B} = -\mu B\cos\theta$, where θ is the angle between the moment and the field. From statistical mechanics, the relative probability of having any angle is $e^{-\text{energy}/kT}$, so angles near zero are more likely than angles near π. Proceeding exactly as we did in Section 11–3, we find that for small magnetic fields $\boldsymbol{M}$ is directed parallel to $\boldsymbol{B}$ and has the magnitude

$$M = \frac{N\mu^2B}{3kT}. \tag{35.9}$$

[See Eq. (11.20).] This approximate formula is correct only for $\mu B/kT$ much less than one.

We find that the induced magnetization—the magnetic moment per unit volume—is proportional to the magnetic field. This is the phenomenon of paramagnetism. You will see that the effect is stronger at lower temperatures and weaker at higher temperatures. When we put a field on a substance, it develops, for small fields, a magnetic moment proportional to the field. The ratio of M to B (for small fields) is called the magnetic *susceptibility*.

Now we want to look at paramagnetism from the point of view of quantum mechanics. We take first the case of an atom with a spin of 1/2. In the absence of

* Ordinary Na vapor is mostly monatomic, although there are also some molecules of Na_2.

a magnetic field the atoms have a certain energy, but in a magnetic field there are two possible energies, one for each value of J_z. For $J_z = +\hbar/2$, the energy is changed by the magnetic field by the amount

$$\Delta U_1 = +g\left(\frac{q_e\hbar}{2m}\right)\cdot\frac{1}{2}\cdot B. \tag{35.10}$$

(The energy shift ΔU is positive for an atom because the electron charge is negative.) For $J_z = -\hbar/2$, the energy is changed by the amount

$$\Delta U_2 = -g\left(\frac{q_e\hbar}{2m}\right)\cdot\frac{1}{2}\cdot B. \tag{35.11}$$

To save writing, let's set

$$\mu_0 = g\left(\frac{q_e\hbar}{2m}\right)\cdot\frac{1}{2}; \tag{35.12}$$

then

$$\Delta U = \pm\mu_0 B. \tag{35.13}$$

The meaning of μ_0 is clear: $-\mu_0$ is the z-component of the magnetic moment in the up-spin case, and $+\mu_0$ is the z-component of the magnetic moment in the down-spin case.

Now statistical mechanics tells us that the probability that an atom is in one state or another is proportional to

$$e^{-(\text{Energy of state})/kT}.$$

With no magnetic field the two states have the same energy; so when there is equilibrium in a magnetic field, the probabilities are proportional to

$$e^{-\Delta U/kT}. \tag{35.14}$$

The number of atoms per unit volume with spin up is

$$N_{\text{up}} = ae^{-\mu_0 B/kt}, \tag{35.15}$$

and the number with spin down is

$$N_{\text{down}} = ae^{+\mu_0 B/kt}. \tag{35.16}$$

The constant a is to be determined so that

$$N_{\text{up}} + N_{\text{down}} = N, \tag{35.17}$$

the total number of atoms per unit volume. So we get that

$$a = \frac{N}{e^{+\mu_0 B/kT} + e^{-\mu_0 B/kT}}. \tag{35.18}$$

What we are interested in is the *average* magnetic moment along the z-axis. The atoms with spin up will contribute a moment of $-\mu_0$, and those with spin down will have a moment of $+\mu_0$; so the average moment is

$$\langle\mu\rangle_{\text{av}} = \frac{N_{\text{up}}(-\mu_0) + N_{\text{down}}(+\mu_0)}{N}. \tag{35.19}$$

The magnetic moment per unit volume M is then $N\langle\mu\rangle_{\text{av}}$. Using Eqs. (35.15), (35.16), and (35.17), we get that

$$M = N\mu_0\,\frac{e^{+\mu_0 B/kT} - e^{-\mu_0 B/kT}}{e^{+\mu_0 B/kT} + e^{-\mu_0 B/kT}}. \tag{35.20}$$

This is the quantum-mechanical formula for M for atoms with $j = 1/2$. Incidentally, this formula can also be written somewhat more concisely in terms of the hyperbolic tangent function:

$$M = N\mu_0 \tanh \frac{\mu_0 B}{kT}. \tag{35.21}$$

A plot of M as a function of B is given in Fig. 35.7. When B gets very large, the hyperbolic tangent approaches 1, and M approaches the limiting value $N\mu_0$. So at high fields, the magnetization *saturates*. We can see why that is; at high enough fields the moments are all lined up in the same direction. In other words, they are all in the spin-down state, and each atom contributes the moment μ_0.

In most normal cases—say, for typical moments, room temperatures, and the fields one can normally get (like 10,000 gauss)—the ratio $\mu_0 B/kT$ is about 0.02. One must go to very low temperatures to see the saturation. For normal temperatures, we can usually replace $\tanh x$ by x, and write

$$M = \frac{N\mu_0^2 B}{kT}. \tag{35.22}$$

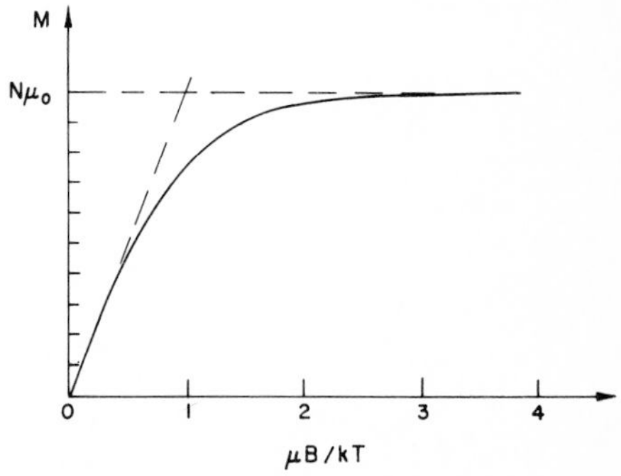

Fig. 35–7. The variation of the paramagnetic magnetization with the magnetic field strength B.

Just as we saw in the classical theory, M is proportional to B. In fact, the formula is almost exactly the same, except that there seems to be a factor of 1/3 missing. But we still need to relate the μ_0 in our quantum formula to the μ that appears in the classical result, Eq. (35.9).

In the classical formula, what appears is $\mu^2 = \boldsymbol{\mu}\cdot\boldsymbol{\mu}$, the square of the vector magnetic moment, or

$$\boldsymbol{\mu}\cdot\boldsymbol{\mu} = \left(g\frac{q_e}{2m}\right)^2 \boldsymbol{J}\cdot\boldsymbol{J}. \tag{35.23}$$

We pointed out in the last chapter that you can very likely get the right answer from a classical calculation by replacing $\boldsymbol{J}\cdot\boldsymbol{J}$ by $j(j+1)\hbar^2$. In our particular example, we have $j = 1/2$, so

$$j(j+1)\hbar^2 = \tfrac{3}{4}\hbar^2.$$

Substituting this for $\boldsymbol{J}\cdot\boldsymbol{J}$ in Eq. (35.23), we get

$$\boldsymbol{\mu}\cdot\boldsymbol{\mu} = \left(g\frac{q_e}{2m}\right)^2 \frac{3\hbar^2}{4},$$

or in terms of μ_0, defined in Eq. (35.12), we get

$$\boldsymbol{\mu}\cdot\boldsymbol{\mu} = 3\mu_0^2.$$

Substituting this for μ^2 in the classical formula, Eq. (35.9), does indeed reproduce the correct quantum formula, Eq. (35.22).

The quantum theory of paramagnetism is easily extended to atoms of any spin j. The low-field magnetization is

$$M = Ng^2 \frac{j(j+1)}{3} \frac{\mu_B^2 B}{kT}, \tag{35.24}$$

where

$$\mu_B = \frac{q_e \hbar}{2m} \tag{35.25}$$

is a combination of constants with the dimensions of a magnetic moment. Most atoms have moments of roughly this size. It is called the *Bohr magneton*. The spin magnetic moment of the electron is almost exactly one Bohr magneton.

35–5 Cooling by adiabatic demagnetization

There is a very interesting special application of paramagnetism. At very low temperatures it is possible to line up the atomic magnets in a strong field. It is then possible to get down to *extremely* low temperatures by a process called *adiabatic demagnetization*. We can take a paramagnetic salt (for example, one containing a number of rare-earth atoms like praseodynium-ammonium-nitrate), and start by cooling it down with liquid helium to one or two degrees absolute in a strong magnetic field. Then the factor $\mu B/kT$ is larger than 1—say more like 2 or 3. Most of the spins are lined up, and the magnetization is nearly saturated. Let's say, to make it easy, that the field is very powerful and the temperature is very low, so that nearly all the atoms are lined up. Then you isolate the salt thermally (say, by removing the liquid helium and leaving a good vacuum) and turn off the magnetic field. The temperature of the salt goes way down.

Now if you were to turn off the field *suddenly*, the jiggling and shaking of the atoms in the crystal lattice would gradually knock all the spins out of alignment. Some of them would be up and some down. But if there is no field (and disregarding the interactions between the atomic magnets, which will make only a slight error), it takes no energy to turn over the atomic magnets. They could randomize their spins without any energy change and, therefore, without any temperature change.

Suppose, however, that while the atomic magnets are being flipped over by the thermal motion there is still some magnetic field present. Then it requires some work to flip them over opposite to the field—*they must do work against the field.* This takes energy from the thermal motions and lowers the temperature. So if the strong magnetic field is not removed too rapidly, the temperature of the salt will decrease—it is cooled by the demagnetization. From the quantum-mechanical view, when the field is strong all the atoms are in the lowest state, because the odds against any being in the upper state are impossibly big. But as the field is lowered, it gets more and more likely that thermal fluctuations will knock an atom into the upper state. When that happens, the atom absorbs the energy $\Delta U = \mu_0 B$. So if the field is turned off slowly, the magnetic transitions can take energy out of the thermal vibrations of the crystal, cooling it off. It is possible in this way to go from a temperature of a few degrees absolute down to a temperature of a few thousandths of a degree.

Would you like to make something even colder than that? It turns out that Nature has provided a way. We have already mentioned that there are also magnetic moments for the atomic nuclei. Our formulas for paramagnetism work just as well for nuclei, except that the moments of nuclei are roughly a *thousand times smaller.* [They are of the order of magnitude of $q\hbar/2m_p$, where m_p is the *proton* mass, so they are smaller by the ratio of the masses of the electron and proton.] With such magnetic moments, even at a temperature of 2°K, the factor $\mu B/kT$ is only a few parts in a thousand. But if we use the paramagnetic demagnetization process to get down to a temperature of a few thousandths of a degree, $\mu B/kT$ becomes a number near 1—at these low temperatures we can begin to saturate the nuclear moments. That is good luck, because we can then use the adiabatic demagnetization of the *nuclear* magnetism to reach still lower temperatures. Thus it is possible to do two stages of magnetic cooling. First we use adiabatic demagnetization of paramagnetic ions to reach a few thousandths of a degree. Then we use the cold paramagnetic salt to cool some material which has a strong nuclear magnetism. Finally, when we remove the magnetic field from this material, its temperature will go down to within a *millionth* of a degree of absolute zero—if we have done everything very carefully.

35–6 Nuclear magnetic resonance

We have said that atomic paramagnetism is very small and that nuclear magnetism is even a thousand times smaller. Yet it is relatively easy to observe the nuclear magnetism by the phenomenon of "nuclear magnetic resonance." Suppose we take a substance like water, in which all of the electron spins are exactly balanced so that their net magnetic moment is zero. The molecules will still have a very, very tiny magnetic moment due to the nuclear magnetic moment of the hydrogen nuclei. Suppose we put a small sample of water in a magnetic field $\boldsymbol{B}$. Since the protons (of the hydrogen) have a spin of 1/2, they will have two possible energy states. If the water is in thermal equilibrium, there will be slightly more protons in the lower energy states—with their moments directed parallel to the field. There is a small net magnetic moment per unit volume. Since the proton moment is only about one-thousandth of an atomic moment, the magnetization which goes as μ^2—using Eq. (35.22)—is only about one-millionth as strong as typical atomic paramagnetism. (That's why we have to pick a material with no atomic magnetism.) If you work it out, the difference between the number of protons with spin up and with spin down is only one part in 10^8, so the effect is indeed very small! It can still be observed, however, in the following way.

Suppose we surround the water sample with a small coil that produces a small horizontal oscillating magnetic field. If this field oscillates at the frequency ω_p, it will induce transitions between the two energy states—just as we described for the Rabi experiment in Section 35–3. When a proton flips from an upper energy state to a lower one, it will give up the energy $\mu_z B$ which, as we have seen, is equal to $\hbar\omega_p$. If it flips from the lower energy state to the upper one, it will *absorb* the energy $\hbar\omega_p$ from the coil. Since there are slightly more protons in the lower state than in the upper one, there will be a net *absorption* of energy from the coil. Although the effect is very small, the slight energy absorption can be seen with a sensitive electronic amplifier.

Just as in the Rabi molecular-beam experiment, the energy absorption will be seen only when the oscillating field is in resonance, that is, when

$$\omega = \omega_p = g\left(\frac{q_e}{2m_p}\right) B.$$

It is often more convenient to search for the resonance by varying B while keeping ω fixed. The energy absorption will evidently appear when

$$B = \frac{2m_p}{g q_e}\,\omega.$$

A typical nuclear magnetic resonance apparatus is shown in Fig. 35–8. A high-frequency oscillator drives a small coil placed between the poles of a large electromagnet. Two small auxiliary coils around the pole tips are driven with a 60-cycle current so that the magnetic field is "wobbled" about its average value by a very small amount. As an example, say that the main current of the magnet is set to give a field of 5000 gauss, and the auxiliary coils produce a variation of ± 1 gauss about this value. If the oscillator is set at 21.2 megacycles per second, it will then be at the proton resonance each time the field sweeps through 5000 gauss [using Eq. (34.13) with $g = 5.58$ for the proton].

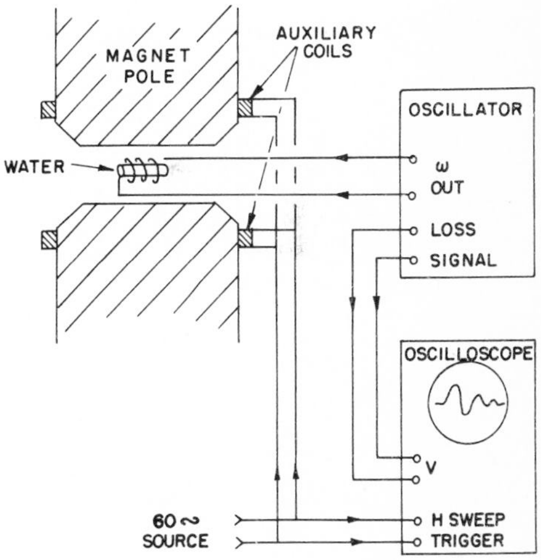

Fig. 35–8. A nuclear magnetic resonance apparatus.

The circuit of the oscillator is arranged to give an additional output signal proportional to any *change* in the power being absorbed from the oscillator. This signal is fed to the vertical deflection amplifier of an oscilloscope. The horizontal sweep of the oscilloscope is triggered once during each cycle of the field-wobbling frequency. (More usually, the horizontal deflection is made to follow in proportion to the wobbling field.)

Before the water sample is placed inside the high-frequency coil, the power drawn from the oscillator is some value. (It doesn't change with the magnetic field.) When a small bottle of water is placed in the coil, however, a signal appears on the oscilloscope, as shown in the figure. We see a picture of the power being absorbed by the flipping over of the protons!

In practice, it is difficult to know how to set the main magnet to exactly 5000 gauss. What one does is to adjust the main magnet current until the resonance signal appears on the oscilloscope. It turns out that this is now the most convenient way to make an accurate measurement of the strength of a magnetic field. Of course, at some time *someone* had to measure accurately the magnetic field and frequency to determine the g-value of the proton. But now that this has been done, a proton resonance apparatus like that of the figure can be used as a "proton resonance magnetometer."

We should say a word about the shape of the signal. If we were to wobble the magnetic field very slowly, we would expect to see a normal resonance curve. The energy absorption would read a maximum when ω_p arrived exactly at the oscillator frequency. There would be some absorption at nearby frequencies because all the protons are not in exactly the same field—and different fields mean slightly different resonant frequencies.

One might wonder, incidentally, whether at the resonance frequency we should see any signal at all. Shouldn't we expect the high-frequency field to equalize the populations of the two states—so that there should be no signal except when the water is first put in? Not exactly, because although we are *trying* to equalize the two populations, the thermal motions on their part are trying to keep the proper ratios for the temperature T. If we sit at the resonance, the power being absorbed by the nuclei is just what is being lost to the thermal motions. There is, however, relatively little "thermal contact" between the proton magnetic moments and the atomic motions. The protons are relatively isolated down in the center of the electron distributions. So in pure water, the resonance signal is, in fact, usually too small to be seen. To increase the absorption, it is necessary to increase the "thermal contact." This is usually done by adding a little iron oxide to the water. The iron atoms are like small magnets; as they jiggle around in their thermal dance, they make tiny jiggling magnetic fields at the protons. These varying fields "couple" the proton magnets to the atomic vibrations and tend to establish thermal equilibrium. It is through this "coupling" that protons in the higher energy states can lose their energy so that they are again capable of absorbing energy from the oscillator.

In practice the output signal of a nuclear resonance apparatus does not look like a normal resonance curve. It is usually a more complicated signal with oscillations—like the one drawn in the figure. Such signal shapes appear because of the changing fields. The explanation should be given in terms of quantum mechanics, but it can be shown that in such experiments the classical ideas of precessing moments always give the correct answer. Classically, we would say that when we arrive at resonance we start driving a lot of the precessing nuclear magnets synchronously. In so doing, we make them precess *together*. These nuclear magnets, all rotating together, will set up an induced emf in the oscillator coil at the frequency ω_p. But because the magnetic field is increasing with time, the precession frequency is increasing also, and the induced voltage is soon at a frequency a little higher than the oscillator frequency. As the induced emf goes alternately in phase and out of phase with the oscillator, the "absorbed" power goes alternately positive and negative. So on the oscilloscope we see the beat note between the proton frequency and the oscillator frequency. Because the proton frequencies are not all identical (different protons are in slightly different fields) and also possibly because of the disturbance from the iron oxide in the water, the freely precessing moments soon get out of phase, and the beat signal disappears.

These phenomena of magnetic resonance have been put to use in many ways as tools for finding out new things about matter—especially in chemistry and nuclear physics. It goes without saying that the numerical values of the magnetic moments of nuclei tell us something about their structure. In chemistry, much has been learned from the structure (or shape) of the resonances. Because of magnetic fields produced by nearby nuclei, the exact position of a nuclear resonance is shifted somewhat, depending on the environment in which any particular nucleus finds itself. Measuring these shifts helps determine which atoms are near which other ones and helps to elucidate the details of the structure of molecules. Equally important is the electron spin resonance of free radicals. Although not present to any very large extent in equilibrium, such radicals are often intermediate states of chemical reactions. A measurement of an electron spin resonance is a delicate test for the presence of free radicals and is often the key to understanding the mechanism of certain chemical reactions.

36

Ferromagnetism

Review: Chapter 10, *Dielectrics*
Chapter 17, *The Laws of Induction*

36–1 Magnetization currents

In this chapter we will discuss some materials in which the net effect of the magnetic moments in the material is much greater than in the case of paramagnetism or diamagnetism. The phenomenon is called *ferromagnetism.* In paramagnetic and diamagnetic materials the induced magnetic moments are usually so weak that we don't have to worry about the additional fields produced by the magnetic moments. For *ferromagnetic* materials, however, the magnetic moments induced by applied magnetic fields are quite enormous and have a great effect on the fields themselves. In fact, the induced moments are so strong that they are often the dominant effect in producing the observed fields. So one of the things we will have to worry about is the mathematical theory of large induced magnetic moments. That is, of course, just a technical question. The real problem is, why are the magnetic moments so strong—how does it all work? We will come to that question in a little while.

Finding the magnetic fields of ferromagnetic materials is something like the problem of finding the electrostatic field in the presence of dielectrics. You will remember that we first described the internal properties of a dielectric in terms of a vector field $\boldsymbol{P}$, the dipole moment per unit volume. Then we figured out that the effects of this polarization are equivalent to a charge density ρ_{pol} given by the divergence of $\boldsymbol{P}$:

$$\rho_{\text{pol}} = -\boldsymbol{\nabla}\cdot\boldsymbol{P}. \tag{36.1}$$

The total charge in any situation can be written as the sum of this polarization charge plus all other charges, whose density we write* ρ_{other}. Then the Maxwell equation which relates the divergence of $\boldsymbol{E}$ to the charge density becomes

$$\boldsymbol{\nabla}\cdot\boldsymbol{E} = \frac{\rho}{\epsilon_0} = \frac{\rho_{\text{pol}} + \rho_{\text{other}}}{\epsilon_0},$$

or

$$\boldsymbol{\nabla}\cdot\boldsymbol{E} = -\frac{\boldsymbol{\nabla}\cdot\boldsymbol{P}}{\epsilon_0} + \frac{\rho_{\text{other}}}{\epsilon_0}.$$

We can then pull out the polarization part of the charge and put it on the other side of the equation, to get the new law

$$\boldsymbol{\nabla}\cdot(\epsilon_0\boldsymbol{E} + \boldsymbol{P}) = \rho_{\text{other}}. \tag{36.2}$$

The new law says the divergence of the quantity $(\epsilon_0\boldsymbol{E} + \boldsymbol{P})$ is equal to the density of the other charges.

Pulling $\boldsymbol{E}$ and $\boldsymbol{P}$ together as in Eq. (36.2), of course, is useful only if we know some relation between them. We have seen that the theory which relates the induced electric dipole moment to the field was a relatively complicated business and can really only be applied to certain simple situations, and even then as an approximation. We would like to remind you of one of the approximate ideas we used. To find the induced dipole moment of an atom inside a dielectric, it is necessary to know the electric field that acts on an individual atom. We made the approximation—which is not too bad in many cases—that the field on the atom

* If all of the "other" charges were on conductors, ρ_{other} would be the same as our ρ_{free} of Chapter 10.

is the same as it would be at the center of the small hole which would be left if we took out the atom (keeping the dipole moments of all the neighboring atoms the same). You will also remember that the electric field in a hole in a polarized dielectric depends on the shape of the hole. We summarize our earlier results in Fig. 36–1. For a thin, disc-shaped hole perpendicular to the polarization, the electric field in the hole is given by

$$\boldsymbol{E}_{\text{hole}} = \boldsymbol{E}_{\text{dielectric}} + \frac{\boldsymbol{P}}{\epsilon_0},$$

which we showed by using Gauss' law. On the other hand, in a needle-shaped slot parallel to the polarization, we showed—by using the fact that the curl of $\boldsymbol{E}$ is zero—that the electric fields inside and outside of the slot are the same. Finally, we found that for a spherical hole the electric field was one-third of the way between the field of the slot and the field of the disc:

$$\boldsymbol{E}_{\text{hole}} = \boldsymbol{E}_{\text{dielectric}} + \frac{1}{3}\frac{\boldsymbol{P}}{\epsilon_0} \text{ (spherical hole).} \tag{36.3}$$

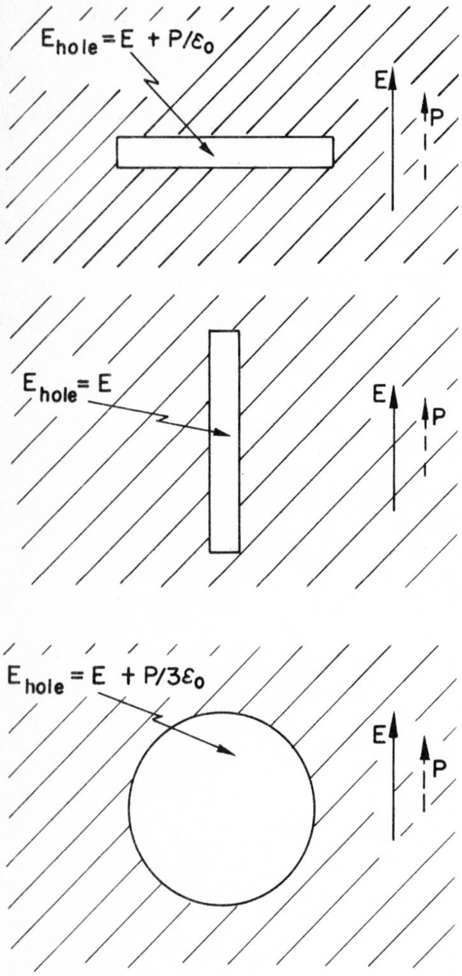

Fig. 36–1. The electric field in a cavity in a dielectric depends on the shape of the cavity.

This was the field we used in thinking about what happens to an atom inside a polarized dielectric.

Now we have to discuss the analog of all this for the case of magnetism. One simple, short-cut way of doing this is to say the $\boldsymbol{M}$, the magnetic moment per unit volume, is just like $\boldsymbol{P}$, the electric dipole moment per unit volume, and that, therefore, the negative of the divergence of $\boldsymbol{M}$ is equivalent to a "magnetic charge density" ρ_m—whatever that may mean. The trouble is, of course, that there isn't any such thing as a "magnetic charge" in the physical world. As we know, the divergence of $\boldsymbol{B}$ is always zero. But that does not stop us from making an artificial *analog* and writing

$$\nabla \cdot \boldsymbol{M} = -\rho_m, \tag{36.4}$$

where it is to be understood that ρ_m is purely mathematical. Then we could make a complete analogy with the electrostatic case and use all our old equations from electrostatics. People have often done something like that. In fact, historically, people even believed that the analogy was right. They believed that the quantity ρ_m represented the density of "magnetic poles." These days, however, we know that the magnetization of materials comes from circulating currents within the atoms—either from the spinning electrons or from the motion of the electrons in the atom. It is therefore nicer from a physical point of view to describe things realistically in terms of the atomic currents, rather than in terms of a density of some mythical "magnetic poles." Incidentally, these currents are sometimes called "Amperian" currents, because Ampere first suggested that the magnetism of matter came from circulating atomic currents.

The actual microscopic current density in magnetized matter is, of course, very complicated. Its value depends on where you look in the atom—it's large in some places and small in others; it goes one way in one part of the atom and the opposite way in another part (just as the microscopic electric field varies enormously inside a dielectric). In many practical problems, however, we are interested only in the fields outside of the matter or in the *average* magnetic field inside of the matter—where we mean an average taken over many, many atoms. It is only for such *macroscopic* problems that it is convenient to describe the magnetic state of the matter in terms of $\boldsymbol{M}$, the average dipole moment per unit volume. What we want to show now is that the atomic currents of magnetized matter can give rise to certain large-scale currents which are related to $\boldsymbol{M}$.

What we are going to do, then, is to separate the current density $\boldsymbol{j}$—which is the real source of the magnetic fields—into various parts: one part to describe the circulating currents of the atomic magnets, and the other parts to describe what other currents there may be. It is usually most convenient to separate the currents into three parts. In Chapter 32 we made a distinction between the currents which flow freely on conductors and the ones which are due to the back and forth motions of the bound charges in dielectrics. In Section 32–2 we wrote

$$\boldsymbol{j} = \boldsymbol{j}_{\text{pol}} + \boldsymbol{j}_{\text{other}},$$

where $\boldsymbol{j}_{\text{pol}}$ represented the currents from the motion of the bound charges in dielectrics and $\boldsymbol{j}_{\text{other}}$ took care of all other currents. Now we want to go further. We want to separate $\boldsymbol{j}_{\text{other}}$ into one part, $\boldsymbol{j}_{\text{mag}}$, which describes the average currents inside of magnetized materials, and an additional term which we can call $\boldsymbol{j}_{\text{cond}}$ for whatever is left over. The last term will generally refer to currents in conductors, but it may also include other currents—for example the currents from charges moving freely through empty space. So we will write for the total current density:

$$\boldsymbol{j} = \boldsymbol{j}_{\text{pol}} + \boldsymbol{j}_{\text{mag}} + \boldsymbol{j}_{\text{cond}}. \tag{36.5}$$

Of course it is this total current which belongs in the Maxwell equation for the curl of B:

$$c^2\nabla \times \boldsymbol{B} = \frac{\boldsymbol{j}}{\epsilon_0} + \frac{\partial \boldsymbol{E}}{\partial t}. \tag{36.6}$$

Now we have to relate the current $\boldsymbol{j}_{\text{mag}}$ to the magnetization vector $\boldsymbol{M}$. So that you can see where we are going, we will tell you that the result is going to be that

$$\boldsymbol{j}_{\text{mag}} = \nabla \times \boldsymbol{M}. \tag{36.7}$$

If we are given the magnetization vector $\boldsymbol{M}$ everywhere in a magnetic material, the circulation current density is given by the curl of $\boldsymbol{M}$. Let's see if we can understand why this is so.

First, let's take the case of a cylindrical rod which has a uniform magnetization parallel to its axis. Physically, we know that such a uniform magnetization really means a uniform density of atomic circulating currents everywhere inside the material. Suppose we try to imagine what the actual currents would look like in a cross section of the material. We would expect to see currents something like those shown in Fig. 36–2. Each atomic current goes around and around in a little circle, with all the circulating currents going around in the same direction. Now what is the effective current of such a thing? Well, in most of the bar there is no effect at all, because right next to each current there is another current going in the opposite direction. If we imagine a small surface—but one still quite a bit larger than a single atom—such as is indicated in Fig. 36–2 by the line $\overline{AB}$, the net current through such a surface is zero. There is no net current anywhere inside the material. Note, however, that at the surface of the material there are atomic currents which are not cancelled by neighboring currents going the other way. At the surface there is a net current always going in the same direction around the rod. Now you see why we said earlier that a uniformly magnetized rod is equivalent to a long solenoid carrying an electric current.

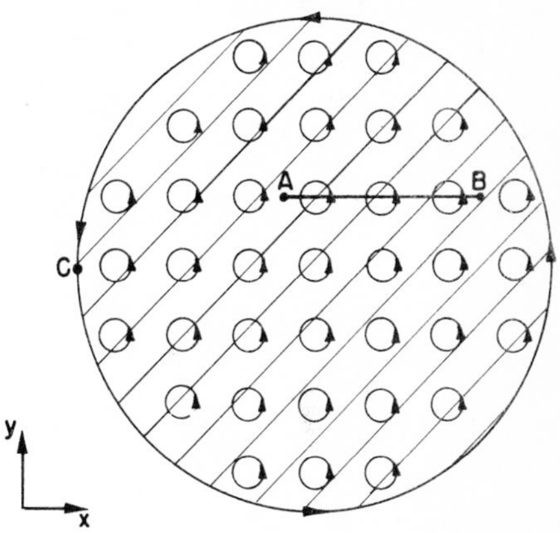

Fig. 36–2. Schematic diagram of the circulating atomic currents as seen in a cross section of an iron rod magnetized in the z-direction.

How does this view fit with Eq. (36.7)? First, inside the material the magnetization $\boldsymbol{M}$ is constant, so all its derivatives are zero. This agrees with our geometric picture. At the surface, however, $\boldsymbol{M}$ is not really constant—it is constant up to the edge and then suddenly collapses to zero. So, right at the surface there are terrific gradients which, according to (36.7), will give a high current density. Suppose we look at what happens near the point C in Fig. 36–2. Taking the x- and y-directions as in the figure, the magnetization $\boldsymbol{M}$ is in the z-direction. Writing out the components of Eq. (36.7), we have

$$\frac{\partial M_z}{\partial y} = (j_{\text{mag}})_x, \qquad -\frac{\partial M_z}{\partial x} = (j_{\text{mag}})_y. \tag{36.8}$$

At the point C, the derivative $\partial M_z/\partial y$ is zero, but $\partial M_z/\partial x$ is large and positive. Equation (36.7) says that there is a large current density in the minus y-direction. This agrees with our picture of a surface current going around the bar.

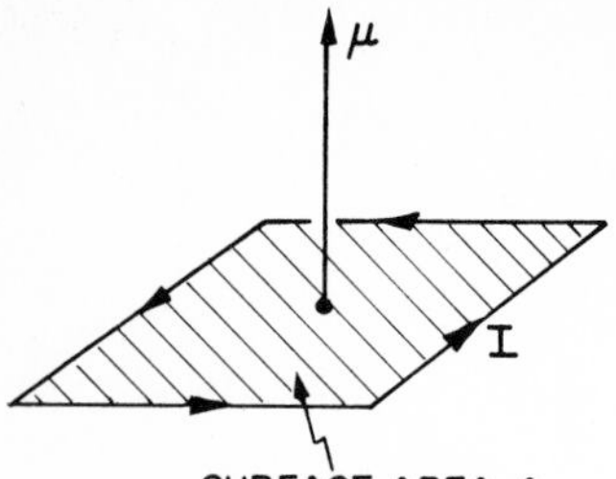

Fig. 36–3. The dipole moment μ of a current loop is IA.

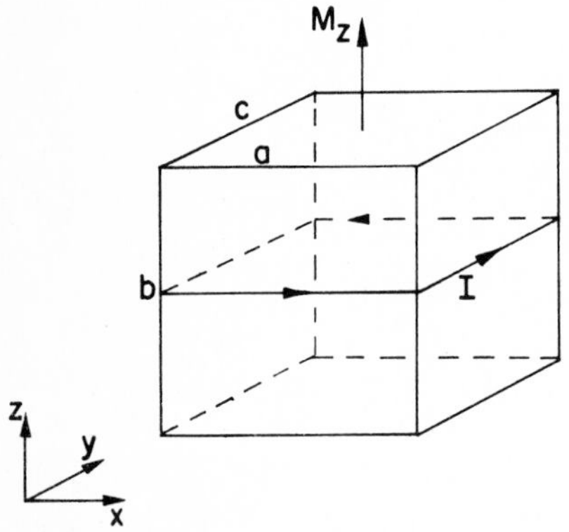

Fig. 36–4. A small magnetized block is equivalent to a circulating surface current.

Now we want to find the current density for a more complicated case in which the magnetization varies from point to point in a material. It is easy to see qualitatively that if the magnetization is different in two neighboring regions, there will not be a perfect cancellation of the circulating currents so that there will be a net current in the volume of the material. It is this effect that we want to work out quantitatively.

First, we need to recall the results of Section 14–5 that a circulating current I has a magnetic moment μ given by

$$\mu = IA, \tag{36.9}$$

where A is the area of the current loop (see Fig. 36–3). Now let's consider a small rectangular block inside of a magnetized material, as sketched in Fig. 36–4. We take the block so small that we can consider that the magnetization is uniform inside it. If this block has a magnetization M_z in the z-direction, the net effect will be the same as a surface current going around on the vertical faces, as shown. We can find the magnitude of these currents from Eq. (36.9). The total magnetic moment of the block is equal to the magnetization times the volume:

$$\mu = M_z(abc),$$

from which we get (remembering that the area of the loop is ac)

$$I = M_z b.$$

In other words, the current per unit length (vertically) on each of the vertical surfaces is equal to M_z.

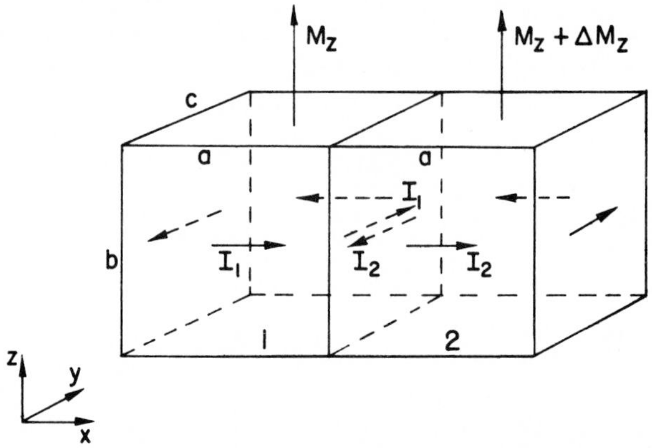

Fig. 36–5. If the magnetization of two neighboring blocks is not the same, there is a net surface current in between.

Now suppose that we imagine two such little blocks next to each other, as shown in Fig. 36–5. Because block 2 is slightly displaced from block 1, it will have a slightly different vertical component of magnetization, which we call $M_z + \Delta M_z$. Now on the surface between the two blocks there will be two contributions to the total current. Block 1 will produce a current I_1 flowing in the positive y-direction, and block 2 will produce a surface current I_2 flowing in the negative y-direction. The total surface current in the positive y-direction is the sum:

$$I = I_1 - I_2 = M_z b - (M_z + \Delta M_z)b$$
$$= -\Delta M_z b.$$

We can write ΔM_z as the derivative of M_z in the x-direction times the displacement from block 1 to block 2, which is just a:

$$\Delta M_z = \frac{\partial M_z}{\partial x}\, a.$$

The current flowing between the two blocks is then

$$I = -\frac{\partial M_z}{\partial x}\, ab.$$

To relate the current I to an average volume current density $\boldsymbol{j}$, we must realize that this current I is really spread over a certain cross-sectional area. If we imagine the whole volume of the material to be filled with such little blocks, one such side face (perpendicular to the x-axis) can be associated with each block.* Then we see that the area to be associated with the current I is just the area ab of one of the front faces. We get the result

$$j_y = \frac{I}{ab} = -\frac{\partial M_z}{\partial x}.$$

We have at least the beginning of the curl of $\boldsymbol{M}$.

There should be another term in j_y from the variation of the x-component of the magnetization with z. This contribution to $\boldsymbol{j}$ will come from the surface between two little blocks stacked one on top of the other, as shown in Fig. 36–6. Using the same arguments we have just made, you can show that this surface will contribute to j_y the amount $\partial M_x/\partial z$. These are the only surfaces which can contribute to the y-component of the current so we have that the total current density in the y-direction is

$$j_y = \frac{\partial M_x}{\partial z} - \frac{\partial M_z}{\partial x}.$$

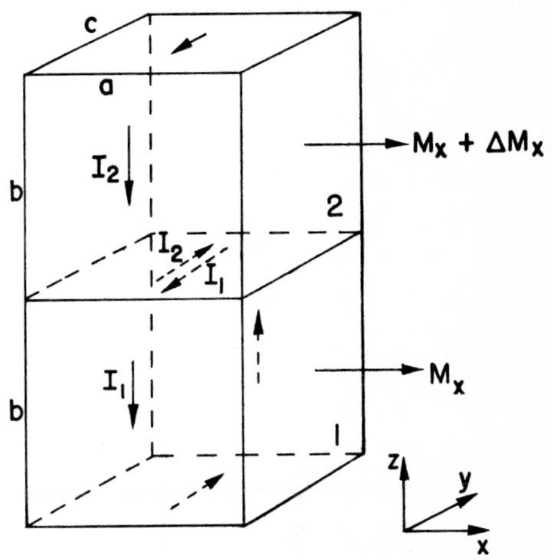

Fig. 36–6. Two blocks, one above the other, may also contribute to j_y.

Working out the currents on the remaining faces of a cube—or using the fact that our z-direction is completely arbitrary—we can conclude that the vector current density is indeed given by the equation

$$\boldsymbol{j} = \boldsymbol{\nabla} \times \boldsymbol{M}.$$

So if we choose to describe the magnetic situation in matter in terms of the average magnetic moment per unit volume $\boldsymbol{M}$, we find that the circulating atomic currents are equivalent to an average current density in matter given by Eq. (36.7). If the material is also a dielectric, there may be, in addition, a polarization current $\boldsymbol{j}_{\text{pol}} = \partial \boldsymbol{P}/\partial t$. And if the material is also a conductor, we may have a conduction current $\boldsymbol{j}_{\text{cond}}$ as well. We can write the total current as

$$\boldsymbol{j} = \boldsymbol{j}_{\text{cond}} + \boldsymbol{\nabla} \times \boldsymbol{M} + \frac{\partial \boldsymbol{P}}{\partial t}. \tag{36.10}$$

36–2 The field $\boldsymbol{H}$

Next, we want to insert the current as written in Eq. (36.10) into Maxwell's equations. We get

$$c^2 \boldsymbol{\nabla} \times \boldsymbol{B} = \frac{\boldsymbol{j}}{\epsilon_0} + \frac{\partial \boldsymbol{E}}{\partial t} = \frac{1}{\epsilon_0}\left(\boldsymbol{j}_{\text{cond}} + \boldsymbol{\nabla} \times \boldsymbol{M} + \frac{\partial \boldsymbol{P}}{\partial t}\right) + \frac{\partial \boldsymbol{E}}{\partial t}.$$

We can move the term in $\boldsymbol{M}$ to the left-hand side:

$$c^2 \boldsymbol{\nabla} \times \left(\boldsymbol{B} - \frac{\boldsymbol{M}}{\epsilon_0 c^2}\right) = \frac{\boldsymbol{j}_{\text{cond}}}{\epsilon_0} + \frac{\partial}{\partial t}\left(\boldsymbol{E} + \frac{\boldsymbol{P}}{\epsilon_0}\right). \tag{36.11}$$

As we remarked in Chapter 32, many people like to write $(\boldsymbol{E} + \boldsymbol{P}/\epsilon_0)$ as a new vector field $\boldsymbol{D}/\epsilon_0$. Similarly, it is often convenient to write $(\boldsymbol{B} - \boldsymbol{M}/\epsilon_0 c^2)$ as a single vector field. *We* choose to define a new vector field $\boldsymbol{H}$ by

$$\boldsymbol{H} = \boldsymbol{B} - \frac{\boldsymbol{M}}{\epsilon_0 c^2}. \tag{36.12}$$

Then Eq. (36.11) becomes

$$\epsilon_0 c^2 \boldsymbol{\nabla} \times \boldsymbol{H} = \boldsymbol{j}_{\text{cond}} + \frac{\partial \boldsymbol{D}}{\partial t}. \tag{36.13}$$

It looks simple, but all the complexity is just hidden in the letters $\boldsymbol{D}$ and $\boldsymbol{H}$.

---

* Or, if you prefer, the current I in each face should be split 50–50 with the blocks on the two sides.

Now we have to give you a warning. Most people who use the mks units have chosen to use a different definition of $\boldsymbol{H}$. Calling *their* field $\boldsymbol{H}'$ (of course, they still call it $\boldsymbol{H}$ without the prime), it is defined by

$$\boldsymbol{H}' = \epsilon_0 c^2 \boldsymbol{B} - \boldsymbol{M}. \tag{36.14}$$

(Also, they usually write $\epsilon_0 c^2$ as a new number $1/\mu_0$; then they have one more constant to keep track of!) With this definition, Eq. (36.13) looks even simpler:

$$\nabla \times \boldsymbol{H}' = \boldsymbol{j}_{\text{cond}} + \frac{\partial \boldsymbol{D}}{\partial t}. \tag{36.15}$$

But the difficulties with this definition of $\boldsymbol{H}'$ are, first, that it doesn't agree with the definition of people who don't use the mks units, and second, that it makes $\boldsymbol{H}'$ and $\boldsymbol{B}$ have different units. We think it is more convenient for $\boldsymbol{H}$ to have the same units as $\boldsymbol{B}$—rather than the units of $\boldsymbol{M}$, as $\boldsymbol{H}'$ does. But if you are going to be an engineer and work on the design of transformers, magnets, and such, you will have to watch out. You will find many books which use for $\boldsymbol{H}$ the definition of Eq. (36.14) rather than our definition of Eq. (36.12), and many other books—especially handbooks about magnetic materials—that relate $\boldsymbol{B}$ and $\boldsymbol{H}$ the way we have done. You'll have to be careful to figure out which convention they are using.

One way to tell is by the units they use. Remember that in the mks system, $\boldsymbol{B}$—and therefore *our* $\boldsymbol{H}$—are measured with the unit: one weber per square meter, equal to 10,000 gauss. In the mks system, a magnetic moment (a current times an area) has the unit: one ampere-meter². The magnetization $\boldsymbol{M}$, then, has the unit: one ampere *per* meter. For $\boldsymbol{H}'$ the units are the same as for $\boldsymbol{M}$. You can see that this also agrees with Eq. (36.15), since ∇ has the dimensions of one over a length. People who are working with electromagnets also get in the habit of calling the unit of $\boldsymbol{H}$ (with the $\boldsymbol{H}'$ definition) "one ampere *turn* per meter"—thinking of the turns of wire on a winding. But a "turn" is really a dimensionless number, so that doesn't need to confuse you. Since our H is equal to $H'/\epsilon_0 c^2$, if you are using the mks system, H (in webers/meter²) is equal to $4\pi \times 10^{-7}$ times H' (in amperes per meter). It is perhaps more convenient to remember that H (in gauss) = $0.0126\, H'$ (in amp/meter).

There is one more horrible thing. Many people who use *our* definition of $\boldsymbol{H}$ have decided to call the units of $\boldsymbol{H}$ and $\boldsymbol{B}$ by *different names*! Even though they have the same dimensions, they call the unit of $\boldsymbol{B}$ one *gauss*, and the unit of $\boldsymbol{H}$ one *oersted* (after Gauss and Oersted, of course). So, in many books you will find graphs with $\boldsymbol{B}$ plotted in gauss and $\boldsymbol{H}$ in oersteds. They are really the same unit—10^{-4} of the mks unit. We have summarized the confusion about magnetic units in Table 36–1.

Table 36–1

Units of magnetic quantities

$[B]$	=	weber/meter² = 10^4 gauss
$[H]$	=	weber/meter² = 10^4 gauss *or* 10^4 oersted
$[M]$	=	ampere/meter
$[H']$	=	ampere/meter

Convenient conversions

B (gauss)	=	$10^4\, B$ (weber/meter²)
H (gauss)	=	H (oersted)
	=	$0.0126\, H'$ (amp/meter)

36–3 The magnetization curve

Now we will look at some simple situations in which the magnetic field is constant, or in which the fields change slowly enough that we can neglect $\partial \boldsymbol{D}/\partial t$ in comparison with $\boldsymbol{j}_{\text{cond}}$. Then the fields obey the equations

$$\nabla \cdot \boldsymbol{B} = 0, \tag{36.16}$$

$$\nabla \times \boldsymbol{H} = \boldsymbol{j}_{\text{cond}}/\epsilon_0 c^2, \tag{36.17}$$

$$\boldsymbol{H} = \boldsymbol{B} - \boldsymbol{M}/\epsilon_0 c^2. \tag{36.18}$$

Suppose we have a torus (a donut) of iron wrapped with a coil of copper wire, as shown in Fig. 36–7(a). A current I flows in the wire. What is the magnetic field? The magnetic field will be mainly inside the iron; there, the lines of $\boldsymbol{B}$ will be circles, as drawn in Fig. 36–7(b). Since the flux of $\boldsymbol{B}$ is continuous, its divergence is zero, and Eq. (36.16) is satisfied. Next, we write Eq. (36.17) in another form by integrating around the closed loop Γ drawn in Fig. 36–7(b). From Stokes's theorem, we have that

$$\oint_\Gamma \boldsymbol{H}\cdot d\boldsymbol{s} = \frac{1}{\epsilon_0 c^2}\int_S \boldsymbol{j}_{\text{cond}}\cdot \boldsymbol{n}\,da, \tag{36.19}$$

where the integral of $\boldsymbol{j}$ is to be carried out over any surface S bounded by Γ. This surface is cut once by each turn of the winding. Each turn contributes the current I to the integral, and, if there are N turns in all, the integral is NI. From the symmetry of our problem, $\boldsymbol{B}$ is the same all around the curve Γ; if we assume that the magnetization, and therefore, the field H is also constant along Γ, Eq. (36.19) becomes

$$Hl = \frac{NI}{\epsilon_0 c^2},$$

where l is the length of the curve Γ. So,

$$H = \frac{1}{\epsilon_0 c^2}\frac{NI}{l}. \tag{36.20}$$

It is because $\boldsymbol{H}$ is directly proportional to the magnetizing current in cases like this one that $\boldsymbol{H}$ is sometimes called the *magnetizing field*.

Now all we need is an equation which relates $\boldsymbol{H}$ to $\boldsymbol{B}$. But there isn't any such equation! There is, of course, Eq. (36.18), but it is no help because there is no direct relation between $\boldsymbol{M}$ and $\boldsymbol{B}$ for a ferromagnetic material like iron. The magnetization $\boldsymbol{M}$ depends on the whole past history of the iron, and not only on what $\boldsymbol{B}$ is at the moment.

All is not lost, though. We can get solutions in certain simple cases. If we start out with unmagnetized iron—let's say with iron that has been annealed at high temperatures—then in the simple geometry of the torus, all the iron will have the same magnetic history. Then we can say something about $\boldsymbol{M}$—and therefore about the relation between $\boldsymbol{B}$ and $\boldsymbol{H}$—from experimental measurements. The field $\boldsymbol{B}$ in the torus is, from Eq. (36.20), given as a constant times the current I in the winding. The field $\boldsymbol{B}$ can be measured by integrating over time the emf in the coil (or in an extra coil wound over the magnetizing coil shown in the figure). This emf is equal to the rate of change of the flux of $\boldsymbol{B}$, so the integral of the emf with time is equal to $\boldsymbol{B}$ times the cross-sectional area of the torus.

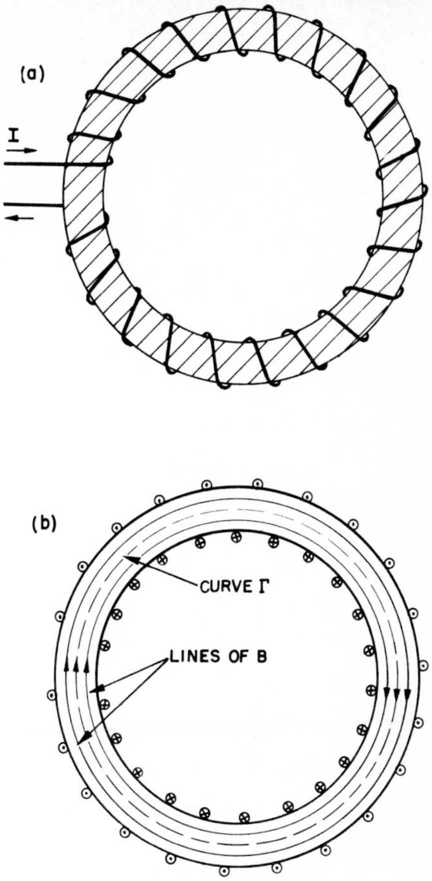

Fig. 36–7. (a) A torus of iron wound with a coil of insulated wire. (b) Cross section of torus showing field lines.

Figure 36–8 shows the relation between $\boldsymbol{B}$ and $\boldsymbol{H}$, observed with a torus of soft iron. When the current is first turned on, $\boldsymbol{B}$ increases with increasing $\boldsymbol{H}$ along the curve a. Note the different scales on $\boldsymbol{B}$ and $\boldsymbol{H}$; initially, it takes only a relatively small $\boldsymbol{H}$ to make a large $\boldsymbol{B}$. Why is $\boldsymbol{B}$ so much larger with the iron than it would be with air? Because there is a large magnetization $\boldsymbol{M}$ which is equivalent to a large surface current on the iron—the field $\boldsymbol{B}$ comes from the *sum* of this current and the conduction current in the winding. Why $\boldsymbol{M}$ should be so large, we will discuss later.

At higher values of $\boldsymbol{H}$, the magnetization curve levels off. We say that the iron *saturates*. With the scales of our figure, the curve appears to become horizontal. Actually, it continues to rise slightly—for large fields, $\boldsymbol{B}$ becomes proportional to $\boldsymbol{H}$, and with a unit slope. There is no further increase of $\boldsymbol{M}$. Incidentally, we should point out that if the torus were made of some nonmagnetic material, $\boldsymbol{M}$ would be zero and $\boldsymbol{B}$ would equal $\boldsymbol{H}$ for all fields.

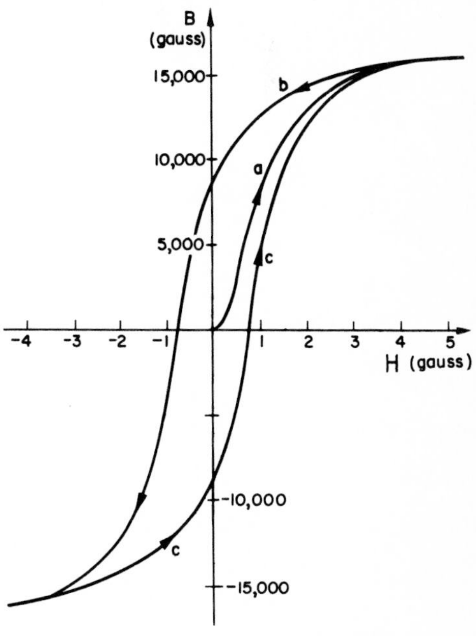

Fig. 36–8. Typical magnetization and hysteresis curves for soft iron.

The first thing we notice is that curve a in Fig. 36–8—which is the so-called *magnetization curve*—is highly nonlinear. But it's worse than that. If, after reaching saturation, we decrease the current in the coil to bring $\boldsymbol{H}$ back to zero, the magnetic field $\boldsymbol{B}$ falls along curve b. When $\boldsymbol{H}$ reaches zero, there is still some $\boldsymbol{B}$ left. Even with no magnetizing current there is a magnetic field in the iron—it has become permanently magnetized. If we now turn on a *negative* current in the coil, the B-H curve continues along b until the iron is saturated in the negative direction. If we then bring the current back to zero again, $\boldsymbol{B}$ goes along curve c. If we alternate the current between large positive and negative values, the B-H curve goes back and forth along very nearly the curves b and c. If we vary $\boldsymbol{H}$ in some arbitrary way, however, we can get more complicated curves which will, in general, lie somewhere between the curves b and c. The loop made by repeated oscillation of the fields is called a *hysteresis* loop of the iron.

We see then that we cannot write a functional relationship like $B = f(H)$, because the value of $\boldsymbol{B}$ at any instant depends not only on what $\boldsymbol{H}$ is at that time, but on its whole past history. Naturally, the magnetization and hysteresis curves are different for different substances. The shape of the curves depends critically on the chemical composition of the material, and also on the details of its preparation and subsequent physical treatment. We will discuss some of the physical explanations for these complications in the next chapter.

36–4 Iron-core inductances

One of the most important applications of magnetic materials is in electrical circuits—for example, in transformers, electric motors, and so on. One reason is that with iron we can control where the magnetic fields go, and also get much larger fields for a given electric current. For example, the typical "toroidal" inductance is made very much like the object shown in Fig. 36–7. For a given inductance, it can be much smaller in volume and use much less copper than an equivalent "air-core" inductance. For a given inductance, we get a much smaller resistance in the winding, so the inductance is more nearly "ideal"—particularly for low frequencies. It is very easy to understand, qualitatively, how such an inductance works. If I is the current in the winding, then the field $\boldsymbol{H}$ which is produced in the inside is proportional to I—as given by Eq. (36.20). The voltage $\mathcal{V}$ across the terminals is related to the magnetic field $\boldsymbol{B}$. Neglecting the resistance of the winding, the voltage $\mathcal{V}$ is proportional to $\partial\boldsymbol{B}/\partial t$. The inductance $\mathfrak{L}$, which is the ratio of $\mathcal{V}$ to dI/dt (see Section 17–7), thus involves the relation between B and H in the iron. Since the $\boldsymbol{B}$ is so much bigger than the $\boldsymbol{H}$, we get a large factor in the inductance. Physically, what happens is that a small current in the coil, which would ordinarily produce a small magnetic field, causes the little "slave" magnets in the iron to line up and produce a tremendously greater "magnetic" current than the external current in the winding. It is as if we had a lot more current going through the coil than we really have. When we reverse the current, all the little magnets flip over—all those internal currents reverse—and we get a much higher induced emf than we would get without the iron. If we want to calculate the inductance, we can do so through the energy—as described in Section 17–8. The *rate* at which energy is delivered from the current source is $I\mathcal{V}$. The voltage $\mathcal{V}$ is the cross-sectional area A of the core, times N, times dB/dt. From Eq. (36.20), $I = (\epsilon_0 c^2 l/N)H$. So we have

$$\frac{dU}{dt} = \mathcal{V}I = (\epsilon_0 c^2 lA)H\,\frac{dB}{dt}\cdot$$

Integrating over time, we have

$$U = (\epsilon_0 c^2 lA)\int H\,dB. \tag{36.21}$$

Notice that lA is the volume of the torus, so we have shown that the energy density $u = U/\text{vol}$ in a magnetic material is given by

$$u = \epsilon_0 c^2 \int H\,dB. \tag{36.22}$$

An interesting feature is involved here. When we use alternating currents, the iron is driven around a hysteresis loop. Since B is not a single-valued function of H, the integral of $\int H\,dB$ around one complete cycle is *not* equal to zero. It is the area enclosed inside the hysteresis curve. Thus, the driving source delivers a certain net energy each cycle—an energy proportional to the area inside the hysteresis loop. And that energy is "lost." It is lost from the electromagnetic goings on, but turns up as heat in the iron. It is called the *hysteresis loss*. To keep such energy losses small, we would like the hysteresis loop to be as narrow as possible. One way to decrease the area of the loop is to reduce the maximum field that is reached during each cycle. For smaller maximum fields, we get a hysteresis curve like the one shown in Fig. 36–9. Also, special materials are designed to have a very narrow loop. The so-called *transformer irons*—which are iron alloys with a small amount of silicon—have been developed to have this property.

When an inductance is run over a small hysteresis loop, the relationship between B and H can be approximated by a linear equation. People usually write

$$B = \mu H. \tag{36.23}$$

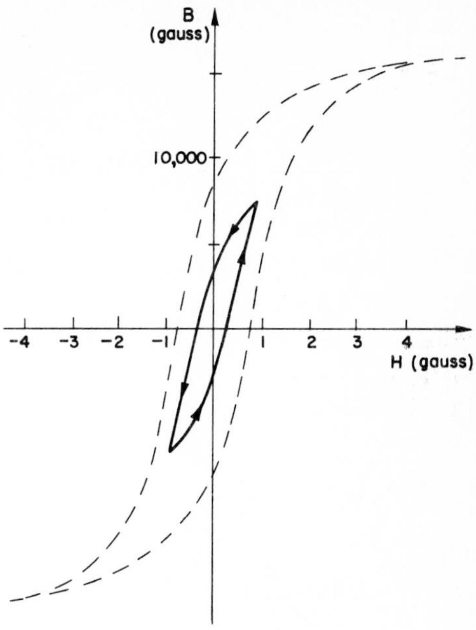

Fig. 36–9. A hysteresis loop that doesn't reach saturation.

The constant μ is *not* the magnetic moment we have used before. It is called the *permeability* of the iron. (It is also sometimes called the "*relative permeability.*") The permeability of ordinary irons is typically several thousand. There are special alloys alike "supermalloy" which can have permeabilities as high as a million.

If we use the approximation that $B = \mu H$ in Eq. (36.21), we can write the energy in a toroidal inductance as

$$U = (\epsilon_0 c^2 lA)\mu \int H\,dH = (\epsilon_0 c^2 lA)\,\frac{\mu H^2}{2}. \tag{36.24}$$

So the energy density is approximately

$$u \approx \frac{\epsilon_0 c^2}{2}\,\mu H^2.$$

We can now set the energy of Eq. (36.24) equal to the energy $\mathcal{L}I^2/2$ of an inductance, and solve for $\mathcal{L}$. We get

$$\mathcal{L} = (\epsilon_0 c^2 lA)\mu\left(\frac{H}{I}\right)^2.$$

Using H/I from Eq. (36.20), we have

$$\mathcal{L} = \frac{\mu N^2 A}{\epsilon_0 c^2 l}. \tag{36.25}$$

The inductance is proportional to μ. If you want inductances for such things as audio amplifiers, you will try to operate them on a hysteresis loop where the *B-H* relationship is as linear as possible. (You will remember that we spoke in Chapter 50, Vol. I, about the generation of harmonics in nonlinear systems.) For such purposes, Eq. (36.23) is a useful approximation. On the other hand, if you *want* to generate harmonics, you may use an inductance which is intentionally operated in a highly nonlinear way. Then you will have to use the complete *B-H* curves, and analyze what happens by graphical or numerical methods.

A "transformer" is often made by putting two coils on the same torus—or *core*—of a magnetic material. (For the larger transformers, the core is made with rectangular proportions for convenience.) Then a varying current in the "primary" winding causes the magnetic field in the core to change, which induces an emf in the "secondary" winding. Since the flux through *each turn* of both windings is the same, the emf's in the two windings are in the same ratio as the number of turns on each. A voltage applied to the primary is transformed to a different voltage at the secondary. Since a certain *net* current around the core is needed to produce the required change in the magnetic field, the *algebraic* sum of the currents in the two windings will be fixed and equal to the required "magnetizing" current. If the current drawn from the secondary increases, the primary current must increase in proportion—there is a "transformation" of currents as well as voltage.

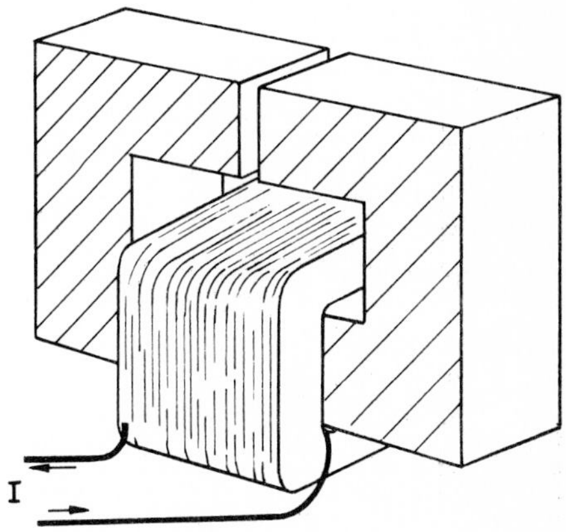

Fig. 36–10. An electromagnet.

36–5 Electromagnets

Now let's discuss a practical situation which is a little more complicated. Suppose we have an electromagnet of the rather standard form shown in Fig. 36–10—there is a "C-shaped" yoke of iron, with a coil of many turns of wire wrapped around the yoke. What is the magnetic field $\boldsymbol{B}$ in the gap?

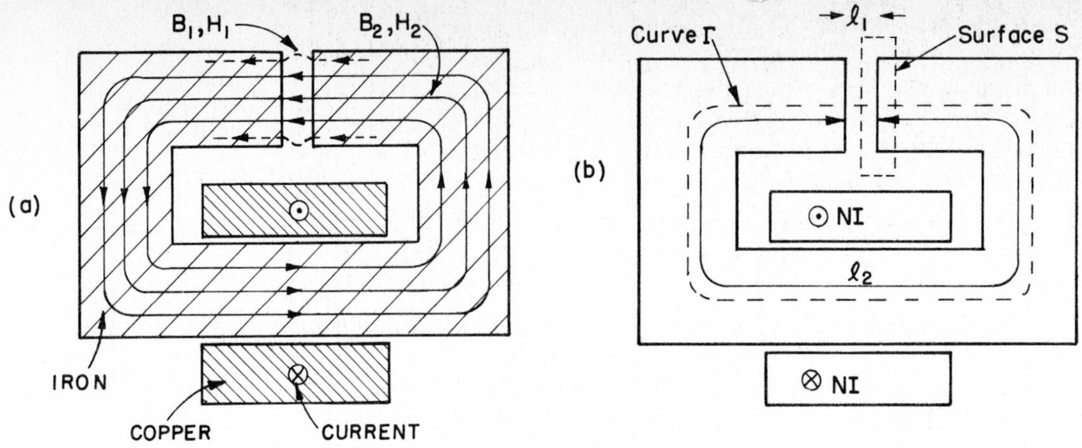

Fig. 36–11. Cross section of an electromagnet.

If the gap thickness is small compared with all the other dimensions, we can, as a first approximation, assume that the lines of $\boldsymbol{B}$ will go around through the loop, just as they did in the torus. They will look more or less as shown in Fig. 36–11(a). They tend to spread out somewhat in the gap, but if the gap is narrow, this will be a small effect. It is a fair approximation to assume that the flux of $\boldsymbol{B}$ through any cross section of the yoke is a constant. If the yoke has a uniform cross-sectional area—and if we neglect any edge effects at the gaps or at the corners—we can say that $\boldsymbol{B}$ is uniform around the yoke.

Also, $\boldsymbol{B}$ will have the same value in the gap. This follows from Eq. (36.16). Imagine the closed surface S, shown in Fig. 36–11(b), which has one face in the gap and the other in the iron. The total flux of $\boldsymbol{B}$ out of this surface must be zero. Calling B_1 the field in the gap and B_2 the field in the iron, we have that

$$B_1A_1 - B_2A_2 = 0.$$

Since $A_1 = A_2$ (to our approximation), it follows that $B_1 = B_2$.

Now let's look at H. We can again use Eq. (36.19), taking the line integral around the curve Γ in Fig. 36–11(b). As before, the right-hand side is NI, the number of turns times the current. Now, however, H will be different in the iron and in the air. Calling H_2 the field in the iron and l_2 the path length around the yoke, this part of the curve will contribute the amount H_2l_2 to the integral. Calling H_1 the field in the gap and l_1 the gap thickness, we get the contribution H_1l_1 from the gap. We have that

$$H_1l_1 + H_2l_2 = \frac{NI}{\epsilon_0c^2}. \tag{36.26}$$

Now we know something else: that in the air gap, the magnetization is negligible, so that $B_1 = H_1$. Since $B_1 = B_2$, Eq. (36.26) becomes

$$B_2l_1 + H_2l_2 = \frac{NI}{\epsilon_0c^2}. \tag{36.27}$$

We still have two unknowns. To find B_2 and H_2, we need another relationship—namely, the one which relates B to H in the iron.

If we can make the approximation that $B_2 = \mu H_2$, we can solve the equation algebraically. However, let's do the general case, in which the magnetization curve of the iron is one like that shown in Fig. 36–8. What we want is the simultaneous solution of this functional relationship together with Eq. (36.27). We can find it by plotting a graph of Eq. (36.27) on the same graph with the magnetization curve, as is done in Fig. 36–12. Where the two curves intersect, we have our solution.

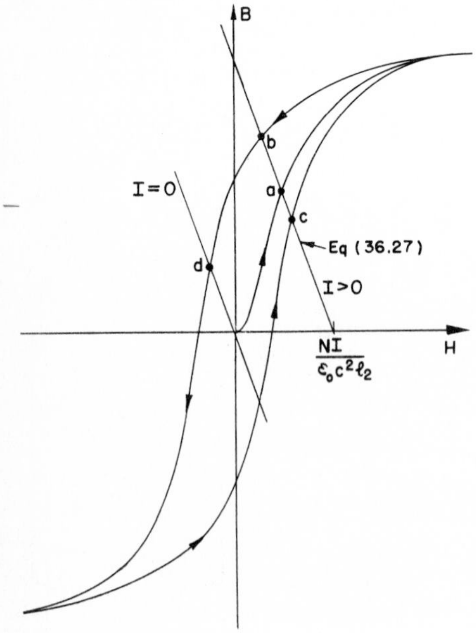

Fig. 36–12. Solving for the field in an electromagnet.

For a given current I, the function (36.27) is the straight line marked $I > 0$ in Fig. 36–12. The line intersects the H-axis ($B_2 = 0$) at $H_2 = NI/\epsilon_0c^2l_2$, and the slope is $-l_2/l_1$. Different currents just shift the line horizontally. From Fig.

36–12, we see that for a given current there are several different solutions, depending on how you got there. If you have just built the magnet and turned the current up to I, the field B_2 (which is also B_1) will have the value given by point a. If you have run the current to some very high value and come down to I, the field will be given by point b. Or, if you have just had a high negative current in the magnet and then come *up* to I, the field is the one at point c. The field in the gap will depend on what you have done in the past.

When the current in the magnet is zero, the relation between B_2 and H_2 in Eq. (36.27) is shown by the line marked $I = 0$ in the figure. There are still various possible solutions. If you have first saturated the iron, there may be a considerable residual field in the magnet as given by point d. You can take the coil off, and you have a permanent magnet. You can see that for a good permanent magnet, you would want a material with a *wide* hysteresis loop. Special alloys, such as Alnico V, have very wide loops.

36–6 Spontaneous magnetization

We now turn to the question of why it is that in ferromagnetic materials a small magnetic field produces such a large magnetization. The magnetization of ferromagnetic materials like iron and nickel comes from the magnetic moment of the electrons in the inner shell of the atom. Each electron has a magnetic moment $\boldsymbol{\mu}$ equal to $q/2m$ times its g-factor, times its angular momentum $\boldsymbol{J}$. For a single electron with no net orbital motion, $g = 2$, and the component of $\boldsymbol{J}$ in any direction—say the z-direction—is $\pm\hbar/2$, so the component of μ along the z-axis is

$$\mu_z = \frac{q\hbar}{2m} = 0.928 \times 10^{-23}\ \text{amp·m}^2. \qquad (36.28)$$

In an iron atom, there are actually two electrons that contribute to the ferromagnetism, so to keep the discussion simpler we will talk about nickel, which is ferromagnetic like iron but which has only one electron in the inner shell. (It is easy to extend the arguments to iron.)

Now the point is that in the presence of an external field $\boldsymbol{B}$, the atomic magnets tend to line up with the field, but are knocked about by thermal motions just as we described for paramagnetic materials. In the last chapter we found out that the balance between a magnetic field trying to line up the atomic magnets and the thermal motions trying to derange them produced the result that the mean magnetic moment per unit volume will end up as

$$M = N\mu \tanh \frac{\mu B_a}{kT}. \qquad (36.29)$$

By $\boldsymbol{B}_a$ we mean the field acting at the atom, and kT is the Boltzmann energy. In the theory of paramagnetism we used for B_a just B itself, neglecting the part of the field at any given atom contributed by the atoms nearby. In the ferromagnetic case, there is a complication. We shouldn't use the average field in the iron for the $\boldsymbol{B}_a$ acting on an individual atom. Instead, we must do as we did in the case of dielectrics—we have to find the *local* field acting at a single atom. For an exact calculation we should add up the fields at the atom in question contributed by all of the other atoms in the crystal lattice. But as we did for dielectrics, we will make the approximation that the field at an atom is the same as we would find in a small spherical hole in the material—assuming that the moments of the atoms in the neighborhood are not changed by the presence of the hole.

Following the arguments we made in Chapter 11, we might think that we could write

$$\boldsymbol{B}_{\text{hole}} = \boldsymbol{B} + \frac{1}{3}\frac{\boldsymbol{M}}{\epsilon_0 c^2} \qquad \text{(wrong!)}.$$

But that is not right. We *can*, however, make use of the results of Chapter 11 if we make a careful comparison of the equations of Chapter 11 with the equations for ferromagnetism in this chapter. Let's put together the corresponding equations. For regions where there are no conduction currents or charges we have:

Electrostatics	Static ferromagnetism
$\nabla\cdot\left(E+\dfrac{P}{\epsilon_0}\right)=0$	$\nabla\cdot B=0$
$\nabla\times E=0$	$\nabla\times\left(B-\dfrac{M}{\epsilon_0c^2}\right)=0$

(36.30)

These two sets of equations can be thought of as analogous if we make the following *purely mathematical* correspondences:

$$E\to B-\frac{M}{\epsilon_0c^2},\qquad E+\frac{P}{\epsilon_0}\to B.$$

This is the same as making the analogy

$$E\to H,\qquad P\to M/c^2. \tag{36.31}$$

In other words, if we write the equations of ferromagnetism as

$$\begin{gathered}\nabla\cdot\left(H+\frac{M}{\epsilon_0c^2}\right)=0,\\ \nabla\times H=0,\end{gathered} \tag{36.32}$$

they *look like* the equations of electrostatics.

This purely algebraic correspondence has led to some confusion in the past. People tended to think that H was "*the* magnetic field." But, as we have seen, B and E are physically the fundamental fields, and H is a derived idea. So although the *equations* are analogous, the *physics* is not analogous. However, that doesn't need to stop us from using the principle that the same equations have the same solutions.

We can use our earlier results for the electric field inside of holes of various shapes in dielectrics—summarized in Fig. 36–1—to find the field H inside of corresponding holes. Knowing H, we can determine B. For instance (using the results we summarized in Section 1), the field H in a needle-shaped hole parallel to M is the same as the H in the material,

$$H_{\text{hole}}=H_{\text{material}}.$$

But since M in the hole is zero, we have

$$B_{\text{hole}}=B_{\text{material}}-\frac{M}{\epsilon_0c^2}. \tag{36.33}$$

On the other hand, for a disc-shaped hole, perpendicular to M, we have

$$E_{\text{hole}}=E_{\text{dielectric}}+\frac{P}{\epsilon_0},$$

which translates into

$$H_{\text{hole}}=H_{\text{material}}+\frac{M}{\epsilon_0c^2}.$$

Or, in terms of B,

$$B_{\text{hole}}=B_{\text{material}}. \tag{36.34}$$

Finally, for a spherical hole, by making our analogy with Eq. (36.3) we would have

$$H_{\text{hole}}=H_{\text{material}}+\frac{M}{3\epsilon_0c^2}$$

or

$$B_{\text{hole}}=B_{\text{material}}-\frac{2}{3}\frac{M}{\epsilon_0c^2}. \tag{36.35}$$

This result is quite different from what we got for E.

It is, of course, possible to get these results in a more physical way, by using the Maxwell equations directly. For example, Eq. (36.34) follows directly from $\nabla \cdot \boldsymbol{B} = 0$. (You use a gaussian surface that is half in the material and half out.) Similarly, you can get Eq. (36.33) by using a line integral along a curve that goes up inside the hole and returns through the material. Physically, the field in the hole is reduced because of the surface currents—which are given by $\nabla \times \boldsymbol{M}$. We will leave it for you to show that Eq. (36.35) can also be obtained by considering the effects of the surface currents on the boundary of the spherical cavity.

In finding the equilibrium magnetization from Eq. (36.29), it turns out to be most convenient to deal with $\boldsymbol{H}$; so write

$$\boldsymbol{B}_{\text{a}} = \boldsymbol{H} + \lambda \frac{\boldsymbol{M}}{\epsilon_0 c^2}. \tag{36.36}$$

In the spherical hole approximation, we would have $\lambda = \frac{1}{3}$, but, as you will see, we will want later to use some other value, so we leave it as an adjustable parameter. Also, we will take all the fields in the same direction so that we won't need to worry about the vector directions. If we were now to substitute Eq. (36.36) into Eq. (36.29), we would have one equation that relates the magnetization M to the magnetizing field H:

$$M = N\mu \tanh\left(\frac{H + \lambda M/\epsilon_0 c^2}{kT}\right).$$

It is, however, an equation that cannot be solved explicitly, so we will do it graphically.

Let's put the problem in a generalized form by writing Eq. (36.29) as

$$\frac{M}{M_{\text{sat}}} = \tanh x, \tag{36.37}$$

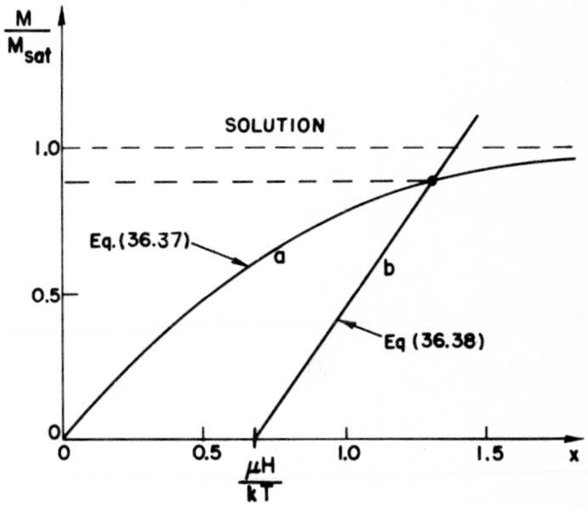

Fig. 36–13. A graphical solution of Eqs. (36.37) and (36.38).

where M_{sat} is the saturation value of the magnetization, namely, $N\mu$, and x represents $\mu B_a/kT$. The dependence of M/M_{sat} on x is shown by curve a in Fig. 36–13. We can also write x as a function of M—using Eq. (36.36) for B_a—as

$$x = \frac{\mu B_a}{kT} = \frac{\mu H}{kT} + \left(\frac{\mu \lambda M_{\text{sat}}}{\epsilon_0 c^2 kT}\right) \frac{M}{M_{\text{sat}}}. \tag{36.38}$$

For any given value of H, this is a straight-line relationship between M/M_{sat} and x. The x intercept is at $x = \mu H/kT$, and the slope is $\epsilon_0 c^2 kT/\mu \lambda M_{\text{sat}}$. For any particular H, we would have a line like the one marked b in Fig. 36–13. The intersection of curves a and b gives us the solution for M/M_{sat}. We have solved the problem.

Let's look at how the solutions will go for various circumstances. We start with $H = 0$. There are two possible situations, shown by the lines b_1 and b_2 in Fig. 36–14. You will notice from Eq. (36.38) that the slope of the line is proportional to the absolute temperature T. So, at *high temperatures* we would have a line like b_1. The solution is $M/M_{\text{sat}} = 0$. When the magnetizing field H is zero, the magnetization is also zero. But at *low temperatures*, we would have a line like b_2, and there are *two solutions* for M/M_{sat}—one with $M/M_{\text{sat}} = 0$ and one with M/M_{sat} near one. It turns out that only the upper solution is stable—as you can see by considering small variations about these solutions.

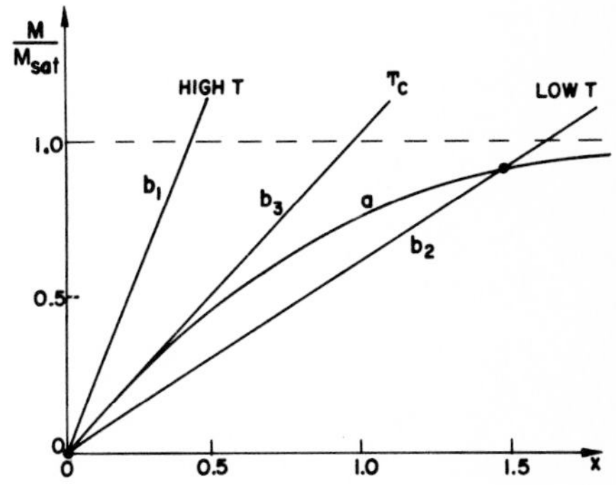

Fig. 36–14. Finding the magnetization when $H = 0$.

According to these ideas, then, a magnetic material should magnetize itself *spontaneously* at sufficiently low temperatures. In short, when the thermal motions are small enough, the coupling between the atomic magnets causes them all to line up parallel to each other—we have a permanently magnetized material analogous to the ferroelectrics we discussed in Chapter 11.

If we start at high temperatures and come down, there is a critical temperature, called the Curie temperature T_c, where the ferromagnetic behavior suddenly sets in. This temperature corresponds to the line b_3 of Fig. 36–14, which is tangent to the curve a, and has, therefore, a slope of 1. The Curie temperature is given by

$$\frac{\epsilon_0 c^2 k T_c}{\mu \lambda M_{\text{sat}}} = 1. \tag{36.39}$$

We can, if we wish, write Eq. (36.38) more simply in terms of T_c as

$$x = \frac{\mu H}{kT} + \frac{T_c}{T}\left(\frac{M}{M_{\text{sat}}}\right). \tag{36.40}$$

Now we want to see what happens for small magnetizing fields H. We can see from Fig. 36–14 how things will go if we shift our straight lines a little to the right. For the low-temperature case, the intersection point will move out a little bit along the low-slope part of curve a, and M will change relatively little. For the high-temperature case, however, the intersection point runs up the steep part of curve a, and M will change relatively rapidly. In fact, we can approximate this part of curve a by a straight line of unit slope, and write:

$$\frac{M}{M_{\text{sat}}} = x = \frac{\mu H}{kT} + \frac{T_c}{T}\left(\frac{M}{M_{\text{sat}}}\right).$$

Now we can solve for M/M_{sat}:

$$\frac{M}{M_{\text{sat}}} = \frac{\mu H}{k(T - T_c)}. \tag{36.41}$$

We have a law that is something like the one we had for paramagnetism. For paramagnetism, we had

$$\frac{M}{M_{\text{sat}}} = \frac{\mu B}{kT}. \tag{36.42}$$

One difference now is that we have the magnetization in terms of H, which includes some of the effects of the interaction of the atomic magnets, but the main difference is that the magnetization is inversely proportional to the *difference* between T and T_c, instead of to the absolute temperate T, alone. Neglecting the interactions between neighboring atoms corresponds to taking $\lambda = 0$, which from Eq. (36.39) means taking $T_c = 0$. Then the results are just what we had in Chapter 35.

We can check our theoretical picture with the experimental data for nickel. It is observed experimentally that the ferromagnetic behavior of nickel disappears when its temperature is raised above 631°K. We can compare this with T_c calculated from Eq. (36.39). Remembering that $M_{\text{sat}} = \mu N$, we have

$$T_c = \lambda\,\frac{N\mu^2}{k\epsilon_0 c^2}.$$

From the density and atomic weight of nickel, we get

$$N = 9.1 \times 10^{28}\ \text{m}^{-3}.$$

Taking μ from Eq. (36.28), and setting $\lambda = \frac{1}{3}$, we get

$$T_c = 0.24°\text{K}.$$

There is a discrepancy of a factor of about 2600! Our theory of ferromagnetism fails completely.

We can try to "patch up" the theory as Weiss did by saying that for some unknown reason λ is not one-third, but $(2600) \times \frac{1}{3}$—or about 900. It turns out that one gets similar values for other ferromagnetic materials like iron. To see what this means, let's go back to Eq. (36.36). We see that a large λ means that B_a, the local field on the atom, appears to be much, much larger than we would think. In fact, writing $H = B - M/\epsilon_0 c^2$, we have

$$B_a = B + \frac{(\lambda - 1)M}{\epsilon_0 c^2}.$$

According to our original idea—with $\lambda = \frac{1}{3}$—the local magnetization M *reduces* the effective field B_a by the amount $-\frac{2}{3}M/\epsilon_0$. Even if our model of a spherical hole were not very good, we would still expect *some* reduction. Instead, to explain the phenomenon of ferromagnetism, we have to imagine that the magnetization of the field *enhances* the local field by some large factor—like one thousand or more. There doesn't seem to be any reasonable way to manufacture such tremendous fields at an atom—nor even fields of the proper sign! Clearly, our "magnetic" theory of ferromagnetism is a dismal failure. We must conclude, then, that ferromagnetism has to do with some *nonmagnetic* interaction between the spinning electrons in neighboring atoms. This interaction must generate a strong tendency for all of the nearby spins to line up in one direction. We will see later that it has to do with quantum mechanics and the Pauli exclusion principle.

Finally, we look at what happens at low temperatures—for $T < T_c$. We have seen that there will then be a spontaneous magnetization—even with $H = 0$—given by the intersection of the curves a and b_2 of Fig. 36–14. If we solve for M for various temperatures—by varying the slope of the line b_2—we get the theoretical curve shown in Fig. 36–15. This curve should be the same for all ferromagnetic materials for which the atomic moment comes from a single electron. The curves for other materials are only slightly different.

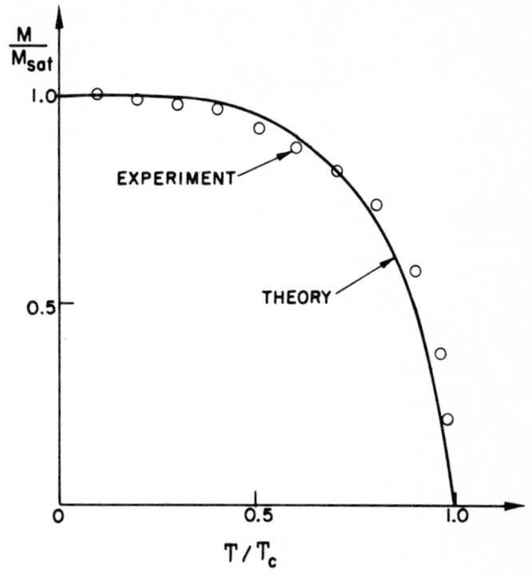

Fig. 36–15. Spontaneous magnetization as a function of temperature for nickel.

In the limit, as T goes to absolute zero, M goes to M_{sat}. As the temperature is increased, the magnetization decreases, falling to zero at the Curie temperature. The points in Fig. 36–15 are the experimental observations for nickel. They fit the theoretical curve fairly well. Even though we don't understand the basic mechanism, the general features of the theory seem to be correct.

Finally, there is one more disturbing discrepancy in our attempt to understand ferromagnetism. We have found that above some temperature the material should behave like a paramagnetic substance with a magnetization M proportional to H (or B), and that below that temperature it should become spontaneously magnetized. But that's not what we found when we measured the magnetization curve for iron. It only became permanently magnetized *after* we had "magnetized" it. According to the ideas just discussed, it would magnetize itself! What is wrong? Well, it turns out that if you look at a *small enough crystal* of iron or nickel, it is indeed completely magnetized! But in large pieces of iron, there are many small regions or "domains" that are magnetized in different directions, so that on a large scale the *average* magnetization appears to be zero. In each small domain, however, the iron has a locked-in magnetization with M nearly equal to M_{sat}. The consequences of this domain structure are that gross properties of large pieces of material are quite different from the microscopic properties that we have really been treating. We will take up in the next lecture the story of the practical behavior of bulk magnetic materials.

37

Magnetic Materials

References: Bozorth, R. M., "Magnetism," *Encyclopaedia Britannica*, Vol. 14, 1957, pp. 636–667.

Kittel, C., *Introduction to Solid State Physics*, John Wiley and Sons, Inc., New York, 2nd ed., 1956.

37–1 Understanding ferromagnetism

In this chapter we will discuss the behavior and peculiarities of ferromagnetic materials and of other strange magnetic materials. Before proceeding to study magnetic materials, however, we will review very quickly some of the things about the general theory of magnets that we learned in the last chapter.

First, we imagine the atomic currents inside the material that are responsible for the magnetism, and then describe them in terms of a volume current density $\boldsymbol{j}_{\text{mag}} = \boldsymbol{\nabla} \times \boldsymbol{M}$. We emphasize that this is not supposed to represent the *actual* currents. When the magnetization is uniform the currents do not *really* cancel out precisely; that is, the whirling currents of one electron in one atom and the whirling currents of an electron in another atom do not overlap in such a way that the sum is exactly zero. Even within a single atom the distribution of magnetism is *not* smooth. For instance, in an iron atom the magnetization is distributed in a more or less spherical shell, not too close to the nucleus and not too far away. Thus, magnetism in matter is quite a complicated thing in its details; it is very irregular. However, we are obliged now to ignore this detailed complexity and discuss phenomena from a gross, average point of view. Then it is true that the *average* current in the interior region, over any finite area that is big compared with an atom, is zero when $\boldsymbol{M} = 0$. So, what we mean by magnetization per unit volume and $\boldsymbol{j}_{\text{mag}}$ and so on, at the level we are now considering, is an average over regions that are large compared with the space occupied by a single atom.

In the last chapter, we also discovered that a ferromagnetic material has the following interesting property: above a certain temperature it is not strongly magnetic, whereas below this temperature it becomes magnetic. This fact is easily demonstrated. A piece of nickel wire at room temperature is attracted by a magnet. However, if we heat it above its Curie temperature with a gas flame, it becomes nonmagnetic and is not attracted toward the magnet—even when brought quite close to the magnet. If we let it lie near the magnet while it cools off, at the instant its temperature falls below the critical temperature it is suddenly attracted again by the magnet!

The general theory of ferromagnetism that we will use supposes that the spin of the electron is responsible for the magnetization. The electron has spin one-half and carries one Bohr magneton of magnetic moment $\mu = \mu_B = q_e\hbar/2m$. The electron spin can be pointed either "up" or "down." Because the electron has a negative charge, when its spin is "up" it has a *negative* moment, and when its spin is "down" it has a *positive* moment. With our usual conventions, the moment $\boldsymbol{\mu}$ of the electron is opposite its spin. We have found that the energy of orientation of a magnetic dipole in a given applied field $\boldsymbol{B}$ is $-\boldsymbol{\mu}\cdot\boldsymbol{B}$, but the energy of the spinning electrons depends on the neighboring spin alignments as well. In iron, if the moment of a nearby atom is "up," there is a very strong tendency that the moment of the one next to it will also be "up." That is what makes iron, cobalt, and nickel so strongly magnetic—the moments all want to be parallel. The first question we have to discuss is *why.*

Soon after the development of quantum mechanics, it was noticed that there is a very strong *apparent* force—not a magnetic force or any other kind of actual force, but only an apparent force—trying to line the spins of nearby electrons *opposite* to one another. These forces are closely related to chemical valence forces. There is a principle in quantum mechanics—called the *exclusion principle*—that two electrons cannot occupy exactly the same state, that they cannot be in exactly the same condition as to location and spin orientation.* For example, if they are at the same point, the only alternative is to have their spins opposite. So, if there is a region of space between atoms where electrons like to congregate (as in a chemical bond) and we want to put another electron on top of one already there, the only way to do it is to have the spin of the second one pointed opposite to the spin of the first one. To have the spins parallel is against the law, unless the electrons stay away from each other. This has the effect that a pair of parallel-spin electrons near to each other have much more energy than a pair of opposite-spin electrons; the net effect is as though there were a force trying to turn the spin over. Sometimes this spin-turning force is called the *exchange force*, but that only makes it more mysterious—it is not a very good term. It is just because of the exclusion principle that electrons have a tendency to make their spins opposite. In fact, that is the explanation of the *lack* of magnetism in almost all substances! The spins of the free electrons on the outside of the atoms have tremendous tendency to balance in opposite directions. The problem is to explain why for materials like iron it is just the reverse of what we should expect.

We have summarized the supposed alignment effect by adding a suitable term in the energy equation, by saying that if the electron magnets in the neighborhood have a mean magnetization M, then the moment of an electron has a strong tendency to be in the same direction as the average magnetization of the atoms in the neighborhood. Thus, we may write for the two possible spin orientations,†

$$\begin{aligned}\text{Spin ``up'' energy} &= +\mu\left(H + \frac{\lambda M}{\epsilon_0 c^2}\right),\\ \text{Spin ``down'' energy} &= -\mu\left(H + \frac{\lambda M}{\epsilon_0 c^2}\right).\end{aligned} \tag{37.1}$$

When it was clear that quantum mechanics could supply a tremendous spin-orientating force—even if, apparently, of the wrong sign—it was suggested that ferromagnetism might have its origin in this same force, that due to the complexities of iron and the large number of electrons involved, the sign of the interaction energy would come out the other way around. Since the time this was thought of—in about 1927 when quantum mechanics was first being understood—many people have been making various estimates and semicalculations, trying to get a theoretical prediction for λ. The most recent calculations of the energy between the two electron spins in iron—assuming that the interaction is a direct one between the two electrons in neighboring atoms—still give the *wrong sign*. The present understanding of this is again to assume that the complexity of the situation is somehow responsible and to hope that the next man who makes the calculation with a more complicated situation will get the right answer!

It is believed that the up-spin of one of the electrons in the inside shell, which is making the magnetism, tends to make the conduction electrons which fly around the outside have the opposite spin. One might expect this to happen because the conduction electrons come into the same region as the "magnetic" electrons. Since they move around, they can carry their prejudice for being upside down over to the next atom; that is, one "magnetic" electron tries to force the conduction electrons to be opposite, and the conduction electron then makes the next "magnetic" electron opposite to *it*. The double interaction is equivalent to an interaction which tries to line up the two "magnetic" electrons. In other words, the tendency to make parallel spins is the result of an intermediary that tends to some extent to be opposite to both. This mechanism does not require that the conduction electrons be completely "upside down." They could just have a slight prejudice to be down, just enough to load the "magnetic" odds the other way. This is the mechanism that

---

* See Chapter 43.

† We write these equations with $H = B - M/\epsilon_0 c^2$ instead of B to agree with the work of the last chapter. You might prefer to write $U = \pm\mu B_a = \pm\mu(B + \lambda' M/\epsilon_0 c^2)$, where $\lambda' = \lambda - 1$. It's the same thing.

the people who have calculated such things now believe is responsible for ferromagnetism. But we must emphasize that to this day nobody can calculate the magnitude of λ simply by knowing that the material is number 26 in the periodic table. In short, we don't thoroughly understand it.

Now let us continue with the theory, and then come back later to discuss a certain error involved in the way we have set it up. If the magnetic moment of a certain electron is "up," energy comes both from the external field and also from the tendency of the spins to be parallel. Since the energy is lower when the spins are parallel, the effect is sometimes thought of as due to an "effective internal field." But remember, it is *not* due to a true *magnetic* force; it is an interaction that is more complicated. In any case, we take Eqs. (37.1) as the formulas for the energies of the two spin states of a "magnetic" electron. At a temperature T, the relative probability of these two states is proportional to $e^{-\text{energy}/kT}$, which we can write as $e^{\pm x}$, with $x = \mu(H + \lambda M/\epsilon_0 c^2)/kT$. Then, if we calculate the mean value of the magnetic moment, we find (as in the last chapter) that it is

$$M = N\mu \tanh x. \tag{37.2}$$

Now we would like to calculate the internal energy of the material. We note that the energy of an electron is exactly proportional to the magnetic moment, so that the calculation of the mean moment and the calculation of the mean energy are the same—except that in place of μ in Eq. (37.2) we would write $-\mu B$, which is $-\mu(H + \lambda M/\epsilon_0 c^2)$. The mean energy is then

$$\langle U \rangle_{\text{av}} = -N\mu\left(H + \frac{\lambda M}{\epsilon_0 c^2}\right) \tanh x.$$

Now this is not quite correct. The term $\lambda M/\epsilon_0 c^2$ represents interactions of all possible *pairs* of atoms, and we must remember to count each pair only *once*. (When we consider the energy of one electron in the field of the rest and then the energy of a second electron in the field of the rest, we have counted part of the first energy once more.) Thus, we must divide the *mutual interaction term* by two, and our formula for the energy then turns out to be

$$\langle U \rangle_{\text{av}} = -N\mu\left(H + \frac{\lambda M}{2\epsilon_0 c^2}\right) \tanh x. \tag{37.3}$$

In the last chapter we discovered an interesting thing—that below a certain temperature the material finds a solution to the equations in which the magnetic moment *is not zero*, even with no external magnetizing field. When we set $H = 0$ in Eq. (37.2), we found that

$$\frac{M}{M_{\text{sat}}} = \tanh\left(\frac{T_c}{T}\frac{M}{M_{\text{sat}}}\right), \tag{37.4}$$

where $M_{\text{sat}} = N\mu$, and $T_c = \mu\lambda M_{\text{sat}}/k\epsilon_0 c^2$. When we solve this equation (graphically or otherwise), we find that the ratio M/M_{sat} as a function of T/T_c is a curve like that labeled "quantum theory" in Fig. 37–1. The dashed curve marked "cobalt, nickel" shows the experimental results for crystals of these elements. The theory and experiment are in reasonably good agreement. The figure also shows the result of the classical theory in which the calculation is carried out assuming that the atomic magnets can have all possible orientations in space. You can see that this assumption gives a prediction that is not even close to the experimental facts.

Even the quantum theory deviates from the observed behavior at both high and low temperatures. The reason for the deviations is that we have made a rather sloppy approximation in the theory: We have assumed that the energy of an atom depends upon the *mean* magnetization of its neighboring atoms. In other words, for each one that is "up" in the neighborhood of a given atom, there will be a contribution of energy due to that quantum mechanical alignment effect. But how many are there pointed "up"? On the average, that is measured by the

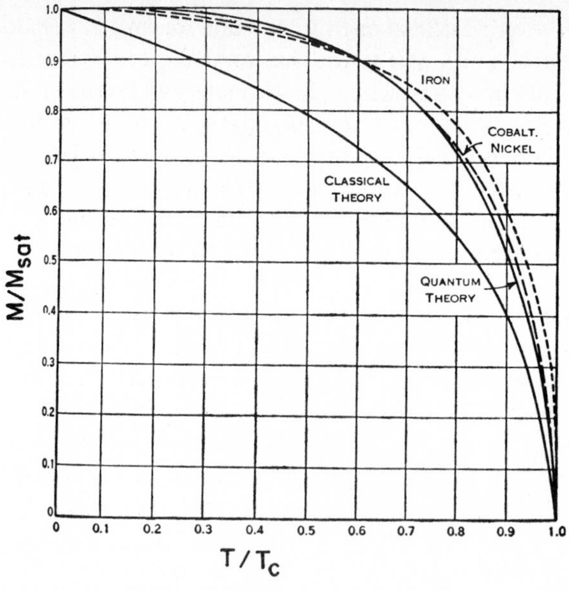

Fig. 37–1. The spontaneous magnetization ($H = 0$) of ferromagnetic crystals as a function of temperature. [Permission from *Encyclopaedia Britannica.*]

magnetization M—but only *on the average*. A particular atom somewhere might find *all* its neighbors "up." Then its energy will be larger than the average. Another one might find some up and some down, perhaps averaging to zero, and it would have *no* energy from that term, and so on. What we ought to do is to use some more complicated kind of average, because the atoms in different places have different environments, and the numbers up and down are different for different ones. Instead of just taking one atom subjected to the average influence, we should take each one in its actual situation, compute its energy, and find the *average energy*. But how do we find out how many are "up" and how many are "down" in the neighborhood? That is, of course, just what we are trying to calculate—the number "up" and "down"—so we have a very complicated interconnected problem of correlations, a problem which has never been solved. It is an intriguing and exciting one which has existed for years and on which some of the greatest names in physics have written papers, but even they have not completely solved it.

It turns out that at low temperatures, when almost all the atomic magnets are "up" and only a few are "down," it is easy to solve; and at high temperatures, far above the Curie temperature T_c when they are almost all random, it is again easy. It is often easy to calculate small departures from some simple, idealized situation, so it is fairly well understood why there are deviations from the simple theory at low temperature. It is also understood physically that for statistical reasons the magnetization *should* deviate at high temperatures. But the exact behavior near the Curie point has never been thoroughly figured out. That's an interesting problem to work out some day if you want a problem that has never been solved.

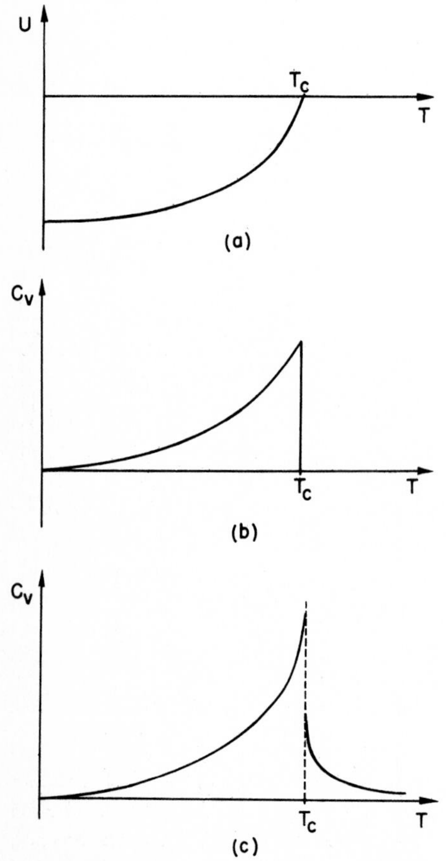

Fig. 37–2. The energy per unit volume and specific heat of a ferromagnetic crystal.

37–2 Thermodynamic properties

In the last chapter we laid the groundwork necessary for calculating the thermodynamic properties of ferromagnetic materials. These are, naturally, related to the internal energy of the crystal, which includes interactions of the various spins, given by Eq. (37.3). For the energy of the spontaneous magnetization below the Curie point, we can set $H = 0$ in Eq. (37.3), and—noticing that $\tanh x = M/M_{\text{sat}}$—we find a mean energy proportional to M^2:

$$\langle U\rangle_{\text{av}} = -\frac{N\mu\lambda M^2}{2\epsilon_0 c^2 M_{\text{sat}}}\cdot \tag{37.5}$$

If we now plot the energy due to the magnetism as a function of temperature, we get a curve which is the negative of the square of the curve of Fig. 37–1, as drawn in Fig. 37–2(a). If we were to measure then the *specific heat* of such a material we would obtain a curve which is the derivative of 37–2(a). It is shown in Fig.

37–2(b). It rises slowly with increasing temperature, but falls suddenly to zero at $T = T_c$. The sharp drop is due to the change in slope of the magnetic energy and is reached right at the Curie point. So without any magnetic measurements at all we could have discovered that something was going on inside of iron or nickel by measuring this thermodynamic property. However, both experiment and improved theory (with fluctuations included) suggest that this simple curve is wrong and that the true situation is really more complicated. The curve goes higher at the peak and falls to zero somewhat slowly. Even if the temperature is high enough to randomize the spins *on the average*, there are still local regions where there is a certain amount of polarization, and in these regions the spins still have a little extra energy of interaction—which only dies out slowly as things get more and more random with further increases in temperature. So the actual curve looks like Fig. 37–2(c). One of the challenges of theoretical physics today is to find an exact theoretical description of the character of the specific heat near the Curie transition—an intriguing problem which has not yet been solved. Naturally, this problem is very closely related to the shape of the magnetization curve in the same region.

Now we want to describe some experiments, other than thermodynamic ones, which show that there is something *right* about our interpretation of magnetism. When the material is magnetized to saturation at low enough temperatures, M is very nearly equal to M_{sat}—nearly all the spins are parallel, as well as their magnetic moments. We can check this by an experiment. Suppose we suspend a bar magnet by a thin fiber and then surround it by a coil so that we can reverse the magnetic field without touching the magnet or putting any torque on it. This is a very difficult experiment because the magnetic forces are so enormous that any irregularities, any lopsidedness, or any lack of perfection in the iron will produce accidental torques. However, the experiment has been done under careful conditions in which such accidental torques are minimized. By means of the magnetic field from a coil that surrounds the bar, we turn all the atomic magnets over at once. When we do this we also change the angular momenta of all the spins from "up" to "down" (see Fig. 37–3). If angular momentum is to be conserved when the spins all turn over, the rest of the bar must have an opposite change in angular momentum. The whole magnet will start to spin. And sure enough, when we do the experiment, we find a slight turning of the magnet. We can measure the total angular momentum given to the whole magnet, and this is simply N times $\hbar$, the change in the angular momentum of each spin. The ratio of angular momentum to magnetic moment measured this way comes out to within about 10 percent of what we calculate. Actually, our calculations assume that the atomic magnets are due purely to the electron spin, but there is, in addition, some orbital motion also in most materials. The orbital motion is not completely free of the lattice and does not contribute much more than a few percent to the magnetism. As a matter of fact, the saturation magnetic field that one gets taking $M_{\text{sat}} = N\mu$ and using the density of iron of 7.9 and the moment μ of the spinning electron is about 20,000 gauss. But according to experiment, it is actually in the neighborhood of 21,500 gauss. This is a typical magnitude of error—5 or 10 percent—due to neglecting the contributions of the orbital moments that have not been included in making the analysis. Thus, a slight discrepancy with the gyromagnetic measurements is quite understandable.

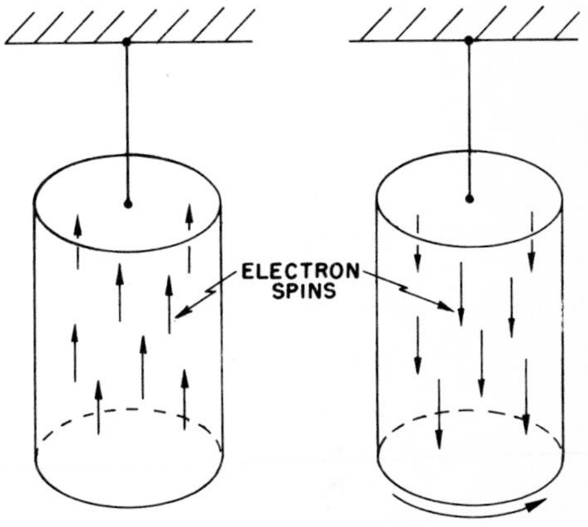

Fig. 37–3. When the magnetization of a bar of iron is reversed, the bar is given some angular velocity.

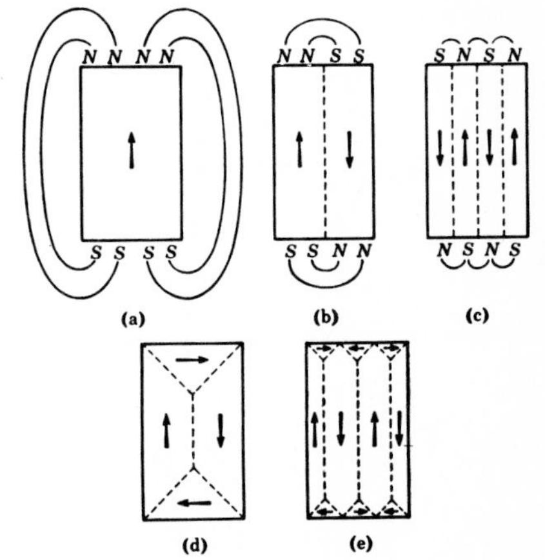

Fig. 37–4. The formation of domains in a single crystal of iron. [From Charles Kittel, *Introduction to Solid State Physics*, John Wiley and Sons, Inc., New York, 2nd ed., 1956.]

37–3 The hysteresis curve

We have concluded from our theoretical analysis that a ferromagnetic material should spontaneously become magnetized below a certain temperature so that all the magnetism would be in the same direction. But we know that this is not true for an ordinary piece of *unmagnetized* iron. Why isn't all iron magnetized? We can explain it with the help of Fig. 37–4. Suppose the iron were all a big single crystal of the shape shown in Fig. 37–4(a) and spontaneously magnetized all in one direction. Then there would be a considerable external magnetic field, which would have a lot of energy. We can reduce that field energy if we arrange that one side of the block is magnetized "up" and the other side magnetized "down," as in Fig. 37–4(b). Then, of course, the fields outside the iron would extend over less volume, so there would be less energy there.

Ah, but wait! In the layer between the two regions we have up-spinning electrons adjacent to down-spinning electrons. But ferromagnetism appears only in those materials for which the energy is *reduced* if the electrons are *parallel* rather than opposite. So, we have added some extra energy along the dotted line in Fig. 37–4(b); this energy is sometimes called *wall energy*. A region having only one direction of magnetization is called a *domain*. At the interface—the "wall"—between two domains, where we have atoms on opposite sides which are spinning in different directions, there is an energy per unit area of the wall. We have described it as though two adjacent atoms were spinning exactly opposite, but it turns out that nature adjusts things so that the transition is more gradual. But we don't need to worry about such fine details at this point.

Now the question is: When is it better or worse to make a wall? The answer is that it depends on the *size* of the domains. Suppose that we were to scale up a block so that the whole thing was twice as big. The volume in the space outside filled with a given magnetic field strength would be *eight* times bigger, and the energy in the magnetic field, which is proportional to the volume, would also be eight times greater. But the *surface* area between two domains, which will give the wall energy, would be only *four* times as big. Therefore, if the piece of iron is big enough, it will pay to split it into more domains. This is why only the very tiny crystals can have but a single domain. Any large object—one more than about a hundredth of a millimeter in size—will have at least one domain wall; and any ordinary, "centimeter-size" object will be split into many domains, as shown in the figure. Splitting into domains goes on *until the energy needed to put in one extra wall is as large as the energy decrease in the magnetic field outside the crystal.*

Actually nature has discovered still another way to lower the energy: It is not necessary to have the field go outside at all, if a little triangular region is magnetized *sideways*, as in Fig. 37–4(d).* Then with the arrangement of Fig. 37–4(d) we see that there is *no* external field, but instead only a little more domain wall.

But that introduces a new kind of problem. It turns out that when a single crystal of iron is magnetized, it changes its length in the direction of magnetization, so an "ideal" cube with its magnetization, say, "up," is no longer a perfect cube. The "vertical" dimension will be different from the "horizontal" dimension. This effect is called *magnetostriction*. Because of such geometric changes, the little triangular pieces of Fig. 37–4(d) do not, so to speak, "fit" into the available space anymore—the crystal has got too long one way and too short the other way. Of course, it *does* fit, really, but only by being squashed in; and this involves some mechanical stresses. So, this arrangement *also* introduces an extra energy. It is the balance of all these various energies which determines how the domains finally arrange themselves in their complicated fashion in a piece of unmagnetized iron.

Now, what happens when we put on an external magnetic field? To take a simple case, consider a crystal whose domains are as shown in Fig. 37–4(d). If we apply an external magnetic field in the upward direction, in what manner does the crystal become magnetized? First, the middle domain wall can *move over sideways* (to the right) and reduce the energy. It moves over so that the region which is "up" becomes bigger than the region which is "down". There are more elementary magnets lined up with the field, and this gives a lower energy. So, for a piece of iron in weak fields—at the very beginning of magnetization—the domain walls begin to move and eat into the regions which are magnetized opposite to the field. As the field continues to increase, a whole crystal shifts gradually into a single

* You may be wondering how spins that have to be either "up" or "down" can also be "sideways"! That's a good question, but we won't worry about it right now. We'll simply adopt the classical point of view, thinking of the atomic magnets as classical dipoles which can be polarized sideways. Quantum mechanics requires considerable expertness to understand how things can be quantized both "up-and-down," and "right-and-left," all at the same time.

large domain which the external field helps to keep lined up. In a strong field the crystal "likes" to be all one way *just because* its energy in the applied field is reduced —it is no longer merely the crystal's own external field which matters.

What if the geometry is not so simple? What if the axes of the crystal and its spontaneous magnetization are in one direction, but we apply the magnetic field in *some other direction*—say at 45°? We might think that domains would reform themselves with their magnetization parallel to the field, and then as before, they could all grow into one domain. But this is not easy for the iron to do, *for the energy needed to magnetize a crystal depends on the direction of magnetization relative to the crystal axis.* It is relatively easy to magnetize iron in a direction parallel to the crystal axes, but it takes *more* energy to magnetize it in some other direction—like 45° with respect to one of the axes. Therefore, if we apply a magnetic field in such a direction, what happens first is that the domains which point along one of the preferred directions which is *near* to the applied field grow until the magnetization is all along one of these directions. Then *with much stronger fields*, the magnetization is gradually pulled around parallel to the field, as sketched in Fig. 37–5.

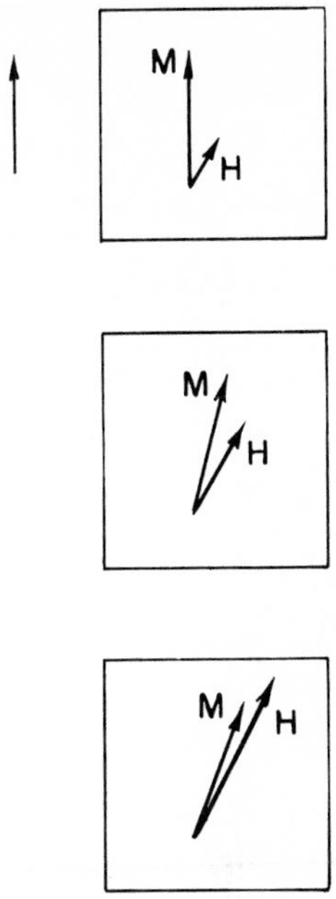

Fig. 37–5. A magnetizing field H at an angle with respect to the crystal axis will gradually change the direction of the magnetization without changing its magnitude.

In Fig. 37–6 are shown some observations of the magnetization curves of single crystals of iron. To understand them, we must first explain something about the notation that is used in describing directions in a crystal. There are many ways in which a crystal can be sliced so as to produce a face which is a plane of atoms. Everyone who has driven past an orchard or vineyard knows this—it is fascinating to watch. If you look one way, you see lines of trees—if you look another way, you see different lines of trees, and so on. In a similar way, a crystal has definite families of planes that hold many atoms, and the planes have this important characteristic (we consider a cubic crystal to make it easier): If we observe where the planes intersect the three coordinate axes—we find that the *reciprocals* of the three distances from the origin are in the ratio of simple whole numbers. These three whole numbers are taken as the definition of the planes. For example, in Fig. 37–7(a), a plane parallel to the yz-plane is shown. This is called a [100] plane; the reciprocals of its intersection of the y- and z-axes are both zero. The direction perpendicular to such a plane (in a cubic crystal) is given the same set of numbers. It is easy to understand the idea in a cubic crystal, for then the indices [100] mean a vector which has a unit component in the x-direction and none in the y- or z-directions. The [110] direction is in a direction 45° from the x- and y-axes, as in Fig. 37–7(b); and the [111] direction is in the direction of the cube diagonal, as in Fig. 37–7(c).

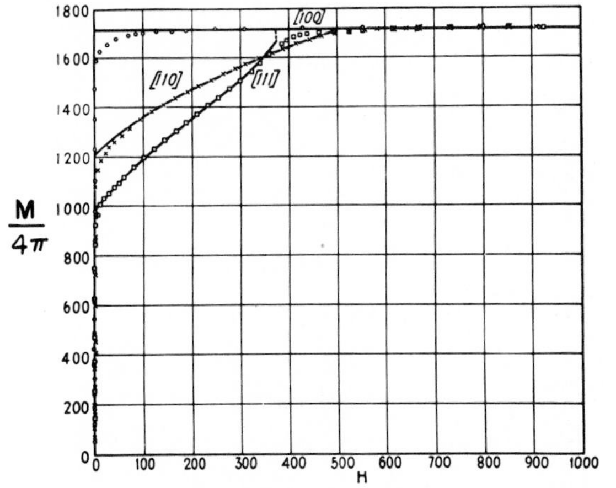

Fig. 37–6. The component of ***M*** parallel to ***H***, for different directions of ***H*** (with respect to the crystal axes). [From F. Bitter, *Introduction to Ferromagnetism*, McGraw-Hill Book Co., Inc., 1937.]

Returning now to Fig. 37–6, we see the magnetization curves of a single crystal of iron for various directions. First, note that for very tiny fields—so weak that it is hard to see them on the scale at all—the magnetization increases extremely rapidly to quite large values. If the field is in the [100] direction—namely along one of those nice, easy directions of magnetization—the curve goes up to a high value, curves around a little, and then is saturated. What happened is that the

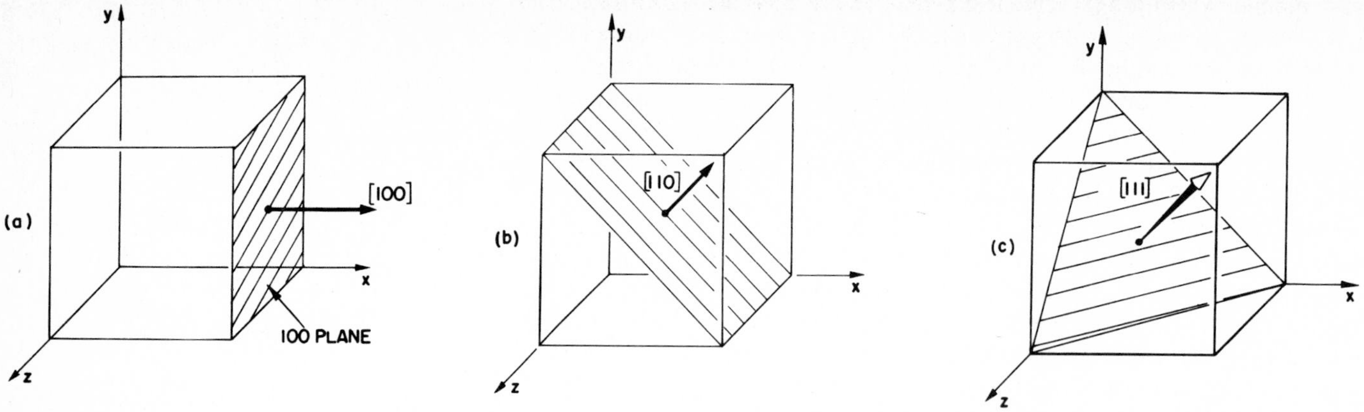

Fig. 37–7. The way the crystal planes are labeled.

domains which were already there are very easily removed. Only a small field is required to make the domain walls move and eat up all of the "wrong-way" domains. Single crystals of iron are enormously permeable (magnetic sense), much more so than ordinary polycrystalline iron. A perfect crystal magnetizes extremely easily. Why is it curved at all? Why doesn't it just go right up to saturation? We are not sure. You might study that some day. We do understand why it is flat for high fields. When the whole block is a single domain, the extra magnetic field cannot make any more magnetization—it is already at M_{sat}, with all the electrons lines up.

Now, if we try to do the same thing in the [110] direction—which is at 45° to the crystal axes—what will happen? We turn on a little bit of field and the magnetization leaps up as the domains grow. Then as we increase the field some more, we find that it takes quite a lot of field to get up to saturation, because now the magnetization is *turning away* from an "easy" direction. If this explanation is correct, the point at which the [110] curve extrapolates back to the vertical axis should be at $1/\sqrt{2}$ of the saturation value. It turns out, in fact, to be very, very close to $1/\sqrt{2}$. Similarly, in the [111] direction—which is along the cube diagonal—we find, as we would expect, that the curve extrapolates back to nearly $1/\sqrt{3}$ of saturation.

Figure 37–8 shows the corresponding situation for two other materials, nickel and cobalt. Nickel is different from iron. In nickel, it turns out that the [111] direction is the easy direction of magnetization. Cobalt has a hexagonal crystal form, and people have botched up the system of nomenclature for this case. They want to have three axes on the bottom of the hexagon and one perpendicular to these, so they have used four indices. The [0001] direction is the direction of the axis of the hexagon, and [1010] is perpendicular to that axis. We see that crystals of different metals behave in different ways.

Now we must discuss a polycrystalline material, such as an ordinary piece of iron. Inside such materials there are many, many little crystals with their crystalline axes pointing every which way. *These are not the same as domains.* Remember that the domains were all part of a *single crystal*, but in a piece of iron there are

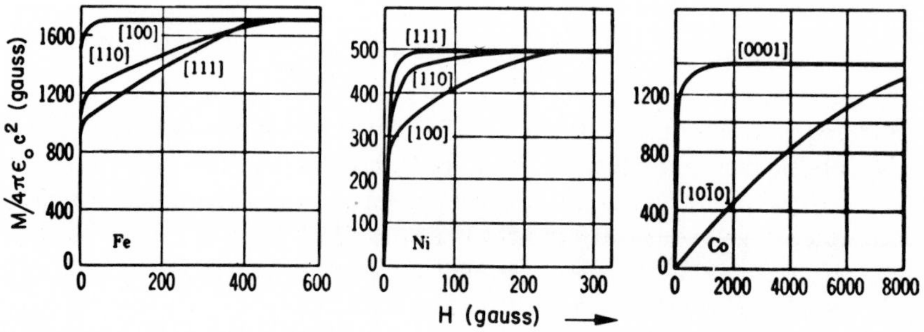

Fig. 37–8. Magnetization curves for single crystals of iron, nickel, and cobalt. [From Charles Kittel, *Introduction to Solid State Physics*, John Wiley and Sons, Inc., New York, 2nd ed., 1956.]

many *different crystals* with axes at different orientations, as shown in Fig. 37–9. Within each of these crystals, there will also generally be some domains. When we apply a *small* magnetic field to a piece of polycrystalline material, what happens is that the domain walls begin to move, and the domains which have a favorable direction of easy magnetization grow larger. This growth is reversible so long as the field stays very small—if we turn the field off, the magnetization will return to zero. This part of the magnetization curve is marked a in Fig. 37–10.

For larger fields—in the region b of the magnetization curve shown—things get much more complicated. In every small crystal of the material, there are strains and dislocations; there are impurities, dirt, and imperfections. And at all but the smallest fields, the domain wall, in moving, gets stuck on these. There is an interaction energy between the domain wall and a dislocation, or a grain boundary, or an impurity. So when the wall gets to one of them, it gets stuck; it sticks there at a certain field. But then if the field is raised some more, the wall suddenly snaps past. So the motion of the domain wall is not smooth the way it is in a perfect crystal—it gets hung up every once in a while and moves in jerks. If we were to look at the magnetization on a microscopic scale, we would see something like the insert of Fig. 37–10.

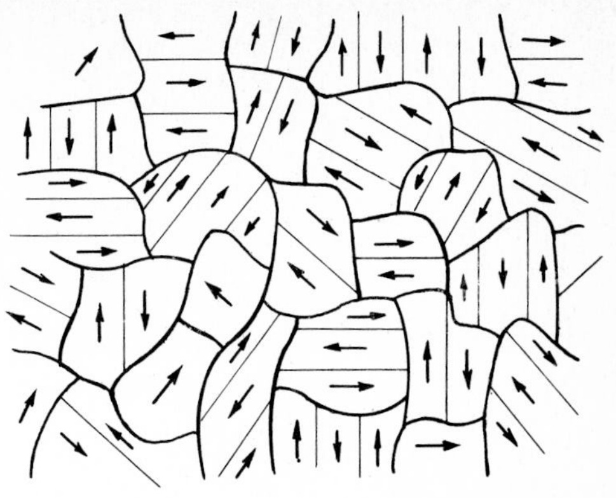

Fig. 37–9. The microscopic structure of an unmagnetized ferromagnetic material. Each crystal grain has an easy direction of magnetization and is broken up into domains which are spontaneously magnetized (usually) parallel to this direction.

Now the important thing is that these jerks in the magnetization can cause an energy loss. In the first place, when a boundary finally slips past an impediment, it moves very quickly to the next one, since the field is already above what would be required for the unimpeded motion. The rapid motion means that there are rapidly changing magnetic fields which produce eddy currents in the crystal. These currents loose energy in heating the metal. A second effect is that when a domain suddenly changes, part of the crystal changes its dimensions from the magnetostriction. Each sudden shift of a domain wall sets up a little sound wave that carries away energy. Because of such effects, the second part of magnetization curve is *irreversible*, and *there is energy being lost*. This is the origin of the hysteresis effect, because to move a boundary wall forward—snap—and then to move it backward—snap—produces a different result. It's like "jerky" friction, and it takes energy.

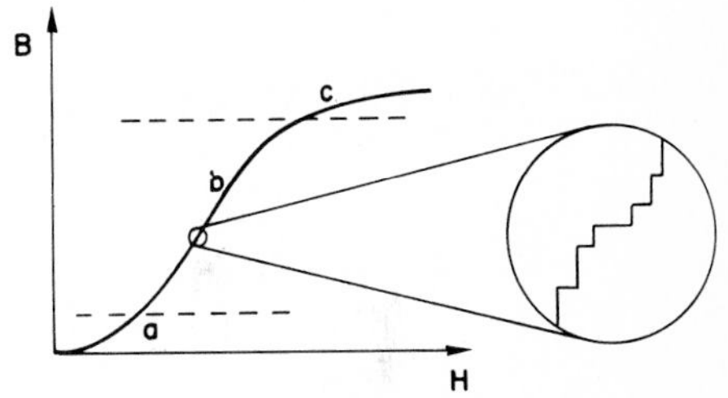

Fig. 37–10. The magnetization curve for polycrystalline iron.

Eventually, for high enough fields, when we have moved all the domain walls and magnetized each crystal in its best direction, there are still some crystallites which happen to have their easy directions of magnetization not in the direction of our external magnetic field. Then it takes a lot of extra field to turn those magnetic moments around. So the magnetization increases slowly, but smoothly, for high fields—namely in the region marked c in the figure. The magnetization does not come sharply to its saturation value, because in the last part of the curve the atomic magnets are *turning* in the strong field. So we see why the magnetization curve of an ordinary polycrystalline materials, such as the one shown in Fig. 37–10, rises a little bit and *reversibly* at first, then rises irreversibly, and then curves over slowly. Of course, there is no sharp break-point between the three regions—they blend smoothly, one into the other.

It is not hard to show that the magnetization process in the middle part of the magnetization curve is jerky—that the domain walls jerk and snap as they shift. All you need is a coil of wire—with many thousands of turns—connected to an amplifier and a loudspeaker, as shown in Fig. 37–11. If you put a few silicon steel sheets (of the type used in transformers) at the center of the coil and bring a bar magnet slowly near the stack, the sudden changes in magnetization will produce impulses of emf in the coil, which are heard as distinct clicks in the loudspeaker. As you move the magnet nearer to the iron you will hear a whole rush of clicks that sound something like the noise of sand grains falling over each other as a can of sand is tilted. The domain walls are jumping, snapping, and jiggling as the field is increased. This phenomenon is called the *Barkhausen effect*.

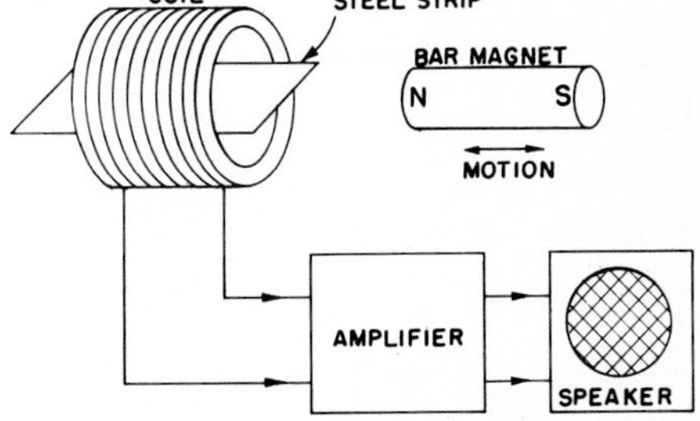

Fig. 37–11. The sudden changes in the magnetization of the steel strip are heard as clicks in the loudspeaker.

As you move the magnet even closer to the iron sheets, the noise grows louder and louder for a while but then there is relatively little noise when the magnet gets very close. Why? Because nearly all the domain walls have moved as far as they can go. Any greater field is merely *turning* the magnetization in each domain, which is a smooth process.

If you now withdraw the magnet, so as to come back on the downward branch of the hysteresis loop, the domains all try to get back to low energy again, and you hear another rush of backward-going jerks. You can also note that if you bring the magnet to a given place and move it back and forth a little bit, there is relatively little noise. It is again like tilting a can of sand—once the grains shift into place, small movements of the can don't disturb them. In the iron the small variations in the magnetic field aren't enough to move any boundaries over any of the "humps."

37–4 Ferromagnetic materials

Now we would like to talk about the various kinds of magnetic materials that there are in the technical world and to consider some of the problems involved in designing magnetic materials for different purposes. First, the term "the magnetic properties of iron," which one often hears, is a misnomer—there is no such thing. "Iron" is not a well-defined material—the properties of iron depend critically on the amount of impurities and also on *how* the iron is formed. You can appreciate that the magnetic properties will depend on how easily the domain walls move and that this is a *gross* property, not a property of the individual atoms. So practical ferromagnetism is not really a property of an iron *atom*—it is a property of *solid iron* in a *certain form*. For example, iron can take on two different crystalline forms. The common form has a body-centered cubic lattice, but it can also have a face-centered cubic lattice, which is, however, stable only at temperatures above 1100°C. Of course, at that temperature the body-centered cubic structure is already past the Curie point. However, by alloying chromium and nickel with the iron (one possible mixture is 18 percent chromium and 8 percent nickel) we can get what is called stainless steel, which, although it is mainly iron, retains the face-centered lattice even at low temperatures. Because its crystal structure is different, it has completely different magnetic properties. Most kinds of stainless steel are not magnetic to any appreciable degree, although there are some kinds which are somewhat magnetic—it depends on the composition of the alloy. Even when such an alloy is magnetic, it is not *ferro*magnetic like ordinary iron—even though it is mostly just iron.

We would like now to describe a few of the special materials which have been developed for their particular magnetic properties. First, if we want to make a *permanent* magnet, we would like material with an enormously *wide* hysteresis loop so that, when we turn the current off and come down to zero magnetizing field, the magnetization will remain large. For such materials the domain boundaries should be "frozen" in place as much as possible. One such material is the remarkable alloy "Alnico V" (51% Fe, 8% Al, 14% Ni, 24% Co, 3% Cu). (The rather complex composition of this alloy is indicative of the kind of detailed effort that has gone into making good magnets. What patience it takes to mix five things together and test them until you find the most ideal substance!) When Alnico solidifies, there is a "second phase" which precipitates out, making many tiny grains and very high internal strains. In this material, the domain boundaries have a hard time moving at all. In addition to having a precise composition, Alnico is mechanically "worked" in a way that makes the crystals appear in the form of long grains along the direction in which the magnetization is going to be. Then the magnetization will have a natural tendency to be lined up in these directions and will be held there from the anisotropic effects. Furthermore, the material is even cooled in an external magnetic field when it is manufactured, so that the grains will grow with the right crystal orientation. The hysteresis loop of Alnico V is shown in Fig. 37–12. You see that it is about 500 times wider than the hysteresis curve for soft iron that we showed in the last chapter in Fig. 36–8.

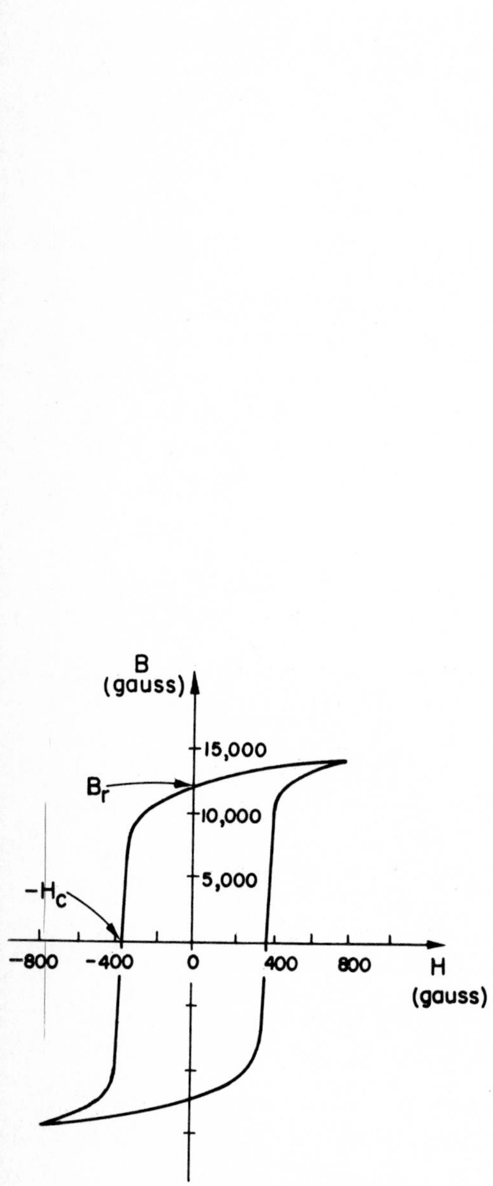

Fig. 37–12. The hysteresis curve of Alnico V.

Let's turn now to a different kind of material. For building transformers and motors, we want a material which is magnetically "soft"—one in which the magnetism is easily changed so that an enormous amount of magnetization results from a very small applied field. To arrange this, we need pure, well-annealed material which will have very few dislocations and impurities so that the domain walls can move easily. It would also be nice if we could make the anisotropy small. Then, even if a grain of the material sits at the wrong angle with respect to the field, it will still magnetize easily. Now we have said that iron prefers to magnetize along the [100] direction, whereas nickel prefers the [111] direction; so if we mix iron and nickel in various proportions, we might hope to find that with just the right proportions the alloy wouldn't prefer *any* direction—the [100] and [111] directions would be equivalent. It turns out that this happens with a mixture of 70 percent nickel and 30 percent iron. In addition—possibly by luck or maybe because of some physical relationship between the anisotropy and the magnetostriction effects—it turns out that the *magnetostriction* of iron and nickel has the opposite sign. And in an alloy of the two metals, this property goes through zero at about 80 percent nickel. So somewhere between 70 and 80 percent nickel we get very "soft" magnetic materials—alloys that are very easy to magnetize. They are called the *permalloys*. Permalloys are useful for high-quality transformers (at low signal levels), but they would be no good at all for permanent magnets. Permalloys must be very carefully made and handled. The magnetic properties of a piece of permalloy are drastically changed if it is stressed beyond its elastic limit—it mustn't be bent. Then, its permeability is reduced because of the dislocations, slip bands, and so on, which are produced by the mechanical deformations. The domain boundaries are no longer easy to move. The high permeability can, however, be restored by annealing at high temperatures.

It is often convenient to have some numbers to characterize the various magnetic materials. Two useful numbers are the intercepts of the hysteresis loop with the B- and H-axes, as indicated in Fig. 37–12. These intercepts are called the *remanent magnetic field* B_r and the *coercive force* H_c. In Table 37–1 we list these numbers for a few magnetic materials.

Table 37–1

Properties of some ferromagnetic materials

Material	B_r Residual magnetic field (gauss)	H_c Coercive force (gauss)
Supermalloy	(≈ 5000)	0.004
Silicon steel (transformer)	12,000	0.05
Armco iron	4000	0.6
Alnico V	13,000	550.

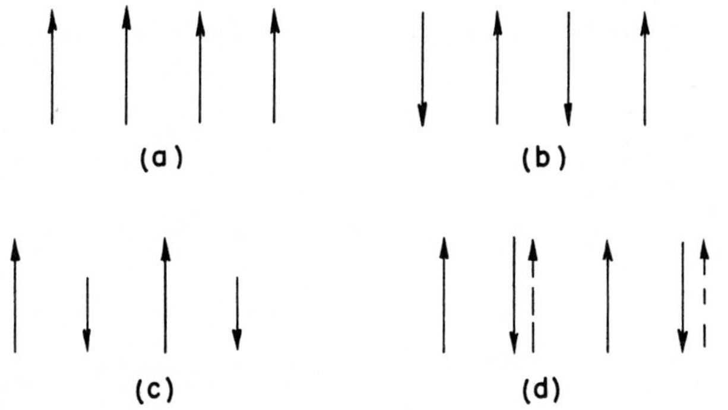

Fig. 37–13. Relative orientation of electron spins in various materials: (a) ferromagnetic, (b) antiferromagnetic, (c) ferrite, (d) yttrium-iron alloy. (Broken arrows show direction of total angular momentum, including orbital motion.)

37–5 Extraordinary magnetic materials

We would now like to discuss some of the more exotic magnetic materials. There are many elements in the periodic table which have incomplete inner electron shells and hence have atomic magnetic moments. For instance, right next to the ferromagnetic elements iron, nickel, and cobalt you will find chromium and manganese. Why aren't *they* ferromagnetic? The answer is that the λ term in Eq. (37.1) has the *opposite sign* for these elements. In the chromium lattice, for example, the spins of the chromium atoms alternate *atom by atom*, as shown in Fig. 37–13(b). So chromium *is* "magnetic" from its own point of view, but it is not technically interesting because there are no *external* magnetic effects. Chromium, then, is an example of a material in which quantum mechanical effects make the spins alternate. Such a material is called *antiferromagnetic*. The alignment in antiferromagnetic materials is also temperature dependent. Below a critical temperature, all the spins are lined up in the alternating array, but when the material is heated above a certain temperature—which is again called the Curie temperature—the spins suddenly become random. There is, internally, a sudden transition. This transition can be seen in the specific heat curve. Also it shows up in some special "magnetic" effects. For instance, the existence of the alternating spins can be verified by scattering neutrons from a crystal of chromium. Because a neutron itself has a spin (and a magnetic moment), it has a different amplitude to be scattered, depending on whether its spin is parallel or opposite to the spin of the scatterer. Thus, we get a different interference pattern when the spins in a crystal are alternating than we do when they have a random distribution.

There is another kind of substance in which quantum mechanical effects make the electron spins alternate, but which is nevertheless *ferromagnetic*—that is, the crystal has a permanent net magnetization. The idea behind such materials is shown in Fig. 37–14. The figure shows the crystal structure of *spinel*, a magnesium-aluminum oxide, which—as it is shown—is *not* magnetic. The oxide has two kinds of metal atoms: magnesium and aluminum. Now if we replace the magnesium and the aluminum by two magnetic elements like iron and zinc, or by zinc and manganese—in other words, if we put in *magnetic* atoms instead of the nonmagnetic ones—an interesting thing happens. Let's call one kind of metal atom a and the other kind of metal atom b; then the following combination of forces must be considered. There is an a-b interaction which tries to make the a atoms and the b atoms have opposite spins—because quantum mechanics always gives the opposite sign (except for the mysterious crystals of iron, nickel, and cobalt). Then, there is a direct a-a interaction which tries to make the a's opposite, and also a b-b interaction which tries to make the b's opposite. Now, of course we cannot have everything opposite everything else—a opposite b, a opposite a, and b opposite b. Presumably because of the distances between the a's and the presence of the oxygen (although we really don't know why), it turns out that the a-b interaction is stronger than the a-a or the b-b. So the solution that nature uses in this case is to make all the a's *parallel to each other*, and all the b's *parallel to each other*, but the two systems *opposite*. That gives the lowest energy because of the stronger a-b interaction. The result: all the a's are spinning up and all the b's are spinning down—or vice versa, of course. But if the *magnetic moments* of the a-type atom and the b-type atom *are not equal*, we can get the situation shown in Fig. 37–13(c), and there can be a net magnetization in the material. The material will then be ferromagnetic—although somewhat weak. Such materials are called *ferrites.* They do not have as high a saturation magnetization as iron—for obvious reasons—so they are only useful for smaller fields. But they have a very important difference—they are insulators; the ferrites are *ferromagnetic insulators.* In high-frequency fields, they will have very small eddy currents and so can be used, for example, in microwave systems. The microwave fields will be able to get inside such an insulating material, whereas they would be kept out by the eddy currents in a conductor like iron.

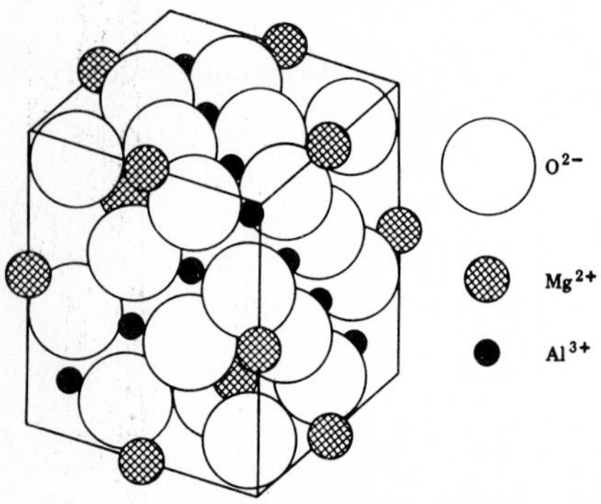

Fig. 37–14. Crystal structure of the mineral spinel ($MgAl_2O_4$); the Mg^{+2} ions occupy tetrahedral sites, each surrounded by four oxygen ions; the Al^{+3} ions occupy octahedral sites, each surrounded by six oxygen ions. [From Charles Kittel, *Introduction to Solid State Physics*, John Wiley and Sons, Inc., New York, 2nd ed., 1956.]

There is another class of magnetic materials which has only recently been discovered—members of the family of the orthosilicates called *garnets.* They are again crystals in which the lattice contains two kinds of metallic atoms, and we have again a situation in which two kinds of atoms can be substituted almost at will. Among the many compounds of interest there is one which is completely ferromagnetic. It has yttrium and iron in the garnet structure, and the reason it is ferromagnetic is very curious. Here again quantum mechanics is making the neighboring spins opposite, so that there is a locked-in system of spins with the electron spins of the iron one way and the electron spins of the yttrium the opposite way. But the yttrium atom is complicated. It is a rare-earth element and gets a large contribution to its magnetic moment from *orbital* motion of the electrons. For yttrium, the orbital motion contribution is *opposite* that of the spin and also is bigger. Thus, although quantum mechanics, working through the exclusion principle, makes the *spins* of the yttrium opposite those of the iron, it makes the *total* magnetic moment of the yttrium atom *parallel* to the iron because of the orbital effect—as sketched in Fig. 37–13(d). The compound is therefore a regular ferromagnet.

Another interesting example of ferromagnetism occurs in some of the rare-earth elements. It has to do with a still more peculiar arrangement of the spins. The material is not ferromagnetic in the sense that the spins are all parallel, nor is it antiferromagnetic in the sense that every atom is opposite. In these crystals all of the spins *in one layer* are parallel and lie in the plane of the layer. In the next layer all spins are again parallel to each other, but point in a somewhat different direction. In the following layer they are in still another direction, and so on. The result is that the local magnetization vector varies in the form of a spiral—the magnetic moments of the successive layers rotate as we proceed along a line perpendicular to the layers. It is interesting to try to analyze what happens when a field is applied to such a spiral—all the twistings and turnings that must go on in all those atomic magnets. (Some people *like* to amuse themselves with the theory of these things!) Not only are there cases of "flat" spirals, but there are also cases in which the directions the magnetic moments of successive layers map out a cone, so that it has a spiral component and also a uniform ferromagnetic component in one direction!

The magnetic properties of materials, worked out on a more advanced level than we have been able to do here, have fascinated physicists of all kinds. In the first place, there are those practical people who love to work out ways of making things in a better way—they love to design better and more interesting magnetic materials. The discovery of things like ferrites, or their application, immediately delights people who like to see clever new ways of doing things. Besides this, there are those who find a fascination in the terrible complexity that nature can produce using a few basic laws. Starting with one and the same general idea, nature goes from the ferromagnetism of iron and its domains, to the antiferromagnetism of chromium, to the magnetism of ferrites and garnets, to the spiral structure of the rare earth elements, and on, and on. It is fascinating to discover experimentally all the strange things that go on in these special substances. Then, to the theoretical physicists, ferromagnetism presents a number of very interesting, unsolved, and beautiful challenges. One challenge is to understand why it exists at all. Another is to predict the statistics of the interacting spins in an ideal lattice. Even neglecting any possible extraneous complications, this problem has, so far, defied full understanding. The reason that it is so interesting is that it is such an easily stated problem: Given a lot of electron spins in a regular lattice, interacting with such-and-such a law, what do they do? It is simply stated, but it has defied complete analysis for years. Although it has been analyzed rather carefully for temperatures not too close to the Curie point, the theory of the sudden transition at the Curie point still needs to be completed.

Finally, the whole subject of the system of spinning atomic magnets—in ferromagnetic, or in paramagnetic materials and in nuclear magnetism, has also been a fascinating thing to advanced students in physics. The system of spins can be pushed on and pulled on with external magnetic fields, so one can do many tricks with resonances, with relaxation effects, with spin-echoes, and with other effects. It serves as a prototype of many complicated thermodynamic systems. But in paramagnetic materials the situation is often fairly simple, and people have been delighted both to do experiments and to explain the phenomena theoretically.

We now close our study of electricity and magnetism. In the first chapter, we spoke of the great strides that have been made since the early Greek observation of the strange behaviors of amber and of lodestone. Yet in all our long and involved discussion we have never explained *why it is that when we rub a piece of amber we get a charge on it*, nor have we explained *why a lodestone is magnetized!* You may say, "Oh, we just didn't get the right sign." No, it is worse than that. Even if we *did* get the right sign, we would still have the question: Why is the piece of lodestone in the ground magnetized? There is the earth's magnetic field, of course, but *where does the earth's field come from*? Nobody really knows—there have only been some good guesses. So you see, this physics of ours is a lot of fakery—we start out with the phenomena of lodestone and amber, and we end up not understanding either of them very well. But we *have* learned a tremendous amount of very exciting and very practical information in the process!

38

Elasticity

Review: Chapter 47, Vol. I, *Sound; the Wave Equation.*

38–1 Hooke's law

The subject of elasticity deals with the behavior of those substances which have the property of recovering their size and shape when the forces producing deformations are removed. We find this elastic property to some extent in all solid bodies. If we had the time to deal with the subject at length, we would want to look into many things: the behavior of materials, the general laws of elasticity, the general theory of elasticity, the atomic machinery that determine the elastic properties, and finally the limitations of elastic laws when the forces become so great that plastic flow and fracture occur. It would take more time than we have to cover all these subjects in detail, so we will have to leave out some things. For example, we will not discuss plasticity or the limitations of the elastic laws. (We touched on these subjects briefly when we were talking about dislocations in metals.) Also, we will not be able to discuss the internal mechanisms of elasticity—so our treatment will not have the completeness we have tried to achieve in the earlier chapters. Our aim is mainly to give you an acquaintance with some of the ways of dealing with such practical problems as the bending of beams.

When you push on a piece of material, it "gives"—the material is deformed. If the force is small enough, the relative displacements of the various points in the material are proportional to the force—we say the behavior is *elastic*. We will discuss only the elastic behavior. First, we will write down the fundamental laws of elasticity, and then we will apply them to a number of different situations.

Suppose we take a rectangular block of material of length l, width w, and height h, as shown in Fig. 38–1. If we pull on the ends with a force F, then the length increases by an amount Δl. We will suppose in all cases that the change in length is a small fraction of the original length. As a matter of fact, for materials like wood and steel, the material will break if the change in length is more than a few percent of the original length. For a large number of materials, experiments show that for sufficiently small extensions the force is proportional to the extension

$$F \propto \Delta l. \tag{38.1}$$

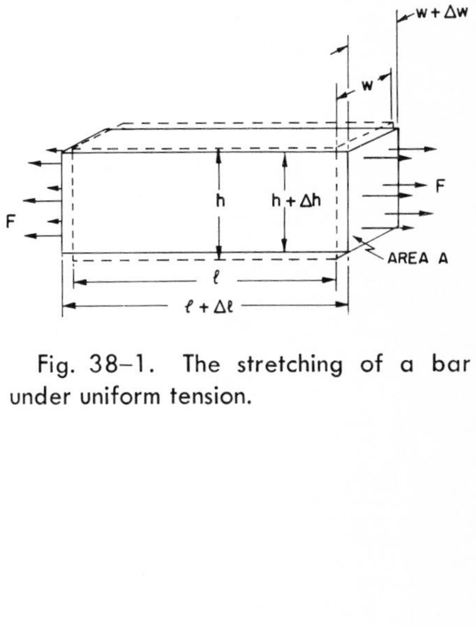

Fig. 38–1. The stretching of a bar under uniform tension.

This relation is known as *Hooke's law*.

The lengthening Δl of the bar will also depend on its length. We can figure out how by the following argument. If we cement two identical blocks together, end to end, the same forces act on each block; each will stretch by Δl. Thus, the stretch of a block of length $2l$ would be twice as big as a block of the same cross section, but of length l. In order to get a number more characteristic of the material, and less of any particular shape, we choose to deal with the ratio $\Delta l/l$ of the extension to the original length. This ratio is proportional to the force but independent of l:

$$F \propto \frac{\Delta l}{l}\cdot \tag{38.2}$$

The force F will also depend on the area of the block. Suppose that we put two blocks side by side. Then for a given stretch Δl we would have the force F on each block, or twice as much on the combination of the two blocks. The force, for a given amount of stretch, must be proportional to the cross-sectional area A of the block. To obtain a law in which the coefficient of proportionality is independent of the dimensions of the body, we write Hooke's law for a rectangular block in the form

$$F = YA\frac{\Delta l}{l}. \tag{38.3}$$

The constant Y is a property only of the nature of the material; it is known as *Young's modulus.* (Usually you will see Young's modulus called E. But we've used E for electric fields, energy, and emf's, so we prefer to use a different letter.)

The *force per unit area* is called the *stress*, and the stretch per unit length—the *fractional* stretch—is called the *strain*. Equation (38.3) can therefore be rewritten in the following way:

$$\frac{F}{A} = Y \times \frac{\Delta l}{l}, \tag{38.4}$$

$$\text{Stress} = (\text{Young's modulus}) \times (\text{Strain}).$$

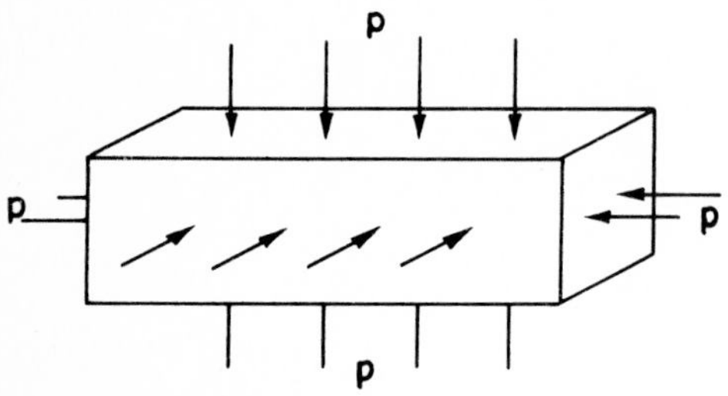

Fig. 38–2. A bar under uniform hydrostatic pressure.

There is another part to Hooke's law: When you *stretch* a block of material in one direction it *contracts* at right angles to the stretch. The contraction in width is proportional to the width w and also to $\Delta l/l$. The sideways contraction is in the same proportion for both width and height, and is usually written

$$\frac{\Delta w}{w} = \frac{\Delta h}{h} = -\sigma\frac{\Delta l}{l}, \tag{38.5}$$

where the constant σ is another property of the material called *Poisson's ratio.* It is always positive in sign and is a number less than 1/2. (It is "reasonable" that σ should be generally positive, but it is not quite clear that it *must* be so.)

The two constants Y and σ specify completely the elastic properties of a *homogeneous' isotropic* (that is, noncrystalline) material. In crystalline materials the stretches and contractions can be different in different directions, so there can be many more elastic constants. We will restrict our discussion temporarily to homogeneous' isotropic materials whose properties can be described by Y and σ. As usual there are different ways of describing things—some people like to describe the elastic properties of materials by different constants. It always takes two, and they can be related to σ and Y.

The last general law we need is the principle of superposition. Since the two laws (38.4) and (38.5) are linear in the forces and in the displacements, superposition will work. If you have one set of forces and get some displacements, and then you add a new set of forces and get some additional displacements, the resulting displacements will be the sum of the ones you would get with the two sets of forces acting independently.

Now we have all the general principles—the superposition principle and Eqs. (38.4) and (38.5)—and that's all there is to elasticity. But that is like saying that once you have Newton's laws that's all there is to mechanics. Or, given Maxwell's equations, that's all there is to electricity. It is, of course, true that with these principles you have a great deal, because with your present mathematical ability you could go a long way. We will, however, work out a few special applications.

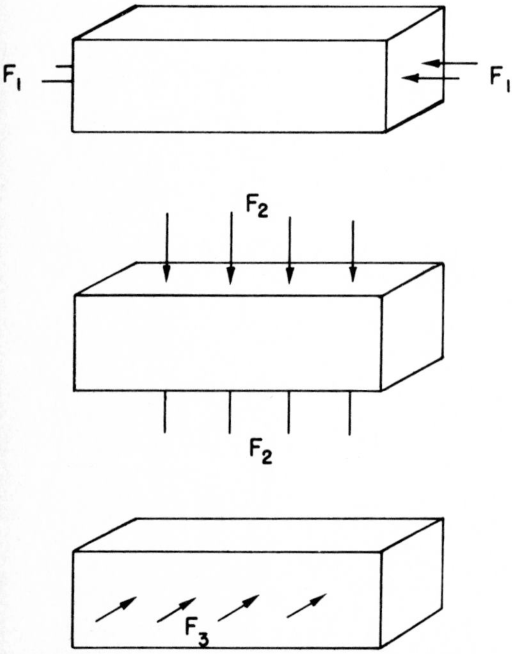

Fig. 38–3. Hydrostatic pressure is the superposition of three longitudinal compressions.

38–2 Uniform strains

As our first example let's find out what happens to a rectangular block under uniform hydrostatic pressure. Let's put a block under water in a pressure tank. Then there will be a force acting inward on every face of the block proportional to the area (see Fig. 38–2). Since the hydrostatic pressure is uniform, the *stress* (force per unit area) on each face of the block is the same. We will work out first the change in the length. The change in length of the block can be thought of as the sum of changes in length that would occur in the three independent problems which are sketched in Fig. 38–3.

Problem 1. If we push on the ends of the block with a pressure p, the compressional strain is p/Y, and it is negative,

$$\frac{\Delta l_1}{l} = -\frac{p}{Y}.$$

Problem 2. If we push on the two sides of the block with pressure p, the compressional strain is again p/Y, but now we want the lengthwise strain. We can get that from the sideways strain multiplied by $-\sigma$. The sideways strain is

$$\frac{\Delta w}{w} = -\frac{p}{Y};$$

so

$$\frac{\Delta l_2}{l} = +\sigma\frac{p}{Y}.$$

Problem 3. If we push on the top of the block, the compressional strain is once more p/Y, and the corresponding strain in the sideways direction is again $-\sigma p/Y$. We get

$$\frac{\Delta l_3}{l} = +\sigma\frac{p}{Y}.$$

Combining the results of the three problems—that is, taking $\Delta l = \Delta l_1 + \Delta l_2 + \Delta l_3$—we get

$$\frac{\Delta l}{l} = -\frac{p}{Y}(1 - 2\sigma). \tag{38.6}$$

The problem is, of course, symmetrical in all three directions; it follows that

$$\frac{\Delta w}{w} = \frac{\Delta h}{h} = -\frac{p}{Y}(1 - 2\sigma). \tag{38.7}$$

The change in the *volume* under hydrostatic pressure is also of some interest. Since $V = lwh$, we can write, for small displacements,

$$\frac{\Delta V}{V} = \frac{\Delta l}{l} + \frac{\Delta w}{w} + \frac{\Delta h}{h}.$$

Using (38.6) and (38.7), we have

$$\frac{\Delta V}{V} = -3\frac{p}{Y}(1 - 2\sigma). \tag{38.8}$$

People like to call $\Delta V/V$ the *volume strain* and write

$$p = -K\frac{\Delta V}{V}.$$

The *volume stress* p is proportional to the volume strain—Hooke's law once more. The coefficient K is called the *bulk modulus;* it is related to the other constants by

$$K = \frac{Y}{3(1 - 2\sigma)}. \tag{38.9}$$

Since K is of some practical interest, many handbooks give Y and K instead of Y and σ. If you want σ you can always get it from Eq. (38.9). We can also see from Eq. (38.9) that Poisson's ratio, σ, must be less than one-half. If it were not, the bulk modulus K would be negative, and the material would expand under increasing pressure. That would allow us to get mechanical energy *out of* any old block—it would mean that the block was in unstable equilibrium. If it started to expand it would continue by itself with a release of energy.

Now we want to consider what happens when you put a "shear" strain on something. By shear strain we mean the kind of distortion shown in Fig. 38–4. As a preliminary to this, let us look at the strains in a *cube* of material subjected to the forces shown in Fig. 38–5. Again we can break it up into two problems: the vertical

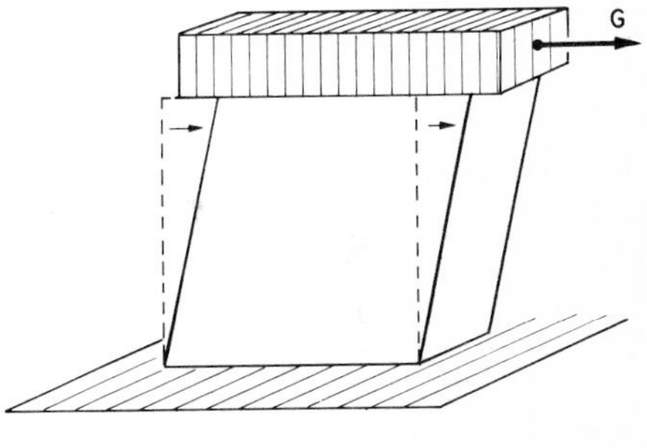

Fig. 38–4. A cube in uniform shear.

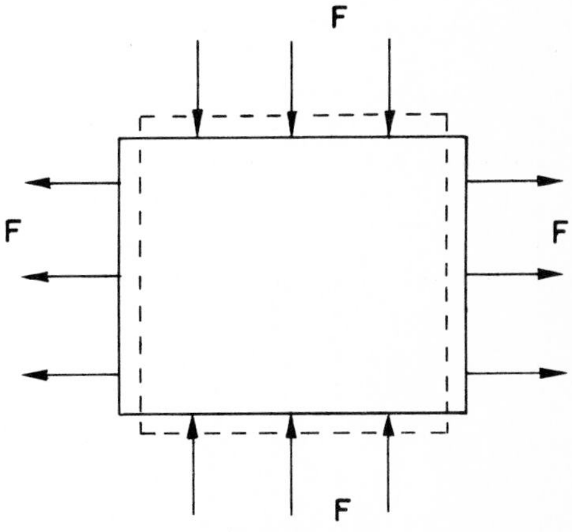

Fig. 38–5. A cube with compressing forces on top and bottom and equal stretching forces on two sides.

pushes, and the horizontal pulls. Calling A the area of the cube face, we have for the change in horizontal length

$$\frac{\Delta l}{l} = \frac{1}{Y}\frac{F}{A} + \sigma\frac{1}{Y}\frac{F}{A} = \frac{1+\sigma}{Y}\frac{F}{A}. \tag{38.10}$$

The change in the vertical height is just the negative of this.

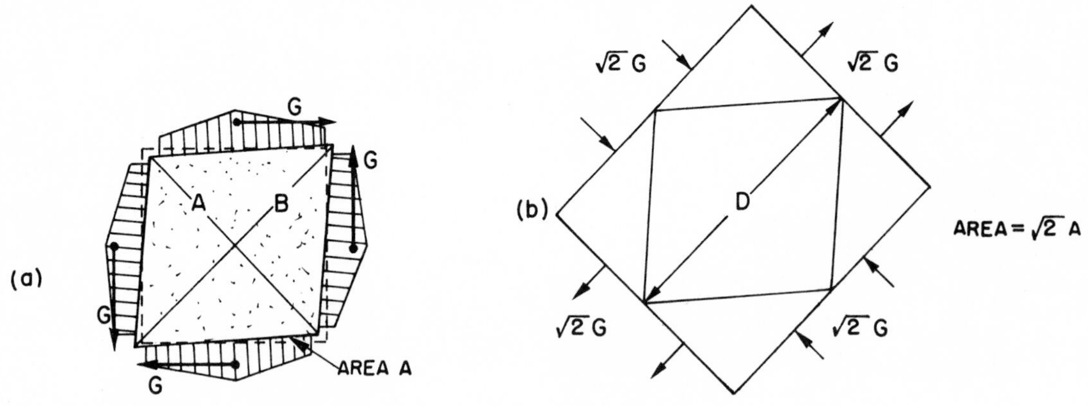

Fig. 38–6. The two pairs of shear forces in (a) produce the same stress as the compressing and stretching forces of (b).

Now suppose we have the same cube and subject it to the shearing forces shown in Fig. 38–6(a). Note that all the forces have to be equal if there are to be no net torques and the cube is to be in equilibrium. (Similar forces must also exist in Fig. 38–4, since the block is in equilibrium. They are provided through the "glue" that holds the block to the table.) The cube is then said to be in a state of pure shear. But note that if we cut the cube by a plane at 45°—say along the diagonal A in the figure—the total force acting across the plane is *normal* to plane and is equal to $\sqrt{2}G$. The area over which this force acts is $\sqrt{2}A$; therefore, the tensile stress normal to this plane is simply G/A. Similarly, if we examine a plane at an angle of 45° the other way—the diagonal B in the figure—we see that there is a compressional stress normal to this plane of $-G/A$. From this, we see that the *stress* in a "pure shear" is equivalent to a combination of tension and compression stresses of equal strength and at right angles to each other, and at 45° to the original faces of the cube. The internal stresses and strains are the same as we would find in the larger block of material with the forces shown in Fig. 38–6(b). But this is the problem we have already solved. The change in length of the diagonal is given by Eq. (38.10),

$$\frac{\Delta D}{D} = \frac{1+\sigma}{Y}\frac{G}{A}. \tag{38.11}$$

(One diagonal is shortened; the other is elongated.)

It is often convenient to express a shear strain in terms of the angle by which the cube is twisted—the angle θ in Fig. 38–7. From the geometry of the figure you can see that the horizontal shift δ of the top edge is equal to $\sqrt{2}\,\Delta D$. So

$$\theta = \frac{\delta}{l} = \frac{\sqrt{2}\,\Delta D}{l} = 2\,\frac{\Delta D}{D}. \tag{38.12}$$

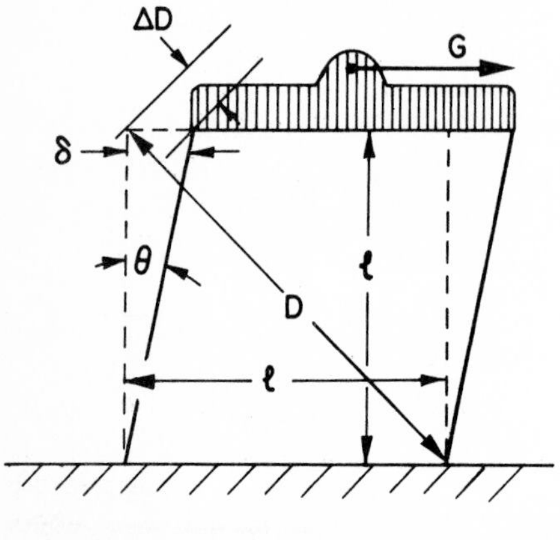

Fig. 38–7. The shear strain θ is $2\,\Delta D/D$.

The shear stress g is defined as the tangential force on one face divided by the area, $g = G/A$. Using Eq. (38.11) in (38.12), we get

$$\theta = 2\,\frac{1+\sigma}{Y}\,g.$$

Or, writing this in the form "stress = constant times strain,"

$$g = \mu\theta. \tag{38.13}$$

The proportionality coefficient μ is called the *shear modulus* (or, sometimes, the coefficient of rigidity). It is given in terms of Y and σ by

$$\mu = \frac{Y}{2(1+\sigma)}. \tag{38.14}$$

Incidentally, the shear modulus must be positive—otherwise you could get work out of a self-shearing block. From Eq. (38.14), σ must be greater than -1. We know, then, that σ must be between -1 and $+\frac{1}{2}$; in practice, however, it is always greater than zero.

As a last example of the type of situation where the stresses are uniform through the material, let's consider the problem of a block which is stretched, while it is at the same time *constrained* so that no lateral contraction can take place. (Technically, it's a little easier to compress it while keeping the sides from bulging out—but it's the same problem.) What happens? Well, there must be sideways forces which keep it from changing its thickness—forces we don't know off-hand but will have to calculate. It's the same kind of problem we have already done, only with a little different algebra. We imagine forces on all three sides, as shown in Fig. 38–8; we calculate the changes in dimensions, and we choose the transverse forces to make the width and height remain constant. Following the usual arguments, we get for the three strains:

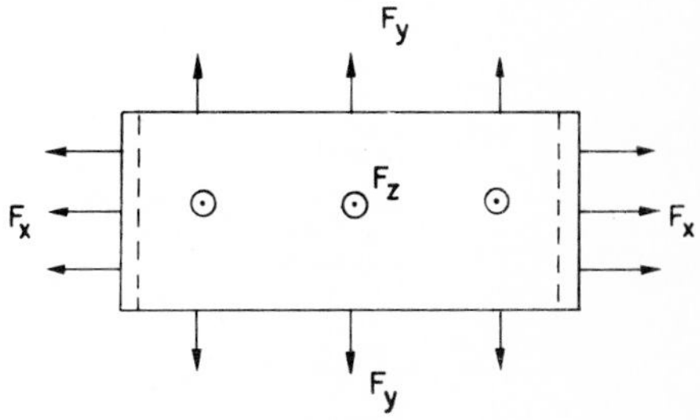

Fig. 38–8. Stretching without lateral contraction.

$$\frac{\Delta l_x}{l_x} = \frac{1}{Y}\frac{F_x}{A_x} - \frac{\sigma}{Y}\frac{F_y}{A_y} - \frac{\sigma}{Y}\frac{F_z}{A_z} = \frac{1}{Y}\left[\frac{F_x}{A_x} - \sigma\left(\frac{F_y}{A_y} + \frac{F_z}{A_z}\right)\right], \tag{38.15}$$

$$\frac{\Delta l_y}{l_y} = \frac{1}{Y}\left[\frac{F_y}{A_y} - \sigma\left(\frac{F_x}{A_x} + \frac{F_z}{A_z}\right)\right], \tag{38.16}$$

$$\frac{\Delta l_z}{l_z} = \frac{1}{Y}\left[\frac{F_z}{A_z} - \sigma\left(\frac{F_z}{A_x} + \frac{F_y}{A_y}\right)\right]. \tag{38.17}$$

Now since Δl_y and Δl_z are supposed to be zero, Eqs. (38.16) and (38.17) give two equations relating F_y and F_z to F_x. Solving them together, we get that

$$\frac{F_y}{A_y} = \frac{F_z}{A_z} = \frac{\sigma}{1-\sigma}\frac{F_x}{A_x}. \tag{38.18}$$

Substituting in (38.15), we have

$$\frac{\Delta l_x}{l_x} = \frac{1}{Y}\left(1 - \frac{2\sigma^2}{1-\sigma}\right)\frac{F_x}{A_x} = \frac{1}{Y}\left(\frac{1-\sigma-2\sigma^2}{1-\sigma}\right)\frac{F_x}{A_x}. \tag{38.19}$$

Often, you will see this turned around, and with the quadratic in σ factored out, it is then written

$$\frac{F}{A} = \frac{1-\sigma}{(1+\sigma)(1-2\sigma)}\,Y\frac{\Delta l}{l}. \tag{38.20}$$

When we constrain the sides, Young's modulus gets multiplied by a complicated function of σ. As you can most easily see from Eq. (38.19), the factor in front of Y is always greater than 1. It is harder to stretch the block when the sides are held—which also means that a block is *stronger* when the sides are held than when they are not.

38–3 The torsion bar; shear waves

Let's now turn our attention to an example which is more complicated because different parts of the material are stressed by different amounts. We consider a twisted rod such as you would find in a drive shaft of some machinery, or in a quartz fiber suspension used in a delicate instrument. As you probably know from experiments with the torsion pendulum, the *torque* on a twisted rod is proportional to the *angle*—the constant of proportionality obviously depending upon the length of the rod, on the radius of the rod, and on the properties of the material. The question is: In what way? We are now in a position to answer this question; it's just a matter of working out some geometry.

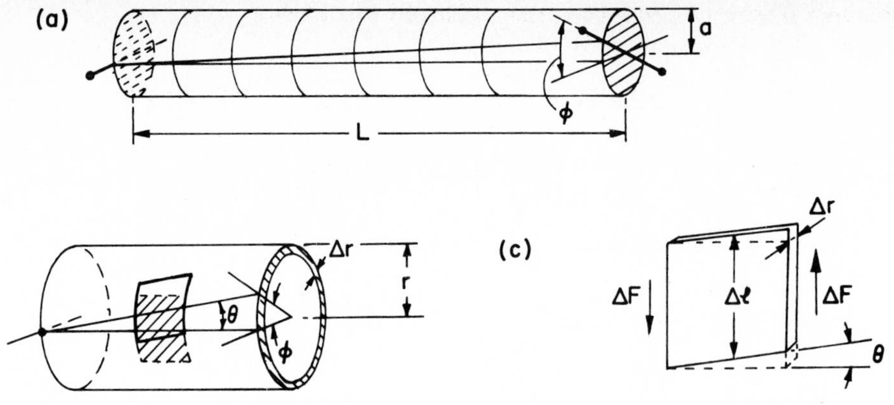

Fig. 38–9. (a) A cylindrical bar in torsion. (b) A cylindrical shell in torsion. (c) Each small piece of the shell is in shear.

Fig. 38–9(a) shows a cylindrical rod of length L, and radius a, with one end twisted by the angle ϕ with respect to the other. If we want to relate the strains to what we already know, we can think of the rod as being made up of many cylindrical shells and work out separately what happens to each shell. We start by looking at a thin, short cylinder of radius r (less than a) and thickness Δr—as drawn in Fig. 38–9(b). Now if we look at a piece of this cylinder that was originally a small square, we see that it has been distorted into a parallelogram. Each such element of the cylinder is in shear, and the shear angle θ is

$$\theta = \frac{r\phi}{L}\cdot$$

The shear stress g in the material is, therefore [from Eq. (38.13)],

$$g = \mu\theta = \mu\,\frac{r\phi}{L}\cdot \tag{38.21}$$

The shear stress is the tangential force ΔF on the end of the square divided by the area $\Delta l\,\Delta r$ of the end [see Fig. 38–9(c)]

$$g = \frac{\Delta F}{\Delta l \Delta r}\cdot$$

The force ΔF on the end of such a square contributes a torque $\Delta\tau$ around the axis of the rod equal to

$$\Delta\tau = r\,\Delta F = rg\,\Delta l\,\Delta r. \tag{38.22}$$

The total torque τ is the sum of such torques around a complete circumference of the cylinder. So putting together enough pieces so that the Δl's add up to $2\pi r$, we find that the total torque, for a *hollow tube*, is

$$rg(2\pi r)\,\Delta r. \tag{38.23}$$

Or, using (38.21),

$$\tau = 2\pi\mu\,\frac{r^3\,\Delta r\phi}{L}\cdot \tag{38.24}$$

We get that the rotational stiffness, τ/ϕ, of a hollow tube is proportional to the cube of the radius r and to the thickness Δr, and inversely proportional to the length L.

We can now imagine a solid rod to be made up of a series of concentric tubes, each twisted by the same angle ϕ (although the internal *stresses* are different for each tube). The total torque is the sum of the torques required to rotate each shell; for the *solid* rod

$$\tau = 2\pi\mu\,\frac{\phi}{L}\int r^3\,dr,$$

where the integral goes from $r = 0$ to $r = a$, the radius of the rod. Integrating, we have

$$\tau = \mu \frac{\pi a^4}{2L} \phi. \tag{38.25}$$

For a rod in torsion, the torque is proportional to the angle and is proportional to the *fourth power* of the diameter—a rod twice as thick is sixteen times as stiff for torsion.

Before leaving the subject of torsion, let us apply what we have just learned to an interesting problem: torsional waves. If you take a long rod and suddenly twist one end, a wave of twist works it way along the rod, as sketched in Fig. 38–10(a). That's a little more exciting than a steady twist—let's see whether we can work out what happens.

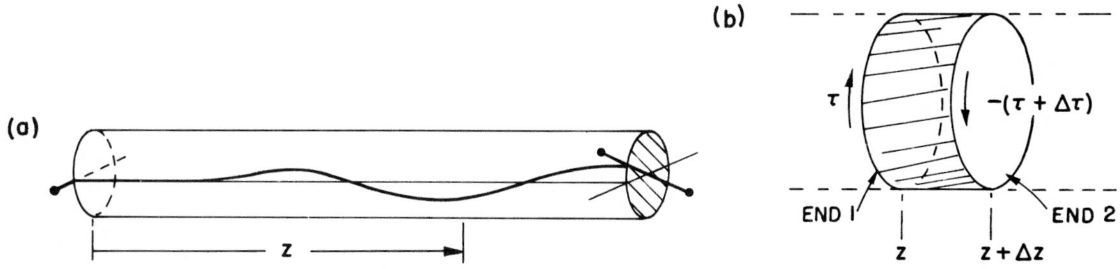

Fig. 38–10. (a) A torsional wave on a rod. (b) A volume element of the rod.

Let z be the distance to some point down the rod. For a static torsion the torque is the same everywhere along the rod, and is proportional to ϕ/L, the total torsion angle over the total length. What matters to the material is the local torsional strain, which is, you will appreciate, $\partial\phi/\partial z$. When the torsion along the rod is not uniform, we should replace Eq. (38.25) by

$$\tau(z) = \mu \frac{\pi a^4}{2} \frac{\partial \phi}{\partial z}\cdot \tag{38.26}$$

Now let's look at what happens to an element of length Δz shown magnified in Fig. 38–10(b). There is a torque $\tau(z)$ at end 1 of the little hunk of rod, and a different torque $\tau(z + \Delta z)$ at end 2. If Δz is small enough, we can use a Taylor expansion and write

$$\tau(z + \Delta z) = \tau(z) + \left(\frac{\partial \tau}{\partial z}\right) \Delta z. \tag{38.27}$$

The net torque $\Delta\tau$ acting *on* the little piece of rod between z and $z + \Delta z$ is clearly the difference between $\tau(z)$ and $\tau(z + \Delta z)$, or $\Delta\tau = (\partial\tau/\partial z)\,\Delta z$. Differentiating Eq. (38.26), we get

$$\Delta\tau = \mu \frac{\pi a^4}{2} \frac{\partial^2 \phi}{\partial z^2} \Delta z. \tag{38.28}$$

The effect of this net torque is to give an angular acceleration to the little slice of the rod. The mass of the slice is

$$\Delta M = (\pi a^2\,\Delta z)\rho,$$

where ρ is the density of the material. We worked out in Chapter 19, Vol. I, that the moment of inertia of a circular cylinder is $mr^2/2$; calling the moment of inertia of our piece ΔI, we have

$$\Delta I = \frac{\pi}{2} \rho a^4\,\Delta z. \tag{38.29}$$

Newton's law says the torque is equal to the moment of inertia times the angular acceleration, or

$$\Delta\tau = \Delta I \frac{\partial^2 \phi}{\partial t^2}\cdot \tag{38.30}$$

Pulling everything together, we get

$$\mu \frac{\pi a^4}{2} \frac{\partial^2 \phi}{\partial z^2} \Delta z = \frac{\pi}{2} \rho a^4 \Delta z \frac{\partial^2 \phi}{\partial t^2},$$

or

$$\frac{\partial^2 \phi}{\partial z^2} - \frac{\rho}{\mu} \frac{\partial^2 \phi}{\partial t^2} = 0. \tag{38.31}$$

You will recognize this as the one-dimensional wave equation. We have found that waves of torsion will propagate down the rod with the speed

$$C_{\text{shear}} = \sqrt{\frac{\mu}{\rho}}. \tag{38.32}$$

The *denser* the rod—for the same stiffness—the *slower* the waves; and the *stiffer* the rod, the quicker the waves work their way down. The speed does *not* depend upon the diameter of the rod.

Torsional waves are a special example of *shear waves.* In general, shear waves are those in which the strains do not change the *volume* of any part of the material. In torsional waves, we have a particular distribution of such shear stresses—namely, distributed on a circle. But for any arrangement of shear stresses, waves will propagate with the same speed—the one given in Eq. (38.32). For example, the seismologists find such shear waves travelling in the interior of the earth.

We can have another kind of a wave in the elastic world inside a solid material. If you push something, you can start "longitudinal" waves—also called "compressional" waves. They are like the sound waves in air or in water—the displacements are in the same direction as the wave propagation. (At the surfaces of an elastic body there can also be other types of waves—called "Rayleigh waves" or "Love waves." In them, the strains are neither purely longitudinal nor purely transverse. We will not have time to study them.)

While we're on the subject of waves, what is the velocity of the pure compressional waves in a *large* solid body like the earth? We say "large" because the speed of sound in a thick body is different from what it is, for instance, along a thin rod. By a "thick" body we mean one in which the transverse dimensions are much larger than the wavelength of the sound. Then, when we push on the object, it cannot expand sideways—it can only compress in one dimension. Fortunately, we have already worked out the special case of the compression of a constrained elastic material. We have also worked out in Chapter 47, Vol. I, the speed of sound waves in a gas. Following the same arguments you can see that the speed of sound in a solid is equal to $\sqrt{Y'/\rho}$, where Y' is the "longitudinal modulus"—or pressure divided by the relative change in length—for the constrained case. This is just the ratio of $\Delta l/l$ to F/A we got in Eq. (38.20). So the speed of the longitudinal waves is given by

$$C_{\text{long}}^2 = \frac{Y'}{\rho} = \frac{1-\sigma}{(1+\sigma)(1-2\sigma)} \frac{Y}{\rho}. \tag{38.33}$$

So long as σ is between zero and 1/2, the shear modulus μ is less than Young's modulus Y, and also Y' is greater than Y, so

$$\mu < Y < Y'.$$

This means that longitudinal waves travel faster than shear waves. One of the most precise ways of measuring the elastic constants of a substance is by measuring the density of the material and the speeds of the two kinds of waves. From this information one can get both Y and σ. It is, incidentally, by measuring the difference in the arrival times of the two kinds of waves from an earthquake that a seismologist can estimate—even from the signals at only one station—the distance to the quake.

38–4 The bent beam

We want now to look at another practical matter—the *bending* of a rod or a beam. What are the forces when we bend a bar of some arbitrary cross section? We will work it out thinking of a bar with a circular cross section, but our answer will be good for any shape. To save time, however, we will cut some corners, so our theory we will work out is only approximate. Our results will be correct only when the radius of the bend is much larger than the thickness of the beam.

Suppose you grab the two ends of a straight bar and bend it into some curve like the one shown in Fig. 38–11. What goes on inside the bar? Well, if it is curved, that means that the material on the inside of the curve is compressed and the material on the outside is stretched. There is some surface which goes along more or less parallel to the axis of the bar that is neither stretched nor compressed. This is called the *neutral* surface. You would expect this surface to be near the "middle" of the cross section. It can be shown (but we won't do it here) that, for small bending of simple beams, the neutral surface goes through the "center of gravity" of the cross section. This is true only for "pure" bending—if you are not stretching or compressing the beam at the same time.

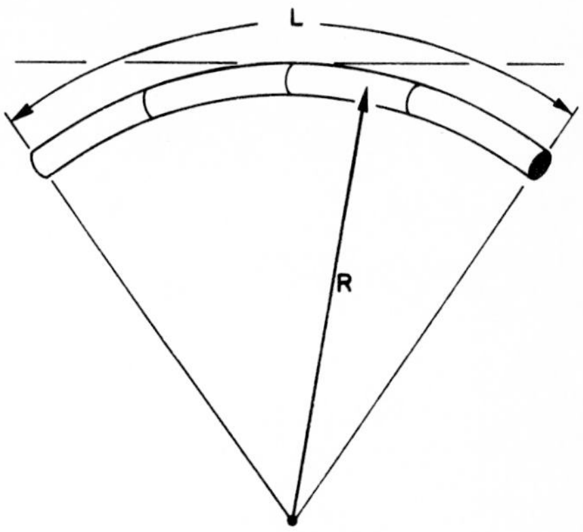

Fig. 38–11. A bent beam.

For pure bending, then, a thin transverse slice of the bar is distorted as shown in Fig. 38–12(a). The material below the neutral surface has a compressional strain which is *proportional to the distance* from the neutral surface; and the material above is stretched, also in proportion to its distance from the neutral surface. So the longitudinal *stretch* Δl is proportional to the height y. The constant of proportionality is just l over the radius of curvature of the bar—see Fig. 38–12:

$$\frac{\Delta l}{l} = \frac{y}{R}.$$

So the force per unit area—the stress—in a small strip at y is also proportional to the distance from the neutral surface

$$\frac{\Delta F}{\Delta A} = Y\frac{y}{R}. \tag{38.34}$$

Now let's look at the *forces* that would produce such a strain. The forces acting on the little segment drawn in Fig. 38–12 are shown in the figure. If we think of any transverse cut, the forces acting across it are one way above the neutral surface and the other way below. They come in pairs to make a "bending moment" $\mathfrak{M}$—by which we mean the torque about the neutral line. We can compute the total moment by integrating the force times the distance from the neutral surface for one of the faces of the segment of Fig. 38–12:

$$\mathfrak{M} = \int_{\substack{\text{cross}\\\text{sect}}} y\,dF. \tag{38.35}$$

From Eq. (38.34), $dF = Yy/R\,dA$, so

$$\mathfrak{M} = \frac{Y}{R}\int y^2\,dA.$$

The integral of $y^2\,dA$ is what we can call the "moment of inertia" of the geometric cross section about a horizontal axis through its "center of mass";* we will call it I:

$$\mathfrak{M} = \frac{YI}{R} \tag{38.36}$$

$$I = \int y^2\,dA. \tag{38.37}$$

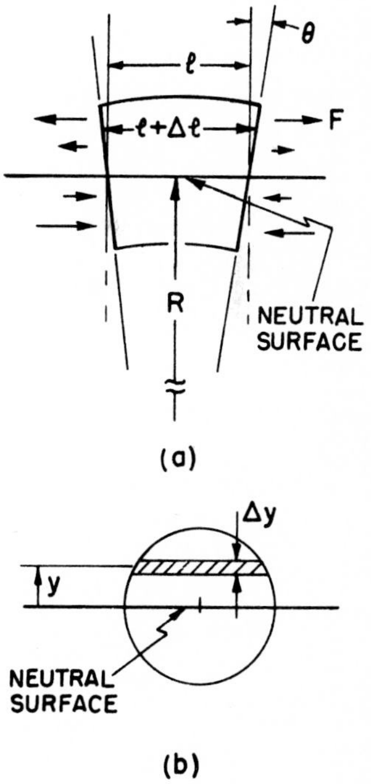

Fig. 38–12. (a) Small segment of a bent beam. (b) Cross section of the beam.

* It is, of course, really the moment of inertia of a slice with unit mass per unit area.

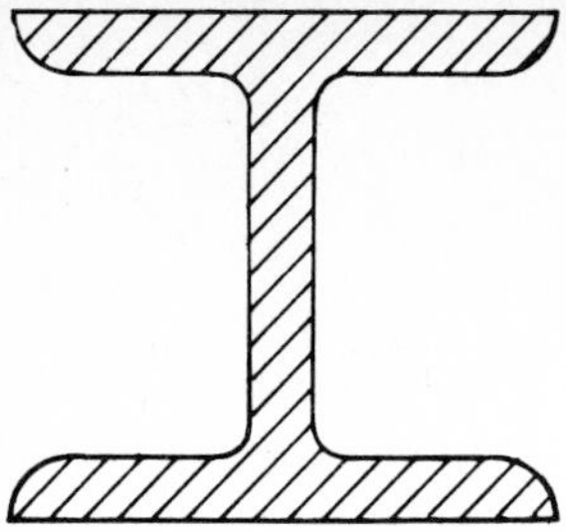

Fig. 38–13. An "I" beam.

Equation (38.36), then, gives us the relation between the bending moment $\mathfrak{M}$ and the curvature $1/R$ of the beam. The "stiffness" of the beam is proportional to Y and to the moment of inertia I. In other words, if you want the stiffest possible beam with a given amount of, say, aluminum, you want to put as much of it as possible as far as you can from the neutral surface, to make a large moment of inertia. You can't carry this to an extreme, however, because then the thing will not curve as we have supposed—it will buckle or twist and become weaker again. But now you see why structural beams are made in the form of an I or an H—as shown in Fig. 38–13.

As an example of the use of our beam equation (38.36), let's work out the deflection of a cantilevered beam with a concentrated force W acting at the free end, as sketched in Fig. 38–14. (By "cantilevered" we simply mean that the beam is supported in such a way that both the position *and* the slope are fixed at one end—it is stuck into a cement wall.) What is the shape of the beam? Let's call the deflection at the distance x from the fixed end z; we want to know $z(x)$. We'll work it out only for small deflections. We will also assume that the beam is long in comparison with its cross section. Now, as you know from your mathematics courses, the curvature $1/R$ of any curve $z(x)$ is given by

$$\frac{1}{R} = \frac{d^2z/dx^2}{[1 + (dz/dx)^2]^{3/2}}. \tag{38.38}$$

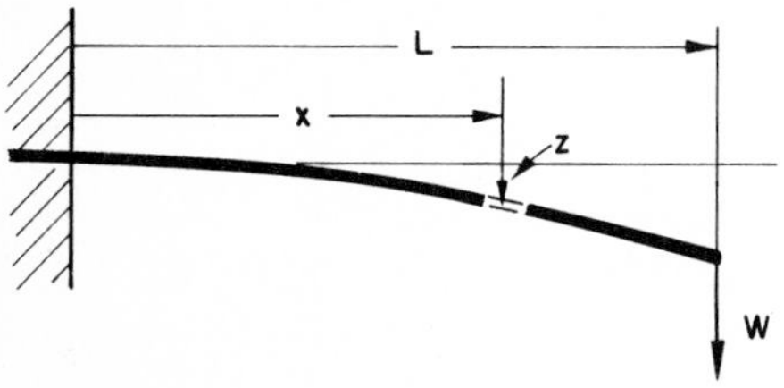

Fig. 38–14. A cantilevered beam with a weight at one end.

Since we are interested only in small slopes—this is usually the case in engineering structures—we neglect $(dz/dx)^2$ in comparison with 1, and take

$$\frac{1}{R} = \frac{d^2z}{dx^2}. \tag{38.39}$$

We also need to know the bending moment $\mathfrak{M}$. It is a function of x because it is equal to the torque about the neutral axis of any cross section. Let's neglect the weight of the beam and take only the downward force W at the end of the beam. (You can put in the beam weight yourself if you want.) Then the bending moment at x is

$$\mathfrak{M}(x) = W(L - x),$$

because that is the torque about the point at x, exerted by the weight W—the torque which the beam must support of x. We get

$$W(L - x) = \frac{YI}{R} = YI\frac{d^2z}{dx^2}$$

or

$$\frac{d^2z}{dx^2} = \frac{W}{YI}(L - x). \tag{38.40}$$

This one we can integrate without any tricks; we get

$$z = \frac{W}{YI}\left(\frac{Lx^2}{2} - \frac{x^3}{6}\right), \tag{38.41}$$

using our assumptions that $z(0) = 0$ and that dz/dx is also zero at $x = 0$. That is the shape of the beam. The displacement of the end is

$$z(L) = \frac{W}{YI}\frac{L^3}{3}; \tag{38.42}$$

the displacement of the end of a beam increases as the cube of the length.

In deriving our approximate beam theory, we have assumed that the cross section of the beam did not change when the beam was bent. When the thickness of the beam is small compared to the radius of curvature, the cross section changes very little and our result is O.K. In general, however, this effect cannot be neglected, as you can easily demonstrate for yourselves by bending a soft-rubber eraser in your fingers. If the cross section was originally rectangular, you will find that when it is bent it bulges at the bottom (see Fig. 38–15). This happens because when we compress the bottom, the material expands sideways—as described by Poisson's ratio. Rubber is easy to bend or stretch, but it is somewhat like a liquid in that it's hard to change the *volume*—as shows up nicely when you bend the eraser. For an incompressible material, Poisson's ratio would be exactly 1/2—for rubber it is nearly that.

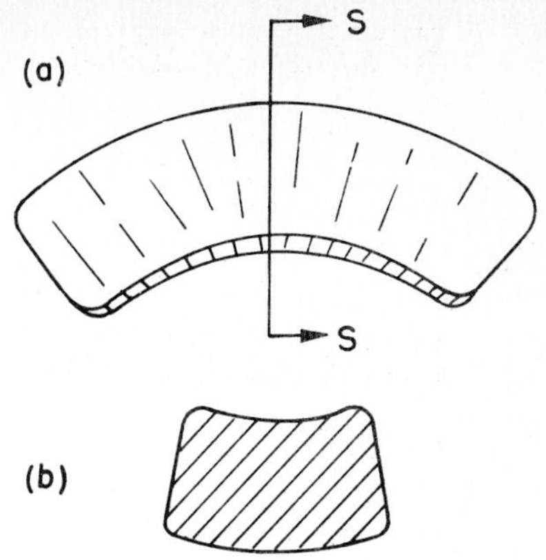

Fig. 38–15. (a) A bent eraser; (b) cross section.

38–5 Buckling

We want now to use our beam theory to understand the theory of the "buckling" of beams, or columns, or rods. Consider the situation sketched in Fig. 38–16 in which a rod that would normally be straight is held in its bent shape by two opposite forces that push on the ends of the rod. We would like to calculate the shape of the rod and the *magnitude of the forces* on the ends.

Let the deflection of the rod from the straight line between the ends be $y(x)$, where x is the distance from one end. The bending moment $\mathfrak{M}$ at the point P in the figure is equal to the force F multiplied by the moment arm, which is the perpendicular distance y,

$$\mathfrak{M}(x) = Fy. \tag{38.43}$$

Using the beam equation (38.36), we have

$$\frac{YI}{R} = Fy. \tag{38.44}$$

For small deflections, we can take $1/R = -d^2y/dx^2$ (the minus sign because the curvature is downward). We get

$$\frac{d^2y}{dx^2} = -\frac{F}{YI}\,y, \tag{38.45}$$

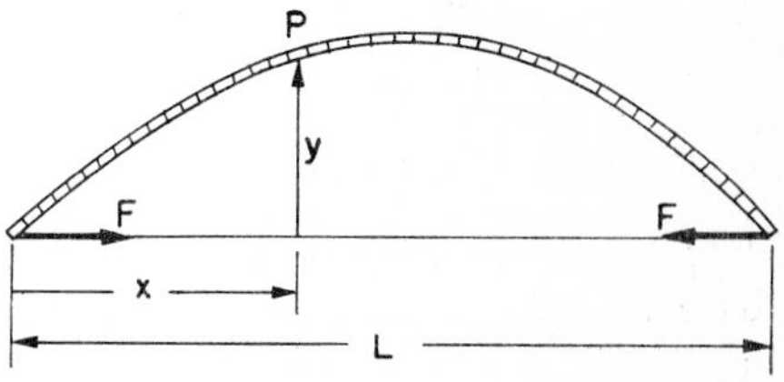

Fig. 38–16. A buckled beam.

which is the differential equation of a sine wave. So for *small* deflections, the curve of such a bent beam is a sine curve. The "wavelength" λ of the sine wave is twice the distance L between the ends. If the bending is small, this is just twice the unbent length of the rod. So the curve is

$$y = K\sin \pi x/L.$$

Taking the second derivative, we get

$$\frac{d^2y}{dx^2} = -\frac{\pi^2}{L^2}\,y.$$

Comparing this to Eq. (38.45), we see that the force is

$$F = \pi^2\frac{YI}{L^2}. \tag{38.46}$$

For small bendings the force is *independent of the bending displacement y!*

We have, then, the following thing physically. If the force is less than the F given in Eq. (38.46), there will be no bending at all. But if it is slightly *greater* than this force, the material will suddenly bend a large amount—that is, for forces above the critical force $\pi^2 YI/L^2$ (often called the "Euler force") the beam will "buckle." If the loading on the second floor of a building exceeds the Euler force for the supporting columns, the building will collapse. Another place where the buckling force is most important is in space rockets. On one hand, the rocket must be able to hold its own weight on the launching pad and endure the stresses during acceleration; on the other hand, it is important to keep the weight of the structure to a minimum, so that the payload and fuel capacity may be made as large as possible.

Actually a beam will not necessarily collapse completely when the force exceeds the Euler force. When the displacements get large, the force is larger than

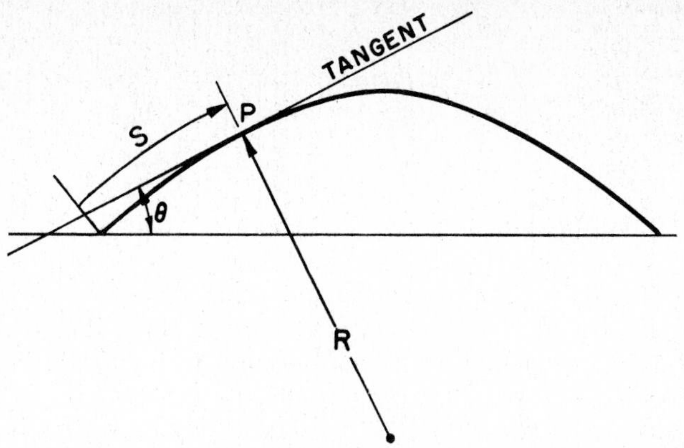

Fig. 38–17. The coordinates S and θ for the curve of a bent beam.

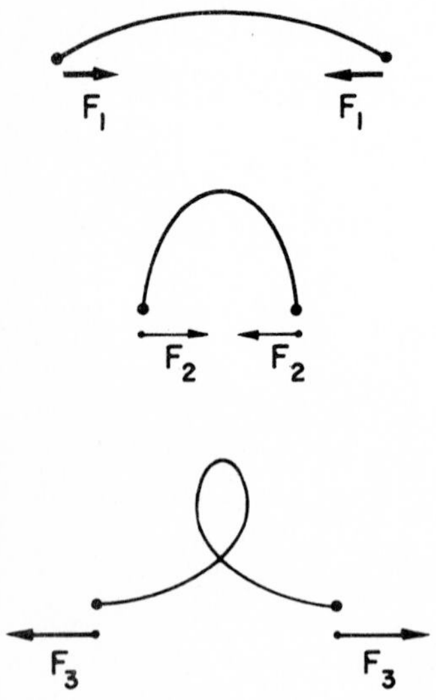

Fig. 38–18. Curves of a bent rod

what we have found because of the terms in $1/R$ in Eq. (38.38) that we have neglected. To find the forces for a large bending of the beam, we have to go back to the exact equation, Eq. (38.44), which we had before we used the approximate relation between R and y. Equation (38.44) has a rather simple geometrical property.* It's a little complicated to work out, but rather interesting. Instead of describing the curve in terms of x and y, we can use two new variables: S, the distance along the curve, and θ the slope of the tangent to the curve. See Fig. 38–17. The curvature is the rate of change of angle with distance:

$$\frac{1}{R} = \frac{d\theta}{dS}.$$

We can, therefore write the exact equation (38.44) as

$$\frac{d\theta}{dS} = -\frac{F}{YI}\,y.$$

If we take the derivative of this equation with respect to S and replace dy/dS by $\sin\theta$, we get

$$\frac{d^2\theta}{dS^2} = -\frac{F}{YI}\sin\theta. \tag{38.47}$$

[If θ is small, we get back Eq. (38.45). Everything is O.K.]

Now it may or may not delight you to know that Eq. (38.47) is exactly the same one you get for the large amplitude oscillations of a pendulum—with F/YI replaced by another constant, of course. We learned way back in Chapter 9, Vol. I, how to find the solution of such an equation by a numerical calculation.† The answers you get are some fascinating curves—known as the curves of the "Elastica." Figure 38–18 shows three curves for different values of F/YI.

* The same equation appears, incidentally, in other physical situations—for example, the meniscus at the surface of a liquid contained between parallel planes—and the same geometrical solution can be used.

† The solutions can also be expressed in terms of some functions, called the "Jacobian elliptic functions," that someone else has already computed.

39

Elastic Materials

Reference: C. Kittel, *Introduction to Solid State Physics*, John Wiley and Sons, Inc., New York, 2nd ed., 1956.

39–1 The tensor of strain

In the last chapter we talked about the distortions of particular elastic objects. In this chapter we want to look at what can happen *in general* inside an elastic material. We would like to be able to describe the conditions of stress and strain inside some big glob of jello which is twisted and squashed in some complicated way. To do this, we need to be able to describe the *local strain* at every point in an elastic body; we can do it by giving a set of six numbers—which are the components of a symmetric tensor—for each point. Earlier, we spoke of the stress tensor (Chapter 31); now we need the tensor of strain.

Imagine that we start with the material initially unstrained and watch the motion of a small speck of "dirt" embedded in the material when the strain is applied. A speck that was at the point P located at $\boldsymbol{r} = (x, y, z)$ moves to a new position P' at $\boldsymbol{r}' = (x', y', z')$ as shown in Fig. 39–1. We will call $\boldsymbol{u}$ the vector displacements from P to P'. Then

$$\boldsymbol{u} = \boldsymbol{r}' - \boldsymbol{r}. \tag{39.1}$$

The displacement $\boldsymbol{u}$ depends, of course, on which point P we start with, so $\boldsymbol{u}$ is a vector function of $\boldsymbol{r}$—or, if you prefer, of (x, y, z).

Let's look first at a simple situation in which the strain is constant over the material—so we have what is called a *homogeneous strain.* Suppose, for instance, that we have a block of material and we stretch it uniformly. We just change its dimensions uniformly in one direction—say, in the x-direction, as shown in Fig. 39–2. The motion u_x of a speck at x is proportional to x. In fact,

$$\frac{u_x}{x} = \frac{\Delta l}{l}.$$

We will write u_x this way:

$$u_x = e_{xx}x.$$

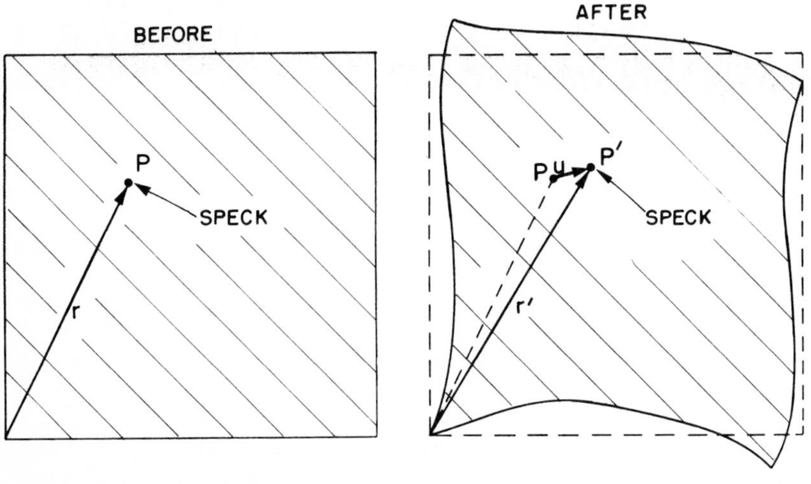

Fig. 39–1. A speck of the material at the point P in an unstrained block moves to P' where the block is strained.

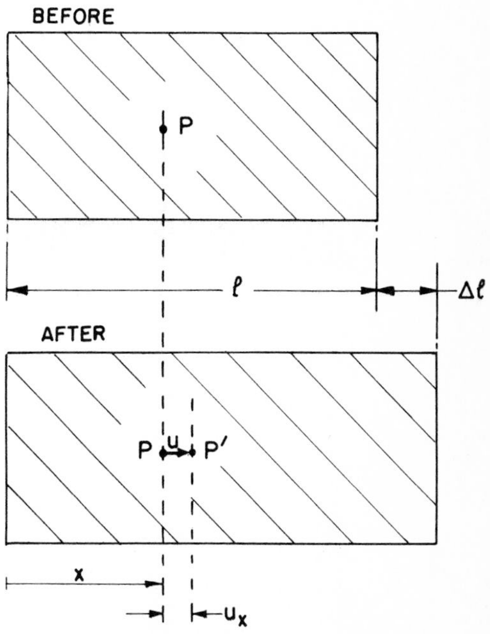

Fig. 39–2. A homogeneous stretch-type strain.

The proportionality constant e_{xx} is, of course, the same thing as $\Delta l/l$. (You will see shortly why we use a double subscript.)

If the strain is not uniform, the relation between u_x and x will vary from place to place in the material. For the general situation, we define the e_{xx} by a kind of local $\Delta l/l$, namely by

$$e_{xx} = \partial u_x/\partial x. \tag{39.2}$$

This number—which is now a function of x, y, and z—describes the amount of stretching in the x-direction throughout the hunk of jello. There may, of course, also be stretching in the y- and z-directions. We describe them by the numbers

$$e_{yy} = \frac{\partial u_y}{\partial y}, \qquad e_{zz} = \frac{\partial u_z}{\partial z}. \tag{39.3}$$

We need to be able to describe also the shear-type strains. Suppose we imagine a little cube marked out in the initially undisturbed jello. When the jello is pushed out of shape, this cube may get changed into a parallelogram, as sketched in Fig. 39–3.* In this kind of a strain, the x-motion of each particle is proportional to its y-coordinate,

$$u_x = \frac{\theta}{2}\,y. \tag{39.4}$$

And there is also a y-motion proportional to x,

$$u_y = \frac{\theta}{2}\,x. \tag{39.5}$$

So we can describe such a shear-type strain by writing

$$u_x = e_{xy}y, \qquad u_y = e_{yx}x$$

with

$$e_{xy} = e_{yx} = \frac{\theta}{2}.$$

Now you might think that when the strains are not homogeneous we could describe the generalized shear strains by defining the quantities e_{xy} and e_{yx} by

$$e_{xy} = \frac{\partial u_x}{\partial y}, \qquad e_{yx} = \frac{\partial u_y}{\partial x}. \tag{39.6}$$

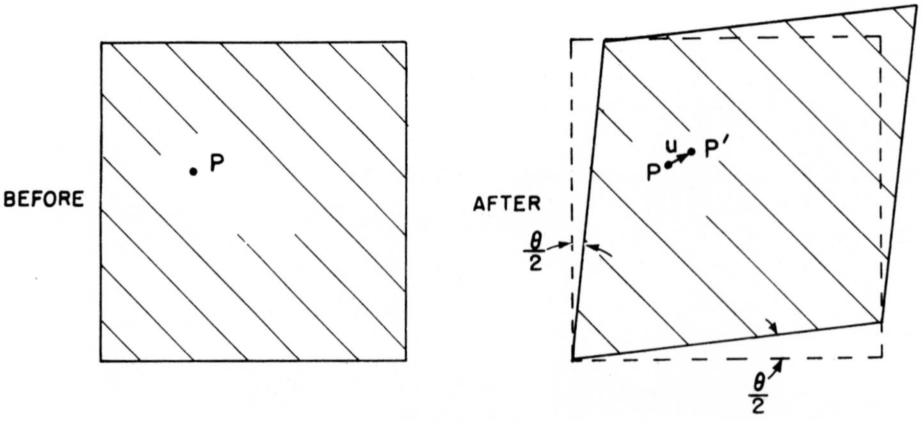

Fig. 39–3. A homogeneous shear strain.

But there is one difficulty. Suppose that the displacements u_x and u_y were given by

$$u_x = \frac{\theta}{2}\,y, \qquad u_y = -\frac{\theta}{2}.$$

---

* We choose for the moment to split the total shear angle θ into two equal parts and make the strain symmetric with respect to x and y.

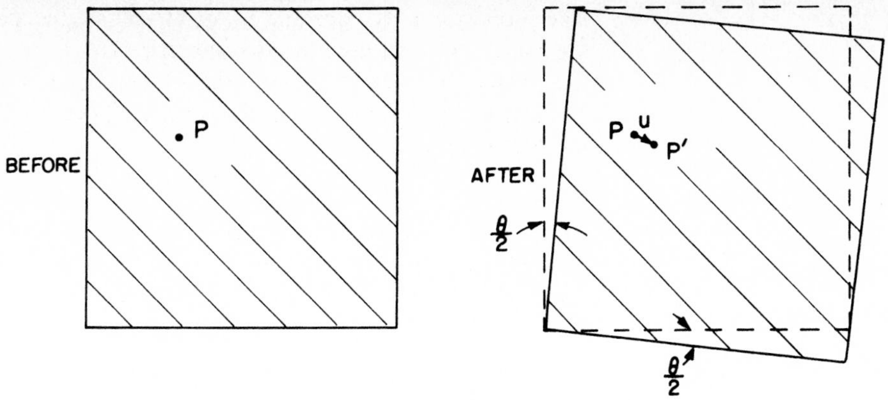

Fig. 39–4. A homogeneous rotation—there is no strain.

They are like Eqs. (39.4) and (39.5) except that the sign of u_y is reversed. With these displacements a little cube in the jello simply gets shifted by the angle $\theta/2$, as shown in Fig. 39–4. There is no strain at all—just a rotation in space. There is no distortion of the material; the *relative* positions of all the atoms are not changed at all. We must somehow make our definitions so that pure rotations are not included in our definitions of a shear strain. The key point is that if $\partial u_y/\partial x$ and $\partial u_x/\partial y$ are equal and opposite, there is no strain; so we can fix things up by *defining*

$$e_{xy} = e_{yx} = \tfrac{1}{2}(\partial u_y/\partial x + \partial u_x/\partial y).$$

For a pure rotation they are both zero, but for a pure shear we get that e_{xy} is equal to e_{yx}, as we would like.

In the most general distortion—which may include stretching or compression as well as shear—we *define* the state of strain by giving the nine numbers

$$\begin{aligned} e_{xx} &= \frac{\partial u_x}{\partial x}, \\ e_{yy} &= \frac{\partial u_y}{\partial y}, \\ &\;\vdots \\ e_{xy} &= \tfrac{1}{2}(\partial u_y/\partial x + \partial u_x/\partial y), \\ &\;\vdots \end{aligned} \tag{39.7}$$

These are the terms of a *tensor of strain*. Because it is a *symmetric tensor*—our definitions make $e_{xy} = e_{yx}$, always—there are really only six different numbers. You remember (see Chapter 31) that the general characteristic of a tensor is that the terms transform like the products of the components of two vectors. (If $\boldsymbol{A}$ and $\boldsymbol{B}$ are vectors, $C_{ij} = A_iB_j$ is a tensor.) Each term of e_{ij} is a product (or the sum of such products) of the components of the vector $\boldsymbol{u} = (u_x, u_y, u_z)$, and of the operator $\boldsymbol{\nabla} = (\partial/\partial x, \partial/\partial y, \partial/\partial z)$, which we know transforms like a vector. Let's let x_1, x_2, and x_3 stand for x, y, and z and u_1, u_2, and u_3 stand for u_x, u_y, and u_z; then we can write the general term e_{ij} of the strain tensor as

$$e_{ij} = \tfrac{1}{2}(\partial u_j/\partial x_i + \partial u_i/\partial x_j), \tag{39.8}$$

where i and j can be 1, 2, or 3.

When we have a homogeneous strain—which may include both stretching and shear—all of the e_{ij} are constants, and we can write

$$u_x = e_{xx}x + e_{xy}y + e_{xz}z. \tag{39.9}$$

(We choose our origin of x, y, z at the point where $\boldsymbol{u}$ is zero.) In this case, the strain tensor e_{ij} gives the relationship between two vectors: the coordinate vector $\boldsymbol{r} = (x, y, z)$ and the displacement vector $\boldsymbol{u} = (u_x, u_y, u_z)$.

When the strains are not homogeneous, any piece of the jello may also get somewhat twisted—there will be a local rotation. If the distortions are all small, we would have

$$\Delta u_i = \sum_j (e_{ij} - \omega_{ij})\,\Delta x_j, \tag{39.10}$$

where ω_{ij} is an *antisymmetric* tensor,

$$\omega_{ij} = \tfrac{1}{2}(\partial u_j/\partial x_i - \partial u_i/\partial x_j), \tag{39.11}$$

which describes the rotation. We will, however, not worry any more about rotations, but only about the strains described by the symmetric tensor e_{ij}.

39–2 The tensor of elasticity

Now that we have described the strains, we want to relate them to the internal forces—the stresses in the material. For each small piece of the material, we assume Hooke's law holds and write that the stresses are proportional to the strains. In Chapter 31 we defined the stress tensor S_{ij} as the ith component of the force across a unit area perpendicular to the j-axis. Hooke's law says that each component of S_{ij} is linearly related to *each* of the components of strain. Since S and e each have nine components, there are $9 \times 9 = 81$ possible coefficients which describe the elastic properties of the material. They are constants if the material itself is homogeneous. We write these coefficients as C_{ijkl} and define them by the equation

$$S_{ij} = \sum_{k,l} C_{ijkl} e_{kl}, \tag{39.12}$$

where i, j, k, l all take on the values 1, 2, or 3. Since the coefficients C_{ijkl} relate one tensor to another, they also form a tensor—a tensor of the *fourth rank.* We can call it the *tensor of elasticity.*

Suppose that all the C's are known and that you put a complicated force on an object of some peculiar shape. There will be all kinds of distortion, and the thing will settle down with some twisted shape. What are the displacements? You can see that it is a complicated problem. If you knew the strains, you could find the stresses from Eq. (39.12)—or vice versa. But the stresses and strains you end up with at any point depend on what happens in all the rest of the material.

The easiest way to get at the problem is by thinking of the energy. When there is a force F proportional to a displacement x, say $F = kx$, the work required for any displacement x is $kx^2/2$. In a similar way, the work w that goes into *each unit volume* of a distorted material turns out to be

$$w = \tfrac{1}{2} \sum_{ijkl} C_{ijkl} e_{ij} e_{kl}. \tag{39.13}$$

The total work W done in distorting the body is the integral of w over its volume:

$$W = \int \tfrac{1}{2} \sum_{ijkl} C_{ijkl} e_{ij} e_{kl}\, d\,\mathrm{Vol}. \tag{39.14}$$

This is then the potential energy stored in the internal stresses of the material. Now when a body is in equilibrium, this internal energy must be *at a minimum.* So the problem of finding the strains in a body can be solved by finding the set of displacements $\boldsymbol{u}$ throughout the body which will make W a minimum. In Chapter 19 we gave some of the general ideas of the calculus of variations that are used in tackling minimization problems like this. We cannot go into the problem in any more detail here.

What we are mainly interested in now is what we can say about the general properties of the tensor of elasticity. First, it is clear that there are *not* really 81 *different* terms in C_{ijkl}. Since both S_{ij} and e_{ij} are symmetric tensors, each with only six different terms, there can be at most 36 different terms in C_{ijkl}. There are, however, usually many fewer than this.

Let's look at the special case of a cubic crystal. In it, the energy density w starts out like this:

$$\begin{aligned} w = \tfrac{1}{2}\{C_{xxxx}e_{xx}^2 &+ C_{xxxy}e_{xx}e_{xy} + C_{xxxz}e_{xx}e_{xz} \\ &+ C_{xxyx}e_{xx}e_{xy} + C_{xxyy}e_{xx}e_{yy} \ldots \text{etc} \ldots \\ &+ C_{yyyy}e_{yy}^2 + \ldots \text{etc} \ldots \text{etc} \ldots\}, \end{aligned} \tag{39.15}$$

with 81 terms in all! Now a cubic crystal has certain symmetries. In particular, if the crystal is rotated 90°, it has the same physical properties. It has the same stiffness for stretching in the y-direction as for stretching in the x-direction. Therefore, if we change our definition of the coordinate directions x and y in Eq. (39.15), the energy wouldn't change. It must be that for a cubic crystal

$$C_{xxxx} = C_{yyyy} = C_{zzzz}. \tag{39.16}$$

Next we can show that the terms like C_{xxxy} must be zero. A cubic crystal has the property that it is symmetric under a *reflection* about any plane perpendicular to one of the axes. If we replace y by $-y$, nothing is different. But changing y to $-y$ changes e_{xy} to $-e_{xy}$—a displacement which was toward $+y$ is now toward $-y$. If the energy is not to change, C_{xxxy} must go into $-C_{xxxy}$ when we make a reflection. But a reflected crystal is the same as before, so C_{xxxy} must be the *same* as $-C_{xxxy}$. This can happen only if both are zero.

You say, "But the same argument will make $C_{yyyy} = 0$!" No, because there are *four* y's. The sign changes once for each y, and four minuses make a plus. If there are *two* or *four* y's, the term does not have to be zero. It is zero only when there is *one*, or *three*. So, for a cubic crystal, any nonzero term of C will have only an *even number* of identical subscripts. (The arguments we have made for y obviously hold also for x and z.) We might then have terms like C_{xxyy}, C_{xyxy}, C_{xyyx}, and so on. We have already shown, however, that if we change all x's to y's and *vice versa* (or all z's and x's, and so on) we must get—for a cubic crystal—the same number. This means that there are *only three different* nonzero possibilities:

$$\begin{aligned} &C_{xxxx}\ (= C_{yyyy} = C_{zzzz}), \\ &C_{xxyy}\ (= C_{yyxx} = C_{xxzz}, \text{etc.}), \\ &C_{xyxy}\ (= C_{yxyx} = C_{xzxz}, \text{etc.}). \end{aligned} \tag{39.17}$$

For a cubic crystal, then, the energy density will look like this:

$$\begin{aligned} w = \tfrac{1}{2}\{&C_{xxxx}(e_{xx}^2 + e_{yy}^2 + e_{zz}^2) \\ &+2C_{xxyy}(e_{xx}e_{yy} + e_{yy}e_{zz} + e_{zz}e_{xx}) \\ &+4C_{xyxy}(e_{xy}^2 + e_{yz}^2 + e_{zx}^2)\}. \end{aligned} \tag{39.18}$$

For an isotropic—that is, noncrystalline—material, the symmetry is still higher. The C's must be the same for *any* choice of the coordinate system. Then it turns out that there is another relation among the C's, namely, that

$$C_{xxxx} = C_{xxyy} + C_{xyxy}. \tag{39.19}$$

We can see that this is so by the following general argument. The stress tensor S_{ij} has to be related to e_{ij} in a way that doesn't depend at all on the coordinate directions—it must be related only by *scalar* quantities. "That's easy," you say. "The only way to obtain S_{ij} from e_{ij} is by multiplication by a scalar constant. It's just Hooke's law. It must be that $S_{ij} = (\text{const})e_{ij}$." But that's not quite right; there could also be the *unit tensor* δ_{ij} multiplied by some scalar, linearly related to e_{ij}. The only invariant you can make that is linear in the e's is $\sum e_{ii}$. (It transforms like $x^2 + y^2 + z^2$, which is a scalar.) So the most general form for the equation relating S_{ij} to e_{ij}—for isotropic materials—is

$$S_{ij} = 2\mu e_{ij} + \lambda\left(\sum_k e_{kk}\right)\delta_{ij}. \tag{39.20}$$

(The first constant is usually written as *two* times μ; then the coefficient μ is equal to the shear modulus we defined in the last chapter.) The constants μ and λ are called the Lamé elastic constants. Comparing Eq. (39.20) with Eq. (39.12), you see that

$$\begin{aligned} C_{xxyy} &= \lambda, \\ C_{xyxy} &= 2\mu, \\ C_{xxxx} &= 2\mu + \lambda. \end{aligned} \tag{39.21}$$

So we have proved that Eq. (39.19) is indeed true. You also see that the elastic properties of an isotropic material are completely given by two constants, as we said in the last chapter.

The C's can be put in terms of any two of the elastic constants we have used earlier—for instance, in terms of Young's modulus Y and Poisson's ratio σ. We will leave it for you to show that

$$\begin{aligned} C_{xxxx} &= \frac{Y}{1+\sigma}\left(1 + \frac{\sigma}{1-2\sigma}\right), \\ C_{xxyy} &= \frac{Y}{1+\sigma}\left(\frac{\sigma}{1-2\sigma}\right), \\ C_{xyxy} &= \frac{Y}{(1+\sigma)}. \end{aligned} \tag{39.22}$$

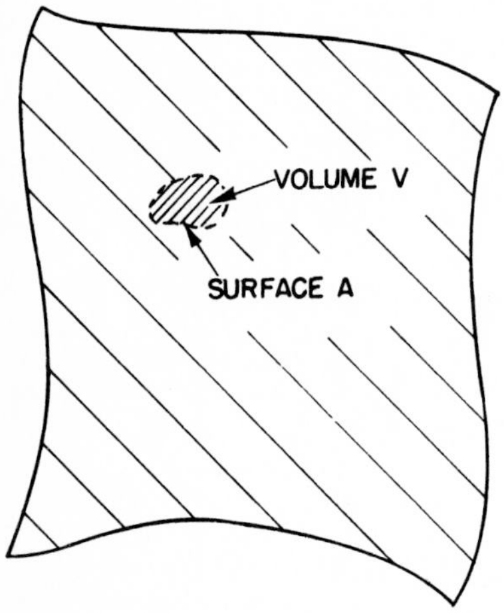

Fig. 39–5. A small volume element V bounded by the surface A.

39–3 The motions in an elastic body

We have pointed out that for an elastic body *in equilibrium* the internal stresses adjust themselves to make the energy a minimum. Now we take a look at what happens when the internal forces are *not* in equilibrium. Let's say we have a small piece of the material inside some surface A. See Fig. 39–5. If the piece is in equilibrium, the total force $\boldsymbol{F}$ acting on it must be zero. We can think of this force as being made up of two parts. There could be one part due to "external" forces like gravity, which act from a distance on the matter in the piece to produce a *force per unit* volume $\boldsymbol{f}_{\text{ext}}$. The total external force $\boldsymbol{F}_{\text{ext}}$ is the integral of $\boldsymbol{f}_{\text{ext}}$ over the volume of the piece:

$$\boldsymbol{F}_{\text{ext}} = \int \boldsymbol{f}_{\text{ext}}\, dV. \tag{39.23}$$

In equilibrium, this force would be balanced by the total force $\boldsymbol{F}_{\text{int}}$ from the neighboring material which acts across the surface A. When the piece is *not* in equilibrium—if it is moving—the sum of the internal and external forces is equal to the mass times the acceleration. We would have

$$\boldsymbol{F}_{\text{ext}} + \boldsymbol{F}_{\text{int}} = \int \rho\ddot{\boldsymbol{r}}\, dV, \tag{39.24}$$

where ρ is the density of the material, and $\ddot{\boldsymbol{r}}$ is its acceleration. We can now combine Eqs. (39.23) and (39.24), writing

$$\boldsymbol{F}_{\text{int}} = \int_v (-\boldsymbol{f}_{\text{ext}} + \rho\ddot{\boldsymbol{r}})\, dV. \tag{39.25}$$

We will simplify our writing by defining

$$\boldsymbol{f} = -\boldsymbol{f}_{\text{ext}} + \rho\ddot{\boldsymbol{r}}. \tag{39.26}$$

Then Eq. (39.25) is written

$$\boldsymbol{F}_{\text{int}} = \int_v \boldsymbol{f}\, dV. \tag{39.27}$$

What we have called $\boldsymbol{F}_{\text{int}}$ is related to the stresses in the material. The stress tensor S_{ij} was defined (Chapter 31) so that the x-component of the force dF across a surface element da, whose unit normal is $\boldsymbol{n}$, is given by

$$dF_x = (S_{xx}n_x + S_{xy}n_y + S_{xz}n_z)\, da. \tag{39.28}$$

The x-component of $\boldsymbol{F}_{\text{int}}$ on our little piece is then the integral of dF_x over the surface. Substituting this into the x-component of Eq. (39.27), we get

$$\int_A (S_{xx}n_x + S_{xy}n_y + S_{xz}n_z)\,da = \int_v f_x\,dV. \tag{39.29}$$

We have a surface integral related to a volume integral—and that reminds us of something we learned in electricity. Note that if you ignore the first subscript x on each of the S's in the left-hand side of Eq. (39.29), it looks just like the integral of a quantity "$\boldsymbol{S}$" $\cdot\, \boldsymbol{n}$—that is, the normal component of a vector—over the surface. It would be the flux of "$\boldsymbol{S}$" out of the volume. And this could be written, using Gauss law, as the volume integral of the divergence of "$\boldsymbol{S}$". It is, in fact, true whether the x-subscript is there or not—it is just a mathematical theorem you get by integrating by parts. In other words, we can change Eq. (39.29) into

$$\int_v \left(\frac{\partial S_{xx}}{\partial x} + \frac{\partial S_{xy}}{\partial y} + \frac{\partial S_{xz}}{\partial z}\right) dV = \int_v f_x\,dV. \tag{39.30}$$

Now we can leave off the volume integrals and write the differential equation for the general component of $\boldsymbol{f}$ as

$$f_i = \sum_j \frac{\partial S_{ij}}{\partial x_j}. \tag{39.31}$$

This tells us how the force per unit volume is related to the stress tensor S_{ij}.

The theory of the motions inside a solid works this way. If we start out knowing the initial displacements—given by, say, $\boldsymbol{u}$—we can work out the strains e_{ij}. From the strains we can get the stresses from Eq. (39.12). From the stresses we can get the force density $\boldsymbol{f}$ in Eq. (39.31). Knowing $\boldsymbol{f}$, we can get, from Eq. (39.26), the acceleration $\ddot{\boldsymbol{r}}$ of the material, which tells us how the displacements will be changing. Putting everything together, we get the horrible equation of motion for an elastic solid. We will just write down the results that come out for an isotropic material. If you use (39.20) for S_{ij}, and write the e_{ij} as $\frac{1}{2}\partial u_i/\partial x_j + \partial u_j/\partial x_i$, you end up with the vector equation

$$\boldsymbol{f} = (\lambda + \mu)\,\nabla(\nabla\cdot\boldsymbol{u}) + \mu\,\nabla^2\boldsymbol{u}. \tag{39.32}$$

You can, in fact, see that the equation relating $\boldsymbol{f}$ and $\boldsymbol{u}$ *must* have this form. The force must depend on the second derivatives of the displacements $\boldsymbol{u}$. What second derivatives of $\boldsymbol{u}$ are there that are vectors? One is $\nabla(\nabla\cdot\boldsymbol{u})$; that's a true vector. The only other one is $\nabla^2\boldsymbol{u}$. So the most general form is

$$\boldsymbol{f} = a\,\nabla(\nabla\cdot\boldsymbol{u}) + b\,\nabla^2\boldsymbol{u},$$

which is just (39.32) with a different definition of the constants. You may be wondering why we don't have a third term using $\nabla\times\nabla\times\boldsymbol{u}$, which is also a vector. But remember that $\nabla\times\nabla\times\boldsymbol{u}$ is the same thing as $\nabla^2\boldsymbol{u} - \nabla(\nabla\cdot\boldsymbol{u})$, so it is a linear combination of the two terms we have. Adding it would add nothing new. We have proved once more that isotropic material has only two elastic constants.

For the equation of motion of the material, we can set (39.32) equal to $\rho\,\partial^2\boldsymbol{u}/\partial t^2$—neglecting for now any body forces like gravity—and get

$$\rho\,\frac{\partial^2\boldsymbol{u}}{\partial t^2} = (\lambda + \mu)\,\nabla(\nabla\cdot\boldsymbol{u}) + \mu\,\nabla^2\boldsymbol{u}. \tag{39.33}$$

It looks something like the wave equation we had in electromagnetism, except that there is an additional complicating term. For materials whose elastic properties are everywhere the same we can see what the general solutions look like in the following way. You will remember that any vector field can be written as the sum of two vectors: one whose divergence is zero, and the other whose curl is zero. In other words, we can put

$$\boldsymbol{u} = \boldsymbol{u}_1 + \boldsymbol{u}_2, \tag{39.34}$$

where

$$\nabla \cdot \boldsymbol{u}_1 = 0, \qquad \nabla \times \boldsymbol{u}_2 = 0. \tag{39.35}$$

Substituting $u_1 + u_2$ for u in (39.33), we get

$$\rho\, \partial^2/\partial t^2[\boldsymbol{u}_1 + \boldsymbol{u}_2] = (\lambda + \mu)\, \nabla(\nabla \cdot \boldsymbol{u}_2) + \mu\nabla^2(\boldsymbol{u}_1 + \boldsymbol{u}_2). \tag{39.36}$$

We can eliminate u_1 by taking the divergence of this equation,

$$\rho\, \partial^2/\partial t^2(\nabla \cdot \boldsymbol{u}_2) = (\lambda + \mu)\, \nabla^2(\nabla \cdot \boldsymbol{u}_2) + \mu\nabla \cdot \nabla^2\boldsymbol{u}_2.$$

Since the operators (∇^2) and $(\nabla\cdot)$ can be interchanged, we can factor out the divergence to get

$$\nabla \cdot \{\rho\, \partial^2\boldsymbol{u}_2/\partial t^2 - (\lambda + 2\mu)\, \nabla^2\boldsymbol{u}_2\} = 0. \tag{39.37}$$

Since $\nabla \times \boldsymbol{u}_2$ is zero by definition, the curl of the bracket $\{\}$ is also zero; so the bracket itself is identically zero, and

$$\rho\, \partial^2\boldsymbol{u}_2/\partial t^2 = (\lambda + 2\mu)\, \nabla^2\boldsymbol{u}_2. \tag{39.38}$$

This is the vector wave equation for waves which move at the speed $C_2 = \sqrt{(\lambda + 2\mu)/\rho}$. Since the curl of $\boldsymbol{u}_2$ is zero, there is no shearing associated with this wave; this wave is just the compressional—sound-type—wave we discussed in the last chapter, and the velocity is just what we found for C_{long}.

In a similar way—by taking the curl of Eq. (39.36)—we can show that $\boldsymbol{u}_1$ satisfies the equation

$$\rho\, \partial^2\boldsymbol{u}_1/\partial t^2 = \mu\, \nabla^2\boldsymbol{u}_1. \tag{39.39}$$

This is again a vector wave equation for waves with the speed $C_2 = \sqrt{\mu/\rho}$. Since $\nabla \cdot \boldsymbol{u}_1$ is zero, $\boldsymbol{u}_1$ produces no changes in density; the vector $\boldsymbol{u}_1$ corresponds to the transverse, or shear-type, wave we saw in the last chapter, and $C_2 = C_{\text{shear}}$.

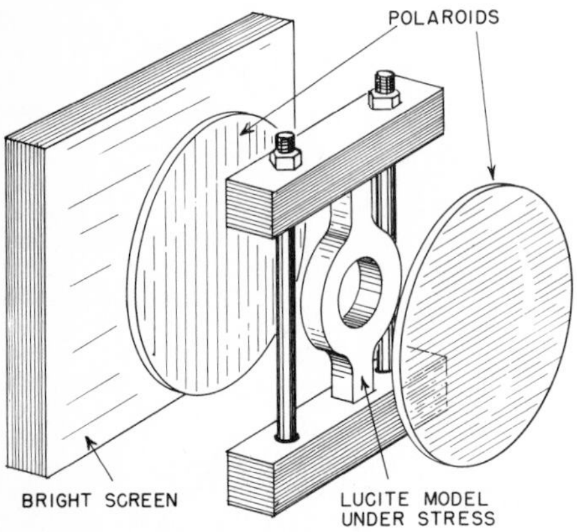

Fig. 39–6. Measuring internal stresses with polarized light.

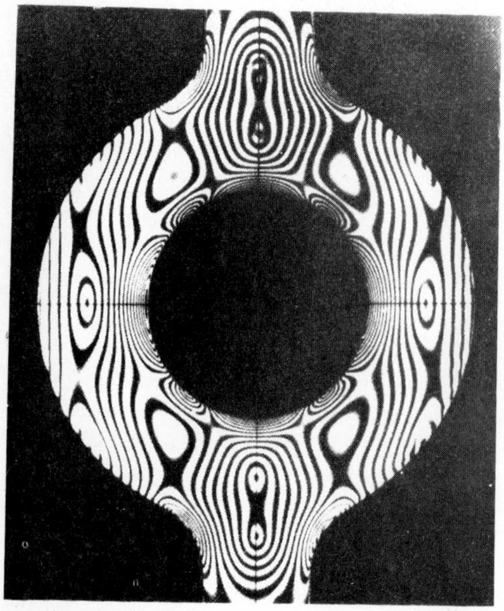

Fig. 39–7. A stressed plastic model as seen between crossed polaroids. [From F. W. Sears, *Optics*, Addison-Wesley Publishing Co., Reading, Mass., 1949.]

If we wished to know the static stresses in an isotropic material, we could, in principle, find them by solving Eq. (39.32) with f equal to zero—or equal to the static body forces from gravity such as ρg—under certain conditions which are related to the forces acting on the surfaces of our large block of material. This is somewhat more difficult to do than the corresponding problems in electromagnetism. It is more difficult, first, because the equations are a little more difficult to handle, and second, because the shape of the elastic bodies we are likely to be interested in are usually much more complicated. In electromagnetism, we are often interested in solving Maxwell's equations around relatively simple geometric shapes such as cylinders, spheres, and so on, since these are convenient shapes for electrical devices. In elasticity, the objects we would like to analyze may have quite complicated shapes—like a crane hook, or an automobile crankshaft, or the rotor of a gas turbine. Such problems can sometimes be worked out approximately by numerical methods, using the minimum energy principle we mentioned earlier. Another way is to use a model of the object and measure the internal strains experimentally, using polarized light.

It works this way: When a transparent isotropic material—for example, a clear plastic like lucite—is put under stress, it becomes birefringent. If you put polarized light through it, the plane of polarization will be rotated by an amount related to the stress: by measuring the rotation, you can measure the stress. Figure 39–6 shows how such a setup might look. Figure 39–7 is a photograph of a photoelastic model of a complicated shape under stress.

39–4 Nonelastic behavior

In all that has been said so far, we have assumed that stress is proportional to strain; in general, that is *not* true. Figure 39–8 shows a typical stress-strain curve for a ductile material. For small strains, the stress is proportional to the strain. Eventually, however, after a certain point, the relationship between stress and strain begins to deviate from a straight line. For many materials—the ones we would call "brittle" —the object breaks for strains only a little above the point where the curve starts to bend over. In general, there are other complications in the stress-strain relationship. For example, if you strain an object, the stresses may be high at first, but decrease slowly with time. Also if you go to high stresses, but still not to the "breaking" point, when you lower the strain the stress will return along a different curve. There is a small hysteresis effect (like the one we saw between B and H in magnetic materials).

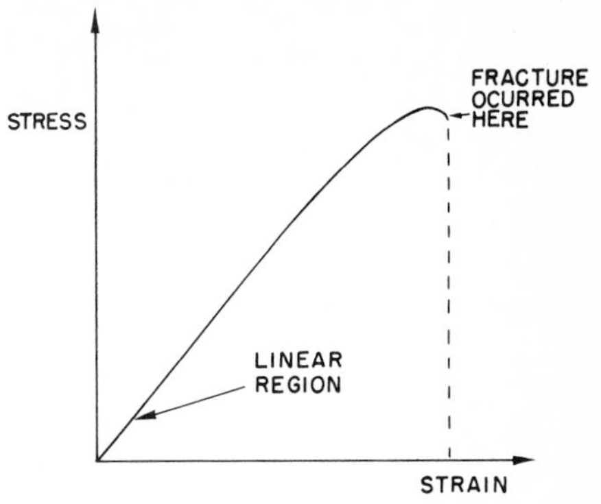

Fig. 39–8. A typical stress-strain relation for large strains.

The stress at which a material will break varies widely from one material to another. Some materials will break when the maximum *tensile* stress reaches a certain value. Other materials will fail when the maximum *shear* stress reaches a certain value. Chalk is an example of a material which is much weaker in tension than in shear. If you pull on the ends of a piece of blackboard chalk, the chalk will break perpendicular to the direction of the applied stress, as shown in Fig. 39–9(a). It breaks perpendicular to the applied force because it is only a bunch of particles packed together which are easily pulled apart. The material is, however, much harder to shear, because the particles get in each other's way. Now you will remember that when we had a rod in torsion there was a shear all around it. Also, we showed that a shear was equivalent to a combination of a tension and compression at 45°. For these reasons, if you *twist* a piece of blackboard chalk, it will break along a complicated surface which starts out at 45° to the axis. A photograph of a piece of chalk broken in this way is shown in Fig. 39–9(b). The chalk breaks where the material is in maximum tension.

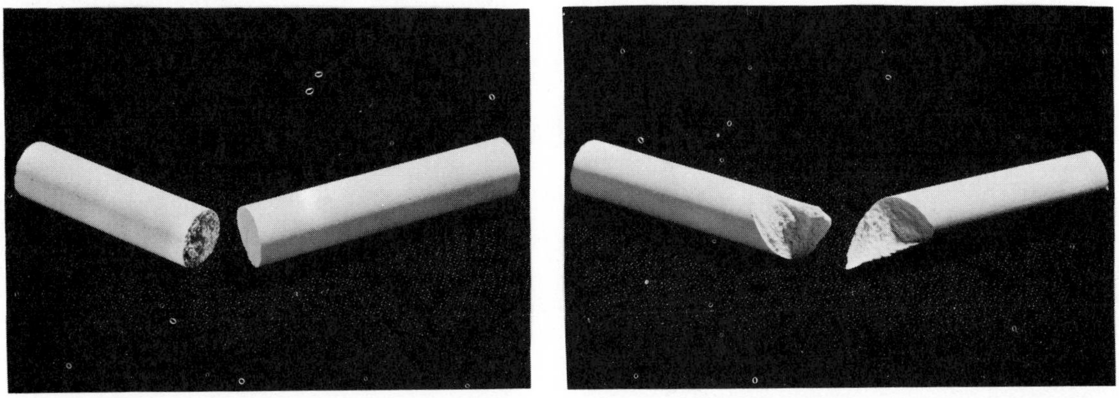

(a) (b)

Fig. 39–9. (a) A piece of chalk broken by pulling on the ends; (b) a piece broken by twisting.

Other materials behave in strange and complicated ways. The more complicated the materials are, the more interesting their behavior. If we take a sheet of "Saran-Wrap" and crumple it up into a ball and throw it on the table, it slowly unfolds itself and returns toward its original flat form. At first sight, we might be tempted to think that it is inertia which prevents it from returning to its original form. However, a simple calculation shows that the inertia is several orders of magnitude too small to account for the effect. There appear to be two important competing effects: "something" inside the material "remembers" the shape it had initially and "tries" to get back there, but something else "prefers" the new shape and "resists" the return to the old shape.

We will not attempt to describe the mechanism at play in the Saran plastic, but you can get an idea of how such an effect might come about from the following *model.* Suppose you imagine a material made of long, flexible, but strong, fibers mixed together with some hollow cells filled with a viscous liquid. Imagine also that there are narrow pathways from one cell to the next so the liquid can leak slowly from a cell to its neighbor. When we crumple a sheet of this stuff, we distort the long fibers, squeezing the liquid out of the cells in one place and forcing it into other cells which are being stretched. When we let go, the long fibers try to return to their original shape. But to do this, they have to force the liquid back to its original location—which will happen relatively slowly because of the viscosity. The forces we apply in crumpling the sheet are much larger than the forces exerted by the fibers. We can crumple the sheet quickly, but it will return more slowly. It is undoubtedly a combination of large stiff molecules and smaller, movable ones in the Saran-Wrap that is responsible for its behavior. This idea also fits with the fact that the material returns more quickly to its original shape when it's warmed up than when it's cold—the heat increases the mobility (decreases the viscosity) of the smaller molecules.

Although we have been discussing how Hooke's law breaks down, the remarkable thing is perhaps not that Hooke's law breaks down for large strains but that it should be so generally true. We can get some idea of why this might be by looking at the strain energy in a material. To say that the stress is proportional to the strain is the same thing as saying that the strain energy varies as the square of the strain. Suppose we have a rod and we twist it through a small angle θ. If Hooke's law holds, the strain energy should be proportional to the square of θ. Suppose we were to assume that the energy were some arbitrary function of the angle; we could write it as a Taylor expansion about zero angle

$$U(\theta) = U(0) + U'(0)\theta + \tfrac{1}{2}U''(0)\theta^2 + \tfrac{1}{6}U'''(\theta)\theta^3 \ldots \tag{39.40}$$

The torque τ is the derivative of U with respect to angle; we would have

$$t(\theta) = U'(0) + U''(0)\theta + \tfrac{1}{2}U'''(0)\theta^2 + \cdots \tag{39.41}$$

Now if we measure our angles from the *equilibrium* position, the first term is zero. So the first remaining term is proportional to θ; and for small enough angles, it will dominate the term in θ^2. [Actually, materials are sufficiently symmetric internally so that $\tau(\theta) = -\tau(-\theta)$; the term in θ^2 will be zero, and the departures from linearity would come only from the θ^3 term. There is, however, no reason why this should be true for compressions and tensions.] The thing we have not explained is why materials usually break soon after the higher-order terms become significant.

39–5 Calculating the elastic constants

As our last topic on elasticity we would like to show how one could try to calculate the elastic constants of a material, starting with some knowledge of the properties of the atoms which make up the material. We will take only the simple case of an *ionic* cubic crystal like sodium chloride. When a crystal is strained, its volume or its shape is changed. Such changes result in an increase in the potential energy of the crystal. To calculate the change in strain energy, we have to know where each atom goes. In complicated crystals, the atoms will rearrange themselves in the lattice in very complicated ways to make the total energy as small as possible. This makes the computation of the strain energy rather difficult. In the case of a simple cubic crystal, however, it is easy to see what will happen. The distortions inside the crystal will be geometrically similar to the distortions of the outside boundaries of the crystal.

We can calculate the elastic constants for a cubic crystal in the following way. First, we assume some force law between each pair of atoms in the crystal. Then, we calculate the change in the internal energy of the crystal when it is distorted from its equilibrium shape. This gives us a relation between the energy and the strains which is quadratic in all the strains. Comparing the energy obtained this way with Eq. (39.13), we can identify the coefficient of each term with the elastic constants C_{ijkl}.

For our example we will assume a simple force law: that the force between neighboring atoms is a *central* force, by which we mean that it acts along the line between the two atoms. We would expect the forces in ionic crystals to be like this, since they are just primarily Coulomb forces. (The forces of covalent bonds are usually more complicated, since they can exert a sideways push on a nearby atom; we will leave out this complication.) We are also going to include only the forces between each atom and its *nearest* and *next-nearest* neighbors. In other words, we will make an approximation which neglects all forces beyond the next-nearest neighbor. The forces we will include are shown for the xy-plane in Fig. 39–10(a). The corresponding forces in the yz- and zx-planes also have to be included.

Since we are only interested in the elastic coefficients which apply to small strains, and therefore only want the terms in the energy which vary quadratically with the strains, we can imagine that the force between each atom pair varies linearly with the displacements. We can then imagine that each pair of atoms is joined by a linear spring, as drawn in Fig. 39–10(b). All of the springs between a sodium atom and a chlorine atom should have the same spring constant, say k_1. The springs between two sodiums and between two chlorines could have different constants, but we will make our discussion simpler by taking them equal; we call them k_2. (We could come back later and make them different after we have seen how the calculations go.)

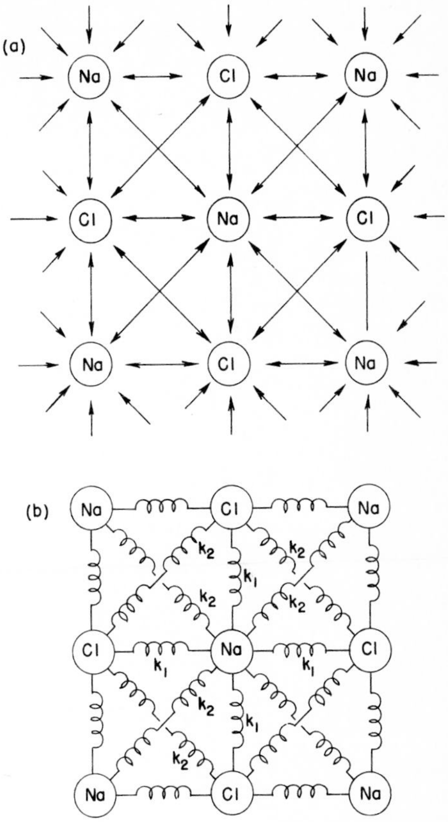

Fig. 39–10. (a) The interatomic forces we are taking into account; (b) a model in which the atoms are connected by springs.

Now we assume that the crystal is distorted by a homogeneous strain described by the strain tensor e_{ij}. In general, it will have components involving x, y, and z; but we will consider now only a strain with the three components e_{xx}, e_{xy}, and e_{yy} so that it will be easy to visualize. If we pick one atom as our origin, the displacement of every other atom is given by equations like Eq. (39.9):

$$\begin{aligned} u_x &= e_{xx}x + e_{xy}y, \\ u_y &= e_{xy}x + e_{yy}y. \end{aligned} \tag{39.42}$$

Suppose we call the atom at $x = y = 0$ "atom 1" and number its neighbors in the xy-plane as shown in Fig. 39–11. Calling the lattice constant a, we get the x and y displacements u_x and u_y listed in Table 39–1.

Now we can calculate the energy stored in the springs, which is $k^2/2$ times the square of the extension for each spring. For example, the energy in the horizontal spring between atom 1 and atom 2 is

$$\frac{k_1(e_{xx}a)^2}{2}. \tag{39.43}$$

Note that to first order, the y-displacement of atom 2 does not change the length of the spring between atom 1 and atom 2. To get the strain energy in a diagonal spring, such as that to atom 3, however, we need to calculate the change in length due to both the horizontal and vertical displacements. For small displacements from the

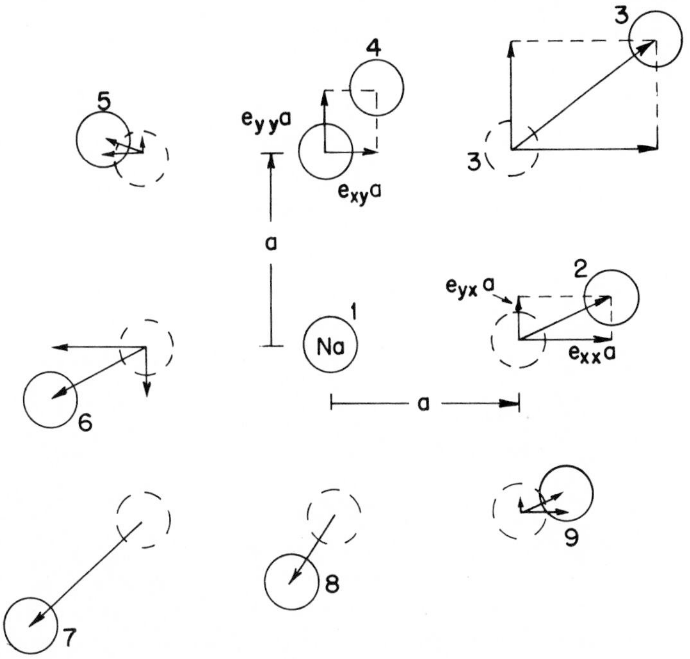

Fig. 39–11. The displacements of the nearest and next-nearest neighbors of atom 1 (exaggerated).

Table 39–1

Atom	Location x, y	u_x	u_y	k
1	$0, a$	0	0	—
2	$a, 0$	$e_{xx}a$	$e_{yx}a$	k_1
3	a, a	$(e_{xx} + e_{xy})a$	$(e_{yx} + e_{yy})a$	k_2
4	$0, a$	$e_{xy}a$	$e_{yy}a$	k_1
5	$-a, a$	$(-e_{xx} + e_{xy})a$	$(-e_{yx} + e_{yy})a$	k_2
6	$-a, 0$	$-e_{xx}a$	$-e_{yx}a$	k_1
7	$-a, -a$	$-(e_{xx} + e_{xy})a$	$-(e_{yx} + e_{yy})a$	k_2
8	$0, -a$	$-e_{xy}a$	$-e_{yy}a$	k_1
9	$a, -a$	$(e_{xx} - e_{xy})a$	$(e_{yx} - e_{yy})a$	k_2

original cube, we can write the change in the distance to atom 3 as the sum of the components of u_x and u_y in the diagonal direction, namely as

$$\frac{1}{\sqrt{2}}\,(u_x + u_y).$$

Using the values of u_x and u_y from the table, we get the energy

$$\frac{k_2}{2}\left(\frac{u_x + u_y}{\sqrt{2}}\right)^2 = \frac{k_2a^2}{4}\,(e_{xx} + e_{yx} + e_{xy} + e_{yy})^2. \tag{39.44}$$

For the total energy for all the springs in the xy-plane, we need the sum of eight terms like (39.43) and (39.44). Calling this energy U_0, we get

$$\begin{aligned} U_0 = \frac{a^2}{2}\Big\{&k_1e_{xx}^2 + \frac{k_2}{2}\,(e_{xx} + e_{yx} + e_{xy} + e_{yy})^2 \\ &+ k_1e_{yy}^2 + \frac{k_2}{2}\,(e_{xx} - e_{yx} - e_{xy} + e_{yy})^2 \\ &+ k_1e_{xx}^2 + \frac{k_2}{2}\,(e_{xx} + e_{yx} + e_{xy} + e_{yy})^2 \\ &+ k_1e_{yy}^2 + \frac{k_2}{2}\,(e_{xx} - e_{yx} - e_{xy} + e_{yy})^2\Big\}\cdot \end{aligned} \tag{39.45}$$

To get the total energy of all the springs connected to atom 1, we must make one addition to the energy in Eq. (39.45). Even though we have only x- and y-components of the strain, there are still some energies associated with the next-nearest neighbors off the xy-plane. This additional energy is

$$k_2(e_{xx}^2a^2 + e_{yy}^2a^2). \tag{39.46}$$

The elastic constants are related to the energy density w by Eq. (39.13). The energy we have calculated is the energy associated with one atom, or rather, it is *twice* the energy per atom, since one-half of the energy of each spring should be assigned to each of the two atoms it joins. Since there are $1/a^3$ atoms per unit volume, w and U_0 are related by

$$w = \frac{U_0}{2a^3}\cdot$$

To find the elastic constants C_{ijkl}, we need only to expand out the squares in Eq. (39.45)—adding the terms of (39.46)—and compare the coefficients of $e_{ij}e_{kl}$ with the corresponding coefficient in Eq. (39.13). For example, collecting the terms in e_{xx}^2 and in e_{yy}^2, we get the factor

$$(k_1 + 2k_2)a^2,$$

so

$$C_{xxxx} = C_{yyyy} = \frac{k_1 + 2k_2}{a}.$$

For the remaining terms, there is a slight complication. Since we cannot distinguish the product of two terms like $e_{xx}e_{yy}$ from $e_{yy}e_{xx}$, the coefficient of such terms in our energy is equal to the sum of two terms in Eq. (39.13). The coefficient of $e_{xx}e_{yy}$ in Eq. (39.45) is $2k_2$, so we have that

$$(C_{xxyy} + C_{yyxx}) = \frac{2k_2}{a}.$$

But because of the symmetry in our crystal, $C_{xxyy} = C_{yyxx}$, so we have that

$$C_{xxyy} = C_{yyxx} = \frac{k_2}{a}.$$

By a similar process, we can also get

$$C_{xyxy} = C_{yxyx} = \frac{k_2}{a}.$$

Finally, you will notice that any term which involves either x or y only once is zero—as we concluded earlier from symmetry arguments. Summarizing our results:

$$\begin{aligned}
C_{xxxx} &= C_{yyyy} = \frac{k_1 + 2k_2}{a}, \\
C_{xyxy} &= C_{yxyx} = \frac{k_2}{a}, \\
C_{xxyy} &= C_{yyxx} = C_{xyyx} = C_{yxxy} = \frac{k_2}{a}, \\
C_{xxxy} &= C_{xyyy} = \text{etc.} = 0.
\end{aligned} \tag{39.47}$$

We have been able to relate the bulk elastic constants to the atomic properties which appear in the constants k_1 and k_2. In our particular case, $C_{xyxy} = C_{xxyy}$. It turns out—as you can perhaps see from the way the calculations went—that these terms are *always* equal for a cubic crystal, no matter how many force terms are taken into account, *provided* only that the forces act along the line joining each pair of atoms—that is, so long as the forces between atoms are like springs and don't have a sideways part such as you might get from a cantilevered beam (and you do get in covalent bonds).

We can check this conclusion with the experimental measurements of the elastic constants. In Table 39–2 we give the observed values of the three elastic coefficients for several cubic crystals.* You will notice that C_{xxyy} and C_{xyxy} are, in general, not equal. The reason is that in metals like sodium and potassium the interatomic forces are not along the line joining the atoms, as we assumed in our model. Diamond does not obey the law either, because the forces in diamond are covalent forces and have some directional properties—the bonds would prefer to be at the tetrahedral angle. The ionic crystals like lithium fluoride, sodium chloride, and so on, do have nearly all the physical properties assumed in our model, and the table shows that the constants C_{xxyy} and C_{xyxy} are almost equal. It is not clear why silver chloride should not satisfy the condition that $C_{xxyy} = C_{xyxy}$.

Table 39–2*

Elastic Moduli of Cubic Crystals in 10^{12} dynes·cm^2

	C_{xxxx}	C_{xxyy}	C_{xyxy}
Na	0.055	0.042	0.049
K	0.046	0.037	0.026
Fe	2.37	1.41	1.16
Diamond	10.76	1.25	5.76
Al	1.08	0.62	0.28
LiF	1.19	0.54	0.53
NaCl	0.486	0.127	0.128
KCl	0.40	0.062	0.062
NaBr	0.33	0.13	0.13
KI	0.27	0.043	0.042
AgCl	0.60	0.36	0.062

* From C. Kittel, *Introduction to Solid State Physics*, John Wiley and Sons, Inc., New York, 2nd. ed., 1956, p. 93.

* In the literature you will often find that a different notation is used. For instance, people usually write $C_{xxxx} = C_{11}$, $C_{xxyy} = C_{12}$, and $C_{xyxy} = C_{44}$.

40

The Flow of Dry Water

40–1 Hydrostatics

The subject of the flow of fluids, and particularly of water, fascinates everybody. We can all remember, as children, playing in the bathtub or in mud puddles with the strange stuff. As we get older, we watch streams, waterfalls, and whirlpools, and we are fascinated by this substance which seems almost alive relative to solids. The behavior of fluids is in many ways very unexpected and interesting—it is the subject of this chapter and the next. The efforts of a child trying to dam a small stream flowing in the street and his surprise at the strange way the water works its way out has its analog in our attempts over the years to understand the flow of fluids. We have tried to dam the water up—in our understanding—by getting the laws and the equations that describe the flow. We will describe these attempts in this chapter. In the next chapter, we will describe the unique way in which water has broken through the dam and escaped our attempts to understand it.

We suppose that the elementary properties of water are already known to you. The main property that distinguishes a fluid from a solid is that a fluid cannot *maintain* a shear stress for any length of time. If a shear is applied to a fluid, it will move under the shear. Thicker liquids like honey move less easily than fluids like air or water. The measure of the ease with which a fluid yields is its viscosity. In this chapter we will consider only situations in which the viscous effects can be ignored. The effects of viscosity will be taken up in the next chapter.

We begin by considering *hydrostatics*, the theory of liquids at rest. When liquids are at rest, there are no shear forces (even for viscous liquids). The law of hydrostatics, therefore, is that the stresses are always normal to any surface inside the fluid. The normal force per unit area is called the *pressure*. From the fact that there is no shear in a static fluid it follows that the pressure stress is the same in all directions (Fig. 40–1). We will let you entertain yourself by proving that if there is no shear on any plane in a fluid, the pressure must be the same in any direction.

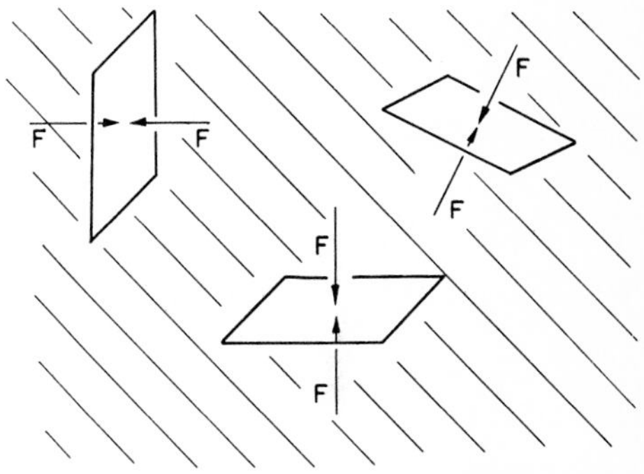

Fig. 40–1. In a static fluid the force per unit area across any surface is normal to the surface and is the same for all orientations of the surface.

The pressure in a fluid may vary from place to place. For example, in a static fluid at the earth's surface the pressure will vary with height because of the weight of the fluid. If the density ρ of the fluid is considered constant, and if the pressure at some arbitrary zero level is called p_0 (Fig. 40–2), then the pressure at a height h above this point is $p = p_0 - \rho g h$, where g is the gravitational force per unit mass. The combination

$$p + \rho g h$$

is, therefore, a constant in the static fluid. This relation is familiar to you, but we will now derive a more general result of which it is a special case.

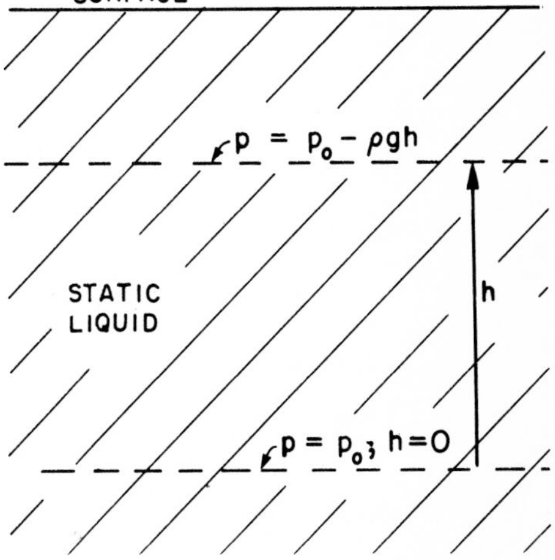

Fig. 40–2. The pressure in a static liquid.

If we take a small cube of water, what is the net force on it from the pressure? Since the pressure at any place is the same in all directions, there can be a net force per unit volume only because the pressure varies from one point to another. Suppose that the pressure is varying in the x-direction—and we take the coordinate directions parallel to the cube edges. The pressure on the face at x gives the force $p\,\Delta y\,\Delta z$ (Fig. 40–3), and the pressure on the face at $x + \Delta x$ gives the force $-[p + (\partial p/\partial x)\,\Delta x]\,\Delta y\,\Delta z$, so that the resultant force is $-(\partial p/\partial x)\,\Delta x\,\Delta y\,\Delta z$. If we take the remaining pairs of faces of the cube, we easily see that the pressure force per unit volume is $-\boldsymbol{\nabla} p$. If there are other forces in addition—such as gravity—then the pressure must balance them to give equilibrium.

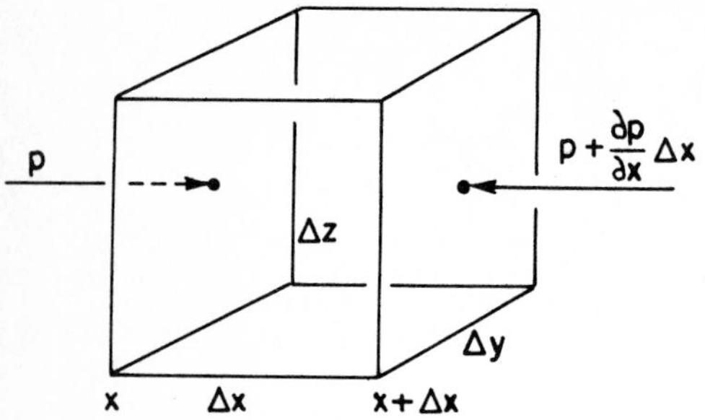

Fig. 40–3. The net pressure force on a cube is $-\nabla p$ per unit volume.

Let's take a circumstance in which such an additional force can be described by a potential energy, as would be true in the case of gravitation; we will let ϕ stand for the potential energy per unit mass. (For gravity, for instance, ϕ is just gz.) The force per unit mass is given in terms of the potential by $-\nabla\phi$, and if ρ is the density of the fluid, the force per unit volume is $-\rho\,\nabla\phi$. For equilibrium this force per unit volume added to the pressure force per unit volume must give zero:

$$-\nabla p - \rho\,\nabla\phi = 0. \tag{40.1}$$

Equation (40.1) is the equation of hydrostatics. In *general*, it *has no solution.* If the density varies in space in an arbitrary way, there is no way for the forces to be in balance, and the fluid cannot be in static equilibrium. Convection currents will start up. We can see this from the equation since the pressure term is a pure gradient, whereas for variable ρ the other term is not. Only when ρ is a constant is the potential term a pure gradient. Then the equation has a solution

$$p + \rho\phi = \text{const.}$$

Another possibility which allows hydrostatic equilibrium is for ρ to be a function only of p. However, we will leave the subject of hydrostatics because it is not nearly so interesting as the situation when fluids are in motion.

40–2 The equations of motion

First, we will discuss fluid motions in a purely abstract, theoretical way and then consider special examples. To describe the motion of a fluid, we must give its properties at every point. For example, at different places, the water (let us call the fluid "water") is moving with different *velocities.* To specify the character of the flow, therefore, we must give the three components of velocity at every point and for any time. If we can find the equations that determine the velocity, then we would know how the liquid moves at all times. The velocity, however, is not the only property that the fluid has which varies from point to point. We have just discussed the variation of the *pressure* from point to point. And there are still other variables. There may also be a variation of *density* from point to point. In addition, the fluid may be a conductor and carry an electric *current* whose density $\boldsymbol{j}$ varies from point to point in magnitude and direction. There may be a *temperature* which varies from point to point, or a *magnetic field*, and so on. So the number of fields needed to describe the complete situation will depend on how complicated the problem is. There are interesting phenomena when currents and magnetism play a dominant part in determining the behavior of the fluid; the subject is called *magnetohydrodynamics*, and great attention is being paid to it at the present time. However, we are not going to consider these more complicated situations because there are already interesting phenomena at a lower level of complexity, and even the more elementary level will be complicated enough.

We will take the situation where there is no magnetic field and no conductivity, and we will not worry about the temperature because we will suppose that the density and pressure determine in a unique manner the temperature at any point. As a matter of fact, we will reduce the complexity of our work by making the assumption that the density is a constant—we imagine that the fluid is essentially incompressible. Putting it another way, we are supposing that the variations of pressure are so small that the changes in density produced thereby are negligible. If that is not the case, we would encounter phenomena additional to the ones we will be discussing here—for example, the propagation of sound or of shock waves. We have already discussed the propagation of sound and shocks to some extent, so we will now isolate our consideration of hydrodynamics from these other phenomena by making the approximation that the density ρ is a constant. It is easy to determine when the approximation of constant ρ is a good one. We can say that if the velocities of flow are much less than the speed of a sound wave in the fluid, we do not have to worry about variations in density. The escape that water makes in our attempts to understand it is not related to the approximation of constant density. The complications that do permit the escape will be discussed in the next chapter.

In the general theory of fluids one must begin with an *equation of state* for the fluid which connects the pressure to the density. In our approximation this equation of state is simply

$$\rho = \text{const.}$$

This then is the first relation for our variables. The next relation expresses the conservation of matter—if matter flows away from a point, there must be a decrease in the amount left behind. If the fluid velocity is $\boldsymbol{v}$, then the mass which flows in a unit time across a unit area of surface is the component of $\rho\boldsymbol{v}$ normal to the surface. We have had a similar relation in electricity. We also know from electricity that the divergence of such a quantity gives the rate of decrease of the density per unit time. In the same way, the equation

$$\boldsymbol{\nabla}\cdot(\rho\boldsymbol{v}) = -\frac{\partial\rho}{\partial t} \tag{40.2}$$

expresses the conservation of mass for a fluid; it is the hydrodynamic *equation of continuity*. In our approximation, which is the incompressible fluid approximation, ρ is a constant, and the equation of continuity is simply

$$\boldsymbol{\nabla}\cdot\boldsymbol{v} = 0. \tag{40.3}$$

The fluid velocity $\boldsymbol{v}$—like the magnetic field $\boldsymbol{B}$—has zero divergence. (The hydrodynamic equations are often closely analogous to the electrodynamic equations; that's why we studied electrodynamics first. Some people argue the other way; they think that one should study hydrodynamics first so that it will be easier to understand electricity afterwards. But electrodynamics is really much easier than hydrodynamics.)

We will get our next equation from Newton's law which tells us how the velocity changes because of the forces. The mass of an element of volume of the fluid times its acceleration must be equal to the force on the element. Taking an element of unit volume, and writing the force per unit volume as $\boldsymbol{f}$, we have

$$\rho \times (\text{acceleration}) = \boldsymbol{f}.$$

We will write the force density as the sum of three terms. We have already considered the pressure force per unit volume, $-\boldsymbol{\nabla}p$. Then there are the "external" forces which act at a distance—like gravity or electricity. When they are conservative forces with a potential per unit mass, ϕ, they give a force density $-\rho\boldsymbol{\nabla}\phi$. (If the external forces are not conservative, we would have to write $\boldsymbol{f}_{\text{ext}}$ for the external force per unit volume.) Then there is another "internal" force per unit volume, which is due to the fact that in a *flowing* fluid there can also be a shearing stress. This is called the viscous force, which we will write $\boldsymbol{f}_{\text{visc}}$. Our equation of motion is

$$\rho \times (\text{acceleration}) = -\boldsymbol{\nabla}p - \rho\,\boldsymbol{\nabla}\phi + \boldsymbol{f}_{\text{visc}}. \tag{40.4}$$

For this chapter we are going to suppose that the liquid is "thin" in the sense that the viscosity is unimportant, so we will omit $\boldsymbol{f}_{\text{visc}}$. When we drop the viscosity term, we will be making an approximation which describes some ideal stuff rather than real water. John von Neumann was well aware of the tremendous difference between what happens when you don't have the viscous terms and when you do, and he was also aware that, during most of the development of hydrodynamics until about 1900, almost the main interest was in solving beautiful *mathematical* problems with this approximation which had almost nothing to do with real fluids. He characterized the theorist who made such analyses as a man who studied "dry water." Such analyses leave out an *essential* property of the fluid. It is because we are leaving this property out of our calculations in this chapter that we have given it the title "The Flow of Dry Water." We are postponing a discussion of *real* water to the next chapter.

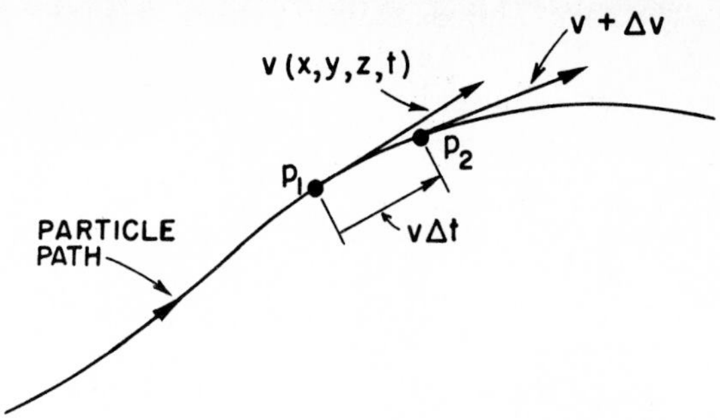

Fig. 40–4. The acceleration of a fluid particle.

If we leave out f_{visc}, we have in Eq. (40.4) everything we need except an expression for the acceleration. You might think that the formula for the acceleration of a fluid particle would be very simple, for it seems obvious that if $\boldsymbol{v}$ is the velocity of a fluid particle at some place in the fluid, the acceleration would just be $\partial\boldsymbol{v}/\partial t$. *It is not*—and for a rather subtle reason. The derivative $\partial\boldsymbol{v}/\partial t$, is the rate at which the velocity $\boldsymbol{v}(x, y, z, t)$ changes at a *fixed point* in space. What we need is how fast the velocity changes for a *particular piece* of fluid. Imagine that we mark one of the drops of water with a colored speck so we can watch it. In a small interval of time Δt, this drop will move to a different location. If the drop is moving along some path as sketched in Fig. 40–4, it might in Δt move from P_1 to P_2. In fact, it will move in the x-direction by an amount $v_x\,\Delta t$, in the y-direction by the amount $v_y\,\Delta t$, and in the z-direction by the amount $v_z\,\Delta t$. We see that, if $\boldsymbol{v}(x, y, z, t)$ is the velocity of the fluid particle which is at (x, y, z) at the time t, then the velocity of the *same* particle at the time $t + \Delta t$ is given by $\boldsymbol{v}(x + \Delta x,\ y + \Delta y,\ z + \Delta z,\ t + \Delta t)$—with

$$\Delta x = v_x\,\Delta t, \qquad \Delta y = v_y\,\Delta t, \qquad \text{and} \qquad \Delta z = v_z\,\Delta t.$$

From the definition of the partial derivatives—recall Eq. (2.7)—we have, to first order, that

$$\boldsymbol{v}(x + v_x\,\Delta t,\ y + v_y\,\Delta t,\ z + v_z\,\Delta t,\ t + \Delta t)$$
$$= \boldsymbol{v}(x, y, z, t) + \frac{\partial\boldsymbol{v}}{\partial x}\,v_x\,\Delta t + \frac{\partial\boldsymbol{v}}{\partial y}\,v_y\,\Delta t + \frac{\partial\boldsymbol{v}}{\partial z}\,v_z\,\Delta t + \frac{\partial\boldsymbol{v}}{\partial t}\,\Delta t.$$

The acceleration $\Delta\boldsymbol{v}/\Delta t$ is

$$v_x\,\frac{\partial\boldsymbol{v}}{\partial x} + v_y\,\frac{\partial\boldsymbol{v}}{\partial y} + v_z\,\frac{\partial\boldsymbol{v}}{\partial z} + \frac{\partial\boldsymbol{v}}{\partial t}\cdot$$

We can write this symbolically——treating $\boldsymbol{\nabla}$ as a vector—as

$$(\boldsymbol{v}\cdot\boldsymbol{\nabla})\boldsymbol{v} + \frac{\partial\boldsymbol{v}}{\partial t}\cdot \tag{40.5}$$

Note that there can be an acceleration even though $\partial\boldsymbol{v}/\partial t = 0$ so that velocity *at a given point* is not changing. As an example, water flowing in a circle at a constant speed is accelerating even though the velocity at a given point is not changing. The reason is, of course, that the velocity of a particular piece of water which is initially at one point on the circle has a different direction a moment later; there is a centripetal acceleration.

The rest of our theory is just mathematical—finding solutions of the equation of motion we get by putting the acceleration (40.5) into Eq. (40.4). We get

$$\frac{\partial\boldsymbol{v}}{\partial t} + (\boldsymbol{v}\cdot\boldsymbol{\nabla})\boldsymbol{v} = -\frac{\boldsymbol{\nabla}p}{\rho} - \boldsymbol{\nabla}\phi, \tag{40.6}$$

where viscosity has been omitted. We can rearrange this equation by using the following identity from vector analysis:

$$(\boldsymbol{v}\cdot\boldsymbol{\nabla})\boldsymbol{v} = (\boldsymbol{\nabla}\times\boldsymbol{v})\times\boldsymbol{v} + \tfrac{1}{2}\boldsymbol{\nabla}(\boldsymbol{v}\cdot\boldsymbol{v}).$$

If we now *define* a new *vector field* $\boldsymbol{\Omega}$, as the curl of $\boldsymbol{v}$,

$$\boldsymbol{\Omega} = \nabla \times \boldsymbol{v}, \tag{40.7}$$

the vector identity can be written as

$$(\boldsymbol{v} \cdot \nabla)\boldsymbol{v} = \boldsymbol{\Omega} \times \boldsymbol{v} + \tfrac{1}{2}\nabla v^2,$$

and our equation of motion (40.6) becomes

$$\frac{\partial \boldsymbol{v}}{\partial t} + \boldsymbol{\Omega} \times \boldsymbol{v} + \frac{1}{2}\nabla v^2 = -\frac{\nabla p}{\rho} - \nabla \phi. \tag{40.8}$$

You can verify that Eqs. (40.6) and (40.8) are equivalent by checking that the components of the two sides of the equation are equal—and making use of (40.7).

The vector field $\boldsymbol{\Omega}$ is called the *vorticity*. If the vorticity is zero everywhere, we say that the flow is *irrotational*. We have already defined in Section 3–5 a thing called the *circulation* of a vector field. The circulation around any closed loop in a fluid is the line integral of the fluid velocity, at a given instant of time, around that loop:

$$(\text{Circulation}) = \oint \boldsymbol{v} \cdot d\boldsymbol{s}.$$

The circulation *per unit area* for an infinitesimal loop is then—using Stokes' theorem—equal to $\nabla \times \boldsymbol{v}$. So the vorticity $\boldsymbol{\Omega}$ is the circulation around a unit area (perpendicular to the direction of $\boldsymbol{\Omega}$). It also follows that if you put a little piece of dirt—*not* an infinitesimal point—at any place in the liquid it will rotate with the angular velocity $\boldsymbol{\Omega}/2$. Try to see if you can prove that. You can also check it out that for a bucket of water on a turntable, $\boldsymbol{\Omega}$ is equal to twice the local angular velocity of the water.

If we are interested only in the velocity field, we can eliminate the pressure from our equations. Taking the curl of both sides of Eq. (40.8), remembering that ρ is a constant and that the curl of any gradient is zero, and using Eq. (40.3), we get

$$\frac{\partial \boldsymbol{\Omega}}{\partial t} + \nabla \times (\boldsymbol{\Omega} \times \boldsymbol{v}) = 0. \tag{40.9}$$

This equation, together with the equations

$$\boldsymbol{\Omega} = \nabla \times \boldsymbol{v} \tag{40.10}$$

and

$$\nabla \cdot \boldsymbol{v} = 0, \tag{40.11}$$

describes completely the velocity field $\boldsymbol{v}$. Mathematically speaking, if we know $\boldsymbol{\Omega}$ at some time, then we know the curl of the velocity vector, and we also know that its divergence is zero, so given the physical situation we have all we need to determine $\boldsymbol{v}$ everywhere. (It is just like the situation in magnetism where we had $\nabla \cdot \boldsymbol{B} = 0$ and $\nabla \times \boldsymbol{B} = \boldsymbol{j}/\epsilon_0 c^2$.) Thus, a given $\boldsymbol{\Omega}$ determines $\boldsymbol{v}$ just as a given $\boldsymbol{j}$ determines $\boldsymbol{B}$. Then, knowing $\boldsymbol{v}$, Eq. (40.9) tells us the rate of change of $\boldsymbol{\Omega}$ from which we can get the new $\boldsymbol{\Omega}$ for the next instant. Using Eq. (40.10), again we find the new $\boldsymbol{v}$, and so on. You see how these equations contain all the machinery for calculating the flow. Note, however, that this procedure gives the velocity field only; we have lost all information about the pressure.

We point out one special consequence of our equation. If $\boldsymbol{\Omega} = 0$ everywhere at any time t, $\partial \boldsymbol{\Omega}/\partial t$ also vanishes, so that $\boldsymbol{\Omega}$ is still zero everywhere at $t + \Delta t$. We have a solution to the equation; the flow is permanently irrotational. If a flow was started with zero rotation, it would always have zero rotation. The equations to be solved then are

$$\nabla \cdot \boldsymbol{v} = 0, \qquad \nabla \times \boldsymbol{v} = 0.$$

They are just like the equations for the electrostatic or magnetostatic fields in free space. We will come back to them and look at some special problems later.

40–3 Steady flow—Bernoulli's theorem

Now we want to return to the equation of motion, Eq. (40.8), but limit ourselves to situations in which the flow is "steady." By steady flow we mean that at any one place in the fluid the velocity never changes. The fluid at any point is always replaced by new fluid moving in exactly the same way. The velocity picture always looks the same—$\boldsymbol{v}$ is a static vector field. In the same way that we drew "field lines" in magnetostatics, we can now draw lines which are always tangent to the fluid velocity as shown in Fig. 40–5. These lines are called *streamlines.* For steady flow, they are evidently the actual paths of fluid particles. (In unsteady flow the streamline pattern changes in time, and the streamline pattern at any instant does not represent the path of a fluid particle.)

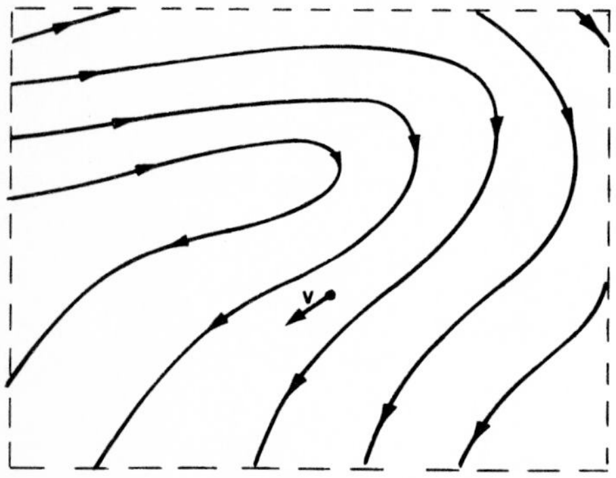

Fig. 40–5. Streamlines in steady fluid flow.

A steady flow does not mean that nothing is happening—atoms in the fluid are moving and changing their velocities. It only means that $\partial\boldsymbol{v}/\partial t = 0$. Then if we take the dot product of $\boldsymbol{v}$ into the equation of motion, the term $\boldsymbol{v}\cdot(\boldsymbol{\Omega}\times\boldsymbol{v})$ drops out, and we are left with

$$\boldsymbol{v}\cdot\boldsymbol{\nabla}\left\{\frac{p}{\rho}+\phi+\frac{1}{2}v^2\right\}=0. \tag{40.12}$$

This equation says that *for a small displacement in the direction of the fluid velocity* the quantity inside the brackets doesn't change. Now in steady flow all displacements are along streamlines, so Eq. (40.12) tells us that *for all the points along a streamline*, we can write

$$\frac{p}{\rho}+\frac{1}{2}v^2+\phi=\text{const (streamline)}. \tag{40.13}$$

This is *Bernoulli's theorem.* The constant may in general be different for different streamlines; all we know is that the left-hand side of Eq. (40.13) is the same all along a *given streamline.* Incidentally, we may notice that for steady *irrotational* motion for which $\boldsymbol{\Omega}=0$, the equation of motion (40.8) gives us the relation

$$\nabla\left\{\frac{p}{\rho}+\frac{1}{2}v^2+\phi\right\}=0,$$

so that

$$\frac{p}{\rho}+\frac{1}{2}v^2+\phi=\text{const (everywhere)}. \tag{40.14}$$

It's just like Eq. (40.13) *except* that *now* the constant has the *same value throughout the fluid.*

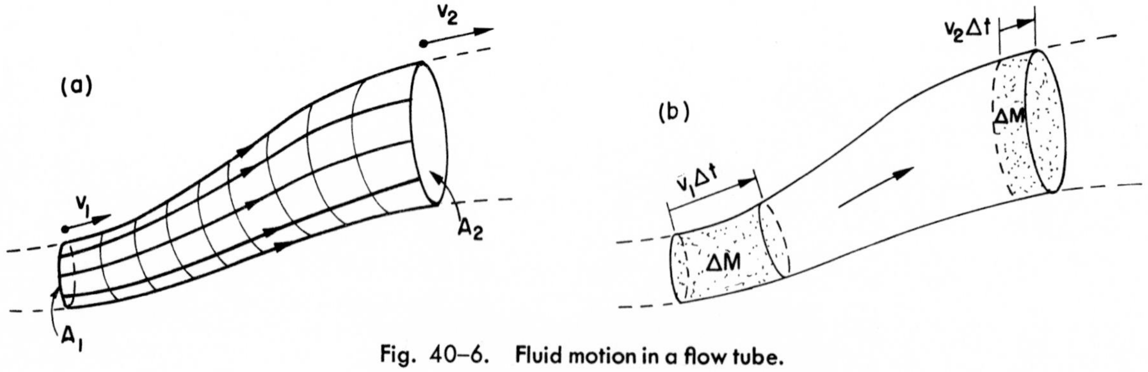

Fig. 40–6. Fluid motion in a flow tube.

The theorem of Bernoulli is in fact nothing more than a statement of the conservation of energy. A conservation theorem such as this gives us a lot of information about a flow without our actually having to solve the detailed equations. Bernoulli's theorem is so important and so simple that we would like to show you how it can be derived in a way that is different from the formal calculations we have just used. Imagine a bundle of adjacent streamlines which form a stream tube as sketched in Fig. 40–6. Since the walls of the tube consist of streamlines, no fluid flows out through the wall. Let's call the area at one end of the stream tube A_1, the fluid velocity there v_1, the density of the fluid ρ_1, and the potential energy ϕ_1. At the other end of the tube, we have the corresponding quantities A_2, v_2, ρ_2, and ϕ_2. Now after a short interval of time Δt, the fluid at A_1 has moved a distance $v_1\,\Delta t$, and the fluid at A_2 has moved a distance $v_2\,\Delta t$ [Fig. 40–6(b)]. The conservation of *mass* requires that the mass which enters through A_1 must be equal to the mass which leaves through A_2. These masses at these two ends must be the same:

$$\Delta M = \rho_1 A_1 v_1\,\Delta t = \rho_2 A_2 v_2\,\Delta t.$$

So we have the equality

$$\rho_1 A_1 v_1 = \rho_2 A_2 v_2. \tag{40.15}$$

This equation tells us that the velocity varies inversely with the area of the stream tube if ρ is constant.

Now we calculate the work done by the fluid pressure. The work done on the fluid entering at A_1 is $p_1 A_1 v_1\,\Delta t$, and the work given up at A_2 is $p_2 A_2 v_2\,\Delta t$. The net work on the fluid between A_1 and A_2 is, therefore,

$$p_1 A_1 v_1\,\Delta t - p_2 A_2 v_2\,\Delta t,$$

which must equal the increase in the energy of a mass ΔM of fluid in going from A_1 to A_2. In other words,

$$p_1 A_1 v_1\,\Delta t - p_2 A_2 v_2\,\Delta t = \Delta M(E_2 - E_1), \tag{40.16}$$

where E_1 is the energy per unit mass of fluid at A_1, and E_2 is the energy per unit mass at A_2. The energy per unit mass of the fluid can be written as

$$E = \tfrac{1}{2}v^2 + \phi + U,$$

where $\frac{1}{2}v^2$ is the kinetic energy per unit mass, ϕ is the potential energy per unit mass, and U is an additional term which represents the internal energy per unit mass of fluid. The internal energy might correspond, for example, to the thermal energy in a compressible fluid, or to chemical energy. All these quantities can vary from point to point. Using this form for the energies in (40.16), we have

$$\frac{p_1 A_1 v_1\,\Delta t}{\Delta M} - \frac{p_2 A_2 v_2\,\Delta t}{\Delta M} = \frac{1}{2}\,v_2^2 + \phi_2 + U_2 - \frac{1}{2}\,v_1^2 - \phi_1 - U_1.$$

But we have seen that $\Delta M = \rho A v\,\Delta t$, so we get

$$\frac{p_1}{\rho_1} + \frac{1}{2}\,v_1^2 + \phi_1 + U_1 = \frac{p_2}{\rho_2} + \frac{1}{2}\,v_2^2 + \phi_2 + U_2, \tag{40.17}$$

which is the Bernoulli result with an additional term for the internal energy. If the fluid is incompressible, the internal energy term is the same on both sides, and we get again that Eq. (40.14) holds along any streamline.

We consider now some simple examples in which the Bernoulli integral gives us a description of the flow. Suppose we have water flowing out of a hole near the bottom of a tank, as drawn in Fig. 40–7. We take a situation in which the flow speed v_{out} at the hole is much larger than the flow speed near the top of the tank; in other words, we imagine that the diameter of the tank is so large that we can neglect the drop in the liquid level. (We could make a more accurate calculation if we wished.) At the top of the tank the pressure is p_0, the atmospheric pressure, and the pressure at the sides of the jet is also p_0. Now we write our Bernoulli equation for a streamline, such as the one shown in the figure. At the top of the tank, we take v equal to zero and we also take the gravity potential ϕ to be zero. At the speed v_{out}, and $\phi = -gh$, so that

$$p_0 = p_0 + \tfrac{1}{2}\rho v_{\text{out}}^2 - \rho g h,$$

or

$$v_{\text{out}} = \sqrt{2gh}. \tag{40.18}$$

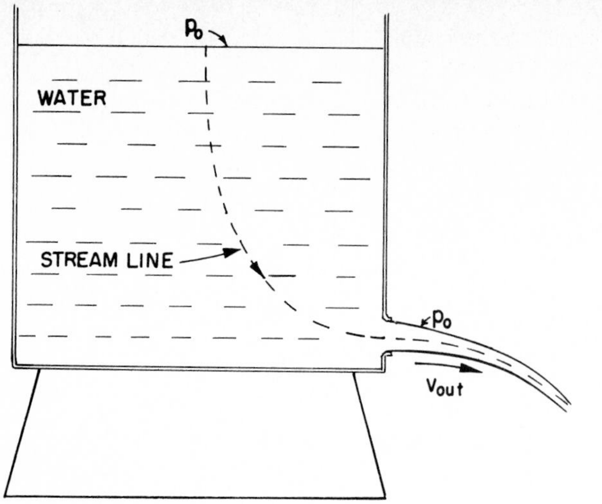

Fig. 40–7. Flow from a tank.

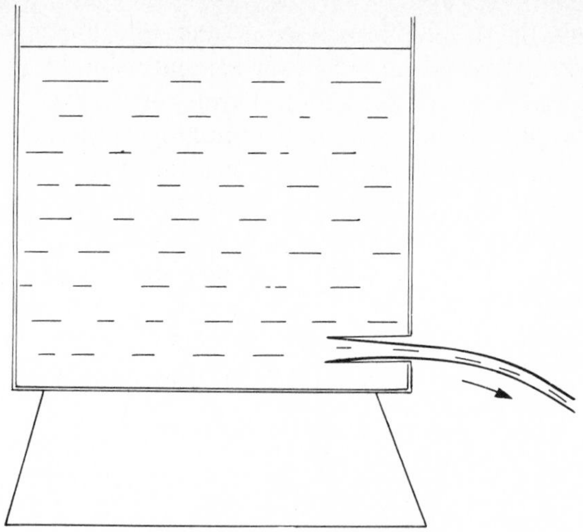

Fig. 40–8. With a re-entrant discharge tube, the stream contracts to one-half the area of the opening.

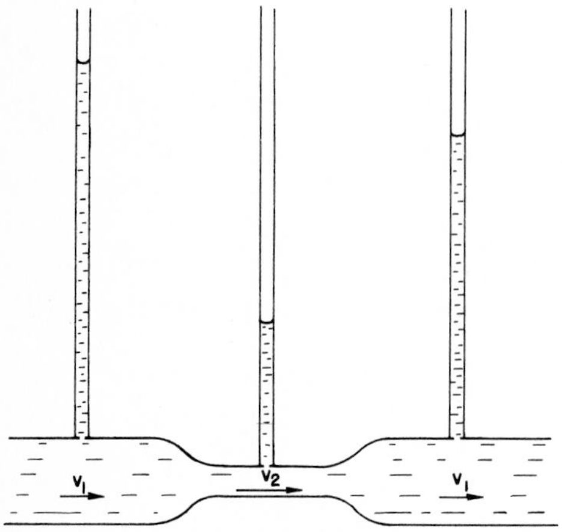

Fig. 40–9. The pressure is lowest where the velocity is highest.

This velocity is just what we would get for something which falls the distance h. It is not too surprising, since the water at the exit gains kinetic energy at the expense of the potential energy of the water at the top. Do not get the idea, however, that you can figure out the rate that the fluid flows out of the tank by multiplying this velocity by the area of the hole. The fluid velocities as the jet leaves the hole are not all parallel to each other but have components inward toward the center of the stream—the jet is converging. After the jet has gone a little way, the contraction stops and the velocities do become parallel. So the total flow is the velocity times the area *at that point*. In fact, if we have a discharge opening which is just a round hole with a sharp edge, the jet contracts to 62 percent of the area of the hole. The reduced effective area of the discharge varies for different shapes of discharge tubes, and experimental contractions are available as tables of *efflux coefficients*.

If the discharge tube is re-entrant, as shown in Fig. 40–8, it is possible to prove in a most beautiful way that the efflux coefficient is exactly 50 percent. We will give just a hint of how the proof goes. We have used the conservation of energy to get the velocity, Eq. (40.18), but there is also momentum conservation to consider. Since there is an outflow of momentum in the discharge jet, there must be a force applied over the cross section of the discharge tube. Where does the force come from? The force must come from the pressure on the walls. As long as the efflux hole is small and away from the walls, the fluid velocity near the walls of the tank will be very small. Therefore, the pressure on every face is almost exactly the same as the static pressure in a fluid at rest—from Eq. (30.14). Then the static pressure at any point on the side of the tank must be matched by an equal pressure at the point on the opposite wall, *except* at the points on the wall opposite the charge tube. If we calculate the momentum poured out through the jet by this pressure, we can show that the efflux coefficient is $1/2$. We cannot use this method for a discharge hole like that shown in Fig. 40–7, however, because the velocity increase along the wall right near the discharge area gives a pressure fall which we are not able to calculate.

Let's look at another example—a horizontal pipe with changing cross section, as shown in Fig. 40–9, with water flowing in one end and out the other. The conservation of energy, namely Bernoulli's formula, says that the pressure is lower in the constricted area where the velocity is higher. We can easily demonstrate this effect by measuring the pressure at different cross sections with small vertical columns of water attached to the flow tube through holes small enough so that they do not disturb the flow. The pressure is then measured by the height of water in these vertical columns. The pressure is found to be less at the constriction than it is on either side. If the area beyond the constriction comes back to the same value it had before the constriction, the pressure rises again.

Bernoulli's formula would predict that the pressure downstream of the constriction should be the same as it was upstream, but actually it is noticeably less. The reason that our prediction is wrong is that we have neglected the frictional, viscous forces which cause a pressure drop along the tube. Despite this pressure drop the pressure is definitely lower at the constriction (because of the increased speed) than it is on either side of it—as predicted by Bernoulli. The speed v_2 must certainly exceed v_1 to get the same amount of water through the narrower tube. So the water accelerates in going from the wide to the narrow part. The force that gives this acceleration comes from the drop in pressure.

We can check our results with another simple demonstration. Suppose we have on a tank a discharge tube which throws a jet of water upward as shown in Fig. 40–10. If the efflux velocity were exactly $\sqrt{2gh}$, the discharge water should rise to a level even with the surface of the water in the tank. Experimentally, it falls somewhat short. Our prediction is roughly right, but again viscous friction which has not been included in our energy conservation formula has resulted in a loss of energy.

Have you ever held two pieces of paper close together and tried to blow them apart? Try it! They come *together*. The reason, of course, is that the air has a higher speed going through the constricted space between the sheets than it does when it gets outside. The pressure between the sheets is *lower* than atmospheric pressure, so they come together rather than separating.

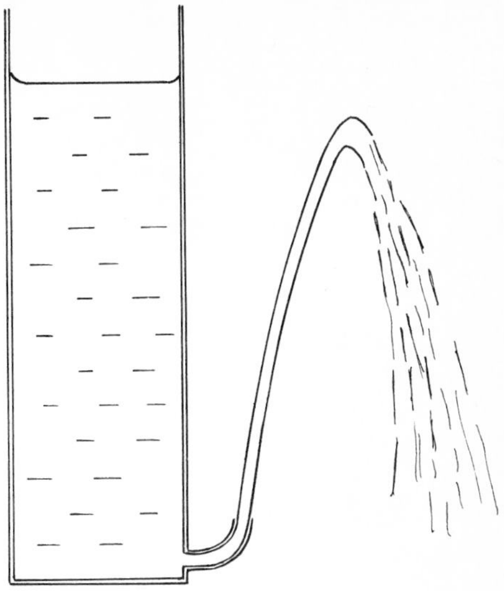

Fig. 40–10. Proof that v is not equal to $\sqrt{2gh}$.

40–4 Circulation

We saw at the beginning of the last section that if we have an incompressible fluid with no circulation, the flow satisfies the following two equations:

$$\nabla \cdot \boldsymbol{v} = 0, \qquad \nabla \times \boldsymbol{v} = 0. \tag{40.19}$$

They are the same as the equations of electrostatics or magnetostatics in empty space. The divergence of the electric field is zero when there are no charges, and the curl of the electrostatic field is always zero. The curl of the magnetic field is zero if there are no currents, and the divergence of the magnetic field is always zero. Therefore, Eqs. (40.19) have the same solutions as the equations for $\boldsymbol{E}$ in electrostatics or for $\boldsymbol{B}$ in magnetostatics. As a matter of fact, we have already solved the problem of the flow of a fluid past a sphere, as an electrostatic analogy, in Section 12–5. The electrostatic analog is a uniform electric field plus a dipole field. The dipole field is so adjusted that the flow velocity normal to the surface of the sphere is zero. The same problem for the flow past a cylinder can be worked out in a similar way by using a suitable line dipole with a uniform flow field. This solution holds for a situation in which the fluid velocity at large distances is constant—both in magnitude and direction. The solution is sketched in Fig. 40–11(a).

There is another solution for the flow around a cylinder when the conditions are such that the fluid at large distances moves in circles around the cylinder. The flow is, then, circular everywhere, as in Fig. 40–11(b). Such a flow has a circulation around the cylinder, although $\nabla \times \boldsymbol{v}$ is still zero *in the fluid.* How can there be circulation without a curl? We have a circulation around the cylinder because the line integral of $\boldsymbol{v}$ around any loop *enclosing* the cylinder is not zero. At the same time, the line integral of $\boldsymbol{v}$ around any closed path which does *not* include the cylinder is zero. We saw the same thing when we found the magnetic field around a wire. The curl of $\boldsymbol{B}$ was zero outside of the wire, although a line integral of $\boldsymbol{B}$ around a path which encloses the wire did not vanish. The velocity field in an irrotational circulation around a cylinder is precisely the same as the magnetic field around a wire. For a circular path with its center at the center of the cylinder, the line integral of the velocity is

$$\oint \boldsymbol{v} \cdot d\boldsymbol{s} = 2\pi r v.$$

For irrotational flow the integral must be independent of r. Let's call the constant

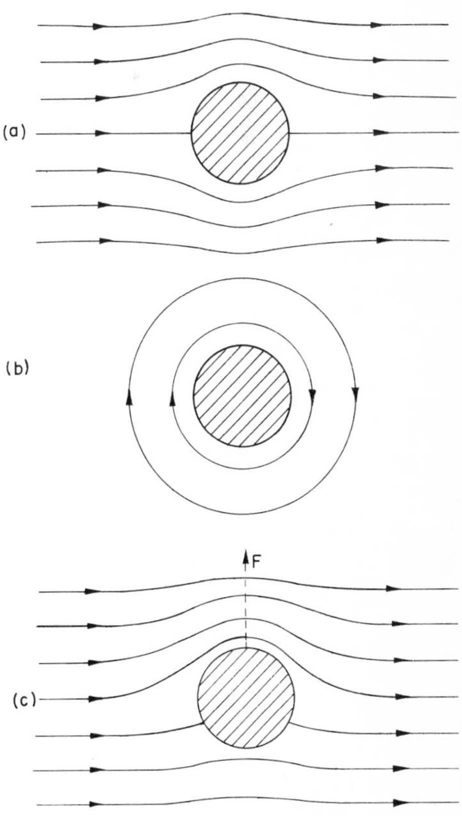

Fig. 40–11. (a) Ideal fluid flow past a cylinder. (b) Circulation around a cylinder. (c) The superposition of (a) and (b).

value C, then we have that

$$v = \frac{C}{2\pi r}, \tag{40.20}$$

where v is the tangential velocity, and r is the distance from the axis.

There is a nice demonstration of a fluid circulating around a hole. You take a transparent cylindrical tank with a drain hole in the center of the bottom. You fill it with water, stir up some circulation with a stick, and pull the drain plug. You get the pretty effect shown in Fig. 40–12. (You've seen a similar thing many times in the bathtub!) Although you put in some $\boldsymbol{\omega}$ at beginning, it soon dies down because of viscosity and the flow becomes irrotational—although still with some circulation around the hole.

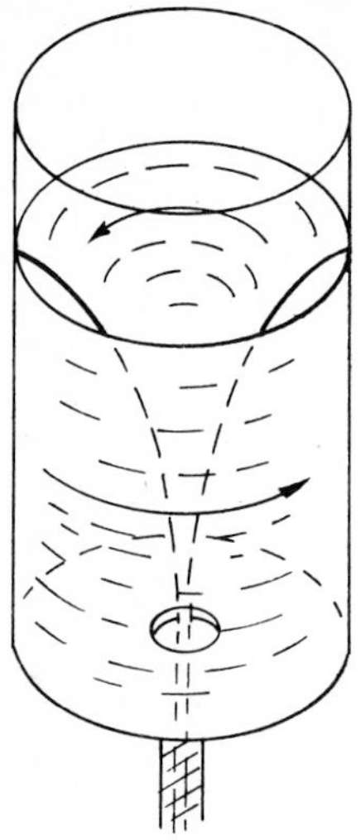

Fig. 40–12. Water with circulation draining from a tank.

From the theory, we can calculate the shape of the inner surface of the water. As a particle of the water moves inward it picks up speed. From Eq. (40.20) the tangential velocity goes as $1/r$—it's just from the conservation of angular momentum, like the skater pulling in her arms. Also the radial velocity goes as $1/r$. Ignoring the tangential motion, we have water going radially inward toward a hole; from $\nabla \cdot \boldsymbol{v} = 0$, it follows that the radial velocity is proportional to $1/r$. So the total velocity also increases as $1/r$, and the water goes in along Archimedean spirals. The air-water surface is all at atmospheric pressure, so it must have—from Eq. (40.14)—the property that

$$gz + \tfrac{1}{2}mv^2 = \text{const.}$$

But v is proportional to $1/r$, so the shape of the surface is

$$(z - z_0) = \frac{k}{r^2}.$$

An interesting point—which *is not true in general* but is true for incompressible, irrotational flow—is that if we have one solution and a second solution, then the sum is also a solution. This is true because the equations in (40.19) are linear. The complete equations of hydrodynamics, Eqs. (40.8), (40.9), and (40.10), are not linear, which makes a vast difference. For the irrotational flow about the cylinder, however, we can superpose the flow of Fig. 40–11(a) on the flow of Fig. 40–11(b) and get the new flow pattern shown in Fig. 40–11(c). This flow is of special interest. The flow velocity is higher on the upper side of the cylinder than on the lower side. The pressures are therefore *lower* on the *upper* side than on the lower side. So when we have a combination of a circulation around a cylinder *and* a net horizontal flow, there is a net *vertical force* on the cylinder—it is called a *lift force*. Of course, if there is no circulation, there is no net force on any body according to our theory of "dry" water.

40–5 Vortex lines

We have already written down the general equations for the flow of an incompressible fluid when there may be vorticity. They are

$$\text{I.}\quad \nabla \cdot \boldsymbol{v} = 0,$$

$$\text{II.}\quad \boldsymbol{\Omega} = \nabla \times \boldsymbol{v},$$

$$\text{III.}\quad \frac{\partial \boldsymbol{\Omega}}{\partial t} + \nabla \times (\boldsymbol{\Omega} \times \boldsymbol{v}) = 0.$$

The physical content of these equations has been described in words by Helmholtz in terms of three theorems. First, imagine that in the fluid we were to draw *vortex lines* rather than streamlines. By vortex lines we mean field lines that have the direction of $\boldsymbol{\Omega}$ and have a density in any region proportional to the magnitude of $\boldsymbol{\Omega}$. From II the divergence of $\boldsymbol{\Omega}$ is *always* zero (remember—Section 3–7—that the divergence of a curl is always zero). So vortex lines are like lines of $\boldsymbol{B}$—they never start or stop, and will tend to go in closed loops. Now Helmholtz described III in words by the following statement: the vortex lines *move with the fluid.* This means that if you were to mark the fluid particles along some vortex lines—by coloring them with ink, for example—then as the fluid moves and carries those particles along, they will always mark the new positions of the vortex lines. In whatever way the atoms of the liquid move, the vortex lines move with them. That is one way to describe the laws.

It also suggests a method for solving any problems. Given the initial flow pattern—say $\boldsymbol{v}$ everywhere—then you can calculate $\boldsymbol{\Omega}$. From the $\boldsymbol{v}$ you can also tell where the vortex lines are going to be a little later—they move with the speed $\boldsymbol{v}$. With the new $\boldsymbol{\Omega}$ you can use I and II to find the new $\boldsymbol{v}$. (That's just like the problem of finding $\boldsymbol{B}$, given the currents.) If we are given the flow pattern at one instant we can in principle calculate it for all subsequent times. We have the general solution for nonviscous flow.

We would like to show how Helmholtz's statement—and, therefore, III—can be at least partly understood. It is really just the law of conservation of angular momentum applied to the fluid. Suppose we imagine a small cylinder of the liquid whose axis is parallel to the vortex lines, as in Fig. 40–13(a). At some time later, this *same* piece of fluid will be somewhere else. Generally it will occupy a cylinder with a different diameter and be in a different place. It may also have a different orientation, say as in Fig. 40–13(b). If the diameter has changed, however, the length will have increased to keep the volume constant (since we are assuming an incompressible fluid). Also, since the vortex lines are stuck with the material, their density will go up as the cross-sectional area goes down. The product of the vorticity $\boldsymbol{\Omega}$ and area A of the cylinder will remain constant, so according to Helmholtz, we should have

$$\Omega_2 A_2 = \Omega_1 A_1. \tag{40.21}$$

Now notice that with zero viscosity all the forces on the surface of the cylindrical volume (or *any* volume, for that matter) are perpendicular to the surface. The pressure forces can cause the volume to be moved from place to place, or can cause it to change shape; but with no *tangential* forces the magnitude of the *angular momentum of the material inside* cannot change. The angular momentum of the liquid in the little cylinder is its moment of inertia I times the angular velocity of the liquid, which is proportional to the vorticity Ω. For a cylinder, the moment of inertia is proportional to mr^2. So from the conservation of angular momentum, we would conclude that

$$(M_1 R_1^2)\Omega_1 = (M_2 R_2^2)\Omega_2.$$

But the mass is the same, $M_1 = M_2$, and the areas are proportional to R^2, so we get again just Eq. (40.21). Helmholtz's statement—which is equivalent to III—is just a consequence of the fact that in the absence of viscosity the angular momentum of an element of the fluid cannot change.

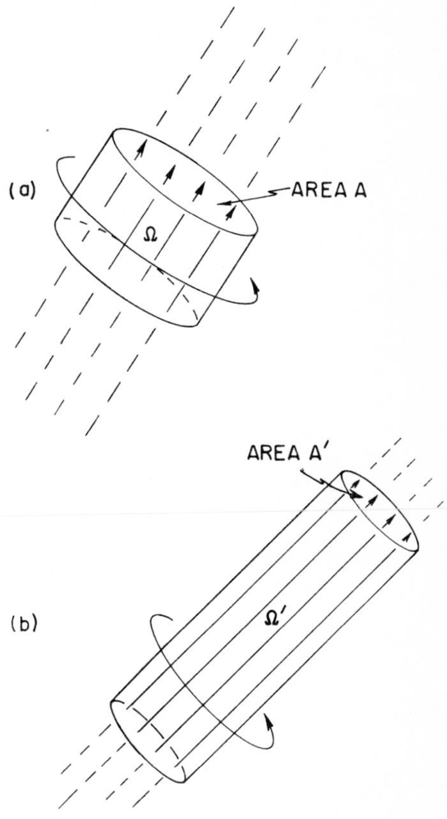

Fig. 40–13. (a) A group of vortex lines at t; (b) the same lines at a later time t'.

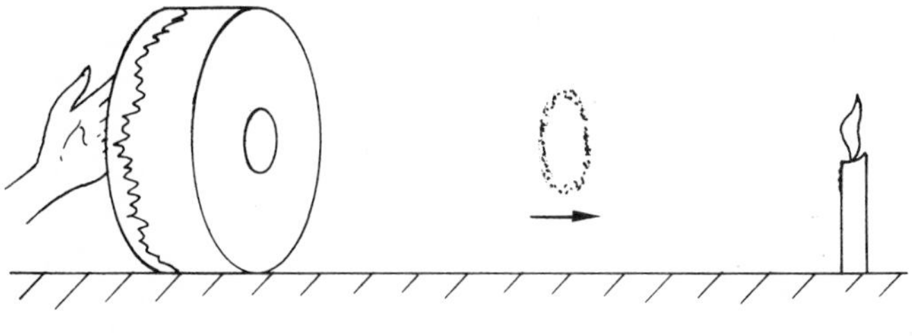

Fig. 40–14. Making a travelling vortex ring.

There is a nice demonstration of a moving vortex which is made with the simple apparatus of Fig. 40–14. It is a "drum" two feet in diameter and two feet long made by stretching a thick rubber sheet over the open end of a cylindrical "box." The "bottom"—the drum is tipped on its side—is solid except for a 3-inch diameter hole. If you give a sharp blow on the rubber diaphragm with your hand, a vortex ring is projected out of the hole. Although the vortex is invisible, you can tell it's there because it will blow out a candle 10 to 20 feet away. By the delay in

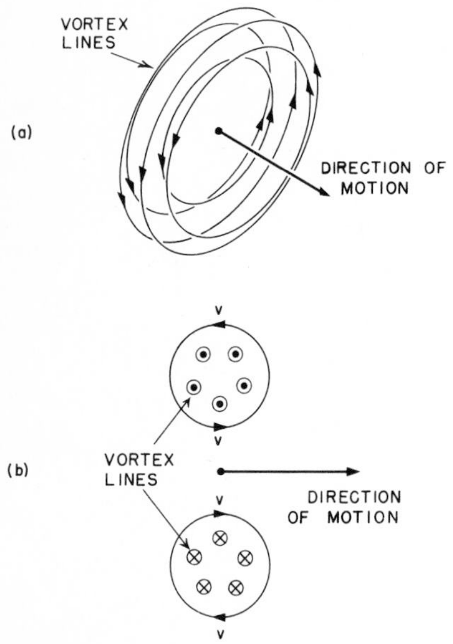

Fig. 40–15. A moving vortex ring (a smoke ring). (a) The vortex lines. (b) A cross section of the ring.

the effect, you can tell that "something" is travelling at a finite speed. You can see better what is going on if you first blow some smoke into the box. Then you see the vortex as a beautiful round "smoke ring."

The smoke ring is a torus-shaped bundle of vortex lines, as shown in Fig. 40–15(a). Since $\boldsymbol{\Omega} = \nabla \times \boldsymbol{v}$, these vortex lines represent also a circulation of $\boldsymbol{v}$ as shown in part (b) of the figure. We can understand the forward motion of the ring in the following way: The circulating velocity around the *bottom* of the ring extends up to the top of the ring, having there a forward motion. Since the lines of $\boldsymbol{\Omega}$ move with the fluid, they also move ahead with the velocity $\boldsymbol{v}$. (Of course, the circulation of $\boldsymbol{v}$ around the top part of the ring is responsible for the forward motion of the vortex lines at the bottom.)

We must now mention a serious difficulty. We have already noted that Eq. (40.9) says that, if $\boldsymbol{\Omega}$ is initially zero, it will always be zero. This result is a great failure of the theory of "dry" water, because it means that once $\boldsymbol{\Omega}$ is zero it is *always* zero—it is impossible to *produce* any vorticity under any circumstance. Yet, in our simple demonstration with the drum, we can generate a vortex ring starting with air which was initially at rest. (Certainly, $\boldsymbol{v} = 0$, $\boldsymbol{\Omega} = 0$ everywhere in the box before we hit it.) Also, we all know that we can start some vorticity in a lake with a paddle. Clearly, we must go to a theory of "wet" water to get a complete understanding of the behavior of a fluid.

Another feature of the dry water theory which is incorrect is the supposition we make regarding the flow at the boundary between it and the surface of a solid. When we discussed the flow past a cylinder—as in Fig. 40–11, for example—we permitted the fluid to slide along the surface of the solid. In our theory, the velocity at a solid surface could have any value depending on how it got started, and we did not consider any "friction" between the fluid and the solid. It is an experimental fact, however, that the velocity of a real fluid always goes to zero at the surface of a solid object. Therefore, our solution for the cylinder, with or without circulation, is wrong—as is our result regarding the generation of vorticity. We will tell you about the more correct theories in the next chapter.

41

The Flow of Wet Water

41–1 Viscosity

In the last chapter we discussed the behavior of water, disregarding the phenomenon of viscosity. Now we would like to discuss the phenomena of the flow of fluids, *including* the effects of viscosity. We want to look at the *real behavior* of fluids. We will describe qualitatively the actual behavior of the fluids under various different circumstances so that you will get some feel for the subject. Although you will see some complicated equations and hear about some complicated things, it is not our purpose that you should learn all these things. This is, in a sense, a "cultural" chapter which will give you some idea of the way the world is. There is only one item which is worth learning, and that is the simple definition of viscosity which we will come to in a moment. The rest is only for your entertainment.

In the last chapter we found that the laws of motion of a fluid are contained in the equation

$$\frac{\partial \boldsymbol{v}}{\partial t} + (\boldsymbol{v} \cdot \nabla)\boldsymbol{v} = -\frac{\nabla p}{\rho} - \nabla \phi + \frac{\boldsymbol{f}_{\text{visc}}}{\rho}. \tag{41.1}$$

In our "dry" water approximation we left out the last term, so we were neglecting all viscous effects. Also, we sometimes made an additional approximation by considering the fluid as incompressible; then we had the additional equation

$$\nabla \cdot \boldsymbol{v} = 0.$$

This last approximation is often quite good—particularly when flow speeds are much slower than the speed of sound. But in real fluids it is almost never true that we can neglect the internal friction that we call viscosity; most of the interesting things that happen come from it in one way or another. For example, we saw that in "dry" water the circulation never changes—if there is none to start out with, there will never be any. Yet, circulation in fluids is an everyday occurrence. We must fix up our theory.

We begin with an important experimental fact. When we worked out the flow of "dry" water around or past a cylinder—the so-called "potential flow"—we had no reason not to permit the water to have a velocity tangent to the surface; only the normal component had to be zero. We took no account of the possibility that there might be a shear force between the liquid and the solid. It turns out—although it is not at all self-evident—that in all circumstances where it has been experimentally checked, the *velocity of a fluid is exactly zero at the surface of a solid.* You have noticed, no doubt, that the blade of a fan will collect a thin layer of dust—and that it is still there after the fan has been churning up the air. You can see the same effect even on the great fan of a wind tunnel. Why isn't the dust blown off by the air? In spite of the fact that the fan blade is moving at high speed through the air, the speed of the air relative to the fan blade goes to zero right at the surface. So the very smallest dust particles are not disturbed.* We must modify the theory to agree with the experimental fact that in all ordinary fluids, the molecules next to a solid surface have zero velocity (relative to the surface).†

---

* You *can* blow *large* dust particles from a table top, but *not* the very finest ones. The large ones stick up into the breeze.

† You can imagine circumstances when it is not true: glass is theoretically a "liquid," but it can certainly be made to slide along a steel surface. So our assertion must break down somewhere.

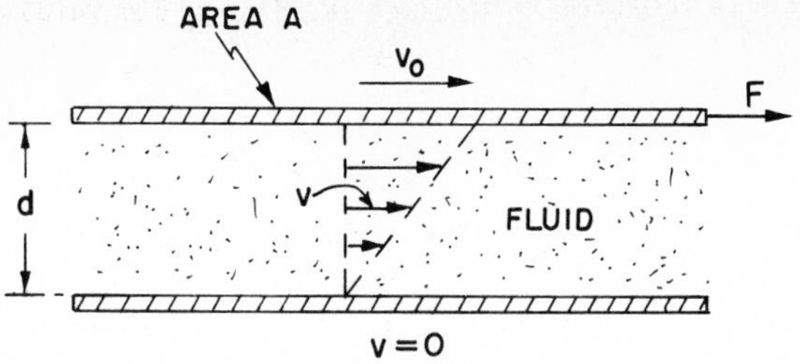

Fig. 41–1. Viscous drag between two parallel plates.

We originally characterized a liquid by the fact that if you put a shearing stress on it—no matter how small—it would give way. It flows. In static situations, there are no shear stresses. But before equilibrium is reached—as long as you still push on it—there can be shear forces. *Viscosity* describes these shear forces which exist in a moving fluid. To get a measure of the shear forces during the motion of a fluid, we consider the following kind of experiment. Suppose that we have two solid plane surfaces with water between them, as in Fig. 41–1, and we keep one stationary while moving the other parallel to it at the slow speed v_0. If you measure the force required to keep the upper plate moving, you find that it is proportional to the area of the plates and to v_0/d, where d is the distance between the plates. So the shear stress F/A is proportional to v_0/d:

$$\frac{F}{A} = \eta\,\frac{v_0}{d}.$$

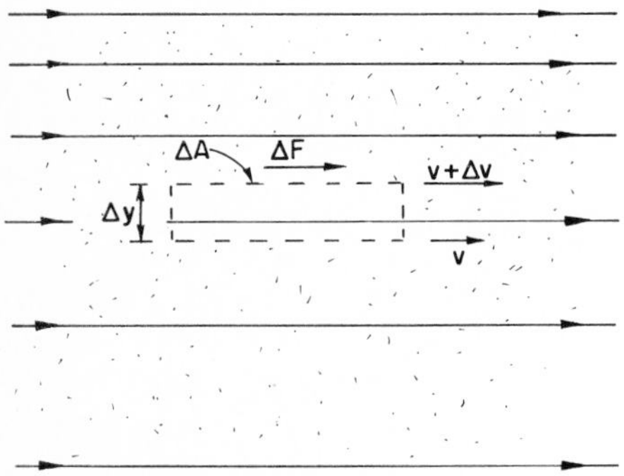

Fig. 41–2. The shear stress in a viscous fluid.

The constant of proportionality η is called the *coefficient of viscosity.*

If we have a more complicated situation, we can always consider a little, flat, rectangular cell in the water with its faces parallel to the flow, as in Fig. 41–2. The shear force across this cell is given by

$$\frac{\Delta F}{\Delta A} = \eta\,\frac{\Delta v_x}{\Delta y} = \eta\,\frac{\partial v_x}{\partial y}. \tag{41.2}$$

Now, $\partial v_x/\partial y$ is the *rate of change* of the shear strain we defined in Chapter 38, so for a liquid, the shear stress is proportional to the *rate of change* of the shear strain.

In the general case we write

$$S_{xy} = \eta\left(\frac{\partial v_y}{\partial x} + \frac{\partial v_x}{\partial y}\right). \tag{41.3}$$

If there is a uniform rotation of the fluid, $\partial v_x/\partial y$ is the negative of $\partial v_y/\partial x$ and S_{xy} is zero—as it should be since there are no stresses in a uniformly rotating fluid. (We did a similar thing in defining e_{xy} in Chapter 39.) There are, of course, the corresponding expressions for S_{yz} and S_{zx}.

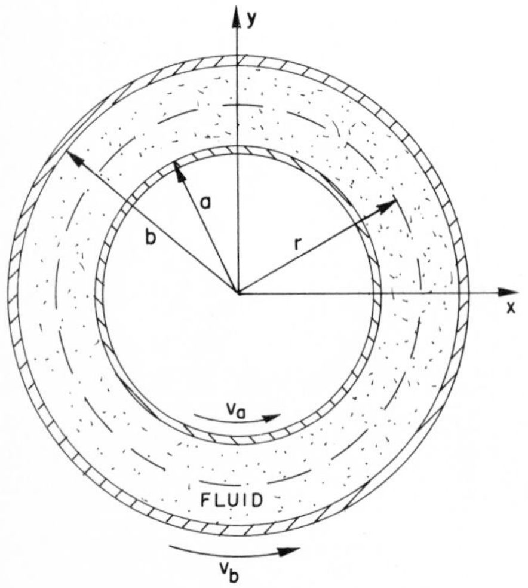

Fig. 41–3. The flow in a fluid between two concentric cylinders rotating at different angular velocities.

As an example of the application of these ideas, we consider the motion of a fluid between two coaxial cylinders. Let the inner one have the radius a and the peripheral velocity v_a, and let the outer one have radius b and velocity v_b. See Fig. 41–3. We might ask, what is the velocity distribution between the cylinders? To answer this question, we begin by finding a formula for the viscous shear in the fluid at a distance r from the axis. From the symmetry of the problem, we can assume that the flow is always tangential and that its magnitude depends only on r; $v = v(r)$. If we watch a speck in the water at the radius r, its coordinates as a function of time are

$$x = r\cos\omega t, \qquad y = r\sin\omega t,$$

where $\omega = v/r$. Then the x- and y-components of velocity are

$$v_x = -r\omega\sin\omega t = -\omega y \quad\text{and}\quad v_y = r\omega\cos\omega t = \omega x. \tag{41.4}$$

From Eq. (41.3), we have

$$S_{xy} = \eta\left[\frac{\partial}{\partial x}(x\omega) - \frac{\partial}{\partial y}(y\omega)\right] = \eta\left[x\,\frac{\partial\omega}{\partial x} - y\,\frac{\partial\omega}{\partial y}\right]. \tag{41.5}$$

For a point at $y = 0$, $\partial\omega/\partial y = 0$, and $x\,\partial\omega/\partial x$ is the same as $r\,d\omega/dr$. So at that point

$$(S_{xy})_{y=0} = \eta r \frac{d\omega}{dr}. \tag{41.6}$$

(It is reasonable that S should depend on $\partial\omega/\partial r$; when there is no change in ω with r, the liquid is in uniform rotation and there are no stresses.)

The stress we have calculated is the tangential shear which is the same all around the cylinder. We can get the *torque* acting *across a cylindrical surface* at the radius r by multiplying the shear stress by the moment arm r and the area $2\pi rl$. We get

$$\tau = 2\pi r^2 l (S_{xy})_{y=0} = 2\pi\eta l r^3 \frac{d\omega}{dr}. \tag{41.7}$$

Since the motion of the water is steady—there is no angular acceleration—the net torque on the cylindrical shell of water between r and $r + dr$ must be zero; that is, the torque at r must be balanced by an equal and opposite torque at $r + dr$, so that τ must be independent of r. In other words, $r^3\,d\omega/dr$ is equal to some constant, say A, and

$$\frac{d\omega}{dr} = \frac{A}{r^3}. \tag{41.8}$$

Integrating, we find that ω varies with r as

$$\omega = -\frac{A}{2r^2} + B. \tag{41.9}$$

The constants A and B are to be determined to fit the conditions that $\omega = \omega_a$ at $r = a$, and $\omega = \omega_b$ at $r = b$. We get that

$$A = \frac{2a^2b^2}{b^2 - a^2}(\omega_b - \omega_a),$$
$$B = \frac{b^2\omega_b - a^2\omega_a}{b^2 - a^2}. \tag{41.10}$$

So we know ω as a function of r, and from it $v = \omega r$.

If we want the torque, we can get it from Eqs. (41.7) and (41.8):

$$\tau = 2\pi\eta l A$$

or

$$\tau = \frac{4\pi\eta l a^2 b^2}{b^2 - a^2}(\omega_b - \omega_a). \tag{41.11}$$

It is proportional to the relative angular velocities of the two cylinders. One standard apparatus for measuring the coefficients of viscosity is built this way. One cylinder—say the outer one—is on pivots but is held stationary by a spring balance which measures the torque on it, while the inner one is rotated at a constant angular velocity. The coefficient of viscosity is then determined from Eq. (41.11).

From its definition, you see that the units of η are newton-sec/m^2. For water at 20°C,

$$\eta = 10^3 \text{ newton-sec/m}^2.$$

It is usually more convenient to use the *specific viscosity*, which is η divided by the density ρ. The values for water and air are then comparable:

$$\begin{aligned} &\text{water at 20°C}, && \eta/\rho = 10^{-6}\text{ m}^2\text{/sec}, \\ &\text{air at 20°C}, && \eta/\rho = 15 \times 10^{-6}\text{ m}^2\text{/sec}. \end{aligned} \tag{41.12}$$

Viscosities usually depend strongly on temperature. For instance, for water just above the freezing point, η/ρ is 1.8 times larger than it is at 20°C.

41-2 Viscous flow

We now go to a general theory of viscous flow—at least in the most general form known to man. We already understand that the shear stress components are proportional to the spatial derivatives of the various velocity components such as $\partial v_x/\partial y$ or $\partial v_y/\partial x$. However, in the general case of a *compressible* fluid there is another term in the stress which depends on other derivatives of the velocity. The general expression is

$$S_{ij} = \eta\left(\frac{\partial v_i}{\partial x_j} + \frac{\partial v_j}{\partial x_i}\right) + \eta'\,\delta_{ij}(\boldsymbol{\nabla}\cdot\boldsymbol{v}), \tag{41.13}$$

where x_i is any one of the rectangular coordinates x, y, or z, and v_i is any one of the rectangular coordinates of the velocity. (The symbol δ_{ij} is the Kronecker delta which is 1 when $i = j$ and 0 for $i \neq j$.) The additional term adds $\eta'\boldsymbol{\nabla}\cdot\boldsymbol{v}$ to all the diagonal elements S_{ii} of the stress tensor. If the liquid is incompressible $\boldsymbol{\nabla}\cdot\boldsymbol{v} = 0$, and this extra term doesn't appear. So it has to do with internal forces during compression. So two constants are required to describe the liquid, just as we had two constants to describe a homogeneous elastic solid. The coefficient η is the "ordinary" coefficient of viscosity which we have already encountered. It is also called the *first coefficient of viscosity* or the "shear viscosity coefficient," and the new coefficient η' is called the *second coefficient of viscosity.*

Now we want to determine the viscous force per unit volume, $\boldsymbol{f}_{\text{visc}}$, so we can put it into Eq. (41.1) to get the equation of motion for a real fluid. The force on a small cubical volume element of a fluid is the resultant of the forces on all the six faces. Taking them two at a time, we will get differences that depend on the derivatives of the stresses, and, therefore, on the second derivatives of the velocity. This is nice because it will get us back to a vector equation. The component of the viscous force per unit volume in the direction of the rectangular coordinate x_i is

$$\begin{aligned}(f_{\text{visc}})_i &= \sum_{j=1}^{3}\frac{\partial S_{ij}}{\partial x_j}\\ &= \eta\sum_{j=1}^{3}\frac{\partial}{\partial x_j}\left\{\eta\left(\frac{\partial v_i}{\partial x_j} + \frac{\partial v_j}{\partial x_i}\right)\right\} + \frac{\partial}{\partial x_i}(\eta'\boldsymbol{\nabla}\cdot\boldsymbol{v}).\end{aligned} \tag{41.14}$$

Usually, the variation of the viscosity coefficients with position is not significant and can be neglected. Then, the viscous force per unit volume contains only second derivatives of the velocity. We saw in Chapter 39 that the most general form of second derivatives that can occur in a vector equation is the sum of a term in the Laplacian ($\boldsymbol{\nabla}\cdot\boldsymbol{\nabla}\boldsymbol{v} = \nabla^2\boldsymbol{v}$), and a term in the gradient of the divergence $(\boldsymbol{\nabla}(\boldsymbol{\nabla}\cdot\boldsymbol{v}))$. Equation (41.14) is just such a sum with the coefficients η and $(\eta + \eta')$. We get

$$\boldsymbol{f}_{\text{visc}} = \eta\,\nabla^2\boldsymbol{v} + (\eta + \eta')\,\boldsymbol{\nabla}(\boldsymbol{\nabla}\cdot\boldsymbol{v}). \tag{41.15}$$

In the incompressible case, $\boldsymbol{\nabla}\cdot\boldsymbol{v} = 0$, and the viscous force per unit volume is just $\eta\,\nabla^2\boldsymbol{v}$. That is all that many people use; however, if you should want to calculate the absorption of sound in a fluid, you would need the second term.

We can now complete our general equation of motion for a real fluid. Substituting Eq. (41.15) into Eq. (41.1), we get

$$\rho\left\{\frac{\partial\boldsymbol{v}}{\partial t} + (\boldsymbol{v}\cdot\boldsymbol{\nabla})\boldsymbol{v}\right\} = -\boldsymbol{\nabla}p - \rho\,\boldsymbol{\nabla}\phi + \eta\,\nabla^2\boldsymbol{v} + (\eta + \eta')\,\boldsymbol{\nabla}(\boldsymbol{\nabla}\cdot\boldsymbol{v}).$$

It's complicated. But that's the way nature is.

If we introduce the vorticity $\boldsymbol{\Omega} = \boldsymbol{\nabla}\times\boldsymbol{v}$, as we did before, we can write our equation as

$$\begin{aligned}\rho\left\{\frac{\partial\boldsymbol{v}}{\partial t} + \boldsymbol{\Omega}\times\boldsymbol{v} + \frac{1}{2}\,\boldsymbol{\nabla}v^2\right\} &= -\boldsymbol{\nabla}p - \rho\,\boldsymbol{\nabla}\phi + \eta\,\nabla^2\boldsymbol{v}\\ &\quad + (\eta + \eta')\,\boldsymbol{\nabla}(\boldsymbol{\nabla}\cdot\boldsymbol{v}).\end{aligned} \tag{41.16}$$

We are supposing again that the only body forces acting are conservative forces like gravity. To see what the new term means, let's look at the incompressible fluid case. Then, if we take the curl of Eq. (41.16), we get

$$\frac{\partial \boldsymbol{\Omega}}{\partial t} + \boldsymbol{\nabla} \times (\boldsymbol{\Omega} \times \boldsymbol{v}) = \frac{\eta}{\rho} \nabla^2 \boldsymbol{\Omega}. \tag{41.17}$$

This is like Eq. (40.9) except for the new term on the right-hand side. When the right-hand side was zero, we had the Helmholtz theorem that the vorticity stays with the fluid. Now, we have the rather complicated nonzero term on the right-hand side which, however, has straightforward physical consequences. If we disregard for the moment the term $\boldsymbol{\nabla} \times (\boldsymbol{\Omega} \times \boldsymbol{v})$, we have a *diffusion equation.* The new term means that the vorticity $\boldsymbol{\Omega}$ *diffuses* through the fluid. If there is a large gradient in the vorticity, it will spread out into the neighboring fluid.

This is the term that causes the smoke ring to get thicker as it goes along. Also, it shows up nicely if you send a "clean" vortex (a "smokeless" ring made by the apparatus described in the last chapter) through a cloud of smoke. When it comes out of the cloud, it will have picked up some smoke, and you will see a hollow shell of a smoke ring. Some of the $\boldsymbol{\Omega}$ diffuses outward into the smoke, while still maintaining its forward motion with the vortex.

41–3 The Reynolds number

We will now describe the changes which are made in the character of fluid flow as a consequence of the new viscosity term. We will look at two problems in some detail. The first of these is the flow of a fluid past a cylinder—a flow which we tried to calculate in the previous chapter using the theory for nonviscous flow. It turns out that the viscous equations can be solved by man today only for a few special cases. So some of what we will tell you is based on experimental measurements—assuming that the experimental model satisfies Eq. (41.17).

The mathematical problem is this: We would like the solution for the flow of an incompressible, viscous fluid past a long cylinder of diameter D. The flow should be given by Eq. (41.17) and by

$$\boldsymbol{\Omega} = \boldsymbol{\nabla} \times \boldsymbol{v} \tag{41.18}$$

with the conditions that the velocity at large distances is some constant velocity, say V (parallel to the x-axis), and at the surface of the cylinder is zero. That is,

$$v_x = v_y = v_z = 0 \tag{41.19}$$

for

$$x^2 + y^2 = \frac{D^2}{4}.$$

That specifies completely the mathematical problem.

If you look at the equations, you see that there are four different parameters to the problem: η, ρ, D, and V. You might think that we would have to give a whole series of cases for different V's, different D's, and so on. However, that is not the case. All the different possible solutions correspond to different values of *one parameter.* This is the most important general thing we can say about viscous flow. To see why this is so, notice first that the viscosity and density appear only in the ratio η/ρ—the *specific* viscosity. That reduces the number of independent parameters to three. Now suppose we measure all distances in the only length that appears in the problem, the diameter D of the cylinder; that is, we substitute for x, y, z, the new variables x', y', z' with

$$x = x'D, \qquad y = y'D, \qquad z = z'D.$$

Then D disappears from (41.19). In the same way, if we measure all velocities in terms of V—that is, we set $v = v'V$—we get rid of the V, and v' is just equal to 1 at large distances. Since we have fixed our units of length and velocity, our unit of time is now D/V; so we should set

$$t = t' \frac{D}{V}. \tag{41.20}$$

With our new variables, the derivatives in Eq. (41.18) get changed from $\partial/\partial x$ to $(1/D)\, \partial/\partial x'$, and so on; so Eq. (41.18) becomes

$$\boldsymbol{\Omega} = \boldsymbol{\nabla} \times \boldsymbol{v} = \frac{V}{D} \boldsymbol{\nabla}' \times \boldsymbol{v}' = \frac{V}{D} \boldsymbol{\Omega}'. \tag{41.21}$$

Our main equation (41.17) then reads

$$\frac{\partial \boldsymbol{\Omega}'}{\partial t'} + \boldsymbol{\nabla}' \times (\boldsymbol{\Omega}' \times \boldsymbol{v}') = \frac{\eta}{\rho V D} \nabla^2 \boldsymbol{\Omega}'.$$

All the constants condense into one factor which we write, following tradition, as $1/\mathcal{R}$:

$$\mathcal{R} = \frac{\rho}{\eta} VD. \tag{41.22}$$

If we just remember that all of our equations are to be written with all quantities in the new units, we can omit all the primes. Our equations for the flow are then

$$\frac{\partial \boldsymbol{\Omega}}{\partial t} + \boldsymbol{\nabla} \times (\boldsymbol{\Omega} \times \boldsymbol{v}) = \frac{1}{\mathcal{R}} \nabla^2 \boldsymbol{\Omega} \tag{41.23}$$

and

$$\boldsymbol{\Omega} = \boldsymbol{\nabla} \times \boldsymbol{v}$$

with the conditions

$$\boldsymbol{v} = 0$$

for

$$x^2 + y^2 = 1/4 \tag{41.24}$$

and

$$v_x = 1, \qquad v_y = v_z = 0$$

for

$$x^2 + y^2 + z^2 \gg 1.$$

What this all means physically is very interesting. It means, for example, that if we solve the problem of the flow for one velocity V_1 and a certain cylinder diameter D_1, and then ask about the flow for a different diameter D_2 and a different fluid, the flow will be the same for the velocity V_2 which gives the same Reynolds number—that is, when

$$\mathcal{R}_1 = \frac{\rho_1}{\eta_1} V_1 D_1 = \mathcal{R}_2 = \frac{\rho_2}{\eta_2} V_2 D_2. \tag{41.25}$$

For any two situations which have the same Reynolds number, the flows will "look" the same—in terms of the appropriate scaled x', y', z', and t'. This is an important proposition because it means that we can determine what the behavior of the flow of air past an airplane wing will be without having to build an airplane and try it. We can, instead, make a model and make measurements using a velocity that gives the same Reynolds number. This is the principle which allows us to apply the results of "wind-tunnel" measurements on small-scale airplanes, or "model-basin" results on scale model boats, to the full-scale objects. Remember, however, that we can only do this provided the compressibility of the fluid can be neglected. Otherwise, a new quantity enters—the speed of sound. And different situations will really correspond to each other only if the ratio of V to the sound speed is also the same. This latter ratio is called the *Mach number*. So, for velocities near the speed of sound or above, the flows are the same in two situations if *both* the *Mach number* and the *Reynolds number* are the same for both situations.

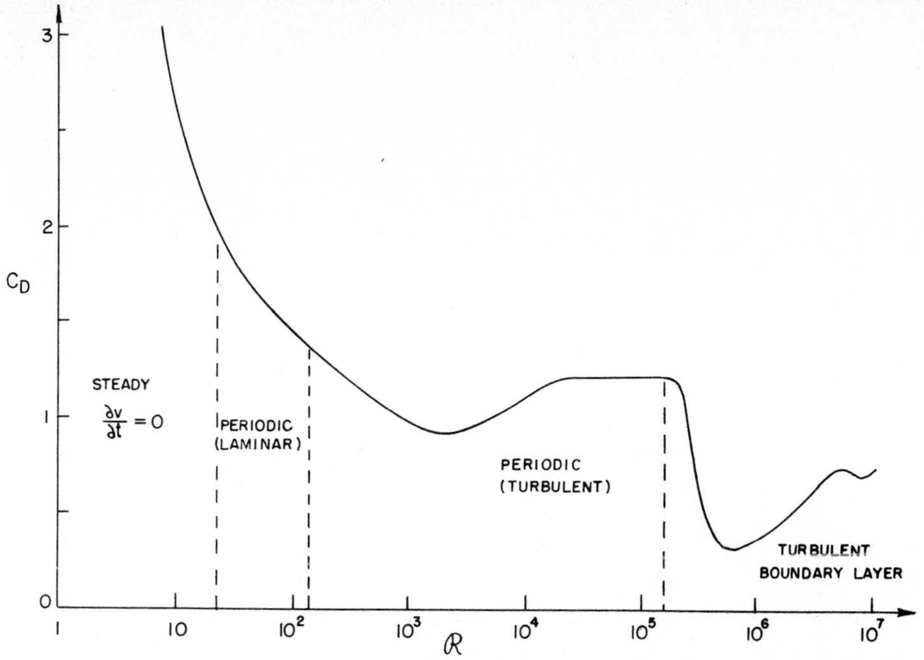

Fig. 41–4. The drag coefficient C_D of a circular cylinder as a function of the Reynolds number.

41–4 Flow past a circular cylinder

Let's go back to the problem of low-speed (nearly incompressible) flow over the cylinder. We will give a qualitative description of the flow of a real fluid. There are many things we might want to know about such a flow—for instance, what is the drag force on the cylinder? The drag force on a cylinder is plotted in Fig. 41–4 as a function of $\mathcal{R}$—which is proportional to the air speed V if everything else is held fixed. What is actually plotted is the so-called *drag coefficient* C_D, which is a dimensionless number equal to the force divided by $\frac{1}{2}\rho V^2 Dl$, where D is the diameter, l is the length of the cylinder, and ρ is the density of the liquid:

$$C_D = \frac{F}{\frac{1}{2}\rho V^2 Dl}.$$

The coefficient of drag varies in a rather complicated way, giving us a pre-hint that something rather interesting and complicated is happening in the flow. We will now describe the nature of flow for the different ranges of the Reynolds number. First, when the Reynolds number is very small, the flow is quite steady; that is, the velocity is constant at any place, and the flow goes around the cylinder. The actual distribution of the flow lines is, however, not like it is in potential flow. They are solutions of a somewhat different equation. When the velocity is very low or, what is equivalent, when the viscosity is very high so the stuff is like honey, then the inertial terms are negligible and the flow is described by the equation

$$\nabla^2\boldsymbol{\Omega} = 0.$$

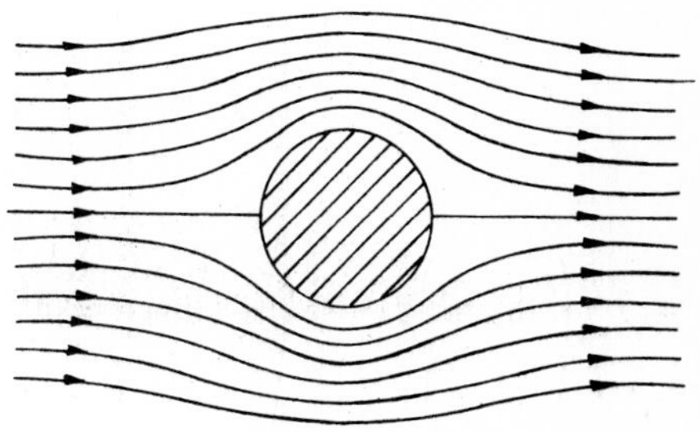

Fig. 41–5. Viscous flow (low velocities) around a circular cylinder.

This equation was first solved by Stokes. He also solved the same problem for a sphere. If you have a small sphere moving under such conditions of low Reynolds number, the force needed to drag it is equal to $6\pi\eta aV$, where a is the radius of the sphere and V is its velocity. This is a very useful formula because it tells the speed at which tiny grains of dirt (or other particles which can be approximated as spheres) move through a fluid under a given force—as, for instance, in a centrifuge, or in sedimentation, or diffusion. In the low Reynolds number region—for $\mathcal{R}$ less than 1—the lines of v around a *cylinder* are as drawn in Fig. 41–5.

If we now increase the fluid speed to get a Reynolds number somewhat greater than 1, we find that the flow is different. There is a circulation behind the sphere, as shown in Fig. 41–6(b). It is still an open question as to whether there is always

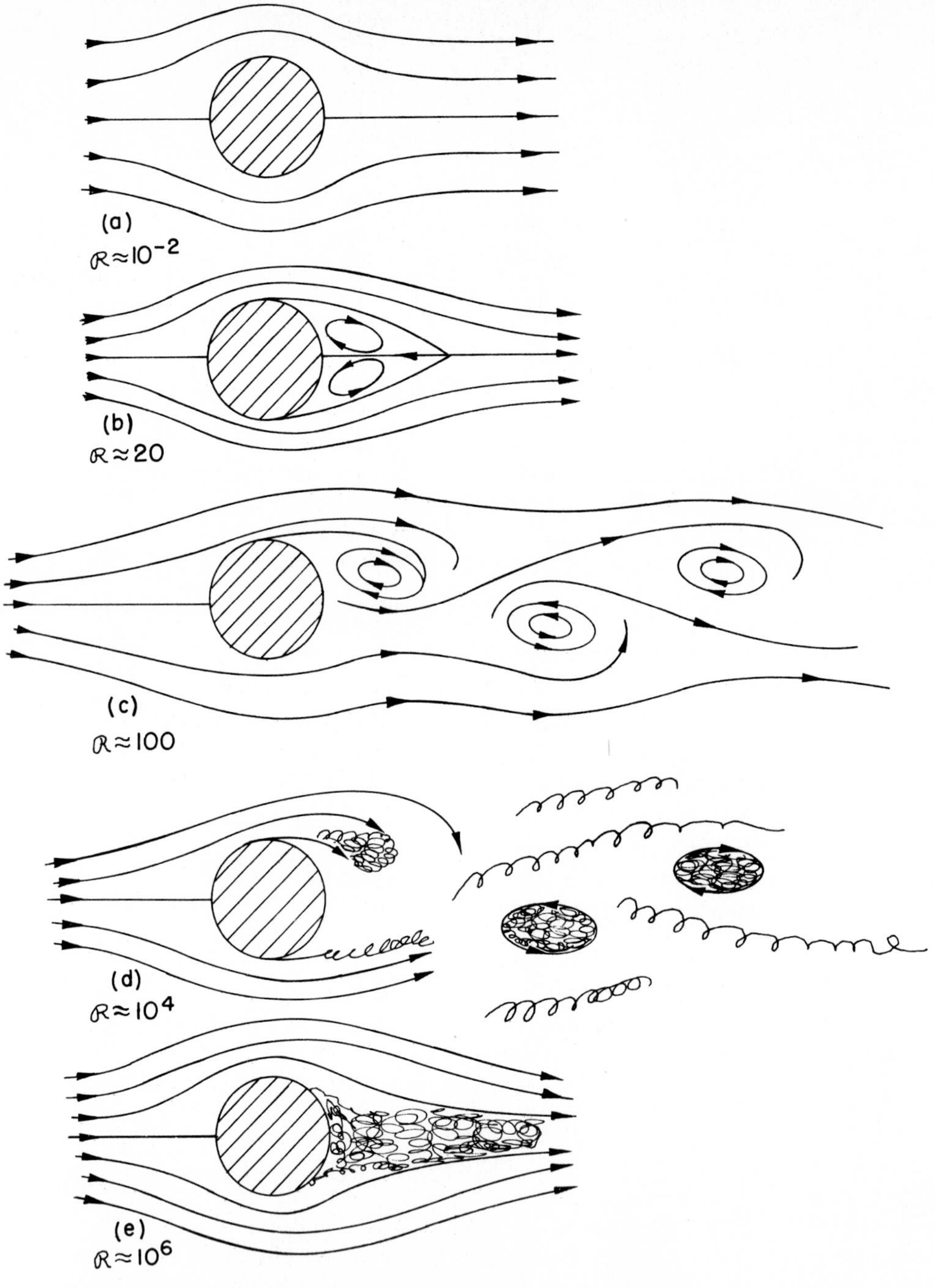

Fig. 41–6. Flow past a cylinder for various Reynolds numbers.

a circulation there even at the smallest Reynolds number or whether things suddenly change at a certain Reynolds number. It used to be thought that the circulation grew continuously. But it is now thought that it appears suddenly, and it is certain that the circulation increases with $\mathcal{R}$. In any case, there is a different character to the flow for $\mathcal{R}$ in the region from about 10 to 30. There is a pair of vortices behind the cylinder.

The flow changes again by the time we get to a number of 40 or so. There is suddenly a complete change in the character of the motion. What happens is that one of the vortices behind the cylinder gets so long that it breaks off and travels downstream with the fluid. Then the fluid curls around behind the cylinder and makes a new vortex. The vortices peel off alternately on each side, so an instantaneous view of the flow looks roughly as sketched in Fig. 41–6(c). The stream of

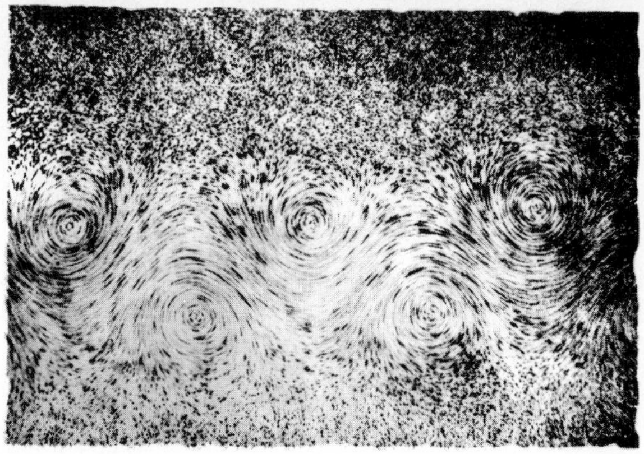

Fig. 41–7. Photograph by Ludwig Prandtl of the "vortex street" in the flow behind a cylinder.

vortices is called a "Kármán vortex street." They always appear for $\mathcal{R} > 40$. We show a photograph of such a flow in Fig. 41–7.

The difference between the two flows in Fig. 41–6(c) and 41–6(b) or 41–6(a) is almost a complete difference in regime. In Fig. 41–6(a) or (b), the velocity is constant, whereas in Fig. 41–6(c), the velocity at any point varies with time. There is no steady solution above $\mathcal{R} = 40$—which we have marked on Fig. 41–4 by a dashed line. For these higher Reynolds numbers, the flow varies with time but in a *regular*, cyclic fashion.

We can get a physical idea of how these vortices are produced. We know that the fluid velocity must be zero at the surface of the cylinder and that it also increases rapidly away from that surface. Vorticity is created by this large local variation in fluid velocity. Now when the main stream velocity is low enough, there is sufficient time for this vorticity to diffuse out of the thin region near the solid surface where it is produced and to grow into a large region of vorticity. This physical picture should help to prepare us for the next change in the nature of the flow as the main stream velocity, or $\mathcal{R}$, is increased still more.

As the velocity gets higher and higher, there is less and less time for the vorticity to diffuse into a larger region of fluid. By the time we reach a Reynolds number of several hundred, the vorticity begins to fill in a thin band, as shown in Fig. 41–6(d). In this layer the flow is chaotic and irregular. The region is called the *boundary layer* and this irregular flow region works its way farther and farther upstream as $\mathcal{R}$ is increased. In the turbulent region, the velocities are very irregular and "noisy"; also the flow is no longer two-dimensional but twists and turns in all three dimensions. There is still a regular alternating motion superimposed on the turbulent one.

As the Reynolds number is increased further, the turbulent region works its way forward until it reaches the point where the flow lines leave the cylinder—for flows somewhat above $\mathcal{R} = 10^5$. The flow is as shown in Fig. 41–6(e), and we have what is called a "turbulent boundary layer." Also, there is a drastic change in the drag force; it drops by a large factor, as shown in Fig. 41–4. In this speed region, the drag force actually *decreases* with increasing speed. There seems to be little evidence of periodicity.

What happens for still larger Reynolds numbers? As we increase the speed further, the wake increases in size again and the drag increases. The latest experiments—which go up to $\mathcal{R} = 10^7$ or so—indicate that a new periodicity appears in the wake, either because the whole wake is oscillating back and forth in a gross motion or because some new kind of vortex is occurring together with an irregular noisy motion. The details are as yet not entirely clear, and are still being studied experimentally.

41–5 The limit of zero viscosity

We would like to point out that none of the flows we have described are anything like the potential flow solution we found in the preceding chapter. This is, at first sight, quite surprising. After all, $\mathcal{R}$ is proportional to $1/\eta$. So η going to zero is equivalent to $\mathcal{R}$ going to infinity. And if we take the limit of large $\mathcal{R}$ in

Eq. (41.23), we get rid of the right-hand side and get just the equations of the last chapter. Yet, you would find it hard to believe that the highly turbulent flow at $\mathcal{R} = 10^7$ was approaching the smooth flow computed from the equations of "dry" water. How can it be that as we approach $\mathcal{R} = \infty$, the flow described by Eq. (41.23) gives a completely different solution from the one we obtained taking $\eta = 0$ to start out with? The answer is very interesting. Note that the right-hand term of Eq. (41.23) has $1/\mathcal{R}$ times a *second derivative*. It is a higher derivative than any other derivative in the equation. What happens is that although the coefficient $1/\mathcal{R}$ is small, there are very rapid variations of $\mathbf{\Omega}$ in the space near the surface. These rapid variations compensate for the small coefficient, and the product *does not go to zero* with increasing $\mathcal{R}$. The solutions do not approach the limiting case as the coefficient of $\nabla^2\mathbf{\Omega}$ goes to zero.

You may be wondering, "What is the fine-grain turbulence and how does it maintain itself? How can the vorticity which is made somewhere at the edge of the cylinder generate so much noise in the background?" The answer is again interesting. Vorticity has a tendency to amplify itself. If we forget for a moment about the diffusion of vorticity which causes a loss, the laws of flow say (as we have seen) that the vortex lines are carried along with the fluid, at the velocity $\boldsymbol{v}$. We can imagine a certain number of lines of $\mathbf{\Omega}$ which are being distorted and twisted by the complicated flow pattern of $\boldsymbol{v}$. This pulls the lines closer together and mixes them all up. Lines that were simple before will get knotted and pulled close together. They will be longer and tighter together. The strength of the vorticity will increase and its irregularities—the pluses and minuses—will, in general, increase. So the magnitude of vorticity in three dimensions increases as we twist the fluid about.

You might well ask, "When is the potential flow a satisfactory theory at all?" In the first place, it is satisfactory outside the turbulent region where the vorticity has not entered appreciably by diffusion. By making special streamlined bodies, we can keep the turbulent region as small as possible; the flow around airplane wings—which are carefully designed—is almost entirely true potential flow.

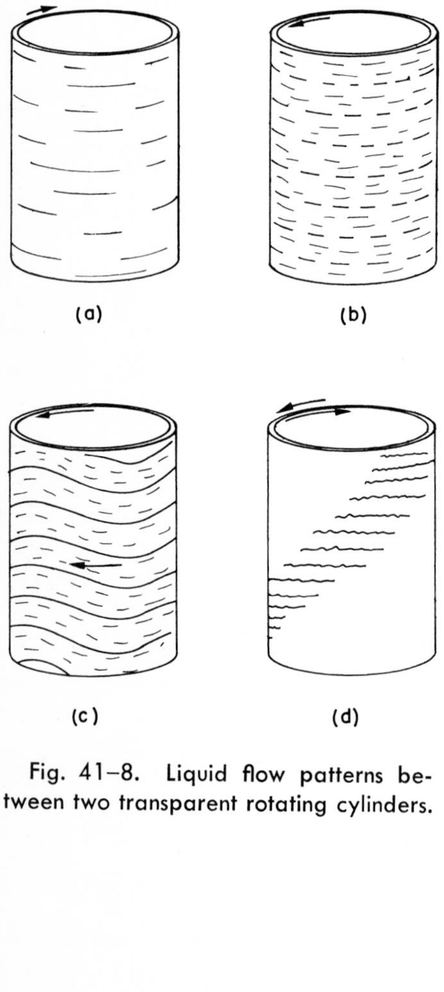

Fig. 41–8. Liquid flow patterns between two transparent rotating cylinders.

41–6 Couette flow

It is possible to demonstrate that the complex and shifting character of the flow past a cylinder is not special but that the great variety of flow possibilities occurs generally. We have worked out in Section 1 a solution for the viscous flow between two cylinders, and we can compare the results with what actually happens. If we take two concentric cylinders with an oil in the space between them and put a fine aluminum powder as a suspension in the oil, the flow is easy to see. Now if we turn the outer cylinder slowly, nothing unexpected happens; see Fig. 41–8(a). Alternatively, if we turn the inner cylinder slowly, nothing very striking occurs. However, if we turn the inner cylinder at a higher rate, we get a surprise. The fluid breaks into horizontal bands, as indicated in Fig. 41–8(b). When the outer cylinder rotates at a similar rate with the inner one at rest, no such effect occurs. How can it be that there is a difference between rotating the inner or the out cylinder? After all, the flow pattern we derived in Section 1 depended only on $\omega_b - \omega_a$. We can get the answer by looking at the cross sections shown in Fig. 41–9. When the inner layers of the fluid are moving more rapidly than the outer ones, they tend to move *outward*—the centrifugal force is larger than the pressure holding them in place. A whole layer cannot move out uniformly because the outer layers are in the way. It must break into cells and circulate, as shown in Fig. 41–9(b). It is like the convection currents in a room which has hot air at the bottom. When the inner cylinder is at rest and the outer cylinder has a high velocity, the centrifugal forces build up a pressure gradient which keeps everything in equilibrium—see Fig. 41–9(c) (as in a room with hot air at the top).

Now let's speed up the inner cylinder. At first, the number of bands increases. Then suddenly you see the bands become wavy, as in Fig. 41–8(c), and the waves travel around the cylinder. The speed of these waves is easily measured. For high rotation speeds they approach 1/3 the speed of the inner cylinder. And no one

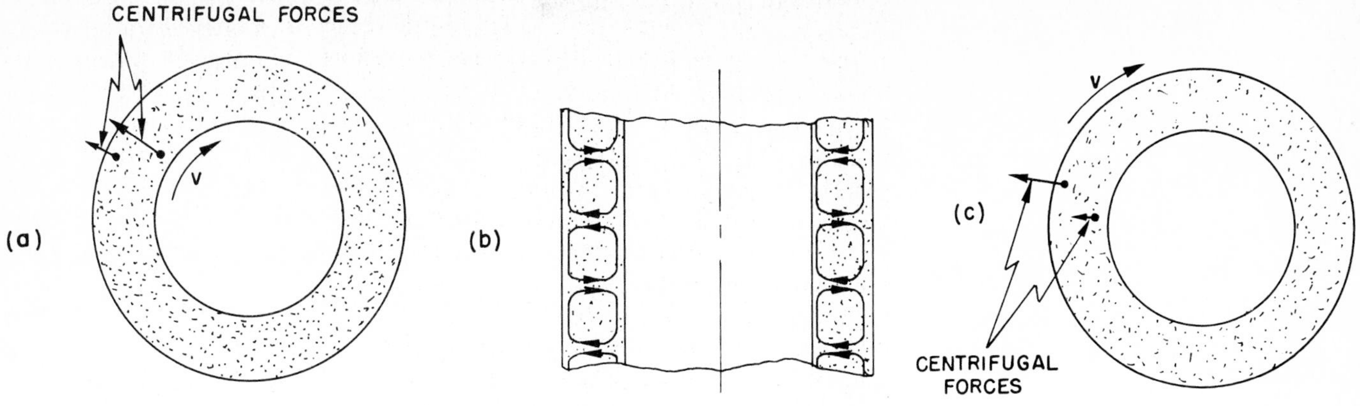

Fig. 41–9. Why the flow breaks up into bands.

knows why! There's a challenge. A simple number like 1/3, and no explanation. In fact, the whole mechanism of the wave formation is not very well understood; yet it is steady laminar flow.

If we now start rotating the outer cylinder also—but in the opposite direction—the flow pattern starts to break up. We get wavy regions alternating with apparently quiet regions, as sketched in Fig. 41–8(d), making a spiral pattern. In these "quiet" regions, however, we can see that the flow is really quite irregular; it is, in fact completely turbulent. The wavy regions also begin to show irregular turbulent flow. If the cylinders are rotated still more rapidly, the whole flow becomes chaotically turbulent.

In this simple experiment we see many interesting regimes of flow which are quite different, and yet which are all contained in our simple equation for various values of the one parameter $\mathcal{R}$. With our rotating cylinders, we can see many of the effects which occur in the flow past a cylinder: first, there is a steady flow; second, a flow sets in which varies in time but in a regular, smooth way; finally, the flow becomes completely irregular. You have all seen the same effects in the column of smoke rising from a cigarette in quiet air. There is a smooth steady column followed by a series of twistings as the stream of smoke begins to break up, ending finally in an irregular churning cloud of smoke.

The main lesson to be learned from all of this is that a tremendous variety of behavior is hidden in the simple set of equations in (41.23). All the solutions are for the same equations, only with different values of $\mathcal{R}$. We have no reason to think that there are any terms missing from these equations. The only difficulty is that we do not have the mathematical power today to analyze them except for very small Reynolds numbers—that is, in the completely viscous case. That we have written an equation does not remove from the flow of fluids its charm or mystery or its surprise.

If such variety is possible in a simple equation with only one parameter, how much more is possible with more complex equations! Perhaps the fundamental equation that describes the swirling nebulae and the condensing, revolving, and exploding stars and galaxies is just a simple equation for the hydrodynamic behavior of nearly pure hydrogen gas. Often, people in some unjustified fear of physics say you can't write an equation for life. Well, perhaps we can. As a matter of fact, we very possibly already have the equation to a sufficient approximation when we write the equation of quantum mechanics:

$$H\psi = -\frac{\hbar}{i}\frac{\partial\psi}{\partial t}.$$

We have just seen that the complexities of things can so easily and dramatically escape the simplicity of the equations which describe them. Unaware of the scope of simple equations, man has often concluded that nothing short of God, not mere equations, is required to explain the complexities of the world.

We have written the equations of water flow. From experiment, we find a set of concepts and approximations to use to discuss the solution—vortex streets, turbulent wakes, boundary layers. When we have similar equations in a less familiar situation, and one for which we cannot yet experiment, we try to solve the equations in a primitive, halting, and confused way to try to determine what new qualitative features may come out, or what new qualitative forms are a consequence of the equations. Our equations for the sun, for example, as a ball of hydrogen gas, describe a sun without sunspots, without the rice-grain structure of the surface, without prominences, without coronas. Yet, all of these are really in the equations; we just haven't found the way to get them out.

There are those who are going to be disappointed when no life is found on other planets. Not I—I want to be reminded and delighted and surprised once again, through interplanetary exploration, with the infinite variety and novelty of phenomena that can be generated from such simple principles. The test of science is its ability to predict. Had you never visited the earth, could you predict the thunderstorms, the volcanos, the ocean waves, the auroras, and the colorful sunset? A salutary lesson it will be when we learn of all that goes on on each of those dead planets—those eight or ten balls, each agglomerated from the same dust cloud and each obeying exactly the same laws of physics.

The next great era of awakening of human intellect may well produce a method of understanding the *qualitative* content of equations. Today we cannot. Today we cannot see that the water flow equations contain such things as the barber pole structure of turbulence that one sees between rotating cylinders. Today we cannot see whether Schrödinger's equation contains frogs, musical composers, or morality—or whether it does not. We cannot say whether something beyond it like God is needed, or not. And so we can all hold strong opinions either way.

42

Curved Space

42–1 Curved spaces with two dimensions

According to Newton everything attracts everything else with a force inversely proportional to the square of the distance from it, and objects respond to forces with accelerations proportional to the forces. They are Newton's laws of universal gravitation and of motion. As you know, they account for the motions of balls, planets, satellites, galaxies, and so forth.

Einstein had a different interpretation of the law of gravitation. According to him, space and time—which must be put together as space-time—are *curved* near heavy masses. And it is the attempt of things to go along "straight lines" in this curved space-time which makes them move the way they do. Now that is a complex idea—very complex. It is the idea we want to explain in this chapter.

Our subject has three parts. One involves the effects of gravitation. Another involves the ideas of space-time which we already studied. The third involves the idea of curved space-time. We will simplify our subject in the beginning by not worrying about gravity and by leaving out the time—discussing just curved space. We will talk later about the other parts, but we will concentrate now on the idea of curved space—what is meant by curved space, and, more specifically, what is meant by curved space in this application of Einstein. Now even that much turns out to be somewhat difficult in three dimensions. So we will first reduce the problem still further and talk about what is meant by the words "curved space" in two dimensions.

In order to understand this idea of curved space in two dimensions you really have to appreciate the limited point of view of the character who lives in such a space. Suppose we imagine a bug with no eyes who lives on a plane, as shown in Fig. 42–1. He can move only on the plane, and he has no way of knowing that there is any way to discover any "outside world." (He hasn't got your imagination.) We are, of course, going to argue by analogy. *We* live in a three-dimensional world, and we don't have any imagination about going off our three-dimensional world in a new direction; so we have to think the thing out by analogy. It is as though we were bugs living on a plane, and there was a space in another direction. That's why we will first work with the bug, remembering that he must live on his surface and can't get out.

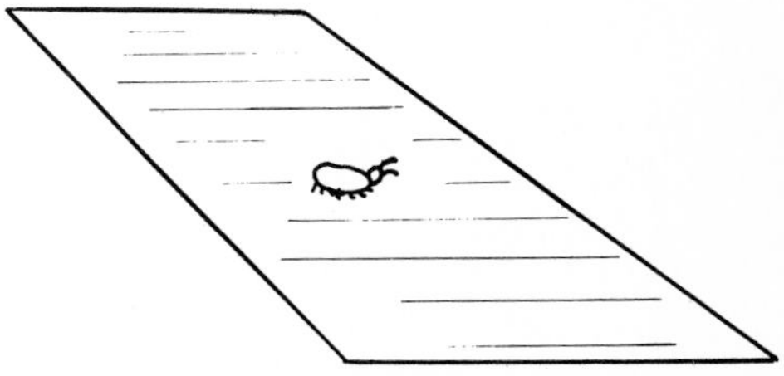

Fig. 42–1. A bug on a plane surface.

As another example of a bug living in two dimensions, let's imagine one who lives on a sphere. We imagine that he can walk around on the surface of the sphere, as in Fig. 42–2 but that he can't look "up," or "down," or "out."

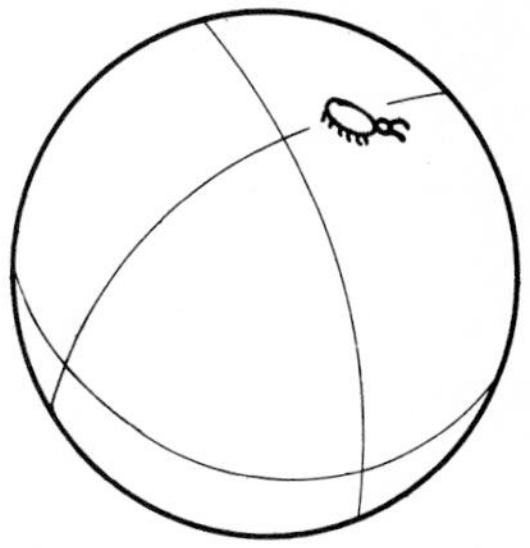

Fig. 42–2. A bug on a sphere.

Now we want to consider still a *third* kind of creature. He is also a bug like the others, and also lives on a plane, as our first bug did, but this time the plane is peculiar. The temperature is different at different places. Also, the bug and any rulers he uses are all made of the same material which expands when it is heated. Whenever he puts a ruler somewhere to measure something the ruler expands immediately to the proper length for the temperature at that place. Wherever he puts any object—himself, a ruler, a triangle, or anything—the thing stretches itself because of the thermal expansion. Everything is longer in the hot places than it is in the cold places, and everything has the same coefficient of expansion. We will call the home of our third bug a "hot plate," although we will particularly want to think of a special kind of hot plate that is cold in the center and gets hotter as we go out toward the edges (Fig. 42–3).

Now we are going to imagine that our bugs begin to study geometry. Although we imagine that they are blind so that they can't see any "outside" world,

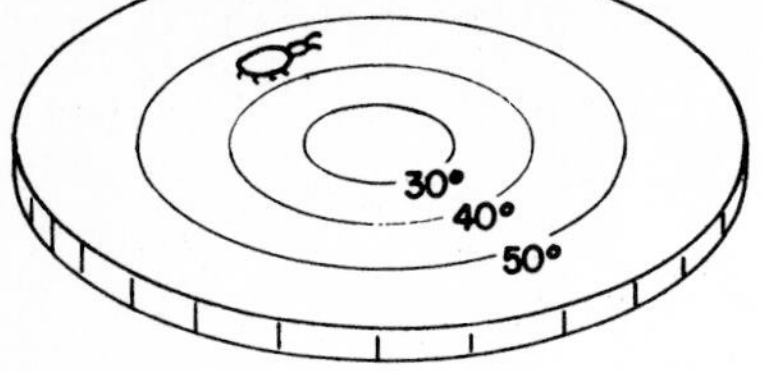

Fig. 42–3. A bug on a hot plate.

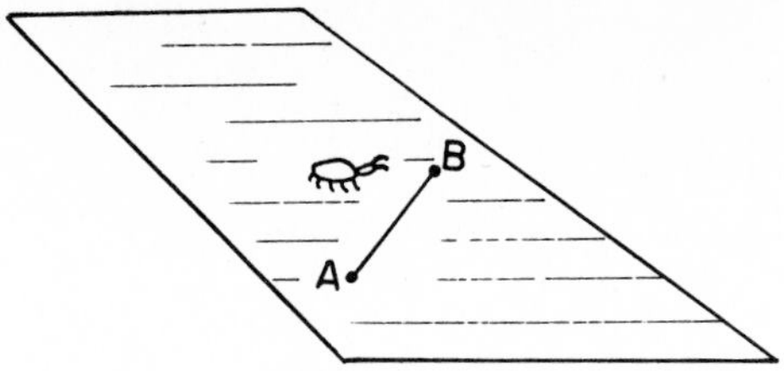

Fig. 42–4. Making a "straight line" on a plane.

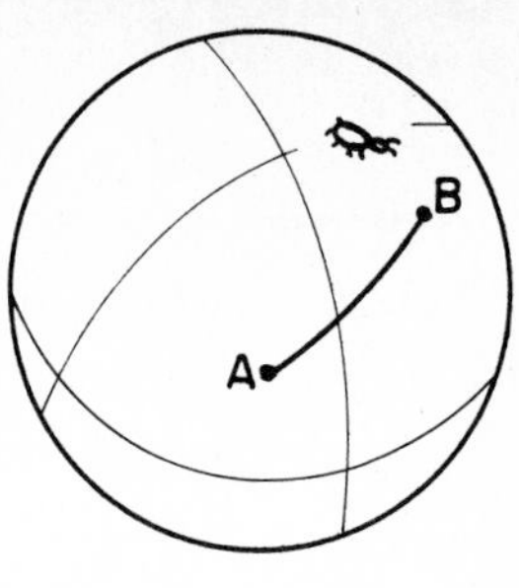

Fig. 42–5. Making a "straight line" on a sphere.

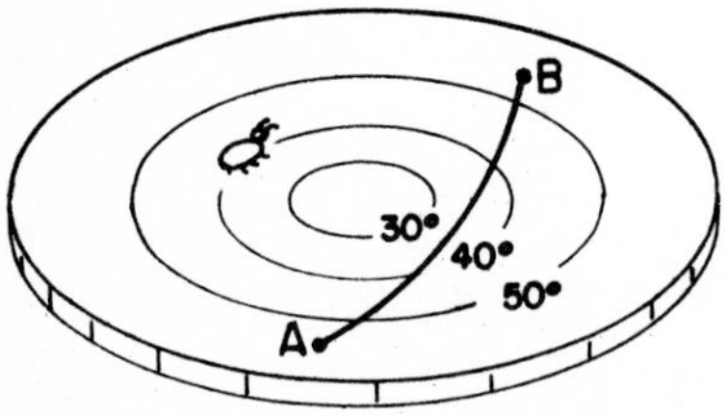

Fig. 42–6. Making a "straight line" on the hot plate.

they can do a lot with their legs and feelers. They can draw lines, and they can make rulers, and measure off lengths. First, let's suppose that they start with the simplest idea in geometry. They learn how to make a straight line—defined as the shortest line between two points. Our first bug—see Fig. 42–4—learns to make very good lines. But what happens to the bug on the sphere? He draws his straight line as the shortest distance—*for him*—between two points, as in Fig. 42–5. It may look like a curve to us, but he has no way of getting off the sphere and finding out that there is "really" a shorter line. He just knows that if he tries any other path *in his world* it is always longer than his straight line. So we will let him have his straight line as the shortest arc between two points. (It is, of course an arc of a great circle.)

Finally, our third bug—the one in Fig. 42–3—will also draw "straight lines" that look like curves to us. For instance, the shortest distance between A and B in Fig. 42–6 would be on a curve like the one shown. Why? Because when his line curves out toward the warmer parts of his hot plate, the rulers get longer (from our omniscient point of view) and it takes fewer "yardsticks" laid end-to-end to get from A to B. So *for him* the line is straight—he has no way of knowing that there could be someone out in a strange three-dimensional world who would call a different line "straight."

We think you get the idea now that all the rest of the analysis will always be from the point of view of the creatures on the particular surfaces and not from *our* point of view. With that in mind let's see what the rest of their geometries looks like. Let's assume that the bugs have all learned how to make two lines intersect at right angles. (You can figure out how they could do it.) Then our first bug (the one on the normal plane) finds an interesting fact. If he starts at the point A and makes a line 100 inches long, then makes a right angle and marks off another 100 inches, then makes another right angle and goes another 100 inches, then makes a third right angle and a fourth line 100 inches long, he ends up right at the starting point as shown in Fig. 42–7(a). It is a property of his world—one of the facts of his "geometry."

Then he discovers another interesting thing. If he makes a triangle—a figure with three straight lines—the sum of the angles is equal to 180°, that is, to the sum of two right angles. See Fig. 42–7(b).

Then he invents the circle. What's a circle? A circle is made this way: You rush off on straight lines in many many directions from a single point, and lay out a lot of dots that are all the same distance from that point. See Fig. 42–7(c). (We have to be careful how we define these things because we've got to be able to make the analogs for the other fellows.) Of course, its equivalent to the curve you can make by swinging a ruler around a point. Anyway, our bug learns how to make circles. Then one day he thinks of measuring the distance around a circle. He measures several circles and finds a neat relationship: The distance around is always the same number times the radius r (which is, of course, the distance from the center out to the curve). The circumference and the radius always have the same ratio—approximately 6.283—independent of the size of the circle.

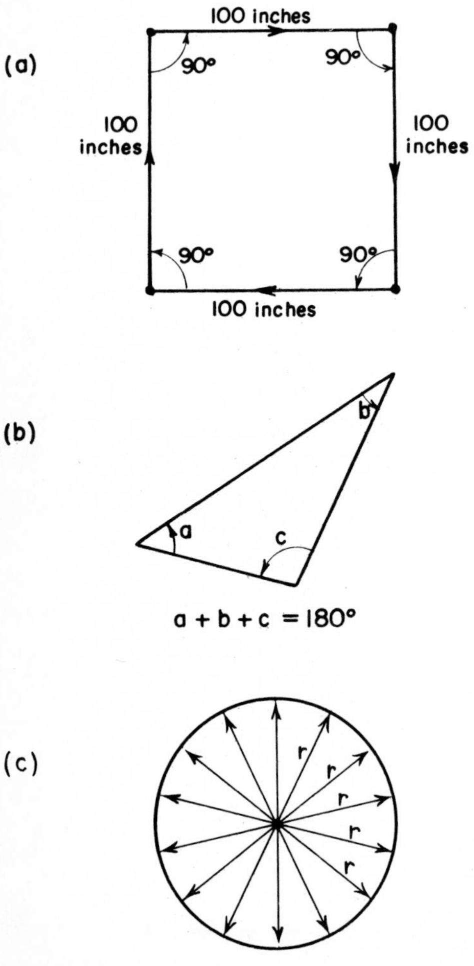

Fig. 42–7. A square, triangle, and circle in a flat space.

Now let's see what our other bugs have been finding out about *their* geometries. First, what happens to the bug on the sphere when he tries to make a "square"?

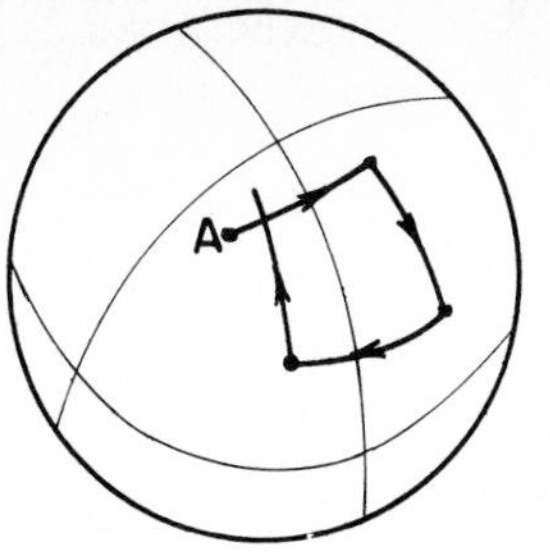

Fig. 42–8. Trying to make a "square" on a sphere.

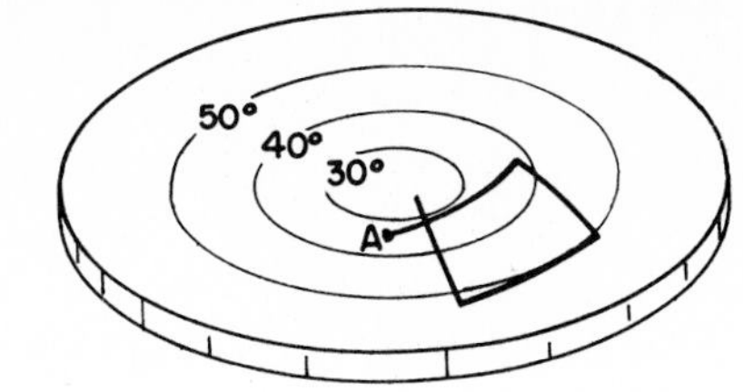

Fig. 42–9. Trying to make a "square" on the hot plate.

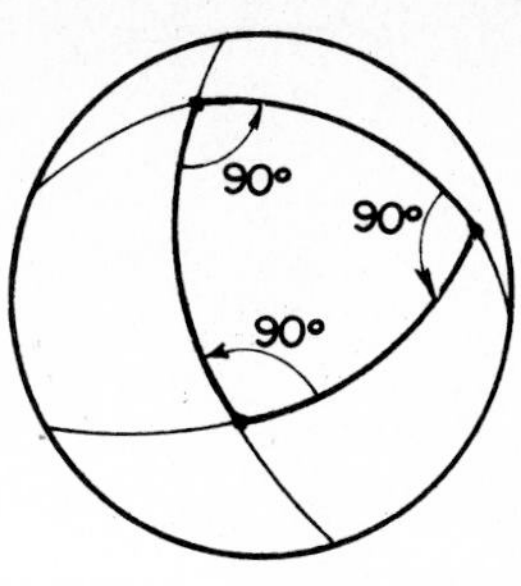

Fig. 42–10. On a sphere a "triangle" can have three 90° angles.

If he follows the prescription we gave above, he would probably think that the result was hardly worth the trouble. He gets a figure like the one shown in Fig. 42–8. His endpoint B isn't on top of the starting point A. It doesn't work out to a closed figure at all. Get a sphere and try it. A similar thing would happen to our friend on the hot plate. If he lays out four straight lines of equal length—as measured with his expanding rulers—joined by right angles he gets a picture like the one in Fig. 42–9.

Now suppose that our bugs had each had their own Euclid who had told them what geometry "should" be like, and that they had checked him out roughly by making crude measurements on a *small* scale. Then as they tried to make accurate squares on a larger scale they would discover that something was wrong. The point is, that just by *geometrical measurements* they would discover that something was the matter with their space. We define a *curved space* to be a space in which the geometry is not what we expect for a plane. The geometry of the bugs on the sphere or on the hot plate is the geometry of a curved space. The rules of Euclidian geometry fail. And it isn't necessary to be able to lift yourself out of the plane in order to find out that the world that you live in is curved. It isn't necessary to circumnavigate the globe in order to find out that it is a ball. You can find out that you live on a ball by laying out a square. If the square is very small you will need a lot of accuracy, but if the square is large the measurement can be done more crudely.

Let's take the case of a triangle on a plane. The sum of the angles is 180 degrees. Our friend on the sphere can find triangles that are very peculiar. He can, for example, find triangles which have *three right angles.* Yes indeed! One is shown in Fig. 42–10. Suppose our bug starts at the north pole and makes a straight line all the way down to the equator. Then he makes a right angle and another perfect straight line the same length. Then he does it again. For the very special length he has chosen he gets right back to his starting point, and also meets the first line with a right angle. So there is no doubt that for him this triangle has three right angles, or 270 degrees in the sum. It turns out that for him the sum of the angles of the triangle is *always* greater than 180 degrees. In fact, the excess (for the special case shown, the extra 90 degrees) is proportional to how much area the triangle has. If a triangle on a sphere is very small, its angles add up to very nearly 180 degrees, only a little bit over. As the triangle gets bigger the discrepancy goes up. The bugs on the hot plate would discover similar difficulties with their triangles.

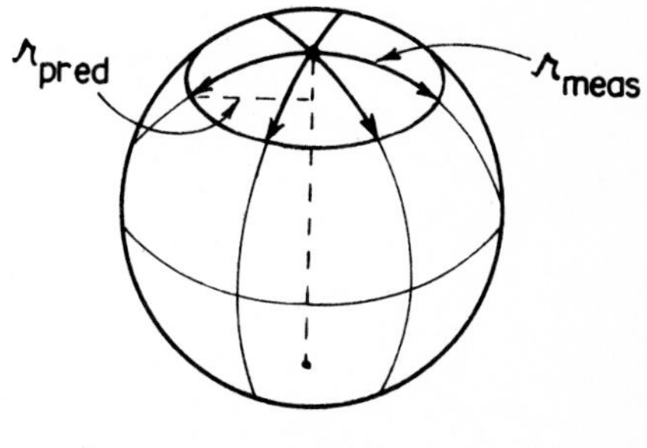

Fig. 42–11. Making a circle on a sphere.

Let's look next at what our other bugs find out about circles. They make circles and measure their circumferences. For example, the bug on the sphere might make a circle like the one shown in Fig. 42–11. And he would discover that the circumference is *less* than 2π times the radius. (You can see that because from the wisdom of our three-dimensional view it is obvious that what he calls the "radius" is a curve which is *longer* than the true radius of the circle.) Suppose that the bug on the sphere had read Euclid, and decided to predict a radius by dividing the circumference C by 2π, taking

$$r_{\text{pred}} = \frac{C}{2\pi}. \tag{42.1}$$

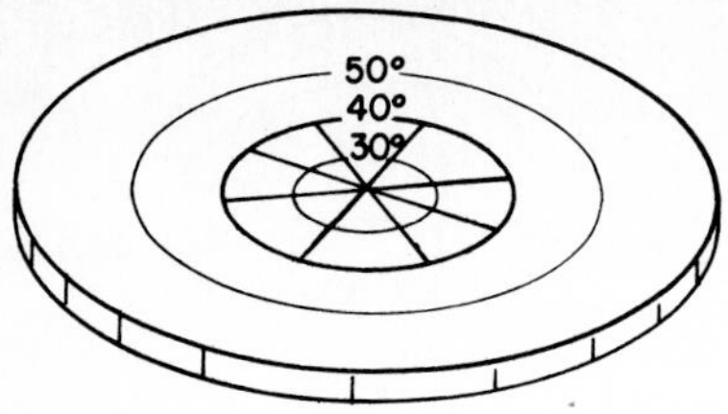

Fig. 42–12. Making a circle on the hot plate.

Then he would find that the measured radius was larger than the predicted radius. Pursuing the subject, he might define the difference to be the "excess radius," and write

$$r_{\text{meas}} - r_{\text{pred}} = r_{\text{excess}}, \tag{42.2}$$

and study how the excess radius effect depended on the size of the circle.

Our bug on the hot plate would discover a similar phenomenon. Suppose he was to draw a circle centered at the cold spot on the plate as in Fig. 42–12. If we were to watch him as he makes the circle we would notice that his rulers are short near the center and get longer as they are moved outward—although the bug doesn't know it, of course. When he measures the circumference the ruler is long all the time, so he, too, finds out that the measured radius is longer than the predicted radius, $C/2\pi$. The hot-plate bug also finds an "excess radius effect." And again the size of the effect depends on the radius of the circle.

We will *define* a "curved space" as one in which these types of geometrical errors occur: The sum of the angles of a triangle is different from 180 degrees; the circumference of a circle divided by 2π is not equal to the radius; the rule for making a square doesn't give a closed figure. You can think of others.

We have given two different examples of curved space: the sphere and the hot plate. But it is interesting that if we choose the right temperature variation as a function of distance on the hot plate, the two *geometries* will be exactly the same. It is rather amusing. We can make the bug on the hot plate get exactly the same answers as the bug on the ball. For those who like geometry and geometrical problems we'll tell you how it can be done. If you assume that the length of the rulers (as determined by the temperature) goes in proportion to one plus some constant times the square of the distance away from the origin, then you will find that the geometry of that hot plate is exactly the same in all details† as the geometry of the sphere.

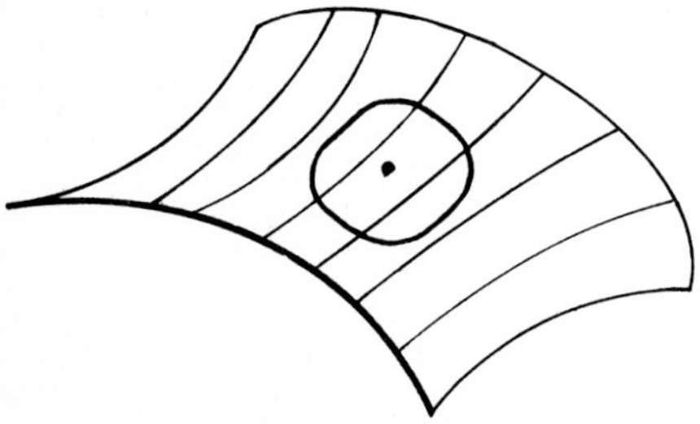

Fig. 42–13. A "circle" on a saddle-shaped surface.

There are, of course, other kinds of geometry. We could ask about the geometry of a bug who lived on a pear, namely something which has a sharper curvature in one place and a weaker curvature in the other place, so that the excess in angles in triangles is more severe when he makes little triangles in one part of his world than when he makes them in another part. In other words, the curvature of a space can vary from place to place. That's just a generalization of the idea. It can also be imitated by a suitable distribution of temperature on a hot plate.

We may also point out that the results could come out with the opposite kind of discrepancies. You could find out, for example, that all triangles when they are made too large have the sum of their angles *less* than 180 degrees. That may sound impossible, but it isn't at all. First of all, we could have a hot plate with the temperature decreasing with the distance from the center. Then all the effects would be reversed. But we can also do it purely geometrically by looking at the two-dimensional geometry of the surface of a saddle. Imagine a saddle-shaped surface like the one sketched in Fig. 42–13. Now draw a "circle" on the surface, defined as the locus of all points the same distance from a center. This circle is a curve that oscillates up and down with a scallop effect. So its circumference is larger than you would expect from calculating $2\pi r$. So $C/2\pi$ is now less than r. The "excess radius" would be negative.

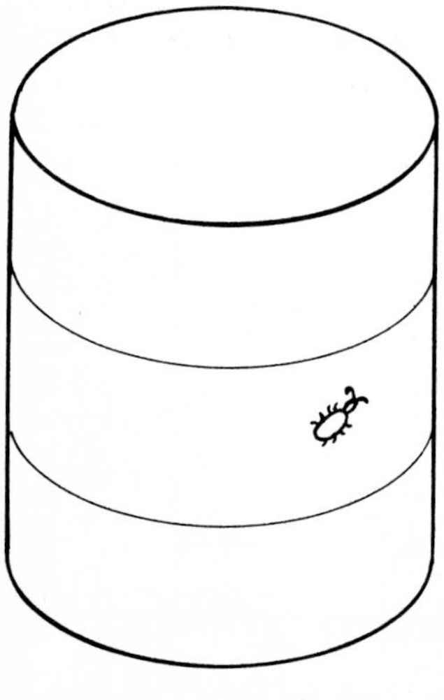

Fig. 42–14. A two-dimensional space with zero intrinsic curvature.

Spheres and pears and such are all surfaces of *positive* curvatures; and the others are called surfaces of *negative* curvature. In general, a two-dimensional world will have a curvature which varies from place to place and may be positive in some places and negative in other places. In general, we mean by a curved space simply one in which the rules of Euclidean geometry break down with one sign of discrepancy or the other. The amount of curvature—defined, say, by the excess radius—may vary from place to place.

We might point out that, from our definition of curvature, a cylinder is, surprisingly enough, not curved. If a bug lived on a cylinder, as shown in Fig. 42–14, he would find out that triangles, squares, and circles would all have the

† Except for the one point at infinity.

same behavior they have on a plane. This is easy to see, by just thinking about how all the figures will look if the cylinder is unrolled onto a plane. Then all the geometrical figures can be made to correspond exactly to the way they are in a plane. So there is no way for a bug living on a cylinder (assuming that he doesn't go all the way around, but just makes local measurements) to discover that his space is curved. In our technical sense, then, we consider that his space is *not* curved. What we want to talk about is more precisely called *intrinsic* curvature; that is, a curvature which can be found by measurements only in a local region. (A cylinder has no intrinsic curvature.) This was the sense intended by Einstein when he said that our space is curved. But we as yet only have defined a curved space in two dimensions; we must go onward to see what the idea might mean in three dimensions.

42–2 Curvature in three-dimensional space

We live in three-dimensional space and we are going to consider the idea that three-dimensional space is curved. You say, "But how can you imagine it being bent in any direction?" Well, we can't imagine space being bent in any direction because our imagination isn't good enough. (Perhaps it's just as well that we can't imagine too much, so that we don't get too free of the real world.) But we can still *define* a curvature without getting out of our three-dimensional world. All we have been talking about in two dimensions was simply an exercise to show how we could get a definition of curvature which didn't require that we be able to "look in" from the outside.

We can determine whether our world is curved or not in a way quite analogous to the one used by the gentlemen who live on the sphere and on the hot plate. We may not be able to distinguish between two such cases but we certainly can distinguish those cases from the flat space, the ordinary plane. How? Easy enough: We lay out a triangle and measure the angles. Or we make a great big circle and measure the circumference and the radius. Or we try to lay out some accurate squares, or try to make a cube. In each case we test whether the laws of geometry work. If they don't work, we say that our space is curved. If we lay out a big triangle and the sum of its angles exceeds 180 degrees, we can say our space is curved. Or if the measured radius of a circle is not equal to its circumference over 2π, we can say our space is curved.

You will notice that in three dimensions the situation can be much more complicated than in two. At any one place in two dimensions there is a certain amount of curvature. But in three dimensions there can be *several components* to the curvature. If we lay out a triangle in some plane, we may get a different answer than if we orient the plane of the triangle in a different way. Or take the example of a circle. Suppose we draw a circle and measure the radius and it doesn't check with $C/2\pi$ so that there is some excess radius. Now we draw another circle at right angles—as in Fig. 42–15. There's no need for the excess to be exactly the same for both circles. In fact, there might be a positive excess for a circle in one plane, and a defect (negative excess) for a circle in the other plane.

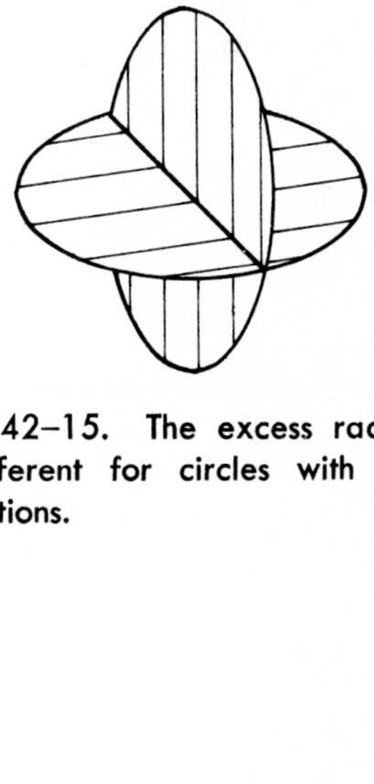

Fig. 42–15. The excess radius may be different for circles with different orientations.

Perhaps you are thinking of a better idea: Can't we get around all of these components by using a *sphere* in three dimensions? We can specify a sphere by taking all the points that are the same distance from a given point in space. Then we can measure the surface area by laying out a fine scale rectangular grid on the surface of the sphere and adding up all the bits of area. According to Euclid the total area A is supposed to be 4π times the square of the radius; so we can define a "predicted radius" as $\sqrt{A/4\pi}$. But we can also measure the radius directly by digging a hole to the center and measuring the distance. Again, we can take the measured radius minus the predicted radius and call the difference the radius excess,

$$r_{\text{excess}} = r_{\text{meas}} - \left(\frac{\text{measured area}}{4\pi}\right)^{1/2},$$

which would be a perfectly satisfactory measure of the curvature. It has the great advantage that it doesn't depend upon how we orient a triangle or a circle.

But the excess radius of a sphere also has a disadvantage; it doesn't completely characterize the space. It gives what is called the *mean curvature* of the three-dimensional world, since there is an averaging effect over the various curvatures. Since it is an average, however, it does not solve completely the problem of defining the geometry. If you know only this number you can't predict all properties of the geometry of the space, because you can't tell what would happen with circles of different orientation. The complete definition requires the specification of six "curvature numbers" at each point. Of course the mathematicians know how to write all those numbers. You can read someday in a mathematics book how to write them all in a high-class and elegant form, but it is first a good idea to know in a rough way what it is that you are trying to write about. For most of our purposes the average curvature will be enough.†

42–3 Our space is curved

Now comes the main question. Is it true? That is, is the actual physical three-dimensional space we live in curved? Once we have enough imagination to realize the possibility that space might be curved, the human mind naturally gets curious about whether the real world is curved or not. People have made direct geometrical measurements to try to find out, and haven't found any deviations. On the other hand, by arguments about gravitation, Einstein discovered that space *is* curved, and we'd like to tell you what Einstein's law is for the amount of curvature, and also tell you a little bit about how he found out about it.

Einstein said that space is curved and that matter is the source of the curvature. (Matter is also the source of gravitation, so gravity is related to the curvature—but that will come later in the chapter.) Let us suppose, to make things a little easier, that the matter is distributed continuously with some density, which may vary, however, as much as you want from place to place.‡ The rule that Einstein gave for the curvature is the following: If there is a region of space with matter in it and we take a sphere small enough that the density ρ of matter inside it is effectively constant, then the *radius excess* for the sphere is proportional to the mass inside the sphere. Using the definition of excess radius, we have

$$\text{Radius excess} = \sqrt{\frac{A}{4\pi}} - r_{\text{meas}} = \frac{G}{3c^2} \cdot M. \tag{42.3}$$

Here, G is the gravitational constant (of Newton's theory), c is the velocity of light, and $M = 4\pi\rho r^3/3$ is the mass of the matter inside the sphere. This is Einstein's law for the mean curvature of space.

Suppose we take the earth as an example and forget that the density varies from point to point—so we won't have to do any integrals. Suppose we were to measure the surface of the earth very carefully, and then dig a hole to the center and measure the radius. From the surface area we could calculate the predicted radius we would get from setting the area equal to $4\pi r^2$. When we compared the predicted radius with the actual radius, we would find that the actual radius exceeded the predicted radius by the amount given in Eq. (42.3). The constant $G/3c^2$ is about 2.5×10^{-29} cm per gram, so for each gram of material the measured radius is off by 2.5×10^{-29} cm. Putting in the mass of the earth, which is about 6×10^{27} grams, it turns out that the earth has 1.5 millimeters more radius

---

† We should mention one additional point for completeness. If you want to carry the hot-plate model of curved space over into three dimensions you must imagine that the length of the ruler depends not only on where you put it, but also on which orientation the ruler has when it is laid down. It is a generalization of the simple case in which the length of the ruler depends on where it is, but is the same if set north-south, or east-west, or up-down. This generalization is needed if you want to represent a three-dimensional space with any arbitrary geometry with such a model, although it happens not to be necessary for two dimensions.

‡ Nobody—not even Einstein—knows how to do it if mass comes concentrated at points.

than it should have for its surface area.† Doing the same calculation for the sun, you find that the sun's radius is one-half a kilometer too long.

You should note that the law says that the *average* curvature *above* the surface area of the earth is zero. But that does *not* mean that all the components of the curvature are zero. There may still be—and, in fact, there is—some curvature above the earth. For a circle in a plane there will be an excess radius of one sign for some orientations and of the opposite sign for other orientations. It just turns out that the average over a sphere is zero when there is no mass *inside* it. Incidentally, it turns out that there is a relation between the various components of the curvature and the *variation* of the average curvature from place to place. So if you know the average curvature everywhere, you can figure out the details of the curvature at each place. The average curvature above the earth varies with altitude, so the space there is curved. And it is that curvature that we see as a gravitational force.

Suppose we have a bug on a plane, and suppose that the "plane" has little pimples in the surface. Wherever there is a pimple the bug would conclude that his space had little local regions of curvature. We have the same thing in three dimensions. Wherever there is a lump of matter, our three-dimensional space has a local curvature—a kind of three-dimensional pimple.

If we make a lot of bumps on a plane there might be an overall curvature besides all the pimples—the surface might become like a ball. It would be interesting to know whether our space has a net average curvature as well as the local pimples due to the lumps of matter like the earth and the sun. The astrophysicists have been trying to answer that question by making measurements of galaxies at very large distances. For example, if the number of galaxies we see in a spherical shell at a large distance is different from what we would expect from our knowledge of the radius of the shell, we would have a measure of the excess radius of a tremendously large sphere. From such measurements it is hoped to find out whether our whole universe is flat on the average, or round—whether it is "closed," like a sphere, or "open" like a plane. You may have heard about the debates that are going on about this subject. There are debates because the atronomical measurements are still completely inconclusive; the experimental data are not precise enough to give a definite answer. Unfortunately, we don't have the slightest idea about the overall curvature of our universe on a large scale.

42–4 Geometry in space-time

Now we have to talk about time. As you know from the special theory of relativity, measurements of space and measurements of time are interrelated. And it would be kind of crazy to have something happening to the space, without the time being involved in the same thing. You will remember that the measurement of time depends on the speed at which you move. For instance, if we watch a guy going by in a spaceship we see that things happen more slowly for him than for us. Let's say he takes off on a trip and returns in 100 seconds flat *by our watches*; his watch might say that he had been gone for only 95 seconds. In comparison with ours, his watch—and all other processes, like his heart beat—have been running slow.

Now let's consider an interesting problem. Suppose you are the one in the spaceship. We ask you to start off at a given signal and return to your starting place just in time to catch a later signal—at, say, exactly 100 seconds later according to *our* clock. And you are also asked to make the trip in such a way that *your* watch will show the *longest possible* elapsed time. How should you move? You should stand still. If you move at all your watch will read less than 100 sec when you get back.

Suppose, however, we change the problem a little. Suppose we ask you to start at Point A on a given signal and go to point B (both fixed relative to us), and to do it in such a way that you arrive back just at the time of a second signal

† Approximately, because the density is not independent of radius as we are assuming.

(say 100 seconds later according to our fixed clock). Again you are asked to make the trip in the way that lets you arrive with the latest possible reading on your watch. How would you do it? For which path and schedule will *your* watch show the greatest elapsed time when you arrive? The answer is that you will spend the longest time from *your* point of view if you make the trip by going at a uniform speed along a straight line. Reason: Any extra motions and any extra-high speeds will make your clock go slower. (Since the time deviations depend on the *square* of the velocity, what you lose by going extra fast at one place you can never make up by going extra slowly in another place.)

The point of all this is that we can use the idea to define "a straight line" in space-time. The analog of a straight line in space is for space-time a *motion* at uniform velocity in a constant direction.

The curve of shortest distance in space corresponds in space-time not to the path of shortest time, but to the one of *longest* time, because of the funny things that happen to signs of the t-terms in relativity. "Straight-line" motion—the analog of "uniform velocity along a straight line"—is then that motion which takes a watch from one place at one time to another place at another time in the way that gives the longest time reading for the watch. This will be our definition for the analog of a straight line in space-time.

42–5 Gravity and the principle of equivalence

Now we are ready to discuss the laws of gravitation. Einstein was trying to generate a theory of gravitation that would fit with the relativity theory that he had developed earlier. He was struggling along until he latched onto one important principle which guided him into getting the correct laws. That principle is based on the idea that when a thing is falling freely everything inside it seems weightless. For example, a satellite in orbit is falling freely in the earth's gravity, and an astronaut in it feels weightless. This idea, when stated with greater precision, is called *Einstein's principle of equivalence.* It depends on the fact that all objects fall with exactly the same acceleration no matter what their mass, or what they are made of. If we have a spaceship that is "coasting"—so it's in a free fall—and there is a man inside, then the laws governing the fall of the man and the ship are the same. So if he puts himself in the middle of the ship he will stay there. He doesn't fall *with respect to the ship.* That's what we mean when we say he is "weightless."

Now suppose you are in a rocket ship which is accelerating. Accelerating with respect to what? Let's just say that its engines are on and generating a thrust so that it is not coasting in a free fall. Also imagine that you are way out in empty space so that there are practically no gravitational forces on the ship. If the ship is accelerating with "lg" you will be able to stand on the "floor" and will feel your normal weight. Also if you let go of a ball, it will "fall" toward the floor. Why? Because the ship is accelerating "upward," but the ball has no forces on it, so it will not accelerate; it will get left behind. Inside the ship the ball will appear to have a downward acceleration of "lg."

Now let's compare that with the situation in a spaceship sitting at rest on the surface of the earth. *Everything is the same!* You would be pressed toward the floor, a ball would fall with an acceleration of lg, and so on. In fact, how could you tell inside a space ship whether you are sitting on the earth or are accelerating in free space? According to Einstein's equivalence principle there is no way to tell if you only make measurements of what happens to things inside!

To be strictly correct, that is true only for one point inside the ship. The gravitational field of the earth is not precisely uniform, so a freely falling ball has a slightly different acceleration at different places—the direction changes and the magnitude changes. But if we imagine a strictly uniform gravitational field, it is completely imitated in every respect by a system with a constant acceleration. That is the basis of the principle of equivalence.

42-6 The speed of clocks in a gravitational field

Now we want to use the principle of equivalence for figuring out a strange thing that happens in a gravitational field. We'll show you something that happens in a rocket ship which you probably wouldn't have expected to happen in a gravitational field. Suppose we put a clock at the "head" of the rocket ship—that is, at the "front" end—and we put another identical clock at the "tail," as in Fig. 42–16. Let's call the two clocks A and B. If we compare these two clocks when the ship is accelerating, the clock at the head seems to run fast relative to the one at the tail. To see that, imagine that the front clock emits a flash of light each second, and that you are sitting at the tail comparing the arrival of the light flashes with the ticks of clock B. Let's say that the rocket is in the position a of Fig. 42–17 when clock A emits a flash, and at the position b when the flash arrives at clock B. Later on the ship will be at position c when the clock A emits its next flash, and at position d when you see it arrive at clock B.

The first flash travels the distance L_1 and the second flash travels the shorter distance L_2. It is a shorter distance because the ship is accelerating and has a higher speed at the time of the second flash. You can see, then, that if the two flashes were emitted from clock A one second apart, they would arrive at clock B with a separation somewhat less than one second, since the second flash doesn't spend as much time on the way. The same thing will also happen for all the later flashes. So if you were sitting in the tail you would conclude that clock A was running faster than clock B. If you were to do the same thing in reverse—letting clock B emit light and observing it at clock A—you would conclude that B was running *slower* than A. Everything fits together and there is nothing mysterious about it all.

But now let's think of the rocket ship at rest in the earth's gravity. *The same thing happens.* If you sit on the floor with one clock and watch another one which is sitting on a high shelf, it will appear to run faster than the one on the floor! You say, "But that is wrong. The times should be the same. With no acceleration there's no reason for the clocks to appear to be out of step." But they must if the principle of equivalence is right. And Einstein insisted that the principle *was* right, and went courageously and correctly ahead. He proposed that clocks at different places in a gravitational field must appear to run at different speeds. But if one always *appears* to be running at a different speed with respect to the other, then so far as the first is concerned the other *is* running at a different rate.

But now you see we have the analog for clocks of the hot ruler we were talking about earlier, when we had the bug on a hot plate. We imagined that rulers and bugs and everything changed lengths in the same way at various temperatures so they could never tell that their measuring sticks were changing as they moved around on the hot plate. It's the same with clocks in a gravitational field. Every clock we put at a higher level is seen to go faster. Heartbeats go faster, all processes run faster.

If they didn't you would be able to tell the difference between a gravitational field and an accelerating reference system. The idea that time can vary from place to place is a difficult one, but it is the idea Einstein used, and it is correct—believe it or not.

Using the principle of equivalence we can figure out how much the speed of a clock changes with height in a gravitational field. We just work out the apparent discrepancy between the two clocks in the accelerating rocket ship. The easiest way to do this is to use the result we found in Chapter 34 of Vol. 1 for the Doppler effect. There, we found—see Eq. (34.14)—that if v is the *relative* velocity of a source and a receiver, the *received* frequency ω is related to the *emitted* frequency ω_0 by

$$\omega = \omega_0 \frac{1 + v/c}{\sqrt{1 - v^2/c^2}}. \tag{42.4}$$

Now if we think of the accelerating rocket ship in Fig. 42–17 the emitter and re-

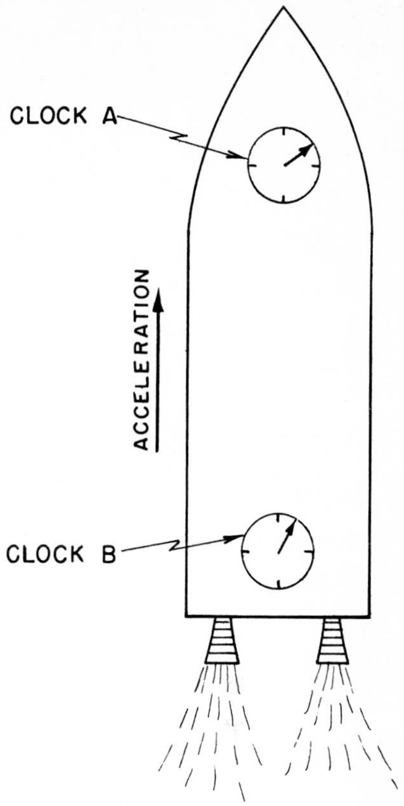

Fig. 42–16. An accelerating rocket ship with two clocks.

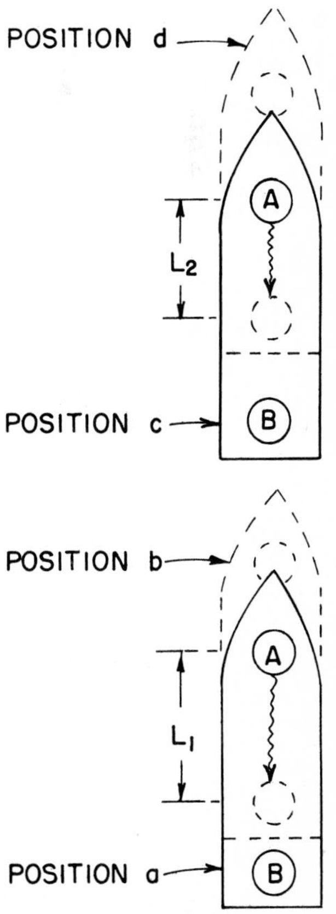

Fig. 42–17. A clock at the head of an accelerating rocket ship appears to run faster than a clock at the tail.

ceiver are moving with equal velocities at any one instant. But in the time that it takes the light signals to go from clock A to clock B the ship has accelerated. It has, in fact, picked up the additional velocity gt, where g is the acceleration and t is time it takes light to travel the distance H from A to B. This time is very nearly H/c. So when the signals arrive at B, the ship has increased its velocity by gH/c. The receiver always has this velocity *with respect to the emitter* at the instant the signal left it. So this is the velocity we should use in the Doppler shift formula, Eq. (42.4). Assuming that the acceleration and the length of the ship are small enough that this velocity is much smaller than c, we can neglect the term in v^2/c^2. We have that

$$\omega = \omega_0\left(1 + \frac{gH}{c^2}\right). \tag{42.5}$$

So for the two clocks in the spaceship we have the relation

$$(\text{Rate at the receiver}) = (\text{Rate of emission})\left(1 + \frac{gH}{c^2}\right), \tag{42.6}$$

where H is the height of the emitter *above* the receiver.

From the equivalence principle the same result must hold for two clocks separated by the height H in a gravitational field with the free fall acceleration g.

This is such an important idea we would like to demonstrate that it also follows from another law of physics—from the conservation of energy. We know that the gravitational force on an object is proportional to its mass M, which is related to its total internal energy E by $M = E/c^2$. For instance, the masses of nuclei determined from the *energies* of nuclear reactions which transmute one nucleus into another agree with the masses obtained from atomic *weights*.

Now think of an atom which has a lowest energy state of total energy E_0 and a higher energy state E_1, and which can go from the state E_1 to the state E_0 by emitting light. The frequency ω of the light will be given by

$$\hbar\omega = E_1 - E_0. \tag{42.7}$$

Now suppose we have such an atom in the state E_1 sitting on the floor, and we carry it from the floor to the height H. To do that we must do some work in carrying the mass $m_1 = E_1/c^2$ up against the gravitational force. The amount of work done is

$$\frac{E_1}{c^2}\,gH. \tag{42.8}$$

Then we let the atom emit a photon and go into the lower energy state E_0. Afterward we carry the atom back to the floor. On the return trip the mass is E_0/c^2; we get back the energy

$$\frac{E_0}{c^2}\,gH, \tag{42.9}$$

so we have done a net amount of work equal to

$$\Delta U = \frac{E_1 - E_0}{c^2}\,gH. \tag{42.10}$$

When the atom emitted the photon it gave up the energy $E_1 - E_0$. Now suppose that the photon happened to go down to the floor and be absorbed. How much energy would it deliver there? You might at first think that it would deliver just the energy $E_1 - E_0$. But that can't be right if energy is conserved, as you can see from the following argument. We started with the energy E_1 at the floor. When we finish, the energy at the floor level is the energy E_0 of the atom in its lower state plus the energy E_{ph} received from the photon. In the meantime we have had to supply the additional energy ΔU of Eq. (42.10). If energy is conserved, the energy we end up with at the floor must be greater than we started with by just the work we have done. Namely, we must have that

$$E_{\text{ph}} + E_0 = E_1 + \Delta U,$$

or (42.11)

$$E_{\text{ph}} = (E_1 - E_0) + \Delta U.$$

It must be that the photon does *not* arrive at the floor with just the energy $E_1 - E_0$ it started with, but with a *little more energy*. Otherwise some energy would have been lost. If we substitute in Eq. (42.11) the ΔU we got in Eq. (42.10) we get that the photon arrives at the floor with the energy

$$E_{\text{ph}} = (E_1 - E_0)\left(1 + \frac{gH}{c^2}\right). \tag{42.12}$$

But a photon of energy E_{ph} has the frequency $\omega = E_{\text{ph}}/\hbar$. Calling the frequency of the *emitted* photon ω_0—which is by Eq. (42.7) equal to $(E_1 - E_0)/\hbar$—our result in Eq. (42.12) gives again the relation of (42.5) between the frequency of the photon when it is absorbed on the floor and the frequency with which it was emitted.

The same result can be obtained in still another way. A photon of frequency ω_0 has the energy $E_0 = \hbar\omega_0$. Since the energy E_0 has the gravitational mass E_0/c^2 the photon has a mass (*not* rest mass) $\hbar\omega_0/c^2$, and is "attracted" by the earth. In falling the distance H it will gain an additional energy $(\hbar\omega_0/c^2)gH$, so it arrives with the energy

$$E = \hbar\omega_0\left(1 + \frac{gH}{c^2}\right).$$

But its frequency after the fall is $E/\hbar$, giving again the result in Eq. (42.5). Our ideas about relativity, quantum physics, and energy conservation all fit together only if Einstein's predictions about clocks in a gravitational field are right. The frequency changes we are talking about are normally very small. For instance, for an altitude difference of 20 meters at the earth's surface the frequency difference is only about two parts in 10^{15}. However, just such a change has recently been found experimentally using the Mössbauer effect.† Einstein was perfectly correct.

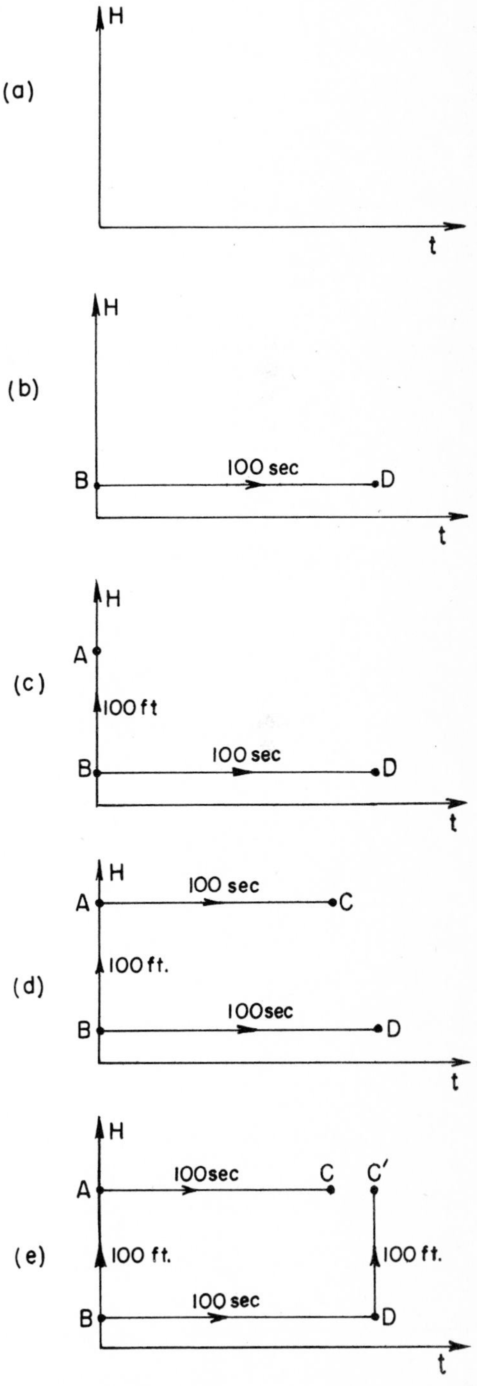

Fig. 42–18. Trying to make a rectangle in space-time.

42–7 The curvature of space-time

Now we want to relate what we have just been talking about to the idea of curved space-time. We have already pointed out that if the time goes at different rates in different places, it is analogous to the curved space of the hot plate. But it is more than an analogy; it means that space-time *is* curved. Let's try to do some geometry in space-time. That may at first sound peculiar, but we have often made diagrams of space-time with distance plotted along one axis and time along the other. Suppose we try to make a rectangle in space-time. We begin by plotting a graph of height H versus t as in Fig. 42–18(a). To make the base of our rectangle we take an object which is *at rest* at the height H_1 and follow its world line for 100 seconds. We get the line BD in part (b) of the figure which is parallel to the t-axis. Now let's take another object which is 100 feet above the first one at $t = 0$. It starts at the point A in Fig. 42–18(c). Now we follow its world line for 100 seconds as measured by a clock at A. The object goes from A to C, as shown in part (d) of the figure. But notice that since time goes at a different rate at the two heights—we are assuming that there is a gravitational field—the two points C and D are not simultaneous. If we try to complete the square by drawing a line to the point C' which is 100 feet above D at the same time, as in Fig. 42–18(e), the pieces don't fit. And that's what we mean when we say that space-time is curved.

---

† R. V. Pound and G. A. Rebka, Jr., *Physical Review Letters* Vol. **4,** p. 337 (1960).

42–8 Motion in curved space-time

Let's consider an interesting little puzzle. We have two identical clocks, A and B, sitting together on the surface of the earth as in Fig. 42–19. Now we lift clock A to some height H, hold it there awhile, and return it to the ground so that it arrives at just the instant when clock B has advanced by 100 seconds. Then clock A will read something like 107 seconds, because it was running faster when it was up in the air. Now here is the puzzle. How should we move clock A so that it reads the latest possible time—always assuming that it returns when B reads 100 seconds? You say, "That's easy. Just take A as high as you can. Then it will run as fast as possible, and be the latest when you return." Wrong. You forgot something—we've only got 100 seconds to go up and back. If we go very high, we have to go very fast to get there and back in 100 seconds. And you mustn't forget the effect of special relativity which causes moving clocks to *slow down* by the factor $\sqrt{1 - v^2/c^2}$. This relativity effect works in the direction of making clock A read *less time* than clock B. You see that we have a kind of game. If we stand still with clock A we get 100 seconds. If we go up slowly to a small height and come down slowly we can get a little more than 100 seconds. If we go a little higher, maybe we can gain a little more. But if we go too high we have to move fast to get there, and we may slow down the clock enough that we end up with less than 100 seconds. What program of height versus time—how high to go and with what speed to get there, carefully adjusted to bring us back to clock B when it has increased by 100 seconds—will give us the largest possible time reading on clock A?

Answer: Find out how fast you have to throw a ball up into the air so that it will fall back to earth in exactly 100 seconds. The ball's motion—rising fast, slowing down, stopping, and coming back down—is exactly the right motion to make the time the maximum on a wrist watch strapped to the ball.

Now consider a slightly different game. We have two points A and B both on the earth's surface at some distance from one another. We play the same game that we did earlier to find what we call the straight line. We ask how we should go from A to B so that the time on our moving watch will be the longest—assuming we start at A on a given signal and arrive at B on another signal at B which we will say is 100 seconds later by a fixed clock. Now you say, "Well we found out before that the thing to do is to coast along a straight line at a uniform speed chosen so that we arrive at B exactly 100 seconds later. If we don't go along a straight line it takes more speed, and our watch is slowed down." But wait! That was before we took gravity into account. Isn't it better to curve upward a little bit and then come down? Then during part of the time we are higher up and our watch will run a little faster? It is, indeed. If you solve the mathematical problem of adjusting the curve of the motion so that the elapsed time of the moving watch is the most it can possibly be, you will find that the motion is a parabola—the same curve followed by something that moves on a free ballistic path in the gravitational field, as in Fig. 42–19. Therefore the law of motion in a gravitational field can also be stated: *An object always moves from one place to another so that a clock carried on it gives a longer time than it would on any other possible trajectory*—with, of course, the same starting and finishing conditions. The time measured by a moving clock is often called its "proper time." In free fall, the trajectory makes the proper time of an object a maximum.

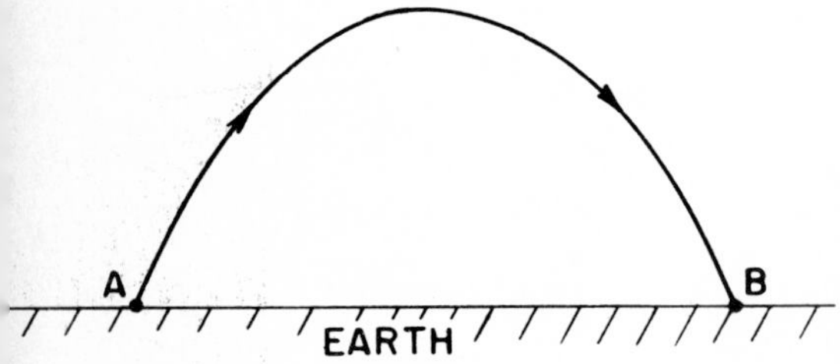

Fig. 42–19. In a uniform gravitational field the trajectory with the maximum proper time for a fixed elapsed time is a parabola.

Let's see how this all works out. We begin with Eq. (42.5) which says that the *excess* rate of the moving watch is

$$\frac{\omega_0 gH}{c^2}. \tag{42.13}$$

Besides this, we have to remember that there is a correction of the opposite sign for the speed. For this effect we know that

$$\omega = \omega_0\sqrt{1 - v^2/c^2}.$$

Although the principle is valid for any speed, we take an example in which the speeds are always much less than c. Then we can write this equation as

$$\omega = \omega_0(1 - v^2/2c^2),$$

and the defect in the rate of our clock is

$$-\omega_0 \frac{v^2}{2c^2}. \tag{42.14}$$

Combining the two terms in (42.13) and (42.14) we have that

$$\Delta\omega = \frac{\omega_0}{c^2}\left(gH - \frac{v^2}{2}\right). \tag{42.15}$$

Such a frequency shift of our moving clock means that if we measure a time dt on a fixed clock, the moving clock will register the time

$$dt\left[1 + \left(\frac{gH}{c^2} - \frac{v^2}{2c^2}\right)\right], \tag{42.16}$$

The total time excess over the trajectory is the integral of the extra term with respect to time, namely

$$\frac{1}{c^2}\int\left(gH - \frac{v^2}{2}\right)dt, \tag{42.17}$$

which is supposed to be a maximum.

The term gH is just the gravitational potential ϕ. Suppose we multiply the whole thing by a constant factor $-mc^2$, where m is the mass of the object. The constant won't change the condition for the maximum, but the minus sign will just change the maximum to a minimum. Equation (42.16) then says that the object will move so that

$$\int\left(\frac{mv^2}{2} - m\phi\right)dt = \text{a minimum}. \tag{42.18}$$

But now the integrand is just the difference of the kinetic and potential energies. And if you look in Chapter 19 of Volume II you will see that when we discussed the principle of least action we showed that Newton's laws for an object in any potential could be written exactly in the form of Eq. (42.18).

42–9 Einstein's theory of gravitation

Einstein's form of the equations of motion—that the proper time should be a maximum in curved space-time—gives the same results as Newton's laws for low velocities. As he was circling around the earth, Gordon Cooper's watch was reading later than it would have in any other path you could have imagined for his satellite.†

So the law of gravitation can be stated in terms of the ideas of the geometry of space-time in this remarkable way. The particles always take the longest proper time—in space-time a quantity analogous to the "shortest distance." That's the law of motion in a gravitational field. The great advantage of putting it this way is that the law doesn't depend on any coordinates, or any other way of defining the situation.

Now let's summarize what we have done. We have given you two laws for gravity:

(1) How the geometry of space-time changes when matter is present—namely, that the curvature expressed in terms of the excess radius is proportional to the mass inside a sphere, Eq. (42.3).

---

† Strictly speaking it is only a *local* maximum. We should have said that the proper time is larger than for any *nearby* path. For example, the proper time on an elliptical orbit around the earth need not be longer than on a ballistic path of an object which is shot to a great height and falls back down.

(2) How objects move if there are only gravitational forces—namely, that objects move so that their proper time between two end conditions is a maximum.

Those two laws correspond to similar pairs of laws we have seen earlier. We originally described motion in a gravitational field in terms of Newton's inverse square law of gravitation and his laws of motion. Now laws (1) and (2) take their places. Our new pair of laws also correspond to what we have seen in electrodynamics. There we had our law—the set of Maxwell's equations—which determines the fields produced by charges. It tells how the character of "space" is changed by the presence of charged matter, which is what law (1) does for gravity. In addition, we had a law about how particles move in the given fields—$d(m\boldsymbol{v})/dt = q(\mathbf{E} + \boldsymbol{v} \times \mathbf{B})$. This, for gravity, is done by law (2).

In the laws (1) and (2) you have a precise statement of Einstein's theory of gravitation—although you will usually find it stated in a more complicated mathematical form. We should, however, make one further addition. Just as time scales change from place to place in a gravitational field, so do also the length scales. Rulers change lengths as you move around. It is impossible with space and time so intimately mixed to have something happen with time that isn't in some way reflected in space. Take even the simplest example: You are riding past the earth. What is "*time*" from *your* point of view is partly space from *our* point of view. So there must also be changes in space. It is the entire *space-time* which is distorted by the presence of matter, and this is more complicated than a change only in time scale. However, the rule that we gave in Eq. (42–3) is enough to determine completely all the laws of gravitation, provided that it is understood that this rule about the curvature of space applies not only from one man's point of view but is true for everybody. Somebody riding by a mass of material sees a different mass content because of the kinetic energy he calculates for its motion past him, and he must include the mass corresponding to that energy. The theory must be arranged so that everybody—no matter how he moves—will, when he draws a sphere, find that the excess radius is $G/3c^2$ times the total mass (or, better, $G/3c^4$ times the total energy content) inside the sphere. That this law —law (1)—should be true in any moving system is one of the great laws of gravitation, called *Einstein's field equation*. The other great law is (2)—that things must move so that the proper time is a maximum—and is called *Einstein's equation of motion.*

To write these laws in a complete algebraic form, to compare them with Newton's laws, or to relate them to electrodynamics is difficult mathematically. But it is the way our most complete laws of the physics of gravity look today.

Although they gave a result in agreement with Newton's mechanics for the simple example we considered, they do not always do so. The three discrepancies first derived by Einstein have been experimentally confirmed: The orbit of Mercury is not a fixed ellipse; starlight passing near the sun is deflected twice as much as you would think; and the rates of clocks depend on their location in a gravitational field. Whenever the predictions of Einstein have been found to differ from the ideas of Newtonian mechanics Nature has chosen Einstein's.

Let's summarize everything that we have said in the following way. First, time and distance rates depend on the place in space you measure them and on the time. This is equivalent to the statement that space-time is curved. From the measured area of a sphere we can define a predicted radius, $\sqrt{A/4\pi}$, but the actual measured radius will have an excess over this which is proportional (the constant is G/c^2) to the total mass contained inside the sphere. This fixes the exact degree of the curvature of space-time. And the curvature must be the same no matter who is looking at the matter or how it is moving. Second, particles move on "straight lines" (trajectories of maximum proper time) in this curved space-time. This is the content of Einstein's formulation of the laws of gravitation.

Index